33
Topics in Medicinal Chemistry

Aims and Scope

Topics in Medicinal Chemistry (TMC) covers all relevant aspects of medicinal chemistry research, e.g. pathobiochemistry of diseases, identification and validation of (emerging) drug targets, structural biology, drugability of targets, drug design approaches, chemogenomics, synthetic chemistry including combinatorial methods, bioorganic chemistry, natural compounds, high-throughput screening, pharmacological in vitro and in vivo investigations, drug-receptor interactions on the molecular level, structure-activity relationships, drug absorption, distribution, metabolism, elimination, toxicology and pharmacogenomics. Drug research requires interdisciplinary team-work at the interface between chemistry, biology and medicine. To fulfil this need, TMC is intended for researchers and experts working in academia and in the pharmaceutical industry, and also for graduates that look for a carefully selected collection of high quality review articles on their respective field of expertise.

Medicinal chemistry is both science and art. The science of medicinal chemistry offers mankind one of its best hopes for improving the quality of life. The art of medicinal chemistry continues to challenge its practitioners with the need for both intuition and experience to discover new drugs. Hence sharing the experience of drug research is uniquely beneficial to the field of medicinal chemistry.

All chapters from Topics in Medicinal Chemistry are published OnlineFirst with an individual DOI. In references, Topics in Medicinal Chemistry is abbreviated as Top Med Chem and cited as a journal.

More information about this series at http://www.springer.com/series/7355

Antonello Mai
Editor

Chemical Epigenetics

With contributions by

L. Altucci · P. B. Arimondo · R. Belle · M. Berdasco ·
P. E. Brennan · F. Cao · M. Conte · F. J. Dekker ·
C. Dell'Aversana · M. Esteller · J. Fernandez · A. Ganesan ·
C. Giorgio · Q. Han · D. Hanly · J. A. Johnson ·
A. Kawamura · M. Lahtela-Kakkonen · C. Luise ·
N. I. Martin · J. Melesina · C. Moreno-Yruela ·
C. A. Olsen · W. C. K. Pomerantz · M. Rahnasto-Rilla ·
D. Robaa · D. Rotili · F. Sarno · G. Sbardella · C. L. Seiler ·
W. Sippl · G. Stazi · T. Suzuki · N. Y. Tretyakova ·
D. Trisciuoglio · J. Tyni · S. Valente · M. J. van Haren ·
M. Wright · P. D. Ycas · C. Zwergel ·
M. R. H. Zwinderman

Springer

Editor
Antonello Mai
Sapienza University of Rome
Rome, Italy

ISSN 1862-2461 ISSN 1862-247X (electronic)
Topics in Medicinal Chemistry
ISBN 978-3-030-42981-2 ISBN 978-3-030-42982-9 (eBook)
https://doi.org/10.1007/978-3-030-42982-9

This Springer imprint is published by the registered company Springer Nature Switzerland AG.
The registered company address is: Gewerbestrasse 11, 6330 Cham, Switzerland

Preface

> Epigenetics doesn't change the genetic code, it changes how that's read. Perfectly normal genes can result in cancer or death. Vice-versa, in the right environment, mutant genes won't be expressed. Genes are equivalent to blueprints; epigenetics is the contractor. They change the assembly, the Structure.
> —Bruce Lipton

Nowadays, science is proving that our body possesses incredible self-healing and self-repairing ability. These mechanisms are markedly influenced by our lifestyle, environmental factors and also our beliefs, thoughts, emotions and intentions. A change in the aforementioned factors can affect or even alter completely the tendency for expression.

During this process, our enormous code base, the DNA, will be read and the manifestation which will be expressed is heavily influenced by epigenetic marks. These marks are either written, read or modified. So we are based not only on plain code, but we are the modifiers of the code through readers, writers and erasers. So we have a profound vibrational effect on our continually evolving genetic code. We are the programmers of the code. DNA activation is our "software upgrade".

The specific scientific term "epigenetics" for these code reading, writing and erasing has been first defined in 1942 by C.H. Waddington. This term has been used in various contexts. Etymologically speaking, epigenetics deals with a precise branch of genetics as the Greek prefix epi means "after", "post" or "additionally". Today, in the molecular realm, all three meanings of epi are somewhat proven in the rapidly growing body of literature especially in the last decade dealing with fundamental processes in a living cell which are outside of the classical genetic processes and sources of genetic information like the DNA base pair sequence.

Today researches try to link an observed phenomenon or a disease down to the molecular level. Besides the longer known epigenetic targets such as histone deacetylases or DNA methyltransferases, a whole bunch of new enzymes and enzyme complexes have been discovered in the last 15 years. This book mainly discusses the recent advances in the drug development of epigenetic modulators from a medicinal chemist's viewpoint. Modern techniques in biology, biochemistry and chemical biology allow researchers faster than ever to describe and discover new epigenetic players as well as novel functions of old and known ones. Medicinal chemistry plays a fundamental role in the discovery process as it provides not only tools to better understand the function of an epigenetic player but also novel therapy options where aberrant epigenetic mechanisms are involved.

The book comprises 16 chapters. Each chapter includes a short introduction for a single epigenetic target or a target family ranging from structural biology aspects to cell biology and biochemistry. Most of the space is devoted to target modulation, either inhibition or activation. The authors give an insight into the discovery and development of mainly small organic molecules and also peptides influencing epigenetic pathways. Modern aspects of drug design such as new methodologies, ranging from computational approaches, crystallography to structural biology are presented with hands-on examples.

In the last 2 years, I have had the privilege to first convene leading experts in the growing field of epigenetics to contribute for the present book. It was an honour to work with them and I think all the contributors can be proud of the results achieved. The present book can be a facile start for everyone to dive into the field of epigenetic drug discovery providing a sound base of knowledge as a strong focus has been put on the most recent and up to date literature.

I would like to express my gratitude to Clemens Zwergel for his enormous support and useful comments in managing this book. I also wish to thank Sofia Costa and Shanti Ramamoorthy from Springer for their valuable assistance during the whole project from the first idea to the final book.

Rome, Italy
February 2019

Antonello Mai

Contents

Targeting the Zinc-Dependent Histone Deacetylases (HDACs) for Drug Discovery 1
A. Ganesan

Hydroxamic Acid-Containing Peptides in the Study of Histone Deacetylases 29
Carlos Moreno-Yruela and Christian A. Olsen

Sirtuin Inhibitors and Activators 55
Minna Rahnasto-Rilla, Jonna Tyni, and Maija Lahtela-Kakkonen

Histone Acetyltransferase Enzymes: From Biological Implications to Most Relevant Inhibitors 93
Daniela Trisciuoglio and Dante Rotili

Lysine Methyltransferases and Their Inhibitors 123
Giulia Stazi, Clemens Zwergel, and Sergio Valente

PRMT Inhibitors 159
Matthijs J. van Haren and Nathaniel I. Martin

Lysine-Specific Histone Demethylases 1/2 (LSD1/2) and Their Inhibitors 197
Takayoshi Suzuki

Inhibitors of JmjC-Containing Histone Demethylases 221
Miranda Wright, Paul E. Brennan, and Akane Kawamura

Chemical Compounds Targeting DNA Methylation and Hydroxymethylation 255
Roman Belle, Akane Kawamura, and Paola B. Arimondo

Applied Biophysics for Bromodomain Drug Discovery 287
William C. K. Pomerantz, Jorden A. Johnson, and Peter D. Ycas

Methyl-Readers and Inhibitors 339
Gianluca Sbardella

Altered Long Non-coding RNA Expression in Cancer: Potential Biomarkers and Therapeutic Targets? 401
David Hanly, Manel Esteller, and María Berdasco

Acetylation and Methylation in Asthma, COPD, and Lung Cancer 429
Martijn R. H. Zwinderman, Fangyuan Cao, and Frank J. Dekker

Structure-Based Design of Epigenetic Inhibitors 455
Dina Robaa, Jelena Melesina, Chiara Luise, and Wolfgang Sippl

Experimental Methodologies for Detection and Mapping of Epigenetic DNA Marks 487
Christopher L. Seiler, Jenna Fernandez, Qiyuan Han, and Natalia Y. Tretyakova

Advanced Assays in Epigenetics 523
Carmela Dell'Aversana, Federica Sarno, Mariarosaria Conte, Cristina Giorgio, and Lucia Altucci

Index 561

About the Editor

Antonello Mai is a Full Professor of Medicinal Chemistry at the Faculty of Pharmacy and Medicine, Sapienza University of Rome, since 2011. He completed his Pharmacy degree in 1984, and his PhD in Medicinal Chemistry and Pharmaceutical Sciences in 1992, both at Sapienza. In 1998 he was appointed as Associate Professor of Medicinal Chemistry at Sapienza's Faculty of Pharmacy. From 2012 to 2018, he was the coordinator of the PhD course in Pharmaceutical Science, and from 2013 to 2019, he was President of the Degree Course in Pharmaceutical Chemistry and Technologies.

From 2004 to 2009, Prof. Mai was a member of the Executive Committee of the Medicinal Chemistry Division of the Italian Chemical Society, and from 2009 to 2014, he was a member of the Scientific Committee of the European School of Medicinal Chemistry. He is an Editor of the journal *ChemMedChem* (Wiley), Regional Editor of *Mini Reviews in Medicinal Chemistry* and *Medicinal Chemistry* (Bentham), and Associate Editor of *Clinical Epigenetics* (Springer) and the *Journal of Enzyme Inhibition and Medicinal Chemistry* (Taylor and Francis).

His career has largely focused on the synthesis of new potential bioactive compounds, particularly in the field of epigenetics (since 1999: inhibitors of HDACs, HATs, sirtuins, PRMTs, HKMTs, KDMs, and DNMTs). Prof. Mai is the author of more than 280 publications in high-impact international journals [h-index = 52] and three book chapters.

Top Med Chem (2020) 33: 1–28
DOI: 10.1007/7355_2019_68

Published online: 14 January 2020

Targeting the Zinc-Dependent Histone Deacetylases (HDACs) for Drug Discovery

A. Ganesan

Contents

1 The Reversible Acetylation of Lysine Residues in Proteins 2
2 The Zinc-Dependent HDACs 4
3 Hydroxamic Acid HDAC Inhibitors 7
4 Thiol and Benzamide HDAC Inhibitors 12
5 Other HDAC Inhibitors 14
6 Therapeutic Applications Targeting Human HDACs 18
7 Therapeutic Applications Targeting Nonhuman HDACs 20
8 Summary 21
References 22

Abstract In humans, the zinc-dependent histone deacetylases (HDACs) are a family of 11 nonredundant isoforms that catalyze the dynamic reversal of posttranslationally modified acyl-lysine residues back to lysine. At the epigenetic level, HDACs have a critical gene silencing effect, promoting the compaction of histone tails with DNA to prevent transcription. In addition, HDACs deacylate many nonhistone substrates in diverse cellular compartments to profoundly influence protein structure and function. While the action of HDACs is indispensable to normal physiology, their abnormal overexpression is linked to the majority of human diseases. Consequently, the inhibition of HDACs has become a valuable target for therapeutic applications. Numerous potent small molecules are known, of both natural product and synthetic origin, that inhibit HDACs, primarily by reversibly interacting with the zinc cation within the enzyme active site. At the present time, five such HDAC inhibitors have received regulatory approval for the treatment of hematological cancers. This review focuses on the typical zinc-binding groups employed in HDAC inhibitors and the major advances within each class in terms of potency, isoform selectivity, and clinical applications.

A. Ganesan (✉)
School of Pharmacy, University of East Anglia, Norwich, UK
e-mail: a.ganesan@uea.ac.uk

Keywords Anticancer drugs, Epigenetics, Histone deacetylases, Zinc metalloenzymes

Abbreviations

BCL2	B-cell lymphoma 2
CoREST	Corepressor RE1 silencing transcription factor
EGFR	Epidermal growth factor receptor
EMA	European medicines agency
FDA	Food and drug administration
HAT	Histone acetyltransferase
HDAC	Histone deacetylase
HMG-CoA	3-Hydroxy-3-methyl-glutaryl-coenzyme A
MiDAC	Mitotic deacetylase complex
NcoR	Nuclear receptor corepressor
NODE	Nanog and Oct4-associated deacetylase
NuRD	Nucleosome remodeling deacetylase
PI3K	Phosphatidylinositol 3-kinase
SAHA	Suberoylanilide hydroxamic acid
Sin3	Septation initiation network transcriptional regulatory protein 3
SMRT	Silencing mediator of retinoid and thyroid receptors

1 The Reversible Acetylation of Lysine Residues in Proteins

Fifty-five years ago, Philipps demonstrated that a significant proportion of lysine residues in histone proteins are acetylated [1]. Shortly thereafter, Allfrey provided evidence that acetylation is rapidly introduced in minutes after protein translation and suggested that it is a mechanism for activating gene transcription [2]. Evidence was soon found for the existence of enzymes that perform histone acetylation and its reversal through deacetylation [3, 4]. These early studies laid the foundation for our present understanding of histone acetylation as a regulator of eukaryotic gene transcription. In the forward direction, the family of histone acetyltransferases (HATs) transfers the acetyl group from the biological acyl donor acetyl coenzyme A to histone proteins (Fig. 1) [5]. In addition, the reactive thioester group in acetyl coenzyme A can undergo direct nonenzymatic transfer to protein substrates [6]. Conversely, the histone deacetylases (HDACs) hydrolyse acetyllysine back to lysine, and it is the balance between these two dynamic processes that determines the transcriptional state of eukaryotic cells.

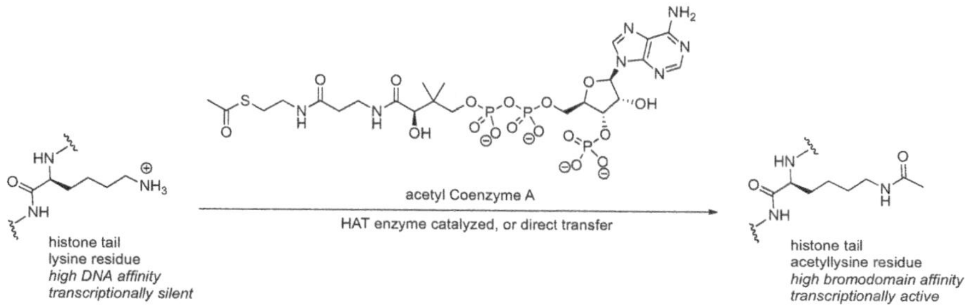

Fig. 1 The acetylation of histone tail lysine residues

The acetylation of lysine residues profoundly alters the physicochemical properties of this sidechain. While lysine bears a positive charge at physiological pH, acetyllysine is neutral, larger in molecular size, and more hydrophobic. These changes have two major consequences for eukaryotic gene regulation: (1) the loss of charge leads to a reduced affinity between the negatively charged DNA and the histone tails that relaxes chromatin and enables transcription by RNA pol II to commence; (2) acetyllysine binding to bromodomain containing proteins recruits transcription factors that are transcriptionally activating [7]. Advances in single cell fluorescence spectroscopy have recently enabled these effects to be observed on a timescale of seconds [8].

While the view of histone acetylation as an on/off transcriptional switch in eukaryotes is fundamentally correct, the details need some refinement. Firstly, although lysine acetylation was historically detected in histones due to their relative abundance, it is by no means restricted to these proteins. In fact, lysine acetylation is a widespread posttranslational modification found in all life forms ranging from prokaryotes to man [9]. In human cells, proteomic studies have identified more than a thousand proteins that undergo lysine acetylation, often at multiple sites. Thus, drugs that modulate acetylation will perturb diverse cellular proteins and physiological pathways in addition to their classical epigenetic effect on histones. It is an open question whether such promiscuity is beneficial for the therapeutic benefits of HDAC inhibitors, the cause of undesirable side effects or, as is more likely, a complex combination of the two. Secondly, the posttranslational acylation of lysine residues is not restricted to acetylation but encompasses a variety of low molecular weight acyl groups (Fig. 2) [10]. Presumably, the degree to which these species are loaded is related to the cell's metabolic state which determines the relative concentration of acyl coenzyme A donors. It is tempting to speculate that each acyl group has its own recognition domain, enabling a cell to go down alternative phenotypic fates through feedback loops that link metabolism and epigenetic regulation [11].

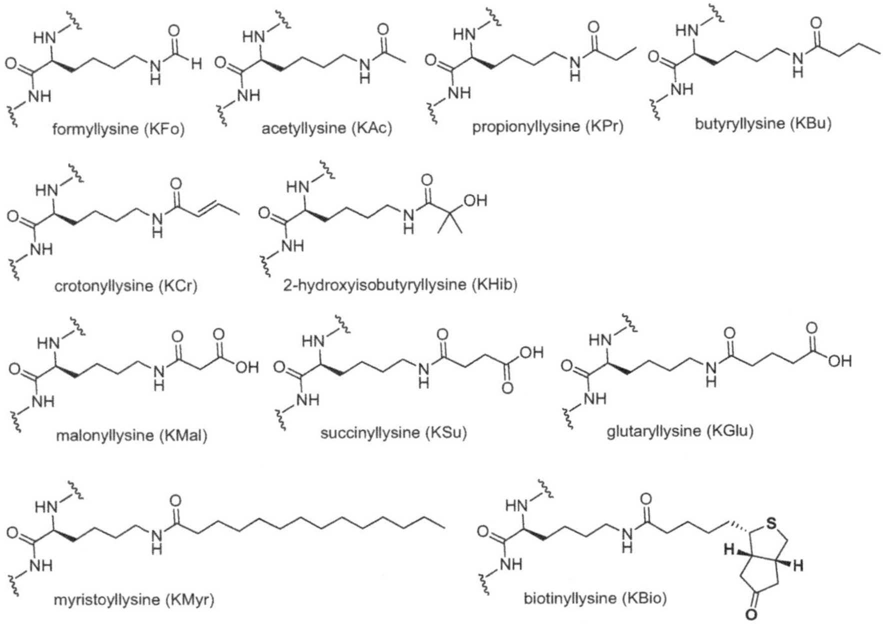

Fig. 2 Examples of lysine posttranslational acylations found in histone proteins

2 The Zinc-Dependent HDACs

The deacylation of lysine residues is carried out by amide bond hydrolases grouped together as HDACs. Within this family, there are two distinct catalytic mechanisms in operation: the zinc-dependent HDACs contain an active site zinc(II) cation and are the focus of this chapter, while the sirtuins are a distinct class of deacetylases that transfer the acetyl group to the ribose sugar of their NAD^+ cofactor [12, 13]. X-ray studies are available for a number of HDACs and HDAC-inhibitor complexes and indicate a catalytic mechanism similar to that of other amide hydrolyzing metalloenzymes [14–16]. The acetyl-lysine sidechain enters a narrow vertical tunnel ~11 Å long at the bottom of which sits the active site zinc(II) cation. The zinc functions as a Lewis acid, simultaneously increasing the electrophilicity of the scissile carbonyl group and the reactivity of the water nucleophile (Fig. 3).

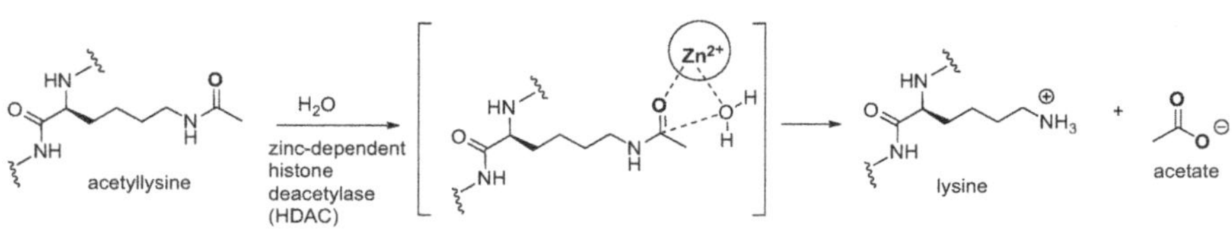

Fig. 3 HDAC-mediated acetyllysine hydrolysis

After hydrolysis, the acetate is released through an ~14 Å long exit channel perpendicular to the acetyl-lysine binding tunnel. In terms of substrate scope, there are two variables for HDACs, the nature of the acyl group and the lysine. For the majority of zinc-dependent HDACs, the acyl group is confined to the shorter chains up to butyryl and crotonyl (Fig. 2), with the exception of HDAC11 that hydrolyses longer acyl chains of 8–18 carbons [17]. As for the lysine, the endogenous substrates are macromolecular proteins, but HDACs will competently hydrolyze single amino acid lysine derivatives under cell-free conditions. Thus, HDACs predominantly recognize the acyl-lysine sidechain and are not particularly sequence selective or sensitive to the nature of the adjacent protein residues. Any specificity is imposed by the cellular compartment within which the HDAC resides and its recruitment to specific substrates by the multiprotein complexes with which it interacts.

In mammals, there are 11 HDAC isoforms, HDAC1–11, that are classified on the basis of sequence homology, size, and cellular localization (Table 1). The class I HDACs are ubiquitous and nuclear in localization with histones as a major substrate and also acting upon additional nuclear proteins. HDAC1 and HDAC2 are present in multiprotein transcriptional complexes such as Sin3, NuRD, CoREST, NODE, and

Table 1 The 11 mammalian HDAC isoforms

Isoform	Localization	Amino acids	Mouse knockout phenotype	Substrate examples
Class I				
HDAC1	Nucleus	482	Embryonic lethal	Histones, RelA, AR, Eg5, NF-KB
HDAC2	Nucleus	488	Perinatal lethal	Histones, GR, NF-KB
HDAC3	Nucleus	428	Embryonic lethal	Histones, RelA, NF-KB, PCAF, STAT1
HDAC8	Nucleus/ cytoplasm	377	Perinatal lethal	Histones, p53, ERR-α, SMC3, ARID1A
Class IIa				
HDAC4	Nucleus/ cytoplasm	1,084	Postnatal lethal	Histones, HIF1α, p53, Runx2, DNAJB8
HDAC5	Nucleus/ cytoplasm	1,122	Cardiac defects	GATA-2, GCMa
HDAC7	Nucleus/ cytoplasm	952	Embryonic lethal	PLAG1
HDAC9	Nucleus/ cytoplasm	1,011	Cardiac defects	ATDC
Class IIb				
HDAC6	Mainly cytoplasm	1,215	Viable	Hsp90, cortactin, α-tubulin, β-catenin, Prx1
HDAC10	Mainly cytoplasm	673	Viable	Hsp70, PP1
Class IV				
HDAC11	Mainly nucleus	347	Viable	Histones, Cdc25A

MiDAC, while HDAC3 is part of NcoR and SMRT complexes, and HDAC8 can function independently of multiprotein complexes. Compared to class I, the class IIa isoforms are tissue-specific, larger in size, and shuttle between the nucleus and cytoplasm upon activation. The class IIa enzymes have weak deacetylase activity in vitro, and relatively little is known about their substrates. It is possible that their primary function is in fact not enzymatic and they function as a scaffold to bring together multiprotein complexes [18]. The class IIa isoforms are found in NCoR and SMRT protein complexes, and the region involved in protein-protein interaction with the HDACs has been identified [19]. The class IIb isoforms HDAC6 and HDAC10 are primarily cytoplasmic in location, as reflected in their known substrates. HDAC6 has two catalytic domains although clear catalytic activity was demonstrated with only one of these [20]. Finally, HDAC11 is placed distinctly in class IV as it has similarities to both the class I and class II isoforms [21]. Although the 11 HDAC isoforms overlap with one another in terms of localization and substrates, they clearly play nonredundant physiological roles as evidenced by the fact that only HDAC6, HDAC10, and HDAC11 knockouts produce a viable phenotype in mice.

The transition state for HDAC catalysis and its similarity to that for previously well-established metalloproteinase drug discovery targets suggests a rational approach to inhibitor design (Fig. 4). Although the first HDAC inhibitors were discovered and optimized without this working model, with hindsight, we can observe its presence in all potent HDAC inhibitors to this day [22–24]. The relatively simple three-point pharmacophore contains three elements: a zinc-binding group, a linker, and a cap. The zinc-binding moiety is a polar functional group that coordinates to the cation in either a mono- or bidentate fashion. This interaction is by far the most dominant for enzyme-inhibitor affinity, and replacement of the zinc-binding group by a weaker coordinator or excising it altogether results in a significant loss of potency. Next is the linker that needs to occupy the narrow channel normally occupied by the acetyllysine sidechain. The linker is a linear moiety that is typically hydrophobic in nature but may incorporate some polar functionality. Finally, there is the cap that forms binding interactions with the enzyme surface as well as protrudes into the solvent-exposed exterior. As the enzyme does not recognize much of the

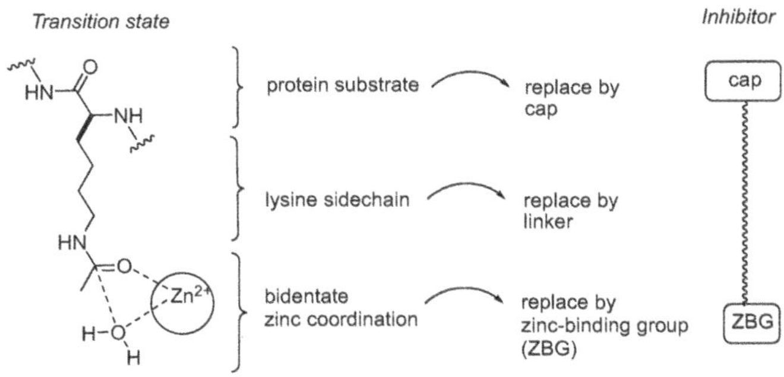

Fig. 4 From HDAC transition state to inhibitor

protein substrate's surface beyond the acetyllysine, a small-sized cap is sufficient and allows potent inhibition with low MW compounds that fit guidelines for oral bioavailability. Furthermore, as the cap is not involved in major interactions with the active site, large variations are possible in this region including modifications introduced to improve pharmacokinetics rather than influence binding.

3 Hydroxamic Acid HDAC Inhibitors

Hydroxamic acids are an effective bidentate metal chelating functional group that are widely used by microorganisms in iron-binding siderophores [25]. Their coordinating ability is also used to advantage in natural products and synthetic compounds that are inhibitors of metalloenzymes including carbonic anhydrase, matrix metalloproteinases, ribonucleotide reductase, and urease [26]. The potent antifungal natural product trichostatin A (**1**, Fig. 5) was the first hydroxamic acid to be identified as a HDAC inhibitor by Yoshida in 1990 [27]. Independently, Breslow's studies on the cellular differentiation caused by dimethyl sulfoxide led to a hydroxamic acid series with suberoylanilide hydroxamic acid (SAHA, **2**) as the optimized clinical candidate. The similarity between these two structures led Breslow to surmise that SAHA was a HDAC inhibitor, as was confirmed by biochemical profiling [28]. Both trichostatin A and SAHA clearly conform to the model for HDAC inhibitor design based on the transition state, as illustrated for vorinostat (**3**). While trichostatin A is too toxic for therapeutic applications, it continues to be widely used as a chemical probe. Meanwhile SAHA, now known by the drug name vorinostat (Zolinza™), would progress to become the first HDAC inhibitor to enter clinical trials and receive FDA approval for the treatment of cutaneous T-cell lymphoma [29].

Subsequent to the identification of the mechanism of action of trichostatin A and vorinostat, hydroxamic acids have become the most popular choice of zinc-binding

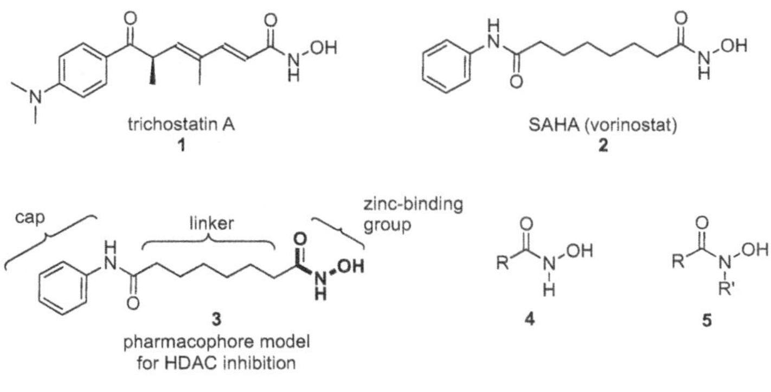

Fig. 5 The natural product trichostatin A and the synthetic compound vorinostat, exemplars of hydroxamic acid HDAC inhibitors

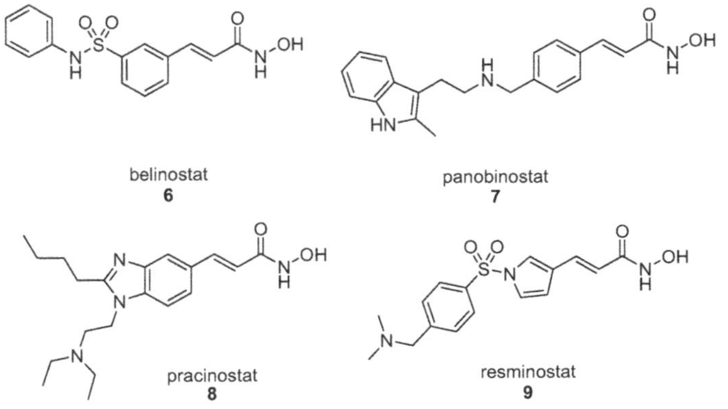

Fig. 6 Examples of HDAC inhibitors containing cinnamoyl linkers

group for the rational design of HDAC inhibitors. However, not all hydroxamic acids are equal: the geometry of the HDAC active site dictates the use of monosubstituted compounds of type **4** rather than N-substituted examples **5**. On the other hand, zinc-binding groups such as sulfonamides or carboxylic acids that are widely found in metal-binding inhibitors for other drug targets rarely achieve high potency against HDACs [30]. Such differences in enzyme topology help reduce the potential for side effects arising from the nonspecific binding of hydroxamic acid HDAC inhibitors to other metalloenzymes. Nevertheless, the pharmacokinetic limitations of hydroxamic acids should not be forgotten – phase I glucuronidation accelerates drug clearance, while Lossen-type rearrangements at alkaline pH can lead to the creation of mutagenic species [31].

A number of second-generation hydroxamic acids have reached clinical trials, among which belinostat (Belodaq™, **6**, Fig. 6) and panobinostat (Farydak™, **7**) have received regulatory approval [32, 33]. All three approved drugs potently inhibit the class I nuclear HDAC isoforms (Table 2). HDAC8 is the least inhibited, consistent with its being an outlier that is the least homologous to the other class I enzymes [34]. The drugs also potently inhibit class IIb and class IV isoforms, whereas HDAC4 and HDAC7 are poorly inhibited among the tissue-specific class IIa isoforms. Compared to vorinostat, belinostat and panobinostat feature a more rigid cinnamoyl linker that is also present in the clinical candidates pracinostat (**8**) and resminostat (**9**) [35, 36]. These cinnamates are superior to vorinostat in their metabolic half-life, while the addition of polar functionality improves oral bioavailability.

In contrast to the HDAC inhibitors containing a saturated hydrocarbon linker (e.g., vorinostat) or unsaturated alkenyl linker (e.g., trichostatin A and the compounds in Fig. 6), even more rigid benzoyl linkers are present in the clinical candidates givinostat (**10**) [37], abexinostat (**11**) [38], and AR-42 (**12**) [39] (Fig. 7). Alternatively, a heteroaromatic pyrimidine linker is employed in quisinostat (**13**) and nanatinostat (**14**) [40, 41]. The more recent compounds feature further refinements in pharmacodynamic or pharmacokinetic properties compared to

Table 2 The IC_{50} values of approved HDAC inhibitors against individual isoforms, based on data provided in reference [33]

Isoform	Vorinostat IC_{50} (nM)	Belinostat IC_{50} (nM)	Panobinostat IC_{50} (nM)
HDAC1	76	18	3
HDAC2	360	34	13
HDAC3	58	21	2
HDAC8	>1,000	160	280
HDAC4	>1,000	>1,000	200
HDAC5	160	76	8
HDAC7	>1,000	600	530
HDAC9	78	44	6
HDAC6	27	15	11
HDAC10	88	31	2
HDAC11	110	44	3

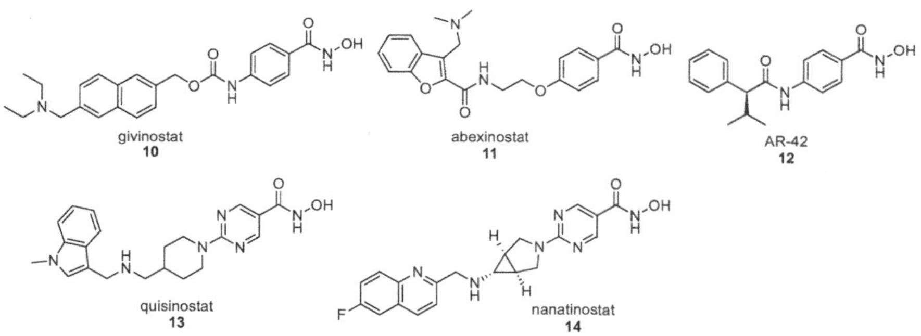

Fig. 7 Examples of HDAC inhibitors in clinical development with aryl or heteroaryl linkers

vorinostat, as illustrated for quisinostat (Table 3). As can be seen, there are differences for the potency of the reference vorinostat between Tables 2 and 3. This is a common phenomenon in the field of HDAC inhibitor discovery, and absolute IC_{50} values, or even the pattern of isoform selectivity, should not be rigidly compared between publications due to differences in assay methodology or batch to batch variation of enzyme preparations.

While the majority of clinical candidate hydroxamic acids show relatively low selectivity between HDAC isoforms, ricolinostat (**15**, Fig. 8) and its chlorinated version citarinostat (**16**) are two exceptions that are HDAC6 selective inhibitors [42, 43]. Additional HDAC6 selective inhibitors at a preclinical stage were reported by academic groups, among others by Ganesan (**17**) with a flexible linker [44], and by Kozikowski (**18**) [45] and Mahboobi (**19**) [46] with more rigid benzoyl linkers. Hydroxamic acids with a different isoform profile include selective HDAC8 inhibitors such as compounds **20** (Fig. 9) and **21** [47, 48]. Meanwhile, cyclopropane **22** is a rare example of a class IIa selective hydroxamic acid [49]. It potently inhibits all the class IIa isoforms at a nanomolar level, whereas class I isoforms are weakly

Table 3 The IC_{50} values of quisinostat against individual HDAC isoforms with vorinostat as a reference, based on data provided in reference [40]

Isoform	Vorinostat IC_{50} (nM)	Quisinostat IC_{50} (nM)
HDAC1	62	0.1
HDAC2	240	0.3
HDAC3	260	5.0
HDAC8	210	4.0
HDAC4	150	0.6
HDAC5	100	3.7
HDAC7	>1,000	120
HDAC9	180	32
HDAC6	20	77
HDAC10	150	0.5
HDAC11	100	0.4

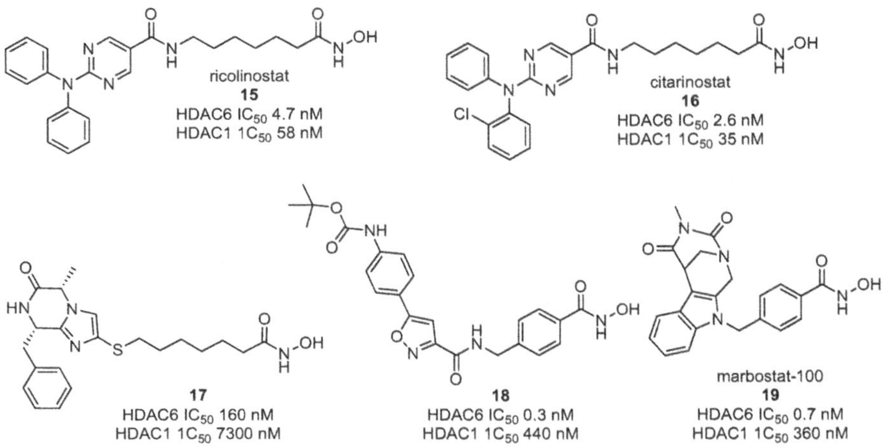

Fig. 8 Examples of HDAC6 selective hydroxamic acid inhibitors

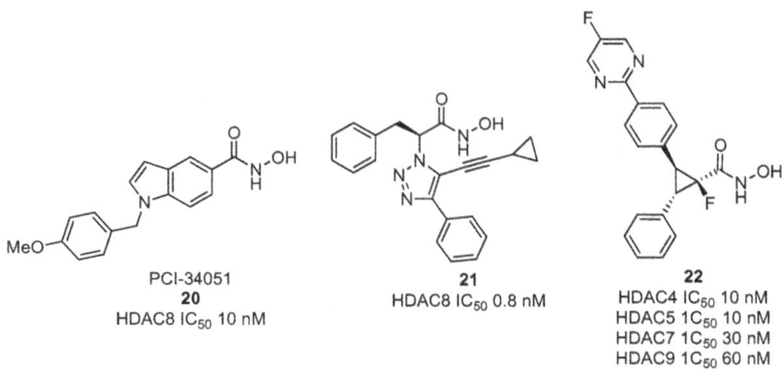

Fig. 9 Examples of HDAC8 and class IIa selective HDAC inhibitors

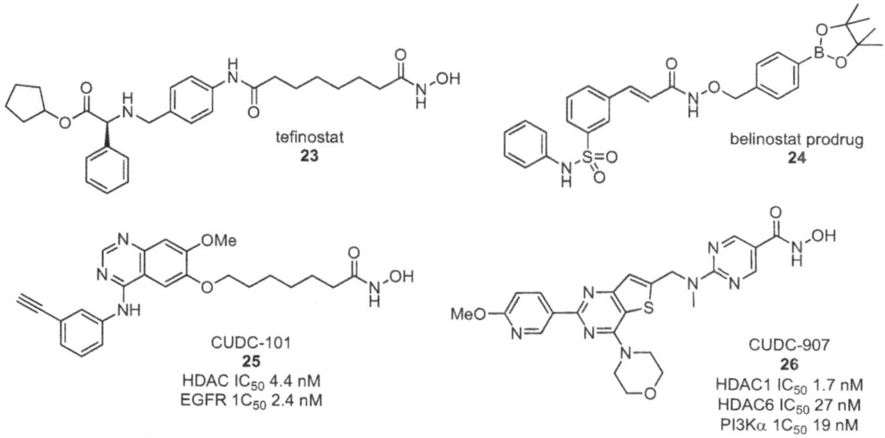

Fig. 10 Examples of prodrug and dual mechanism hydroxamic acid HDAC inhibitors

inhibited at a micromolar level. The compound design takes advantage of a "lower pocket" present in the class IIa enzymes that accommodates the phenyl ring, and this was corroborated by X-ray structures of enzyme-inhibitor complexes.

Besides isoform discrimination, two alternative approaches for HDAC drug discovery involve the use of prodrugs or multitarget inhibitors with a dual mechanism of action. Chroma's tefinostat (**23**, Fig. 10) is relatively nonselective but accumulates less in normal cells due to secretion through efflux pumps. In the liver, human carboxylesterase-1 hydrolyses the cyclopentyl ester to give a poorly diffusible ionized carboxylic acid, enabling selective targeting to liver cancers for which the compound is currently undergoing clinical trials [50]. The hydroxamic acid itself can be converted to a prodrug form to improve bioavailability. A recent example with belinostat (**24**) was shown to possess higher activity compared to the parent drug in a MCF-7 breast cancer tumor xenograft animal model [51]. Meanwhile, the simplicity of the HDAC pharmacophore suggests the "cap" can incorporate the pharmacophore for a second orthogonal target. Inhibitors have been successfully developed that are dual inhibitors of HDACs and kinase inhibitors as well as other enzymes such as HMG-CoA reductase, phosphodiesterase type 5, DNA topoisomerase, and receptors including the vitamin D receptor, retinoid X receptor, and estrogen receptor [52, 53]. Two examples from Curis are currently in clinical trials: CUDC-101 (**25**) for dual HDAC/EGFR inhibition and CUDC-907 (**26**) for dual HDAC/PI3K inhibition [54, 55].

4 Thiol and Benzamide HDAC Inhibitors

In 1998, Yoshida identified the natural product FK228 (**27**, Fig. 11), a bicyclic depsipeptide isolated from the bacteria *Chromobacterium violaceum* [56], as a HDAC inhibitor [57]. At first sight, FK228 does not follow the HDAC pharmacophore due to the absence of a zinc-binding group. However, Yoshida's experiments suggested the disulfide bridge undergoes reductive cleavage to release a thiol sidechain that can then coordinate to the active site zinc [58]. Subsequently, related bacterial natural product prodrugs such as largazole (**28**) and thailandepsin A (**29**) were discovered [59, 60]. All contain an identical zinc-binding group and benefit from a large macrocyclic cap that aids in binding to the enzyme surface more tightly than the minimal phenyl ring present in hydroxamic acid inhibitors such as trichostatin A or vorinostat [61, 62]. While the free zinc-binding thiol has relatively poor cell permeability and may be subject to nonspecific binding, the natural products mask this functionality as a cell-permeable disulfide or ester prodrug that is metabolically labile. In their active form, the natural products are highly potent HDAC inhibitors, especially against the class I isoforms HDAC1, HDAC2, and HDAC3 and relatively weak against the cytoplasmic isoform HDAC6 (Table 4). FK228 progressed to clinical trials sponsored by the NCI and became the second HDAC inhibitor to receive FDA approval for the treatment of cutaneous T-cell lymphoma and is now known as romidepsin (Istodax™) [63].

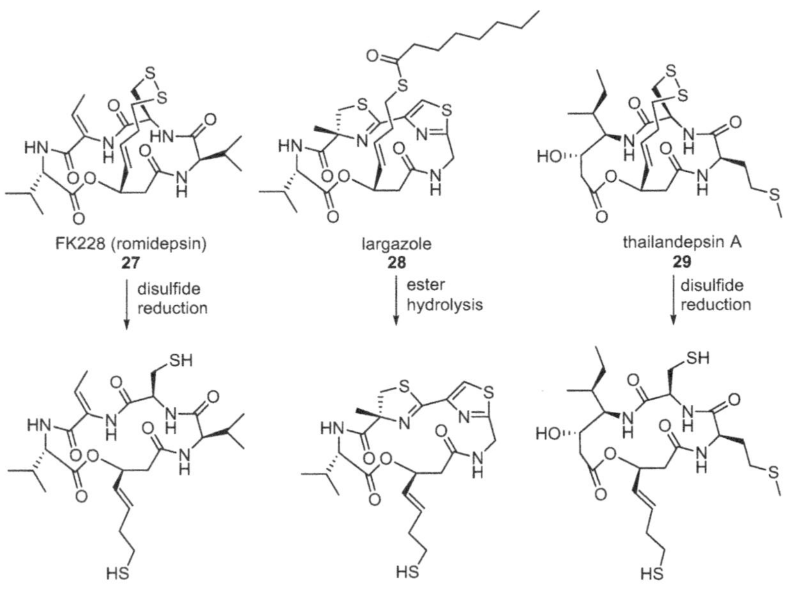

Fig. 11 Examples of depsipeptide natural product HDAC inhibitors containing a protected zinc-binding thiol

Table 4 The IC_{50} values of the active forms of depsipeptide HDAC inhibitors against individual isoforms, based on data taken from reference [61]

Isoform	Romidepsin IC_{50} (nM)	Largazole IC_{50} (nM)	Thailandepsin A IC_{50} (nM)
HDAC1	0.8	0.4	14
HDAC2	1.0	0.9	3.5
HDAC3	1.3	0.7	4.8
HDAC8	26	100	>1,000
HDAC4	470	>1,000	>1,000
HDAC5	>1,000	>1,000	
HDAC7	>1,000	>1,000	>1,000
HDAC9	>1,000	>1,000	>1,000
HDAC6	330	42	380
HDAC10	0.9	0.5	
HDAC11	0.3	>1,000	

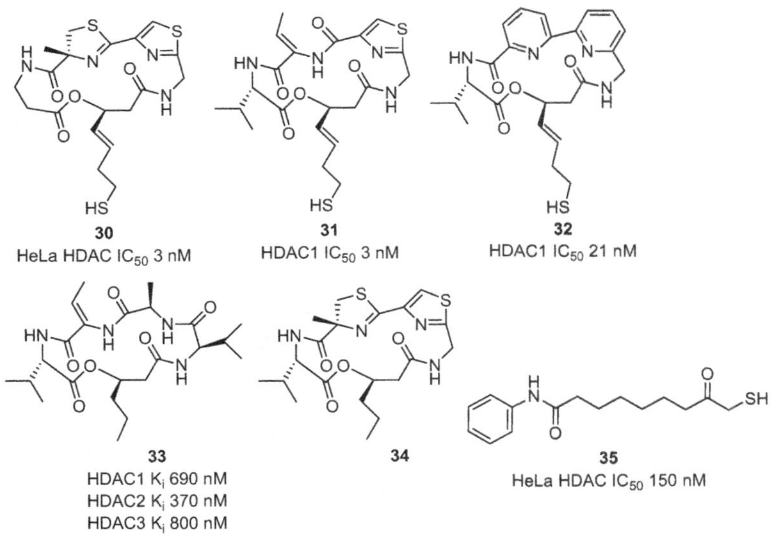

Fig. 12 Examples of synthetic HDAC inhibitors based on a thiol zinc-binding group

A number of academic groups have completed total syntheses of the depsipeptide natural products including a multigram scale preparation of largazole [64, 65]. These routes have been applied to the preparation of synthetic analogues that have illuminated the SAR within this series [66]. While the zinc-bearing thiol and its stereochemistry are necessary for potent enzyme inhibition, the remaining amino acid sidechains can be varied, as illustrated by the incorporation of a β-amino acid to give a ring-expanded largazole analogue **30** (Fig. 12) [67], the romidepsin-largazole hybrid **31** [68], and the bipyridyl analogue **32** [69]. An intriguing example is Olsen's romidepsin analogue **33** and largazole analogue **34** in which the zinc-binding thiol

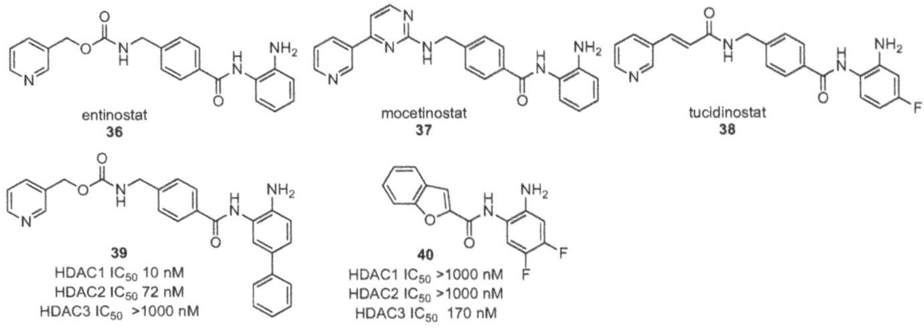

Fig. 13 Examples of benzamide HDAC inhibitors

was excised [70]. These compounds no longer obey the classical three-point HDAC pharmacophore, and while **34** was inactive at the tested concentrations, **33** inhibits HDACs at a micromolar level. Although this represents a substantial drop in activity relative to the approved drug romidepsin, the analogue avoids the pharmacokinetic liabilities associated with zinc-binding groups and might be amenable to further optimization to produce potent compounds. An alternative approach to increasing potency with synthetic thiols is the introduction of a second functional group to enable bidentate coordination. The α-mercaptoketone vorinostat analogue **35**, for example, was more active in HDAC inhibition compared to vorinostat [71].

In addition to hydroxamic acids and thiols, the other zinc-binding group that has led to an approved HDAC inhibitor is the benzamide or *ortho*-anilinoamide. X-ray studies indicate coordination of the amine to the zinc cation, as well as weak binding to the carbonyl oxygen. Thus, the benzamides, like the hydroxamic acids, are capable of bidentate chelation. Two examples in clinical development are entinostat (**36**, Fig. 13) and mocetinostat (**37**) [72, 73]. Analogue-based drug discovery at Chipscreen Biosciences with the benzamide scaffold led to chidamide (now known as tucidinostat, **38**, Epidaza™) that received regulatory approval in China for the treatment of relapsed or refractory peripheral T-cell lymphoma [74]. These clinical candidate benzamides are primarily selective against class I isoforms (Table 5) in their biological activity [75]. By taking advantage of an internal cavity present in the HDAC1 and HDAC2 active site but not in HDAC3, attachment of an aryl ring to entinostat produced the more selective compound **39** [76]. On the other hand, analogue **40** displays the opposite selectivity between the class I isoforms, being most active against HDAC3 [77].

5 Other HDAC Inhibitors

Although hydroxamic acids, thiols, and benzamides have successfully yielded multiple clinical candidates and five approvals to date, the first HDAC inhibitor to be identified was in fact sodium butyrate with a carboxylic acid zinc-binding

Table 5 The IC_{50} values of entinostat and tucidinostat against individual isoforms, based on data provided in reference [75]

Isoform	Entinostat IC_{50} (nM)	Tucidinostat IC_{50} (nM)
HDAC1	260	100
HDAC2	310	160
HDAC3	500	70
HDAC8	>1,000	730
HDAC4	>1,000	>1,000
HDAC5	>1,000	>1,000
HDAC7	>1,000	>1,000
HDAC9	>1,000	>1,000
HDAC6	>1,000	>1,000
HDAC10	250	80
HDAC11	650	430

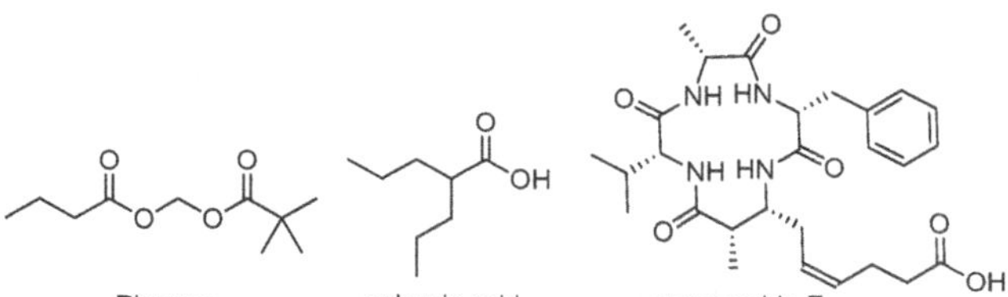

Fig. 14 Examples of carboxylic acid HDAC inhibitors

group [78]. As butyrate suffers from a short in vivo half-life [79], clinical trials have focused on prodrugs such as Pivanex (**41**, Fig. 14) or the repurposing of the well-established antiepileptic drug valproic acid (**42**). However, the low potency of such monodentate short chain carboxylic acids (millimolar IC_{50} against HDAC enzymes) compared to other zinc-binding groups is a challenge for therapeutic applications, and none have received regulatory approval. At the preclinical stage, the marine depsipeptide natural product azumamide E (**43**) is a rare example of a carboxylic acid HDAC inhibitor with submicromolar potency [80]. In this case, the active site zinc coordination is presumably augmented by additional interactions from the macrocyclic cap to improve binding affinity.

Hydroxamic acids, thiols, benzamides, and carboxylic acids are the only four zinc-binding groups incorporated into HDAC inhibitors that have progressed to clinical investigation. Besides these motifs, ketones and derivatives thereof are an important zinc-binding group at the preclinical stage. Their presence was first observed in a family of fungal toxin cyclic tetrapeptide natural products such as trapoxin A (**44**, Fig. 15), AS1387392 (**45**) and apicidin (**46**) that are potent inhibitors of class I HDAC isoforms. Trapoxin A was originally believed to be an irreversible inhibitor that undergoes epoxide ring opening by a nucleophilic residue in the active site. However, a recent X-ray structure of the natural product bound to HDAC8

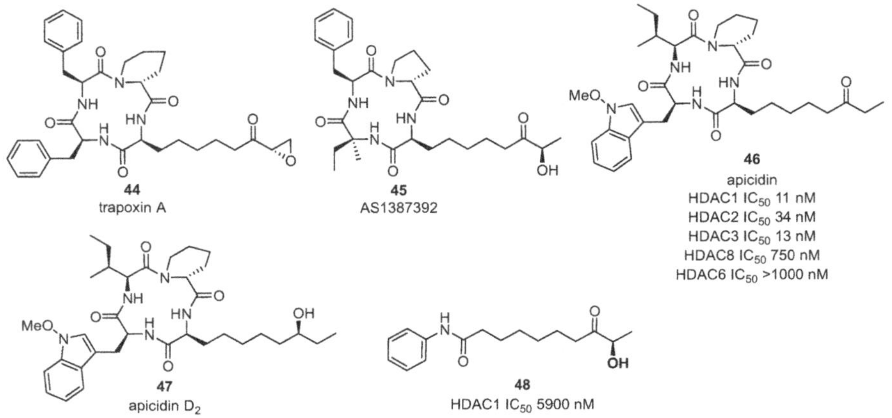

Fig. 15 Examples of HDAC inhibitors with ketone or alcohol zinc-binding groups

reveals that although inhibition is irreversible, covalent modification is not involved, and the epoxide is intact [81]. Instead, the ketone is nucleophilically attacked by the active site water to produce a diolate that coordinates in monodentate fashion to the zinc cation. Presumably, similar gem-diol forms of the ketone in AS1387392 and apicidin are engaged in monodentate coordination, and this explains why **47**, another member of the apicidin family, retains HDAC inhibitory activity with only an alcohol as zinc-binding group (albeit, weaker in potency than apicidin itself). Extensive SAR studies have been performed on the cyclic tetrapeptide natural products including variation of the amino acid residues and replacement of the ketone by other zinc-binding groups such as hydroxamic acids and thiols as well as modifications of the peptidic backbone [82, 83]. Although some analogues have shown promise in animal models, none has progressed further as yet. Incorporation of the α-hydroxy ketone warhead within the much simpler vorinostat scaffold afforded analogue **48** with micromolar potency against HDAC1 [84].

In addition to the traditional zinc-binding groups, there are isolated cases where other motifs were successfully employed for HDAC inhibition. The scope and limitations of these rarer zinc-binding groups are difficult to evaluate until they are more widely adopted and SAR studies appear from multiple investigators. Nevertheless, there are reported examples that have achieved a combination of high potency and isoform selectivity. Novartis disclosed a phenylalanine derivative (**49**, Fig. 16) for which X-ray crystallography shows bidentate coordination to zinc through the carbonyl and amine groups [85]. The dichlorophenyl aromatic ring forms π-π interactions with the acetate exit channel and results in isoform selectivity for HDAC8. The rest of the molecule lies within the substrate-binding tunnel, meaning that the compound effectively has no cap. Thus, it illustrates that active site interactions alone are sufficient to provide potent and selective inhibitors. Tempero's TMP269 (**50**) displays pronounced class IIa isoform selectivity, and in the X-ray cocrystal with HDAC7, the zinc cation is 2.7 Å away from one of the

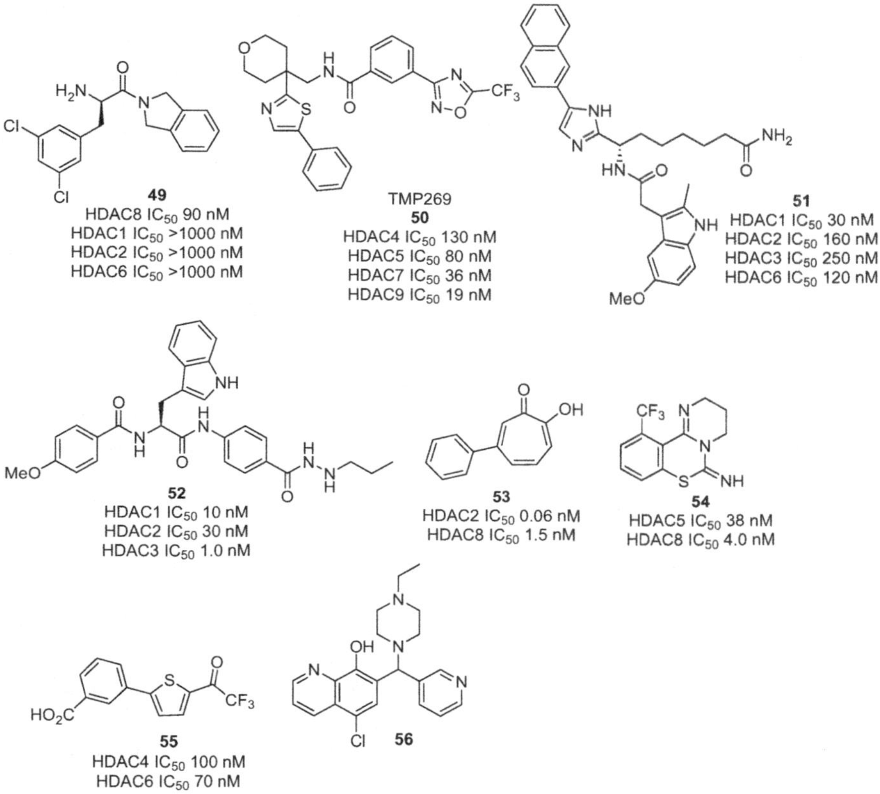

Fig. 16 Examples of atypical HDAC inhibitors

fluorines and 3.0 Å away from the oxygen of the oxadiazole ring [86]. With these relatively large distances, metal binding involves weaker electrostatic interactions, and the Tempero example shows high affinity and selectivity can still be achieved compared to the direct bidentate coordination typical of hydroxamic acids.

Besides inhibitors **49** and **50**, there are compounds with putative zinc-binding groups whose binding modes are not yet confirmed through X-ray crystallographic evidence. In the acetyl-lysine mimic **51**, a primary amide is the likely zinc binder and the compound inhibited HDAC1, HDAC2, HDAC3, and HDAC6 at a submicromolar level [87]. Class I isoform selectivity was also observed with the tryptophan derivative **52**, in which the more common benzamide zinc-binding group was replaced by a metabolically more stable acyl hydrazide [88]. The metal-binding properties of tropolones is well-known, and the β-phenyl derivative **53** was reported to be a remarkably selective subnanomolar HDAC2 inhibitor [89]. A pyrimido-benzothiazine framework has led to selective HDAC8 inhibitors such as **54** [90]. Thiophene **55** is predicted to bind the lipophilic substrate binding tunnel and was selective for HDAC4 and HDAC6 inhibition [91]. A radically different mechanism is involved for the hydroxyquinoline **56**, a micromolar HDAC5 and

HDAC9 inhibitor [92]. Upon coordination to the zinc, the piperazine moiety is ejected to form a reactive quinone methide that covalently binds to a cysteine residue in the enzyme. Finally, HDAC inhibition does not have to rely on active site occupancy [93]. In principle, allosteric inhibitors or disruptors of protein-protein interactions that recruit HDACs to their cellular multiprotein complexes are among the alternative strategies that are likely to be developed in the future and offer unique opportunities for isoform specificity.

6 Therapeutic Applications Targeting Human HDACs

The first HDAC inhibitors were identified on the basis of phenotypic effects in cell culture such as differentiation or growth inhibition of eukaryotic cells. With this background, cancer was an obvious therapeutic indication for HDAC inhibitors, and the rationale is supported by the increased levels of HDAC expression observed in many cancers. Indeed, HDAC inhibitors are potent antiproliferative agents against human cancer cell lines, and, barring pharmacokinetic liabilities, the level of activity correlates with that measured in mechanism-based enzyme assays. Evidence for cellular target engagement is typically obtained through Western blotting to show increased acetylation levels of substrate proteins, e.g., histones for nuclear HDACs and tubulin for HDAC6, and the induction of downstream proteins such as the cyclin-dependent kinase inhibitor p21 (CIP1/WAF1) [94]. In addition, HDAC inhibitors typically cause morphological changes that can be observed microscopically.

The hallmarks of proliferating cells are multiple mutations that promote division while silencing alternative fates such as repair, differentiation, and apoptosis. Cancer chemotherapy covers a broad spectrum of drugs ranging from blunt instruments that are general cytotoxic agents such as cisplatin to highly specific mechanism-based agents such as the antibody rituximab with HDAC inhibitors sitting somewhere in between. While normal cells are relatively tolerant of HDAC inhibition, transformed cells respond by reactivating pathways leading to cell death, cell cycle arrest, senescence, differentiation, or tumor immunogenicity [95]. Microarray experiments with cells treated with HDAC inhibitors indicate significant changes in expression levels of the BCL2 family of proteins that regulate apoptosis [96]. Based on their potent activity in cells and tumor xenograft animal models, the first candidates, vorinostat and romidepsin, entered clinical trials in the USA, from which the most promising results were seen in the treatment of T-cell lymphomas. Vorinostat received FDA approval in 2006 for the treatment of cutaneous T-cell lymphoma, and romidepsin followed suit in 2009. Romidepsin received additional approval for the treatment of peripheral T-cell lymphoma in 2011, while belinostat and tucidinostat were approved for the same indication in 2014 and 2015, respectively. The T-cell lymphomas appear particularly sensitive to HDAC inhibition as this result in the downregulation of expression of immunosuppressive cytokines such as IL-10 [97]. Nevertheless, the rarity of these T-cell lymphomas, combined with the fact that the HDAC inhibitors are not first-line therapies, has limited the commercial success of these four compounds. Meanwhile, the Novartis drug panobinostat has

Table 6 Non-cancer clinical trials with approved HDAC inhibitors

Drug	Indication
Givinostat	Arthritis
Givinostat	Crohn's disease
Givinostat	Autoinflammatory disease
Vorinostat	Pruritus
Vorinostat Panobinostat	Sickle cell disease
Givinostat	Muscular dystrophy
Vorinostat	Niemann-Pick disease
Vorinostat Romidepsin Panobinostat	Graft vs host disease
Vorinostat Romidepsin Tucidinostat	HIV
Vorinostat	Alzheimer's disease
Ricolinostat	Neuropathic pain
Vorinostat	Schizophrenia

Based on data from https://clinicaltrials.gov/

tapped into a more lucrative market, receiving FDA and EMA approval in 2015 for use against multiple myeloma.

In addition to the hematological indications, multiple clinical trials are investigating the approved HDAC inhibitors in other cancer types including solid tumors. Besides monotherapy, combinations of HDAC inhibitors with a second agent are also being explored. These avenues of investigation may result in further approvals for the established drugs, provided a therapeutic window for efficacy and tolerability can be identified.

The history of HDAC inhibitors as anticancer agents suggests utility against other conditions that have a proliferative or immunological profile [98, 99]. Indeed, the approved drugs are undergoing evaluation as both monotherapy and in combinations against a number of inflammatory diseases (Table 6), genetic syndromes, and autoimmune disorders. Early stage trials are exploring vorinostat in neurological diseases that will need to address the additional challenge of brain penetration. Outside cancer, perhaps the most exciting potential application of HDAC inhibitors lies in antiretroviral therapy. Here, the transcriptional activation by HDAC inhibitors facilitates the expression of latent reservoirs of the HIV virus that are then susceptible to conventional antiviral agents [100]. Preliminary indications suggest the strategy can be applied to other viruses such as hepatitis B and Epstein-Barr [101].

Moving away from the already approved HDAC inhibitors, further therapeutic advances are impending with the second-generation clinical candidates. The newer inhibitors have the advantage of superior pharmacokinetics that may enable successful application against solid tumors. Meanwhile, at the preclinical stage, HDAC inhibitors have shown promise in animal models against many disease conditions apart from cancer (Table 7). Since some of these studies were performed with relatively nonspecific older inhibitors, the recent availability of isoform selective

Table 7 Examples of non-cancer indications in which HDAC inhibitors were tested in animal models

Indication	Compound and reference
Alcohol use disorder	Vorinostat [102]
Autism	Romidepsin [103]
Cardiac hypertrophy	Trichostatin A [104]
Colitis	Entinostat [105]
Contact hypersensitivity	Ricolinostat [106]
Cued fear extinction	RGFP963 [107]
Cystic fibrosis	Vorinostat [108]
Diabetes	BRD3308 [109]
Huntington's disease	RGFP966 [110]
Hypertension	Trichostatin A [111]
Kidney fibrosis	Tubastatin A [112]
Obesity	Entinostat [113]
Parkinson's disease	K560 [114]
Pulmonary fibrosis	Romidepsin [115]
Retinal disease	Trichostatin A [116]
Sepsis	Tubastatin A [117]

compounds suggests further progress will be possible. Ultimately, our predictive power should enable the matching of a given therapeutic indication with the optimum inhibitor profile and thereby reduce the incidence of side effects.

7 Therapeutic Applications Targeting Nonhuman HDACs

Although the therapeutic targeting of HDACs has predominantly concentrated on the human enzymes, it can be extrapolated to other species. The natural product trichostatin A (**1**), for example, was first identified due to its potent antifungal activity. MethylGene have reported the compound MGCD290 (structure undisclosed) as an inhibitor of the fungal HDAC Hos2 [118]. Although a phase II trial in acute vulvovaginal candidiasis with a combination of MGCD290 and fluconazole did not demonstrate superior efficacy over fluconazole alone, fungal HDACs remain an interesting target for drug discovery [119]. Meanwhile, there is extensive literature on the inhibition of parasitic HDACs for the treatment of malaria and other neglected diseases [120]. The cancer clinical candidate pracinostat, for example, was active upon oral administration in a *Plasmodium berghei*-infected malaria mouse model while romidepsin inhibited adult worm pairing and egg production in *Schistosoma mansoni* [121, 122]. Medicinal chemistry efforts have aimed to selectively target the parasite HDACs. For example, compound **57** (Fig. 17) with an amide zinc-binding group inhibited the *P. falciparum* HDAC1 with more than 50-fold selectivity over the human enzyme [123], while hydroxamic acid **58** displays some selectivity for the *S. mansoni* HDAC8 over the human isoform [124]. Finally, although bacteria do not contain histones, they do have histone deacetylase-like proteins. Interestingly, the vorinostat analogue **59** with a perfluorinated linker showed selectivity against *Pseudomonas aeruginosa* HDAC-like enzymes over the human enzymes and is potentially a lead for antibacterial agents [125].

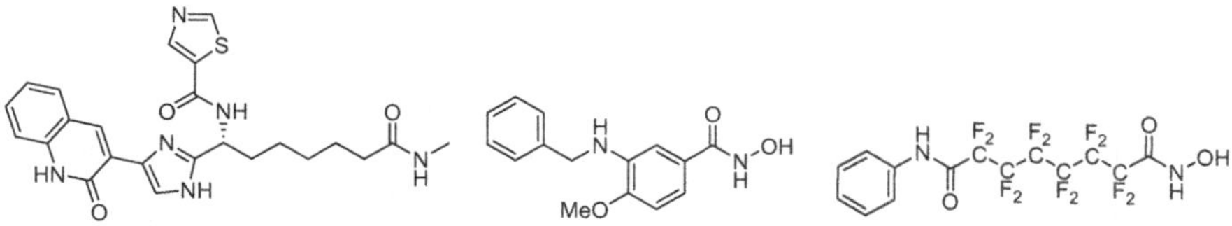

57
PfHDAC1 IC_{50} 500 nM
hHDAC1 IC_{50} >25,000 nM

58
smHDAC8 IC_{50} 120 nM
hHDAC8 IC_{50} 510 nM

59
PA0321 IC_{50} 700 nM
PA1409 IC_{50} >1000 nM
PA3774 IC_{50} 22 nM
hHDAC1 IC_{50} >10000 nM
hHDAC6 IC_{50} >10000 nM

Fig. 17 Examples of inhibitors of microbial HDACs and HDAC-like proteins

Table 8 Examples of highly isoform selective HDAC inhibitors

Compound	Selectivity
39	HDAC1, HDAC2
40	HDAC3
21	HDAC8
22	HDAC4, HDAC5, HDAC7, HDAC9
50	HDAC4, HDAC5, HDAC7, HDAC9
18	HDAC6

8 Summary

Four decades have passed since the first report of an enzyme that catalyzes the deacetylation of acetyllysine protein residues. Within that time, tremendous progress has been made in understanding the biological functions of HDACs and the potential applications of HDAC inhibitors. Many such compounds have been discovered through a combination of phenotypic screening and enzyme-based assays. While the classical pharmacophore for HDAC inhibition consisting of a zinc-binding group, linker, and cap largely holds true, it is by no means obligatory. There are now inhibitors where zinc coordination is replaced by electrostatic attractions or completely removed, while there are others that lack a cap. Structure-based design has played an important role and aided the development of selective inhibitors by taking advantage of differences in enzyme architecture between isoforms. Potent inhibitors are now available with a high degree of isoform selectivity (Table 8), and these will play a critical role in target validation and optimization of the target product profile for a given indication.

Compliance with Ethical Standards

Conflict of Interest: Author declares that he has no conflict of interest.

Ethical approval: Not applicable.

References

1. Philipps DMP (1963) The presence of acetyl groups in histones. Biochem J 87:258–263
2. Allfrey VG, Faulkner R, Mirsky AE (1964) Acetylation and methylation of histones and their possible role in the regulation of RNA synthesis. Proc Natl Acad Sci U S A 51:786–794
3. Gallwitz D, Sekeris CE (1969) The acetylation of histones of rat liver nuclei in vitro by acetyl-CoA. Z Physiol Chem 350:150–154
4. Inoue A, Fujimoto D (1969) Enzymatic deacetylation of histone. Biochem Biophys Res Commun 36:146–150
5. He M, Han Z, Liu L et al (2018) Chemical biology approaches for investigating the functions of lysine acetyltransferases. Angew Chem Int Ed 57:1162–1184
6. Harmel R, Fiedler D (2018) Features and regulation of non-enzymatic posttranslational modifications. Nat Chem Biol 14:244–252
7. Pérez-Salvia M, Esteller M (2017) Bromodomain inhibitors and cancer therapy: from structures to applications. Epigenetics 12:323–339
8. Lyon K, Stasevich TJ (2017) Imaging translational and post-translational gene regulatory dynamics in living cells with antibody-based probes. Trends Genet 33:322–335
9. Drazic A, Myklebust LM, Ree R et al (2016) The world of protein acetylation. Biochim Biophys Acta 1864:1372–1401
10. Jiang H, Zhang X, Chen X et al (2018) Protein lipidation: occurrence, mechanisms, biological functions, and enabling technologies. Chem Rev 118:919–988
11. Reid MA, Dai Z, Locasale JW (2017) The impact of cellular metabolism on chromatin dynamics and epigenetics. Nat Cell Biol 19:1298–1306
12. Yoshida M, Kudo N, Kosono S et al (2017) Chemical and structural biology of protein lysine deacetylases. Proc Jpn Acad Ser B 93:297–321
13. Seto E, Yoshida M (2014) Erasers of histone acetylation: the histone deacetylase enzymes. Cold Spring Harb Perspect Biol 6:a018713
14. Bertrand P (2010) Inside HDAC with HDAC inhibitors. Eur J Med Chem 45:2095–2116
15. Micelli C, Rastelli G (2015) Histone deacetylases: structural determinants of inhibitor selectivity. Drug Discov Today 20:718–735
16. Roche J, Bertrand P (2016) Inside HDACs with more selective HDAC inhibitors. Eur J Med Chem 121:451–483
17. Kutil Z, Novakova Z, Meleshin M et al (2018) Histone deacetylase 11 is a fatty-acid deacylase. ACS Chem Biol 13:685–693
18. Di Giorgio E, Brancolini C (2016) Regulation of class IIa HDAC activities: it is not only matter of subcellular localization. Epigenomics 8:251–269
19. Desravines DC, Serna Martin I, Schneider R et al (2017) Structural characterization of the SMRT corepressor interacting with histone deacetylase. Sci Rep 7:3678
20. Hai Y, Christianson DW (2016) Histone deacetylase 6 structure and molecular basis of catalysis and inhibition. Nat Chem Biol 12:741–747
21. Yanginlar C, Logie C (2018) HDAC11 is a regulator of diverse immune functions. Biochem Biophys Acta 1861:54–59
22. Zwergel C, Stazi G, Valente S et al (2016) Histone deacetylase inhibitors: updated studies in various epigenetic-related diseases. J Clin Epigenetics 2:1–7
23. Manal M, Chandrasekar MJN, Priya JG et al (2016) Inhibitors of histone deacetylase as antitumor agents: a critical review. Bioorg Chem 67:18–42
24. Ganesan A (2018) Epigenetic drug discovery: a success story for cofactor interference. Philos Trans R Soc B 373:20170069
25. Khan A, Singh P, Srivastava K (2018) Synthesis, nature and utility of universal iron chelator – siderophore: a review. Microbiol Res 212:103–111
26. Codd R (2008) Traversing the coordination chemistry and chemical biology of hydroxamic acids. Coord Chem Rev 252:1387–1408

27. Yoshida M, Kijima M, Akita M et al (1990) Potent and specific inhibition of mammalian histone deacetylase both in vivo and in vitro by trichostatin A. J Biol Chem 265:17174–17179
28. Richon VM, Emiliani S, Verdin E et al (1998) A class of hybrid polar inducers of transformed cell differentiation inhibits histone deacetylases. Proc Natl Acad Sci U S A 95:3003–3007
29. Marks PA, Breslow R (2007) Dimethyl sulfoxide to vorinostat: development of this histone deacetylase inhibitor as an anticancer drug. Nat Biotechnol 25:84–90
30. Kawai K, Nagata N (2012) Metal–ligand interactions: an analysis of zinc binding groups using the Protein Data Bank. Eur J Med Chem 51:271–276
31. Shen S, Kozikowski AP (2016) Why hydroxamates may not be the best histone deacetylase inhibitors – what some may have forgotten or would rather forget? ChemMedChem 11:15–21
32. Finn PW, Loza E, Carstensten E (2016) The discovery and development of belinostat. In: Fischer J, Childers WE (eds) Successful drug discovery, vol 2. Wiley-VCH, Weinheim, pp 31–57
33. Atadja P, Perez L (2016) Discovery and development of Farydak (NVP-LBH589, panobinostat) as an anticancer drug. In: Fischer J, Childers WE (eds) Successful drug discovery, vol 2. Wiley-VCH, Weinheim, pp 59–88
34. Maolanon AR, Madsen AS, Olsen CA (2016) Innovative strategies for selective inhibition of histone deacetylases. Cell Chem Biol 23:759–768
35. Novotny-Diermayr V, Sangthongpitag K, Hu CY et al (2010) SB939, a novel potent and orally active histone deacetylase inhibitor with high tumor exposure and efficacy in mouse models of colorectal cancer. Mol Cancer Ther 9:642–652
36. Mandl-Weber S, Meinel FG, Jankowsky R et al (2010) The novel inhibitor of histone deacetylase resminostat (RAS2410) inhibits proliferation and induces apoptosis in multiple myeloma (MM) cells. Br J Haematol 149:518–528
37. Leoni F, Fossati G, Lewis EC et al (2005) The histone deacetylase inhibitor ITF2357 reduces production of pro-inflammatory cytokines in vitro and systemic inflammation in vivo. Mol Med 11:1–15
38. Buggy JJ, Cao ZA, Bass KE et al (2006) CRA-024781: a novel synthetic inhibitor of histone deacetylase enzymes with antitumor activity in vitro and in vivo. Mol Cancer Ther 5:1309–1317
39. Lu Q, Wang DS, Chen CS et al (2005) Structure-based optimization of phenylbutyrate-derived histone deacetylase inhibitors. J Med Chem 48:5530–5535
40. Arts J, King P, Mariën A et al (2009) JNJ-26481585, a novel "second-generation" oral histone deacetylase inhibitor, shows broad-spectrum preclinical antitumoral activity. Clin Cancer Res 15:6841–6851
41. Moffat D, Patel S, Day F et al (2010) Discovery of 2-(6-{[(6-fluoroquinolin-2-yl)methyl] amino}bicyclo[3.1.0]hex-3-yl)-*N*-hydroxypyrimidine-5-carboxamide (CHR-3996), a class I selective orally active histone deacetylase inhibitor. J Med Chem 53:8663–8678
42. Santo L, Hideshima T, Kung AL et al (2012) Preclinical activity, pharmacodynamic, and pharmacokinetic properties of a selective HDAC6 inhibitor, ACY-1215, in combination with bortezomib in multiple myeloma. Blood 119:2579–2589
43. Huang P, Almeciga-Pinto I, Jarpe M et al (2017) Selective HDAC inhibition by ACY-241 enhances the activity of paclitaxel in solid tumor models. Oncotarget 8:2694–2707
44. Lecointre B, Narozny R, Borrello MT et al (2018) Isoform-selective HDAC1/6/8 inhibitors with an imidazo-ketopiperazine cap containing stereochemical diversity. Philos Trans R Soc B 373:20170364
45. Gaisina IN, Tueckmantel W, Ugolkov A et al (2016) Identification of HDAC6-selective inhibitors of low cancer cell cytotoxicity. ChemMedChem 11:81–92
46. Sellmer A, Stangl H, Beyer M et al (2018) Marbostat-100 defines a new class of potent and selective antiinflammatory and antirheumatic histone deacetylase 6 inhibitors. J Med Chem 61:3454–3477
47. Balasubramanian S, Ramos J, Luo W et al (2008) A novel histone deacetylase 8 (HDAC8)-specific inhibitor PCI-34051 induces apoptosis in T-cell lymphomas. Leukemia 22:1026–1034

48. Ingham OJ, Paranal RM, Smith WB et al (2016) Development of a potent and selective HDAC8 inhibitor. ACS Med Chem Lett 7:929–932
49. Luckhurst CA, Breccia P, Stott AJ et al (2015) Potent, selective, and CNS-penetrant tetrasubstituted cyclopropane class IIa histone deacetylase (HDAC) inhibitors. ACS Med Chem Lett 7:34–39
50. Ossenkoppele GJ, Lowenberg B, Zachee P et al (2013) A phase I first-in-human study with tefinostat – a monocyte/macrophage targeted histone deacetylase inhibitor – in patients with advanced haematological malignancies. Br J Haematol 162:191–201
51. Zheng S, Guo S, Zhong Q et al (2018) Biocompatible boron-containing prodrugs of belinostat for the potential treatment of solid tumors. ACS Med Chem Lett 9:149–154
52. Ganesan A (2016) Multitarget drugs: an epigenetic epiphany. ChemMedChem 11:1227–1241
53. de Lera AR, Ganesan A (2016) Epigenetic polypharmacology: from combination therapy to multitargeted drugs. Clin Epigenetics 8:105
54. Cai X, Zhai HX, Wang J et al (2010) Discovery of 7-(4-(3-ethynylphenylamino)-7-methoxyquinazolin-6-yloxy)-*N*-hydroxyheptanamide (CUDC-101) as a potent multi-acting HDAC, EGFR, and HER2 inhibitor for the treatment of cancer. J Med Chem 53:2000–2009
55. Qian C, Lai CJ, Bao R et al (2012) Cancer network disruption by a single molecule inhibitor targeting both histone deacetylase activity and phosphatidylinositol 3-kinase signaling. Clin Cancer Res 18:4104–4113
56. Shigematsu N, Ueda H, Takase S et al (1994) FR901228, a novel antitumor bicyclic depsipeptide produced by *Chromobacterium violaceum* No. 968. II. Structure determination. J Antibiot 47:311–314
57. Nakajima H, Kim YB, Terano H et al (1998) FR901228, a potent antitumor antibiotic, is a novel histone deacetylase inhibitor. Exp Cell Res 241:126–133
58. Furumai R, Matsuyama A, Kobashi N et al (2002) FK228 (depsipeptide) as a natural prodrug that inhibits class I histone deacetylases. Cancer Res 62:4916–4921
59. Taori K, Paul VJ, Luesch H (2008) Structure and activity of largazole, a potent antiproliferative agent from the Floridian marine cyanobacterium *Symploca* sp. J Am Chem Soc 130:1806–1807
60. Wang C, Henkes LM, Doughty LB et al (2011) Thailandepsins: bacterial products with potent histone deacetylase inhibitory activities and broad-spectrum antiproliferative activities. J Nat Prod 74:2031–2038
61. Ganesan A (2015) Macrocyclic inhibitors of zinc-dependent histone deacetylases (HDACs). In: Levin J (ed) Macrocycles in drug discovery. RSC, Cambridge, pp 109–140
62. Maolanon AR, Kristensen HM, Leman LJ et al (2017) Natural and synthetic macrocyclic inhibitors of the histone deacetylase enzymes. Chembiochem 18:5–49
63. Ganesan A (2016) Romidepsin and the zinc-binding thiol family of natural product HDAC inhibitors. In: Fischer J, Childers WE (eds) Successful drug discovery, vol 2. Wiley-VCH, Weinheim, pp 13–30
64. Stolze SC, Kaiser M (2013) Case studies of the synthesis of bioactive cyclodepsipeptide natural products. Molecules 18:1337–1367
65. Chen QY, Chaturvedi PR, Luesch H (2018) Process development and scale-up total synthesis of largazole, a potent class I histone deacetylase inhibitor. Org Process Res Dev 22:190–199
66. Poli G, di Fabio R, Ferrante L et al (2017) Largazole analogues as histone deacetylase inhibitors and anticancer agents: an overview of structure–activity relationships. ChemMedChem 12:1917–1926
67. Benelkebir H, Marie S, Hayden A et al (2011) Total synthesis of largazole and analogues: HDAC inhibition, antiproliferative activity and metabolic stability. Bioorg Med Chem 19:3650–3658
68. Yao Y, Tu Z, Liao C et al (2015) Discovery of novel class I histone deacetylase inhibitors with promising in vitro and in vivo antitumor activities. J Med Chem 58:7672–7680

69. Almaliti J, Al-Hamashi AA, Negmeldin AT et al (2016) Largazole analogues embodying radical changes in the depsipeptide ring: development of a more selective and highly potent analogue. J Med Chem 59:10642–10660
70. Kitir B, Maolanon AR, Ohm RG et al (2017) Chemical editing of macrocyclic natural products and kinetic profiling reveal slow, tight-binding histone deacetylase inhibitors with picomolar affinities. Biochemistry 56:5134–5146
71. Gu W, Nusinzon I, Smith RD Jr et al (2006) Carbonyl- and sulfur-containing analogs of suberoylanilide hydroxamic acid: potent inhibition of histone deacetylases. Bioorg Med Chem 14:3320–3329
72. Saito A, Yamashita T, Mariko Y et al (1999) A synthetic inhibitor of histone deacetylase, MS-27-275, with marked in vivo antitumor activity against human tumors. Proc Natl Acad Sci U S A 96:4592–4597
73. Fournel M, Bonfils C, Hou Y et al (2008) MGCD0103, a novel isotype-selective histone deacetylase inhibitor, has broad spectrum antitumor activity in vitro and in vivo. Mol Cancer Ther 7:759–768
74. Lu XP, Ning ZQ, Li ZB et al (2016) Discovery and development of HDAC subtype selective inhibitor chidamide: potential immunomodulatory activity against cancers. In: Fischer J, Childers WE (eds) Successful drug discovery, vol 2. Wiley-VCH, Weinheim, pp 89–114
75. Ning ZQ, Li ZB, Newman MJ et al (2012) Chidamide (CS055/HBI-8000): a new histone deacetylase inhibitor of the benzamide class with antitumor activity and the ability to enhance immune cell-mediated tumor cell cytotoxicity. Cancer Chemother Pharmacol 69:901–909
76. Witter DJ, Harrington P, Wilson KJ et al (2008) Optimization of biaryl selective HDAC1&2 inhibitors (SHI-1:2). Bioorg Med Chem Lett 18:726–731
77. McClure JJ, Inks ES, Zhang C et al (2017) Comparison of the deacylase and deacetylase activity of zinc-dependent HDACs. ACS Chem Biol 12:1644–1655
78. Candido EP, Reeves R, Davie JR (1978) Sodium butyrate inhibits histone deacetylation in cultured cells. Cell 14:105–113
79. Kim SW, Hooker JM, Otto N et al (2013) Whole-body pharmacokinetics of HDAC inhibitor drugs, butyric acid, valproic acid and 4-phenylbutyric acid measured with carbon-11 labeled analogs by PET. Nucl Med Biol 40:912–918
80. Wen S, Carey KL, Nakao Y et al (2007) Total synthesis of azumamide A and azumamide E, evaluation as histone deacetylase inhibitors, and design of a more potent analogue. Org Lett 9:1105–1108
81. Porter NJ, Christianson DW (2017) Binding of the microbial cyclic tetrapeptide trapoxin A to the class I histone deacetylase HDAC8. ACS Chem Biol 12:2281–2286
82. Olsen CA, Montero A, Leman LJ et al (2012) Macrocyclic peptoid–peptide hybrids as inhibitors of class I histone deacetylases. ACS Med Chem Lett 3:749–753
83. Islam MN, Islam MS, Hoque MA et al (2014) Bicyclic tetrapeptides as potent HDAC inhibitors: effect of aliphatic loop position and hydrophobicity on inhibitory activity. Bioorg Med Chem 22:3862–3870
84. Traoré MDM, Zwick V, Simões-Pires CA et al (2017) Hydroxyl ketone-based histone deacetylase inhibitors to gain insight into class I HDAC selectivity versus that of HDAC6. ACS Omega 2:1550–1562
85. Whitehead L, Dobler MR, Radetich B et al (2011) Human HDAC isoform selectivity achieved via exploitation of the acetate release channel with structurally unique small molecule inhibitors. Bioorg Med Chem 19:4626–4634
86. Lobera M, Madauss KP, Pohlhaus DT et al (2013) Selective class IIa histone deacetylase inhibition via a nonchelating zinc-binding group. Nat Chem Biol 9:319–325
87. Attenni B, Ontoria JM, Cruz JC et al (2009) Histone deacetylase inhibitors with a primary amide zinc binding group display antitumor activity in xenograft model. Bioorg Med Chem Lett 19:3081–3084

88. Li X, Peterson YK, Inks ES et al (2018) Class I HDAC inhibitors display different antitumor mechanism in leukemia and prostatic cancer cells depending on their p53 status. J Med Chem 61:2589–2603
89. Ononye SN, VanHeyst MD, Oblak EZ et al (2013) Tropolones as lead-like natural products: the development of potent and selective histone deacetylase inhibitors. ACS Med Chem Lett 4:757–761
90. Kleinschek A, Meyners C, Digiorgio E et al (2016) Potent and selective non-hydroxamate histone deacetylase 8 inhibitors. ChemMedChem 11:2598–2606
91. Ontoria JM, Altamura S, Di Marco A et al (2009) Identification of novel, selective, and stable inhibitors of class II histone deacetylases. Validation studies of the inhibition of the enzymatic activity of HDAC4 by small molecules as a novel approach for cancer therapy. J Med Chem 52:6782–6789
92. Boskovic ZV, Kemp MM, Freedy AM et al (2016) Inhibition of zinc-dependent histone deacetylases with a chemically triggered electrophile. ACS Chem Biol 11:1844–1851
93. Millard CJ, Watson PJ, Fairall L et al (2017) Targeting class I histone deacetylases in a "complex" environment. Trends Pharmacol Sci 38:363–377
94. Villagra A, Sahakian E, Seto E (2016) Preparation and biochemical analysis of classical histone deacetylases. Methods Enzymol 573:161–181
95. Newbold A, Falkenberg KJ, Prince HM et al (2016) How do tumor cells respond to HDAC inhibition? FEBS J 283:4032–4046
96. Bolden JE, Shi W, Jankowski K et al (2013) HDAC inhibitors induce tumor-cell-selective pro-apoptotic transcriptional responses. Cell Death Dis 4:e519
97. Tiffon C, Adams J, van der Fits L et al (2011) The histone deacetylase inhibitors vorinostat and romidepsin downmodulate IL-10 expression in cutaneous T-cell lymphoma cells. Br J Pharmacol 162:1590–1602
98. Das Gupta K, Shakespear MR, Iyer A et al (2016) Histone deacetylases in monocyte/macrophage development, activation and metabolism: refining HDAC targets for inflammatory and infectious diseases. Clin Trans Immunol 5:e62
99. Zhang Q, Dai Y, Cai Z et al (2018) HDAC inhibitors: novel immunosuppressants for Allo- and Xeno- transplantation. Chem Select 3:176–187
100. Thorlund K, Horwitz MS, Fife BT et al (2017) Landscape review of current HIV 'kick and kill' cure research – some kicking, not enough killing. BMC Infect Dis 17:595
101. Ghosh SK, Perrine SP, Williams RM et al (2012) Histone deacetylase inhibitors are potent inducers of gene expression in latent EBV and sensitize lymphoma cells to nucleoside antiviral agents. Blood 119:1008–1017
102. Warnault V, Darcq E, Levine A et al (2013) Chromatin remodeling – a novel strategy to control excessive alcohol drinking. Transl Psychiatry 3:e231
103. Qin L, Ma K, Wang ZJ et al (2018) Social deficits in *Shank3*-deficient mouse models of autism are rescued by histone deacetylase (HDAC) inhibition. Nat Neurosci 21:564–575
104. Ooi JY, Tuano NK, Rafehi H et al (2015) HDAC inhibition attenuates cardiac hypertrophy by acetylation and deacetylation of target genes. Epigenetics 10:418–430
105. Akimova T, Beier UH, Liu Y et al (2012) Histone/protein deacetylases and T-cell immune responses. Blood 119:2443–2451
106. Tsuji G, Okiyama N, Villarroel VA et al (2015) Histone deacetylase 6 inhibition impairs effector CD8 T-cell functions during skin inflammation. J Allergy Clin Immunol 135:1228–1239
107. Bowers ME, Xia B, Carreiro S et al (2015) The class I HDAC inhibitor RGFP963 enhances consolidation of cued fear extinction. Learn Mem 22:225–231
108. Bodas M, Mazur S, Min T et al (2018) Inhibition of histone-deacetylase activity rescues inflammatory cystic fibrosis lung disease by modulating innate and adaptive immune responses. Respir Res 19:2

109. Dirice E, Ng RWS, Martinez R et al (2017) Isoform-selective inhibitor of histone deacetylase 3 (HDAC3) limits pancreatic islet infiltration and protects female nonobese diabetic mice from diabetes. J Biol Chem 292:17598–17608
110. Jia H, Wang Y, Morris CD et al (2016) The effects of pharmacological inhibition of histone deacetylase 3 (HDAC3) in Huntington's disease mice. PLoS One 11:e0152498
111. Usui T, Okada M, Mizuno W et al (2012) HDAC4 mediates development of hypertension via vascular inflammation in spontaneous hypertensive rats. Am J Physiol Heart Circ Physiol 302: H1894–H1904
112. Choi SY, Ryu Y, Kee HJ et al (2015) Tubastatin A suppresses renal fibrosis via regulation of epigenetic histone modification and Smad3-dependent fibrotic genes. Vascul Pharmacol 72:130–140
113. Ferrari A, Fiorino E, Longo R et al (2017) Attenuation of diet-induced obesity and induction of white fat browning with a chemical inhibitor of histone deacetylases. Int J Obes (Lond) 41:289–298
114. Choong CJ, Sasaki T, Hayakawa H et al (2016) A novel histone deacetylase 1 and 2 isoform-specific inhibitor alleviates experimental Parkinson's disease. Neurobiol Aging 37:103–116
115. Conforti F, Davies ER, Calderwood CJ et al (2017) The histone deacetylase inhibitor, romidepsin, as a potential treatment for pulmonary fibrosis. Oncotarget 8:48737–48754
116. Trifunović D, Arango-Gonzalez B, Comitato A et al (2016) HDAC inhibition in the *cpfl1* mouse protects degenerating cone photoreceptors in vivo. Hum Mol Genet 25:4462–4472
117. Li Y, Zhao T, Liu B et al (2015) Inhibition of histone deacetylase 6 improves long-term survival in a lethal septic model. J Trauma Acute Care Surg 78:378–385
118. Pfaller MA, Messer SA, Georgopapadakou N et al (2009) Activity of MGCD290, a Hos2 histone deacetylase inhibitor, in combination with azole antifungals against opportunistic fungal pathogens. J Clin Microbiol 47:3797–3804
119. Kuchler K, Jenull S, Shivarathri R et al (2016) Fungal KATs/KDACs: a new highway to better antifungal drugs? PLoS Pathog 12:e1005938
120. Hailu GS, Robaa D, Forgione M et al (2017) Lysine deacetylase inhibitors in parasites: past, present, and future perspectives. J Med Chem 60:4780–4804
121. Sumanadasa SD, Goodman CD, Lucke AJ et al (2012) Antimalarial activity of the anticancer histone deacetylase inhibitor SB939. Antimicrob Agents Chemother 56:3849–3856
122. Chua MJ, Arnold MS, Xu W et al (2017) Effect of clinically approved HDAC inhibitors on *Plasmodium*, *Leishmania* and *Schistosoma* parasite growth. Int J Parasitol Drugs Drug Resist 7:42–50
123. Ontoria JM, Paonessa G, Ponzi S et al (2016) Discovery of a selective series of inhibitors of *Plasmodium falciparum* HDACs. ACS Med Chem Lett 7:454–459
124. Heimburg T, Chakrabarti A, Lancelot J et al (2016) Structure-based design and synthesis of novel inhibitors targeting HDAC8 from *Schistosoma mansoni* for the treatment of schistosomiasis. J Med Chem 59:2423–2435
125. Meyners C, Wolff B, Kleinschek A et al (2017) Perfluorinated hydroxamic acids are potent and selective inhibitors of HDAC-like enzymes from *Pseudomonas aeruginosa*. Bioorg Med Chem Lett 27:1508–1512

Top Med Chem (2020) 33: 29–54
DOI: 10.1007/7355_2019_69

Published online: 28 November 2019

Hydroxamic Acid-Containing Peptides in the Study of Histone Deacetylases

Carlos Moreno-Yruela and Christian A. Olsen

Contents

1 Introduction 30
2 Hydroxamic Acid: A Transition State Mimic 33
3 Synthesis of Hydroxamic Acid-Containing Peptides 35
4 Structure-Activity Relationship of Macrocyclic HDAC Inhibitors 36
4.1 Biological Activity of Selected Inhibitors 43
5 Chemical Tools for the Study of HDAC Structure and Function 44
6 Conclusions and Future Perspectives 47
References 49

Abstract Histone deacetylases (HDACs) are ubiquitous enzymes that remove ε-*N*-acetyl-lysine posttranslational modifications (PTMs) on histone tails. The resulting PTM landscape affects chromatin packing and recruitment of transcription factors, in turn playing an indirect role in regulation of gene expression. Deregulation of the activity of these hydrolases has been associated with several complex diseases. Thus, HDAC inhibitors have been approved for cancer treatment and are being studied against inflammation, neurodegeneration, and autoimmune disorders among others. The role of each of the 11 Zn^{2+}-dependent HDACs has not yet been elucidated, mainly due to their structural similarity and, in part, due to the absence of isotype-selective probes. Such selectivity may be achievable by targeting features outside of the active site pocket, which is highly conserved. Peptides, which may cover larger areas than small molecules, may become useful chemical tools able to reach unexplored areas of the protein surface to achieve selectivity. In addition, by incorporating hydroxamic acid-containing lysine mimics in their structure, strong binding to the catalytic cavity is achieved. Furthermore, such molecules present similarities to the native substrates, which could be exploited for determining targets of their deacetylase activity. Therefore, hydroxamic acid-containing peptides have

C. Moreno-Yruela and C. A. Olsen (✉)
Department of Drug Design and Pharmacology, Faculty of Health and Medical Sciences, University of Copenhagen, Copenhagen, Denmark
e-mail: cao@sund.ku.dk

potential for investigating HDAC function. Several examples of the application of these chemotypes are discussed in this book chapter.

Keywords Cyclic peptide, Epigenetics, HDAC, Histone deacetylases, Hydroxamic acid, Peptide probe

Abbreviations

Api	Apicidin
Asu	L-α-Aminosuberic acid
Asuha	L-α-Aminosuberic hydroxamic acid
Azu	Azumamide
CHAP	Cyclic hydroxamic acid-containing peptide
Chlam	Chlamydocin
HCtx	HC-toxin
HDAC	Histone deacetylase
Kac	ε-*N*-Acetyl-lysine
KDAC	Lysine deacylase
NMR	Nuclear magnetic resonance
PTM	Posttranslational modification
SAHA	Suberoylanilide hydroxamic acid
SAR	Structure-activity relationship
SPPS	Solid-phase peptide synthesis
Tpx	Trapoxin
TSA	Trichostatin A

1 Introduction

The first mammalian histone deacetylase (HDAC) was isolated and cloned in 1996 [1], identifying one of the key enzymes in Vincent Allfrey's hypothesis that RNA synthesis is regulated by reversible posttranslational histone modification [2, 3]. Twenty years later, the superfamily of HDACs has been extensively characterized and counts 18 proteins: 11 zinc-dependent enzymes divided into classes I, II, and IV by sequence similarity [4] and 7 structurally distinct NAD^+-dependent enzymes known as the sirtuins [5].

The conventional enzymatic activity of these biomolecules is the removal of the acetyl posttranslational modification (PTM) from ε-*N*-acetyllysine (Kac) residues. Although the entire protein superfamily is referred to as "histone deacetylases," mainly class I members catalyze Kac hydrolysis at the tails of histone proteins [6]. Thus, class I HDACs are present in the nucleus of human cells, where their substrates are modified histones involved in DNA packing by the formation of nucleosome supramolecular assemblies [6–9]. In addition, HDAC1–HDAC3 are known to interact with nuclear proteins and to be recruited to multiprotein complexes involved in chromatin remodeling, cell cycle progression, as well as DNA replication and transcription. HDAC1 and HDAC2 are associated with several monomeric and homo- and heteropolymeric complexes, such as CoREST (co-repressor of RE1-silencing transcription factor), MiDAC (mitotic deacetylase complex), NuRD (nucleosome remodeling and deacetylase), and SIN3 (switch-independent 3); and HDAC3 is recruited to the SMRT/NCoR (nuclear receptor co-repressor 2) complex. These protein partners appear to be essential for the deacetylase enzymatic activity, as well as for directing HDAC function through substrate recognition [8]. HDAC8 is also part of the class I, but it is not involved in such protein complexes and thus thought to prefer biological targets other than histones [10]. Furthermore, HDAC8 is not able to cleave ε-*N*-crotonyllysine (Kcr) PTMs as the other three class I human HDACs, which reflects its functional distinction from HDAC1–HDAC3 [11, 12].

Similar to class I HDACs, members of class IIa are also localized in the nucleus, and they have been found to interact with some of the aforementioned protein partners involved in epigenetic regulation. However, their biological mechanism is still under discussion, since they lack a key tyrosine residue in the active site, which appears important for the deacetylase activity and is characteristic of class I HDACs [13]. Recombinant class IIa HDACs show very poor deacetylase activity in vitro, whereas they do cleave other non-biologically relevant substrates [13, 14]. HDAC11, the sole human enzyme member of class IV HDACs, was also controversial due to poor in vitro deacetylase activity. Interestingly, recent studies indicate that this isotype is responsible for the removal of ε-*N*-myristoyllysine (Kmyr) PTMs [15–17], which was an activity only associated with the sirtuins previously. Finally, HDAC6 and HDAC10, members of class IIb, are mainly present in the cytosol and therefore do not directly target chromatin. HDAC6 acts on α-tubulin among other acetylated proteins in human cells, whereas HDAC10 has recently been proposed to be a polyamine deacetylase [6, 18].

Initially, scientific interest in HDACs was raised by the observation that some compounds that induce differentiation and inhibit proliferation of cancer cells also resulted in the accumulation of hyperacetylated histones. This was the case for *n*-butyrate [19], for dimethyl sulfoxide (DMSO) [20], and for the natural products trichostatin A (TSA, Fig. 1 compound **1.1**) [21] and trapoxin A (TpxA) [22]. Parallel to these findings, Ronald Breslow, Paul A. Marks, and coworkers developed a potent inducer of murine erythroleukemia cell differentiation, suberoylanilide hydroxamic acid (SAHA, Fig. 1 compound **1.2**) [23]. All these compounds were discovered to inhibit the deacetylation of histones by targeting HDACs, and SAHA became a powerful probe for the study of their function, as well as the first of a handful of HDAC inhibitors approved for cancer treatment [23]. Moreover, targeting HDACs has not only found relevance against several types of cancer but also against neurodegenerative and inflammatory diseases [9, 24].

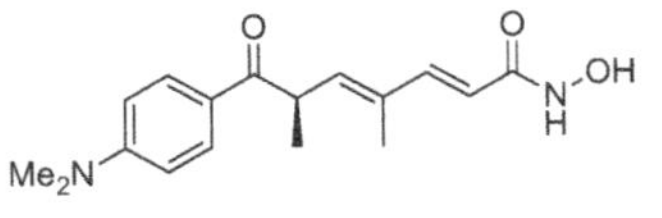

1.1, Trichostatin A (TSA)

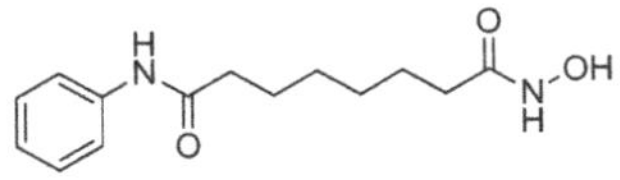

1.2, Vorinostat
Suberoylanilide hydroxamic acid (SAHA)

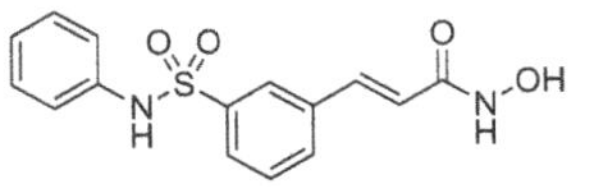

1.3, Belinostat

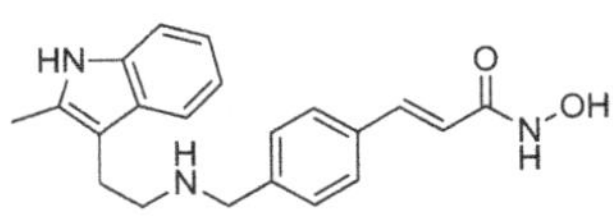

1.4, Panobinostat

Fig. 1 Chemical structures of a naturally occurring HDAC inhibitor, trichostatin A (**1.1**), and three synthetic HDAC inhibitors approved for cancer treatment by the FDA: vorinostat (**1.2**), belinostat (**1.3**), and panobinostat (**1.4**)

Early in the development of HDAC inhibitors, a common pharmacophore was proposed consisting of three elements: a binding region, which provides isotype selectivity (also referred to as "capping group"), a spacer, and an enzyme-inhibiting group [25]. The "capping group" of the pharmacophore interacts with the rim of the active site, which is the same area of the enzyme that binds to the peptide backbone of the protein substrate and, therefore, directs the enzymatic activity toward the right target. Not surprisingly, this region exhibits structural variability across HDAC isotypes, which may be exploited for the design of isotype-selective inhibitors [26]. Since small molecules interact mainly with the active site pocket and only a small portion of the rim, isotype selectivity has proven elusive for this chemotype. Thus, peptides have arisen as chemical tools able to cover a larger area of the protein surface and to establish interactions similar to those of the substrate. In the case of the enzyme-inhibiting group, it is now known that it drives coordination of the inhibitor to the catalytic zinc (Zn^{2+}) ion and, therefore, it has been renamed as Zn^{2+}-binding group. Numerous different chemical moieties have been studied for this purpose, but there is one that stands out: the hydroxamic acid, present in the natural product **1.1** and in three out of five HDAC inhibitors approved for cancer treatment including compound **1.2** (Fig. 1) [24].

Potent inhibitors have been created as a combination of a hydroxamic acid Zn^{2+}-binding group, which provides strong binding to the catalytic site, and a cyclic peptide capping group that adds both potency and selectivity via specific interactions with the enzyme surface. Also, beyond inhibitors, the fact that HDACs are able to bind Kac and aliphatic hydroxamic acids likewise offers the possibility of studying these enzymes with peptides that mimic the interaction with the substrate. Here, we present a summary of different tools developed for studying and targeting HDACs, which combine the hydroxamic acid functionality with peptide scaffolds.

2 Hydroxamic Acid: A Transition State Mimic

Zn^{2+}-dependent HDACs catalyze deacetylation through the following mechanism: first, Kac is accommodated in the catalytic pocket favoring coordination of the PTM to the Zn^{2+} ion, then, a water molecule present in the active site performs a nucleophilic attack on the carbonyl group, and, finally, acetate is released [27]. The reaction is possible because coordination to Zn^{2+} enhances electrophilicity of the acetamido group, which favors the nucleophilic attack, and it also stabilizes the hydrated transition state of the reaction via chelation to the metal ion. In addition, the interactions between the PTM and Zn^{2+} ensure the right conformation for the reaction with the water molecule. All this has been proposed with the support of kinetic studies and several crystal structures of inactive HDAC8 mutants (Fig. 2a, b) [27, 28]. Interestingly, the hydroxamic acid group of HDAC inhibitors binds to the active site in analogy to the transition state of the reaction, forming two oxygen-metal bonds with Zn^{2+} and, presumably, interacting with the water molecule (as shown in a crystal structure of HDAC2 with compound **1.2**, PDB code: 4LXZ). Hydrolysis, though, is not favorable for this moiety due to the electronic characteristics of the carbonyl group. Moreover, the nitrogen and two oxygen atoms, and their spatial distribution, allow for additional interactions with the catalytic pocket that further stabilize the interaction (Fig. 2d) [29, 30].

In general, aliphatic hydroxamic acids are neutral at physiological pH, since their pK_a is 9.4 in aqueous solution. However, it has been argued whether coordination to Zn^{2+} in HDACs can lead to deprotonation due to a decrease in pK_a influenced by the biological environment. Likewise, a monodentate or bidentate character of the interaction between this functional group and the metal ion has been debated. Inhibitors with a bulky warhead such as phenylhydroxamic acids have been shown to prefer a monodentate hydroxamate-Zn^{2+} interaction in the active site of zebrafish HDAC6 [31]. On the other hand, quantitative structure-activity relationship (QSAR) studies, together with X-ray crystal structures, have demonstrated that the protonated and more stable form of the hydroxamic acid generally coordinates in a bidentate fashion to the catalytic Zn^{2+} in class I HDACs [32–39]. The chelating effect, together with multiple hydrogen bonds established with side chains and, sometimes, the water molecule in the active site, explains why this moiety contributes so significantly to the binding affinity of HDAC inhibitors, especially when compared to other Zn^{2+}-binding groups.

Many of its properties support hydroxamic acid as the ideal warhead for the design of HDAC inhibitors, but its versatility has been discussed in terms of two major disadvantages: promiscuity and mutagenicity. The fact that inhibitors bearing strong chelating groups are able to target most metalloenzymes is widespread, and it has been argued as an explanation for the amount of side effects associated with them [40]. HDAC inhibitors such as SAHA frequently present such promiscuity toward several HDAC isotypes in vitro [24]. On the other hand, a number of in vitro studies have supported the inability of SAHA to inhibit other metalloenzymes than HDACs [41, 42], as well as to alter the concentration of free Zn^{2+} in living cells

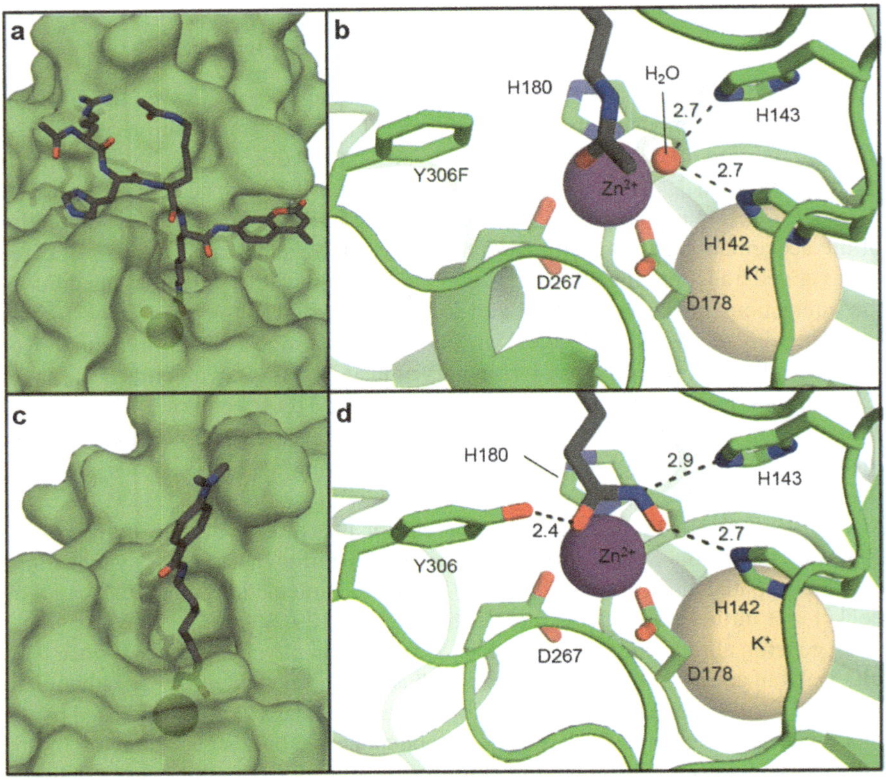

Fig. 2 Crystal structure of (**a**) Kac-containing peptide interacting with the catalytic pocket of an inactive HDAC8 mutant (Y306F), where the acetamido group is coordinated to the Zn^{2+} atom (surface representation) and (**b**) key residues and their interactions with a water molecule in the catalytic site (PDB code: 5D1C) [28]. Crystal structure of (**c**) hydroxamic acid-containing inhibitor 3-(1-methyl-4-phenylacetyl-1*H*-2-pyrrolyl)-*N*-hydroxy-2-propenamide (APHA) in complex with the active site of HDAC8 (surface representation) and (**d**) key interactions with residues in the catalytic pocket (PDB code: 3EW8) [29]. Potassium (K^+) satisfies a structural function and is not involved in the enzymatic activity. Distances are expressed in angstroms (Å)

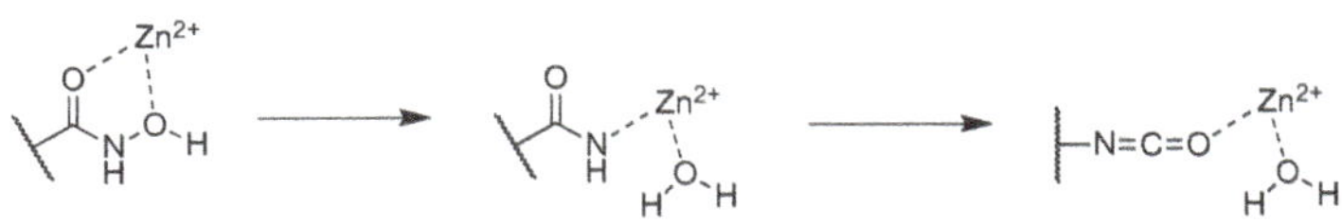

Fig. 3 Proposed Zn^{2+}-assisted Lossen rearrangement of hydroxamic acids [44]

[43]. In terms of mutagenicity, it is known that hydroxamic acids undergo Lossen rearrangement in the presence of metal ions, although no in vivo interrogation has been reported to date (Fig. 3) [44].

3 Synthesis of Hydroxamic Acid-Containing Peptides

Cyclic peptides have gained interest as a chemotype for HDAC inhibition. Two strategies have been followed in order to introduce a hydroxamic acid-containing side chain in the structure for improving inhibitory properties (Fig. 4). The more widespread route consists of the synthesis of a linear peptide, either in solution or by solid-phase peptide synthesis (SPPS), containing protected L-α-aminosuberic acid (Asu). Then, appropriate protecting group manipulation allows for selective head-to-tail cyclization of the linear peptide, followed by deprotection of the carboxylic acid side chain and functionalization with hydroxylamine (Fig. 4a) [45–49]. On the other hand, a fully solid supported synthesis is also possible, either by displacement from oxime resin using hydroxylamine as nucleophile [50] or with various hydroxylamine-functionalized resins which generate a hydroxamic acid moiety upon acidic cleavage [51–59]. For example, hydroxylamine-functionalized 2-chlorotrityl resin was employed by the group of M. Reza Ghadiri for the synthesis of macrocyclic HDAC inhibitors (Fig. 4b) [60, 61].

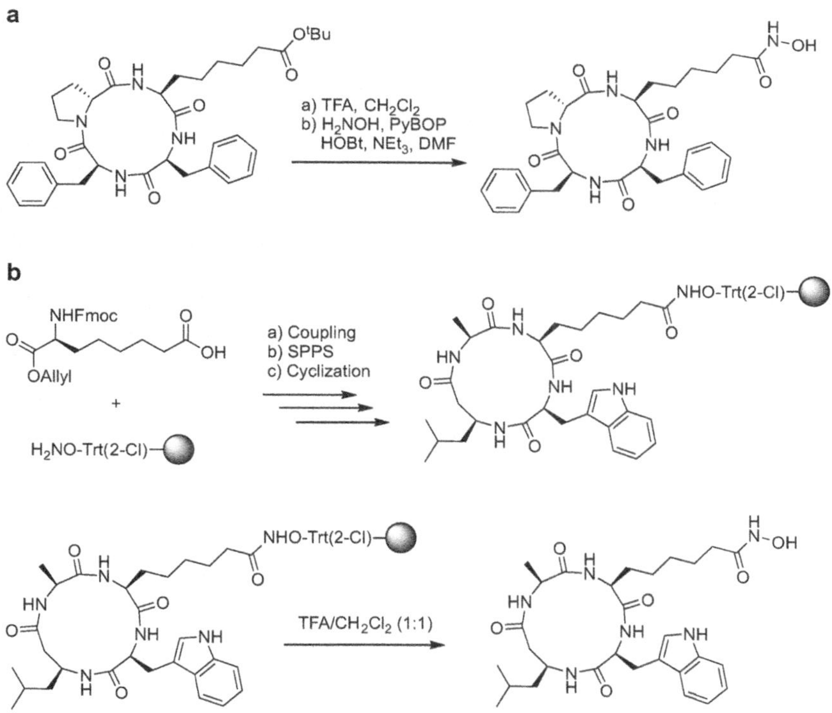

Fig. 4 Synthesis of hydroxamic acid-containing cyclic peptides by functionalization of L-α-aminosuberic acid (Asu) (**a**) in solution or (**b**) on solid support [49, 60]

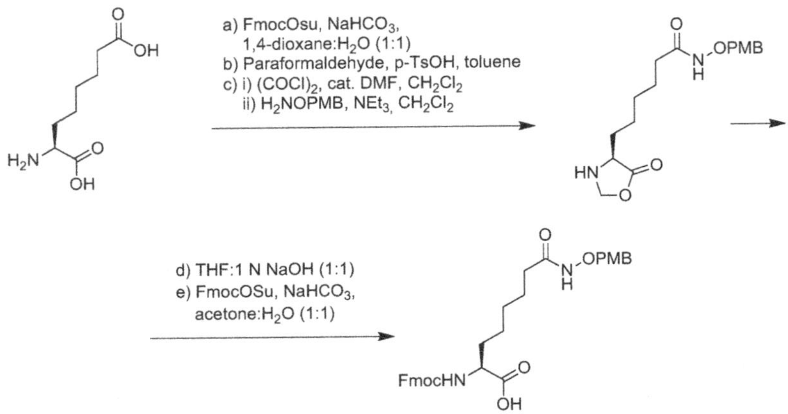

Fig. 5 Synthesis of protected L-α-aminosuberic hydroxamic acid (Asuha) building block for SPPS [63]

In the case of linear peptides, Schwabe and coworkers also utilized such a resin for immobilization of protected Asu through its side chain carboxylate. Then, orthogonal deprotection of the C- and N-terminal groups of the building block allowed for bidirectional elongation of the peptide, followed by acidic cleavage to yield the hydroxamic acid-containing peptides [62]. Even though this is a successful strategy, efficient and versatile synthesis of long peptides benefits from availability of a protected L-α-aminosuberic hydroxamic acid (Asuha) building block compatible with standard SPPS. Such a building block has been prepared by a number of groups using Asu as starting material and already presented as potential tool for the investigation of HDACs using substrate mimics [63]. Standard Fmoc SPPS was used for the synthesis of a histone 2B fragment analogue containing a surrogate of lysine 12 (H2BK12Asuha), and the hydroxamic acid group of Asuha was protected with a 4-methoxybenzyl (PMB) group during SPPS (Fig. 5). A similar synthetic strategy was followed by Schwarzer and coworkers employing *tert*-butyl as protecting group, although its deprotection required longer acidic treatment than common groups used for SPPS [64].

4 Structure-Activity Relationship of Macrocyclic HDAC Inhibitors

Trapoxin A (TpxA, Fig. 6 compound **6.18**) was the first of a series of naturally occurring cyclic peptides described to inhibit histone deacetylation by targeting HDACs [22]. The structural characteristics of these compounds are interesting for the design of HDAC inhibitors, since their macrocyclic scaffold acts as the "capping group" of the pharmacophore model and is able to interact with a large surface area

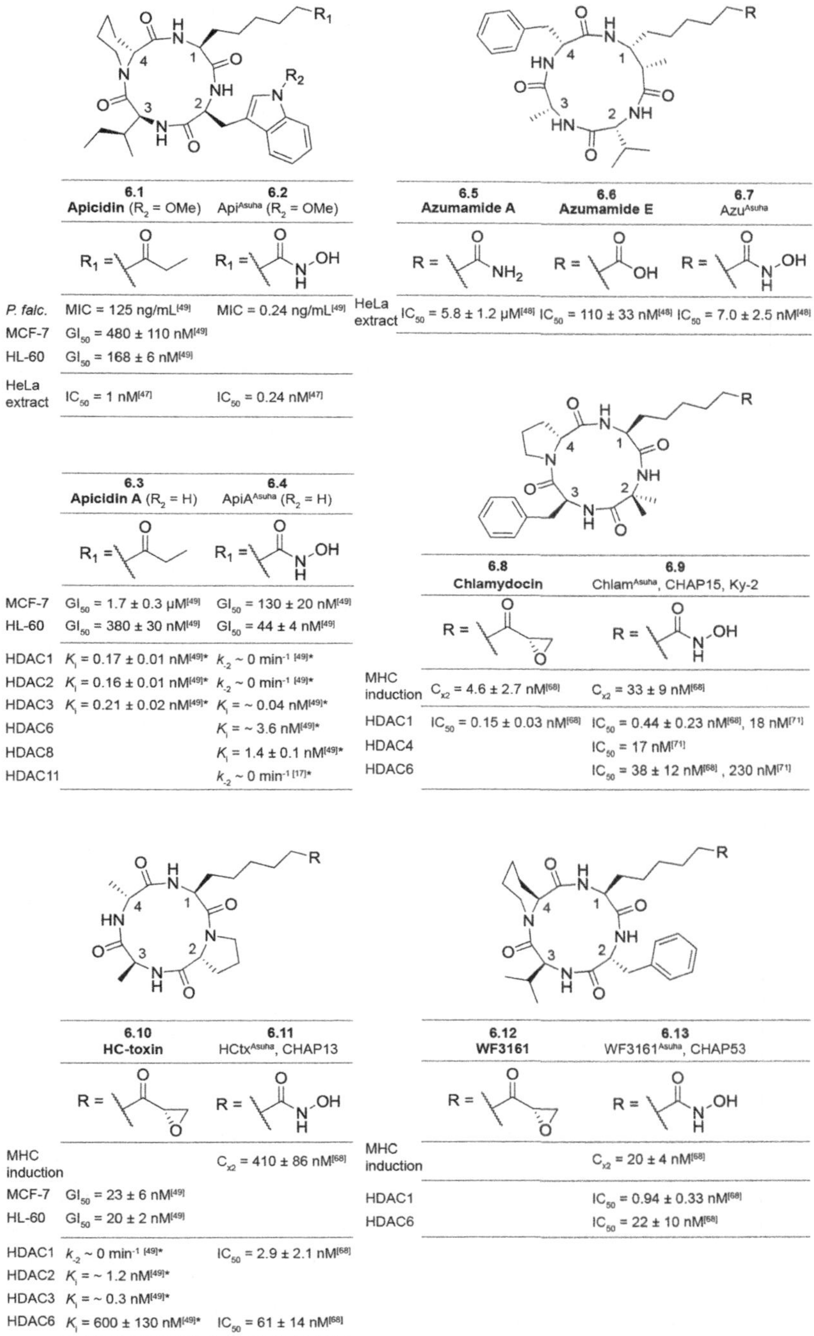

Fig. 6 Naturally occurring cyclic peptides reported as HDAC inhibitors and their hydroxamic acid-containing derivatives, with potencies measured in cell-based assays (*P. falc.* whole red blood cells

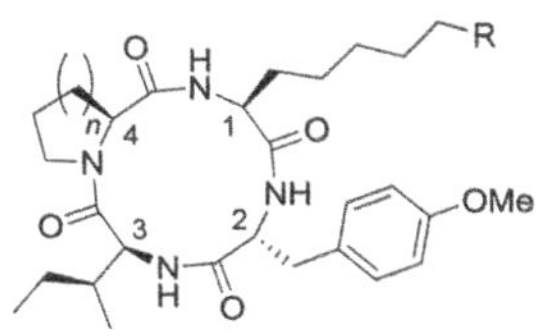

	6.14 Cyl-1 (n = 1)	**6.15** Cyl-1^{Asuha}, CHAP30 (n = 1)
	R = 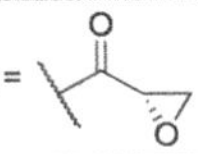	R = 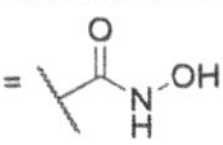
MHC induction		C_{x2} = 17 ± 8 nM[68]
HDAC1		IC_{50} = 4.4 ± 1.8 nM[68]
HDAC6		IC_{50} = 110 ± 84 nM[68]

	6.16 Cyl-2 (n = 2)	**6.17** Cyl-2^{Asuha}, CHAP49 (n = 2)
	R = 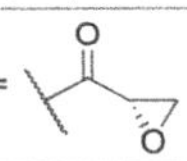	R = 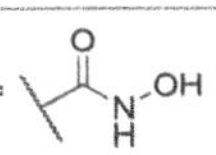
MHC induction	C_{x2} = 10 ± 7 nM[68]	C_{x2} = 5.3 ± 2.2 nM[68]
HDAC1	IC_{50} = 0.70 ± 0.45 nM[68]	IC_{50} = 1.2 ± 0.7 nM[68]
HDAC6	IC_{50} = 40 ± 11 μM[68]	IC_{50} = 36 ± 17 nM[68]

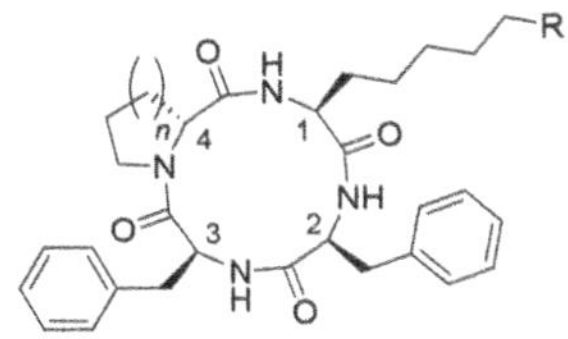

	6.18 Trapoxin A (n = 2)	**6.19** TpxA^{Asuha}, CHAP56 (n = 2)
	R = 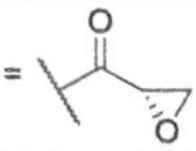	R = 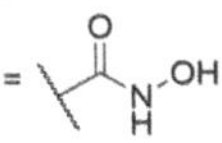
MHC induction	C_{x2} = 3.5 ± 0.6 nM[68]	C_{x2} = 30 ± 5 nM[68]
MCF-7	GI_{50} = 9.5 ± 0.1 nM[49]	
HL-60	GI_{50} = 1.9 ± 0.7 nM[49]	
HDAC1	IC_{50} = 0.82 ± 0.29 nM[68]	IC_{50} = 6.1 ± 1.4 nM[68]
HDAC3	k_{-2} ~ 0 min^{-1} [49]*	
HDAC6	IC_{50} = 524 ± 240 nM[68]	IC_{50} = 150 ± 84 nM[68]
HDAC11	K_i = 24 ± 2 nM[17]	

	6.20 Trapoxin B (n = 1)	**6.21** TpxB^{Asuha}, CHAP1 (n = 1)
	R = 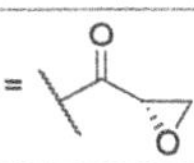	R = 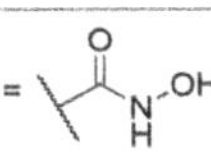
MHC induction	C_{x2} = 1.2 ± 0.9 nM[68]	C_{x2} = 98 ± 23 nM[68]
MCF-7		GI_{50} = 180 ± 30 nM[49]
HL-60		GI_{50} = 170 ± 20 nM[49]
HDAC1	IC_{50} = 0.11 ± 0.01 nM[68]	k_{-2} ~ 0 min^{-1} [49]*
HDAC2		k_{-2} ~ 0 min^{-1} [49]*
HDAC3		K_i = ~ 0.02 nM[49]*
HDAC4	IC_{50} = 0.30 ± 0.03 nM[68]	IC_{50} = 2.7 ± 1.3 nM[68]
HDAC6	IC_{50} = 360 ± 160 nM[68]	K_i = ~ 1.2 nM[49]*
HDAC8		K_i = 3.5 ± 0.9 nM[49]*
HDAC11		K_i = 24 ± 1 nM[17]*

Fig. 6 (continued) infected with *Plasmodium falciparum*, *MCF-7* human breast cancer, *HL-60* human promyelocytic leukemia, *MHC* class I major histocompatibility complex molecules, induced in B16/BL6 cells) and in vitro HDAC inhibition assays. *Data obtained in continuous assays. K_i values could not be determined for compounds exhibiting k_{-2} rates approaching 0

around the catalytic pocket. Potentially, this can be used for modulating isotype selectivity. In addition, cyclic peptides have attracted interest as drug candidates due to their oral bioavailability and stability in biological environments [65, 66]. Taking advantage of the latest techniques in peptide synthesis and the functional group versatility provided by canonical and noncanonical amino acids, efforts have been pursued toward the development of potent macrocyclic HDAC inhibitors exhibiting different selectivity profiles [67].

Early work carried out by the group of Yoshida was predicated on combining the structures of several naturally occurring cyclic tetrapeptides with that of TSA (**1.1**)

[68]. In particular, incorporation of the hydroxamic acid Zn^{2+}-binding group onto the macrocyclic structure of trapoxin B (TpxB, **6.20**) yielded a compound with cross-class activity and enhanced half-life in blood. This compound was named cyclic hydroxamic acid-containing peptide (CHAP) 1 (**6.21**). The optimal separation between the macrocycle and the chelating group was found to be 5 carbon atoms, which was further applied for the design of subsequent inhibitors inspired by compound **6.18**, chlamydocin (**6.8**), cyl-1 (**6.14**) and cyl-2 (**6.16**), WF3161 (**6.12**), and HC-toxin (**6.10**) [68, 69]. The same strategy has been followed by others, to the extent that several naturally occurring macrocyclic HDAC inhibitors have been modified with the hydroxamic acid moiety, with various effects on in vitro activity and selectivity (Fig. 6). Modified apicidin (**6.1**) had already been published by Merck Research Laboratories few months before Yoshida's study, showing slight improvement in in vitro anti-deacetylase activity for the derivative with respect to the natural compound [47]. Later, in 2007, Ganesan and coworkers synthesized both the natural and hydroxamic acid-containing version of azumamide E (**6.6**), which also exhibited the same trends in potency [48].

In 1993, when the natural product **6.18** was identified as an HDAC inhibitor, in vitro experiments suggested "slow-binding" kinetics, with decreasing, nonlinear rates measured over the course of incubation with the inhibitor, only reaching equilibrium after 1 h incubation with the enzyme. In addition, it was argued that the epoxide moiety acts as a covalent handle, since HDAC activity was not recovered by dialysis [22]. This experiment was also performed with analogue **6.21**, and restoration of the deacetylase activity indicated that the hydroxamic acid group interacts in a non-covalent manner with HDACs as expected [68, 69]. However, it was not investigated whether the binding kinetics of trapoxin-inspired CHAPs also followed slow-binding profiles. Recent results from Olsen and coworkers demonstrate that hydroxamic acid-containing macrocycles may in fact be able to exhibit in vitro slow, tight-binding behavior. In particular, compounds **6.4** and **6.21** displayed delayed equilibria in the inhibition of HDAC isotypes 1–3 and 6 [49]. It is possible to calculate the equilibrium constants (K_i) from kinetic parameters. For example, estimated K_i values for HDAC3 inhibition were reported to be 40 pM and 20 pM for macrocycles **6.4** and **6.21**, respectively. In the cases of HDAC1 and HDAC2 in which the complex dissociation rate is close to 0 (tight binding), estimation of K_i was not possible however. HDAC8 was also potently inhibited by both hydroxamic acid-containing macrocycles, but, in this case, the mechanism of inhibition was observed to be fast-on/fast-off (**6.4**: $K_i = 3.5 \pm 0.9$ nM, **6.21**: $K_i = 1.4 \pm 0.1$ nM) [49]. Moreover, the recently reported demyristoylase activity of HDAC11 was inhibited by compound **6.4** in a slow, tight-binding manner and by compound **6.21** with a fast-on/fast-off profile ($K_i = 24 \pm 1$ nM) [17]. This opens the door for the development of new inhibitors exhibiting isotype-selective inhibition mechanisms. Further results, including slow, tight-binding inhibition data for macrocycles **6.10** and **6.18**, are included in Fig. 6. It is important to mention that the performance of slow, tight-binding inhibitors such as **6.4** and **6.21** in conventional end-point experiments yielded apparent K_i values up to 30 times higher than those obtained from continuous assays. These differences result from the fact that end-point assays rely on a linear behavior of the inhibitor over time, which is not the case for slow, tight-binding molecules. Thus, preincubation times and the duration of the assay can introduce bias in the measurement, especially when compared to fast-on/fast-off inhibitors.

Kinetic evidence highlights a possible ambiguity in the characterization of natural and hydroxamic acid-containing peptide inhibitors. Previous in vitro studies with compounds inspired by the structures of **6.3**, **6.10**, **6.18**, and **6.20** (such as those concerning compound **6.21**) thus call for reevaluation, since only end-point inhibition data was reported. This also means that the differences observed between in vitro HDAC inhibition and cellular antiproliferase activity for some compounds might be derived not only from differences in membrane permeability but also resulting from differences in mechanism. It has now been shown that potencies measured in continuous assays can differ from other in vitro experiments and that the mechanism of inhibition is not only class-dependent but, sometimes, even isotype-dependent. Thus, investigation of the mechanism of action should be taken into account in future studies.

Elaborating on the modification of naturally occurring macrocyclic HDAC inhibitors, further studies involved modification of the tetrapeptide scaffold. Yoshida's group carried out thorough structure-activity relationship (SAR) studies of the side chain configuration of trapoxin and cyl derivatives, which interestingly revealed that, in each case, the LDLD diastereomers were superior in cell-based assays. It was notable that most members of the SAR study exhibited in vitro inhibitory activities in the nanomolar range against partially purified HDACs, but only the mentioned LDLD diastereomers maintained this degree of activity against B16/BL6 cells (Table 1). Thus, the leading explanation for those observations was argued to be associated with permeability of the cyclic peptide, although binding kinetics and X-ray or NMR structures were not taken into account.

The position of the proline (Pro) residue in the cycle was also investigated for analogues of macrocycles **6.14** and **6.20**, for which the original position 4 was

Table 1 In vitro activity comparison of hydroxamic acid-containing tetrapeptides with LDLD configuration [70]

			In vitro HDAC inhibition	MHC induction	B16/BL6 growth inhibition
Name	Scaffold	Configuration	IC_{50} (nM)	C_{x2} (nM)	GI_{50} (nM)
6.15	Cyl-1	LDLL	3.3 ± 0.3	17 ± 8	112 ± 3
T1.1, CHAP31	Cyl-1	LDLD	3 ± 1.5	1.4 ± 0.5	5.4 ± 0.4
6.17	Cyl-2	LDLL	5 ± 1.7	5 ± 2.2	70 ± 14
T1.2, CHAP50	Cyl-2	LDLD	4 ± 1.2	1.4 ± 0.4	5.4 ± 0.4
6.19	TpxA	LLLD	4.8 ± 0.5	30 ± 5	90 ± 13
T1.3, CHAP57	TpxA	LDLD	2.9 ± 0.8	3 ± 0.5	34 ± 18
6.21	TpxB	LLLD	6 ± 1.5	100 ± 23	260 ± 35
T1.4, CHAP27	TpxB	LDLD	3.4 ± 0.6	3 ± 1.3	18 ± 2.5

optimal for HDAC inhibition. This correlated with the lower efficiency of CHAP **6.11** (HC-toxin derivative), which does present a favorable LDLD configuration but with the Pro residue in position 2 instead [68, 70]. Peptide **T1.1**, which was the most promising compound from this SAR (cyl-1 derivative), and the chlamydocin-derived compound **6.9** have been the subject of in vivo studies, which are discussed in Sect. 4.1.

Chlamydocin analogues have also been subject of SAR studies at positions 2 and 3 of the macrocycle [45, 71]. In particular, the replacement of 2-aminoisobutyric acid (Aib) at position 2 by more bulky and constrained cyclic amino acids leads to significant improvements in the antiproliferase activity against cancer cells (MCF-7, HeLa, and K562 cells) [71]. Moreover, a separate study on chlamydocin derivatives was taken into account for the design of bicyclic inhibitors with an aliphatic linkage between side chains 2 and 3. Increase in hydrophobicity and, presumably, change of conformation are possible explanations for the improved antiproliferase activity of these compounds against HL-60, K562, and U937 leukemia cells in culture. These compounds presented improved in vitro selectivity toward HDAC1 and HDAC4 when compared to HDAC6, which resembles the behavior of chlamydocin [72, 73].

The cyclic tetrapeptide scaffold, characteristic of most macrocyclic HDAC inhibitors, consists of 12 atoms and is highly conformationally constrained. This is a challenge for the synthesis of analogues, since the cyclization step from linear peptide to 12-member ring is usually low yielding. In this regard, some cyclic peptides have been designed with a larger ring size, which is more prone to cyclization. Together with the original CHAP studies, an octapeptide was prepared as a result of combining two linear precursors of compound **6.15**. However, this compound showed 26-fold loss in in vitro HDAC inhibition and no activity in cell-based assays [70]. Jose et al. designed a substrate-based cyclic hexapeptide inhibitor selective toward HDAC6 [74]. Since acetylated α-tubulin had been reported as an HDAC6 target, a macrocycle containing amino acids 38–43 of α-tubulin was prepared, with K40 replaced by Asuha. The hypothesis was that the Zn^{2+}-binding group would help accommodate the rest of the peptide in a conformation similar to the native substrate. However, the cyclic peptide adopted a different spatial projection of side chains and was inactive [74]. A more successful approach, carried out by the group of M. Reza Ghadiri, was based on introduction of β-amino acid residues in the structure of natural tetrapeptides in order to increase the size of the cycle. The lead structure, apicidin A (**6.3**), was simplified and modified with one or two β-amino acids ($\alpha_3\beta$ or $\alpha_2\beta_2$ scaffolds, respectively) yielding 13- and 14-member ring scaffolds. This resulted in improved synthetic yields but also in a single conformation of the macrocycles as determined by NMR spectroscopy in DMSO-d_6, as opposed to the native compound, which adopts at least three conformations on the NMR timescale. Thus, increasing the ring size by a single methylene group relieved the strain sufficiently to improve the yields of ring closure as well as the conformational flexibility. In terms of HDAC inhibitory activity, the analogues bearing one β-amino acid in position 3 retained potency and isotype selectivity compared to the parent compound. Then, when the native Zn^{2+}-binding group was changed to hydroxamic acid (Fig. 7, compounds **7.1** and **7.2**), the overall HDAC

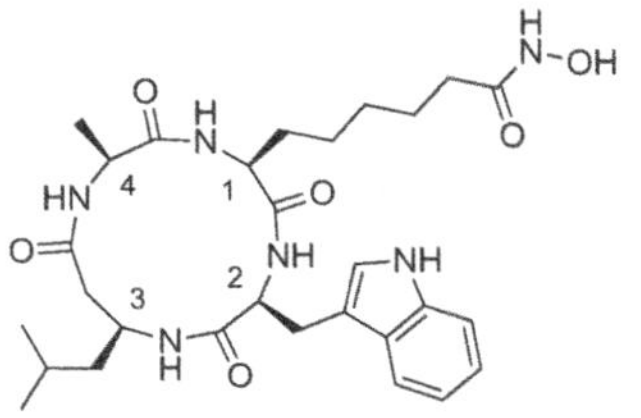

7.1

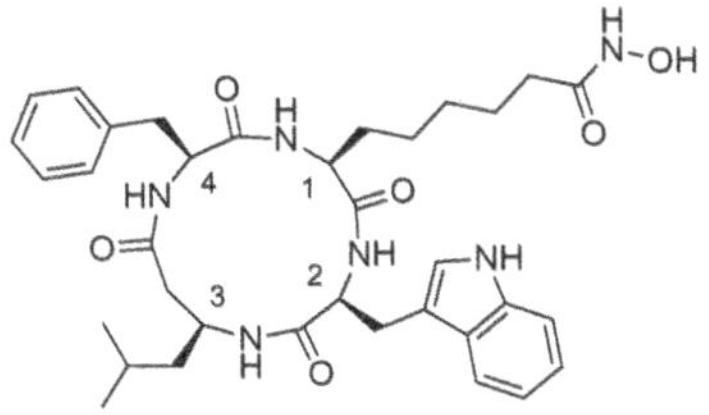

7.2

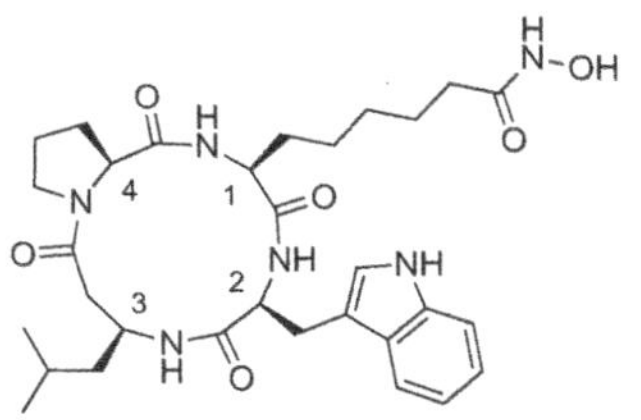

7.3

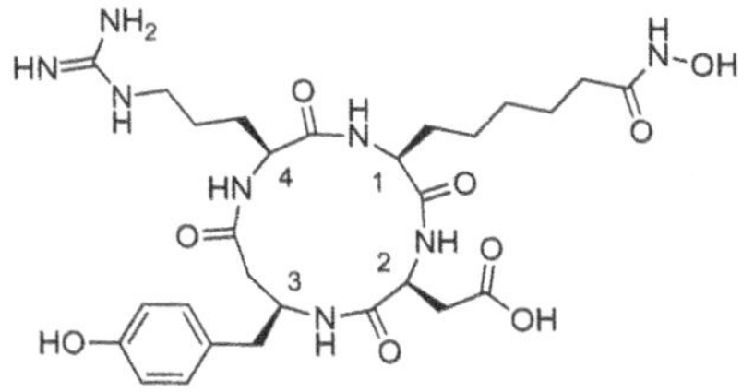

7.4

Fig. 7 Chemical structures of hydroxamic acid-containing $\alpha_3\beta$ cyclic HDAC inhibitors

inhibition was improved as well as expanded to HDAC6, in agreement with previous results for hydroxamic acid-containing macrocycles. Compound **7.1** exhibited growth inhibition activity similar to apicidin toward HeLa, K-562, KYO-1, and Molt-3 human cancer cells in culture, slight improvement against MCF-7 (breast cancer), and novel inhibition of Huh-7 growth (hepatocarcinoma) [60]. Interestingly, a family of naturally occurring HDAC inhibitors with $\alpha_3\beta$ scaffold was discovered and reported parallel to this study, the azumamides (Fig. 6, compounds **6.5** and **6.6**) [75, 76]. Combinatorial libraries of cyclic $\alpha_3\beta$-tetrapeptides furnished two novel inhibitor scaffolds (compounds **7.3** and **7.4**, Fig. 7) showing improved selectivity toward HDAC6 versus class I HDAC1, HDAC3, and HDAC8. However, growth inhibition activity against various cancer cell lines was lost for compound **7.4** while remained similar for compound **7.3** compared to **7.1** [61].

Three-dimensional structures of naturally occurring and synthetic cyclic peptides have been studied by NMR, including macrocycles **6.1** [77], **6.6** [76, 78], **6.8** [79, 80], and **6.10** [81]. Even though it is known that the measured conformation may vary depending on the solvent employed [67, 80], such structures are highly relevant for explaining potency, selectivity, and membrane permeability of macrocyclic inhibitors. For example, SAR concerning $\alpha_3\beta$ cyclic scaffolds **7.1**, **7.2**, and **7.3** correlated with a very specific positioning of the Trp and the Zn^{2+}-binding group-containing side chains, which also correlated with the apicidin scaffold and not with that of azumamide, the only natural $\alpha_3\beta$ tetrapeptide [60, 76, 78]. It was also shown that alternative β-amino acid replacements induced rigid conformations deviating from the HDAC-binding pharmacophore model, which explains the loss in potency with respect to the lead compound [60]. Potent analogues of compounds **6.16** and **6.20** have been shown to also adopt three-dimensional structures similar to that of

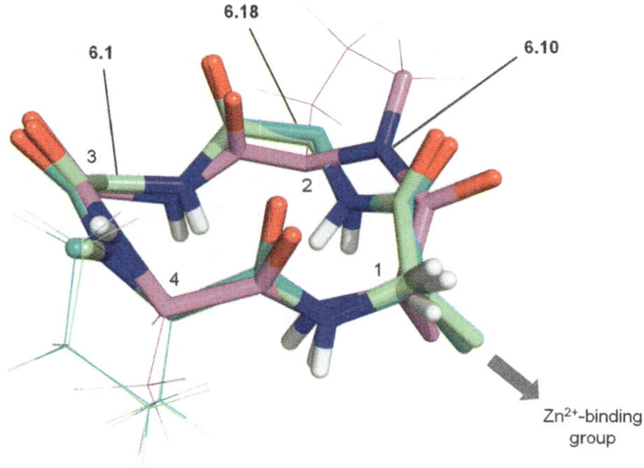

Fig. 8 Crystal structure of the scaffold of apicidin (**6.1**, green, CCDC code: 274844) overlayed with the crystal structures of those of HC-toxin from a complex with zebrafish HDAC6 (**6.10**, pink, PDB code: 5EFJ) and trapoxin A from a complex with human HDAC8 (**6.18**, cyan, PDB code: 5VI6)

compound **6.1** at positions 1 and 2 of the macrocycle, further supporting a common pharmacophore [49].

X-ray crystal structures are available for examples of these macrocyclic inhibitors. In particular, the crystal structure of compound **6.1** was solved in the absence of a biological target. Interestingly, inhibitors co-crystallized with HDACs maintain the same conformational features observed in the absence of protein, as shown in Fig. 8. The three-dimensional distribution of the 12-membered ring is almost identical for isolated **6.1** and co-crystallized compound **6.18**, being both of them potent inhibitors of several HDAC isotypes. Peptide **6.10**, however, exhibits conformational differences at position 2 of the cycle, which is also in agreement with a lower inhibitory potency of this scaffold. It is reasonable to say that the conformation of these cyclic peptides is highly optimized for binding to a region of the HDAC surface, which is common among several HDACs and particularly isotypes 1–3.

More extensive structural studies including, for example, the LDLD diastereomeric derivatives mentioned above would also be useful for rationalizing membrane permeability and establishing a more detailed pharmacophore model.

4.1 Biological Activity of Selected Inhibitors

Three macrocyclic hydroxamic acid-containing HDAC inhibitors have been studied in more detail biologically, including effects in mice and patients. Compound **T1.2**, a hydroxamic acid-containing derivative of cyl-1 bearing LDLD configuration (Table 1), is the best characterized to date. This compound enhances transgene and not endogenous gene expression in rat 2 fibroblasts (DLD1 and MKN7 cell lines) and human keratinocytes (HaCaT, NHEK, and NHDF cell lines), measured by β-galactosidase and green fluorescent protein (GFP) expression [82, 83]. This effect was observed after 72 h of exposure, and it lasted up to 7 days in the case of rat fibroblasts (similar to the behavior of romidepsin, a depsipeptide HDAC inhibitor approved for the treatment of cutaneous T-cell lymphoma). Histone hyperacetylation and mRNA expression presented similar patterns, supporting the hypothesis that expression is induced at the transcription level by inhibition of HDAC activity. Transgene expression was also induced in mice (especially in the liver) and in cultured epidermal sheets [82, 83]. Cancer growth suppression has also been evaluated for peptide **T1.2**. Preliminary GI_{50} values of 13 nM and 14 nM against human esophageal cancer cell lines (T.Tn and TE2, respectively) and T.Tn cancer growth suppression measured in mice encouraged further studies. The in vitro mechanism of action involved induction of apoptosis, which was further investigated and shown to occur through the intrinsic pathway, i.e., upregulation of the Bax/Bcl-2 ratio in a p53-independent manner [84]. Furthermore, esophageal squamous cell carcinoma (ESCC) patients who were treated with compound **T1.2** presented upregulation of *miR-375*. The tumor-suppressive microRNA produced was identified to target *LDHB*, which, upon knockdown, showed tumor suppression. This gene, *miR-375*, is present in an area on chromosome 2q35, accessible by nucleosome disruption, therefore supporting HDAC-mediated effect of macrocycle **T1.2** in this context [85].

Chlamydocin derivative **6.9** (Fig. 6) was reported as cytotoxic against mouse myeloma cell lines HS-72 and P3U1, but not against healthy spleen B and T cells. This effect occurred through a mitochondrial intrinsic apoptotic pathway, mediated by caspase-3 and caspase-9, which was completely inhibited by transfecting HS-72 cells with a *bcl-2* expression plasmid. Other effects of exposure to compound **6.9** were accumulation of hyperacetylated histone 3 and downregulation of the expression of HDAC1 and HDAC2 [86].

Compound **7.1** was tested in two different disease models in vitro: first, as a molecular chaperone for the recovery of the function of ΔF508-cystic fibrosis transmembrane conductance regulator (CFTR), where HDAC inhibitors with Zn^{2+}-binding groups other than the hydroxamic acid were more promising, most likely due to toxicity associated with the high potency of hydroxamic acid-containing chemotypes [87], and, second, as activator of the survival of motor neuron 2 (SMN2) gene, which can overcome the loss of SMN1 in spinal muscular atrophy, for which this compound proved more promising. However, the effects of various macrocyclic inhibitors in the last-mentioned study did not correlate with their in vitro HDAC inhibitory profiles, and further experiments would be needed in order to define such activity [88].

5 Chemical Tools for the Study of HDAC Structure and Function

Linear hydroxamic acid-containing peptides have not attracted as much attention for the design of HDAC inhibitors as their cyclic counterparts. Such molecules present lower membrane permeability than macrocycles due to N- and C-terminal electrical charges and, generally, smaller hydrophobic surface, as well as being more sensitive to proteolytic degradation. Amino acid and peptoid derivatives of SAHA have been studied as small molecule inhibitors with significant improvements in in vitro and cancer antiproliferase activities [89–93]. However, linear peptides with a length of three or more amino acids have found more powerful application as substrate mimics for the study of HDAC structure and function.

As mentioned, HDAC1–HDAC3 are recruited to multicomponent nuclear complexes, which enhance and direct their deacetylase activity toward the desired target [8]. HDAC3, in particular, is part of the SMRT/NCoR co-repressor complex involved in repression of gene expression. This interaction has been investigated by the research group of Schwabe, who solved the X-ray crystal structure of HDAC3 in complex with a domain of the SMRT protein. This crystal structure included a molecule of inositol tetraphosphate (Ins(1,4,5,6)P_4) bound to the interface between the two proteins. Further investigation revealed that interaction with inositol tetraphosphate enhanced HDAC3 deacetylase activity beyond sole activation by SMRT [94]. A similar binding site, this time occupied by sulfate ions from the buffer, was later observed in a co-crystal structure of HDAC1:MTA1 from the NuRD complex. These observations raised the hypothesis that inositol phosphates could play an overall regulatory role in HDAC function and assembly to multiprotein complexes [95]. In this regard, a revealing crystal structure was reported by the same group in 2016, in which inositol hexaphosphate ($InsP_6$) was accommodated between HDAC1 and MTA1, in clear analogy to the HDAC3:SMRT interface already reported [62]. Subsequent experiments described allosteric activation of the deacetylase activity mediated by different inositol phosphates, together with a conformational cross talk between the HDAC active site and the protein-protein interface where these molecules bind.

This latter mentioned crystal structure presented an additional feature, particularly interesting for the present review, since a hydroxamic acid-containing peptide inhibitor was also bound in the catalytic pocket of HDAC1. The design of this inhibitor was inspired by the tail of histone 4 (H4), where a surrogate of the HDAC1 targeted K16 ac residue was replaced by Asuha in order to ensure binding to the active site and avoid catalytic turnover (Fig. 9a). In the crystal structure, the hydroxamic acid moiety is chelating to Zn^{2+} in analogy to previous inhibitor-enzyme structures. Moreover, residues 14–18 of H4 were resolved for this histone tail analogue. The binding pose of the peptide reveals possible interactions that HDAC1 establishes with the histone and, at the same time, provides important insight into the conformational changes that the enzyme undergo upon substrate binding (such as for residue D99 at the rim of the protein). These are key features for understanding differences among the HDAC isotypes, although investigation with several substrates would be required in order to map more possible interactions around the active site rim. It is also notable that this peptide ligand adopts a quasi-cyclic conformation upon binding, which could inspire future macrocyclic inhibitor design (Fig. 9b). In terms of inhibitory activity, the IC_{50} of the peptide was 336 nM against HDAC1:MTA1 in in vitro end-point assays [62].

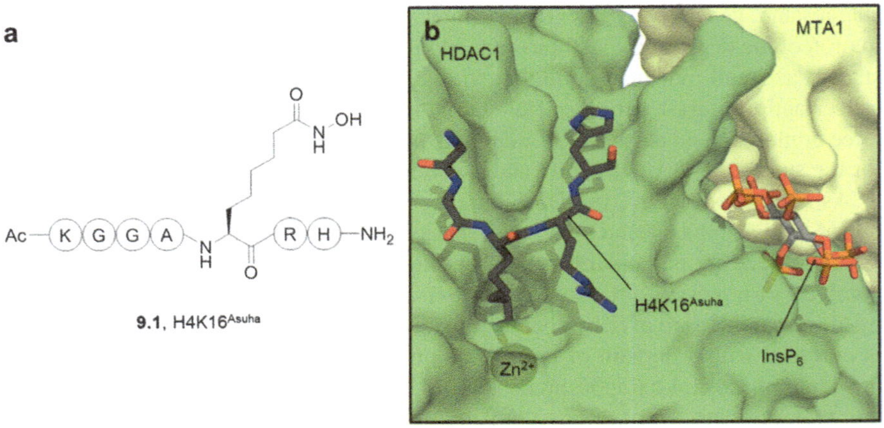

Fig. 9 (**a**) Chemical structure of a hydroxamic acid-containing peptide HDAC inhibitor (**9.1**, H4K16Asuha) and (**b**) crystal structure of the HDAC1:MTA1 interaction, assisted by a molecule of inositol hexaphosphate ($InsP_6$) and in complex with such inhibitor (PDB code: 5ICN) [62]

Substrate-inspired peptides have also been employed by the group of Dirk Schwarzer for affinity capture of HDACs and their interacting partners. Chemoproteomic studies of HDAC-containing protein complexes were first developed at Cellzome, relying on the analysis of cell lysate samples after being exposed to beads functionalized with HDAC inhibitors [96, 97]. The proteomic analysis was performed for samples containing a competing HDAC inhibitor in solution, which would avoid the removal of HDACs and their interacting partners. Such differences in enrichment were measured in the presence of different inhibitors in order to evaluate their potency as well as to identify their selectivity toward specific isotypes or HDAC-containing complexes. Data from these experiments highlighted the importance of studying HDAC1–HDAC3 in a biological environment rather than isolated, since both affinity and selectivity of the inhibitors differed from previous in vitro studies [96]. Schwarzer and coworkers implemented hydroxamic acid-containing peptides as chemical probes for the enrichment instead of small molecule inhibitors [64, 98]. First, a set of short peptides (Fig. 10, structure **10.1**) was synthesized and immobilized on agarose resin by reaction of a C-terminal cysteine residue. Some of these peptides, especially the one containing Asuha ($n = 4$), were able to enrich class I, IIa, and IIb HDACs, as well as members of the CoREST complex.

Then, two new probes were prepared based on the sequence of the tumor suppressor protein p53 (p53-K382ac, which is a known substrate for HDACs) and the nuclear transport factor 2 (NTF2-K4ac). In both cases, Kac was replaced by Asuha, as shown in Fig. 10 (peptides **10.2** and **10.3**). These new probes were able to enrich HDACs with a different selectivity profile when compared to **10.1**. Enrichment of class IIb HDACs dropped dramatically when using peptide **10.2**, together with HDAC8, whereas HDAC4 and members of NuRD and Sin3 complexes gained

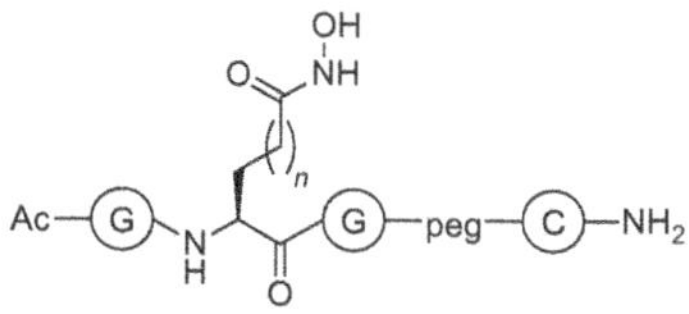

10.1, *n* = 1-4

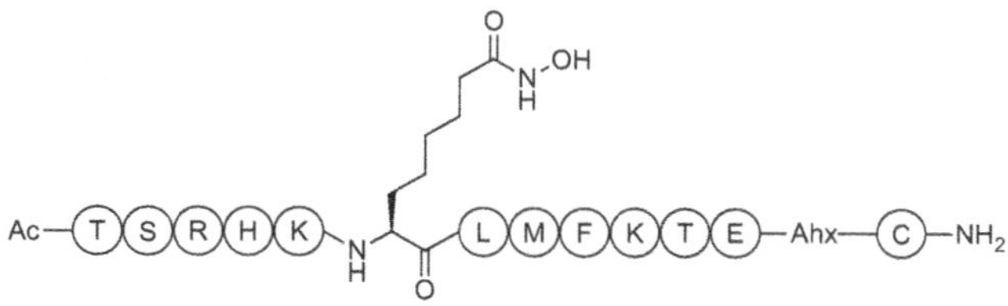

10.2

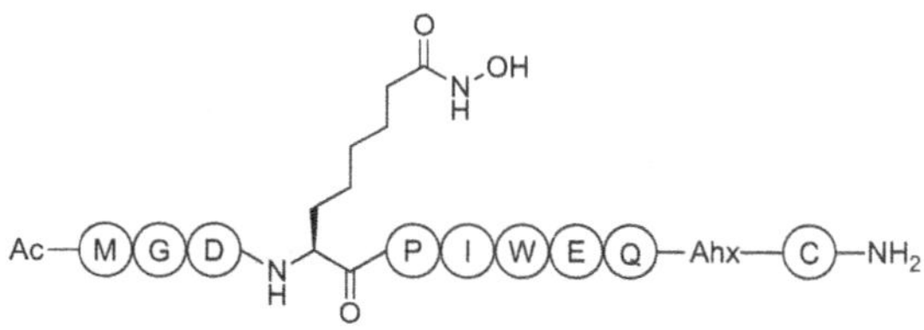

10.3

Fig. 10 Chemical structures of hydroxamic acid-containing peptides used for the pulldown of HDACs and their interacting partners from biological samples. *Peg* polyethylene glycol linker, *Ahx* 6-aminohexanoyl linker

intensity [64]. Conversely, peptide **10.3** presented better selectivity toward HDACs 6 and 10 [64]. This study further confirms the potential of hydroxamic acid-containing peptides for studying HDAC function. The peptide sequence modulated distinct enrichment profiles not only in terms of HDAC isotypes but also of the interacting partners. Therefore, it would be interesting to profile several acetylation sites and to investigate the differences in the recruitment of HDAC-containing complexes. On the other hand, it remains to be seen whether Asuha-containing peptides could provide information about the substrate selectivity of HDACs and HDAC-containing complexes. A study employing nucleosomes with Asuha-containing histone 3 (H3K9Asuha and H3K14Asuha) as inhibitors failed at reproducing preferences in CoREST-mediated deacetylation [99]. Nevertheless, further investigation would be needed in order to extract a definite conclusion.

6 Conclusions and Future Perspectives

Zn^{2+}-dependent HDACs are involved in the epigenetic control of gene expression and multiple other biological pathways. They draw special interest because of their role in complex diseases such as autoimmune response and neurodegeneration, and several HDAC inhibitors are currently used in the clinic as treatment of certain cancers [9]. Among the 11 HDAC isotypes, some exhibit distinct and independent functions, whereas others present overlapping activities and share interacting protein partners [8]. Development of probes for the study of HDACs remains a challenge, and especially isotype selectivity has been found highly elusive. In this regard, peptides bearing a hydroxamic acid functionality are promising chemical tools, as the peptide scaffold is able to interact with a large surface area where the structural differences between isotypes are more substantial. Additionally, the Zn^{2+}-binding group allows a tight binding to the active site of the enzyme.

Cyclic peptide inhibitors containing the hydroxamic acid functionality have achieved sub-nanomolar potencies against HDACs and exhibited efficacy against cancer cell growth in culture [70, 84]. Recent studies have also revealed that these molecules may exhibit slow, tight-binding kinetics, which is dependent on the HDAC isotype targeted [17, 49]. This is interesting for investigating the biology of the enzymatic activity of HDACs and for better understanding the effects of drugs. However, the hydroxamic acid-containing cyclic peptides tested to date are in general broadly cytotoxic, which limits their potential as clinical candidates. In this regard, the incorporation of less potent Zn^{2+}-binding groups could lead to a better therapeutic outcome. In terms of chemical space, there is still room for investigating larger cycles that could target isotype-specific interactions, since the tetrapeptide scaffold seems to be highly optimized for interacting with the conserved HDAC surface and therefore inhibit several enzymes at a time. Macrocycles covering a larger area at the protein surface surrounding the active site or targeting protein-protein interaction regions can potentially avoid this pharmacophore and interact with other non-conserved residues.

Peptides can help generate biologically relevant information about which substrates each HDAC recognizes and how their interactome affects this. For this purpose, the ability of the unnatural amino acid Asuha to bind to the HDAC active site has been exploited, since it mimics the interaction between the enzyme and its native substrate Kac. Linear peptides that incorporate Asuha have been shown to bind to HDACs in a sequence-dependent manner that can, potentially, be indicative of substrate preference. In addition, it appears that the interacting partners of the enzyme can affect such preferences, although this has not yet been studied in depth [64]. Thus far, Asuha has been incorporated in rather short peptides and only in two cases into nucleosomes. Therefore, it could be of potential interest to extend these studies to a larger variety of proteins and to be able to capture transient HDAC-substrate interactions in a biological context. Such information would be highly valuable for understanding the role of each HDAC isotype and improving future drug development.

Compliance with Ethical Standards

Funding: This work was supported by a Ph.D. fellowship funded by the University of Copenhagen.

Conflict of Interest: The authors declare no conflict of interest.

Ethical Approval: No part of the contents of this chapter requires ethical review.

Informed Consent: No part of the contents of this chapter requires informed consent.

References

1. Taunton J, Hassig CA, Schreiber SL (1996) A mammalian histone deacetylase related to the yeast transcriptional regulator Rpd3p. Science 272(5260):408–410
2. Verdin E, Ott M (2015) 50 years of protein acetylation: from gene regulation to epigenetics, metabolism and beyond. Nat Rev Mol Cell Biol 16(4):258–264
3. Allfrey V, Faulkner R, Mirsky A (1964) Acetylation and methylation of histones and their possible role in the regulation of RNA synthesis. PNAS 51(5):786–794
4. Gregoretti IV, Lee Y-M, Goodson HV (2004) Molecular evolution of the histone deacetylase family: functional implications of phylogenetic analysis. J Mol Biol 338(1):17–31
5. Frye RA (2000) Phylogenetic classification of prokaryotic and eukaryotic Sir2-like proteins. Biochem Biophys Res Commun 273(2):793–798
6. Yang X-J, Seto E (2008) The Rpd3/Hda1 family of lysine deacetylases: from bacteria and yeast to mice and men. Nat Rev Mol Cell Biol 9(3):206–218
7. Arrowsmith CH, Bountra C, Fish PV, Lee K, Schapira M (2012) Epigenetic protein families: a new frontier for drug discovery. Nat Rev Drug Discov 11(5):384–400
8. Millard CJ, Watson PJ, Fairall L, Schwabe JWR (2017) Targeting class I histone deacetylases in a "complex" environment. Trends Pharmacol Sci 38(4):363–377
9. Falkenberg KJ, Johnstone RW (2014) Histone deacetylases and their inhibitors in cancer, neurological diseases and immune disorders. Nat Rev Drug Discov 13(9):673–691
10. Chakrabarti A, Oehme I, Witt O, Oliveira G, Sippl W, Romier C, Pierce RJ, Jung M (2015) HDAC8: a multifaceted target for therapeutic interventions. Trends Pharmacol Sci 36(7): 481–492
11. Wei W, Liu X, Chen J, Gao S, Lu L, Zhang H, Ding G, Wang Z, Chen Z, Shi T, Li J, Yu J, Wong J (2017) Class I histone deacetylases are major histone decrotonylases: evidence for critical and broad function of histone crotonylation in transcription. Cell Res 27:898–915
12. Fellows R, Denizot J, Stellato C, Cuomo A, Jain P, Stoyanova E, Balázsi S, Hajnády Z, Liebert A, Kazakevych J, Blackburn H, Correa RO, Fachi JL, Sato FT, Ribeiro WR, Ferreira CM, Peree H, Spagnuolo M, Mattiuz R, Matolcsi C, Guedes J, Clark J, Veldhoen M, Bonaldi T, Ramirez Vinolo MA, Varga-Weisz P (2018) Microbiota derived short chain fatty acids promote histone crotonylation in the colon through histone deacetylases. Nat Commun 9(105):1–15
13. Bradner JE, West N, Grachan ML, Greenberg EF, Haggarty SJ, Warnow T, Mazitschek R (2010) Chemical phylogenetics of histone deacetylases. Nat Chem Biol 6(3):238–243
14. Lahm A, Paolini C, Pallaoro M, Nardi M, Jones P, Neddermann P, Sambucini S, Bottomley M, Surdo PL, Carfı A, Koch U, De Francesco R, Seinkühler C, Gallinari P (2007) Unraveling the hidden catalytic activity of vertebrate class IIa histone deacetylases. PNAS 104(44): 17335–17340
15. Kutil Z, Novakova Z, Meleshin M, Mikesova J, Schutkowski M, Barinka C (2018) Histone deacetylase 11 is a fatty-acid deacylase. ACS Chem Biol 13(3):685–693
16. Cao J, Sun L, Aramsangtienchai P, Spiegelman NA, Zhang X, Seto E, Lin H (2019) HDAC11 regulates type I interferon signaling through defatty-acylation of Shmt2. PNAS 116(12): 5487–5492
17. Moreno-Yruela C, Galleano I, Madsen AS, Olsen CA (2018) Histone deacetylase 11 is an E-N-myristoyllysine hydrolase. Cell Chem Biol 25(7):849–856
18. Hai Y, Shinsky SA, Porter NJ, Christianson DW (2017) Histone deacetylase 10 structure and molecular function as a polyamine deacetylase. Nat Commun 8(15368):1–9
19. Riggs MG, Whittaker RG, Neumann JR, Ingram VM (1977) N-butyrate causes histone modification in HeLa and Friend erythroleukaemia cells. Nature 268(5619):462–464

20. Richon VM, Webb Y, Merger R, Sheppard T, Jursic B, Ngo L, Civoli F, Breslow R, Rifkind RA, Marks PA (1996) Second generation hybrid polar compounds are potent inducers of transformed cell differentiation. PNAS 93(12):5705–5708
21. Yoshida M, Kijima M, Akita M, Beppu T (1990) Potent and specific inhibition of mammalian histone deacetylase both in vivo and in vitro by trichostatin A. J Biol Chem 265(28): 17174–17179
22. Kijima M, Yoshida M, Sugita K, Horinouchi S, Beppu T (1993) Trapoxin, an antitumor cyclic tetrapeptide, is an irreversible inhibitor of mammalian histone deacetylase. J Biol Chem 268(30):22429–22435
23. Marks PA, Breslow R (2007) Dimethyl sulfoxide to vorinostat: development of this histone deacetylase inhibitor as an anticancer drug. Nat Biotechnol 25(1):84–90
24. Zagni C, Floresta G, Monciino G, Rescifina A (2017) The search for potent, small-molecule HDAC is in cancer treatment: a decade after vorinostat. Med Res Rev 37(6):1373–1428
25. Jung M, Hoffmann K, Brosch G, Loidl P (1997) Analogues of trichosтatin A and trapoxin B as histone deacetylase inhibitors. Bioorg Med Chem Lett 7(13):1655–1658
26. Maolanon AR, Madsen AS, Olsen CA (2016) Innovative strategies for selective inhibition of histone deacetylases. Cell Chem Biol 23(7):759–768
27. Gantt SML, Decroos C, Lee MS, Gullett LE, Bowman CM, Christianson DW, Fierke CA (2016) General base–general acid catalysis in human histone deacetylase 8. Biochemistry 55(5):820–832
28. Decroos C, Christianson NH, Gullett LE, Bowman CM, Christianson KE, Deardorff MA, Christianson DW (2015) Biochemical and structural characterization of HDAC8 mutants associated with cornelia de lange syndrome spectrum disorders. Biochemistry 54(42): 6501–6513
29. Dowling DP, Gantt SL, Gattis SG, Fierke CA, Christianson DW (2008) Structural studies of human histone deacetylase 8 and its site-specific variants complexed with substrate and inhibitors. Biochemistry 47(51):13554–13563
30. Lauffer BE, Mintzer R, Fong R, Mukund S, Tam C, Zilberleyb I, Flicke B, Ritscher A, Fedorowicz G, Vallero R, Ortwine DF, Gunzner J, Modrusan Z, Neumann L, Koth CM, Lupardus PJ, Kaminker JS, Heise CE, Steiner P (2013) Histone deacetylase (HDAC) inhibitor kinetic rate constants correlate with cellular histone acetylation but not transcription and cell viability. J Biol Chem 288(37):26926–26943
31. Porter NJ, Mahendran A, Breslow R, Christianson DW (2017) Unusual zinc-binding mode of HDAC6-selective hydroxamate inhibitors. PNAS 114(51):13459–13464
32. Gupta SP (2015) QSAR studies on hydroxamic acids: a fascinating family of chemicals with a wide spectrum of activities. Chem Rev 115(13):6427–6490
33. Chen K, Xu L, Wiest O (2013) Computational exploration of zinc binding groups for HDAC inhibition. J Org Chem 78(10):5051–5055
34. Wang D, Helquist P, Wiest O (2007) Zinc binding in HDAC inhibitors: a DFT study. J Org Chem 72(14):5446–5449
35. Wang D-F, Wiest O, Helquist P, Lan-Hargest H-Y, Wiech NL (2004) QSAR studies of PC-3 cell line inhibition activity of TSA and SAHA-like hydroxamic acids. Bioorg Med Chem Lett 14(3):707–711
36. Kalyaanamoorthy S, Chen Y-PP (2013) Quantum polarized ligand docking investigation to understand the significance of protonation states in histone deacetylase inhibitors. J Mol Graph Model 44:44–53
37. Wu R, Lu Z, Cao Z, Zhang Y (2011) Zinc chelation with hydroxamate in histone deacetylases modulated by water access to the linker binding channel. J Am Chem Soc 133(16):6110–6113
38. Cross JB, Duca JS, Kaminski JJ, Madison VS (2002) The active site of a zinc-dependent metalloproteinase influences the computed P K a of ligands coordinated to the catalytic zinc ion. J Am Chem Soc 124(37):11004–11007
39. Gong W, Wu R, Zhang Y (2015) Thiol versus hydroxamate as zinc binding group in HDAC inhibition: an Ab initio QM/MM molecular dynamics study. J Comput Chem 36(30): 2228–2235

40. Mann BS, Johnson JR, He K, Sridhara R, Abraham S, Booth BP, Verbois L, Morse DE, Jee JM, Pope S, Harapanhalli RS, Dagher R, Farrell A, Justice R, Pazdur R (2007) Vorinostat for treatment of cutaneous manifestations of advanced primary cutaneous T-cell lymphoma. Clin Cancer Res 13(8):2318–2322
41. Day JA, Cohen SM (2013) Investigating the selectivity of metalloenzyme inhibitors. J Med Chem 56(20):7997–8007
42. Chen Y, Cohen SM (2015) Investigating the selectivity of metalloenzyme inhibitors in the presence of competing metalloproteins. ChemMedChem 10(10):1733–1738
43. Chen Y, Lai B, Zhang Z, Cohen SM (2017) The effect of metalloprotein inhibitors on cellular metal ion content and distribution. Metallomics 9(3):250–257
44. Shen S, Kozikowski AP (2016) Why hydroxamates may not be the best histone deacetylase inhibitors—what some may have forgotten or would rather forget? ChemMedChem 11(1): 15–21
45. Nishino N, Jose B, Shinta R, Kato T, Komatsu Y, Yoshida M (2004) Chlamydocin–hydroxamic acid analogues as histone deacetylase inhibitors. Biorg Med Chem 12(22):5777–5784
46. Islam N, Islam S, Hoque A, Kato T, Nishino N (2015) Synthetic strategy for bicyclic tetrapeptides HDAC inhibitors using ring closing metathesis. J Chem Sci 127(9):1563–1569
47. Meinke PT, Colletti SL, Ayer MB, Darkin-Rattray SJ, Myers RW, Schmatz DM, Wyvratt MJ, Fisher MH (2000) Synthesis of side chain modified apicidin derivatives: potent mechanism-based histone deacetylase inhibitors. Tetrahedron Lett 41(41):7831–7835
48. Wen S, Carey KL, Nakao Y, Fusetani N, Packham G, Ganesan A (2007) Total synthesis of azumamide A and azumamide E, evaluation as histone deacetylase inhibitors, and design of a more potent analogue. Org Lett 9(6):1105–1108
49. Kitir B, Maolanon AR, Ohm RG, Colaço AR, Fristrup P, Madsen AS, Olsen CA (2017) Chemical editing of macrocyclic natural products provides HDAC inhibitors with pico-molar affinities. Biochemistry 56(38):5134–5146
50. Thouin E, Lubell WD (2000) Effective synthesis of enantiopure hydroxamates by displacement of resin-bound esters with hydroxylamine. Tetrahedron Lett 41(4):457–460
51. Floyd CD, Lewis CN, Patel SR, Whittaker M (1996) A method for the synthesis of hydroxamic acids on solid phase. Tetrahedron Lett 37(44):8045–8048
52. Ngu K, Patel DV (1997) A new and efficient solid phase synthesis of hydroxamic acids. J Org Chem 62(21):7088–7089
53. Gordeev MF, Hui HC, Gordon EM, Patel DV (1997) A general and efficient solid phase synthesis of quinazoline-2, 4-diones. Tetrahedron Lett 38(10):1729–1732
54. Chen JJ, Spatola AF (1997) Solid phase synthesis of peptide hydroxamic acids. Tetrahedron Lett 38(9):1511–1514
55. Richter LS, Desai MC (1997) A TFA-cleavable linkage for solid-phase synthesis of hydroxamic acids. Tetrahedron Lett 38(3):321–322
56. Barlaam B, Koza P, Berriot J (1999) Solid-phase synthesis of hydroxamic acid based TNF-α convertase inhibitors. Tetrahedron 55(23):7221–7232
57. Mellor SL, McGuire C, Chan WC (1997) N-Fmoc-aminooxy-2-chlorotrityl polystyrene resin: a facile solid-phase methodology for the synthesis of hydroxamic acids. Tetrahedron Lett 38(18): 3311–3314
58. Bauer U, Ho W-B, Koskinen AM (1997) A novel linkage for the solid-phase synthesis of hydroxamic acids. Tetrahedron Lett 38(41):7233–7236
59. Bang CG, Jensen JF, O'Hanlon Cohrt E, Olsen LB, Siyum SG, Mortensen KT, Skovgaard T, Berthelsen J, Yang L, Givskov M (2017) A linker for the solid-phase synthesis of hydroxamic acids and identification of HDAC6 inhibitors. ACS Comb Sci 19(10):657–669
60. Montero A, Beierle JM, Olsen CA, Ghadiri MR (2009) Design, synthesis, biological evaluation, and structural characterization of potent histone deacetylase inhibitors based on cyclic A/B-tetrapeptide architectures. J Am Chem Soc 131(8):3033–3041
61. Olsen CA, Ghadiri MR (2009) Discovery of potent and selective histone deacetylase inhibitors via focused combinatorial libraries of cyclic A3β-tetrapeptides. J Med Chem 52(23):7836–7846

62. Watson PJ, Millard CJ, Riley AM, Robertson NS, Wright LC, Godage HY, Cowley SM, Jamieson AG, Potter BVL, Schwabe JWR (2016) Insights into the activation mechanism of class I HDAC complexes by inositol phosphates. Nat Commun 7(11262):1–13
63. Wilson DM, Silverman LN, Bergauer M, Keshari KR (2013) Solid phase synthesis of hydroxamate peptides for histone deacetylase inhibition. Tetrahedron Lett 54(2):151–153
64. Dose A, Sindlinger J, Bierlmeier J, Bakirbas A, Schulze-Osthoff K, Einsele-Scholz S, Hartl M, Essmann F, Finkemeier I, Schwarzer D (2016) Interrogating substrate selectivity and composition of endogenous histone deacetylase complexes with chemical probes. Angew Chem Int Ed 55(3):1192–1195
65. Nielsen DS, Shepherd NE, Xu W, Lucke AJ, Stoermer MJ, Fairlie DP (2017) Orally absorbed cyclic peptides. Chem Rev 117(12):8094–8128
66. Craik DJ, Fairlie DP, Liras S, Price D (2013) The future of peptide-based drugs. Chem Biol Drug Des 81(1):136–147
67. Maolanon A, Kristensen H, Leman L, Ghadiri R, Olsen CA (2017) Natural and synthetic macrocyclic inhibitors of the histone deacetylase enzymes. ChemBioChem 18(1):5–49
68. Furumai R, Komatsu Y, Nishino N, Khochbin S, Yoshida M, Horinouchi S (2001) Potent histone deacetylase inhibitors built from trichostatin A and cyclic tetrapeptide antibiotics including trapoxin. PNAS 98(1):87–92
69. Yoshida M, Furumai R, Nishiyama M, Komatsu Y, Nishino N, Horinouchi S (2001) Histone deacetylase as a new target for cancer chemotherapy. Cancer Chemother Pharmacol 48: S20–S26
70. Komatsu Y, Tomizaki K-Y, Tsukamoto M, Kato T, Nishino N, Sato S, Yamori T, Tsuruo T, Furumai R, Yoshida M, Horinouchi S, Hayashi H (2001) Cyclic hydroxamic-acid-containing peptide 31, a potent synthetic histone deacetylase inhibitor with antitumor activity. Cancer Res 61(11):4459–4466
71. Wang S, Li X, Wei Y, Xiu Z, Nishino N (2014) Discovery of potent HDAC inhibitors based on chlamydocin with inhibitory effects on cell migration. ChemMedChem 9(3):627–637
72. Islam NM, Kato T, Nishino N, Kim H-J, Ito A, Yoshida M (2010) Bicyclic peptides as potent inhibitors of histone deacetylases: optimization of alkyl loop length. Bioorg Med Chem Lett 20(3):997–999
73. Li X-H, Huang M-L, Wang S-M, Wang Q (2013) Selective inhibition of bicyclic tetrapeptide histone deacetylase inhibitor on HDAC4 and K562 leukemia cell. Asian Pac J Cancer Prev 14(12):7095–7100
74. Jose B, Okamura S, Kato T, Nishino N, Sumida Y, Yoshida M (2004) Toward an HDAC6 inhibitor: synthesis and conformational analysis of cyclic hexapeptide hydroxamic acid designed from α-tubulin sequence. Biorg Med Chem 12(6):1351–1356
75. Nakao Y, Yoshida S, Matsunaga S, Shindoh N, Terada Y, Nagai K, Yamashita JK, Ganesan A, van Soest RWM, Fusetani N (2006) Azumamides A–E: histone deacetylase inhibitory cyclic tetrapeptides from the marine sponge mycale izuensis. Angew Chem Int Ed 45(45):7553–7557
76. Maulucci N, Chini MG, Di Micco S, Izzo I, Cafaro E, Russo A, Gallinari P, Paolini C, Nardi MC, Casapullo A (2007) Molecular insights into azumamide E histone deacetylases inhibitory activity. J Am Chem Soc 129(10):3007–3012
77. Kranz M, Murray PJ, Taylor S, Upton RJ, Clegg W, Elsegood MRJ (2006) Solution, solid phase and computational structures of apicidin and its backbone-reduced analogs. J Pept Sci 12(6):383–388
78. Izzo I, Maulucci N, Bifulco G, De Riccardis F (2006) Total synthesis of azumamides A and E. Angew Chem Int Ed 45(45):7557–7560
79. Kawai M, Jasensky RD, Rich DH (1983) Conformational analysis by NMR spectrometry of the highly substituted cyclic tetrapeptides, chlamydocin and Ala4-chlamydocin. Evidence for a unique amide bond sequence in dimethyl sulfoxide-d6. J Am Chem Soc 105(13):4456–4462
80. Rich DH, Kawai M, Jasensky RD (1983) Conformational studies of cyclic tetrapeptides. Chem Biol Drug Des 21(1):35–42
81. Kawai M, Rich DH, Walton JD (1983) The structure and conformation of HC-toxin. Biochem Biophys Res Commun 111(2):398–403

82. Taura K, Yamamoto Y, Nakajima A, Hata K, Uchinami H, Yonezawa K, Hatano E, Nishino N, Yamaoka Y (2004) Impact of novel histone deacetylase inhibitors, CHAP31 and FR901228 (FK228), on adenovirus-mediated transgene expression. J Gene Med 6(5):526–536
83. Yasukawa K, Sawamura D, Goto M, Nakamura H, Shimizu H (2007) Histone deacetylase inhibitors preferentially augment transient transgene expression in human dermal fibroblasts. Br J Dermatol 157(4):662–669
84. Murakami K, Matsubara H, Hoshino I, Akutsu Y, Miyazawa Y, Matsushita K, Sakata H, Nishimori T, Usui A, Kano M, Nishino N, Yoshida M (2010) CHAP31 induces apoptosis only via the intrinsic pathway in human esophageal cancer cells. Oncology 78(1):62–74
85. Isozaki Y, Hoshino I, Nohata N, Kinoshita T, Akutsu Y, Hanari N, Mori M, Yoneyama Y, Akanuma N, Takeshita N, Maruyama T, Seki N, Nishino N, Yoshida M, Matsubara H (2012) Identification of novel molecular targets regulated by tumor suppressive miR-375 induced by histone acetylation in esophageal squamous cell carcinoma. Int J Oncol 41(3):985–994
86. Fujii S, Okinaga T, Ariyoshi W, Takahashi O, Iwanaga K, Nishino N, Tominaga K, Nishihara T (2013) Mechanisms of G1 cell cycle arrest and apoptosis in myeloma cells induced by hybrid-compound histone deacetylase inhibitor. Biochem Biophys Res Commun 434(3):413–420
87. Hutt DM, Olsen CA, Vickers CJ, Herman D, Chalfant MA, Montero A, Leman LJ, Burkle R, Maryanoff BE, Balch WE, Ghadiri MR (2011) Potential agents for treating cystic fibrosis: cyclic tetrapeptides that restore trafficking and activity of ΔF508-CFTR. ACS Med Chem Lett 2(9):703–707
88. Lai J-I, Leman LJ, Ku S, Vickers CJ, Olsen CA, Montero A, Ghadiri MR, Gottesfeld JM (2017) Cyclic tetrapeptide HDAC inhibitors as potential therapeutics for spinal muscular atrophy: screening with IPSC-derived neuronal cells. Bioorg Med Chem Lett 27(15):3289–3293
89. Kahnberg P, Lucke AJ, Glenn MP, Boyle GM, Tyndall JD, Parsons PG, Fairlie DP (2006) Design, synthesis, potency, and cytoselectivity of anticancer agents derived by parallel synthesis from α-aminosuberic acid. J Med Chem 49(26):7611–7622
90. Andrews KT, Tran TN, Lucke AJ, Kahnberg P, Le GT, Boyle GM, Gardiner DL, Skinner-Adams TS, Fairlie DP (2008) Potent antimalarial activity of histone deacetylase inhibitor analogues. Antimicrob Agents Chemother 52(4):1454–1461
91. Taddei M, Cini E, Giannotti L, Giannini G, Battistuzzi G, Vignola D, Vesci L, Cabri W (2014) Lactam based 7-amino suberoylamide hydroxamic acids as potent HDAC inhibitors. Bioorg Med Chem Lett 24(1):61–64
92. Krieger V, Hamacher A, Gertzen CG, Senger J, Zwinderman MR, Marek M, Romier C, Dekker FJ, Kurz T, Jung M, Gohlke H, Kassack MU, Hansen FK (2017) Design, multicomponent synthesis and anticancer activity of a focused histone deacetylase (HDAC) inhibitor library with peptoid-based cap groups. J Med Chem 60(13):5493–5506
93. Belvedere S, Witter DJ, Yan J, Secrist JP, Richon V, Miller TA (2007) Aminosuberoyl hydroxamic acids (ASHAs): a potent new class of HDAC inhibitors. Bioorg Med Chem Lett 17(14):3969–3971
94. Watson PJ, Fairall L, Santos GM, Schwabe JW (2012) Structure of HDAC3 bound to corepressor and inositol tetraphosphate. Nature 481(7381):335–340
95. Millard CJ, Watson PJ, Celardo I, Gordiyenko Y, Cowley SM, Robinson CV, Fairall L, Schwabe JWR (2013) Class I HDACs share a common mechanism of regulation by inositol phosphates. Mol Cell 51(1):57–67
96. Bantscheff M, Hopf C, Savitski MM, Dittmann A, Grandi P, Michon A-M, Schlegl J, Abraham Y, Becher I, Bergamini G, Boesche M, Delling M, Dümpelfeld B, Eberhard D, Huthmacher C, Mathieson T, Poeckel D, Reader V, Strunk K, Sweetman G, Kruse U, Neubauer G, Ramsden NG, Drewes G (2011) Chemoproteomics profiling of HDAC inhibitors reveals selective targeting of HDAC complexes. Nat Biotechnol 29(3):255–265
97. Becher I, Dittmann A, Savitski MM, Hopf C, Drewes G, Bantscheff M (2014) Chemoproteomics reveals time-dependent binding of histone deacetylase inhibitors to endogenous repressor complexes. ACS Chem Biol 9(8):1736–1746

98. Sindlinger J, Bierlmeier J, Geiger LC, Kramer K, Finkemeier I, Schwarzer D (2016) Probing the structure–activity relationship of endogenous histone deacetylase complexes with immobilized peptide-inhibitors. J Pept Sci 22(5):352–359
99. Wu M, Hayward D, Kalin JH, Song Y, Schwabe JWR, Cole PA (2018) Lysine-14 acetylation of histone H3 in chromatin confers resistance to the deacetylase and demethylase activities of an epigenetic silencing complex. eLife 7:e37231

Top Med Chem (2020) 33: 55–92
DOI: 10.1007/7355_2019_70

Published online: 30 October 2019

Sirtuin Inhibitors and Activators

Minna Rahnasto-Rilla, Jonna Tyni, and Maija Lahtela-Kakkonen

Contents

1 Introduction 57
2 Sirtuin Inhibitors 58
  2.1 Inhibitors Mimicking Substrates or Product 58
  2.2 Substrate-Based Inhibitors 60
  2.3 Small Molecule Inhibitors 64
3 Sirtuin Activators 76
  3.1 Natural Compounds Activating Sirtuins 76
  3.2 Various Scaffolds of Sirtuin Activators 79
4 General Conclusions 81
References 82

Abstract The mammalian family of sirtuins (SIRT1–7) target a large variety of proteins at various subcellular localizations and thus exert regulatory effects on critical biological processes such as gene silencing, DNA repair, and chromosomal stability and longevity. Sirtuins play crucial roles in many signaling pathways and are regarded as potential therapeutic targets in several pathological conditions, such as cancer, metabolic disorders, and also cardiovascular and neurodegenerative diseases. Therefore, the modulation of sirtuin activity by inhibitors or activators could be beneficial for human health, a topic that has been interesting scientists for over 15 years. Researchers have developed novel inhibitors and activators toward sirtuins mainly for human silent information regulator type 1 (SIRT1) because both cellular studies and experiments in animal models have indicated that SIRT1 regulators could be used for the treatment for multiple human diseases. Gradually, an appreciation of the importance of the other members of the sirtuin family has increased, and potent inhibitors and activators are designed for various sirtuins.

Keywords Activation, Deacetylation, Drug design and development, Inhibition, Sirtuin

M. Rahnasto-Rilla, J. Tyni, and M. Lahtela-Kakkonen (✉)
School of Pharmacy, University of Eastern Finland, Kuopio, Finland
e-mail: maija.lahtela-kakkonen@uef.fi

Abbreviations

AD	Alzheimer's disease
ADP	Adenosine diphosphate
AMC	7-Amino-4-methylcoumarin
ATP	Adenosine triphosphate
BIM	Bisindolylmaleimides
CPS1	Carbamoyl phosphate synthetase peptides
$EC_{1.5}$	Concentration of compound required to increase the enzyme activity by 50%
EC_{150}	Effective concentration which increases the enzyme activity to 150%
EC_{50}	Half maximal effective concentration
FOXO3α	Forkhead O transcription factor 3 α
H3K9	Histone H3 lysine 9
HD	Huntington's disease
HDAC	Histone deacetylase enzymes
HTS	High-throughput screening
IC_{50}	Half maximal inhibitory concentration
ICL-SIRT078	3-((2-Methoxynaphthalen-1-yl)methyl)-7-((pyridin-3-ylmethyl) amino)-5,6,7,8-tetrahydrobenzo[4,5]thieno[2,3-d]pyrimidin-4 (3H)-one
K_m	Value of substrate concentration at half maximal velocity
LPS	Lipopolysaccharides
Mn-SOD	Manganese superoxide dismutase
NAD^+	Nicotinamide adenine dinucleotide
NAM	Nicotinamide
NO	Nitric oxide
PCK, PGC-1α	Peroxisome proliferator-activated receptor gamma coactivator 1-alpha Protein kinase C
PD	Parkinson's disease
ROS	Reactive oxygen species
SAR	Structure-activity relationships
SirReal2	Sirtuin-rearranging ligand2
SIRT	Sirtuins
SIRT1	Silent information regulator type 1
STACs	Sirtuin-activating compounds
TAMRA	Tetramethylrhodamine
TNFα	Tumor necrosis factor alpha

1 Introduction

Sirtuins are a family of proteins regarded as nicotinamide adenine dinucleotide (NAD^+)-dependent histone deacetylase enzymes (HDACs) [1–5]. Sirtuins are found in a wide variety of organisms from eukaryotes to humans. In mammals, a total of seven sirtuins have been identified. In addition to the deacetylation activity, some of the sirtuins have also other enzymatic activities; SIRT6 and SIRT4 act as adenosine diphosphate (ADP)-ribosyltransferases [6, 7], SIRT5 has higher demalonylase and desuccinylase activities than deacetylase activity [8], and SIRT6 deacylates long-chain fatty acyl groups [9]. Sirtuins bind to many proteins in various subcellular localizations, and thus they exert regulatory effects on many biological processes.

Sirtuins are regarded as potential therapeutic targets in cancer, metabolic disorders, and cardiovascular and neurodegenerative diseases [10–19]. In the case of cancer, sirtuins have been linked to the initiation and the development of cancer, but their exact role has remained somewhat unclear as sirtuins have both tumor-promoting and tumor-suppressing effects [16, 17]. Therefore, both sirtuin inhibitors and activators have been intensively studied in the therapy of cancer. Multiple factors, such as tissue type, species, age, and the cellular localization of the sirtuins, may be regulating their role in the development of cancer and may lie behind the contradictory observations, e.g., on the tumor-suppressing and tumor-promoting roles [17, 19].

Due to their NAD^+ dependency, sirtuins are activated in the conditions of nutrient depletion, starvation, and cellular stress, thus exerting an important role in many metabolic pathways [20]. The activators of sirtuins could serve as novel treatment strategies for type2 diabetes [10]. In particular, nuclear sirtuins have been linked to inflammatory signaling pathways, and thus sirtuin activators could confer beneficial effects via the downregulation of these pathways [21]. Some sirtuins have protective roles in the development of cardiovascular diseases, e.g., cardiac hypertrophy and atherosclerosis [18, 22]. All in all, sirtuin activators hold great promise in treating cardiovascular and metabolic diseases.

The impact of sirtuins in the development of common neurodegenerative diseases has been extensively studied (e.g., [23–25]); for example, the progression of Alzheimer's disease (AD) and Parkinson's disease (PD) have been shown to be affected by the enzymatic activity of sirtuins [26]. The aging process may decrease SIRT1 activity, and thus specific sirtuin activators could be used as potential treatments for AD [11]. However, it has been postulated that certain isoforms of sirtuins should be inhibited rather than activated in neurodegenerative diseases [27]. In addition, sirtuins have been linked to the progress of Huntington's disease (HD), amyotrophic lateral sclerosis, and spinal and bulbar muscular atrophy [12, 28].

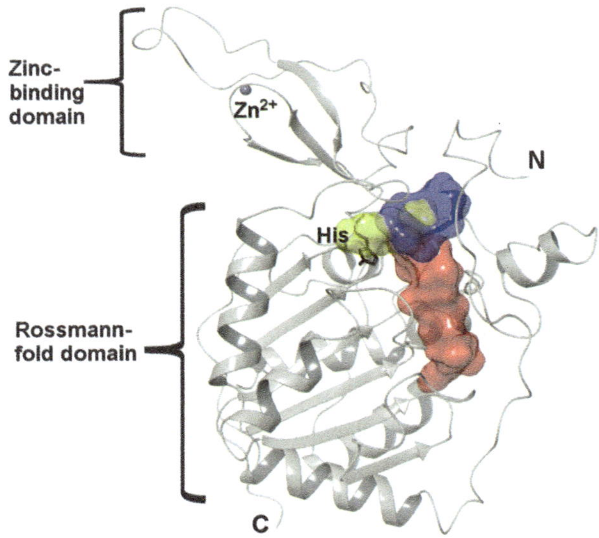

Fig. 1 The crystal structure of SIRT6 [31] showing sirtuin's two domains and the general binding sites for adenine ribose A- and the B-pocket (red) and the NAM moiety: C-pocket (blue). The binding site of the acetylated lysine is marked in yellow with the histidine needed for the reaction (marked with black)

2 Sirtuin Inhibitors

Sirtuins are a family of proteins which share a highly conserved approximately 275 amino acid-long catalytic core [29, 30]. The catalytic core of sirtuins consists of a Rossmann-fold domain, a smaller zinc-binding domain, and several loops connecting these two domains. The N- and C-terminal segments attached to the catalytic core vary in length and sequence in the different sirtuins. Little is known about the role and structure of the N- and C-terminals in sirtuins, but they have been claimed to mediate protein-specific activities [29]. The NAD^+ binding site is divided into three sites: (a) the adenine ribose binding site, (b) the nicotinamide (NAM) ribose binding site, and (c) the NAM moiety binding site (Fig. 1). The acetylated lysine of the substrate binds to the hydrophobic tunnel formed between the small and large domains. Sirtuin inhibitors can be roughly classified into two classes: substrate/product-based inhibitors and small molecule inhibitors.

2.1 *Inhibitors Mimicking Substrates or Product*

The activity of sirtuins can be regulated by changing NAD^+ levels, but it is still unclear whether the physiological fluctuations in NAD^+ could actually regulate sirtuin activity [32, 33]. NADH is also known to inhibit the deacetylation reaction of SIRT1–3, SIRT5, and SIRT6 with the half maximal inhibitory concentration (IC_{50}) values in a range of 1.3–68 μM [34]. Over the years, some sirtuin inhibitors have been developed by making minor modifications to the structure of NAD^+. One

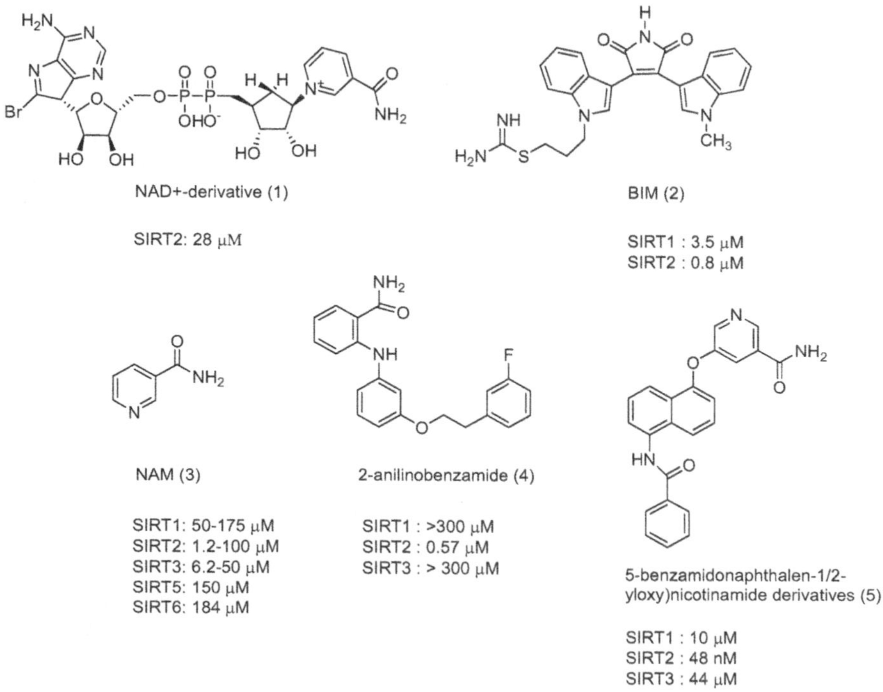

Fig. 2 The structures of substrate and product mimicking sirtuin inhibitors with their IC_{50} values

example of those compounds was carba-NAD where an oxygen atom was replaced with a methylene group [35, 36].

There are some other NAD^+ mimicking derivatives, such as 8-bromo-NAD (compound 1 in Fig. 2), which inhibited SIRT2 with an IC_{50} value in a range of 28–128 μM [37]. Some NAD^+ mimicking compounds such as the bisindolylmaleimides (BIMs) that were originally identified as adenosine triphosphate (ATP)-competitive kinase, protein kinase C (PKC) inhibitors, were developed by modifying the adenosine part of NAD^+. The most potent compound in the BIM family (compound 2 in Fig. 2) displayed inhibition at a low micromolar level for SIRT1 and SIRT2 with IC_{50} values of 3.5 μM and 0.8 μM, respectively [38]. The development of NAD^+-based inhibitors is challenging as it is difficult to avoid unwanted adverse effects, as NAD^+ is a ubiquitous coenzyme.

NAM (compound 3 in Fig. 2) is released from NAD^+ during the deacetylation reaction of sirtuins [39]. NAM is noncompetitive inhibitor, and its IC_{50} values have been reported for various sirtuin isoforms: 120 μM (SIRT1), 100 μM (SIRT2), 50 μM (SIRT3), 150 μM (SIRT5), and 184 μM (SIRT6) [40]. There are some variations in the literature concerning the IC_{50} values of NAM attributable to the various assays used in the determinations, for example, the IC_{50} value for SIRT1 was found to be within a range of 50–175 μM and for SIRT2 in the range of 1.2–100 μM

[41–46]. Interestingly, a recent study revealed that NAM could stimulate SIRT1 activity in cells even though it acted as an inhibitor in vitro [47].

Several NAM analogs were developed as sirtuin inhibitors [48–51]. A series of 2-anilinobenzamide derivatives showed SIRT1 inhibition with IC_{50} values between 52 and 300 μM [48]. Some of these derivatives were even more potent toward SIRT2. The most potent derivative displayed an IC_{50} value of 0.57 μM (compound 4 in Fig. 2) for SIRT2 [50]. Recently, a 2-anilinobenzamide scaffold was combined with a peptide substrate mimicking inhibitor, and a novel type of inhibitor for SIRT2 was identified. This compound, 3′-phenethyloxy-2-anilinobenzamide, was an isoform-selective inhibitor, with an IC_{50} value of 28 μM for SIRT2 [52].

In addition, other NAM derivatives, (5-benzamidonaphthalen-1/2-yloxy)nicotinamides, have been reported to be potent sirtuin inhibitors for several sirtuin isoforms: The IC_{50} values of the derivatives were 0.80–100 μM for SIRT1, 48 nM–1 μM for SIRT2, and 4.4–232 μM for SIRT3. Interestingly, the most potent of these derivatives (compound 5 in Fig. 2) showed excellent selectivity toward SIRT2 with an IC_{50} value of 48 nM, whereas its ability to inhibit the activity of SIRT1 and SIRT3 was at the micromolar level. Based on the kinetic studies, it was postulated that compound 5 acted as a competitive inhibitor against the peptide substrate and in a noncompetitive manner against NAD^+. Compound 5 exhibited also moderate anticancer activity in breast cancer (MCF7), prostate cancer (DU 145), and chronic myelogenous leukemia cell lines [41].

2.2 Substrate-Based Inhibitors

The first substrate-based inhibitor was developed from human p53 which is a SIRT1 substrate by taking the part of its C-terminal (residues 372–389). This peptide-type inhibitor had an IC_{50} value of 2 μM for SIRT1. The inhibition of its truncated analogs such as N^{α}-Fmoc-N^{ε}-thioacetyl-lysine and N^{α}-acetyl-N^{ε}-thioacetyl-lysine was also tested on SIRT1, but they displayed virtually no inhibition at all [53]. However, the inhibition potential of various acetylated and thioacetylated peptides was studied with SIRT1, SIRT2, and SIRT3, and several low micromolar inhibitors were found [54]. The mechanism of thioacetylated peptidic inhibitors was examined, and it was observed that the replacement of acetyl-group with thioacetyl-group formed a covalent and stable 1′-*S*-alkylimidate intermediate in the deacetylation reaction instead of *O*-alkylamidate. This stalled intermediate produced very slowly a deacetylated peptide and 1′-SH-2′-*O*-acetyl-ADP-ribose [55].

The acetyl-lysine in peptide substrate can be replaced with various groups such as thioacetyl- or trifluoroacetyl-group (examples in Fig. 3). Huhtiniemi and coworkers devised a set of acetyl-lysine analogs to evaluate their inhibitory activity on SIRT1 and SIRT2 [56]. N^{ε}-Thioacetyl-lysine (compound 7 in Fig. 3) displayed the best inhibitory activity toward SIRT1, whereas the selenoacetyl moiety (compound 8 in Fig. 3) had the best inhibition toward SIRT2. The study revealed that the substrate binding site can accommodate moieties larger than an acetyl-group such as

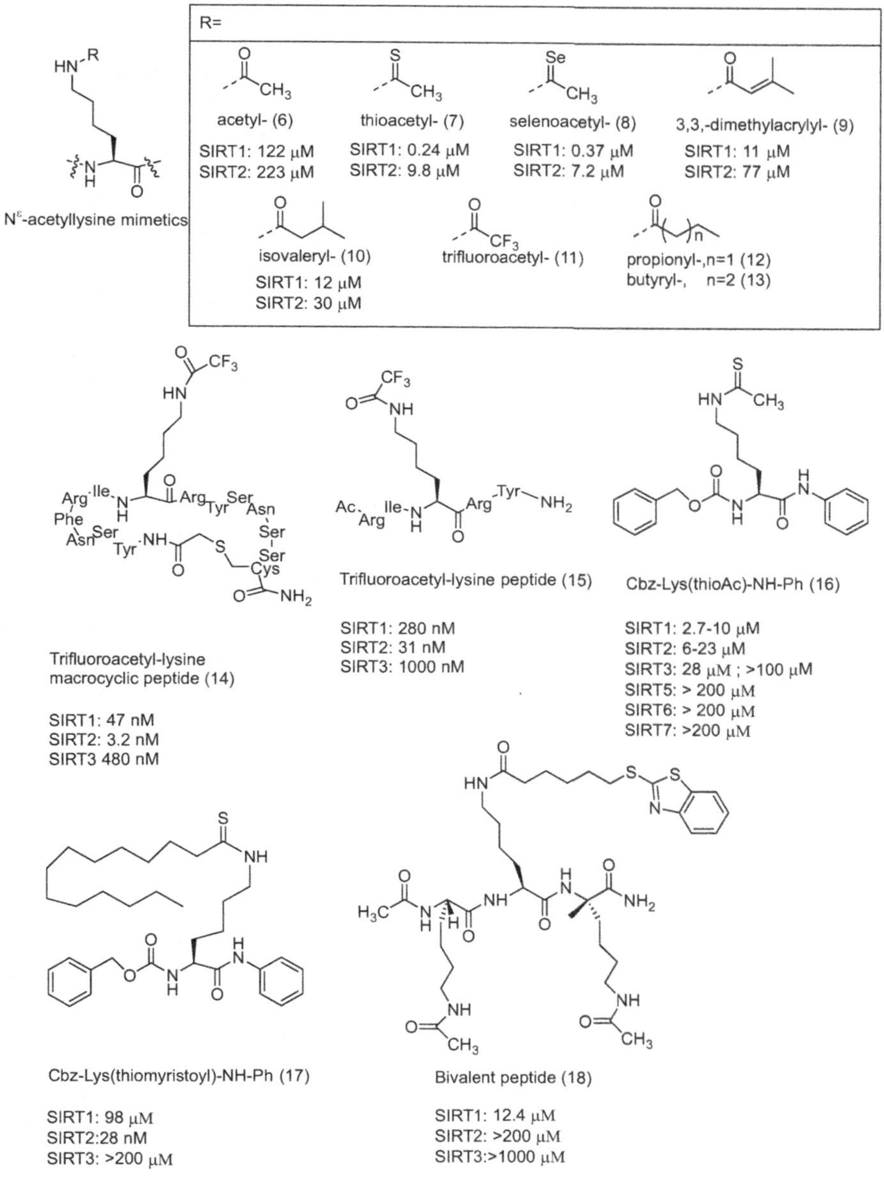

Fig. 3 The structures of substrate-based sirtuin inhibitors and their IC_{50} values

3,3-dimethylacrylyl (compound 9 in Fig. 3) and isovaleryl moieties (compound 10 in Fig. 3). The other acetyl-lysine analogs such as monofluoroacetyl, trifluoroacetyl (compound 11 in Fig. 3), propionyl, and butyryl moiety (compounds 12 and 13 in Fig. 3) increased the affinity toward yeast sirtuins but reduced the rate of the formation of the *O*-alkylamidate intermediate [57].

The substrate-based inhibitors were also designed by incorporating modifications into the peptide scaffold. The thioacetyl-lysine containing tri-, tetra-, and pentapeptides based on α-tubulin (SDK(thioAc)TI) and p53 (HKK(thioAc)LM) were developed, and they inhibited SIRT1 at the nanomolar scale and SIRT2 at a low micromolar scale [58]. It was also observed that the p53 sequence displayed a better inhibition than the α-tubulin sequence. Interestingly, SIRT1 inhibition was not very sensitive to the length of peptide, whereas the pentapeptide (HKK(thioAc)AM) was more potent toward SIRT2 with an IC_{50} value of 3.8 μM. Some of these peptides such as H_2N-KK(thioAc)L-OH showed clear selectivity for SIRT1 over SIRT2 [58]. Subsequently, it has been reported that the simple peptide Ac-Ala-Lys (thioAc)-Ala could achieve SIRT1 and SIRT2 inhibition similar to the longer and more complicated peptides [56]. A set of p53 and α-tubulin mimetic peptides was also screened against SIRT6. The most potent compounds exerted more than 60% inhibition activity at the 200 μM concentration, and subsequently they displayed IC_{50} values below 100 μM [59]. The exploitation of peptides as therapeutic agents has disadvantages due to their delivery, poor bioavailability, and short half-life, and thus, other types of inhibitors have been developed.

The macrocyclic peptide inhibitors had improved potency, cell permeability, and resistance toward proteolytic enzymes compared to the analog linear peptide. However, most of them were pan-SIRT1/2/3 inhibitors [60]. Simple cyclic peptides containing N^{ε}-thioacetyl-lysine showed inhibitory properties in the low nanomolar range for SIRT1, SIRT2, and SIRT3. Interestingly, macrocyclic peptides containing trifluoroacetyl-lysine (e.g., compound 14 in Fig. 3) were able to act as catalytic mechanism-based inhibitors for sirtuins. They have a greater inhibitory potency toward SIRT2 (IC_{50} = 3–4 nM) than toward SIRT1 (IC_{50} = 5–6 nM) and SIRT3 (IC_{50} = 480 nM). However, their linear counterparts were equally potent SIRT2 inhibitors with IC_{50} values of 5–6 nM. The short and linear RIK(Trifluoroacetyl)RY (compound 15 in Fig. 3) was less potent toward SIRT2 (IC_{50} = 31 nM) than the cyclic compound, and the inhibition was even weaker for SIRT1 (IC_{50} = 280 nM) and SIRT3 (IC_{50} = 1 μM) [61]. Investigation of these macrocyclic peptides with SIRT2 revealed that the trifluoroacetyl-lysine moiety was orientated into the catalytic tunnel of the binding site and the macrocyclic structure was stabilized via multiple intramolecular hydrogen bonds [62].

Non-peptide *N*-thioacetyl-lysine analogs represented another attempt to improve the inhibitory activity of the substrate-based inhibitors. The first non-peptide *N*-thioacetyl-lysine analogs were developed by Asaba et al. [63] and Suzuki et al. [64]. The most potent inhibitor of these compounds had a thioacetyl-lysine scaffold with an aniline group attached to the carbonyl terminal and a benzyloxycarbonyl group to the amino terminal (compound 16 in Fig. 3). The compound exhibited low micromolar inhibition for SIRT1 (IC_{50} = 3 μM) and SIRT2 (IC_{50} = 23 μM), but it was selective over SIRT3 (IC_{50} > 100 μM) [64]. Jing and coworkers developed Cbz-Lys(ThioAc)-NH-Phe that showed IC_{50} values of 10 μM for SIRT1, 6 μM for SIRT2, 28 μM for SIRT3, and >200 μM for SIRT5–7 [65]. This inhibitor exhibited also a dose-dependent increase in p53 acetylation in human colon cancer cells [64].

Subsequently, other *N*-thioacetyl-lysine analogs were developed with IC_{50} values of 0.24–23 μM for SIRT1 and 1.8–75 μM for SIRT2 with some of those compounds showing moderate inhibition toward SIRT6 [59, 66, 67]. The most potent compound of these analogs was not isoform-selective as it inhibited also SIRT3 ($IC_{50} = 3.89$ μM); it exhibited an antiproliferative effect in lung carcinoma and breast carcinoma cells [67]. Although *N*-thioacetyl-lysine analogs lack isoform-selectivity, they can be used as a chemical probe to investigate the mechanism of sirtuin deacetylation reaction or as a starting point in the design of novel small-molecule inhibitors.

In attempts to improve the isoform-selectivity of *N*-thioacetyl-lysine analogs, a set of compounds with various aliphatic acyl groups such as thiobutyryl-, thioheptanoyl-, and thiomyristoyl-moieties were developed [65]. Interestingly, it was observed that the length of the aliphatic acyl group exerted an influence on the isoform-selectivity of sirtuins. The selective inhibitor was thiomyristoyl-lysine (compound 17 in Fig. 3) with the IC_{50} values of 28 nM and 98 μM toward SIRT2 and SIRT1, respectively, and it did not inhibit SIRT3 at all. Thiomyristoyl analogs, which mimic a tumor necrosis factor alpha (TNFα) or a histone H3 lysine 9 (H3K9) peptide sequence, were also potent inhibitors of SIRT6. The most potent inhibitor, the H3K9 derivative, inhibited both SIRT6 demyristoylation and deacetylation activity with IC_{50} values of 1.7 μM and 8.2 μM, respectively. This compound also inhibited SIRT6 catalyzed defatty-acylation of TNFα in human embryonic kidney (HEK) 293 T cells. Thiomyristoyl-lysine had a broad anticancer effect on various human cancer cells and mouse models of breast cancer [65].

Inhibitors have been developed also for SIRT5; these have been based on the H3K9 peptide sequence [68]. The 11-residues succinyl-lysine peptide was the most potent inhibitor for SIRT5 having an IC_{50} value of 5 μM. A pentapeptide with thiosuccinyl-lysine at the middle of sequence seemed to increase the potency of the inhibition as compared to compounds with the corresponding group at the C-terminus or N-terminus. Carbamoyl phosphate synthetase peptides (CPS1) containing succinyl-lysine were demonstrated to inhibit specifically SIRT5; these displayed the inhibitory constant (K_i) values from 5 to 100 μM [69]. Zang and coworkers [70] identified a potent thiourea-type linear peptide which contained N^{ε}-carboxyethyl-thiocarbamoyl-lysine (called the "SIRT5 inhibitory warhead"). This compound was selective, but the warhead seemed to result in a metabolically unstable compound. Later, the cyclic pentapeptide-based counterpart was observed to be more stable, cell permeable, and selective toward SIRT5 with an IC_{50} value of 7.5 μM for SIRT5, but, for other sirtuins, it was a very weak inhibitor [71].

Bivalent sirtuin inhibitors were designed to bind into two binding pockets at the same time; the *N*-thioacetyl-lysine part could bind to the peptide substrate binding site, and the other moiety occupied the pocket of the NAM moiety of NAD^+. The most potent bivalent inhibitor exhibited an IC_{50} value of 12 μM for SIRT1, but only weak inhibition toward SIRT2 and SIRT3 (compound 18 in Fig. 3). This bivalent inhibitor was more isoform-selective than the monovalent counterpart and had IC_{50} values of 39 μM, 22 μM, and 54 μM for SIRT1, SIRT2, and SIRT3, respectively [72].

2.3 Small Molecule Inhibitors

Since 2001, a series of small molecular sirtuin inhibitors have been developed that can be classified based on their scaffolds as β-naphthols, indoles, chromanones, sulfobenzoic acid, or others. SIRT1 has been the most widely studied of the sirtuins mainly because it was the first sirtuin linked with aging-associated disorders [10, 13, 73, 74]. The predominance of SIRT1 is also partly attributable to the fact that initially small-molecule inhibitors were tested only for SIRT1 and SIRT2. The importance of other sirtuins increased with time, and today, compounds inhibiting the other sirtuins have been examined, especially those binding to SIRT3 and SIRT5. Currently, the design of sirtuin inhibitors has focused more on identifying selective SIRT2 inhibitors, and several of them have been published since 2012. A few inhibitors have been developed for SIRT6 [59, 75, 76] and SIRT7 [77]. For SIRT4 there are no compounds yet reported to have an IC_{50} value.

2.3.1 β-Naphthol Derivatives

Sirtinol was one of the first identified sirtuin inhibitors (compound 19 in Fig. 4); it showed inhibitory activity in vitro toward both yeast Sir2 ($IC_{50} = 68$ μM) and human various sirtuins [78–80]. It was observed that the 2-hydroxyl-1-napthol scaffold was essential for preventing the deacetylation reaction of sirtuins [78]. Later, the inhibitory activity of sirtinol was reported to lie in a range of IC_{50} values of 37–131 μM for SIRT1 and a range of IC_{50} values of 38–58 μM for SIRT2 and at IC_{50} value of 48.9 μM for SIRT5 [80]. In addition, several analogs of sirtinol have been designed. M15, another β-naphthol derivative, was synthesized, but it was less potent than sirtinol. A set of *ortho-* to *meta-* or *para*-substitution derivatives of sirtinol exhibited two to tenfold improvements in the SIRT1 and SIRT2 inhibition activity [79]. Interestingly, some of these compounds also exhibited anticancer activity [81, 82].

Modifications of sirtinol were used to identify a novel potent SIRT1 inhibitor, JGB-1741 (compound 20 in Fig. 4), that consists of a hydroxyl-naphthol moiety and a thiophene ring. JGB-1741 had an IC_{50} value of 15 μM for SIRT1, whereas it was a weak inhibitor for SIRT2 and SIRT3. JGB-1741 was observed to influence the proliferation of human cancer cells (K562, HepG2, and MDA-MB-231), to increase p53 acetylation, and to induce p53-mediated apoptosis in human breast cancer cells [83].

Salermide (compound 21 in Fig. 4) was designed as analog of sirtinol, with a reverse amide in the meta-position, and it showed SIRT2 inhibition with an IC_{50} value of 25 μM. Salermide was nontoxic to mice, and it induced apoptosis in several cancer cells [82, 84–86]. Salermide showed SIRT2 inhibition with an IC_{50} value of 25 μM. Further modifications of Salermide led to the development of several potent SIRT1 and SIRT2 inhibitors. The most potent inhibitor of these derivatives was 4-(2-phenylpropyl)thio-derivative (compound 22 in Fig. 4). This derivative of

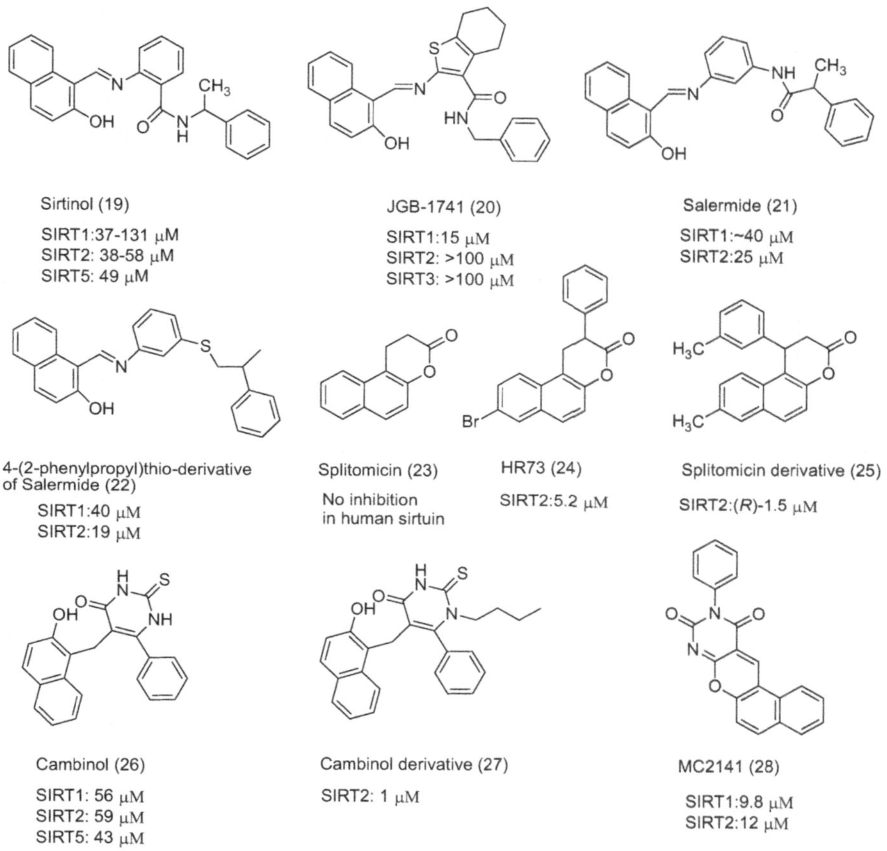

Fig. 4 The structures of some β-naphthol-based sirtuin inhibitors and their IC_{50} values

Salermide also exhibited an antiproliferative effect on human leukemia cells (MOLT4), breast cancer (MDA-MB-231), and colon (RKO) cancer cells. It displayed potency against colorectal carcinoma and glioblastoma multiforme cancer stem cells. Some of these analogs evoked massive apoptosis of human leukemia (U937) cells [87].

Splitomicin (compound 23 in Fig. 4) was discovered in a cell-based phenotypic screening for yeast sirtuin Sir2p [88], but it did not have any impact on the human sirtuin subtypes. However, a series of splitomicin analogs were synthetized and tested against the human sirtuins, and they showed improved inhibitory activity as compared to splitomicin [89]. Structure-activity relationships (SAR)-studies revealed that analogs without a lactone ring were inactive, but the naphthalene moiety could be replaced with smaller aromatic rings without losing the inhibitory activity [90, 91]. A splitomicin derivative, HR73 (compound 24 in Fig. 4), showed inhibition toward human SIRT1 ($IC_{50} < 5$ μM). Interestingly, HR73 significantly decreased HIV transcription. Further modification of splitomicin analogs (e.g., compound 25 in Fig. 4) led to the discovery of novel human SIRT2 inhibitors with IC_{50} values in a range of 0.8–38 μM for SIRT2 [92, 93]. Some splitomicin analogs exerted a weak antiproliferative effect in human breast cancer (MCF7) cells [93], and splitomicin prevented inflammatory responses in human neutrophils [94].

Cambinol (compound 26 in Fig. 4) had a beta-naphthol moiety similar to that of sirtinol and its derivatives. Cambinol inhibited both SIRT1 (IC_{50} = 56 μM) and SIRT2 (IC_{50} = 59 μM) but showed no SIRT3 inhibition [95]. In later studies, cambinol displayed IC_{50} value of 43 μM also toward SIRT5 [80]. The β-naphthol structure was reported to be critical for the inhibition and replacing it with a phenyl ring abolished the activity. The kinetic studies showed that cambinol was a competitive inhibitor with the acetylated substrate but not with NAD^+. Cambinol had low toxicity in vivo, and cambinol was a potent antitumor agent in vitro and in vivo [95]. Recently, cambinol was identified to be ten times more potent in inhibiting neutral sphingomyelinase 2 in brain than its effects on either SIRT1 and SIRT2 [96]. In attempts to improve isoform-selectivity, a *p*-bromo-analog of cambinol was designed; this was a very potent SIRT1 inhibitor with an IC_{50} value of 13 μM, while it exhibited only modest potency toward SIRT2 (IC_{50} > 90 μM). Interestingly, adding a substituent at the N1-position in cambinol derivative (compound 27 in Fig. 4) led to an increase in the activity against SIRT2 (IC_{50} = 1 μM), and a decrease in the inhibition against SIRT1 [97]. Other cambinol analogs were synthesized with improved potencies and isoform-selectivities [98, 99]. These analogs exhibited some cytotoxicity in cell lines derived from Burkitt's lymphoma (Dakiki, Daudi, Mutu, Oku, Ramos, Namalwa) and in colon (HCT116), breast (MCF7), and non-small cell lung carcinoma (NCI-H460).

Various benzodeazaoxaflavin (compound 28 in Fig. 4) derivatives were synthetized with low micromolar level potencies at inhibiting both SIRT1 and SIRT2 [100, 101]. MC2141 exhibited antiproliferative activity in human cancer cell lines (Raji, DLD1, and HeLa) as well as in cancer stem cells of colorectal carcinoma and glioblastoma multiforme. Some analogs of MC2141 were designed showing SIRT1 and SIRT2 inhibition with IC_{50} values in the low micromolar range.

2.3.2 Indole Derivatives as Potent SIRT1 Inhibitors

A high-throughput screening (HTS) of 280,000 compounds led to the discovery of a potent indole-based inhibitor, called Ex-527 [102]. Ex-527 (compound 29 in Fig. 6) was a nanomolar inhibitor for SIRT1 (IC_{50} = 38–100 nM), and this was a 500-fold improvement compared to other sirtuin inhibitors available at that time. Interestingly, the (S)-isomer was the active form, while the (R)-isomer had significantly lower potency. In addition, this new indole compound showed isoform-selectivity over SIRT2 and SIRT3 since their IC_{50} values were in the micromolar range and Ex-527 exerted no inhibition toward SIRT4–7 [102, 103]. This inspired the synthesis of various analogs with an indole scaffold. SAR analysis of these compounds revealed the necessity of carboxamide and its position for the inhibitory activity since changing the carboxamide from 1-position to 2-position led to a 350-fold loss in potency. The substitution of a small nonpolar group such as methyl at the para-position and the expansion of the carbon 6-membered ring to a 7-membered ring retained the inhibition potency. Kinetic studies suggested that Ex-527 was not a competitive inhibitor with the substrate peptide or with NAD^+. Ex-527 was postulated to bind to the protein after NAM release and potentially affects the release of the reaction products. The crystal structure of Ex-527 bound with different sirtuins revealed that Ex-527 was occupying the nicotinamide site and a pocket of the ribose of NAD^+ (Fig. 5) [104, 105]. Ex-527 induced apoptosis in acute myeloid leukemia via p53 [106]. In addition, Ex-527 was assessed as a potential HD therapy as it suppressed the deacetylation of mutant huntingtin protein, i.e., it accelerated the degradation of the acetylated mutant via autophagy [107]. Ex-527, also called selisistat, passed through phase II clinical trials for the treatment of HD [108, 109].

Selisistat was also combined with carprofen, a known anti-inflammatory drug. The most potent analog of this combination increased the acetyl-p53 and acetyl-alpha-tubulin levels and induced apoptosis in human macrophage (U937) cells [110]. A few other indole derivatives were developed as sirtuin inhibitors (compound 30 in Fig. 6). One of these derivatives was called AC-93253 (compound 31 Fig. 6); this compound displayed low micromolar inhibition toward SIRT1, SIRT2, and SIRT3. AC-93253 induced cytotoxicity in several cancer cell lines (DU-145, A549, NCI-H460, MiaPaCa2, HuVEC, PrEC, HMEC) in the submicromolar range [111]. In addition, 3-arylideneindolin-2-ones were identified as novel sirtuin inhibitors. Among these compounds, GW5074 (compound 32 in Fig. 6) was observed to be a potent inhibitor of various sirtuins [112, 113].

Inauhzin (compound 33 in Fig. 6) belongs to indole-based derivatives that showed inhibitory activity toward SIRT1 with an IC_{50} value ranging from 0.7 to 2 μM. Inauhzin evoked p53-dependent apoptosis of human cancer cells without causing any apparently genotoxic stress [114]. Some indole derivatives were developed by combining an indole ring with a substituted triazole; they were found to be

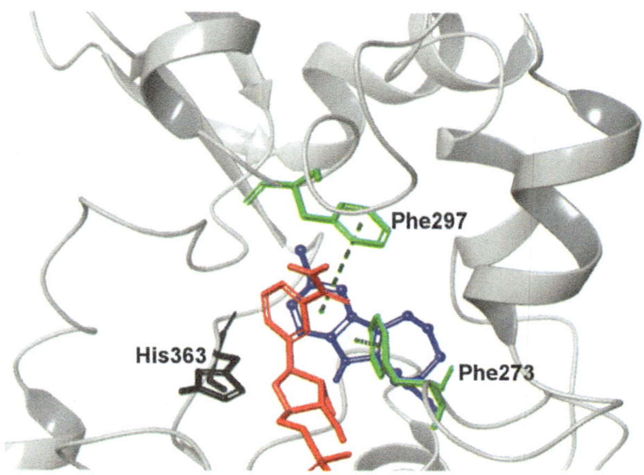

Fig. 5 The crystal structure of SIRT1 together with inhibitor Ex-527 (blue) and ribose-nicotinamide moiety of NAD^+ (red) [104]. Active histidine is marked with black. Residues forming interactions to Ex-527 are marked with green color. Green dashes indicate π-π interactions

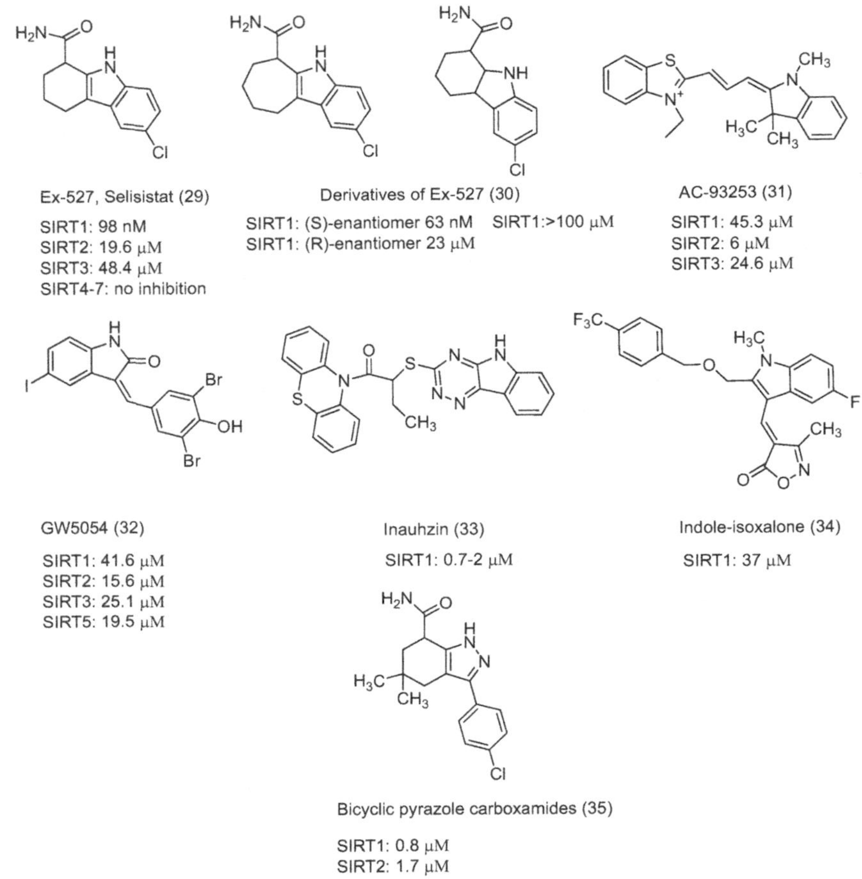

Fig. 6 The structures of indole-based sirtuin inhibitors and their IC_{50} values

SIRT1 inhibitors. These compounds were able to prevent the growth in breast cancer cell lines (MDA-MB-231) and prostate cancer cell line (LnCAP) [115]. A series of indole-isoxazolone derivatives were synthetized, and the most potent inhibitor (compound 34 in Fig. 6) showed IC_{50} value of 37 μM for SIRT1. The compounds were screened against human metastatic breast cancer cells (MDA-MB-231), human breast cancer cells (MCF7), and human colon adenocarcinoma cells (HT-29) to evaluate their in vitro cytotoxic property [116].

A set of bicyclic pyrazole derivatives was synthetized and tested against SIRT1 and SIRT2 [117]. In general, the bicyclic pyrazoles were slightly more potent against SIRT1 over SIRT2. The 4-chlorophenyl pyrazole carboxamide was a micromolar inhibitor of SIRT1 and showed only weak inhibition for SIRT2. The most potent SIRT1 inhibitor was 3-(4-chlorophenyl)-5,5-dimethyl-4,5,6,7-tetrahydro-1H-indazole-7-carboxamide (compound 35 in Fig. 6) showing an IC_{50} value of 0.8 μM

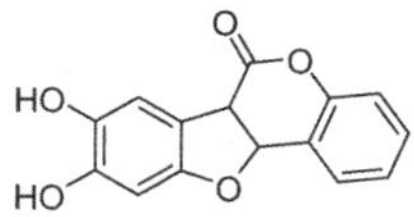

A3 (36)

SIRT2: 45 μM

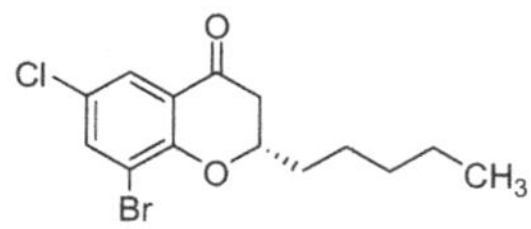

(*S*)-8-bromo-6-chloro-2-pentylchroman-4-one (37)

SIRT2: 1.5 μM

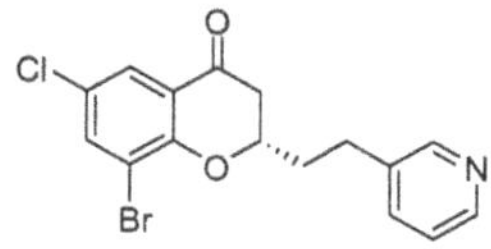

Chroman-4-one derivative (38)

SIRT2: 3.7 μM

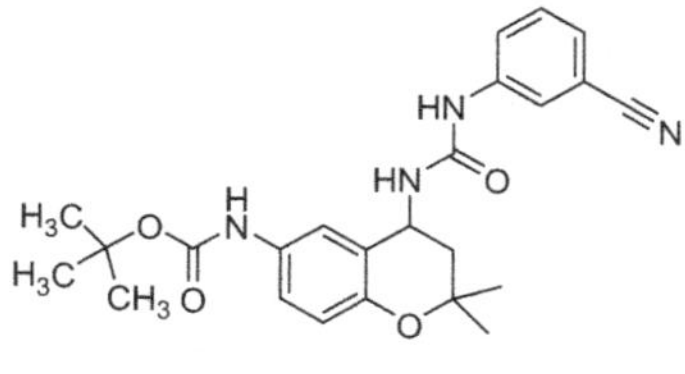

R/S-N-3-cyanophenyl-N'-(6-tert-butoxycarbonylamino-3,4-dihydro-2,2-dimethyl-2H-1-benzopyran-4-yl)urea (39)

SIRT1: 6.1 μM
SIRT2: 4.2 μM

Fig. 7 The structures of chromanone and chroman derived sirtuin inhibitors and their IC_{50} values

[117]. One indole derivative, GW5074 (3-(3,5-dibromo-4-hydroxybenzylidene)-5-iodo-1,3-dihydroindol-2-one), was reported to be a potent inhibitor of the desuccinylation reaction of SIRT5 ($IC_{50} = 19.5$ μM) but not against deacetylation reaction [113].

2.3.3 Chromanone and Chroman Derivatives as SIRT2 Inhibitors

A3 (compound 36 in Fig. 7) was discovered by phenotypic screening already in the year 2001. A3 inhibited in vitro yeast Sir2p and human SIRT2 deacetylase activity with IC_{50} values of 66 μM and 45 μM, respectively [78]. Several chromanone derivatives were developed and reported as selective and potent SIRT2 inhibitors. The most potent compound, (S)-8-bromo-6-chloro-2-pentylchroman-4-one, displayed SIRT2 inhibition with IC_{50} value of 1.5 μM (compound 37 Fig. 7). Compound 37 did not exhibit any inhibitory activity against SIRT1 and SIRT3 [118].

Although compound 37 was potent and selective for SIRT2 over SIRT1, it had poor solubility, and thus a series of analogs were developed. These analogs displayed low micromolar potency similar to that of compound 37. Two analogs showed an antiproliferative effect in human breast cancer (MCF7) and in human lung cancer (A549) cell lines [119]. Recently, the binding of chroman-4-one derivatives (compound 38 in Fig. 7) to SIRT2 was explored by a photoaffinity labeling technique [120].

Various 2,2-dimethylchroman derivatives exhibited an antiproliferative effect in glioma cells. One of the antiproliferative agents, compound 39 (Fig. 7), was also a potent SIRT1 inhibitor ($IC_{50} = 6.1$ μM) and SIRT2 inhibitor ($IC_{50} = 4.2$ μM). This compound displayed the reduction of tumor formation in vivo in zebrafish xenografts [121].

2.3.4 Sirtuin-Rearranging Ligand2 as an Isoform-Selective Inhibitor for SIRT2

Sirtuin-rearranging ligand2 (SirReal2) was identified by in vitro screening campaign (compound 40 in Fig. 8) [122]. SirReal2 was highly isoform-selective compound which exhibited the inhibition of SIRT2 with IC_{50} value of 400 nM, and it showed no activity toward SIRT3–5. SirReal2 displayed low inhibitory activities for SIRT1 (22% inhibition at 100 μM) and for SIRT6 (19% inhibition at 200 μM). The crystal structure of SIRT2 with co-crystallized SirReal2 revealed that SirReal2 bound via hydrophobic interactions into the extended nicotinamide moiety pocket of SIRT2 (Fig. 9) [122]. The extended pocket of nicotinamide moiety had been proposed from a previous molecular modeling study [123].

Several aminothiazole derivatives were developed based on the SirReal2 compound, and they were also SIRT2 selective inhibitors [124, 125]. Interestingly, one derivative (compound 41 in Fig. 8) was generated by combining aminothiazoles with thalidomide; this compound displayed an IC_{50} value of 250 nM for SIRT2 [126].

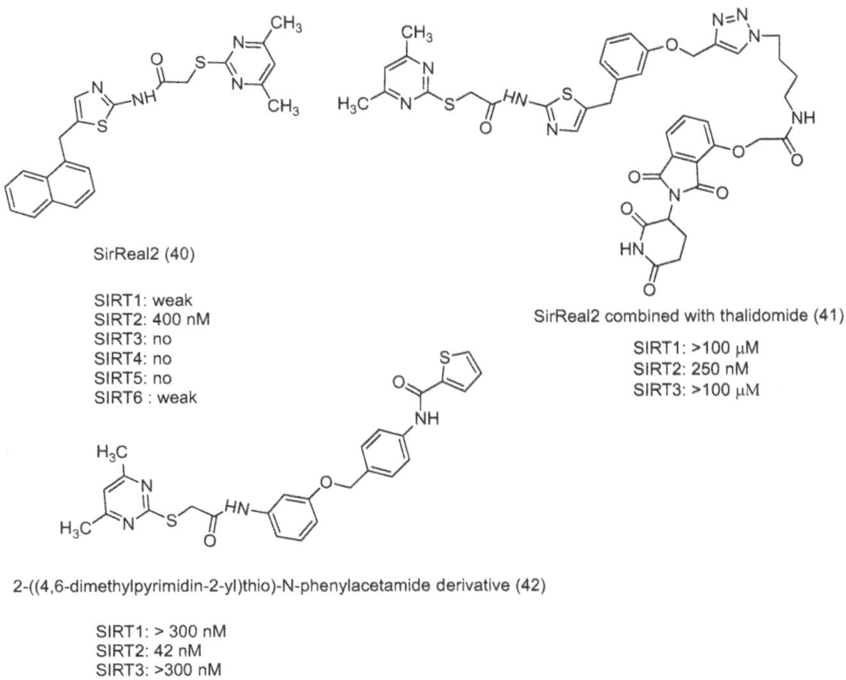

Fig. 8 The structures of SirReal2 and its derivatives and their IC_{50} values

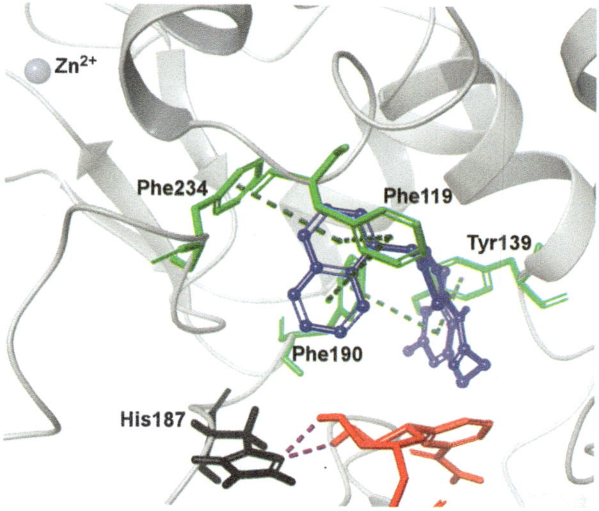

Fig. 9 The crystal structure of SIRT2 together with inhibitor SirReal2 (blue) and NAD^+ (red) [122]. Active histidine is marked with black. Residues making interactions with the inhibitor are marked with green color. Green dashes indicate π-π interactions and magenta dashes indicate hydrogen bonding

Another series of SirReal2 analogs was developed by Yang and coworkers; the most potent inhibitor (compound 42 in Fig. 8) exhibited an IC_{50} value of 42 nM for SIRT2 and displayed an antiproliferative effect in a breast cancer cell line (MCF7) [125].

2.3.5 AGK2 and Sulfobenzoic Acid Derivatives

Based on screening, AGK2 (compound 43 in Fig. 10) was identified as a selective SIRT2 inhibitor exhibiting an IC_{50} value of 3.5 μM [125]. In this screening program, other potent SIRT2 inhibitors were also identified: sulfobenzoic acid derivatives, AK–1 (IC_{50} = 12.5 μM) (compound 44 in Fig. 10) [127] and C2–8 [128]. Interestingly, both AGK2 and AK-1 prevented alpha-synuclein toxicity in cell-based and *Drosophila* models of Parkinson's disease (PD) [129]. They also prevented mutant polyglutamine toxicity in *Drosophila* and *C. elegans* PD models [130] and were also neuroprotective in mouse models [131]. The sulfobenzoic acid derivative C2–8 was neuroprotective in HD mice [132, 133].

Several analogs of AK-1 were developed and tested for inhibitory properties against SIRT1–3. Many of these analogs were more potent inhibitors against SIRT2 displaying IC_{50} values in a range between 4 μM and 10 μM in comparison with their effects on the other sirtuins [134]. 3-(1-Azepanylsulfonyl-)-*N*-(3-bromophenyl) benzamide, also called AK-7 (compound 45 in Fig. 10), showed a dose-dependent and selective inhibition for SIRT2 (IC_{50} = 15.5 μM) [133]. However, AK-7 was neuroprotective in vitro, and it reduced polyglutamine inclusions and cholesterol levels in neurons. AK-7 has also been studied in vivo in both truncated and full-length HD mouse models where it was able to improve the mice's motor function, extend their survival, and reduce the extent of brain atrophy [132]. AK-7 was also claimed to be neuroprotective in PD [135, 136]. In addition, the novel thiazole-containing inhibitors of SIRT2 that are derivatives of 8-nitro-5-R-quinoline and

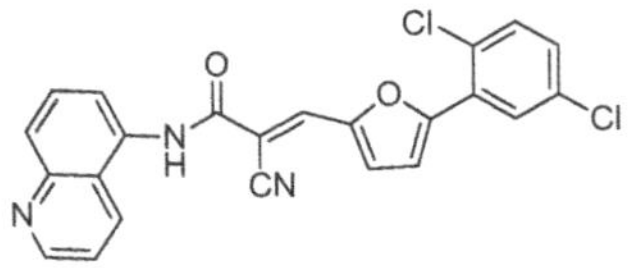

AGK2 (43)

SIRT1: > 50 μM
SIRT2: 3.5 μM
SIRT3: > 50 μM

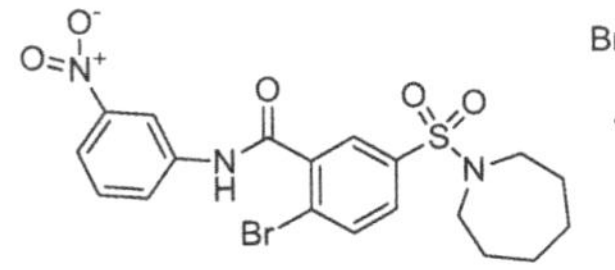

AK-1 (44)

SIRT2: 12.5 μM

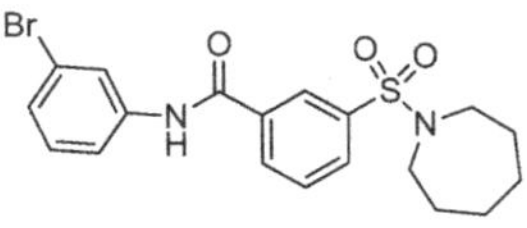

AK-7 (45)

SIRT1: >300 μM
SIRT2: 15.5 μM
SIRT3: >300 μM

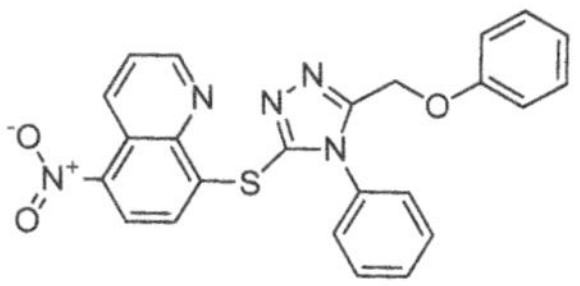

MIND4 (46)

SIRT2: 1.2 μM

Fig. 10 The structures of AGK2 and sulfobenzoic acid-based sirtuin inhibitors and their IC_{50} values

5-nitro-8-R-quinoline have also been proposed as potential treatments for HD. The most potent compound, MIND4 (compound 46 in Fig. 10), displayed SIRT2 inhibitory activity with an IC_{50} value of 1.2 μM. It displayed some neuroprotective activity both in ex vivo brain slices and in *Drosophila* models of HD [137].

2.3.6 Various Small Molecules with Inhibitory Activities Towards Sirtuins

Suramin is an old drug that was discovered to be a potent SIRT1 inhibitor [138]. Suramin is a G-protein antagonist known to target purine binding sites, and it was originally used in the treatment of trypanosomiasis and onchocerciasis. Suramin exhibited IC_{50} values of 297 μM, 1.2 μM, and 22 μM for SIRT1, SIRT2, and SIRT5, respectively. Various suramin analogs have been synthetized with the most potent suramin derivative having an IC_{50} value of 93 nM for SIRT1 and exhibiting selectivity over SIRT2 ($IC_{50} = 2.3$ μM) [139].

N,N′-Bisbenzylidenebenzene-1,4-diamines and *N,N′*-bisbenzylidenenaphthalene-1,4-diamines were identified as sirtuin inhibitors from a virtual screening campaign conducted by Tervo et al. [140]. The most potent compound exhibited inhibition with an IC_{50} value of 58 μM for SIRT2 [141]. Glycine tryptamide was also identified by virtual screening [45]. This scaffold proved to be a starting structure for developing a novel series of sirtuin inhibitors; the best of these SIRT1 inhibitors had an IC_{50} value of 52 μM, and the best SIRT2 inhibitor had an IC_{50} value of 47 μM [142].

A virtual screening identified an oxadiazole-carbonylaminothiourea to be a sirtuin inhibitor. The hit compound displayed a preference toward SIRT2 over SIRT1 with

IC_{50} values of 192 μM and 57 μM for SIRT1 and SIRT2, respectively. Various substituents with an oxadiazole moiety were evaluated, and a low micromolar SIRT1 inhibitor was developed. The study also revealed that the oxadiazole-carbonylaminothiourea scaffold was more potent than an oxadiazole-carbonylaminourea scaffold, similar to the case of thioacetylated peptides and acetylated peptides [143]. Recently, a set of oxadiazole analogs has been designed as selective SIRT2 inhibitors with IC_{50} values in the range of 1.5–20 μM. Two of these analogs displayed antiproliferative effects in leukemia cell lines. Kinetic studies suggested that oxadiazole analogs were uncompetitive inhibitors. The crystal structure of SIRT2 with one of the analogs has been published [144].

Tenovin-1 and Tenovin-6 were identified in a cell-based screening. Tenovin-1 had problems with solubility, but Tenovin-6 was sixfold more water soluble than Tenovin-1. Tenovin-6 (compound 47 in Fig. 11) showed an IC_{50} value of 21 μM for SIRT1 and an IC_{50} value of 10 μM for SIRT2, but it was slightly less potent toward SIRT3. Tenovin-6 decreased tumor growth in vivo [145]. There are also reports that Tenovin-6 was able to prevent cell proliferation in chronic myeloid leukemia cells and an acute promyelocytic leukemia (NB4) cell line [73, 146]. Tenovin-6 has exhibited antitumor effects in various cancer cell lines such as gastric and uveal melanoma cells [147–149]. Tenovin-6 also displayed an antineoplastic effect in vitro and in vivo on various hematopoietic malignancies [150, 151]. A set of analogs of Tenovin-1 and Tenovin-6 was synthetized by adding various substituents with different sizes and electronic properties in the tert-butyl-substituted aromatic ring [152, 153]. The most potent analog was Tenovin-D3 (compound 48 in Fig. 11) with IC_{50} values of 22 μM for SIRT2 and >90 μM for SIRT1 [154].

The thieno[3,2-d]pyrimidine-6-carboxamide scaffold was discovered by screening a 1.2 million compound library using Encoded Library Technology. Several potent SIRT1/SIRT2/SIRT3 inhibitors were identified with a nanomolar level inhibition for sirtuins. Compound 49 in Fig. 11 is an example of a carboxamide-based sirtuin inhibitor, showing equal potency for SIRT1, SIRT2, and SIRT3. The crystal structure was examined; this revealed that thieno[3,2-d]pyrimidine-6-carboxamides (Fig. 12) bound onto the site of the nicotinamide moiety [155].

Various benzimidazole derivatives were synthesized and their SIRT1 and SIRT2 inhibitory activities were evaluated. The most potent compound exhibited an IC_{50} value of 27 μM for SIRT2 and 52 μM for SIRT1. The compound showed also anticancer activity against different cancer cell lines, i.e., colon (HCT-116), breast (MDA-MB-468), and blood-leukemia (CCRF-CEM) [156].

Based on pharmacophore screening, ICL-SIRT078 (compound 50 in Fig. 11) was identified as a potent SIRT2 inhibitor with IC_{50} value of 0.17–3.96 μM. The compound showed no activity toward SIRT1, SIRT3, or SIRT5. Interestingly, the compound exerted a neuroprotective effect in a PD model. A series of analogs was prepared with some of them displaying inhibitory activity in a range of 1.90 μM and 6.52 μM [157].

SIRT3 isoform-selective inhibitor, SDX437 (compound 51 in Fig. 11), was screened from a 100,000 diverse compound library by using a novel substrate GYK(Ac)RGC developed during the study. Several of the hit compounds inhibited

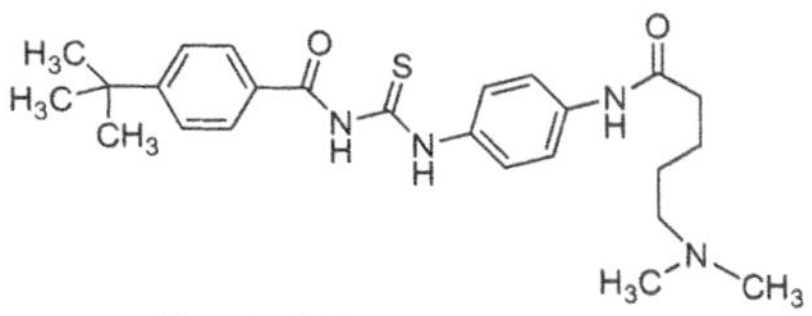

Tenovin-6 (47)

SIRT1: 21 μM
SIRT2: 10 μM
SIRT3: 67 μM

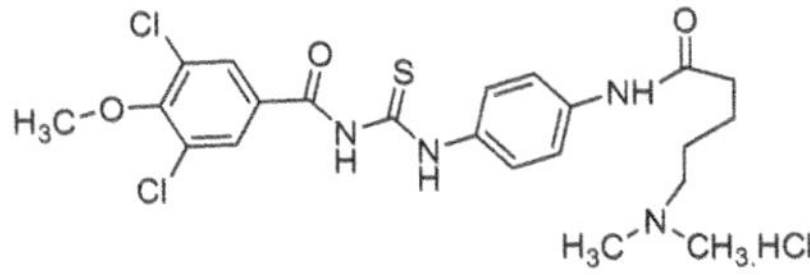

Tenovin-D3 (48)

SIRT1: >90 μM
SIRT2: 22 μM

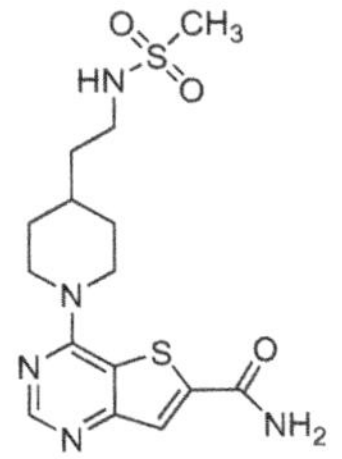

Thieno[3,2-d]pyrimidine-6-carboxamide derivative (49)

SIRT1: 4 nM
SIRT2: 1 nM
SIRT3: 7 nM

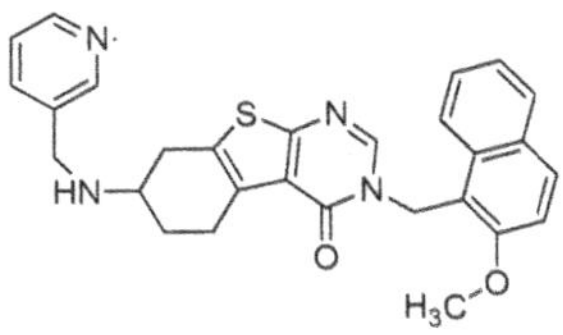

ICL-SIRT078 (50)

SIRT2: 170nm- 4μM

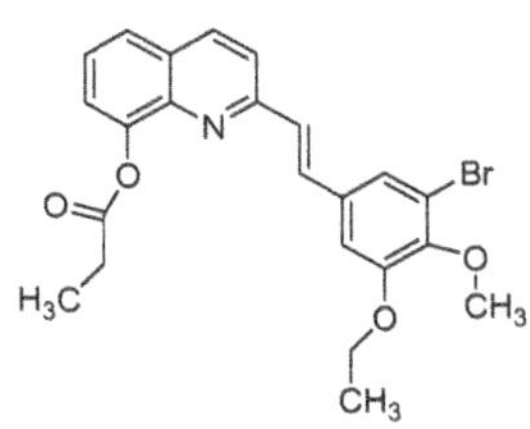

SDX-437 (51)

SIRT1:-
SIRT2: 700 nM

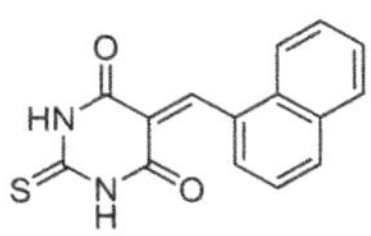

Thiobarbiturate (52)

SIRT1: 13 μM
SIRT2: 9 μM

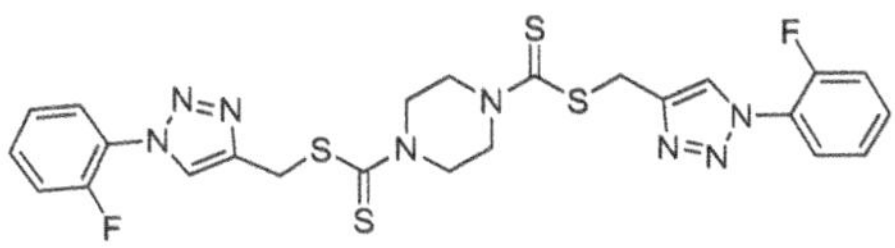

Triazole-dithiocarbate derivates (53)

SIRT1:1.08 μM

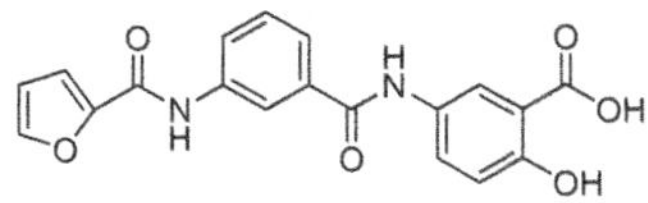

5-(3-(furan-2-carboxamido)benzamido)-2-hydroxybenzoic acid (54)

SIRT1:1578 μM
SIRT2: 751 μM
SIRT6: 89 μM

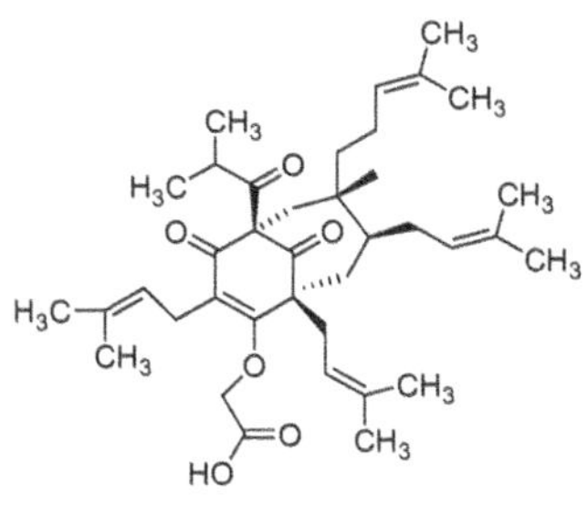

Aristoforin (55)

SIRT1: 7 μM
SIRT2: 21 μM

Fig. 11 The structures of various sirtuin inhibitors and their IC_{50} values

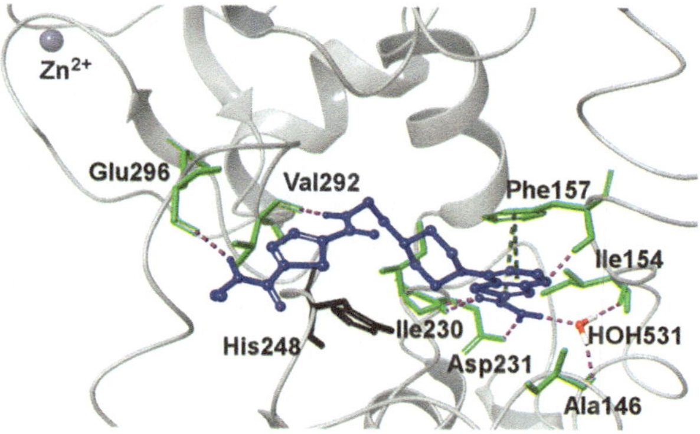

Fig. 12 The crystal structure of SIRT3 together with inhibitor ELT (blue) [155]. Active histidine is marked in black. Residues making interactions with the inhibitor are marked with green color. Green dashes indicate π-π interactions and magenta dashes indicate hydrogen bonding

also SIRT1 in addition to SIRT3; those compounds were filtered out in order to achieve isoform-selectivity. The compound named SDX-437 was observed to exhibit an IC_{50} value of 700 nM for SIRT3, whereas no inhibition was observed against SIRT1 [158].

Thiobarbiturates were reported to be potent inhibitors for SIRT1, SIRT2, and SIRT5 and generally less potent toward SIRT3. The most potent thiobarbiturate inhibitor (compound 52 in Fig. 11) showed IC_{50} values of 13 μM and 9 μM for SIRT1 and SIRT2, respectively [159]. Maurer and coworkers discovered several thiobarbiturate analogs, which displayed inhibition with IC_{50} values in the range of 3.4–89 μM toward SIRT1, in the range of 3.4–20.3 μM for SIRT2, and in the range of 2.3–67.3 μM for SIRT5 [80]. Modifications of the basic skeleton of thiobarbiturate led to the identification of a novel inhibitor scaffold, aurone. Some aurones showed inhibition toward the deacetylation activity of SIRT1, and the most potent of the aurones exhibited an IC_{50} value of 1.0 μM. Aurones showed also antiproliferative properties against two breast cancer cell lines (MDA-MB-231 and MCF7) in vitro [160].

Recently, a set of SIRT1 inhibitors having a 1,4-bispiperazinecarbodithioic acid methyl ester scaffold (compound 53 in Fig. 11) has been identified. The most potent compounds displayed IC_{50} values in a range of 1–11 μM for SIRT1; no IC_{50} values for other sirtuins were reported. The most potent compound with this scaffold inhibited SIRT2 at the cellular level and exerted also an antiproliferative effect in a human gastric cancer cell line MGC-803 [161].

So far, fewer SIRT6 regulators have been detected. In silico screening revealed a set of compounds which displayed inhibitory activity against SIRT6 at the micromolar level [75]. The most potent compound of the screening hits, 5-(3-(furan-2-carboxamido)benzamido)-2-hydroxybenzoic acid (compound 54 in Fig. 11), exhibited IC_{50} values of 89 μM for SIRT6, 1,578 μM for SIRT1, and 751 μM for SIRT2. A study where human primary pancreatic adenocarcinoma (BxPC-3) cells were treated with compound 54 resulted in an increased glucose uptake. Other micromolar SIRT6 inhibitors, quinazolinedione-based derivatives, also increased glucose uptake, and they also sensitized cancer cells to both gemcitabine and olaparib [162].

Many natural compounds possess an ability to inhibit the sirtuins. One of the compounds which was able to inhibit sirtuins was extracted from the bark of *Garcinia cochinchinensis* containing guttiferone G and hyperforin. In particular, a structurally similar but synthetic compound, aristoforin (compound 55 in Fig. 11), displayed IC_{50} values of 7 μM and 21 μM for SIRT1 and SIRT2, respectively, and it displayed an antiproliferative effect on HUVEC cells [163]. The tanikolide dimer from the Madagascar marine cyanobacterium *Lyngbya majuscula* was identified as a potent SIRT2 inhibitor ($IC_{50} = 176$ nM or 2.4 μM depending on assay). Synthetic stereoisomers had equal potency for SIRT1 (IC_{50} = 29–36 μM) and SIRT2 (IC_{50} = 2.4–3.3 μM) [164].

3 Sirtuin Activators

The identification of sirtuin activators has proved to be challenging, and thus far, fewer activators have been reported than inhibitors. Several SIRT1 activators have been published over the years since SIRT1 activation would be desirable in many age-related and metabolic conditions [42, 138, 165–171]. In recent years, there has been an intense debate about the identified SIRT1 activators; it has been claimed that sirtuins are only activated when hydrophobic fluorescent moieties have been attached to the peptide substrates, but not when unmodified peptides or the native protein substrates were used [172, 173].

Subsequently, these compounds were demonstrated to activate SIRT1 catalyzed deacetylation through an allosteric mechanism. This means that the binding of a small molecule to an allosteric site can induce a conformational change subsequently modifying the affinity of the enzyme for its native substrate [166, 171]. The crystal structures with activators [174, 175] revealed that most of the SIRT1 activators were binding to a SIRT1-specific N-terminal domain, which was postulated to induce a closure of the substrate-containing active site. The rational design of sirtuin activators, however, has been lacking and only few activators for other than SIRT1 have been published so far. This might partly be due to the absence of a deep understanding of the kinetics of sirtuin-catalyzed deacylation and especially an elucidation of the activation property has proven to be far more elusive. Nonetheless, recently, some SIRT6 activators have been identified.

3.1 Natural Compounds Activating Sirtuins

Certain plant polyphenols, members of a large and diverse group of plant secondary metabolites, were the first compounds discovered to be able to increase sirtuin activity [138]. The most potent of these compounds was resveratrol (compound 56 in Fig. 13), which was observed to stimulate the catalytic activity of yeast and human sirtuins. Dose-response experiments showed that resveratrol doubled the rate

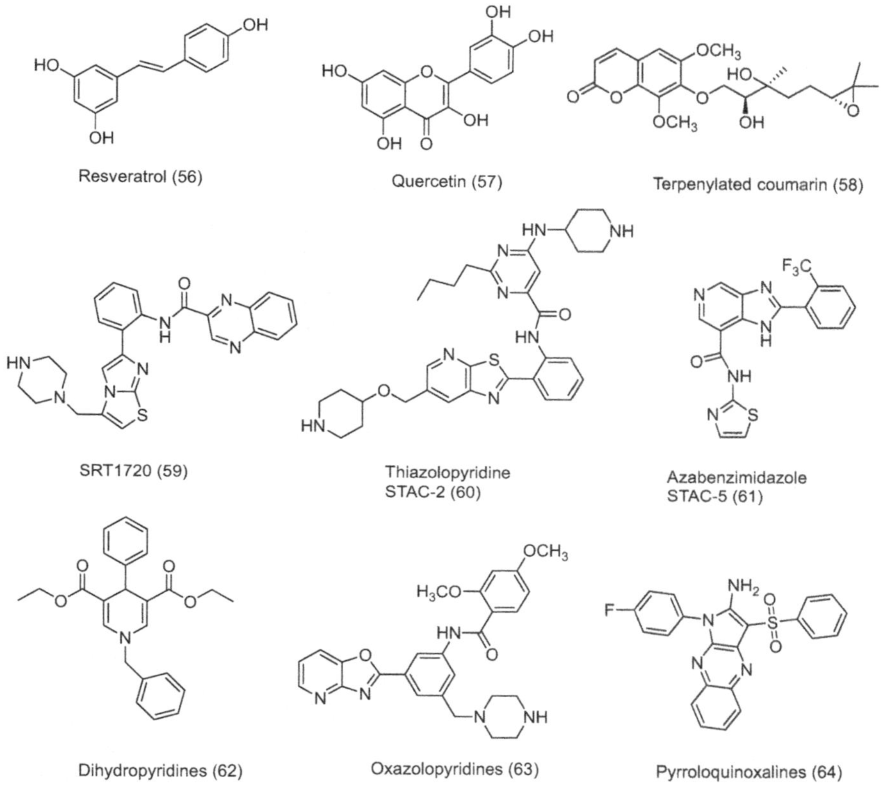

Fig. 13 Molecular structures of some sirtuin activators

of the SIRT1 catalyzed deacetylation reaction at about 11 μM concentration via a K_m (value of substrate concentration at half maximal velocity)-lowering mechanism [138]. Furthermore, resveratrol was also reported to stimulate the deacetylation activity of SIRT5 (by twofold); this was observed with the substrate consisting of fluorescent moieties. Interestingly resveratrol inhibited the deacetylation reaction of SIRT3 and desuccinylase reaction of SIRT5 [176]. Several other synthetic activators called sirtuin-activating compounds (STACs) related to resveratrol have been developed [166, 169, 171, 177].

Cao et al. [174] showed that the N-terminal domain is required for the stimulation of SIRT1's deacetylase activity by resveratrol (Fig. 14); in particular, Glu230 in this domain was demonstrated to be critical for stimulation. In addition, Dai et al. [175] demonstrated that residues 183–229 in the N-terminal domain were involved in STAC binding. The crystal structure of SIRT1 with three resveratrol molecules (Res1-Res3) and a 7-amino-4-methylcoumarin (AMC)-containing peptide revealed that the binding of two of these resveratrols to the SIRT1 N-terminal domain was probably due to a tighter binding between SIRT1 and the peptide substrate. The

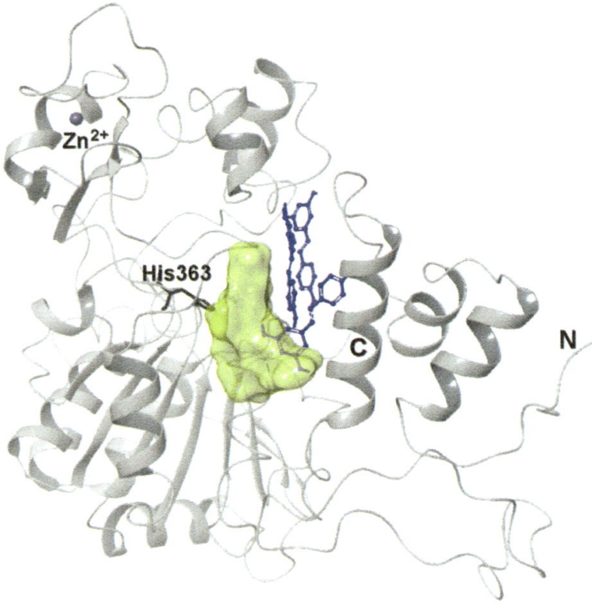

Fig. 14 The crystal structure of SIRT1 together with three resveratrol molecules (blue) and an AMC-containing peptide (yellow). Active histidine is marked with black [174]

contrasting binding modes of resveratrol and its analogs in different sirtuin complexes are speculated to be evidence of distinct binding conformations of the peptides and the local environments in the individual sirtuins. For example, in contrast to SIRT1, SIRT5 does not have an extended N-terminal domain, and thus when the structure of SIRT5 with resveratrol was examined, it was found that the compound bound to a location corresponding to the binding site of Res3 in SIRT1 [176]. Resveratrol exposure has been reported to extend the lifespan of numerous organisms including *C. elegans* and *S. cerevisiae* [178–180]. Resveratrol has displayed also a supportive effect in numerous age-related disease models including those mimicking cancer, type 1 and 2 diabetes, and cardiovascular disease [181].

Other polyphenols, such as chalcones including butein, and flavones including quercetin (compound 57 in Fig. 13) were demonstrated to increase SIRT1 deacetylation activity when a fluorescent AMC moiety was attached to the peptide substrate [168]. Additionally, flavonoids quercetin and luteolin were shown to stimulate SIRT6 at high concentration [182]. They exhibited SIRT6 activation with half maximal effective concentration (EC_{50}) values of 990 μM (Quercetin) and 270 μM (Luteolin). Rahnasto-Rilla et al. [76] observed that other flavonoids including anthocyanidins could stimulate SIRT6 catalyzed deacetylase reaction. Cyanidin was the most potent compound in the class of anthocyanidins, achieving a 55-fold maximal activation.

The terpenylated coumarins (compound 58 in Fig. 13) extracted from the bark of *Ailanthus altissima* were reported to increase SIRT1 activity in vitro and in vivo [167]. Furthermore, some ginsenosides such as ginsenoside Rb2 from *Panax ginseng* have been identified as SIRT1 activators. These ginsenoides increased the ATP content, inhibited the formation of reactive oxygen species (ROS), and enhanced the activity of a mitochondria-localized antioxidant enzyme, manganese superoxide dismutase (Mn-SOD) [183]. Although many studies have revealed that SIRT1 stimulates the expression of Mn-SOD [184–186], interestingly, the catalytic activity of enzyme is regulated by SIRT3-mediated deacetylation, particularly at the Lys68 site [187]. The lignan honokiol (2-(4-hydroxy-3-prop-2-enyl-phenyl)-4-prop-2-enyl-phenol), a natural biphenolic compound derived from the bark of magnolia trees, has been demonstrated to activate SIRT3-dependent deacetylation of MnSOD in vitro as well as enhancing SIRT3 expression in cardiomyocytes [188]. In addition, SIRT3 activator, 7-hydroxy-3-(4′-methoxyphenyl) coumarin, promoted the deacetylation and activation of Mn-SOD [189]. The increased activation of SIRT3 was associated with reduced ROS levels in cardiomyocytes obtained from wild-type mice [188]. Phlorotannin- and fucoidan-rich extracts, from brown macroalgae, namely, *Fucus distichus*, *Fucus vesiculosus*, and *Cystoseira tamariscifolia*, have demonstrated significant activation toward SIRT6. Subsequently the isolated compound fucoidan from *F. dichitus* was observed to stimulate deacetylation activity. The results suggest that the effect was SIRT6 specific and fucoidan did not exert any effects on either SIRT1 or SIRT3 but displayed weak inhibition towards SIRT2 [190].

3.2 *Various Scaffolds of Sirtuin Activators*

3.2.1 Imidazothiazoles

The first synthetic sirtuin activators, imidazothiazole derivatives, were chemically distinct from the polyphenol backbone of resveratrol but demonstrated to activate SIRT1 via the same K_m-lowering mechanism [171]. The experiment was carried out using a fluorescence polarization assay with a carboxytetramethylrhodamine (TAMRA)-tagged substrate and verified using mass spectrometry. The potent activators, SRT1460 and SRT1720 (compound 59 in Fig. 13), displayed EC_{50} values of 2.9 μM and 0.16 μM toward SIRT1, respectively. The activation by these compounds was selective versus SIRT2 and SIRT3 (SIRT2: SRT1460 $EC_{50} > 300$ μM, SRT1720 $EC_{50} = 37$ μM; SIRT3: $EC_{50} > 300$ μM). SRT1720 has been widely studied in various disease models such as type 2 diabetes, inflammation, and fatty liver disease. SRT1720 treatment has shown beneficial metabolic effects by reducing the blood glucose level and improving insulin sensitivity and glucose tolerance. It was able to reduce lipid accumulation in the liver. SRT1720 has displayed anti-inflammatory, anticancer, and cardioprotective properties [171].

Series of other synthetic compounds including thiazolopyridines (STAC-2) (compound 60 in Fig. 13), benzimidazoles (STAC-5) (compound 61 in Fig. 13), and bridged ureas (STAC-9) activated also SIRT1 [166, 169]. Synthetic compounds increased SIRT1 deacetylation when an AMC-tagged peptide was used as the substrate. However, it was demonstrated that the fluorescent moiety on substrates was indeed unnecessary for activation since it could be replaced by hydrophobic amino acids. In addition, specific hydrophobic motifs found in SIRT1 natural substrates such as peroxisome proliferator-activated receptor gamma coactivator 1-alpha (PGC-1α) and forkhead O transcription factor 3α (FOXO3α) were demonstrated to facilitate SIRT1 activation by all of the reported STACs compounds. Thus, SIRT1 activation was demonstrated as a substrate-specific effect with certain native sequences, while no effect or inhibition was observed with others [169].

3.2.2 Pyridine and Related Derivatives as Sirtuin Activators

One group of sirtuin activators, dihydropyridines (compound 62 in Fig. 13), with a benzyl group at the N1 position displayed SIRT1 activation with EC_{150} values (the effective concentration which increases the enzyme activity to 150%) about 1 μM but also a moderate activation for SIRT2 and SIRT3 with EC_{150} values of 15 μM and 50 μM, respectively. Dihydropyridines with a carbethoxy (ester) or carboxy substitution at the 3,5 positions demonstrated the highest activating potencies against SIRT1 when compared to the carboxamide moiety [170]. Valente et al. [191] synthetized a series of dihydropyridine analogs, and the results demonstrated that the replacement of the N1-benzyl moiety with the benzoyl portion, 2-pyrazinoyl, or 2-naphthoyl group led to higher SIRT1 activation. In addition, the substitution of the C4-phenyl ring with mono- or bicyclic heteroaromatic rings increased potency against SIRT1. Modified compounds displayed high nitric oxide (NO) release in human keratinocyte (HaCat) cells and ameliorated skin repair in a mouse model of wound healing [191]. NO release is involved in the maintenance of skin homeostasis as well as in the modulation of inflammatory response.

Another group of pyridines, oxalopyridines (compound 63 in Fig. 13), showed moderate activation of SIRT1 with $EC_{1.5}$ (the concentration of compound required to increase the enzyme activity by 50%) in a range of 6–25 μM. Later, a set of oxalopyridine analogs was synthetized, and the results revealed the importance of the position of the substituents in the phenyl ring. Ortho-substitution in the central phenyl ring gave the most potent activation, $EC_{1.5} = 0.9$ μM. By replacing the bicyclic scaffold to a benzimidazole and its analog, even more potent SIRT1 activators were developed: benzimidazoles and azalbenzimidazoles exhibited $EC_{1.5}$ values of 0.3 μM benzimidazole and 0.5 μM = azalbenzimidasoles, respectively [165].

3.2.3 Other Activators

Feldman et al. [42] reported the activation of SIRT6 by long-chain fatty acids. Free long-chain fatty acids stimulated the deacetylation activity of SIRT6 by factors of 5.6 and 6.2 in the case of oleic and linoleic acid, respectively. Other fatty acids, such as myristic, palmitic, stearic, γ-linolenic, and α-linolenic acids, showed activation factors of 2.5–2.8. The activation mechanism was investigated by steady-state kinetic analyses, and the results suggested that myristic acid could stimulate the deacetylase activity of SIRT6 by increasing the affinity of the enzyme for an acetylated substrate by 35 times. A series of *N*-acylethanolamines also possessed slightly increased SIRT6 activity, with oleoylethanolamide and myristoylethanolamide exhibiting maximal activation, i.e., a doubling [182].

Pyrroloquinoxalines (compound 64 in Fig. 13) were identified as SIRT1 activators displaying EC_{50} values greater than 1 μM for SIRT1. These compounds decreased lipopolysaccharides (LPS)-induced tumor necrosis factor-α (TNF-α) in human THP-1 leukemia cells [192]. You and coworkers screened as SIRT6 activators a series of the pyrrolo[1,2-a]quinoxaline-based compounds [31]. The most potent compound evoked a dose-dependent increase in SIRT6 activity, with a maximum of ≈twofold stimulation and an EC_{50} value of 38 μM. Examination of several X-ray structures with co-crystallized compounds has demonstrated that these compounds bind predominantly through the C-pocket at the acyl channel.

4 General Conclusions

The sirtuins are a family of antiaging enzymes; they are some of the most promising epigenetic targets in drug development for age-related diseases. The human SIRT1–7 play important roles in numerous biological processes. They affect gene transcription, tolerance to stress, adaptive mechanisms of metabolism, inflammatory response, biological rhythms, and cellular senescence, to mention but a few. Human SIRT1 has been one of the most intriguing research targets during the past years. Gradually, the role of other members of the sirtuin family has increased, and it has been speculated that modulation of these sirtuins could exert impacts on cancer, inflammation, and cardiovascular and metabolic diseases.

The regulation of sirtuins has attracted significant research interest, and many drug discovery research groups around the world have been searching for new sirtuin regulators. A large number of new sirtuin inhibitors, belonging to various compound classes, have been identified during the past years. In particular, compounds capable of inhibiting the sirtuins are intriguing targets for neurodegenerative diseases and cancer. There has been a special interest in developing sirtuin activators since they might have beneficial impact on human health. Although a group of synthetic and more potent activators has been described, there are still relatively few activators for most of the sirtuins. Selective and potent sirtuin inhibitors and activators may be beneficial in improving human health and combatting many diseases.

Compliance with Ethical Standards

Funding: M.R-R. was supported by Academy of Finland (grant no. 269341), Finnish Cultural Foundation, and Maud Kuistila Memorial Foundation. J.T was supported by UEF Doctoral School.

M.L-K. was supported by Academy of Finland (grant no 315824).

Conflict of Interest: There is no conflict of interest.

Ethical Approval: This chapter does not contain any studies with human participants or animals performed by any of the authors.

Informed Consent: This chapter does not contain any informed consent material.

References

1. Laurent G, German NJ, Saha AK et al (2013) SIRT4 coordinates the balance between lipid synthesis and catabolism by repressing malonyl CoA decarboxylase. Mol Cell 50(5):686–698. https://doi.org/10.1016/j.molcel.2013.05.012
2. Michishita E, McCord RA, Berber E et al (2008) SIRT6 is a histone H3 lysine 9 deacetylase that modulates telomeric chromatin. Nature 452(7186):492–496. https://doi.org/10.1038/nature06736
3. Nakagawa T, Lomb DJ, Haigis MC et al (2009) SIRT5 deacetylates carbamoyl phosphate synthetase 1 and regulates the urea cycle. Cell 137(3):560–570. https://doi.org/10.1016/j.cell.2009.02.026
4. Schwer B, North BJ, Frye RA et al (2002) The human silent information regulator (Sir)2 homologue hSIRT3 is a mitochondrial nicotinamide adenine dinucleotide-dependent deacetylase. J Cell Biol 158(4):647–657. https://doi.org/10.1083/jcb.200205057
5. Smith JS, Brachmann CB, Celic I et al (2000) A phylogenetically conserved NAD+-dependent protein deacetylase activity in the Sir2 protein family. Proc Natl Acad Sci U S A 97(12):6658–6663
6. Haigis MC, Mostoslavsky R, Haigis KM et al (2006) SIRT4 inhibits glutamate dehydrogenase and opposes the effects of calorie restriction in pancreatic beta cells. Cell 126(5):941–954. https://doi.org/10.1016/j.cell.2006.06.057
7. Kugel S, Mostoslavsky R (2014) Chromatin and beyond: the multitasking roles for SIRT6. Trends Biochem Sci 39(2):72–81. https://doi.org/10.1016/j.tibs.2013.12.002
8. Gertz M, Steegborn C (2010) Function and regulation of the mitochondrial sirtuin isoform Sirt5 in Mammalia. Biochim Biophys Acta 1804(8):1658–1665. https://doi.org/10.1016/j.bbapap.2009.09.011
9. Jiang H, Khan S, Wang Y et al (2013) SIRT6 regulates TNF-α secretion through hydrolysis of long-chain fatty acyl lysine. Nature 496(7443):110–113. https://doi.org/10.1038/nature12038
10. Aditya R, Kiran AR, Varma DS et al (2017) A review on SIRtuins in diabetes. Curr Pharm Des 23(16):2299–2307. https://doi.org/10.2174/1381612823666170125153334
11. Dai H, Sinclair DA, Ellis JL et al (2018) Sirtuin activators and inhibitors: promises, achievements, and challenges. Pharmacol Ther 188:140–154. https://doi.org/10.1016/j.pharmthera.2018.03.004
12. Herskovits AZ, Guarente L (2013) Sirtuin deacetylases in neurodegenerative diseases of aging. Cell Res 23(6):746–758. https://doi.org/10.1038/cr.2013.70
13. Kitada M, Koya D (2013) SIRT1 in type 2 diabetes: mechanisms and therapeutic potential. Diabetes Metab J 37(5):315–325. https://doi.org/10.4093/dmj.2013.37.5.315
14. Lin Z, Fang D (2013) The roles of SIRT1 in cancer. Genes Cancer 4(3–4):97–104. https://doi.org/10.1177/1947601912475079
15. Morris BJ (2013) Seven sirtuins for seven deadly diseases of aging. Free Radic Biol Med 56:133–171. https://doi.org/10.1016/j.freeradbiomed.2012.10.525
16. Roth M, Chen WY (2014) Sorting out functions of sirtuins in cancer. Oncogene 33 (13):1609–1620. https://doi.org/10.1038/onc.2013.120
17. Wilking MJ, Ahmad N (2015) The role of SIRT1 in cancer: the saga continues. Am J Pathol 185(1):26–28. https://doi.org/10.1016/j.ajpath.2014.10.002
18. Winnik S, Auwerx J, Sinclair DA et al (2015) Protective effects of sirtuins in cardiovascular diseases: from bench to bedside. Eur Heart J 36(48):3404–3412. https://doi.org/10.1093/eurheartj/ehv290

19. Yuan H, Su L, Chen WY (2013) The emerging and diverse roles of sirtuins in cancer: a clinical perspective. Onco Targets Ther 6:1399–1416. https://doi.org/10.2147/OTT.S37750
20. Anderson KA, Madsen AS, Olsen CA et al (2017) Metabolic control by sirtuins and other enzymes that sense NAD+, NADH, or their ratio. Biochim Biophys Acta Bioenerg 1858 (12):991–998. https://doi.org/10.1016/j.bbabio.2017.09.005
21. Mendes KL, Lelis DF, Santos SHS (2017) Nuclear sirtuins and inflammatory signaling pathways. Cytokine Growth Factor Rev 38:98–105. https://doi.org/10.1016/j.cytogfr.2017.11.001
22. Luo XY, Qu SL, Tang ZH et al (2014) SIRT1 in cardiovascular aging. Clin Chim Acta 437:106–114. https://doi.org/10.1016/j.cca.2014.07.019
23. Jęśko H, Wencel P, Strosznajder RP et al (2017) Sirtuins and their roles in brain aging and neurodegenerative disorders. Neurochem Res 42(3):876–890. https://doi.org/10.1007/s11064-016-2110-y
24. Tang BL (2017) Sirtuins as modifiers of Parkinson's disease pathology. J Neurosci Res 95(4):930–942. https://doi.org/10.1002/jnr.23806
25. Wong SY, Tang BL (2016) SIRT1 as a therapeutic target for Alzheimer's disease. Rev Neurosci 27(8):813–825. https://doi.org/10.1515/revneuro-2016-0023
26. Rizzi L, Roriz-Cruz M (2018) Sirtuin 1 and Alzheimer's disease: an up-to-date review. Neuropeptides 71:54–60. https://doi.org/10.1016/j.npep.2018.07.001
27. Donmez G, Outeiro TF (2013) SIRT1 and SIRT2: emerging targets in neurodegeneration. EMBO Mol Med 5(3):344–352. https://doi.org/10.1002/emmm.201302451
28. Naia L, Rego AC (2015) Sirtuins: double players in Huntington's disease. Biochim Biophys Acta 1852(10 Pt A):2183–2194. https://doi.org/10.1016/j.bbadis.2015.07.003
29. Pan M, Yuan H, Brent M et al (2012) SIRT1 contains N- and C-terminal regions that potentiate deacetylase activity. J Biol Chem 287(4):2468–2476. https://doi.org/10.1074/jbc.M111.285031
30. Sanders BD, Jackson B, Marmorstein R (2010) Structural basis for sirtuin function: what we know and what we don't. Biochim Biophys Acta 1804(8):1604–1616. https://doi.org/10.1016/j.bbapap.2009.09.009
31. You W, Rotili D, Li TM et al (2017) Structural basis of Sirtuin 6 activation by synthetic small molecules. Angew Chem Int Ed Engl 56(4):1007–1011. https://doi.org/10.1002/anie.201610082
32. Avalos JL, Bever KM, Wolberger C (2005) Mechanism of sirtuin inhibition by nicotinamide: altering the NAD(+) cosubstrate specificity of a Sir2 enzyme. Mol Cell 17(6):855–868. https://doi.org/10.1016/j.molcel.2005.02.022
33. Imai S, Guarente L (2016) It takes two to tango: NAD+ and sirtuins in aging/longevity control. NPJ Aging Mech Dis 2:16017. https://doi.org/10.1038/npjamd.2016.17
34. Madsen AS, Andersen C, Daoud M et al (2016) Investigating the sensitivity of NAD+-dependent sirtuin deacylation activities to NADH. J Biol Chem 291(13):7128–7141. https://doi.org/10.1074/jbc.M115.668699
35. Nguyen GT, Schaefer S, Gertz M et al (2013) Structures of human sirtuin 3 complexes with ADP-ribose and with carba-NAD+ and SRT1720: binding details and inhibition mechanism. Acta Crystallogr D Biol Crystallogr 69(Pt 8):1423–1432. https://doi.org/10.1107/S0907444913015448
36. Szczepankiewicz BG, Dai H, Koppetsch KJ et al (2012) Synthesis of carba-NAD and the structures of its ternary complexes with SIRT3 and SIRT5. J Org Chem 77(17):7319–7329. https://doi.org/10.1021/jo301067e
37. Pesnot T, Kempter J, Schemies J et al (2011) Two-step synthesis of novel, bioactive derivatives of the ubiquitous cofactor nicotinamide adenine dinucleotide (NAD). J Med Chem 54 (10):3492–3499. https://doi.org/10.1021/jm1013852
38. Trapp J, Jochum A, Meier R et al (2006) Adenosine mimetics as inhibitors of NAD+-dependent histone deacetylases, from kinase to sirtuin inhibition. J Med Chem 49 (25):7307–7316. https://doi.org/10.1021/jm060118b

39. Sauve AA, Schramm VL (2003) Sir2 regulation by nicotinamide results from switching between base exchange and deacetylation chemistry. Biochemistry 42(31):9249–9256. https://doi.org/10.1021/bi034959l
40. Hu J, Jing H, Lin H (2014) Sirtuin inhibitors as anticancer agents. Future Med Chem 6(8):945–966. https://doi.org/10.4155/fmc.14.44
41. Cui H, Kamal Z, Ai T et al (2014) Discovery of potent and selective sirtuin 2 (SIRT2) inhibitors using a fragment-based approach. J Med Chem 57(20):8340–8357. https://doi.org/10.1021/jm500777s
42. Feldman JL, Dittenhafer-Reed KE, Kudo N et al (2015) Kinetic and structural basis for acyl-group selectivity and NAD(+) dependence in sirtuin-catalyzed deacylation. Biochemistry 54(19):3037–3050. https://doi.org/10.1021/acs.biochem.5b00150
43. Lawson M, Uciechowska U, Schemies J (2010) Inhibitors to understand molecular mechanisms of NAD(+)-dependent deacetylases (sirtuins). Biochim Biophys Acta 1799 (10–12):726–739. https://doi.org/10.1016/j.bbagrm.2010.06.003
44. Rye PT, Frick LE, Ozbal CC et al (2011) Advances in label-free screening approaches for studying histone acetyltransferases. J Biomol Screen 16(10):1186–1195. https://doi.org/10.1177/1087057111418653
45. Tervo AJ, Suuronen T, Kyrylenko S (2006) Discovering inhibitors of human sirtuin type 2: novel structural scaffolds. J Med Chem 49(24):7239–7241. https://doi.org/10.1021/jm060686r
46. Yang H, Lavu S, Sinclair DA (2006) Nampt/PBEF/Visfatin: a regulator of mammalian health and longevity? Exp Gerontol 41(8):718–726. https://doi.org/10.1016/j.exger.2006.06.003
47. Hwang ES, Song SB (2017) Nicotinamide is an inhibitor of SIRT1 in vitro, but can be a stimulator in cells. Cell Mol Life Sci 74(18):3347–3362. https://doi.org/10.1007/s00018-017-2527-8
48. Suzuki T, Imai K, Nakagawa H et al (2006) 2-Anilinobenzamides as SIRT inhibitors. ChemMedChem 1(10):1059–1062. https://doi.org/10.1002/cmdc.200600162
49. Suzuki T, Imai K, Imai E et al (2009a) Design, synthesis, enzyme inhibition, and tumor cell growth inhibition of 2-anilinobenzamide derivatives as SIRT1 inhibitors. Bioorg Med Chem 17(16):5900–5905. https://doi.org/10.1016/j.bmc.2009.07.001
50. Suzuki T, Khan MN, Sawada H et al (2012) Design, synthesis, and biological activity of a novel series of human sirtuin-2-selective inhibitors. J Med Chem 55(12):5760–5773. https://doi.org/10.1021/jm3002108
51. Tatum PR, Sawada H, Ota Y (2014) Identification of novel SIRT2-selective inhibitors using a click chemistry approach. Bioorg Med Chem Lett 24(8):1871–1874. https://doi.org/10.1016/j.bmcl.2014.03.026
52. Mellini P, Itoh Y, Tsumoto H et al (2017) Potent mechanism-based sirtuin-2-selective inhibition by an *in situ* generated occupant of the substrate-binding site, "selectivity pocket" and NAD+ binding site. Chem Sci 8(9):6400–6408. https://doi.org/10.1039/c7sc02738a
53. Fatkins DG, Monnot AD, Zheng W (2006) Nε-thioacetyllysine: a multi-facet functional probe for enzymatic protein lysine Nε-deacetylation. Bioorg Med Chem Lett 16(14):3651–3656. https://doi.org/10.1016/j.bmcl.2006.04.075
54. Fatkins DG, Zheng W (2008) Substituting Nε-thioacetyl-lysine for Nε-acetyl-lysine in peptide substrates as a general approach to inhibiting human NAD+-dependent protein deacetylases. Int J Mol Sci 9(1):1–11
55. Smith BC, Denu JM (2007b) Mechanism-based inhibition of Sir2 deacetylases by thioacetyl-lysine peptide. Biochemistry 46(50):14478–14486. https://doi.org/10.1021/bi7013294
56. Huhtiniemi T, Suuronen T, Lahtela-Kakkonen M et al (2010) N(epsilon)-modified lysine containing inhibitors for SIRT1 and SIRT2. Bioorg Med Chem 18(15):5616–5625. https://doi.org/10.1016/j.bmc.2010.06.035
57. Smith BC, Denu JM (2007a) Acetyl-lysine analog peptides as mechanistic probes of protein deacetylases. J Biol Chem 282(51):37256–37265. https://doi.org/10.1074/jbc.M707878200

58. Kiviranta PH, Suuronen T, Wallén EAA et al (2009) N(ε)-thioacetyllysine-containing tri-, tetra-, and pentapeptides as SIRT1 and SIRT2 inhibitors. J Med Chem 52(7):2153–2156. https://doi.org/10.1021/jm801401k
59. Kokkonen P, Rahnasto-Rilla M, Kiviranta PH et al (2012) Peptides and pseudopeptides as SIRT6 deacetylation inhibitors. ACS Med Chem Lett 3(12):969–974. https://doi.org/10.1021/ml300139n
60. Huang Y, Liu J, Yan L et al (2016) Simple N(ε)-thioacetyl-lysine-containing cyclic peptides exhibiting highly potent sirtuin inhibition. Bioorg Med Chem Lett 26(6):1612–1617. https://doi.org/10.1016/j.bmcl.2016.01.086
61. Morimoto J, Hayashi Y, Suga H (2012) Discovery of macrocyclic peptides armed with a mechanism-based warhead: isoform-selective inhibition of human deacetylase SIRT2. Angew Chem Int Ed Engl 51(14):3423–3427. https://doi.org/10.1002/anie.201108118
62. Yamagata K, Goto Y, Nishimasu H et al (2014) Structural basis for potent inhibition of SIRT2 deacetylase by a macrocyclic peptide inducing dynamic structural change. Structure 22 (2):345–352. https://doi.org/10.1016/j.str.2013.12.001
63. Asaba T, Suzuki T, Ueda R et al (2009) Inhibition of human sirtuins by in situ generation of an acetylated lysine-ADP-ribose conjugate. J Am Chem Soc 131(20):6989–6996. https://doi.org/10.1021/ja807083y
64. Suzuki T, Asaba T, Imai E et al (2009b) Identification of a cell-active non-peptide sirtuin inhibitor containing N-thioacetyl lysine. Bioorg Med Chem Lett 19(19):5670–5672. https://doi.org/10.1016/j.bmcl.2009.08.028
65. Jing H, Hu J, He B et al (2016) A SIRT2-selective inhibitor promotes c-Myc oncoprotein degradation and exhibits broad anticancer activity. Cancer Cell 29(3):297–310. https://doi.org/10.1016/j.ccell.2016.02.007
66. Huhtiniemi T, Salo HS, Suuronen T et al (2011) Structure-based design of pseudopeptidic inhibitors for SIRT1 and SIRT2. J Med Chem 54(19):6456–6468. https://doi.org/10.1021/jm200590k
67. Mellini P, Kokkola T, Suuronen T et al (2013) Screen of pseudopeptidic inhibitors of human sirtuins 1-3: two lead compounds with antiproliferative effects in cancer cells. J Med Chem 56(17):6681–6695. https://doi.org/10.1021/jm400438k
68. He B, Du J, Lin H (2012) Thiosuccinyl peptides as Sirt5-specific inhibitors. J Am Chem Soc 134(4):1992–1995. https://doi.org/10.1021/ja2090417
69. Roessler C, Nowak T, Pannek M et al (2014) Chemical probing of the human sirtuin 5 active site reveals its substrate acyl specificity and peptide-based inhibitors. Angew Chem Int Ed Engl 53(40):10728–10732. https://doi.org/10.1002/anie.201402679
70. Zang W, Hao Y, Wang Z et al (2015) Novel thiourea-based sirtuin inhibitory warheads. Bioorg Med Chem Lett 25(16):3319–3324. https://doi.org/10.1016/j.bmcl.2015.05.058
71. Liu J, Huang Y, Zheng W (2016) A selective cyclic peptidic human SIRT5 inhibitor. Molecules 21(9):1217. https://doi.org/10.3390/molecules21091217
72. Wang J, Zang W, Liu J et al (2017) Bivalent SIRT1 inhibitors. Bioorg Med Chem Lett 27(2):180–186. https://doi.org/10.1016/j.bmcl.2016.11.082
73. Li L, Wang L, Li L et al (2012) Activation of p53 by SIRT1 inhibition enhances elimination of CML leukemia stem cells in combination with imatinib. Cancer Cell 21(2):266–281. https://doi.org/10.1016/j.ccr.2011.12.020
74. Ng FW, Wijaya L, Tang BL (2015) SIRT1 in the brain-connections with aging-associated disorders and lifespan. Front Cell Neurosci 9:64. https://doi.org/10.3389/fncel.2015.00064
75. Parenti MD, Grozio A, Bauer I et al (2014) Discovery of novel and selective SIRT6 inhibitors. J Med Chem 57(11):4796–4804. https://doi.org/10.1021/jm500487d
76. Rahnasto-Rilla M, Tyni J, Huovinen M et al (2018) Natural polyphenols as sirtuin 6 modulators. Sci Rep 8(1):4163. https://doi.org/10.1038/s41598-018-22388-5
77. Kim JH, Kim D, Cho SJ et al (2018) Identification of a novel SIRT7 inhibitor as anticancer drug candidate. Biochem Biophys Res Commun 508(2):451–457. https://doi.org/10.1016/j.bbrc.2018.11.120

78. Grozinger CM, Chao ED, Blackwell HE et al (2001) Identification of a class of small molecule inhibitors of the sirtuin family of NAD-dependent deacetylases by phenotypic screening. J Biol Chem 276(42):38837–38843. https://doi.org/10.1074/jbc.M106779200
79. Mai A, Massa S, Lavu S et al (2005) Design, synthesis, and biological evaluation of sirtinol analogues as class III histone/protein deacetylase (sirtuin) inhibitors. J Med Chem 48(24):7789–7795. https://doi.org/10.1021/jm050100l
80. Maurer B, Rumpf T, Scharfe M et al (2012) Inhibitors of the NAD(+)-dependent protein desuccinylase and demalonylase sirt5. ACS Med Chem Lett 3(12):1050–1053. https://doi.org/10.1021/ml3002709
81. Ota H, Tokunaga E, Chang K et al (2006) Sirt1 inhibitor, Sirtinol, induces senescence-like growth arrest with attenuated Ras-MAPK signaling in human cancer cells. Oncogene 25(2):176–185. https://doi.org/10.1038/sj.onc.1209049
82. Peck B, Chen CY, Ho KK et al (2010) SIRT inhibitors induce cell death and p53 acetylation through targeting both SIRT1 and SIRT2. Mol Cancer Ther 9(4):844–855. https://doi.org/10.1158/1535-7163.MCT-09-0971
83. Kalle AM, Mallika A, Badiger J et al (2010) Inhibition of SIRT1 by a small molecule induces apoptosis in breast cancer cells. Biochem Biophys Res Commun 401(1):13–19. https://doi.org/10.1016/j.bbrc.2010.08.118
84. Dastjerdi MN, Salahshoor MR, Mardani M et al (2013) The apoptotic effects of sirtuin1 inhibitor on the MCF-7 and MRC-5 cell lines. Res Pharm Sci 8(2):79–89
85. Lara E, Mai A, Calvanese V et al (2009) Salermide, a Sirtuin inhibitor with a strong cancer-specific proapoptotic effect. Oncogene 28(6):781–791. https://doi.org/10.1038/onc.2008.436
86. Liu G, Su L, Hao X et al (2012b) Salermide up-regulates death receptor 5 expression through the ATF4-ATF3-CHOP axis and leads to apoptosis in human cancer cells. J Cell Mol Med 16(7):1618–1628. https://doi.org/10.1111/j.1582-4934.2011.01401.x
87. Rotili D, Tarantino D, Nebbioso A et al (2012b) Discovery of salermide-related sirtuin inhibitors: binding mode studies and antiproliferative effects in cancer cells including cancer stem cells. J Med Chem 55(24):10937–10947. https://doi.org/10.1021/jm3011614
88. Bedalov A, Gatbonton T, Irvine WP et al (2001) Identification of a small molecule inhibitor of Sir2p. Proc Natl Acad Sci U S A 98(26):15113–15118. https://doi.org/10.1073/pnas.261574398
89. Hirao M, Posakony J, Nelson M et al (2003) Identification of selective inhibitors of NAD+-dependent deacetylases using phenotypic screens in yeast. J Biol Chem 278 (52):52773–52782. https://doi.org/10.1074/jbc.M308966200
90. Pagans S, Pedal A, North BJ et al (2005) SIRT1 regulates HIV transcription via Tat deacetylation. PLoS Biol 3(2):e41. https://doi.org/10.1371/journal.pbio.0030041
91. Posakony J, Hirao M, Stevens S et al (2004) Inhibitors of Sir2: evaluation of splitomicin analogues. J Med Chem 47(10):2635–2644. https://doi.org/10.1021/jm030473r
92. Freitag M, Schemies J, Larsen T et al (2011) Synthesis and biological activity of splitomicin analogs targeted at human NAD(+)-dependent histone deacetylases (sirtuins). Bioorg Med Chem 19(12):3669–3677. https://doi.org/10.1016/j.bmc.2011.01.026
93. Neugebauer RC, Uchiechowska U, Meier R et al (2008) Structure-activity studies on splitomicin derivatives as sirtuin inhibitors and computational prediction of binding mode. J Med Chem 51(5):1203–1213. https://doi.org/10.1021/jm700972e
94. Liu FC, Day YJ, Liou JT et al (2012a) Splitomicin inhibits MLP-induced superoxide anion production in human neutrophils by activate cAMP/PKA signaling inhibition of ERK pathway. Eur J Pharmacol 688(1–3):68–75. https://doi.org/10.1016/j.ejphar.2012.05.006
95. Heltweg B, Gatbonton T, Schuler AD et al (2006) Antitumor activity of a small-molecule inhibitor of human silent information regulator 2 enzymes. Cancer Res 66(8):4368–4377. https://doi.org/10.1158/0008-5472.CAN-05-3617
96. Figuera-Losada M, Stathis M, Dorskind JM (2015) Cambinol, a novel inhibitor of neutral sphingomyelinase 2 shows neuroprotective properties. PLoS One 10(5):e0124481. https://doi.org/10.1371/journal.pone.0124481

97. Medda F, Russell RJM, Higgins M et al (2009) Novel cambinol analogs as sirtuin inhibitors: synthesis, biological evaluation, and rationalization of activity. J Med Chem 52 (9):2673–2682. https://doi.org/10.1021/jm8014298
98. Mahajan SS, Scian M, Sripathy S et al (2014) Development of pyrazolone and isoxazol-5-one cambinol analogues as sirtuin inhibitors. J Med Chem 57(8):3283–3294. https://doi.org/10.1021/jm4018064
99. Medda F, Joseph TL, Pirrie L et al (2011) N1-Benzyl substituted cambinol analogues as isozyme selective inhibitors of the sirtuin family of proteins deacetylases. Med Chem Commun 2(7):611–615. https://doi.org/10.1039/C1MD00023C
100. Rotili D, Carafa V, Tarantino D et al (2011) Simplification of the tetracyclic SIRT1-selective inhibitor MC2141: coumarin- and pyrimidine-based SIRT1/2 inhibitors with different selectivity profile. Bioorg Med Chem 19(12):3659–3668. https://doi.org/10.1016/j.bmc.2011.01.025
101. Rotili D, Tarantino D, Carafa V et al (2012a) Benzodeazaoxaflavins as sirtuin inhibitors with antiproliferative properties in cancer stem cells. J Med Chem 55(18):8193–8197. https://doi.org/10.1021/jm301115r
102. Napper AD, Hixon J, McDonagh T et al (2005) Discovery of indoles as potent and selective inhibitors of the deacetylase SIRT1. J Med Chem 48(25):8045–8054. https://doi.org/10.1021/jm050522v
103. Solomon JM, Pasupuleti R, Xu L et al (2006) Inhibition of SIRT1 catalytic activity increases p53 acetylation but does not alter cell survival following DNA damage. Mol Cell Biol 26(1):28–38. https://doi.org/10.1128/MCB.26.1.28-38.2006
104. Zhao X, Allison D, Condon B et al (2013) The 2.5 Å crystal structure of the SIRT1 catalytic domain bound to nicotinamide adenine dinucleotide (NAD+) and an indole (EX527 analogue) reveals a novel mechanism of histone deacetylase inhibition. J Med Chem 56(3):963–969. https://doi.org/10.1021/jm301431y
105. Gertz M, Fischer F, Nguyen GTT (2013) Ex-527 inhibits Sirtuins by exploiting their unique NAD+-dependent deacetylation mechanism. Proc Natl Acad Sci U S A 110(30):E2772–E2781. https://doi.org/10.1073/pnas.1303628110
106. Sasca D, Hähnel PS, Szybinski J et al (2014) SIRT1 prevents genotoxic stress-induced p53 activation in acute myeloid leukemia. Blood 124(1):121–133. https://doi.org/10.1182/blood-2013-11-538819
107. Kim BS, Lee CH, Chang GE et al (2016) A potent and selective small molecule inhibitor of sirtuin 1 promotes differentiation of pluripotent P19 cells into functional neurons. Sci Rep 6:34324. https://doi.org/10.1038/srep34324
108. Süssmuth SD, Haider S, Landwehrmeyer GB et al (2015) An exploratory double-blind, randomized clinical trial with selisistat, a SirT1 inhibitor, in patients with Huntington's disease. Br J Clin Pharmacol 79(3):465–476. https://doi.org/10.1111/bcp.12512
109. Westerberg G, Chiesa JA, Andersen CA et al (2015) Safety, pharmacokinetics, pharmacogenomics and QT concentration-effect modelling of the SirT1 inhibitor selisistat in healthy volunteers. Br J Clin Pharmacol 79(3):477–491. https://doi.org/10.1111/bcp.12513
110. Mellini P, Carafa V, Di Rienzo B et al (2012) Carprofen analogues as sirtuin inhibitors: enzyme and cellular studies. ChemMedChem 7(11):1905–1908. https://doi.org/10.1002/cmdc.201200318
111. Zhang Y, Au Q, Zhang M et al (2009) Identification of a small molecule SIRT2 inhibitor with selective tumor cytotoxicity. Biochem Biophys Res Commun 386(4):729–733. https://doi.org/10.1016/j.bbrc.2009.06.113
112. Huber K, Schemies J, Uciechowska U et al (2010) Novel 3-arylideneindolin-2-ones as inhibitors of NAD+-dependent histone deacetylases (sirtuins). J Med Chem 53(3):1383–1386. https://doi.org/10.1021/jm901055u
113. Suenkel B, Fischer F, Steegborn C (2013) Inhibition of the human deacylase Sirtuin 5 by the indole GW5074. Bioorg Med Chem Lett 23(1):143–146. https://doi.org/10.1016/j.bmcl.2012.10.136

114. Zhang Q, Zeng SX, Zhang Y et al (2012) A small molecule Inauhzin inhibits SIRT1 activity and suppresses tumour growth through activation of p53. EMBO Mol Med 4(4):298–312. https://doi.org/10.1002/emmm.201100211
115. Panathur N, Dalimba U, Koushik PV et al (2013) Identification and characterization of novel indole based small molecules as anticancer agents through SIRT1 inhibition. Eur J Med Chem 69:125–138. https://doi.org/10.1016/j.ejmech.2013.08.018
116. Panathur N, Gokhale N, Dalimba U et al (2015) New indole-isoxazolone derivatives: synthesis, characterisation and in vitro SIRT1 inhibition studies. Bioorg Med Chem Lett 25(14):2768–2772. https://doi.org/10.1016/j.bmcl.2015.05.015
117. Therrien E, Laourche G, Nguyen N et al (2015) Discovery of bicyclic pyrazoles as class III histone deacetylase SIRT1 and SIRT2 inhibitors. Bioorg Med Chem Lett 25(12):2514–2518. https://doi.org/10.1016/j.bmcl.2015.04.068
118. Fridén-Saxin M, Seifert T, Landergren MR (2012) Synthesis and evaluation of substituted chroman-4-one and chromone derivatives as sirtuin 2-selective inhibitors. J Med Chem 55(16):7104–7113. https://doi.org/10.1021/jm3005288
119. Seifert T, Malo M, Kokkola T et al (2014) Chroman-4-one- and chromone-based sirtuin 2 inhibitors with antiproliferative properties in cancer cells. J Med Chem 57(23):9870–9888. https://doi.org/10.1021/jm500930h
120. Seifert T, Malo M, Lengqvist J et al (2016) Identification of the binding site of chroman-4-one-based Sirtuin 2-Sselective inhibitors using photoaffinity labeling in combination with tandem mass spectrometry. J Med Chem 59(23):10794–10799. https://doi.org/10.1021/acs.jmedchem.6b01117
121. Schnekenburger M, Goffin E, Lee JY et al (2017) Discovery and characterization of R/S-N-3-cyanophenyl-N′-(6-tert-butoxycarbonylamino-3,4-dihydro-2,2-dimethyl-2H-1-benzopyran-4-yl)urea, a new histone deacetylase class III inhibitor exerting antiproliferative activity against cancer cell lines. J Med Chem 60(11):4714–4733. https://doi.org/10.1021/acs.jmedchem.7b00533
122. Rumpf T, Schiedel M, Karaman B et al (2015) Selective Sirt2 inhibition by ligand-induced rearrangement of the active site. Nat Commun 6:6263. https://doi.org/10.1038/ncomms7263
123. Kiviranta PH, Salo HS, Leppänen J et al (2008) Characterization of the binding properties of SIRT2 inhibitors with a N-(3-phenylpropenoyl)-glycine tryptamide backbone. Bioorg Med Chem 16(17):8054–8062. https://doi.org/10.1016/j.bmc.2008.07.059
124. Schiedel M, Rumpf T, Karaman B et al (2016) Aminothiazoles as potent and selective Sirt2 inhibitors: a structure-activity relationship study. J Med Chem 59(4):1599–1612. https://doi.org/10.1021/acs.jmedchem.5b01517
125. Yang L, Ma X, Yuan C et al (2017) Discovery of 2-((4,6-dimethylpyrimidin-2-yl)thio)-N-phenylacetamide derivatives as new potent and selective human sirtuin 2 inhibitors. Eur J Med Chem 134:230–241. https://doi.org/10.1016/j.ejmech.2017.04.010
126. Schiedel M, Herp D, Hammelmann S, Swyter S, Lehotzky A, Robaa D, Oláh J, Ovádi J, Sippl W, Jung M (2017) Chemically induced degradation of sirtuin 2 (Sirt2) by a proteolysis targeting chimera (PROTAC) based on sirtuin rearranging ligands (SirReals). J Med Chem 61(2):482–491. https://pubs.acs.org/doi/10.1021/acs.jmedchem.6b01872
127. Outeiro TF, Kontopoulos E, Altmann SM (2007) Sirtuin 2 inhibitors rescue alpha-synuclein-mediated toxicity in models of Parkinson's disease. Science 317(5837):516–519. https://doi.org/10.1126/science.1143780
128. Zhang X, Smith DL, Meriin AB et al (2005) A potent small molecule inhibits polyglutamine aggregation in Huntington's disease neurons and suppresses neurodegeneration in vivo. Proc Natl Acad Sci U S A 102(3):892–897. https://doi.org/10.1073/pnas.0408936102
129. Bodner RA, Outeiro TF, Atlmann S et al (2006) Pharmacological promotion of inclusion formation: a therapeutic approach for Huntington's and Parkinson's diseases. Proc Natl Acad Sci U S A 103(11):4246–4251. https://doi.org/10.1073/pnas.0511256103

130. Luthi-Carter R, Taylor DM, Pallos J et al (2010) SIRT2 inhibition achieves neuroprotection by decreasing sterol biosynthesis. Proc Natl Acad Sci U S A 107(17):7927–7932. https://doi.org/10.1073/pnas.1002924107
131. Spires-Jones TL, Fox LM, Rozkalne A et al (2012) Inhibition of Sirtuin 2 with sulfobenzoic acid derivative AK1 is non-toxic and potentially neuroprotective in a mouse model of frontotemporal dementia. Front Pharmacol 3:42. https://doi.org/10.3389/fphar.2012.00042
132. Chopra V, Quinti L, Kim J et al (2012) The sirtuin 2 inhibitor AK-7 is neuroprotective in Huntington's disease mouse models. Cell Rep 2(6):1492–1497. https://doi.org/10.1016/j.celrep.2012.11.001
133. Taylor DM, Balabadra U, Xiang Z et al (2011) A brain-permeable small molecule reduces neuronal cholesterol by inhibiting activity of sirtuin 2 deacetylase. ACS Chem Biol 6(6):540–546. https://doi.org/10.1021/cb100376q
134. Khanfar MA, Quinti L, Wang H et al (2014) Development and characterization of 3-(Benzylsulfonamido)benzamides as potent and selective SIRT2 inhibitors. Eur J Med Chem 76:414–426. https://doi.org/10.1016/j.ejmech.2014.02.003
135. Guan Q, Wang M, Chen H et al (2016) Aging-related 1-methyl-4-phenyl-1,2,3,6-tetrahydropyridine-induced neurochemical and behavioral deficits and redox dysfunction: improvement by AK-7. Exp Gerontol 82:19–29. https://doi.org/10.1016/j.exger.2016.05.011
136. Szegő ÉM, Gerhardt E, Outeiro TF (2017) Sirtuin 2 enhances dopaminergic differentiation via the AKT/GSK-3β/β-catenin pathway. Neurobiol Aging 56:7–16. https://doi.org/10.1016/j.neurobiolaging.2017.04.001
137. Quinti L, Casale M, Moniot S et al (2016) SIRT2- and NRF2-targeting thiazole-containing compound with therapeutic activity in Huntington's disease models. Cell Chem Biol 23(7):849–861. https://doi.org/10.1016/j.chembiol.2016.05.015
138. Howitz KT, Bitterman KJ, Cohen HY et al (2003) Small molecule activators of sirtuins extend *Saccharomyces cerevisiae* lifespan. Nature 425(6954):191–196. https://doi.org/10.1038/nature01960
139. Trapp J, Meier R, Hongwiset D et al (2007) Structure-activity studies on suramin analogues as inhibitors of NAD+-dependent histone deacetylases (sirtuins). ChemMedChem 2(10):1419–1431. https://doi.org/10.1002/cmdc.200700003
140. Tervo AJ, Kyrylenko S, Niskanen P et al (2004) An in silico approach to discovering novel inhibitors of human sirtuin type 2. J Med Chem 47(25):6292–6298. https://doi.org/10.1021/jm049933m
141. Kiviranta PH, Leppänen J, Kyrylenko S et al (2006) N,N′-Bisbenzylidenebenzene-1,4-diamines and N,N′-bisbenzylidenenaphthalene-1,4-diamines as sirtuin type 2 (SIRT2) inhibitors. J Med Chem 49(26):7907–7911. https://doi.org/10.1021/jm060566j
142. Kiviranta PH, Leppänen J, Rinne VM et al (2007) N-(3-(4-Hydroxyphenyl)-propenoyl)-amino acid tryptamides as SIRT2 inhibitors. Bioorg Med Chem Lett 17(9):2448–2451. https://doi.org/10.1016/j.bmcl.2007.02.023
143. Huhtiniemi T, Suuronen T, Rinne VM et al (2008) Oxadiazole-carbonylaminothioureas as SIRT1 and SIRT2 inhibitors. J Med Chem 51(15):4377–4380. https://doi.org/10.1021/jm800639h
144. Moniot S, Forgione M, Lucidi A et al (2017) Development of 1,2,4-oxadiazoles as potent and selective inhibitors of the human deacetylase sirtuin 2: structure-activity relationship, X-ray crystal structure, and anticancer activity. J Med Chem 60(6):2344–2360. https://doi.org/10.1021/acs.jmedchem.6b01609
145. Lain S, Hollick JJ, Campbell J et al (2008) Discovery, in vivo activity, and mechanism of action of a small-molecule p53 activator. Cancer Cell 13(5):454–463. https://doi.org/10.1016/j.ccr.2008.03.004
146. Sunami Y, Araki M, Hironaka Y et al (2013) Inhibition of the NAD-dependent protein deacetylase SIRT2 induces granulocytic differentiation in human leukemia cells. PLoS One 8(2):e57633. https://doi.org/10.1371/journal.pone.0057633

147. Dai W, Zhou J, Jin B et al (2016b) Class III-specific HDAC inhibitor Tenovin-6 induces apoptosis, suppresses migration and eliminates cancer stem cells in uveal melanoma. Sci Rep 6:22622. https://doi.org/10.1038/srep22622
148. Hirai S, Endo S, Saito R et al (2014) Antitumor effects of a sirtuin inhibitor, tenovin-6, against gastric cancer cells via death receptor 5 up-regulation. PLoS One 9(7):e102831. https://doi.org/10.1371/journal.pone.0102831
149. Ueno T, Endo S, Saito R et al (2013) The sirtuin inhibitor tenovin-6 upregulates death receptor 5 and enhances cytotoxic effects of 5-fluorouracil and oxaliplatin in colon cancer cells. Oncol Res 21(3):155–164. https://doi.org/10.3727/096504013X13854886566598
150. MacCallum SF, Groves MJ, James J et al (2013) Dysregulation of autophagy in chronic lymphocytic leukemia with the small-molecule Sirtuin inhibitor Tenovin-6. Sci Rep 3:1275. https://doi.org/10.1038/srep01275
151. Yuan H, He M, Cheng F et al (2017) Tenovin-6 inhibits proliferation and survival of diffuse large B-cell lymphoma cells by blocking autophagy. Oncotarget 8(9):14912–14924. https://doi.org/10.18632/oncotarget.14741
152. McCarthy AR, Pirrie L, Hollick JJ et al (2012) Synthesis and biological characterisation of sirtuin inhibitors based on the tenovins. Bioorg Med Chem 20(5):1779–1793. https://doi.org/10.1016/j.bmc.2012.01.001
153. Pirrie L, McCarthy AR, Major LL et al (2012) Discovery and validation of SIRT2 inhibitors based on tenovin-6: use of a ^{1}H-NMR method to assess deacetylase activity. Molecules 17 (10):12206–12224. https://doi.org/10.3390/molecules171012206
154. McCarthy AR, Sachweh MCC, Higgins M et al (2013) Tenovin-D3, a novel small-molecule inhibitor of sirtuin SirT2, increases p21 (CDKN1A) expression in a p53-independent manner. Mol Cancer Ther 12(4):352–360. https://doi.org/10.1158/1535-7163.MCT-12-0900
155. Disch JS, Evindar G, Chiu CH et al (2013) Discovery of thieno[3,2-d]pyrimidine-6-carboxamides as potent inhibitors of SIRT1, SIRT2, and SIRT3. J Med Chem 56(9):3666–3679. https://doi.org/10.1021/jm400204k
156. Yoon YK, Ali MA, Wei AC et al (2014) Benzimidazoles as new scaffold of sirtuin inhibitors: green synthesis, in vitro studies, molecular docking analysis and evaluation of their anti-cancer properties. Eur J Med Chem 83:448–454. https://doi.org/10.1016/j.ejmech.2014.06.060
157. Di Fruscia P, Zacharioudakis E, Liu C et al (2015) The discovery of a highly selective 5,6,7,8-tetrahydrobenzo[4,5]thieno[2,3-d]pyrimidin-4(3H)-one SIRT2 inhibitor that is neuroprotective in an in vitro Parkinson's disease model. ChemMedChem 10(1):69–82. https://doi.org/10.1002/cmdc.201402431
158. Patel K, Sherrill J, Mrksich M et al (2015) Discovery of SIRT3 inhibitors using SAMDI mass spectrometry. J Biomol Screen 20(7):842–848. https://doi.org/10.1177/1087057115588512
159. Uciechowska U, Schemies J, Neugebauer RC et al (2008) Thiobarbiturates as sirtuin inhibitors: virtual screening, free-energy calculations, and biological testing. ChemMedChem 3 (12):1965–1976. https://doi.org/10.1002/cmdc.200800104
160. Manjulatha K, Srinivas S, Mulakayala N et al (2012) Ethylenediamine diacetate (EDDA) mediated synthesis of aurones under ultrasound: their evaluation as inhibitors of SIRT1. Bioorg Med Chem Lett 22(19):6160–6165. https://doi.org/10.1016/j.bmcl.2012.08.017
161. Zheng YC, Wang LZ, Zhao LJ et al (2016) 1,2,3-Triazole-dithiocarbamate hybrids, a group of novel cell active SIRT1 inhibitors. Cell Physiol Biochem 38(1):185–193. https://doi.org/10.1159/000438620
162. Sociali G, Galeno L, Parenti MD et al (2015) Quinazolinedione SIRT6 inhibitors sensitize cancer cells to chemotherapeutics. Eur J Med Chem 102:530–539. https://doi.org/10.1016/j.ejmech.2015.08.024
163. Gey C, Kyrylenko S, Hennig L et al (2007) Phloroglucinol derivatives guttiferone G, aristoforin, and hyperforin: inhibitors of human sirtuins SIRT1 and SIRT2. Angew Chem Int Ed Engl 46(27):5219–5222. https://doi.org/10.1002/anie.200605207
164. Gutiérrez M, Andrianasolo EH, Shin WK et al (2009) Structural and synthetic investigations of tanikolide dimer, a SIRT2 selective inhibitor, and tanikolide seco-acid from the Madagascar

marine cyanobacterium *Lyngbya majuscula*. J Org Chem 74(15):5267–5275. https://doi.org/10.1021/jo900578j
165. Bemis JE, Vu CB, Xie R et al (2009) Discovery of oxazolo[4,5-b]pyridines and related heterocyclic analogs as novel SIRT1 activators. Bioorg Med Chem Lett 19(8):2350–2353. https://doi.org/10.1016/j.bmcl.2008.11.106
166. Dai H, Kustigian L, Carney A et al (2010) SIRT1 activation by small molecules: kinetic and biophysical evidence for direct interaction of enzyme and activator. J Biol Chem 285 (43):32695–32703. https://doi.org/10.1074/jbc.M110.133892
167. Dao TT, Tran TL, Kim J et al (2012) Terpenylated coumarins as SIRT1 activators isolated from *Ailanthus altissima*. J Nat Prod 75(7):1332–1338. https://doi.org/10.1021/np300258u
168. de Boer VCJ, de Goffau MC, Arts ICW et al (2006) SIRT1 stimulation by polyphenols is affected by their stability and metabolism. Mech Ageing Dev 127(7):618–627. https://doi.org/10.1016/j.mad.2006.02.007
169. Hubbard BP, Gomes AP, Dai H et al (2013) Evidence for a common mechanism of SIRT1 regulation by allosteric activators. Science 339:1216–1219. https://doi.org/10.1126/science.1231097
170. Mai A, Valente S, Meade S et al (2009) Study of 1,4-dihydropyridine structural scaffold: discovery of novel sirtuin activators and inhibitors. J Med Chem 52(17):5496–5504. https://doi.org/10.1021/jm9008289
171. Milne JC, Lambert PD, Schenk S et al (2007) Small molecule activators of SIRT1 as therapeutics for the treatment of type 2 diabetes. Nature 450(7170):712–716. https://doi.org/10.1038/nature06261
172. Borra MT, Smith BC, Denu JM (2005) Mechanism of human SIRT1 activation by resveratrol. J Biol Chem 280(17):17187–17195. https://doi.org/10.1074/jbc.M501250200
173. Kaeberlein M, McDonagh T, Heltweg B et al (2005) Substrate-specific activation of sirtuins by resveratrol. J Biol Chem 280(17):17038–17045. https://doi.org/10.1074/jbc.M500655200
174. Cao D, Wang M, Qiu X et al (2015) Structural basis for allosteric, substrate-dependent stimulation of SIRT1 activity by resveratrol. Genes Dev 29(12):1316–1325. https://doi.org/10.1101/gad.265462.115
175. Dai H, Case AW, Riera TV et al (2015) Crystallographic structure of a small molecule SIRT1 activator-enzyme complex. Nat Commun 6:7645. https://doi.org/10.1038/ncomms8645
176. Gertz M, Nguyen GTT, Fischer F et al (2012) A molecular mechanism for direct sirtuin activation by resveratrol. PLoS One 7(11):e49761. https://doi.org/10.1371/journal.pone.0049761
177. Dai H, Ellis JL, Sinclair DA et al (2016a) Synthesis and assay of SIRT1-activating compounds. Methods Enzymol 574:213–244. https://doi.org/10.1016/bs.mie.2016.01.012
178. Anderson RM, Bitterman KJ, Wood JG et al (2003) Nicotinamide and PNC1 govern lifespan extension by calorie restriction in *Saccharomyces cerevisiae*. Nature 423(6936):181–185. https://doi.org/10.1038/nature01578
179. Lin SJ, Defossez PA, Guarente L (2000) Requirement of NAD and SIR2 for life-span extension by calorie restriction in *Saccharomyces cerevisiae*. Science 289(5487):2126–2128
180. Lin SJ, Kaeberlein M, Andalis AA et al (2002) Calorie restriction extends *Saccharomyces cerevisiae* lifespan by increasing respiration. Nature 418(6895):344–348. https://doi.org/10.1038/nature00829
181. Hubbard BP, Sinclair DA (2014) Small molecule SIRT1 activators for the treatment of aging and age-related diseases. Trends Pharmacol Sci 35(3):146–154. https://doi.org/10.1016/j.tips.2013.12.004
182. Rahnasto-Rilla M, Kokkola T, Jarho E et al (2016) N-Acylethanolamines bind to SIRT6. ChemBioChem 17(1):77–81. https://doi.org/10.1002/cbic.201500482
183. Wang Y, Liang X, Chen Y et al (2016) Screening SIRT1 activators from medicinal plants as bioactive compounds against oxidative damage in mitochondrial function. Oxid Med Cell Longev 2016:4206392. https://doi.org/10.1155/2016/4206392

184. Daitoku H, Hatta M, Matsuzaki H et al (2004) Silent information regulator 2 potentiates Foxo1-mediated transcription through its deacetylase activity. Proc Natl Acad Sci U S A 101 (27):10042–10047. https://doi.org/10.1073/pnas.0400593101
185. Gan L, Han Y, Bastianetto S et al (2005) FoxO-dependent and -independent mechanisms mediate SirT1 effects on IGFBP-1 gene expression. Biochem Biophys Res Commun 337 (4):1092–1096. https://doi.org/10.1016/j.bbrc.2005.09.169
186. Kobayashi Y, Furukawa-Hibi Y, Chen C et al (2005) SIRT1 is critical regulator of FOXO-mediated transcription in response to oxidative stress. Int J Mol Med 16(2):237–243. https://doi.org/10.3892/ijmm.16.2.237
187. Chen Y, Zhang J, Lin Y et al (2011) Tumour suppressor SIRT3 deacetylates and activates manganese superoxide dismutase to scavenge ROS. EMBO Rep 12(6):534–541. https://doi.org/10.1038/embor.2011.65
188. Pillai VB, Samant S, Sundaresan NR et al (2015) Honokiol blocks and reverses cardiac hypertrophy in mice by activating mitochondrial Sirt3. Nat Commun 6:6656. https://doi.org/10.1038/ncomms7656
189. Lu J, Zhang H, Chen X et al (2017) A small molecule activator of SIRT3 promotes deacetylation and activation of manganese superoxide dismutase. Free Radic Biol Med 112:287–297. https://doi.org/10.1016/j.freeradbiomed.2017.07.012
190. Rahnasto-Rilla M, McLoughlin P, Kulikowicz T et al (2017) The identification of a SIRT6 activator from brown algae *Fucus distichus*. Mar Drugs 15(6):E190. https://doi.org/10.3390/md15060190
191. Valente S, Mellini P, Spallotta F et al (2016) 1,4-Dihydropyridines active on the SIRT1/AMPK pathway ameliorate skin repair and mitochondrial function and exhibit inhibition of proliferation in cancer cells. J Med Chem 59(4):1471–1491. https://doi.org/10.1021/acs.jmedchem.5b01117
192. Nayagam VM, Wang X, Tan YC et al (2006) SIRT1 modulating compounds from high-throughput screening as anti-inflammatory and insulin-sensitizing agents. J Biomol Screen 11 (8):959–967. https://doi.org/10.1177/1087057106294710

Top Med Chem (2020) 33: 93–122
DOI: 10.1007/7355_2019_71

Published online: 28 November 2019

Histone Acetyltransferase Enzymes: From Biological Implications to Most Relevant Inhibitors

Daniela Trisciuoglio and Dante Rotili

Contents

1 Background 94
2 HAT Superfamilies 97
3 HAT and Cancer 99
4 HAT and Other Diseases 102
5 HAT Inhibitors 103
6 Bisubstrate Inhibitors 103
7 Natural Products and Derivatives 106
8 Synthetic Inhibitors 110
9 Conclusions and Perspectives 113
References 115

Abstract The acetylation of lysine residues of histone and nonhistone proteins is a post-translational modification catalysed by the so-called histone acetyltransferases (HATs) that plays a crucial role in several biological settings. The deregulation of this enzymatic activity is implicated in many disease conditions such as cancer and inflammatory and neurological disorders. Despite many histone acetyltransferase inhibitors (HATi) have been identified so far, there is still the need for new, metabolically stable, more potent and selective HATi as potential therapeutic agents and/or as chemical tools for studying HAT biology. In the present chapter, the main features of HAT enzymes and related diseases have been summarized, with a particular focus on HATi, analysing their structure-activity relationships, mechanisms of action and potential therapeutic applications.

D. Trisciuoglio (✉)
Institute of Molecular Biology and Pathology, National Research Council, Rome, Italy

Preclinical Models and New Therapeutic Agents Unit, IRCCS-Regina Elena National Cancer Institute, Rome, Italy
e-mail: daniela.trisciuoglio@uniroma1.it

D. Rotili (✉)
Department of Chemistry and Technologies of Drugs, Sapienza University of Rome, Rome, Italy
e-mail: dante.rotili@uniroma1.it

Keywords Cancer, Chemical probes, Epigenetics, Histone acetyltransferase inhibitors, Structure- and ligand-based drug discovery

Abbreviations

ALL	Acute lymphoid leukaemia
AML	Acute myeloid leukaemia
BRD	Bromodomain
CBP	CREB-binding protein
CSCC	Cutaneous squamous cell carcinoma
DTT	Dithiothreitol
GCN5	General control nonderepressible 5
GNAT	Gcn5-related *N*-acetyltransferase
HATi	Histone acetyltransferase inhibitors
HATs	Histone acetyltransferases
HDACs	Histone deacetylases
LoCAMs	Long chain alkylidene malonates
LOH	Loss of heterozygosity
MSL	Male-specific lethal
MYST	Moz, Ybf2/Sas3, Sas2, Tip60
NCOA	Nuclear receptor coactivator
NSCLC	Non-small-cell lung cancer
NSL	Non-specific lethal
PCAF	p300/CBP-associated factor
SRC-1	Steroid receptor coactivator-1
SRC-3/AIB-1	Steroid receptor coactivator-3/activated in breast cancer-1
TAFII250	TATA box binding protein (TBP)-associated factor
TIF-2	Transcriptional intermediary factor-2
TRAM-1	Thyroid hormone receptor activator molecule-1
α-TAT1	α-Tubulin acetyltransferase 1

1 Background

Histone acetylation is one of most studied post-translational modifications involved in a plethora of cell functions, including the regulation of gene expression [1]. The balance between acetylation and deacetylation of histone proteins is regulated by the action of two protein families: histone acetyltransferases (HATs) and histone deacetylases (HDACs). HATs catalyse the transfer of an acetyl group from acetyl-CoA to the lysine (K) located near the amino terminus of core histone proteins (Fig. 1). The reaction is preferentially carried out on specific lysine residues: for

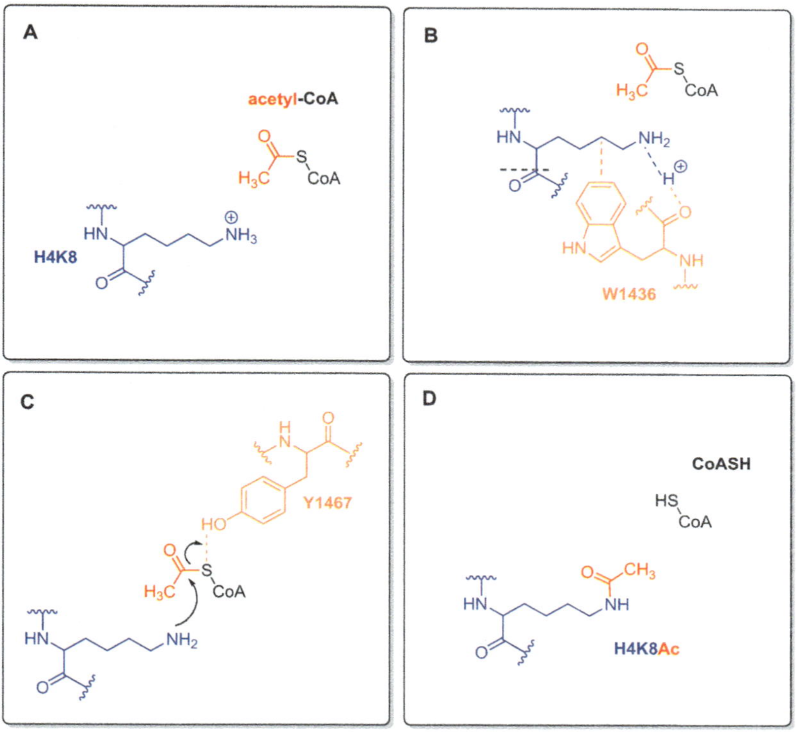

Fig. 1 Proposed four-step catalytic mechanism of the HAT enzyme p300. (**a**) Acetyl-CoA and a specific histone lysine (H4K8) bind in sequence within the active site. (**b**) The W1436 promotes the deprotonation of the charged lysine H4K8 and position it for the nucleophilic attack. (**c**) The free lysine attacks the carbonyl of acetyl-CoA, while Y1467 protonates the leaving CoA performing a general acid catalysis. (**d**) The acetylated lysine product and the CoASH leave in sequence the active site

example, on histone H3, the lysine residues which are mainly acetylated are those in positions 9, 14, 18 and 23, while on histone H4, the preferentially acetylated residues are K5, K8, K12 and K16. This reaction, neutralizing the positive charge of the amino acid, weakens the electrostatic interaction between the histones and the negatively charged DNA. In addition to the electrostatic repulsion that makes the chromatin more lax, the addition of the acetyl group recalls the activation of chromatin remodelling protein complexes (e.g. transcription factors, remodellers, chromatin modifiers), thus leading to acetylation-mediated gene activation.

In the last decade, it has become clear that HAT enzymes can also acetylate numerous nonhistone targets such as transcription factors, nuclear regulators and various cytoplasmic proteins [2, 3]. Thus, the human HATs have been recently classified as lysine acetyltransferases (KATs), considering their ability to acetylate different proteins (Table 1). The HATs are grouped into three main superfamilies: (1) Gcn5-related *N*-acetyltransferase (GNAT), (2) p300 and CREB-binding protein

Table 1 HAT family members: classification and main biological implications

Protein name	Gene names	Cellular localization	Histone targets	Nonhistone targets	Pathology	Ref.
P300/CBP family						
KAT3A	CREBBP, CBP	Nucleus	H2A, H2B	NF-kappaB, *c*-myb, Foxo1	Neurodegenerative diseases: Rubinstein-Taybi syndrome. Cancers: acute myeloid leukaemia; acute lymphoid leukaemia; lung, colon, breast and ovarian carcinomas; nasopharyngeal carcinoma; cutaneous squamous cell carcinoma	[40–45]
KAT3B	EP300, P300	Nucleus	H2A, H2B	NF-kappaB, *c*-myc, p53, STAT3, β-catenin, Foxo1, AR		
GNAT family						
KAT1	HAT1	Nucleus	H3, H4, H2A	–	Neurodegenerative diseases. Cancers: lung, colon, breast, ovarian, hepatic, gastric and oesophageal cancers	[15, 67]
KAT2A	KAT2A, GCN5, GCN5L2	Nucleus	H3, H4, H2A	CDC6, CDK9, cyclin D1, cyclin E1 and E2F1, HDM2, PTEN, *c*-myc		
KAT2B	pCAF	Nucleus	H3	p53, CDK9, *c*-myc, Foxo1, AR		
KAT9	ELP3		H4, H2A, H3	–		
α-TAT1	α-TAT1, MEC17, C6orf134	Cytosol		Tubulin, cortactin		
MYST family						
KAT5	KAT5, TIP60, HTATIP	Nucleus	H4, H2A	ATM, TRRAP, p53, E2F1, *c*-myc	Neurodegenerative diseases: genitopatellar syndrome. Cancers: acute myeloid leukaemia; acute lymphoid leukaemia; colorectal, gastric, breast, ovarian, hepatocellular and renal cell carcinomas	[51–53]
KAT6A	KAT6A, MOZ, MYST3, RUNXBP	Nucleus	H3	–		
KAT6B	KAT6B, MORF, MOZ2, MYST4	Nucleus	–	–		
KAT7	KAT7, HBO1, HBOa, MYST2	Nucleus	H3, H4	–		
KAT8	KAT8, MOF, MYST1		H4			

(p300/CBP) and (3) Moz, Ybf2/Sas3, Sas2, Tip60 (MYST). Other three families have also been classified as HAT for their similarity: (1) the nuclear receptor coactivator (NCOA)-related HAT family, (2) transcription factor-related HAT family and (3) the novel Camello HAT family.

The HATs are predominantly expressed in both the nucleus and cytoplasm, while some of them are also found in cytoplasmatic organelles such as mitochondria and endoplasmic reticulum [2, 3]. The main function shared by all HAT members is the activation of transcription. At least three different mechanisms of transcriptional activation mediated by HATs have been described. In the first, HATs can directly acetylate their targets (histone proteins and transcriptional factors) to facilitate a transcriptional response [4]. The histone acetylation facilitates the unwinding of the chromatin structure, while the acetylation of transcriptional factors can increase their DNA binding affinity, so leading to an increase in transactivation and gene expression [5, 6]. Alternatively, a HAT enzyme can act as a bridge connecting transcription factors to the transcription machinery or as a protein scaffold contributing to the assembly of multi-protein complexes that promote transcriptional activation [7].

Nowadays, it is clear that HAT members do not exclusively control nuclear transcription or transcriptional factors, but they can also act in other cellular compartments thus regulating cellular processes not directly related to transcriptional activation [1]. For their multiple biological functions, HATs play a key role in the pathogenesis of several diseases, including cancer and neurodegenerative disorders. Although a growing body of evidence demonstrates a direct relationship between HATs and cancer and suggests them as new therapeutic targets, HAT inhibitors (HATi) are still in preclinical development [8, 9]. Herein, we summarize the biological functions of the main HAT members, their role in human disorders and the most promising inhibitors identified so far.

2 HAT Superfamilies

The four histone subtypes (histone H2A, H2B, H3 and H4) are the main substrates for the p300/CBP family. The p300/CBP family comprises only two protein members with interchangeable functions, CBP and its paralog p300. They show similar structures and share 86% sequence identity at the HAT domains. The HAT domain of p300/CBP consists of about 500 residues located in the central region of the protein. p300/CBP contain several other protein domains, including a bromodomain (BRD) and three cysteine-histidine-rich domains (TAZ, PHD and ZZ) serving for protein-protein interaction. So far, a plethora of interacting proteins, including cofactors and transcription factors, has been identified thus confirming the relevant role of this family in the transcriptional control. In this context, p300/CBP act as a bridge connecting transcription factors to the transcription machinery, but also directly acetylate histones and/or transcription factors to facilitate a transcriptional response [7]. It has been reported that p300/CBP modulate the activity and cellular localization of different factors producing multiple downstream effects. For example, the acetylation on Stat3 by p300 at K685 enhances its DNA binding and transactivation activities, as well as its nuclear localization [5]. Similarly, CBP acetylates *c*-myb and E2F1 thus increasing their transactivating capacity [6]. Conversely, CBP-mediated acetylation of FoxO1 attenuates its DNA binding activity [10].

The GNAT family consists of 12 proteins with different cellular functions and substrates, including GCN5 (general control nonderepressible 5; KAT2A), PCAF (p300/CBP-associated factor, KAT2B), α-tubulin acetyltransferase 1 (α-TAT1), HAT1, the elongator complex subunit Elp3, the mediator-complex subunit Nut1, Hpa2 and other proteins showing a sequence and structural resemblance with GCN5. GNAT proteins share a domain composed of four A–D conserved sequence motifs and rarely have bromodomains or chromodomains required for binding acetylated or methylated lysine, respectively. In general, GNAT family is involved in cell growth, playing an important role in cell cycle regulation [11]. The two main members GCN5 and PCAF are closely related proteins. GCN5 is needed for normal progression through the G2/M phases and mitotic gene expression. PCAF shares 73% sequence homology with GCN5 and plays a role in transcriptional activation, cell cycle arrest and cell differentiation. Under stress conditions, PCAF is required for the acetylation of histone H3 on p21 promoter, thus stopping cell growth [12]. The GNAT member α-TAT1 is the main responsible for α-tubulin acetylation at K40 in higher organisms [13–15]. α-TAT1 is required for cell migration and invasion [16], and its overexpression in breast cancer cells increases α-tubulin acetylation and enhances formation of microtentacles, flexible cell protrusions enhancing the attachment of circulating cells [17].

The MYST family (acronym for the founding members MOZ, Ybf2, Sas2, TIP60) is the largest but still poorly studied HAT family. Currently, it comprises five human enzymes: Tip60, MOF, MOZ, MORF and HBO1. This family is characterized by the presence of a highly conserved MYST domain containing acetyl-CoA binding and zinc finger motifs [18]. Many MYSTs also contain other domains for recognition of other proteins [19]. Moreover, they are involved in a wide range of cellular processes including regulation of transcription, cell growth, cell cycle and stem cell differentiation [18]. Tip60 is the most studied MYST family member. This enzyme is mainly involved in transcriptional regulation, acting as cofactor for different transcription factor, including *c*-Myc [20], p53 [21] and E2F1 [22]. Beyond its role as transcriptional activator, Tip60 is implicated in multiple cellular pathways, including transcription, DNA damage-induced checkpoint activation and apoptosis [23, 24]. MOF is a well-conserved member of the MYST family [25]. It plays important roles in transcriptional activation by acetylating histone H4 on K16, a prevalent mark associated with chromatin de-condensation. The incorporation of MOF in distinct transcription regulatory complexes, namely, the MSL (male-specific lethal: MSL1, MSL2) and the NSL (non-specific lethal: KANSL1, KANSL3, MCRS1) complexes, is important for its enzymatic activity and target selection [26]. Recently, MOF and MOF-associated complex have been found to be dual transcriptional regulators of nuclear and mitochondrial genomes [27]. MOZ was firstly identified as a protein with a zinc finger and a putative acetyltransferase signature that in the translocation t(8;16)(p11;p13) of acute myeloid leukaemia (AML) is a fusion partner of the CBP.

Subsequently, on the base of its homology to MOZ, another MYST member called MORF was discovered [28, 29]. In cells, MOZ and MORF form similar stable protein complexes with the ING proteins. These complexes possess a HAT activity specific for histone H3 and can function as transcriptional coactivators. The HAT activity of MOZ/MORF complexes is required for normal developmental programmes, including haematopoiesis and skeletogenesis, and for the regulation of various genes, especially the Hox family [19, 30–33]. HBO1 was discovered in a two-hybrid screening looking for new interactor of ORC1, the subunit of the origin recognition complex (ORC) [34]. HBO1 is a major mediator of both histone H3 (K14, K23) and H4 (K5, K8, K12) acetylation. As for other MYST members, the specificity of histone tail modification is finely controlled by the different scaffold subunits of HBO1 complex: for example, JADE proteins are required for H4 acetylation, while the HBO1 complex at the H3 contains BRPF (bromodomain- and PHD finger-containing protein 1) protein [35].

Finally, the basal transcription factor family and the nuclear receptor cofactors family are classified as HATs; nevertheless both these families do not harbour acetylation-related structural motifs. The transcription factor-related HATs include TATA box binding protein (TBP)-associated factor TAFII250 and TFIIIC. Nuclear receptor cofactors family include steroid receptor coactivator-1 (SRC-1), steroid receptor coactivator-3/activated in breast cancer-1 (SRC-3/AIB-1), thyroid hormone receptor activator molecule-1 (TRAM-1), nuclear receptor coactivator-3 (NCOA-3) and transcriptional intermediary factor-2 (TIF-2).

More recently, a new family lacking canonical features of known HATs, being relatively smaller in size, and other associated domains has been identified. This new family has been named Camello protein family and includes functional HATs that show specificity towards histone H4, exhibit perinuclear localization and are essential for zebrafish development [36].

3 HAT and Cancer

Abnormal acetylation patterns can be the result of genetic lesions (e.g. haploinsufficiency or inactivating mutations, resulting in the silencing of target genes), but also of an enhanced HAT activity on the wrong targets (e.g. oncogenes), due to their aberrant recruitment or overexpression. Hence, HAT genetic alterations as well as HAT functional dysregulation are strongly linked to human diseases, in particular to cancer [8]. As mentioned above, alterations of protein acetylation impact on different hallmarks of cancer such as cell motility and invasion, cell cycle, senescence, cell death, differentiation, DNA repair and DNA damage response; thus both hyperacetylation and hypoacetylation can cooperate with other cancerous modifications to promote tumorigenesis and tumour progression (Fig. 2). To date, the genetic alterations of HATs as well as of the opposing enzymes HDACs are reported to be involved in the expression of

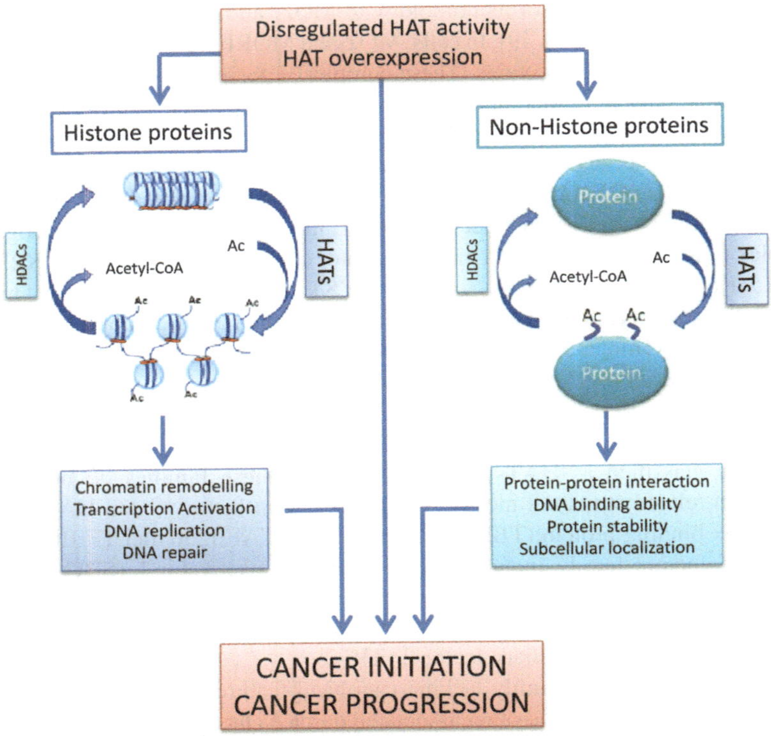

Fig. 2 HAT genetic alterations as well as HAT functional dysregulation are strongly linked to human diseases, in particular to cancer. The acetylation status of histones affects chromatin modelling and transcriptional activity. In fact, histone acetylation by HAT enzymes alters the expression of genes involved in both cancer initiation and progression. During cancer initiation, HATs activate genes which results in uncontrolled cell proliferation, loss of differentiation and inhibition of apoptosis. During cancer progression, HATs activate genes which results in a loss of adhesion, migration, invasion and angiogenesis. The acetylation status of a protein affects various protein functions, thus controlling different cellular processes. Overall, alterations of protein acetylation can cooperate with other cancerous modifications to promote tumorigenesis and tumour progression

malignant phenotypes in cancer. In particular, different HATs are reported to be mutated in both haematological and solid tumours [19, 37] (Table 1). p300/CBP genes are involved in various chromosomal translocation events during haematological malignancies giving rise to tumour-specific hybrid proteins. In AML, chromosome translocations of CBP gene are associated with the development of this neoplasia following chemotherapy for other forms of cancer [38]. In acute lymphoid leukaemia (ALL), mutations of CBP gene impairing its HATs activity have been found in about 18.3% of patients. Moreover, the mutations were often associated with relapsed tumours, indicating that cells with mutated HATs were resistant to first-line therapy [39].

Mutated or deleted p300/CBP genes have been reported in several solid cancers, such as lung, colon, breast and ovarian carcinomas, nasopharyngeal carcinoma and cutaneous squamous cell carcinoma (CSCC), and in most cases these mutations result into the truncation of the proteins [40–45]. Loss of heterozygosity (LOH) at the p300 locus has been observed in numerous cancers, including hepatocellular, colorectal, oral, breast, ovarian, gastric carcinomas and glioblastomas [37]. Notably, p300 is also involved in the regulation of expression and function of several oncoproteins including c-myc [46], androgen receptor [47], tumour suppressor protein BRCA1 [48] and p53 [49].

Like p300/CBP, MYST family members are often mutated in cancer, and chromosomal aberrations involving MOZ and MORF genes can drive leukaemogenesis. In AML, both MOZ and MORF fuse with multiple genes, including CBP and p300 [28, 29]. These hybrid proteins lead to an aberrant acetylation and transcriptional activation generally associated with overexpression of oncogenes. Recently, in vitro and in vivo studies have demonstrated that the depletion of MORF expression enhances cancer growth and aggressiveness of small-cell lung tumours [50].

Different and independent studies have also shown abnormal MOF expression, and its corresponding acetylation mark (H4K16) has been found in certain primary cancer tissues, including breast cancer, medulloblastoma, ovarian cancer, renal cell carcinoma, colorectal carcinoma, gastric cancer as well as non-small-cell lung cancer (NSCLC) [19]. However, the role of MOF in human tumorigenesis is still controversial and deserves more investigation. For example, MOF is frequently downregulated and is a prognostic marker in colorectal, gastric, breast, ovarian, hepatocellular and renal cell carcinomas [51–53]. Conversely, the overexpression of MOF in NSCLC predicts poor prognosis of the disease [54].

The human Tip60 locus is frequently mutated or lost in a variety of tumours including breast and prostate carcinomas [55]. Indeed, in prostate cancer Tip60 is upregulated in clinical specimens, and its expression correlates with disease progression. Mechanistically, Tip60 acetylates the androgen receptor even in a ligand-independent manner, thus inducing the expression of target genes [55]. In breast cancer, Bassi and coauthors have reported an interesting correlation between Tip60 levels and p53 mutations, thus suggesting that Tip60 is a novel breast tumour suppressor gene [56].

Because of their cellular functions, also GNAT family members have been implicated in different types of cancer. GCN5 is found to be upregulated in human glioma, colon and lung cancer [57]. Conversely, PCAF gene is frequently deleted in solid tumours such as ovarian cancer, gastric cancer and oesophageal carcinoma [58]. Recent reports suggest that also α-TAT1 plays a key role in many cellular processes related to cancer dissemination, including cell adhesion, migration and invasion [17, 59, 60]. Notably, α-TAT1 is also associated with pancreatic cancer-initiating cells and breast cancer progression [61].

4 HAT and Other Diseases

Altered or abnormal HAT activity is also related to several other diseases, including cardiac hypertrophy, asthma, AIDS and neurodegenerative disorders. Well-studied are the roles of different HATs in neural development. CBP/p300 are crucial enzymes in development. Indeed, dysfunctions of CBP/p300 activity deregulate gene transcriptions that are prominently linked to Rubinstein-Taybi syndrome, an incurable genetic disorder with combination of mental retardation and physical features. Most of Rubinstein-Taybi syndrome patients carry heterozygous mutation in gene CBP, while only a small percentage of patients show mutations in p300 gene [62]. Several studies in transgenic mice have described a role for CBP/p300 in neural development [63]. Disruption of both p300 and CBP expression leads to embryonic lethality in mice between embryonic day 9 (E9.0) and E11.5, associated with different types of neural tube closure and embryonic vascular and cardiac defects. Surprisingly, p300 +/− mice also manifested considerable embryonic lethality. GCN5 expression and activity are also required for the corrected neural tube closure. Notably, GCN5-null mice also exhibit early embryonic lethality [64], while deletion of PCAF causes no obvious abnormal phenotypes in mice [65, 66]. Despite this, mice deleted of both GCN5 and PCAF show a more severe phenotype than GCN5-null mice, indicating that some PCAF functions are redundant to those of GCN5 in the early stages of embryogenesis [66]. Interestingly, deficits in learning abilities, spatial and recognition memory as well as in both short-term and contextual long-term memory have been observed in adult PCAF null mice [67]. Little is known about the role of MYST members in neural development. Heterozygous mutations of MYST members exhibit no relevant phenotypes, while homozygous mutations of main members (TIP60, HBO1 and MOF) result in early embryonic lethality around the blastocyst stage or post-gastrulation. MORF is highly expressed in the brain, and its mutations are associated with different neurodevelopmental disorders in humans. In particular, mutations of MORF have been found in patients affected by genitopatellar syndrome, a skeletal dysplasia with cerebral and genital anomalies [68].

The role of HATs has been also extensively studied in heart disease. By using a genetic approach, it has been found that HAT members are key factors in the pathological processes of cardiac remodelling, including hypertrophy, contractility and fibrosis [69]. p300/CBP family plays critical roles in physiological and pathological growth of cardiac myocytes. The role of p300 in normal cardiac transcription is demonstrated by p300 null mice showing cardiac structural defects and reduced expression of muscle structural proteins such as β-myosin heavy chain and α-actinin. Moreover, p300/CBP activity is enhanced by signalling pathways that promote cardiac hypertrophy, and ectopic overexpression of both enzymes stimulates cardiac growth, while dominant-negative mutants of p300 block agonist-mediated cardiac growth [70].

Lysine acetylation is also essential in regulating immuno-metabolism, and it has been implicated as both a post-transcriptional and post-translational mechanism in modulating immunological and metabolic pathways and therefore may have a crucial role in maintaining energy homeostasis [71]. To date, the role of HDACs in these processes is well known. In contrast, the functions of HAT members are not as well explored as HDACs.

5 HAT Inhibitors

In the past 20 years, a great amount of evidence about the involvement of aberrant histone protein acetylation in development of different diseases has stimulated a continuous research on HATi as potential therapeutics or at least as chemical probes for better understanding of HAT biology. To date, despite the great efforts made by different research groups into finding small molecule HATi, only a few promising compounds have been identified. By a medicinal chemistry point of view, according to origin and mechanism of action, HATi can be divided into three main groups: bisubstrate inhibitors, natural inhibitors and related (semi)synthetic derivatives and fully synthetic inhibitors (Table 2).

6 Bisubstrate Inhibitors

The bisubstrate inhibitors are HAT substrate mimics that consist of two moieties connected by spacers of variable length: the coenzyme A, resembling the natural co-substrate acetyl-CoA, and a (pseudo)peptide portion, mimicking the lysine-containing histone substrate sequences (Fig. 3).

In the so-called Lys-CoA (**1**), CoA is connected to the ε-amino group of a *N*-acetyl-lysine through a carboxymethylene linker, while the lysine carboxylic acid is converted into a primary amide. Lys-CoA shows activity towards p300 in the sub-micromolar range ($IC_{50} = 0.5$ μM), with a selectivity over PCAF ($IC_{50} = 200$ μM) of 400 times, that can be ascribed to the different catalytic mechanism of the two HATs [72]. In 2008, has been reported the co-crystal structure of **1** bound within the p300 catalytic site, shading light into the interactions that are crucial for the inhibition [73].

The main medicinal chemistry approach pursued to modulate selectivity and affinity of bisubstrate inhibitors has been to vary the sequence and length of the peptide chain bound to Lys-CoA. In this way have been obtained the peptide **2** (H3-CoA-20) that, resembling the K14-containing sequence of histone H3 (the main PCAF substrate), is a sub-micromolar inhibitor of this enzyme($IC_{50} = 0.3$ μM) [72] and the peptide **3** that, mimicking the K16-containing sequence of histone H4 (the main substrate of many MYST family enzymes), is a micromolar inhibitor of Tip60 ($IC_{50} = 17.6$ μM) and p300 ($IC_{50} = 6.62$ μM) [74]. Unfortunately, all these compounds suffer from metabolic instability and low cell permeability, primarily due to their (pseudo)peptidic nature and the presence of phosphate groups within the

Table 2 Most relevant HATi

Compound	Structure	Enzyme inhibitory activity	Cell-based activity	Ref.
1 Lys-CoA	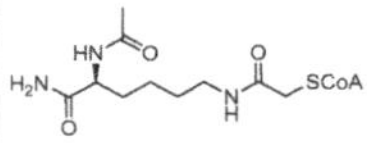	**p300** $IC_{50} = 0.5\ \mu M$ **PCAF** $IC_{50} = 200\ \mu M$	–	[72]
5a Anacardic acid	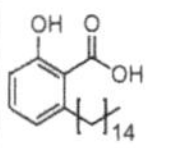	**p300/CBP** $IC_{50} = 5–1{,}000\ \mu M$ **PCAF** $IC_{50} = 5–667\ \mu M$ **Tip60** $IC_{50} = 64–348\ \mu M$ **MOF** $IC_{50} = 43–64\ \mu M$	Inhibition of p65 subunit acetylation and subsequent repression of NF-κB signalling	[74, 77–79]
9a LoCAM (SPV106)	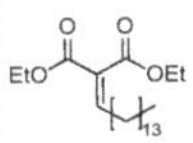	**p300/CBP** 74% inhibition @50 μM **PCAF** 137% activation @100 μM	Apoptotic and antiproliferative effects in human leukaemia U937 cells	[85, 86]
10 Curcumin	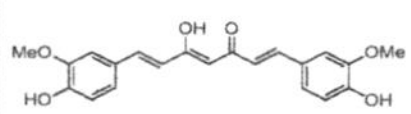	**p300** $IC_{50} = 25\ \mu M$	Reduction of p300-dependent acetylation of histone H3/H4 and p53. Under examination in many clinical trials for various indications	[87–91]
11a Garcinol	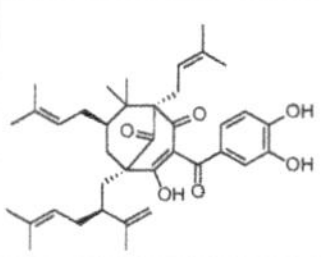	**p300** $IC_{50} = 7\ \mu M$ **PCAF** $IC_{50} = 5\ \mu M$	Inhibition of histone acetylation and induction of apoptosis in several cancer cell lines	[93]
11d EML425	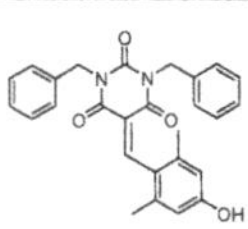	**p300** $IC_{50} = 2.9\ \mu M$ **CBP** $IC_{50} = 1.1\ \mu M$	Reduction of H4K5 and H3K9 acetylation levels and induction of cell cycle arrest in the G0/G1 phase (U937 cells)	[96]
12a NK13650A	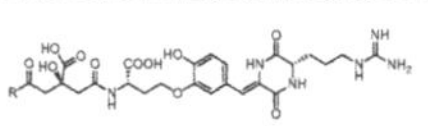 R = NH-aspartic acid	**p300** $IC_{50} = 11\ nM$	Repression of transcription mediated by androgen and oestrogen receptors and antiproliferative effects in various cancer cell lines	[97]

(continued)

Table 2 (continued)

Compound	Structure	Enzyme inhibitory activity	Cell-based activity	Ref.
14a PU139	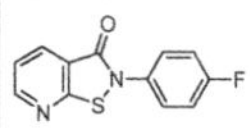	**PCAF** IC_{50} = 1.6–9.7 μM **p300** IC_{50} = 5.3 μM **CBP** IC_{50} = 2.5 μM **Gcn5** IC_{50} = 8.4 μM	Histone hypoacetylation and growth inhibition (**LNCaP** GI_{50} = 36.2 μM; **HL-60** GI_{50} = 3.4 μM) in various cancer lines and healthy mice	[101, 102]
15c CPTH6	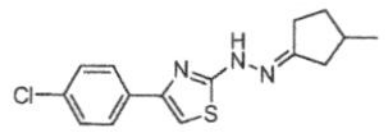	**p300** 27% inhibition @100 μM **PCAF** 40% inhibition @800 μM (phenotypic screen) **Gcn5** 40% inhibition @800 μM (phenotypic screen)	Inhibition of H3/H4 histone and α-tubulin acetylation in various cancer cell lines. Autophagy induction in different tumour cells. Induction of apoptosis in lung cancer stem-like cells (LCSCs) derived from NSCLC patients	[103–107]
16a C646	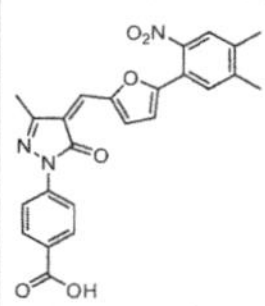	**p300** IC_{50} = 1.6 μM K_i = 460 nM	Repression of proliferation and histone H3/H4 hypoacetylation in melanoma and lung cancer lines. Apoptosis induction in AML1-ETO-positive leukaemia cells	[108, 111]
17 A-485	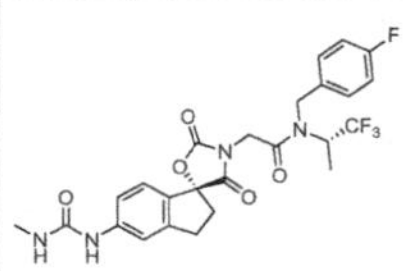	**p300** IC_{50} = 9.8 nM **CBP** IC_{50} = 2.6 nM	H3K18 and H3K27 hypoacetylation in various cancer lines [after 24 h incubation: **PC-3** EC_{50}(H3K18) = 0.46 μM and EC_{50} (H3K27) = 0.16 μM; **LnCaP-FGC** EC_{50} (H3K18) = 0.97 μM and EC_{50} (H3K27) = 0.18 μM]. Antiproliferative activity in lineage-specific haematological and prostate cancer cells. Tumour growth inhibition in a xenograft model of prostate cancer	[113, 114]

(continued)

Table 2 (continued)

Compound	Structure	Enzyme inhibitory activity	Cell-based activity	Ref.
18 NU9056	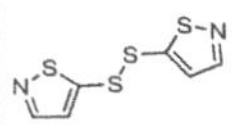	**Tip60** $IC_{50} = 2\ \mu M$	Histone hypoacetylation and apoptosis induction in prostate cancer cells. Cell growth inhibitory activity in various prostate cancer lines (**LnCaP** $GI_{50} = 24\ \mu M$; **PC-3** $GI_{50} = 27\ \mu M$)	[115]

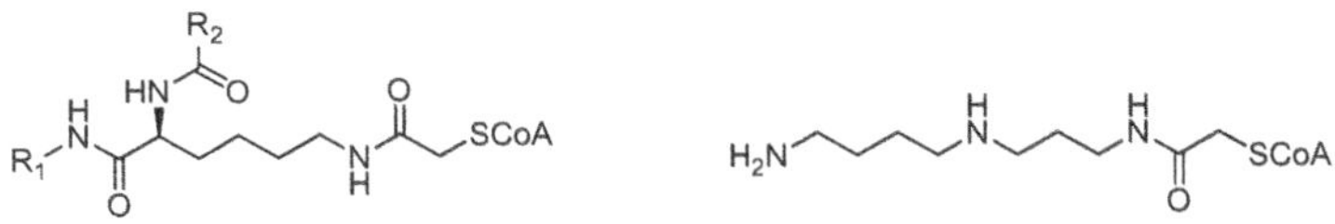

1 Lys-CoA $R_1 = H$
$R_2 = CH_3$

4 Spd(N^1)-CoA

2 H3-CoA-20 R_1 = A-P-R-K-Q-L-OH
R_2 = G-G-T-S-L-R-A-T-Q-K-T-R-A-$NHCH_3$

3 H4-K16-CoA R_1 = K-A-G-G-K-G-L-G-K-G-G-K-G-R-G-S-OCH_3
R_2 = R-H-R-K-NH_2

Fig. 3 Bisubstrate HATi

CoA moiety. Among the various strategies applied to overcome these limitations, particularly effective has been the linking of the polyamine spermidine to CoA by using a carboxymethylene bridge that led to the Spd(N^1)-CoA (**4**). Indeed, despite the presence of phosphate groups and an amide bond, this molecule is endowed with great cell permeability as its spermidine portion is recognized by the specific transport system of polyamines and is able to block the histone acetylation-dependent DNA repair and synthesis, so leading to chemo- and radiosensitization [75, 76].

7 Natural Products and Derivatives

Among the many HATi identified from natural sources, virtually all suffer from poor selectivity for a particular HAT enzyme, display undesired off-target effects that arise from their promiscuous chemical nature (e.g. polyphenols and Michael acceptors) and do not have the necessary physicochemical properties required for a potential candidate drug (Fig. 4).

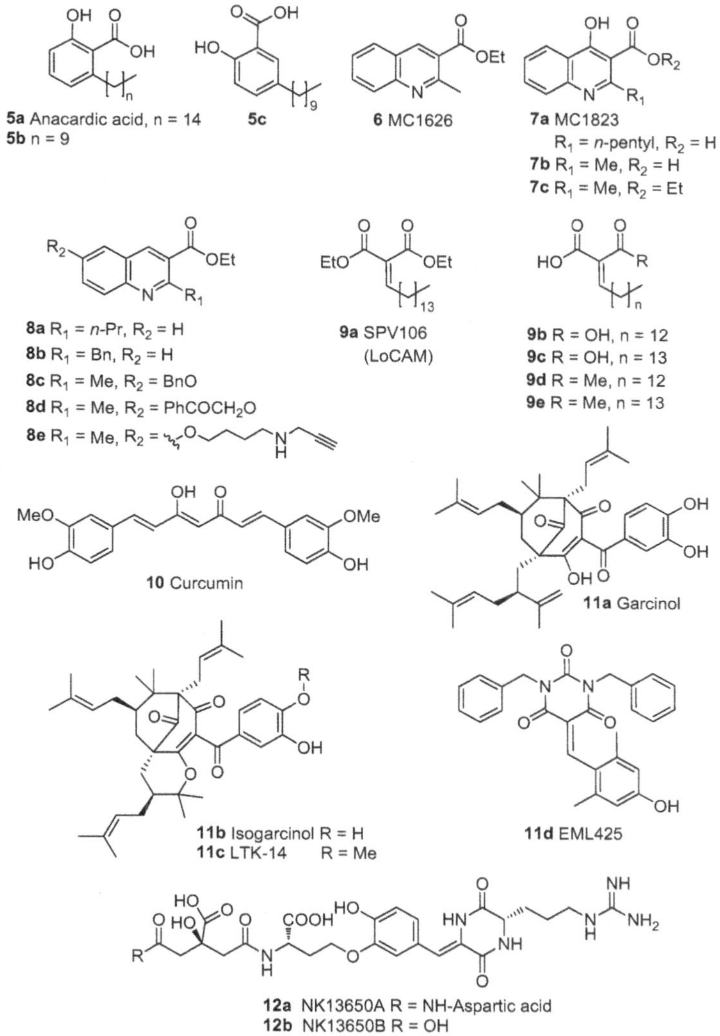

Fig. 4 HATi from natural sources and related compounds

Isolated from the liquid of cashew nut shells, the so-called anacardic acid (**5a**) is a non-competitive inhibitor of various HAT enzymes. Depending on the experimental assay and enzyme purification conditions, its IC_{50} values range between 8.5 [77] and over 1,000 μM [74] for p300/CBP, from 5 [77] to 667.1 μM [74] for PCAF and from 64 [78] to 347.6 μM [74] for Tip60. It is also able to inhibit MOF, with K_i and IC_{50} values of 64 [79] and 43 μM [78], respectively. One of its most relevant cellular effects is the repression of NF-κB signalling as a consequence of the inhibition of the p65 subunit acetylation. Nevertheless, the effects of **5a** in various cellular contexts seem mostly to be the consequence of its promiscuous inhibition of multiple enzymes [80]. For the low cell permeability due to its high lipophilicity, **5a** has been the object of an extensive med-chem optimization effort aimed at generating derivatives with increased solubility in water as well as better inhibitory potency and isoform selectivity. Recently, Wapenaar et al. reported a series of **5a** derivatives differing in regiochemistry, alkyl chain length and chemical composition that were tested against both p300 and MOF [79]. Among them, two derivatives, **5b** and **5c**, showed an inhibitory potency against MOF (IC_{50} = 37 and 57 μM, respectively) better than **5a** (IC_{50} = 64 μM). Since both compounds, despite differing in the regiochemistry, possess a linear decyl aliphatic chain instead of the pentadecyl tail of **5a**, it has been proposed that the hydrophobic contacts due to the aliphatic chain could play a key role in the interactions of this chemotype with MOF [79].

A yeast phenotypic screening where the reduction of *S. cerevisiae* viability was correlated to Gcn5 inhibition led to the identification of the quinoline **6** (MC1626) as a sub-millimolar HATi also effective at inhibiting acetylation and gene transcription mediated by Gcn5 [81]. Starting from **6** a first hit optimization effort led to **7a** (MC1823), a related derivative that is also an analogue of **5a** with the three substituents around the central quinoline core retaining the same relative positions as in **5a**, and with the linear pentadecyl chain of the prototype substituted with a pentyl tail [82]. More potent than **6** and **5a** at inhibiting the total HAT activity of nuclear extracts [82], **7a** was the lead for a new series of anacardic acid analogues based on the 3-carboxy-4-hydroxyquinoline moiety [83]. Among them, the 2-methyl derivatives **7b** and its ethyl ester **7c** resulted to be micromolar inhibitors of both p300 and CBP, more potent than **5a** [83]. Starting from **6** it has also been developed a series of HATi bearing longer alkyl/aryl groups at the quinoline C2 position or additional side chains at the C6 position. Among them, a few derivatives with *n*-propyl (**8a**) or benzyl (**8b**) groups at C2 position or extended side chains (**8c-e**) at C6 position showed twofold to threefold increased potency against p300 compared to **6**, with **8e** being the most potent against p300 (IC_{50} = 57.5 μM) and very effective at decreasing H3/H4 acetylation in human leukaemia cells (U937) [84].

The so-called long chain alkylidene malonates (LoCAMs) are a large class of HAT modulators derived from anacardic acid that includes SPV106 (**9a**) and related compounds. Endowed with an inhibitory potency against p300/CBP comparable to that of anacardic acid, **9a** is also a PCAF activator [85]. Among the LoCAMs, the most potent HATi are represented by the series of the bicarboxylic derivatives, with both the ethyl esters of the prototype hydrolysed, that are strong p300 inhibitors (IC_{50} = 1.3 and 1.1 μM for **9b** and **9c**, respectively) and partially inhibit PCAF, and the acetoacetic derivatives that are good inhibitors of p300 (IC_{50} = 2.4 and 4.7 μM for **9d** and **9e**, respectively), but are also strong activators of PCAF [86]. Interestingly, the corresponding acetylacetone derivatives that have both ethyl esters of **9a** replaced by acetyl groups show no effect on p300 while are pure activators of PCAF. In summary, by a structure-activity relationship point of view, it is possible to state that (1) the hydrolysis of the ester groups of the prototype leads to a significant p300 inhibition but decreases the affinity for PCAF; (2) in contrast, the substitution of the carboxylic moiety for an acetyl group results in a drop of p300 inhibition and a corresponding rise in PCAF activation; and (3) both small variations in the length of the aliphatic chain and the introduction of heteroatoms into it abolish any HAT modulatory activity [85, 86].

The rhizome of *Curcuma longa*, also indicated as turmeric, is a commonly used remedy in both Indian and Chinese traditional medicines. One of its principal components is represented by the natural product curcumin (**10**) that, among various other in vitro activities, has shown the capability to inhibit p300 with a potency in the micromolar range ($IC_{50} = 25$ μM) [87]. The cinnamoyl moieties of **10** are required for p300 inhibition because by functioning as Michael acceptors, they react covalently with a crucial cysteine residue within the active site of the enzyme [88]. Given the thiol reactivity and, more generally, its pleiotropic nature, **10** cannot be considered a selective HATi as it also targets diverse other proteins playing a crucial role in many epigenetic and non-epigenetic cellular networks [89]. Moreover, like other phytochemicals, **10** is also able to disrupt cell membranes, so it cannot be excluded that some of its biological effects are the result of this feature [90]. Presently, curcumin is under evaluation in many clinical trials for various disease conditions, but the results of the first completed placebo-controlled, double-blinded trials (e.g. colon and pancreatic cancer, Alzheimer's disease and radiation dermatitis) have revealed that the translation from in vitro to in vivo conditions leads to a massive drop in the activity of **10**, with high chemical instability, extremely low oral bioavailability and unspecific reactivity as the main causes of inefficacy [91]. Various synthetic derivatives of **10** have been shown to inhibit p300 in vitro with a better potency than the prototype, but all of them, sharing with the lead the main mechanism of action (thiol reactivity), are able to interact with multiple off-targets, and their non-specific nature makes hard and almost useless to draw correlations between observations in vitro and in vivo [92].

The polyisoprenylated benzophenone obtained from *Garcinia indica* called garcinol (**11a**) is a low micromolar inhibitor of p300 and PCAF ($IC_{50} = 7$ and 5 μM, respectively) [93]. Isogarcinol (**11b**), the product of its intramolecular cyclization, is also a low micromolar inhibitor of p300 and PCAF, while the semi-synthetic derivative LTK-14 (**11c**), obtained through the mono-methylation of **11b**, is a selective inhibitor of p300 ($IC_{50} = 5$–7 μM) [94]. Data of isothermal titration calorimetry suggest that the binding mode of **11c** is different from those of **11a** and **11b**. In fact, while the two natural HATi have likely two principal binding sites, with the hydroxyl groups of the catechol motif accommodated by the acetyl-CoA binding pocket, and the isoprenoid chains interacting close to the substrate-binding site region, LTK-14 is a non-competitive inhibitor for both histone and acetyl-CoA substrates that likely binds only to the second binding site of **11a** and **11b**, since the methylation of one of the two catechol hydroxyl groups impedes the interactions with the acetyl-CoA binding pocket [94]. In addition, probably due to differences in specificity, **11c** is much less toxic than its parent compounds in cellular tests (HeLa cell line) [95]. The molecular simplification of garcinol led to the benzylidenebarbituric acid EML425 (**11d**), which is a potent and selective p300/CBP inhibitor ($IC_{50} = 2.9$ and 1.1 μM for p300 and CBP, respectively) [96]. Despite the presence of an α,β-unsaturated carbonyl system would suggest a thiol reactivity and a covalent binding mode, molecular modelling studies performed on the **11d**-p300 interaction show that this inhibitor occupies the same binding pocket as **11c**, with a reversible and non-competitive mode of action. Interestingly, upon test in human leukaemia cells (U937), **11d** reduces the levels of H4K5 and H3K9 acetylation and induces the arrest of the cell cycle in the G0/G1 phase [96].

The two fungal metabolites of *Penicillium* species **12a** (NK13650A) and **12b** (NK13650B) are peptidic compounds that display a very potent and selective p300 inhibition (IC_{50}s = 11 and 22 nM, respectively). Moreover, they are able to repress the transcriptional activation mediated by the oestrogen and androgen receptors and to reduce the cell viability in various tumour cell lines. Their main disadvantage is the low metabolic stability and cell permeability due to the peptidic nature. Anyway, since through proper chemical modifications it is possible to decrease the peptide character, the very high inhibitory potency of both **12a** and **12b** suggests a great potential for their possible future applications in cancer management [97].

8 Synthetic Inhibitors

Over the years, a wide range of synthetic HATi have been reported exploiting strategies such as in silico library screening, high-throughput screening and structure- or ligand-based drug design approaches. Structure-activity relationship studies have provided a good understanding of the structural requirements for a strong HAT inhibition, but only very recently has been reported the first example of a fully synthetic, extremely potent, selective and safe HATi (Fig. 5).

Identified in 2005 by the means of a high-throughput screening, the isothiazolones of general formula **13** are PCAF and p300 inhibitors, able to reduce histone acetylation and block cell proliferation in different cellular settings. Their proposed mechanism of inhibition relies on a covalent binding between the isothiazolone sulphur and a key cysteine residue within the enzyme active site, forming a disulphide bridge. This assumption has been confirmed by the complete loss of HAT inhibition that can be observed after co-incubation with dithiothreitol (DTT) that likely competes for the formation of the disulphide bridge [98]. As for other HATi, the thiol reactivity of this series of inhibitors is their major weakness, since they can also cross-react with many other proteins [99]. A SAR study has highlighted the pivotal role of the substituent on the nitrogen atom in modulating the inhibitory activity of this series of HATi [100]. In fact, *N*-aryl-substituted derivatives are able to inhibit all enzymes tested in the study (p300/CBP, PCAF, Gcn5 and MOF), while *N*-alkyl or *N*-benzyl substituents lead to different patterns of isoform-selective inhibition. In particular, *N*-alkyl derivatives are weak inhibitors of both PCAF and Gcn5, while are still effective at inhibiting p300, CBP and MOF, with a small preference for p300/CBP [100]. An interesting series of related inhibitors is represented by the pyridoisothiazolones **14a** and **14b**. First reported as a low micromolar PCAF inhibitor, the *para*-fluorophenyl derivative PU139 (**14a**) [101], when tested in the above-cited SAR study, revealed to be a non-selective HATi with

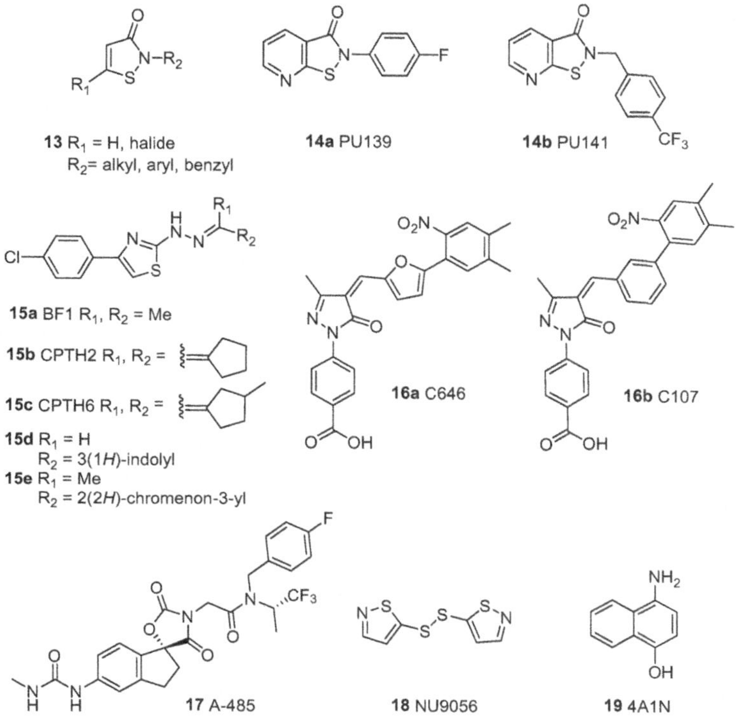

Fig. 5 Synthetic HATi

an inhibitory potency against PCAF ($IC_{50} = 9.7\ \mu M$) even lower than those observed with Gcn5 ($IC_{50} = 8.4\ \mu M$), p300 ($IC_{50} = 5.3\ \mu M$) and CBP ($IC_{50} = 2.5\ \mu M$) [100]. Interestingly, **14a** and its *para*-trifluoromethylbenzyl analogue **14b**, when tested against a panel of cysteine protease to evaluate any potential cross reactivity, did not show significant in vitro inhibition, demonstrating a promising target selectivity despite the covalent mechanism of action [102].

The derivatives BF1 (**15a**), CPTH2 (**15b**) and CPTH6 (**15c**) belong to a series of 4-phenyl-2-thiazolylhydrazones that have been reported first as Gcn5 inhibitors from a yeast phenotypic screening, where their reduction of *S. cerevisiae* cell growth could be correlated to the histone H3 hypoacetylation arising from Gcn5 inhibition [103]. In the following studies, **15a** and **15c** showed the ability to inhibit also p300 and PCAF, respectively, with **15a** being able to induce histone hypoacetylation in glioblastoma and neuroblastoma cell lines and **15c** to reduce histone H3/H4 and α-tubulin acetylation in several leukaemia cell lines [104]. **15b** and **15c** were also characterized for their ability to modulate the autophagic process in various cancer cell lines [105]. **15c** showed the ability to induce apoptosis in lung cancer stem-like cells derived from NSCLC patients, and its growth inhibitory effect correlated to the baseline level of K40-acetylated α-tubulin [106]. In 2014, a library of hydrazone derivatives and analogues was evaluated against p300 and PCAF, and some compounds resulted more potent than **15b** and **15c** at inhibiting p300, with the most active ones, **15d** and **15e**, that retain the 4-(*para*-chlorophenyl)-2-thiazolylhydrazone core of the prototypes and display IC_{50} values of 77.6 and 78.3 μM, respectively [107].

Identified through a structure-based virtual screening carried out on a library of commercially available small molecules that were docked into the binding pocket of Lys-CoA within p300 [108], the pyrazolone C646 (**16a**) is a selective (sub)micromolar inhibitor (K_i = 460 nM and IC_{50} = 1.6 μM) of p300. Various site-directed mutagenesis, molecular docking and SAR studies have allowed to clarify the binding mode and to identify the structural elements required for its strong inhibition. In this regard, the carboxylic acid seems essential for binding, but also crucial is a hydrogen bond acceptor in the position of the nitro group, while the oxidation to carbonyl or reduction of the C646 exo-methylene leads to a drop of inhibitory potency, suggesting that the electronic properties and planarity of the conjugated system are critical [108]. Despite reported as a non-covalent and competitive p300 inhibitor, the presence of an α,β-unsaturated system makes **16a** susceptible to Michael addition. In fact, subsequent studies have demonstrated that C646 reacts with numerous cysteine-containing proteins that could be, at least in part, responsible for its cellular effects [99, 109]. Another weakness of **16a** is its great intrinsic fluorescence which prevents the use in fluorescence-based assays. In order to overcome this problem, the furan ring of the prototype has been replaced with a phenyl moiety, leading to the non-fluorescent p300 inhibitor C107 (**16b**, IC_{50} = 9 μM) [110]. Able to reduce histone H3 and H4 acetylation and block cellular proliferation in lung and melanoma cancer cells [108], when assayed in AML1-ETO-positive leukaemia cells, **16a** inhibits cell growth and colony formation, with a partial cell cycle arrest in G1 phase and a significant apoptosis, associated with reduction of histone H3 acetylation and *bcl-2* and *c-kit* levels [111]. Interestingly, **16a** is also able to inhibit HDACs with K_i values ranging from 7 to 25 μM depending on the specific HDAC subtype, so confirming that it is not selective and that its cellular effects could also arise from the inhibition of enzymes other than p300 [112].

Very recently has been reported the first drug-like catalytic inhibitor of p300/CBP [113]. This spiro-oxazolidinedione compound, indicated as A-485 (**17**), has been identified through a virtual ligand screening performed on open conformation of the docked p300 binding pocket of Lys-CoA, followed by a medicinal chemistry optimization study performed on the first hits [113, 114]. With single-digit nanomolar inhibitory potencies against p300 (IC_{50} = 9.8 nM and K_d = 15 nM) and CBP (IC_{50} = 2.6 nM), **17** is particularly selective over various HAT enzymes (GCN5, PCAF, HAT1, TIP60, MYST3 and MYST4) and more than 150 epigenetic and non-epigenetic targets. Indeed, the co-crystal structure of A-485 bound to the p300 active site, in addition to show the competition with acetyl-CoA for the binding to the same pocket, by highlighting key interactions within the catalytic site, clearly explains the elevated HAT isoform specificity of **17** (Fig. 6). Moreover, A-485 reduces in a dose-dependent way the acetylation of H3K18 and H3K27 in various cell lines so confirming the selective inactivation of p300/CBP, displays selective antiproliferative effects in multiple prostate and haematological tumour cell lines,

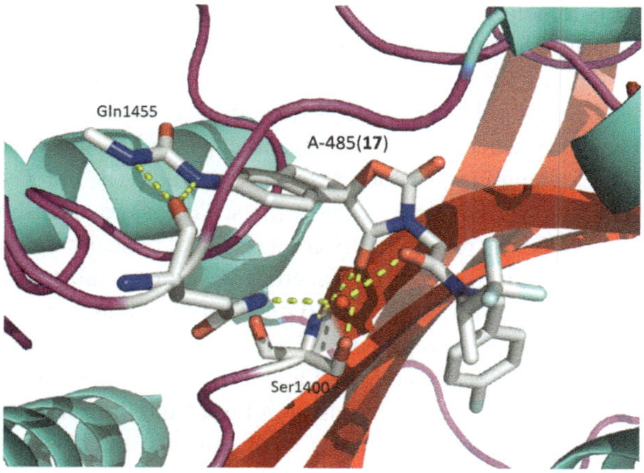

Fig. 6 Binding mode of compound **17** derived from the co-crystal structure with p300

blocks the androgen receptor-mediated transcriptional programmes in both androgen-sensitive and castration-resistant prostate tumour cells and reduces the growth of the prostate cancer in the xenograft model LuCaP-77 [113].

Identified through a high-throughput screening as a single-digit micromolar Tip60 inhibitor ($IC_{50} = 2\ \mu M$), the bis-isothiazole disulphide NU9056 (**18**), when assayed in prostate cancer cells, induced a time- and concentration-dependent decrease in histone acetylation and a significant amount of apoptosis [115].

Recently, by means of a fragment screening approach, has been reported the fragment 4-amino-1-naphthol (**19**, 4A1N) as a micromolar inhibitor of MOF ($IC_{50} = 9.7\ \mu M$), with K_i values that suggest an inhibition occurring mainly through the interaction with the acetylated form of the enzyme [116]. Despite unselective over PCAF ($IC_{50} = 3.6\ \mu M$) and p300 ($IC_{50} = 1.4\ \mu M$), **19** can be considered an interesting starting point for the development of more active and selective MOFi.

9 Conclusions and Perspectives

Despite more than 50 years have passed since the first discoveries about HAT activity and numerous studies over this timeframe have disclosed the structural features of HATs and their involvement in various pathological conditions, with only one very recent exception, no in vivo potent and selective HATi are available so far. In fact, despite virtual screening and structure-based drug design approaches, by using both enzyme and enzyme-(co)substrates crystallographic structures, have been successful in providing some promising HATi, there is still the necessity to develop more potent (particularly in cells and in vivo) and selective (between HAT subtypes and between HAT and other enzymes) inhibitors that could work either as chemical tools to better our knowledge about HAT biology or as potential chemotherapeutics to target the numerous disorders where HATs seem significantly involved.

One of the main problems in HATi development is the structure of the target itself: the highly conserved co-substrate-binding cleft and the shallow substrate-binding pocket make attaining potency and/or selectivity for any specific isoform very demanding. Another serious problem is the issue of the different results obtained from diverse assay formats. The usage of different techniques, that provide IC_{50} values as results of different physical measurements, and the great variability of types of buffer and substrate, enzyme sources, substrate/cofactor concentrations and incubation times often lead to quite inconsistent and even contradictory results. For this reason, the comparison of IC_{50} values, even for similar assay settings, cannot be reliable. The best solution to this issue is the standardization of the assay protocols, the usage of reliable reference inhibitors and the comparison only of K_i values after a precise study of the catalytic mechanism of the diverse HAT enzymes and of their inhibitors' kinetics. Orthogonal assays are also vital to confirm initial results, and counter screens against off-targets should be performed to establish selectivity of hit/lead compounds.

Another problem in HATi development is that quite often their cellular effects do not reflect the enzyme inhibitory activity measured in vitro. A possible explanation of the evidence that potency and selectivity of HATi often vary between cellular and enzymatic assays is that natively HATs are frequently part of multi-protein complexes, and their structural and/or catalytic functions can be modulated by the other proteins within the complex depending on the specific cellular context. Moreover, the majority of the HATi reported so far exhibit low potency, very scarce selectivity and metabolic instability, which may be responsible of their low efficacy in cells and in vivo.

The consequence of this situation is that currently, with the peculiar exception of curcumin, there are no HATi in advanced clinical development. Even curcumin, displaying an inhibiting activity against many other proteins in addition to p300, and showing additional mechanisms of action that likely contribute to the pleiotropic effects observed in cellular and animal models, when translated to human clinical trials, is not as effective as predicted by both in vitro and cellular assays [91].

Nowadays, only two crystallographic structures of HATi in complex with a single HAT enzyme are freely available: the complexes of p300 with **1** [73] and **17** [113]. Resolving this type of structures for other HATs, especially within their native multicomponent protein complexes, will be of vital importance to improve our understanding of crucial enzyme-inhibitor interactions, thereby providing precious information for the development of more potent and selective HATi. In fact, supported by the new crystallographic data, computer-aided methods, in combination with the most recent approaches in the drug discovery field, will likely significantly increase our chances to developing HATi that could be used as potential chemotherapeutics and/or at least as chemical probes for studying HAT biology.

Compliance with Ethical Standards

Conflict of Interest: The authors declare no conflict of interest.

Ethical Approval: This article does not contain any studies with human participants or animals performed by any of the authors.

References

1. Choudhary C, Kumar C, Gnad F, Nielsen ML, Rehman M, Walther TC, Olsen JV, Mann M (2009) Lysine acetylation targets protein complexes and co-regulates major cellular functions. Science 325:834–840. https://doi.org/10.1126/science.1175371
2. Roth SY, Denu JM, Allis CD (2001) Histone acetyltransferases. Annu Rev Biochem 70:81–120
3. Lee KK, Workman JL (2007) Histone acetyltransferase complexes: one size doesn't fit all. Nat Rev Mol Cell Biol 8:284–295
4. Kalkhoven E (2004) CBP and p300: HATs for different occasions. Biochem Pharmacol 68:1145–1155
5. Wang R, Cherukuri P, Luo J (2005) Activation of Stat3 sequence-specific DNA binding and transcription by p300/CREB-binding protein-mediated acetylation. J Biol Chem 280:11528–11534. https://doi.org/10.1074/jbc.M413930200
6. Sano Y, Ishii S (2001) Increased affinity of c-Myb for CREB-binding protein (CBP) after CBP-induced acetylation. J Biol Chem 276:3674–3682. https://doi.org/10.1074/jbc.M006896200
7. Chan HM, La Thangue NB (2001) p300/CBP proteins: HATs for transcriptional bridges and scaffolds. J Cell Sci 114:2363–2373
8. di Martile M, del Bufalo D, Trisciuoglio D (2016) The multifaceted role of lysine acetylation in cancer: prognostic biomarker and therapeutic target. Oncotarget 7:55789–55810. https://doi.org/10.18632/oncotarget.10048
9. Esteller M (2008) Epigenetics in cancer. N Engl J Med 358:1148–1159. https://doi.org/10.1056/NEJMra072067
10. Matsuzaki H, Daitoku H, Hatta M, Aoyama H, Yoshimochi K, Fukamizu A (2005) Acetylation of Foxo1 alters its DNA-binding ability and sensitivity to phosphorylation. Proc Natl Acad Sci U S A 102:11278–11283. https://doi.org/10.1073/pnas.0502738102
11. Dyda F, Klein DC, Hickman AB (2000) GCN5-related N-acetyltransferases: a structural overview. Annu Rev Biophys Biomol Struct 29:81–103
12. Love IM, Sekaric P, Shi D, Grossman SR, Androphy EJ (2012) The histone acetyltransferase PCAF regulates p21 transcription through stress-induced acetylation of histone H3. Cell Cycle 11:2458–2466
13. Kormendi V, Szyk A, Piszczek G, Roll-Mecak A (2012) Crystal structures of tubulin acetyltransferase reveal a conserved catalytic core and the plasticity of the essential N terminus. J Biol Chem 287:41569–41575
14. Taschner M, Vetter M, Lorentzen E (2012) Atomic resolution structure of human α-tubulin acetyltransferase bound to acetyl-CoA. Proc Natl Acad Sci U S A 109:19649–19654
15. Friedmann DR, Aguilar A, Fan J, Nachury MV, Marmorstein R (2012) Structure of the α-tubulin acetyltransferase, αTAT1, and implications for tubulin-specific acetylation. Proc Natl Acad Sci U S A 109:19655–19660
16. Castro-Castro A, Janke C, Montagnac G, Paul-Gilloteaux P, Chavrier P (2012) ATAT1/MEC-17 acetyltransferase and HDAC6 deacetylase control a balance of acetylation of alpha-tubulin and cortactin and regulate MT1-MMP trafficking and breast tumor cell invasion. Eur J Cell Biol 91:950–960

17. Boggs AE, Vitolo MI, Whipple RA, Charpentier MS, Goloubeva OG, Ioffe OB, Tuttle KC, Slovic J, Lu Y, Mills GB, Martin SS (2015) α-Tubulin acetylation elevated in metastatic and basal-like breast cancer cells promotes microtentacle formation, adhesion, and invasive migration. Cancer Res 75:203–215
18. Sapountzi V, Côté J (2011) MYST-family histone acetyltransferases: beyond chromatin. Cell Mol Life Sci 68:1147–1156
19. Avvakumov N, Côté J (2007) The MYST family of histone acetyltransferases and their intimate links to cancer. Oncogene 26:5395–5407
20. Patel JH, Du Y, Ard PG, Phillips C, Carella B, Chen CJ, Rakowski C, Chatterjee C, Lieberman PM, Lane WS, Blobel GA, McMahon SB (2004) The c-MYC oncoprotein is a substrate of the acetyltransferases hGCN5/PCAF and TIP60. Mol Cell Biol 24:10826–10834. https://doi.org/10.1128/MCB.24.24.10826-10834.2004
21. Tang Y, Luo J, Zhang W, Gu W (2006) Tip60-dependent acetylation of p53 modulates the decision between cell-cycle arrest and apoptosis. Mol Cell 24:827–839
22. van den Broeck A, Nissou D, Brambilla E, Eymin B, Gazzeri S (2012) Activation of a Tip60/E2F1/ERCC1 network in human lung adenocarcinoma cells exposed to cisplatin. Carcinogenesis 33:320–325
23. Murr R, Loizou JI, Yang YG, Cuenin C, Li H, Wang ZQ, Herceg Z (2006) Histone acetylation by Trrap-Tip60 modulates loading of repair proteins and repair of DNA double-strand breaks. Nat Cell Biol 8:91–99
24. Sykes SM, Mellert HS, Holbert MA, Li K, Marmorstein R, Lane WS, McMahon SB (2006) Acetylation of the p53 DNA-binding domain regulates apoptosis induction. Mol Cell 24:841–851
25. Su J, Wang F, Cai Y, Jin J (2016) The functional analysis of histone acetyltransferase MOF in tumorigenesis. Int J Mol Sci 17. https://doi.org/10.3390/ijms17010099
26. Taipale M, Rea S, Richter K, Vilar A, Lichter P, Imhof A, Akhtar A (2005) hMOF histone acetyltransferase is required for histone H4 lysine 16 acetylation in mammalian cells. Mol Cell Biol 25:6798–6810. https://doi.org/10.1128/MCB.25.15.6798-6810.2005
27. Chatterjee A, Seyfferth J, Lucci J, Gilsbach R, Preissl S, Bottinger L, Martensson CU, Panhale A, Stehle T, Kretz O, Sahyoun AH, Avilov S, Eimer S, Hein L, Pfanner N, Becker T, Akhtar A (2016) MOF acetyl transferase regulates transcription and respiration in mitochondria. Cell 167:722–738.e23. https://doi.org/10.1016/j.cell.2016.09.052
28. Borrow J, Stanton VP Jr, Andresen JM, Becher R, Behm FG, Chaganti RS, Civin CI, Disteche C, Dubé I, Frischauf AM, Horsman D, Mitelman F, Volinia S, Watmore AE, Housman DE (1996) The translocation t(8;16)(p11;p13) of acute myeloid leukaemia fuses a putative acetyltransferase to the CREB-binding protein. Nat Genet 14:33–41
29. Chaffanet M, Gressin L, Preudhomme C, Soenen-Cornu V, Birnbaum D, Pébusque MJ (2000) MOZ is fused to p300 in an acute monocytic leukemia with t(8;22). Genes Chromosomes Cancer 28:138–144
30. Yang XJ, Ullah M (2007) MOZ and MORF, two large MYSTic HATs in normal and cancer stem cells. Oncogene 26:5408–5419. https://doi.org/10.1038/sj.onc.1210609
31. Perez-Campo FM, Costa G, Lie-a-Ling M, Kouskoff V, Lacaud G (2013) The MYSTerious MOZ, a histone acetyltransferase with a key role in haematopoiesis. Immunology 139:161–165. https://doi.org/10.1111/imm.12072
32. Perez-Campo FM, Borrow J, Kouskoff V, Lacaud G (2009) The histone acetyl transferase activity of monocytic leukemia zinc finger is critical for the proliferation of hematopoietic precursors. Blood 113:4866–4874. https://doi.org/10.1182/blood-2008-04-152017
33. Crump JG, Swartz ME, Eberhart JK, Kimmel CB (2006) Moz-dependent Hox expression controls segment-specific fate maps of skeletal precursors in the face. Development 133:2661–2669. https://doi.org/10.1242/dev.02435
34. Iizuka M, Stillman B (1999) Histone acetyltransferase HBO1 interacts with the ORC1 subunit of the human initiator protein. J Biol Chem 274:23027–23034

35. Lalonde ME, Avvakumov N, Glass KC, Joncas FH, Saksouk N, Holliday M, Paquet E, Yan K, Tong Q, Klein BJ, Tan S, Yang XJ, Kutateladze TG, Cote J (2013) Exchange of associated factors directs a switch in HBO1 acetyltransferase histone tail specificity. Genes Dev 27:2009–2024. https://doi.org/10.1101/gad.223396.113
36. Karmodiya K, Anamika K, Muley V, Pradhan SJ, Bhide Y, Galande S (2014) Camello, a novel family of Histone Acetyltransferases that acetylate histone H4 and is essential for zebrafish development. Sci Rep 4:6076. https://doi.org/10.1038/srep06076
37. Iyer NG, Ozdag H, Caldas C (2004) p300/CBP and cancer. Oncogene 23:4225–4231
38. Sun XJ, Man N, Tan Y, Nimer SD, Wang L (2015) The role of histone acetyltransferases in normal and malignant hematopoiesis. Front Oncol 5:108
39. Mullighan CG, Zhang J, Kasper LH, Lerach S, Payne-Turner D, Phillips LA, Heatley SL, Holmfeldt L, Collins-Underwood JR, Ma J, Buetow KH, Pui CH, Baker SD, Brindle PK, Downing JR (2011) CREBBP mutations in relapsed acute lymphoblastic leukaemia. Nature 471:235–239. https://doi.org/10.1038/nature09727
40. Kishimoto M, Kohno T, Okudela K, Otsuka A, Sasaki H, Tanabe C, Sakiyama T, Hirama C, Kitabayashi I, Minna JD, Takenoshita S, Yokota J (2005) Mutations and deletions of the CBP gene in human lung cancer. Clin Cancer Res 11:512–519
41. Gayther SA, Batley SJ, Linger L, Bannister A, Thorpe K, Chin SF, Daigo Y, Russell P, Wilson A, Sowter HM, Delhanty JD, Ponder BA, Kouzarides T, Caldas C (2000) Mutations truncating the EP300 acetylase in human cancers. Nat Genet 24:300–303
42. Yokomizo C, Yamaguchi K, Itoh Y, Nishimura T, Umemura A, Minami M, Yasui K, Mitsuyoshi H, Fujii H, Tochiki N, Nakajima T, Okanoue T, Yoshikawa T (2011) High expression of p300 in HCC predicts shortened overall survival in association with enhanced epithelial mesenchymal transition of HCC cells. Cancer Lett 310:140–147
43. Xiao XS, Cai MY, Chen JW, Guan XY, Kung HF, Zeng YX, Xie D (2011) High expression of p300 in human breast cancer correlates with tumor recurrence and predicts adverse prognosis. Chin J Cancer Res 23:201–207
44. Chen MK, Cai MY, Luo RZ, Tian X, Liao QM, Zhang XY, Han JD (2015) Overexpression of p300 correlates with poor prognosis in patients with cutaneous squamous cell carcinoma. Br J Dermatol 172:111–119
45. Li M, Luo RZ, Chen JW, Cao Y, Lu JB, He JH, Wu QL, Cai MY (2011) High expression of transcriptional coactivator p300 correlates with aggressive features and poor prognosis of hepatocellular carcinoma. J Transl Med 9:5
46. Vervoorts J, Lüscher-Firzlaff JM, Rottmann S, Lilischkis R, Walsemann G, Dohmann K, Austen M, Lüscher B (2003) Stimulation of c-MYC transcriptional activity and acetylation by recruitment of the cofactor CBP. EMBO Rep 4:484–490
47. Zhong J, Ding L, Bohrer LR, Pan Y, Liu P, Zhang J, Sebo TJ, Karnes RJ, Tindall DJ, van Deursen J, Huang H (2014) p300 acetyltransferase regulates androgen receptor degradation and PTEN-deficient prostate tumorigenesis. Cancer Res 74:1870–1880
48. Pao GM, Janknecht R, Ruffner H, Hunter T, Verma IM (2000) CBP/p300 interact with and function as transcriptional coactivators of BRCA1. Proc Natl Acad Sci U S A 97:1020–1025
49. Teufel DP, Freund SM, Bycroft M, Fersht AR (2007) Four domains of p300 each bind tightly to a sequence spanning both transactivation subdomains of p53. Proc Natl Acad Sci U S A 104:7009–7014
50. Simo-Riudalbas L, Perez-Salvia M, Setien F, Villanueva A, Moutinho C, Martinez-Cardus A, Moran S, Berdasco M, Gomez A, Vidal E, Soler M, Heyn H, Vaquero A, de la Torre C, Barcelo-Batllori S, Vidal A, Roz L, Pastorino U, Szakszon K, Borck G, Moura CS, Carneiro F, Zondervan I, Savola S, Iwakawa R, Kohno T, Yokota J, Esteller M (2015) KAT6B is a tumor suppressor histone H3 lysine 23 acetyltransferase undergoing genomic loss in small cell lung cancer. Cancer Res 75:3936–3945. https://doi.org/10.1158/0008-5472.CAN-14-3702
51. Liu N, Zhang R, Zhao X, Su J, Bian X, Ni J, Yue Y, Cai Y, Jin J (2013) A potential diagnostic marker for ovarian cancer: involvement of the histone acetyltransferase, human males absent on the first. Oncol Lett 6:393–400. https://doi.org/10.3892/ol.2013.1380

52. Wang Y, Zhang R, Wu D, Lu Z, Sun W, Cai Y, Wang C, Jin J (2013) Epigenetic change in kidney tumor: downregulation of histone acetyltransferase MYST1 in human renal cell carcinoma. J Exp Clin Cancer Res 32:8. https://doi.org/10.1186/1756-9966-32-8
53. Cai M, Hu Z, Liu J, Gao J, Tan M, Zhang D, Zhu L, Liu S, Hou R, Lin B (2015) Expression of hMOF in different ovarian tissues and its effects on ovarian cancer prognosis. Oncol Rep 33:685–692. https://doi.org/10.3892/or.2014.3649
54. Chen Z, Ye X, Tang N, Shen S, Li Z, Niu X, Lu S, Xu L (2014) The histone acetylranseferase hMOF acetylates Nrf2 and regulates anti-drug responses in human non-small cell lung cancer. Br J Pharmacol 171:3196–3211. https://doi.org/10.1111/bph.12661
55. Halkidou K, Gnanapragasam VJ, Mehta PB, Logan IR, Brady ME, Cook S, Leung HY, Neal DE, Robson CN (2003) Expression of Tip60, an androgen receptor coactivator, and its role in prostate cancer development. Oncogene 22:2466–2477
56. Bassi C, Li YT, Khu K, Mateo F, Baniasadi PS, Elia A, Mason J, Stambolic V, Pujana MA, Mak TW, Gorrini C (2016) The acetyltransferase Tip60 contributes to mammary tumorigenesis by modulating DNA repair. Cell Death Differ 23:1198–1208
57. Yin YW, Jin HJ, Zhao W, Gao B, Fang J, Wei J, Zhang DD, Zhang J, Fang D (2015) The histone acetyltransferase GCN5 expression is elevated and regulated by c-Myc and E2F1 transcription factors in human colon cancer. Gene Expr 16:187–196
58. Li Q, Liu Z, Xu M, Xue Y, Yao B, Dou C, Jia Y, Wang Y, Tu K, Zheng X, Yao Y (2016) PCAF inhibits hepatocellular carcinoma metastasis by inhibition of epithelial-mesenchymal transition by targeting Gli-1. Cancer Lett 375:190–198
59. Akella JS, Wloga D, Kim J, Starostina NG, Lyons-Abbott S, Morrissette NS, Dougan ST, Kipreos ET, Gaertig J (2010) MEC-17 is an alpha-tubulin acetyltransferase. Nature 467:218–222
60. Aguilar A, Becker L, Tedeschi T, Heller S, Iomini C, Nachury MV (2014) A-tubulin K40 acetylation is required for contact inhibition of proliferation and cell-substrate adhesion. Mol Biol Cell 25:1854–1866
61. Bailey JM, Alsina J, Rasheed ZA, McAllister FM, Fu YY, Plentz R, Zhang H, Pasricha PJ, Bardeesy N, Matsui W, Maitra A, Leach SD (2014) DCLK1 marks a morphologically distinct subpopulation of cells with stem cell properties in preinvasive pancreatic cancer. Gastroenterology 146:245–256
62. Park E, Kim Y, Ryu H, Kowall NW, Lee J, Ryu H (2014) Epigenetic mechanisms of Rubinstein-Taybi syndrome. Neuromolecular Med 16:16–24. https://doi.org/10.1007/s12017-013-8285-3
63. Wood MA, Kaplan MP, Park A, Blanchard EJ, Oliveira AM, Lombardi TL, Abel T (2005) Transgenic mice expressing a truncated form of CREB-binding protein (CBP) exhibit deficits in hippocampal synaptic plasticity and memory storage. Learn Mem 12:111–119. https://doi.org/10.1101/lm.86605
64. Phan HM, Xu AW, Coco C, Srajer G, Wyszomierski S, Evrard YA, Eckner R, Dent SY (2005) GCN5 and p300 share essential functions during early embryogenesis. Dev Dyn 233:1337–1347. https://doi.org/10.1002/dvdy.20445
65. Xu W, Edmondson DG, Evrard YA, Wakamiya M, Behringer RR, Roth SY (2000) Loss of Gcn5l2 leads to increased apoptosis and mesodermal defects during mouse development. Nat Genet 26:229–232. https://doi.org/10.1038/79973
66. Yamauchi T, Yamauchi J, Kuwata T, Tamura T, Yamashita T, Bae N, Westphal H, Ozato K, Nakatani Y (2000) Distinct but overlapping roles of histone acetylase PCAF and of the closely related PCAF-B/GCN5 in mouse embryogenesis. Proc Natl Acad Sci U S A 97:11303–11306. https://doi.org/10.1073/pnas.97.21.11303
67. Maurice T, Duclot F, Meunier J, Naert G, Givalois L, Meffre J, Celerier A, Jacquet C, Copois V, Mechti N, Ozato K, Gongora C (2008) Altered memory capacities and response to stress in p300/CBP-associated factor (PCAF) histone acetylase knockout mice. Neuropsychopharmacology 33:1584–1602. https://doi.org/10.1038/sj.npp.1301551

68. Campeau PM, Kim JC, Lu JT, Schwartzentruber JA, Abdul-Rahman OA, Schlaubitz S, Murdock DM, Jiang MM, Lammer EJ, Enns GM, Rhead WJ, Rowland J, Robertson SP, Cormier-Daire V, Bainbridge MN, Yang XJ, Gingras MC, Gibbs RA, Rosenblatt DS, Majewski J, Lee BH (2012) Mutations in KAT6B, encoding a histone acetyltransferase, cause Genitopatellar syndrome. Am J Hum Genet 90:282–289. https://doi.org/10.1016/j.ajhg.2011.11.023
69. Wang Y, Miao X, Liu Y, Li F, Liu Q, Sun J, Cai L (2014) Dysregulation of histone acetyltransferases and deacetylases in cardiovascular diseases. Oxid Med Cell Longev 2014:641979. https://doi.org/10.1155/2014/641979
70. Yanazume T, Hasegawa K, Morimoto T, Kawamura T, Wada H, Matsumori A, Kawase Y, Hirai M, Kita T (2003) Cardiac p300 is involved in myocyte growth with decompensated heart failure. Mol Cell Biol 23:3593–3606
71. Iyer A, Fairlie DP, Brown L (2012) Lysine acetylation in obesity, diabetes and metabolic disease. Immunol Cell Biol 90:39–46. https://doi.org/10.1038/icb.2011.99
72. Lau OD, Kundu TK, Soccio RE, Ait-Si-Ali S, Khalil EM, Vassilev A, Wolffe AP, Nakatani Y, Roeder RG, Cole PA (2000) HATs off: selective synthetic inhibitors of the histone acetyltransferases p300 and PCAF. Mol Cell 5:589–595
73. Liu X, Wang L, Zhao K, Thompson PR, Hwang Y, Marmorstein R, Cole PA (2008) The structural basis of protein acetylation by the p300/CBP transcriptional coactivator. Nature 451:846–850. https://doi.org/10.1038/nature06546
74. Wu J, Xie N, Wu Z, Zhang Y, Zheng YG (2009) Bisubstrate inhibitors of the MYST HATs Esa1 and Tip60. Bioorg Med Chem 17:1381–1386. https://doi.org/10.1016/j.bmc.2008.12.014
75. Cullis PM, Wolfenden R, Cousens LS, Alberts BM (1982) Inhibition of histone acetylation by N-[2-(S-coenzyme A)acetyl] spermidine amide, a multisubstrate analog. J Biol Chem 257:12165–12169
76. Bandyopadhyay K, Baneres JL, Martin A, Blonski C, Parello J, Gjerset RA (2009) Spermidinyl-CoA-based HAT inhibitors block DNA repair and provide cancer-specific chemo- and radiosensitization. Cell Cycle 8:2779–2788
77. Balasubramanyam K, Swaminathan V, Ranganathan A, Kundu TK (2003) Small molecule modulators of histone acetyltransferase p300. J Biol Chem 278:19134–19140. https://doi.org/10.1074/jbc.M301580200
78. Ghizzoni M, Wu J, Gao T, Haisma HJ, Dekker FJ, George Zheng Y (2012) 6-alkylsalicylates are selective Tip60 inhibitors and target the acetyl-CoA binding site. Eur J Med Chem 47:337–344. https://doi.org/10.1016/j.ejmech.2011.11.001
79. Wapenaar H, van der Wouden PE, Groves MR, Rotili D, Mai A, Dekker FJ (2015) Enzyme kinetics and inhibition of histone acetyltransferase KAT8. Eur J Med Chem 105:289–296. https://doi.org/10.1016/j.ejmech.2015.10.016
80. Hemshekhar M, Sebastin Santhosh M, Kemparaju K, Girish KS (2012) Emerging roles of anacardic acid and its derivatives: a pharmacological overview. Basic Clin Pharmacol Toxicol 110:122–132. https://doi.org/10.1111/j.1742-7843.2011.00833.x
81. Ornaghi P, Rotili D, Sbardella G, Mai A, Filetici P (2005) A novel Gcn5p inhibitor represses cell growth, gene transcription and histone acetylation in budding yeast. Biochem Pharmacol 70:911–917. https://doi.org/10.1016/j.bcp.2005.06.013
82. Mai A, Rotili D, Tarantino D, Ornaghi P, Tosi F, Vicidomini C, Sbardella G, Nebbioso A, Miceli M, Altucci L, Filetici P (2006) Small-molecule inhibitors of histone acetyltransferase activity: identification and biological properties. J Med Chem 49:6897–6907. https://doi.org/10.1021/jm060601m
83. Mai A, Rotili D, Tarantino D, Nebbioso A, Castellano S, Sbardella G, Tini M, Altucci L (2009) Identification of 4-hydroxyquinolines inhibitors of p300/CBP histone acetyltransferases. Bioorg Med Chem Lett 19:1132–1135. https://doi.org/10.1016/j.bmcl.2008.12.097
84. Lenoci A, Tomassi S, Conte M, Benedetti R, Rodriguez V, Carradori S, Secci D, Castellano S, Sbardella G, Filetici P, Novellino E, Altucci L, Rotili D, Mai A (2014) Quinoline-based p300

histone acetyltransferase inhibitors with pro-apoptotic activity in human leukemia U937 cells. ChemMedChem 9:542–548. https://doi.org/10.1002/cmdc.201300536
85. Sbardella G, Castellano S, Vicidomini C, Rotili D, Nebbioso A, Miceli M, Altucci L, Mai A (2008) Identification of long chain alkylidenemalonates as novel small molecule modulators of histone acetyltransferases. Bioorg Med Chem Lett 18:2788–2792. https://doi.org/10.1016/j.bmcl.2008.04.017
86. Castellano S, Milite C, Feoli A, Viviano M, Mai A, Novellino E, Tosco A, Sbardella G (2015) Identification of structural features of 2-alkylidene-1,3-dicarbonyl derivatives that induce inhibition and/or activation of histone acetyltransferases KAT3B/p300 and KAT2B/PCAF. ChemMedChem 10:144–157. https://doi.org/10.1002/cmdc.201402371
87. Balasubramanyam K, Varier RA, Altaf M, Swaminathan V, Siddappa NB, Ranga U, Kundu TK (2004) Curcumin, a novel p300/CREB-binding protein-specific inhibitor of acetyltransferase, represses the acetylation of histone/nonhistone proteins and histone acetyltransferase-dependent chromatin transcription. J Biol Chem 279:51163–51171. https://doi.org/10.1074/jbc.M409024200
88. Marcu MG, Jung YJ, Lee S, Chung EJ, Lee MJ, Trepel J, Neckers L (2006) Curcumin is an inhibitor of p300 histone acetylatransferase. Med Chem 2:169–174
89. Fu S, Kurzrock R (2010) Development of curcumin as an epigenetic agent. Cancer 116:4670–4676. https://doi.org/10.1002/cncr.25414
90. Ingolfsson HI, Thakur P, Herold KF, Hobart EA, Ramsey NB, Periole X, de Jong DH, Zwama M, Yilmaz D, Hall K, Maretzky T, Hemmings HC Jr, Blobel C, Marrink SJ, Kocer A, Sack JT, Andersen OS (2014) Phytochemicals perturb membranes and promiscuously alter protein function. ACS Chem Biol 9:1788–1798. https://doi.org/10.1021/cb500086e
91. Nelson KM, Dahlin JL, Bisson J, Graham J, Pauli GF, Walters MA (2017) The essential medicinal chemistry of curcumin. J Med Chem 60:1620–1637. https://doi.org/10.1021/acs.jmedchem.6b00975
92. Arif M, Vedamurthy BM, Choudhari R, Ostwal YB, Mantelingu K, Kodaganur GS, Kundu TK (2010) Nitric oxide-mediated histone hyperacetylation in oral cancer: target for a water-soluble HAT inhibitor, CTK7A. Chem Biol 17:903–913. https://doi.org/10.1016/j.chembiol.2010.06.014
93. Balasubramanyam K, Altaf M, Varier RA, Swaminathan V, Ravindran A, Sadhale PP, Kundu TK (2004) Polyisoprenylated benzophenone, garcinol, a natural histone acetyltransferase inhibitor, represses chromatin transcription and alters global gene expression. J Biol Chem 279:33716–33726. https://doi.org/10.1074/jbc.M402839200
94. Mantelingu K, Reddy BA, Swaminathan V, Kishore AH, Siddappa NB, Kumar GV, Nagashankar G, Natesh N, Roy S, Sadhale PP, Ranga U, Narayana C, Kundu TK (2007) Specific inhibition of p300-HAT alters global gene expression and represses HIV replication. Chem Biol 14:645–657. https://doi.org/10.1016/j.chembiol.2007.04.011
95. Arif M, Pradhan SK, Thanuja GR, Vedamurthy BM, Agrawal S, Dasgupta D, Kundu TK (2009) Mechanism of p300 specific histone acetyltransferase inhibition by small molecules. J Med Chem 52:267–277. https://doi.org/10.1021/jm800657z
96. Milite C, Feoli A, Sasaki K, La Pietra V, Balzano AL, Marinelli L, Mai A, Novellino E, Castellano S, Tosco A, Sbardella G (2015) A novel cell-permeable, selective, and noncompetitive inhibitor of KAT3 histone acetyltransferases from a combined molecular pruning/classical isosterism approach. J Med Chem 58:2779–2798. https://doi.org/10.1021/jm5019687
97. Tohyama S, Tomura A, Ikeda N, Hatano M, Odanaka J, Kubota Y, Umekita M, Igarashi M, Sawa R, Morino T (2012) Discovery and characterization of NK13650s, naturally occurring p300-selective histone acetyltransferase inhibitors. J Org Chem 77:9044–9052. https://doi.org/10.1021/jo301534b
98. Stimson L, Rowlands MG, Newbatt YM, Smith NF, Raynaud FI, Rogers P, Bavetsias V, Gorsuch S, Jarman M, Bannister A, Kouzarides T, McDonald E, Workman P, Aherne GW

(2005) Isothiazolones as inhibitors of PCAF and p300 histone acetyltransferase activity. Mol Cancer Ther 4:1521–1532. https://doi.org/10.1158/1535-7163.MCT-05-0135

99. Dahlin JL, Nelson KM, Strasser JM, Barsyte-Lovejoy D, Szewczyk MM, Organ S, Cuellar M, Singh G, Shrimp JH, Nguyen N, Meier JL, Arrowsmith CH, Brown PJ, Baell JB, Walters MA (2017) Assay interference and off-target liabilities of reported histone acetyltransferase inhibitors. Nat Commun 8:1527. https://doi.org/10.1038/s41467-017-01657-3
100. Furdas SD, Hoffmann I, Robaa D, Herquel B, Malinka W, Świątek P, Akhtar A, Sippl W, Jung M (2014) Pyrido- and benzisothiazolones as inhibitors of histone acetyltransferases (HATs). Med Chem Commun 5(12):1856
101. Furdas SD, Shekfeh S, Bissinger EM, Wagner JM, Schlimme S, Valkov V, Hendzel M, Jung M, Sippl W (2011) Synthesis and biological testing of novel pyridoisothiazolones as histone acetyltransferase inhibitors. Bioorg Med Chem 19:3678–3689. https://doi.org/10.1016/j.bmc.2011.01.063
102. Gajer JM, Furdas SD, Grunder A, Gothwal M, Heinicke U, Keller K, Colland F, Fulda S, Pahl HL, Fichtner I, Sippl W, Jung M (2015) Histone acetyltransferase inhibitors block neuroblastoma cell growth in vivo. Oncogene 4:e137. https://doi.org/10.1038/oncsis.2014.51
103. Chimenti F, Bizzarri B, Maccioni E, Secci D, Bolasco A, Chimenti P, Fioravanti R, Granese A, Carradori S, Tosi F, Ballario P, Vernarecci S, Filetici P (2009) A novel histone acetyltransferase inhibitor modulating Gcn5 network: cyclopentylidene-[4-(4′-chlorophenyl) thiazol-2-yl]hydrazone. J Med Chem 52:530–536. https://doi.org/10.1021/jm800885d
104. Trisciuoglio D, Ragazzoni Y, Pelosi A, Desideri M, Carradori S, Gabellini C, Maresca G, Nescatelli R, Secci D, Bolasco A, Bizzarri B, Cavaliere C, D'Agnano I, Filetici P, Ricci-Vitiani L, Rizzo MG, del Bufalo D (2012) CPTH6, a thiazole derivative, induces histone hypoacetylation and apoptosis in human leukemia cells. Clin Cancer Res 18:475–486. https://doi.org/10.1158/1078-0432.CCR-11-0579
105. Ragazzoni Y, Desideri M, Gabellini C, de Luca T, Carradori S, Secci D, Nescatelli R, Candiloro A, Condello M, Meschini S, del Bufalo D, Trisciuoglio D (2013) The thiazole derivative CPTH6 impairs autophagy. Cell Death Dis 4:e524. https://doi.org/10.1038/cddis.2013.53
106. di Martile M, Desideri M, de Luca T, Gabellini C, Buglioni S, Eramo A, Sette G, Milella M, Rotili D, Mai A, Carradori S, Secci D, de Maria R, del Bufalo D, Trisciuoglio D (2016) Histone acetyltransferase inhibitor CPTH6 preferentially targets lung cancer stem-like cells. Oncotarget 7:11332–11348. https://doi.org/10.18632/oncotarget.7238
107. Carradori S, Rotili D, de Monte C, Lenoci A, D'Ascenzio M, Rodriguez V, Filetici P, Miceli M, Nebbioso A, Altucci L, Secci D, Mai A (2014) Evaluation of a large library of (thiazol-2-yl)hydrazones and analogues as histone acetyltransferase inhibitors: enzyme and cellular studies. Eur J Med Chem 80:569–578. https://doi.org/10.1016/j.ejmech.2014.04.042
108. Bowers EM, Yan G, Mukherjee C, Orry A, Wang L, Holbert MA, Crump NT, Hazzalin CA, Liszczak G, Yuan H, Larocca C, Saldanha SA, Abagyan R, Sun Y, Meyers DJ, Marmorstein R, Mahadevan LC, Alani RM, Cole PA (2010) Virtual ligand screening of the p300/CBP histone acetyltransferase: identification of a selective small molecule inhibitor. Chem Biol 17:471–482. https://doi.org/10.1016/j.chembiol.2010.03.006
109. Shrimp JH, Sorum AW, Garlick JM, Guasch L, Nicklaus MC, Meier JL (2015) Characterizing the covalent targets of a small molecule inhibitor of the lysine acetyltransferase P300. ACS Med Chem Lett 7:151–155. https://doi.org/10.1021/acsmedchemlett.5b00385
110. Dancy BM, Crump NT, Peterson DJ, Mukherjee C, Bowers EM, Ahn YH, Yoshida M, Zhang J, Mahadevan LC, Meyers DJ, Boeke JD, Cole PA (2012) Live-cell studies of p300/CBP histone acetyltransferase activity and inhibition. Chembiochem 13:2113–2121. https://doi.org/10.1002/cbic.201200381
111. Gao XN, Lin J, Ning QY, Gao L, Yao YS, Zhou JH, Li YH, Wang LL, Yu L (2013) A histone acetyltransferase p300 inhibitor C646 induces cell cycle arrest and apoptosis selectively in AML1-ETO-positive AML cells. PLoS One 8:e55481. https://doi.org/10.1371/journal.pone.0055481

112. van den Bosch T, Boichenko A, Leus NG, Ourailidou ME, Wapenaar H, Rotili D, Mai A, Imhof A, Bischoff R, Haisma HJ, Dekker FJ (2016) The histone acetyltransferase p300 inhibitor C646 reduces pro-inflammatory gene expression and inhibits histone deacetylases. Biochem Pharmacol 102:130–140. https://doi.org/10.1016/j.bcp.2015.12.010
113. Lasko LM, Jakob CG, Edalji RP, Qiu W, Montgomery D, Digiammarino EL, Hansen TM, Risi RM, Frey R, Manaves V, Shaw B, Algire M, Hessler P, Lam LT, Uziel T, Faivre E, Ferguson D, Buchanan FG, Martin RL, Torrent M, Chiang GG, Karukurichi K, Langston JW, Weinert BT, Choudhary C, de Vries P, van Drie JH, McElligott D, Kesicki E, Marmorstein R, Sun C, Cole PA, Rosenberg SH, Michaelides MR, Lai A, Bromberg KD (2017) Discovery of a selective catalytic p300/CBP inhibitor that targets lineage-specific tumours. Nature 550:128–132. https://doi.org/10.1038/nature24028
114. Michaelides MR, Kluge A, Patane M, van Drie JH, Wang C, Hansen TM, Risi RM, Mantei R, Hertel C, Karukurichi K, Nesterov A, McElligott D, de Vries P, Langston JW, Cole PA, Marmorstein R, Liu H, Lasko L, Bromberg KD, Lai A, Kesicki EA (2017) Discovery of spiro oxazolidinediones as selective, orally bioavailable inhibitors of p300/CBP histone acetyltransferases. ACS Med Chem Lett 9:28–33. https://doi.org/10.1021/acsmedchemlett.7b00395
115. Coffey K, Blackburn TJ, Cook S, Golding BT, Griffin RJ, Hardcastle IR, Hewitt L, Huberman K, McNeill HV, Newell DR, Roche C, Ryan-Munden CA, Watson A, Robson CN (2012) Characterisation of a Tip60 specific inhibitor, NU9056, in prostate cancer. PLoS One 7:e45539. https://doi.org/10.1371/journal.pone.0045539
116. Wapenaar H, van den Bosch T, Leus NGJ, van der Wouden PE, Eleftheriadis N, Hermans J, Hailu GS, Rotili D, Mai A, Domling A, Bischoff R, Haisma HJ, Dekker FJ (2017) The relevance of Ki calculation for bi-substrate enzymes illustrated by kinetic evaluation of a novel lysine (K) acetyltransferase 8 inhibitor. Eur J Med Chem 136:480–486. https://doi.org/10.1016/j.ejmech.2017.05.015

Top Med Chem (2020) 33: 123–158
DOI: 10.1007/7355_2019_72

Published online: 28 November 2019

Lysine Methyltransferases and Their Inhibitors

Giulia Stazi, Clemens Zwergel, and Sergio Valente

Contents

1 Introduction 126
1.1 The H3K9 Methyltransferases EHMT1 and EHMT2 (GLP and G9a) 126
1.2 The H3K9 Methyltransferase SETDB1 129
1.3 The H3K9 Methyltransferases SUV39H1 and SUV39H2 130
1.4 The H3K27 Methyltransferases EZH1 and EZH2 130
1.5 The H3K79 Histone Methyltransferase DOT1L 135
1.6 The H3K4 Histone Methyltransferases SET1/MLL Family: MLL1–4 and SETD1A and 1B 138
1.7 The H3K4 Histone Methyltransferase SETD7 140
1.8 The H3K4 Histone Methyltransferase SMYD 141
1.9 The H3K36 Histone Methyltransferase SETD2 143
1.10 The H4K20 Histone Methyltransferases SETD8, SUV420H1, and SUV420H2 . . 144
2 Conclusions 146
References 147

Abstract Since 2000, the histone methyltransferases that catalyze the methylation of a number of histone and nonhistone substrates have been discovered.

A growing body of literature is indicating that lysine methyltransferases (KMTs) play a crucial role for transcriptional regulation and are involved in cancer and various other human diseases, thus being of high interest as potential therapeutic targets.

In this book chapter, we highlight the discovery, characterization, and application of selective KMT inhibitors, useful for dissecting their physiological functions as well as their disease implications.

Over the past decade, there has been an impressive progress regarding the KMT inhibitor discovery, especially conjugating the research interest with the available and novel techniques including new assay methods, high-throughput screening,

G. Stazi, C. Zwergel, and S. Valente (✉)
Department of Drug Chemistry and Technologies, Sapienza University of Rome, Rome, Italy
e-mail: sergio.valente@uniroma1.it

structural biology, and medicinal chemistry approaches. Our goal is to point out herein key advances, challenges, possible future opportunities, and directions, regarding KMT modulation in a preclinical and clinical setting.

Keywords Cancer, Chromatin, DOT1L, Epigenetic modulators, EZH2, G9a, Lysine methyltransferases, SETs, SUVs

Abbreviations

5-HT2A	5-Hydroxytryptamine receptor 2A
AML	Acute myeloid leukemia
ASH2L	ASH2 like histone lysine methyltransferase complex subunit
CARM1	Coactivator-associated arginine methyltransferase 1
DLBCL	Diffuse large B-cell lymphoma
DNMT	DNA methyltransferase
DNMTi	DNA methyltransferase inhibitor
DOT1L	Disruptor of telomeric silencing 1-like
EED	Embryonic ectoderm development
EHMT1	Euchromatin histone methyltransferase 1 (see also GLP)
EHMT2	Euchromatin histone methyltransferase 2 (see also G9a)
ERα	Estrogen receptor alpha
EZH1	Enhancer of zeste homologue 1
EZH2	Enhancer of zeste homologue 2
FP	Fluorescence polarization
G9a	Euchromatic histone-lysine N-methyltransferase 2 (EHMT2)
GLP	G9a-like protein (EHMT1)
GPCRs	G protein-coupled receptor
H1	Histamine receptor 1
H3K9	Lysine 9 of histone 3
H3K9me2	Demethylated lysine 9 of histone 3
HbF	Hemoglobin F
HCT116	Colon colorectal carcinoma cell line
HDACi	Histone deacetylase inhibitors
HEK293T	Human embryonic kidney cell line
HL60	Human leukemia cell line
HSP90	Heat shock protein 90
HTS	High-throughput screening
IUGR	Intrauterine growth restriction
KARPAS-422	Lymphoma cell line
KMTs	Lysine methyltransferases
LnCaP	Prostate adenocarcinoma cells

MAP 3K2	Mitogen-activated protein kinase kinase kinase 2
MCF10A	Non-tumorigenic epithelial cell line
MCF7	Breast cancer cell line
MDA-MB-231	Human breast cancer cell line
MLL	Mixed-lineage leukemia
MOA	Mechanism of action
MV4–11 (MLL-AF4), MOLM-13 (MLL-AF9), and THP1 (MLL-AF9)	MLL-rearranged cell lines
MYND	Myeloid translocation protein-8, Nervy, and DEAF-1
PARP1	Poly(ADP-ribose)-polymerase 1
PC-3	Prostate cancer cell line
PCNA	Proliferating cell nuclear antigen
PK	Physical-chemical (properties)
PKMTs	Protein lysine methyltransferases
PML-NB	Promyelocytic leukemia protein nuclear bodies
PMTs	Protein methyltransferases
PPAR-γ	Peroxisome proliferator-activated receptor gamma
PPI	Protein-protein interaction
PRC2	Polycomb-repressive complex 2
PRMTs	Protein arginine methyltransferases
PWS	Prader-Willi syndrome
Rb	Retinoblastoma
RBBP5	Retinoblastoma-binding protein 5
RE-IIBP	Response element II-binding protein
RNMTs	RNA methyltransferases
SAH	S-adenosyl homocysteine
SAM	S-adenosyl-L-methionine
SAR	Structure-activity relationship
SCD	Sickle cell disease
SCLC	Small cell lung cancer
SET	Nuclear proto-oncogene
SETD2	SET domain containing 2 histone methyltransferase
SETD7	SET domain containing 7 histone methyltransferase
SETD8	SET domain containing 8 histone methyltransferase
SETDB1	SET domain bifurcated 1
SMYD	SET and MYND domain-containing
SUM159	Breast cancer cell line
SUV	Histone-lysine N-methyltransferase
SUZ12	Subunit polycomb-repressive complex 2
U2OS	Osteosarcoma cell line
U937	Histiocytic lymphoma cell line
VEGFR1	Vascular endothelial growth factor receptor 1
WDR5	WD repeat-containing protein 5
WT	Wild type

1 Introduction

First discovered in 2000, histone methyltransferases catalyze the methylation of a number of histone and nonhistone substrates. Protein methylation is a dynamic process, involved in gene expression, transcriptional regulation, and in many key steps in cell fate decision. The two major families of histone methyltransferases are the lysine methyltransferases (KMTs) and the arginine methyltransferases (PRMTs). The KMTs, in turn, can be divided in two different classes: the SET domain- and the non-SET domain-containing KMTs. DOT1L is the sole member of the non-SET KMT family. The SET domain-containing KMTs are classified according to their sequence similarity into five major families including SUV, SET1, SET2, EZ, and RIZ families [1]. Aberrant histone/protein methylation patterns have been associated with various disorders including cancer. In time, there has been a growing interest in elucidating the functions and probing the therapeutic relevance of these enzymes [2]. Hence, the research on small-molecule modulators of PMTs to be used as tool compounds or as drug discovery hits has been very prolific. However, despite the number of studies on several different lysine methyltransferases, relatively few selective inhibitors have been identified so far. In this chapter we will focus on the current progress toward the discovery of small-molecule and peptide-based inhibitors of KMTs.

1.1 The H3K9 Methyltransferases EHMT1 and EHMT2 (GLP and G9a)

Euchromatin histone methyltransferases such as EHMT1 and EHMT2, which are better known as GLP and G9a, are SET domain-containing methyltransferases which mono- or di-methylate H3K9 as well as other nonhistone substrates, such as K373 of the tumor suppressor p53 [3]. EHMTs are responsible for transcriptional repression and activation via the formation of a heterodimeric complex during germ cell formation, embryogenesis, and cardiac morphogenesis [4]. G9a is known to be overexpressed across different cancers, such as leukemia, prostate, lung or liver cancer, where its overexpression was associated with poor prognosis [5]. The reduction of G9a expression diminished cell proliferation, migration, and invasion of lung and breast cancer cells in vitro and suppressed tumor growth and metastasis in vivo [3]. Renneville et al. proved that *G9a* or *GLP* knockdown increased the expression of γ-globin genes leading to an augmented number of cells expressing hemoglobin F (HbF). This finding could be an interesting starting point for a new therapy of sickle cell disease via expressing hemoglobin F (HbF), as increased HbF levels are a well-validated strategy for sickle cell disease (SCD) treatment [6].

1.1.1 Non-SAM-Competitive G9a/GLP Inhibitors

In 2007, BIX01294 (**1**, Fig. 1), possessing three subunits, the quinazoline, the piperidine, and the diazepane one, was identified via a high-throughput screening approach. This compound was the first reported selective small-molecule inhibitor of G9a and GLP, exhibiting an IC_{50} of 1.9 μM for G9a and 0.7 μM for GLP [7]. With the aid of the BIX-GLP crystal structure, an unoccupied tunnel space, normally hosting the substrate histone lysine (H3K9), has been identified, allowing the extension of the molecule into the aforementioned tunnel to achieve more specific and potent inhibitors. Liu et al. extended at first the methyl ether substituent at C-7 to a three-carbon chain terminating with a dimethyl amine, leading to UNC0224 (**2**, Fig. 1), which was fivefold more potent in an enzyme-based assay than the parent compound **1**.

The same research group continued to investigate on the 7-position of the quinazoline moiety, not only by varying the linker size but also by using different aliphatic and alicyclic terminal groups [8]. This study yielded UNC0321 (**3**, Fig. 1) being a nanomolar inhibitor. Compound **1** was found to be rather GLP-specific, while both compounds **2** and **3** possess a slightly higher specificity toward G9a. Although the aforementioned EHMT inhibitors were very potent in enzyme-based assays, they lack desirable drug-like properties, displaying also metabolic issues. To address these problems, Liu et al. continued their med-chem optimization by modifying the C-2, C-4, and C-7 positions with numerous chemically different substituents [9]. The most advanced compound from this study, UNC0638 (**4**, Fig. 1), features at the C-7 a side chain. So far, this compound is the most deeply studied G9a/GLP inhibitor present in the literature, being 600-fold selective over other epigenetic and non-epigenetic targets.

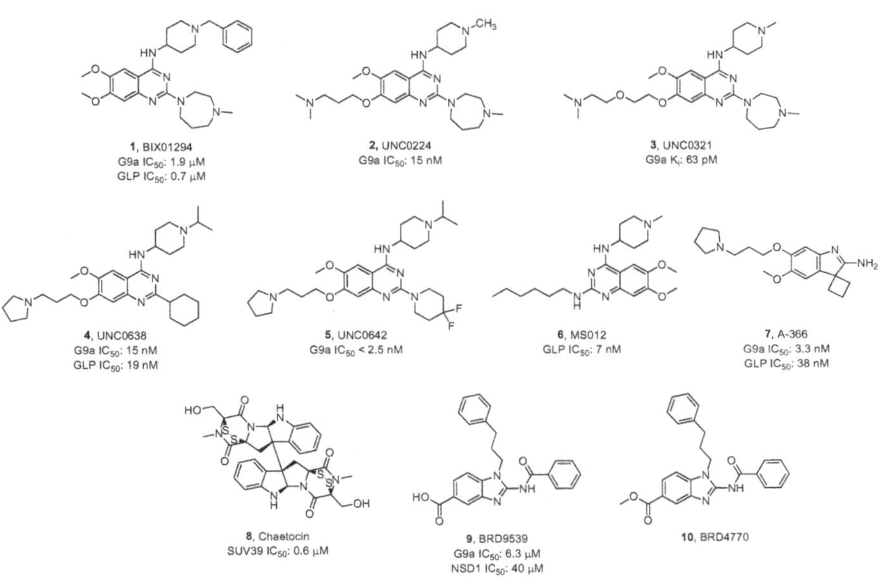

Fig. 1 H3K4 methyltransferases G9a/GLP and SUV39 inhibitors

Displaying enzymatic IC_{50} values of 19 nM and less than 15 nM against GLP and G9a, respectively [10], the inhibitor **4** is more or less equipotent against G9a/GLP. Kinetic studies in the presence of histone and SAM, as well as a crystal structure analysis, confirmed that compound **4** is a substrate-competitive inhibitor [10]. Compound **4** exhibited high cellular potency and low cell toxicity in several cancer and normal cell lines. For example, it lowered global H3K9me2 levels in human breast cancer MDA-MB-231 cells, being significantly more potent (IC_{50}: 81 $\pm$ 9 nM) than compound **1** (IC_{50}: 500 $\pm$ 43 nM) [10]. As mentioned before, compound **4** was less cytotoxic (EC_{50}: 11,000 $\pm$ 710 nM) than compound **1** (EC_{50}: 2,700 $\pm$ 76 nM) in MDA-MB-231 cells, thus offering a better therapeutic window. Additionally, compound **4** has been shown to induce leukemic stem cell differentiation, suppressing the proliferation of primary human AML cells [11]. The potential clinical benefit of such a pharmacological inhibition of G9a in AML treatment was confirmed in G9a-conditional knockout AML cells [11]. The cyclohexyl group present in the 2-position of compound **4** is known to be prone to metabolic modifications, so that a more suitable and stable candidate for in vivo studies was needed [10]. The same research group further investigated on the 2-position by using more stable substituents such as nitrogen-containing saturated heterocycles, leading to excellent biochemical properties together with IC_{50}s against G9 lower than 2.5 nM [12]. UNC0642 (**5**, Fig. 1), possessing a 4,4-difluoropiperidine substitution at C-2, succeeded compound **4**, being a suitable candidate for in vivo studies [12]. Compound **5** retained high in vitro potency for G9a and GLP (IC_{50} < 2.5 nM) and was >20,000-fold selective for G9a/GLP over other methyltransferases (e.g., SETD7, SETD8, SETDB1, PRMTs, SUVs, DOT1L, and DNMT1) and >300-fold selective over other non-epigenetic targets such as kinases. This compound reduced the H3K9me2 mark, without exhibiting a strong cytotoxicity, in both healthy and cancer cell lines. Compound **5** might not only be useful in cancer therapy but also in Prader-Willi syndrome (PWS), where it was proved to reactivate genes from the maternal allele not only in a cellular context but also in a mouse model, leading to a significantly increased growth and lifespan of PWS mouse pups [13]. Recently, MS012 (**6**, Fig. 1), a potent and selective (over PKMTs, PRMTs, DNMTs, and RNMTs) GLP inhibitor has been described [14]. This inhibitor possesses still a quinazoline scaffold and exhibits a 140-fold selectivity for GLP (IC_{50}: 7 $\pm$ 2 nM), over G9a. It should be mentioned that the X-ray structures revealed that this substrate-competitive inhibitor has an almost identical binding mode for GLP and G9a, underlining the challenge for a medicinal chemist to design potent and highly selective inhibitors for such homologous enzymes [14]. A-366 (**7**, Fig. 1), the first potent and selective non-quinazoline G9a inhibitor, has been described by Abbvie (IL, USA) featuring a novel spiro(cyclobutane-1,3′-indol)-2′-amine core preserving the right-hand side of the UNC G9a inhibitors. This compound potently inhibited G9a (IC_{50}: 3.3 nM) and GLP (IC_{50}: 38 nM) with a substrate-competitive and non-SAM-competitive mechanism of action (MOA) [15]. The crystal structure of compound **7** in complex with G9a revealed a comparable binding mode to compound **4** [15]. Like compound **4**, compound **7** at 3 μM reduced by 50% H3K9me2 levels in prostate cancer cells (PC-3), although no effects on cellular proliferation have been observed on a panel of 38 cancer cell lines even at 10 μM.

However, compound **7** when used in an acute myeloid leukemia flank xenograft model at 30 mg/kg/day for 14 days reduced H3K9 methylation, with a 45% reduction of tumor volume [16].

1.1.2 SAM-Competitive G9a/GLP Inhibitors

So far most of the G9a/GLP inhibitors are non-SAM competitive with limited focus on the development of SAM-competitive inhibitors [17]. Chaetocin (**8**, Fig. 1), a fungal epidithiodiketopiperazine alkaloid, will be discussed in more detail among the SUV inhibitors. However, 8 inhibits EHMT1 and EHMT2 as well, being at least tenfold less potent than BIX01294 (**1**). The selectivity of this compound is debated controversially in the literature [18]. In 2012, Yuan et al. screened a library of 2-substituted benzimidazoles mimicking SAM. They discovered the G9a SAM-competitive inhibitors BRD9539 (**9**, Fig. 1) and BRD4770 (**10**, Fig. 1), its methyl ester, as a prodrug [17]. While compound **9** inhibited G9a with an IC_{50} of 6.3 μM, it also inhibited EZH2 with a similar potency and NSD1 with a lower potency (IC_{50}: 40 μM). It was, however, selective over other epigenetic methyltransferases such as SUV39H1, SUV39H2, MLL1, SETD7, SETD8, PRMT1, PRMT3, PRMT5, and DNMT1. Strangely, by now, the inhibition properties of compound **9** versus GLP are not reported in the literature.

There are also further G9a/GLP inhibitors described in recent works [19, 20], but these compounds have been just validated in biochemical assays without extensive further characterization.

1.2 The H3K9 Methyltransferase SETDB1

SET domain bifurcated 1 (SETDB1, also known as ESET or KMT1E) is a HMT specifically catalyzing H3K9 methylation in the nucleus [21]. SETDB1 is responsible of transcriptional repression and gene silencing. It is implicated in PML-NB (nuclear bodies)-associated functions and plays a role in embryonic stem cells. During embryogenesis and postnatal development, SETDB1 is involved in X-chromosome inactivation, proviral silencing, differentiation of osteocytes and chondrocytes [22], and myogenesis. All these different aspects have been recently well reviewed by Karanth et al. [23]. SETDB1 is considered as an oncogene, and it has been found dysregulated already in early stages of cancer development. Its expression increases with cancer progression and invasiveness; indeed SETDB1 overexpression occurs in various cancers such as melanoma, lung cancer, ovarian cancer, hepatocellular carcinoma, and breast cancer [23]. Nevertheless, it is worth to remember that SETDB1 is involved also in adipogenesis and obesity (through PPAR-γ modulation), Huntington's disease, schizophrenia, and autism [23]. So far, no selective SETDB1 inhibitors have been reported; however the interest for targeting SETDB1 in cancer is growing. Currently, non-specific SETDB inhibitors, namely mithramycin, DZNep, and paclitaxel, are used to study the effects of targeting this MT in cancer cells [23].

1.3 The H3K9 Methyltransferases SUV39H1 and SUV39H2

SUV39H1 was the first histone lysine methyltransferase to be discovered. SUV39H1 and SUV39H2 catalyze, via a SAM-dependent mechanism, H3K9 dimethylation (me2) and trimethylation (me3) by preferentially binding to monomethylated (me1) H3K9 [24, 25]. Both enzymes are crucial in several biologically relevant processes [26]. For example, SUV39H1/2 may serve as a tumor suppressor by maintaining the H3K9 trimethylation mark at pericentric heterochromatin [27]. SUV39 knockdown induces higher sensitivity of tumor cells to radiation or chemotherapy agents, while their overexpression promotes malignancy and resistance to these treatments [28]. Furthermore, SUV39H2 might be involved in maintaining HIV silencing [29]. To date, only one SUV39 inhibitor has been published chaetocin (**8**, Fig. 1), that is, a potent SAM-competitive inhibitor with an IC_{50} of 0.6 μM [30]. However, follow-up studies have demonstrated that compound **8** is not very selective [18, 31] as it acts also on EHMT1 and EHMT2 and other targets such as thioredoxin [32]. Nevertheless, when used for AML treatment, compound **8** has been proven to induce apoptosis and to reduce cancer progression in vivo, similar to loss of SUV39H1 alone [32]. However, as off-target effect, upregulation of reactive oxygen species is also believed to be responsible for the response to chaetocin treatment [32]. To date, no small synthetic selective SUV39 inhibitor is known in the literature, underlying the need of the discovery of new specific hit compounds for this target.

1.4 The H3K27 Methyltransferases EZH1 and EZH2

The lysine methyltransferases EZH1 or EZH2 constitute the catalytic subunit of the PRC2 complex. Even though EZH1 and EZH2 possess a high sequence similarity, they exhibit different catalytic efficiencies, distinct chromatin binding properties, and expression patterns [33]. By themselves, EZH1 and EZH2 are not able to catalyze H3K27 methylation, but they require the copresence of at least two other protein subunits in the complex: EED and SUZ12 [34]. EZH2 has been proved to have also PRC2-independent functions, being involved in the methylation of a number of nonhistone substrates, such as the transcription factors GATA4 and PLZF [35]. The non-PRC2 activity of EZH2 could drive to transcriptional activation, rather than repression [36]. The polycomb complex is involved in the regulation of different functions in addition to the *HOX* gene silencing including X-chromosome inactivation, germline development, cell fate decision, cell-cycle regulation, senescence, stem cell pluripotency, and cancer metastasis [37]. In humans, the self-renewal ability of embryonic and tissue-specific stem cells is maintained by the PRC2 activity, and its dynamic regulation is critical for proper development and differentiation [38]. The wide biological role of EZH2 and, specially, its deep involvement in the regulation of cell-cycle progression, as well as its pivotal role in several cellular pathways, well explain why its dysfunction is associated with several solid or hematological cancers as well as its involvement in stem cell maintenance and tumor development [36]. Activating or inactivating somatic EZH2 mutations and deletions and missense, nonsense, and frameshift *EZH2* heterozygous or homozygous mutations have been found in various cancers [36]. Considering that both gain- and loss-of-function EZH2 mutations have been reported in cancers, we could conclude that EZH2 behaves in turn as an oncogene or as a tumor suppressor, based on the context. A balanced EZH2 activity is required to keep homeostasis [39]. In the light of these findings, EZH2 has been considered an attractive target for cancer therapy.

1.4.1 EZH2 Catalytic Inhibitors

The carbocyclic adenosine analogue 3-deazaneplanocin (DZNep, **11**, Fig. 2), a derivative of the natural antibiotic neplanocin A, has been one of the first small molecules to be tested as EZH2 inhibitor [40]. By mechanism, compound **11** is an *S*-adenosyl-L-homocysteine hydrolase inhibitor, affecting all SAM-dependent processes; hence it is an indirect and unselective inhibitor [41]. The poor PK and toxicological profile of compound **11** [42] encouraged the development of novel, potent, and selective inhibitors of EZH2. High-throughput biochemical screenings led to the development of SAM-competitive catalytic EZH2 inhibitors, many of them containing a dimethylpyridone moiety. In 2012, Epizyme Inc. reported EPZ005687 (**12**, Fig. 2) as a potent, selective, and SAM-competitive small-molecule inhibitor of EZH2 with a K_i of 24 nM [43]. Treatment with compound **12** in EZH2-WT and Y641- or A677-mutant lymphoma cells, as well as in other cancer cell lines, including breast and prostate cancers, resulted in dose-dependent ablation of H3K27 methylation. Simultaneously, GlaxoSmithKline (GSK), via a high-throughput biochemical screening, followed by an extensive medicinal chemistry optimization, disclosed GSK126 (**13**, Fig. 2), able to potently (K_i^{app}: 0.5–3 nM) and selectively inhibit WT and mutant EZH2. Compound **13** markedly inhibits the growth of EZH2-mutant diffuse large B-cell lymphoma in xenograft mice [44] and is currently being under evaluation in phase I clinical trials against various types of lymphoma [45]. GSK343 (**14**, Fig. 2) is another indazole-based potent EZH2 inhibitor (IC_{50}: 4 nM) [46]. EI1 (**15**, Fig. 2), a SAM-competitive inhibitor, is effective against WT and mutant EZH2 and displays >10,000-fold selectivity for EZH2 over other methyltransferases and 90-fold selectivity over EZH1. In cell-based studies, EI1 reduced H3K27 methylation levels reactivating PRC2 target genes. Moreover, it was able to decrease proliferation and to induce cell-cycle arrest and apoptosis in Y641 mutant large B-cell lymphoma [47]. Reported as the first orally bioavailable inhibitor in mice, UNC1999 (**16**, Fig. 2) was a dual and highly selective inhibitor of WT and Y641 mutant EZH2, as well as of EZH1. Also compound **16** is a

Fig. 2 An overview on the most widely reported EZH2 inhibitors

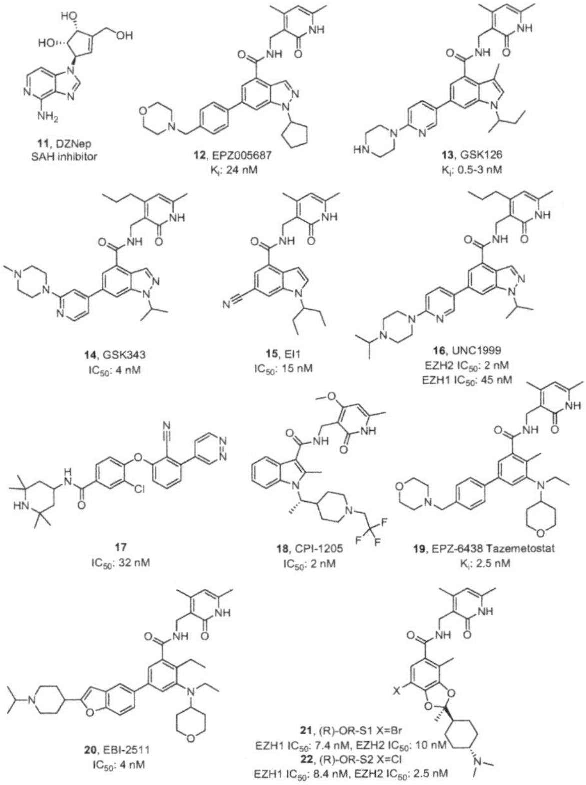

SAM-competitive inhibitor, able to reduce H3K27 methylation levels in cells and to induce apoptosis in Y641N-mutant large B-cell lymphoma [48].

The dual inhibitor **16** was proved to efficiently arrest cell proliferation MLL-rearranged leukemia [49] ex vivo and in vivo, while the selective EZH2 inhibitor compound **13** was not as effective. Importantly, these data raise the question around the possibility and the need of simultaneously targeting EZH1 and EZH2 and call for deeper investigation about the still misty EZH1, its functions, and its relation with EZH2. Interestingly, compound **17** (Fig. 2) is the sole potent EZH2i where the dimethylpyridone was efficiently replaced with a 2,2,6,6-tetramethylpiperidine. Compound **17** proved effective in KARPAS-422 lymphoma cells [50], but its poor PK properties did not allow further in vivo evaluation. CPI-360 showed a good activity profile in the biochemical assays (WT-EZH2 IC_{50}: 0.5 nM) as well as in a KARPAS-422 mouse xenograft model [51]. More recently, an optimization study on this compound led to the identification of CPI-1205 (**18**, Fig. 2), a highly potent (biochemical IC_{50}: 2 nM, cellular EC_{50}: 32 nM) and selective EZH2 inhibitor. Vaswani et al. also disclosed the co-crystal structure of compound **18** bound to the human PRC2 complex. This compound showed once again a potent antitumor activity in a KARPAS-422 mouse xenograft model. Compound **18** is currently in phase I clinical trials for B-cell lymphoma treatment [52]. Tazemetostat (EPZ6438, **19**, Fig. 2) was developed by means of a follow-up optimization of the Epizyme scaffold (EPZ005687, **12**). It has greater potency and better PK profile than compound **12**, including good oral bioavailability [53]. Compound **19** recently entered the clinical arena and is currently studied for a number of lymphomas in phase II but also for solid tumors in phase I [54–56]. Very recently Lu et al. published an optimization study of tazemetostat leading to compound EBI-2511 (**20**) demonstrating excellent in vivo efficacy in Pfeiffer tumor xenograft mouse [57]. Honma et al. recently described (R)-OR-S1 **21** and (R)-OR-S2 **22** (Fig. 2) as dual inhibitors of EZH1/2 suppressing trimethylation of histone H3K27 in cells more than EZH2 selective inhibitors. They also showed greater antitumor efficacy than EZH2 selective inhibitor in vitro and in vivo against diffuse large B-cell lymphoma as well as solid cancers, without exhibiting severe toxicity in rats, thus indicating the possibility of EZH1/2 dual inhibitors for clinical applications [58].

1.4.2 PRC2 Disruptors

Even though EZH2 knockdown studies showed that EZH2 is crucial in many tumor types, small-molecules EZH2 inhibitors proved to be effective in a smaller range of cancers. This phenomenon could be explained considering the need to target both the catalytic and the scaffolding activity of EZH2/PRC2 [59] and/or thinking about selectively targeting PRC2-dependent functions. Thus, both issues call for an alternative way of targeting the PRC2. In this regard, targeting protein-protein interactions has been considered an interesting strategy to be applied. The first PRC2 disruptors to be described have been stabilized α-helix of EZH2 peptides (also called SAH-EZH2). Kim et al. applied all-hydrocarbon stapling to stabilize α-helical structures of their peptides, leading to an enhanced target binding affinity, protease resistance, and membrane penetration. Upon treatment with SAH-EZH2, MLL-AF9 leukemia cells undergo growth arrest and monocyte-macrophage differentiation [60]. High-throughput screenings led to the identification of the first small-molecule EZH2-EED disruptors: astemizole (**23**, Fig. 3), an old antihistamine drug [61], and wedelolactone (**24**, Fig. 3), a natural compound [62]. Between December 2016 and March 2017, a series of publications proved the growing interest and the huge effort done in developing EED-targeting agents as an alternative strategy to inhibit the PRC2 functions [63–68]. In December 2016, Novartis reported the discovery of five structurally distinct EED binder hits **25–29** (Fig. 3), identified by a high-throughput screen [69]. Compounds **25–29** displayed low micromolar activities against PRC2, with similar values in presence of EZH1 or EZH2, and they were quite selective in inhibition with respect to other HMTs ($IC_{50} > 100$ μM). The co-crystal structure of compounds **25–29** in complex with EED has been resolved and studied in detail [63]. This work was followed by three other publications describing the subsequent development of some of the newly reported scaffolds. An optimization study of compound **28**, through X-ray crystallography-guided fragmentation and regrowth,

Fig. 3 PRC2 disruptors

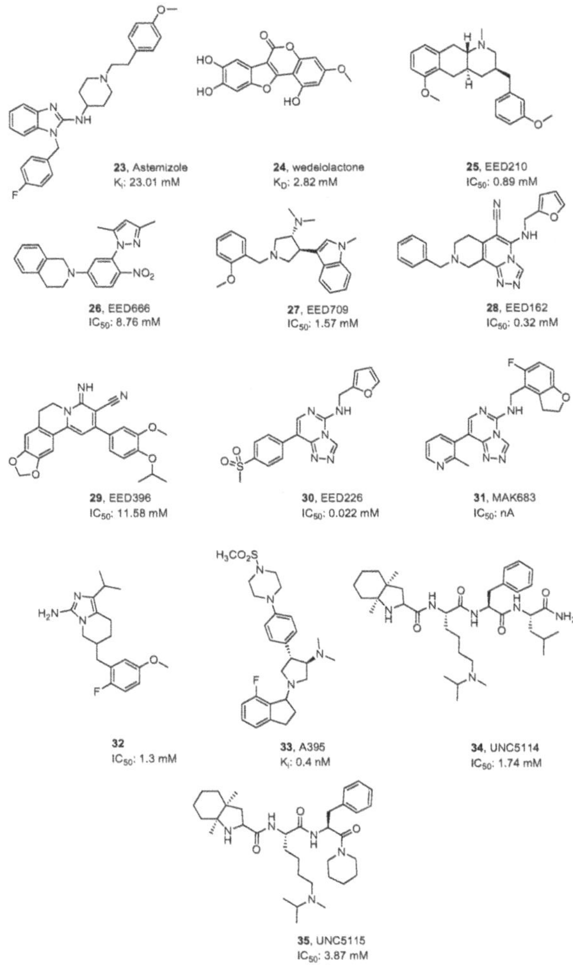

yielded compound EED226 (**30**, Fig. 3) [65], a potent and selective PRC2-EZH2/PRC2-EZH1 disruptor, reducing global H3K27me3 in cells, able to kill selectively cells with a heterozygous Y641N mutation. This compound showed optimal PK properties, encouraging and enabling extensive preclinical studies, demonstrating an optimal tolerability in various species, also inducing regression of tumor xenografts in vivo [65]. Notably, this compound is even effective in cell lines with acquired resistance to SAM-competitive EZH2 inhibitors and shows a synergistic effect when combined with EZH2 inhibitors [67]. Compound **31** (Fig. 3), MAK-683, structurally related to compound **30**, is currently in phase I/II clinical trials for diffuse large B-cell lymphoma nasopharyngeal carcinoma [56]. Compound **31** has been described in a patent in 2017 without giving details on its precise biological activity [70] in contrast to compound **30**, which has been described in various contexts [65, 67].

Applying the same deconstruction and regrowth strategy, the Novartis researchers also developed compound **32** (Fig. 3) from the HTS hit **25** [64]. In the same period, also AbbVie Inc. reported the identification of a novel potent and selective EED binder, compound A-395 (**33**, Fig. 3), able to bind EED in the H3K27me3 binding pocket and to inhibit H3K27 methylation in vitro via inhibition of PRC2 activation. When tested in cell-based assays, compound **33** strongly reduced the H3K27 methylation and inhibited cancer cell growth and proved to be effective also in cell lines with acquired resistance to SAM-competitive EZH2 inhibitors. Finally, the novel AbbVie compound was efficacious in a diffuse large B-cell lymphoma (DLBCL) xenograft mouse model (using cell lines containing an activating heterozygous EZH2 A677G mutation) [66].

Lastly, Barnash et al. optimized a low-affinity methylated Jarid2 peptide to give the first peptidomimetic EED ligands UNC5114 (**34**) and UNC5115 (**35**) (Fig. 3), using a coupled combinatorial and structure-based design approach. The binding mode and the exact mechanism of PRC2 inhibition are not yet fully understood. However, it is confirmed that the novel peptidomimetic binder is an allosteric inhibitor, targeting the methyl-lysine reader function [68].

Recently, EZH2 inhibitors have been used in combination with known or novel epi-drugs or standard drugs, for example, DZNep was combined with HDACi and/or DNMTi resulting in an effective and synergistic treatment in AML than each single agent alone [71], but also the pyrimidine-based inhibitors with HDACi displayed antitumor activity in glioblastoma [72].

1.5 The H3K79 Histone Methyltransferase DOT1L

DOT1L (disruptor of telomeric silencing 1-like), also known as KMT4, is a histone methyltransferase responsible for mono-, di-, and trimethylation of H3K79 [73], being associated with active gene transcription. H3K79 methylation is involved in the expression of genes involved in cell cycle progression, DNA damage response, as well as embryonic development, hematopoiesis, and cardiac function [74]. DOT1L was believed to be the sole H3K79 methyltransferase, but recent findings support the idea that MMSET isoform RE-IIBP (interleukin-5 response element II-binding protein) methylates this mark as well [75]. DOT1L possesses no SET domain, different from the other identified human PKMTs [76]; instead it exhibits its enzymatic activity via a non-SET catalytic domain, similar to those observed in PRMTs and DNMTs [77, 78]. The turnover of the modifications catalyzed by DOT1L is rather slow, and no demethylase capable to remove this mark has been described to date [79]. DOT1L has been studied mainly for its hematopoietic regulation properties [74], being described as an attractive epigenetic target in AML with mixed-lineage leukemia (MLL) gene translocations. In vitro studies in such cancer cell lines confirmed that DOT1L pharmacological inhibition or its genetic ablation led to differentiation and apoptosis [80]. Furthermore, it has been discovered as a potential promising target also in breast cancer [81].

In 2011, Epizyme Inc. described the first selective DOT1L inhibitor EPZ004777 (**36**, Fig. 4), based on the cofactor SAM and the crystal structure of the enzyme active site. This compound possesses a high picomolar potency in an enzyme-based assay (IC_{50}: 400 ± 100 pM) [82]. It shows a remarkable selectivity (1,000-fold selective) over other SAM-dependent methyltransferases despite the structural similarity of the cofactor binding pocket. As expected, binding studies and crystal structures of this inhibitor within DOT1L confirmed it as a SAM competitor [83]. Reduced levels of H3K79me2 were found in several leukemia cell lines treated with compound **36**, without affecting other histone methylation marks, thus underlining the specificity of this compound [82]. Furthermore, compound **36** led to a reduced expression of the MLL-rearranged leukemia hallmarks *HOXA9* and *MEIS1* [82], being able to selectively kill at low micromolar potencies MLL-rearranged leukemia cells, while having little effect on non-MLL-translocated cells, and prolonging survival in a mouse model of MLL-rearranged leukemia [84]. These effects can be taken as a proof of concept for the pharmacological inhibition of DOT1L in specific cancer therapy [82]. However, despite these first promising cell-based results, further clinical development was not pursued as the pharmacokinetic properties of this compound were not ideal. In 2013, Epizyme Inc. published the second generation of an improved derivative of compound **36**, known as EPZ-5676 or pinometostat (**37**, Fig. 4) [85]. In this compound the ribose moiety was replaced with a cyclobutyl ring to enhance its pharmacokinetic properties as well as its inhibition activity against DOT1L (K_i < 0.08 nM; EPZ004777 K_i: 0.3 nM) [85]. Compound **37** retained the same binding mode as its parent compound. Interestingly, the selectivity over other protein methyltransferases has been further increased up to 37,000-fold [85]. The new analogue displayed a nanomolar antiproliferative activity against most of the other MLL-rearranged cell lines such as MV4–11 (MLL-AF4), MOLM-13 (MLL-AF9), and THP1 (MLL-AF9), with little effect on leukemia cells lacking the MLL translocation [82, 85]. Despite improved pharmacokinetic profile [86], compound **37** is still poorly orally bioavailable [87]. To summarize, compound **37** is a major breakthrough in the protein methyltransferase (PMT) inhibitor field, which entered as the first PMT inhibitor in the clinical arena: phase I clinical trials for this inhibitor were recently completed for the treatment of patients with MLL-r, a genetically defined type of acute leukemia [56]. However, the outcome of this study has not yet been published. Furthermore, there are also several other DOT1L inhibitors possessing a deazadenosine core reported in the literature. Using the crystal structure of the DOT1L-**36** complex, Yu et al. developed a chemical probe called SGC0946 (**38**, Fig. 4), just showing an additional bromo atom at the 7-position of the deazadenosine ring of **36**, with enhanced in vitro (IC_{50}: 0.3 nM) and in vivo potency. This compound was also able to reduce H3K79 methylation levels (IC_{50}: 8.8 ± 1.6 nM) tenfold more potently than the lead **36** (IC_{50}: 84 ± 20 nM) in MCF10A cells while maintaining a good selectivity profile [83]. Yao et al. described compound **39** (Fig. 4) as a selective, covalent, and potent DOT1L inhibitor (IC_{50}: 38 nM), possessing greater than 29-fold selectivity for DOT1L over other methyltransferases such as CARM1, PRMT1, G9a, and SUV39H1 [88]. The authors proposed, additionally to the hydrogen bond interactions within the SET domain, an intramolecular cyclization of compound **39**

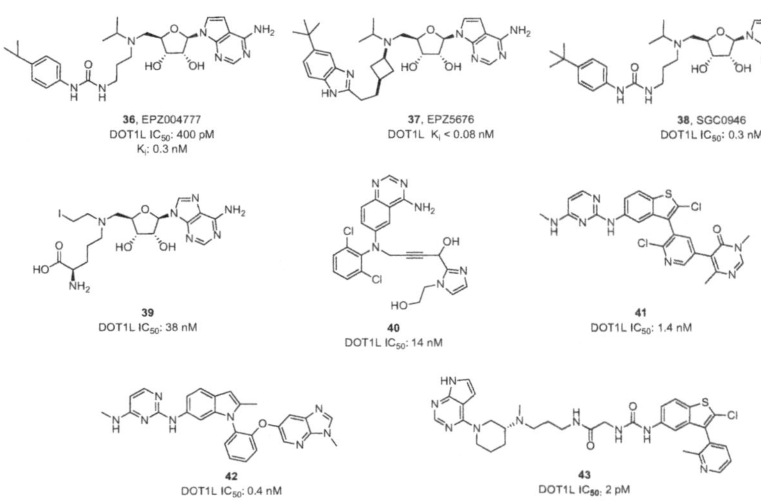

Fig. 4 H3K79 methyltransferase DOT1L inhibitors

to form a reactive aziridinium intermediate, able to react with the ε-NH_2 group of lysine 79 forming a covalent bond to H3K79. This compound has not yet been evaluated in cell-based assays. Scheufler et al. published for the first time in 2016 the non-nucleoside DOT1L inhibitor **40** (Fig. 4) via a weak fragment-based screening being chemically different from other previously published SAM-based DOT1L inhibitors [89]. This novel compound turned out to be potent in a biochemical assay, exhibiting an IC_{50} of 14 nM. It is interesting to note that compound **40** was binding to an induced binding pocket different from the SAM binding site of the classical DOT1L. The inhibitor **40** was confirmed to be SAM-competitive in X-ray crystallography studies, as upon binding of the compound **40**, a conformational change of the enzyme did not allow anymore the binding of the SAM cofactor essential for the functionality of DOT1L. In a subsequent paper, the same research group continued the abovementioned fragment-based approach, leading to the discovery of compounds **41** and **42** (Fig. 4), retaining the same binding mode of **40** and acting on the same induced binding pocket [90]. In a biochemical screen, both compounds turned out to be potent and selective DOT1L inhibitors with IC_{50} values of 1.4 and 0.4 nM, respectively. Cellular studies revealed that compounds **41** and **42** are able to decrease H3K79me2 levels with IC_{50} in the nanomolar range (23 and 16 nM, respectively) and to inhibit the proliferation of the MLL-rearranged MV4–11 leukemia cells in a nanomolar range as well. Furthermore, extensive PK studies on compound **41** in rats highlighted good pharmacokinetic properties, including a moderate half-life and a good oral bioavailability. Recently, the aforementioned research group carried on working with other fragments resulting in the discovery of a highly potent (IC_{50}: 2 pM) and selective DOT1L inhibitor **43** (Fig. 4) [91]. Impressively, inhibitor **43** was threefold more potent than compound **37** (IC_{50}: 5 nM and

15 nM, respectively) in proliferation arrest of MV4–11 cells. Overall, **40–43** are all exciting new non-nucleoside-based potent and selective DOT1L inhibitors, being useful tools for cellular and in vivo studies with high potential to achieve a clinical trial setting.

1.6 The H3K4 Histone Methyltransferases SET1/MLL Family: MLL1–4 and SETD1A and 1B

The human SET1 family includes MLL1 (KMT2A), MLL2 (KMT2D), MLL3 (KMT2C), MLL4 (KMT2B), SET1A (KMT2F), and SET1B (KMT2G). MLL1 (also known as KMT2A or TRX1) is a multidomain histone methyltransferase specific for mono-, di-, and trimethylation of H3K4 [92]. MLL1 undergoes several types of chromosomal rearrangements, and all of them have been associated with different leukemias [92]. A number of nonsense, frameshift, or coding mutations in the MLL protein family have been identified, but their physiological role is not yet fully understood [92]. The crystal structure of MLL1 SET domain, together with biochemical data, showed that MLL1 itself displays a poor methyltransferase activity [93]. Indeed, an optimal catalytic activity is achieved within the MLL complex including at least three evolutionary conserved subunits: WDR5 (WD repeat-containing protein 5), RBBP5 (retinoblastoma-binding protein 5), and ASH2L (ASH2 like histone lysine methyltransferase complex subunit) [93]. Selective MLL inhibitors have not been described so far; however, considering the relevance of MLL interactions with its partners [94, 95], a number of PPI disruptors have been identified. These compounds can be divided in two classes: on the one hand the WDR5 inhibitors and on the other the menin-MLL PPI disruptors.

1.6.1 WDR5 Inhibitors

In 2013, Karatas et al. reported novel peptidomimetic antagonists of WDR5-MLL interaction, among them MM-102 (**44**, IC_{50}: 400 nM, Fig. 5) was the most potent and effective in cells, and it became a starting point for the development of cyclic peptidomimetic derivatives. First, in 2014, Cao et al. reported compound MM-401 (**45**, Fig. 5). Despite the cyclization, compound 45 proved to be a tight binder (K_D: 0.9 nM) and potent (IC_{50}: 320 nM) and selective inhibitor [96]. Interestingly, in MLL-AF9 cells, compound **45** (at 20 μM) specifically reduced H3K4me2 and H3K4me3 across 5′ HoxA loci, after 48 h of treatment, while selectively inducing cell-cycle arrest, apoptosis, and differentiation in murine leukemia cells MLL1-AF9, MLL1-ENL, and MLL1-AF1. Moreover, it efficiently inhibited cell growth in human blasts derived from AML patients with MLL1 rearrangements [96]. In 2017, in a follow-up study on the same scaffold, the authors determined the optimal linker length and discovered a number of promising analogues, being MM-589 (**46**,

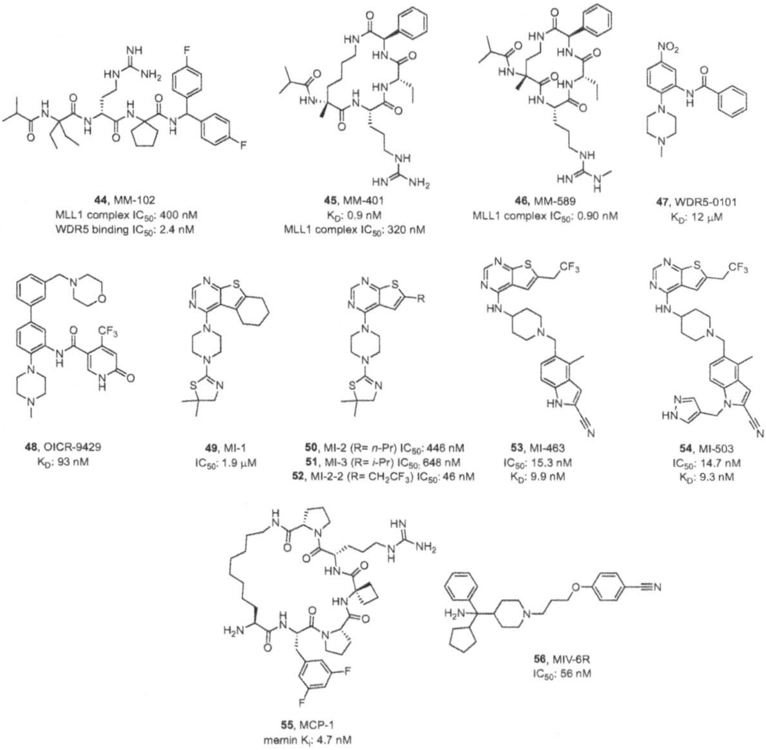

Fig. 5 H3K4 methyltransferase MLL1 modulators: WDR5 inhibitors and MLL-menin disruptors

Fig. 5) the most potent (IC_{50}: 0.90 nM). Compound **46** was surprisingly selective in growth inhibition of MOLM13 and MV4–11 cells carrying MLL translocation over HL60 leukemia cell line. Additionally, compound **46** showed also a good PK profile and good metabolic stability, being a promising tool for further optimization and in vivo studies [97]. A number of studies allowed the discovery of WDR5-0101 (**47**, Fig. 5) with the aim to optimize the potency in cells of this scaffold. In 2015, Grebien and coworkers reported the potent (K_D 93 ± 28 nM) and highly selective WDR5 binder OICR-9429 (**48**, Fig. 5), proved to bind the MLL1 WIN motif-binding pocket of WDR5 [98]. The efficacy and target selectivity of the novel compound were also confirmed in two different systems in which oncogenic transcription factors (p-30 and p-53) drive cell growth in a WDR5-MLL-dependent manner [98, 99].

1.6.2 MLL-Menin Disruptors

Another druggable interaction is the MLL-menin one. The thienopyrimidine-based compound MI-1 (**49**, Fig. 5) was the first small molecule to be identified as antagonist of menin-MLL1 interaction (K_D: 1.9 μM). Compound **49** was selected from a library of 49,000 compounds through a fluorescent polarization-based screening. Following structural optimizations on compound **49** yielded in order MI-2 (**50**) and MI-3 (**51**) and MI-2-2 (**52**) (IC_{50} of 446, 648, and 46 nM, respectively) (Fig. 5). In 2015, MI-463 (**53**) and MI-503 (**54**) (Fig. 5) were developed as orally bioavailable analogues of compound **52** and proved to be effective in xenografted leukemia models [100]. Starting from a linear octapeptide (MLL1 residues 6–13; RWRFPARP), Zhou et al. developed macrocyclic peptidomimetic antagonists of the menin-MLL interaction. Among them MCP-1 (**55**, Fig. 5) was the most efficient (K_i: 4.7 nM), being >600 times more potent than the acyclic analogue and 15 times more potent than the initial linear hit [101]. Displaying a reduced molecular weight and peptidic features compared to the original linear peptide, compound **55** is a promising starting point for the development of cell-permeable menin-MLL disruptors. FP-based screening of a 288,000-compound library, followed by hit optimization, resulted in the identification of MIV-6R (**56**) (IC_{50}: 56 nM), able to inhibit proliferation and induce differentiation in different leukemia cell lines expressing MLL fusion proteins [102].

1.7 The H3K4 Histone Methyltransferase SETD7

SETD7 (also known as SET9, SET7/9, or KMT7) was one of the first lysine methyltransferases to be reported, and it was originally described as H3K4 monomethyltransferase [103]. SETD7 can be localized both in the cytoplasm and in the nucleus, and its localization is regulated in an unusual way involving also other cellular factors. Even if SETD7 displays a robust activity on H3K4 N-terminal peptides in vitro, a little or null activity is observed on H3 substrate. Accordingly, SETD7 knockdown or depletion does not affect the H3K4me levels in living cells [103]. In time, a number of different studies highlighted the role of SETD7 as modulator of various transcriptional regulators [103]. SETD7 is also involved in the regulation of DNMT1 stability [103], DNA repair, cell differentiation, and cell-cycle control. The number and the variety of its substrates place SETD7 as player in several molecular pathways involved in cancer, metabolism (e.g., diabetes), inflammation, and viral infections (e.g., HIV, HCV infections) [103] (Fig. 6).

In 2014, the first SETD7 inhibitor, (R)-PFI-2 (**57**, Fig. 6), was reported as the result of an HTS screening on a library of 150,000 compounds followed by lead optimization [104]. Compound **57** proved to be highly potent (IC_{50} 2.0 ± 0.2 nM and Morrison K_i 0.33 ± 0.04 nM) and selective for SETD7 over a panel of 18 MTs and 134 GPCRs, ion channels, and other enzyme targets. Interestingly, its enantiomer (S)-

Fig. 6 H3K4 methyltransferase SETD7 inhibitors

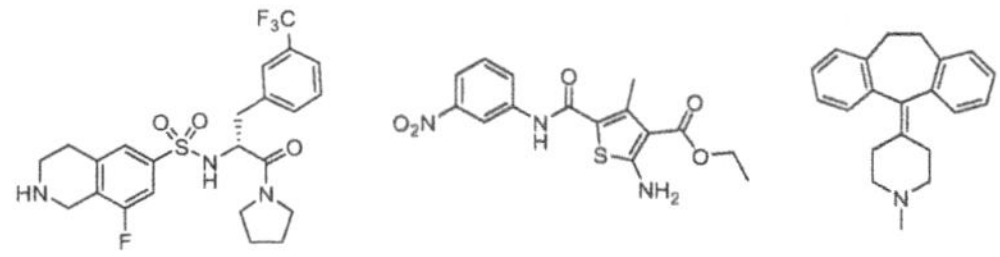

57, (R)-PFI-2
SETD7 IC_{50}: 2 nM

58, DC-S239
SETD7 IC_{50}: 4.6 μM

59, cyproheptadine
SETD7 IC_{50}: 1 μM

PFI-2 was about 500-fold less potent. More recently, in in-depth studies, Niu et al. elucidated the molecular and thermodynamic basis of the different interactions of the two enantiomers of **57** with SETD7, giving an explanation to the difference in activity [105]. This compound displayed no cytotoxicity up to 50 μM in various cell lines, phenocopying the effects of SETD7 genetic deletion and displaying a good PK profile [104]. To date, both enantiomers of **57** are the most potent and selective SETD7 inhibitors endowed with a good cellular activity. By the way, efforts aimed to the discovery of novel and potent SETD7 inhibitors have been done in the last years [106–109]. In 2015, a pharmacophore- and docking-based virtual screening, followed by SAR optimization studies, led Meng et al. to the identification of DC-S239 (**58**, Fig. 6) as novel selective SETD7 inhibitor (IC_{50}: 4.6 μM) [108]. When tested in HCT116 and DHL4 cells, compound **58** displayed no cytotoxicity but dose-dependently reduced proliferation in MCF7, HL60, and MV4–11 cells. The known histamine H1 and serotonin 5-HT2A receptors antagonist cyproheptadine (**59**, Fig. 6), in use as anti-allergy drug, has been recently proved to be also a SETD7 inhibitor (IC_{50}: 1.0 μM) [109]. This drug was shown to be selective for SETD7 over SETD8, G9a, SUV39H1, and DOT1L, displaying a substrate-competitive MOA. Similar to SETD7 knockdown, this compound dose- and time-dependently reduced ERα expression in MCF7 cells. This effect was proved to be independent from H1 or 5-HT2A receptors inhibition [109]. However, the application of compound **59** is limited to cells that are not expressing H1 and 5-HT2A receptors. Moreover, chemical manipulation and SAR investigation may be possible to dissect these different activities.

1.8 The H3K4 Histone Methyltransferase SMYD

The SMYD (SET and MYND domain-containing) family of proteins includes five conserved members (SMYD1 to SMYD5). The SMYD proteins contain a unique SET domain that is divided in two parts by the MYND (myeloid translocation protein-8, Nervy, and DEAF-1) domain, a cysteine-rich zinc finger motif primarily involved in protein-protein interaction [110]. SMYD proteins have been shown to methylate a number of histone and nonhistone targets, thus being involved in several processes including chromatin remodeling, transcription, signal transduction, and cell-cycle control. SMYD1–3 are classically known to methylate H3K4 [110]. Moreover, other histone substrates have been identified for SMYD2 (in vitro H3K36) and SMYD3 (H4K5). SMYD proteins methylate several nonhistone targets such as p53,

Rb, PARP1, HSP90, and ERα (SMYD2) [111] and the kinases MAP 3K2 and VEGFR1 (SMYD3) [110].

SMYD2 and SMYD3 overexpression has been associated with various cancers and correlated with bad prognosis [110]. Hence, the development of specific inhibitors became a point of interest.

SMYD2 inhibitors were developed first in 2011, as the result of an HTS campaign. AZ-505 (**60**, Fig. 7) was reported as the first substrate-competitive and selective SMYD2 inhibitor (IC_{50}: 0.12 μM) [112]. The authors also solved the co-crystal structure of SMYD2 in complex with compound **60**, confirming the mechanism of action of the inhibitor and giving interesting tips for further optimization [112]. Later on, in 2015, an in-depth SAR study on compound **59** scaffold led to the identification of A-893 (**61**, Fig. 7) [113], where the insertion of the hydroxyl substituent on the lysine-channel binding moiety yielded an impressive improvement in inhibition potency (IC_{50}: 2.8 nM). Compound **61** was confirmed to be a substrate-competitive inhibitor endowed with a good selectivity for SMYD2 over a panel of 30 MTs. In A549 lung cancer cells, compound **61** reduced p53 methylation levels by 42%, without affecting protein expression [113]. Thanks to the principles provided by Ferguson et al. [112], Nguyen et al. designed the potent ($IC_{50} < 15$ nM), selective, and substrate-competitive SMYD2 inhibitor LLY-507 (**62**, Fig. 7) [114]. This compound reduced p53 methylation in various cell lines and inhibited proliferation in cancer cells overexpressing SMYD2 [114]. In 2016, aminopyrazoline-based small-molecule (S)-BAY-598 (**63**, Fig. 7) was reported as potent (IC_{50}: 27 ± 7 nM), selective, substrate-competitive (K_i: 8 ± 1 nM), and cell-active inhibitor of SMYD2, resulting from an optimization of HTS hits [115]. In HEK293T cells, compound **63** decreased p53 methylation in a dose-dependent way (IC_{50}: 58 nM), without altering protein levels. Tested against a panel of 240 cancer cell lines, compound **63** displayed antiproliferative effects only on a limited number of cell lines, but was capable to reduce methylation at SMYD2 target sites when tested in in vivo models (mice xenografted with tumor tissues derived from the SMYD2-overexpressing KYSE-150 cell line) at doses starting from 30 mg/kg, with most significant effects at 100 mg/kg dose [115]. The initial in vivo effects were only moderate, but the collected data encourage for further investigation.

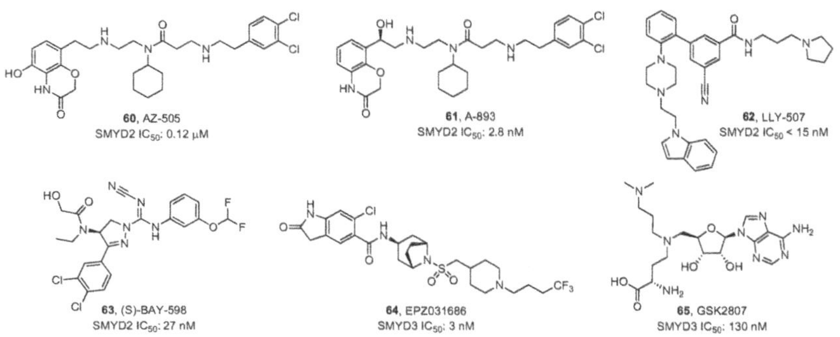

Fig. 7 H3K4 methyltransferase SMYD2/3 inhibitors

Also SMYD3 is a promising target in cancer [116]; however SMYD3 inhibitors appeared in literature just recently. As the result of an HTS followed by structure-guided optimization, in 2016, Mitchell et al. described EPZ031686 (**64**, Fig. 7) as the first in class SMYD3 inhibitor [117]. Compound **64** proved very potent (biochemical IC_{50}: 3 nM, cell IC_{50}: 36 nM) and selective SMYD3 inhibitor. Compound **64**, selected for in vivo PK evaluation, proved to be orally bioavailable and eligible for in vivo efficacy studies. In the same year, starting from structural studies on the SMYD3 and a MAP 3K2 peptide co-crystal structure, bi-substrate-type inhibitors have been designed by Van Aller et al. [118]. This study led to the identification of GSK2807 (**65**, Fig. 7), a SAH analogue with a dimethylaminopropyl side chain, that proved to be a potent (IC_{50}: 130 nM) and selective inhibitor. X-ray crystallography and kinetic studies elucidated its binding mode and MOA. Due to its poor cell permeability, compound **65** was not suitable for in-cell evaluation [118].

1.9 The H3K36 Histone Methyltransferase SETD2

H3K36 trimethylation is specifically catalyzed by SETD2 (KMT3A or SET2), and this mark is associated with transcription elongation and RNA splicing, DNA repair, and tumor suppression [119]. SETD2 mutations have been reported in different cancers [119], and aberrant SETD2-H3K36 trimethylation patterns seem to be involved in leukemia development [120].

In 2012, Zheng et al. designed *N*-alkyl sinefungins as bi-substrate-type PKMT inhibitors aimed to resemble certain transition-state features. More in detail, the *N*-alkyl sinefungins should locate their secondary amines at the substrate-cofactor interface and the *N*-alkyl chains in the lysine-binding pocket, so to achieve high affinity for specific methyltransferases [121]. Pr-SNF (**66**) and Bn-SNF (**67**) (Fig. 8) were identified as SAM-competitive inhibitors of human SETD2 exhibiting an IC_{50} of 0.8 ± 0.2 and 0.48 ± 0.06 μM, respectively. These two compounds displayed medium to high selectivity for SETD2 [121]. However, Pr-SNF/Bn-SNF activity proved to be influenced also by the peptide substrate. Due to their poor membrane permeability, Pr-SNF and Bn-SNF could not be successfully tested in cell-based or in vivo assays [121].

Fig. 8 SETD2 methyltransferase inhibitors

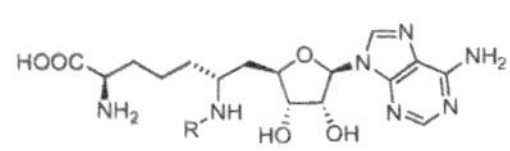

66, Pr-SNF (R= *n*-Pr) SETD2 IC_{50}: 0.8 μM
67, Bn-SNF (R= CH_2Ph) SETD2 IC_{50}: 0.48 μM

1.10 The H4K20 Histone Methyltransferases SETD8, SUV420H1, and SUV420H2

Methylation of H4K20 is catalyzed by the SETD8, SUV420H1, and SUV420H2, protein methyltransferases in humans [122].

Among the SET domain-containing methyltransferases, SETD8 (also known as PR-SET7, SET8, or KMT5A) is the sole enzyme responsible for H4K20 monomethylation [123]. H4K20 methylation has been found to play a key role in processes ranging from DNA damage repair, DNA replication, mitotic condensation, to the regulation of gene expression [123]. In addition to H4K20, SETD8 has many nonhistone substrates, including PCNA (PCNAK248me1) and the tumor suppressor p53 (p53K382me1) [123]. Even though H4K20me is now broadly studied, its function in gene expression is still not well defined, being associated with both gene expression and repression [123]. The outcome of H4K20me may vary in relation to the cellular and genomic context. SETD8 is overexpressed across a number of cancers including bladder cancer, non-small cell and small cell lung carcinoma, chronic myelogenous leukemia, hepatocellular carcinoma, and pancreatic cancers [123], and it is playing a role in invasiveness and metastasis [123]. Additionally, SETD8 has been reported to be involved in other processes including erythroid maturation regulation [123], maintenance of adult skin and epidermal proliferation, adipogenesis and metabolism, and intrauterine growth restriction (IUGR) process [123]. In 2007, Reinberg et al. reported the first two SETD8 inhibitors: the dye-like compounds H acid (**68**) and thymolphthalein (**69**) (Fig. 9) [124]. Identified through the screening of a focused library against four MTs (SETD8, G9a, SETD7, PRMT1), these compounds both displayed dual inhibition against SETD8 (IC_{50}: 3.8 and 9.0 μM, respectively) and EZH2 (IC_{50}: 3.0 and 25.2 μM, respectively). The cellular activity of compound **68** could not be appreciated due to cell permeability issues. Conversely, when tested in HeLa cells, compound **69** induced a dose-dependent reduction of cell viability together with a marked global H4K20 demethylation, without affecting other histone marks (not even H3K27me3). Additionally, this compound dose- and time-dependently enriched the mitotic population, while abrogating the DNA stimulatory effect on SETD8 methyltransferase activity [124].

As the result of different studies aimed to identify epigenetic multiple ligands, Mai et al. in 2012 reported bis(bromomethoxyphenol)- and bis(dibromomethoxyphenol)-containing derivatives as SETD8 inhibitors [125]. Interestingly, compounds MC1946 (**70**), MC1947 (**71**), MC1948 (**72**), and MC2569 (**73**) (Fig. 9) displayed SETD8 selective inhibition with IC_{50} values ranging from 2.6 to 10.2 μM. All these compounds reduced H4K20me1 levels in U937 cells, after 24 h treatment at 50 μM, and compound **71** induced 28% of granulocytic differentiation and massive cell death in the same cell line [125].

One year later, the polyketide nahuoic acid A (**74**, Fig. 9), isolated from *Streptomyces* sp. cultures, was shown to inhibit SETD8 (IC_{50}: 6.5 ± 0.5 μM) competing with SAM cofactor (K_i: 2 ± 0.3 μM) and without any significant effect on other tested MTs [126]. More recently, compound **74** and its penta-acetate derivative have been shown to inhibit cancer cell (U2OS, SUM159, MDA-MB-436) proliferation [127]. Further structural manipulation of compound **74**, as in its analogues nahuoic

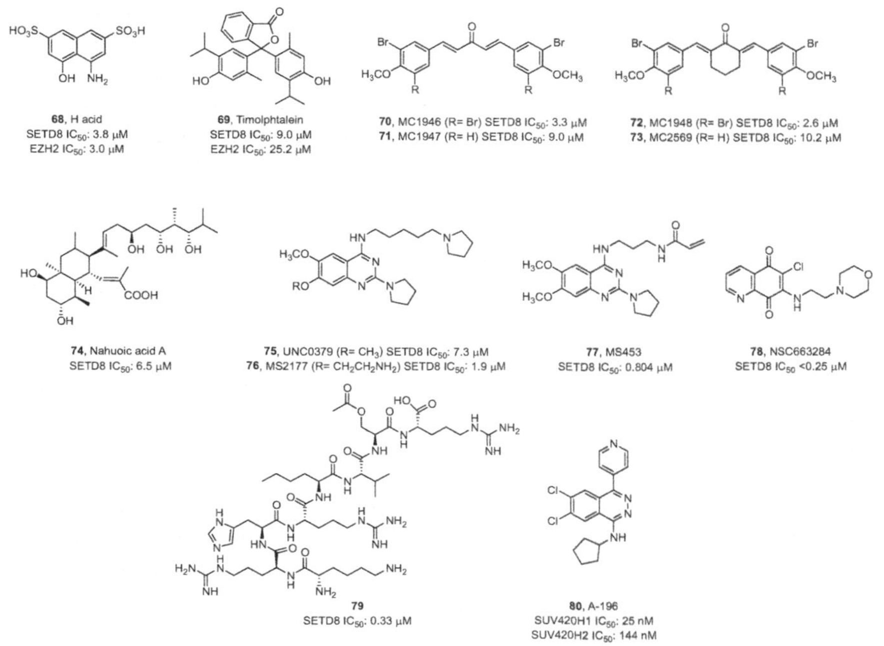

Fig. 9 H4K20 methyltransferases SETD8, SUV420H1, and SUV420H2 inhibitors

acids B–E, did not significantly improve the potency against SETD8 [127]. The quinazoline-based SETD8 selective substrate-competitive inhibitor UNC0379 (**75**, Fig. 9) (IC_{50}: 7.3 ± 1.0 μM) was reported in 2014 by Ma et al. [128, 129]. The novel quinazoline-based inhibitor proved selective for SETD8 over 15 other methyltransferases. Recently, it was proved that pharmacological inhibition of SETD8 by compound **75** phenocopied the effects of SETD8 knockdown and inhibited the proliferation of neuroblastoma cells [130]. In 2016, among various substitutions on the scaffold compound **75**, the insertion of an aminoethyl chain to the 7-position yielded MS2177 (**76**, Fig. 9) as a more potent SETD8 inhibitor (IC_{50}: 1.9 μM) [131]. ITC confirmed the target binding (K_D: 1.3 μM), while kinetic studies showed that compound **76** is a substrate-competitive inhibitor. Thanks to the resolution of the co-crystal structure of SETD8 in complex with compound **76**, Cys311 was identified as a possible site to target with covalent inhibitors. Indeed, MS453 (**77**, Fig. 9) was developed as covalent analogue of compound **76**, containing an acrylamide that could specifically react with Cys311. The authors, using MS techniques in presence of WT and mutant SETD8, demonstrated that compound **77** specifically links to Cys311, displaying an IC_{50} of 0.8 μM and a time-dependent inhibition activity. Additionally, the novel covalent inhibitor proved selective for SETD8 over 29 other methyltransferases in biochemical assays. Despite its nice biochemical profile, compound **77** displayed poor PK properties, not being suitable for further evaluation in cell-based assays.

Via a high-throughput screen, NSC663284 (**78**, Fig. 9) has been identified and validated as potent SETD8 inhibitor by suppressing the H4K20me1 mark of SETD8 mark at single doses of 1–5 μM in HEK293T cells leading to characteristic S/G2/M-phase cell-cycle defects as observed for RNAi-mediated SETD8 knockdown [132]. More recently, a completely different approach based on the use of an H4 substrate peptide (amino acids 16–23 of histone H4) as a starting point for molecular design led to the discovery of a potent peptide-based inhibitor of SETD8. In detail, Judge et al. replaced the K20 with more hydrophobic nonnatural amino acids. The replacement of K20 with *nor*-leucine led to the identification of the novel substrate-competitive inhibitor **79** (Fig. 9, IC_{50}: 0.33 μM). The new inhibitor showed also a good selectivity profile for SETD8 over a panel of 32 methyltransferases [133]. Due to its peptide nature, this compound is not cell permeable, thus not suitable for in-cell investigation.

SUV420H1 and SUV420H2 are two highly homologous methyltransferases that di- and trimethylate H4K20 have recently emerged as important regulators of genomic integrity, being important in the regulation of proliferation, cell cycle, and chromatin remodeling [134]. Aberrant H4K20 methylation has been associated with human cancer; thus modulation of the methylation status might be beneficial in cancer [135]. Indeed, A-196 (**80**, Fig. 9) has been recently described as the first potent, selective inhibitor of SUV420H1 and SUV420H2 [134], exhibiting IC_{50} values of 25 and 144 nM, respectively. Reduced levels of H4K20me2 and H4K20me3 and increased levels of H4K20me1 were found after treatment with this inhibitor throughout the cell cycle in U2OS osteosarcoma cells and LnCaP prostate adenocarcinoma cells. Compound **80** was not exhibiting significant toxic effects in cells being a valuable lead compound that can serve as a starting point for med-chem optimization but also as a chemical probe to further study the biological functions of SUV4–20H1/2 [134].

2 Conclusions

In this book chapter, we highlighted the discovery, characterization, and application of selective KMTs inhibitors, useful for dissecting their physiological functions as well as their disease implications. A growing body of literature is indicating that KMTs play a crucial role for transcriptional regulation and involved in cancer and various other human diseases, thus being of high interest as potential therapeutic targets.

Over the past decade, there has been an impressive progress regarding the PMT inhibitors discovery, especially conjugating the research interest with the available and novel techniques including new assay methods, high-throughput screening, structural biology, and medicinal chemistry approaches. Our goal was to point out key advances, challenges, possible future opportunities, and directions, regarding KMT modulation.

In the last years, the resolution of KMT crystal structures provided rational bases for designing highly selective inhibitors. Here, we have summarized the discovery and validation of selective inhibitors of KMTs focusing on the most potent, selective, and well-characterized small molecules with robust on-target activities in cells.

Some of the molecules presented are now in a more advanced stage of development such as the first inhibitors of DOT1L and EZH2 that entered the clinical arena for diseases such as leukemia, lymphoma, and SCLC.

We have seen significant progress, in the KMT inhibitor field; however for some KMTs, such as the MMSET (NSD-2) or the MLL family, the biological function in both physiological and pathological processes is not yet fully understood, partly due to the lack of selective inhibitors of these PMTs.

Furthermore, a better understanding of the dynamic interactions of the aforementioned proteins is a very desirable goal, possibly using selective inhibitors able to dissect highly homologous KMTs.

In the years to come, an amazing development can be expected as a lot of research groups are actively developing valuable chemical tools to better understand the biological functions of KMTs and test therapeutic hypotheses regarding these proteins.

Compliance with Ethical Standards

Conflict of Interest: The authors declare no conflict of interest.

Ethical Approval: This article does not contain any studies with human participants or animals performed by any of the authors.

References

1. Bennett RL, Licht JD (2018) Targeting epigenetics in cancer. Annu Rev Pharmacol Toxicol 58:187–207. https://doi.org/10.1146/annurev-pharmtox-010716-105106
2. Ribich S, Harvey D, Copeland RA (2017) Drug discovery and chemical biology of cancer epigenetics. Cell Chem Biol 24(9):1120–1147. https://doi.org/10.1016/j.chembiol.2017.08.020
3. Huang J, Dorsey J, Chuikov S, Perez-Burgos L, Zhang X, Jenuwein T, Reinberg D, Berger SL (2010) G9a and Glp methylate lysine 373 in the tumor suppressor p53. J Biol Chem 285(13): 9636–9641. https://doi.org/10.1074/jbc.M109.062588
4. Inagawa M, Nakajima K, Makino T, Ogawa S, Kojima M, Ito S, Ikenishi A, Hayashi T, Schwartz RJ, Nakamura K, Obayashi T, Tachibana M, Shinkai Y, Maeda K, Miyagawa-Tomita S, Takeuchi T (2013) Histone H3 lysine 9 methyltransferases, G9a and GLP are essential for cardiac morphogenesis. Mech Dev 130(11–12):519–531. https://doi.org/10.1016/j.mod.2013.07.002
5. Artal-Martinez de Narvajas A, Gomez TS, Zhang JS, Mann AO, Taoda Y, Gorman JA, Herreros-Villanueva M, Gress TM, Ellenrieder V, Bujanda L, Kim DH, Kozikowski AP,

Koenig A, Billadeau DD (2013) Epigenetic regulation of autophagy by the methyltransferase G9a. Mol Cell Biol 33(20):3983–3993. https://doi.org/10.1128/MCB.00813-13
6. Renneville A, Van Galen P, Canver MC, McConkey M, Krill-Burger JM, Dorfman DM, Holson EB, Bernstein BE, Orkin SH, Bauer DE, Ebert BL (2015) EHMT1 and EHMT2 inhibition induces fetal hemoglobin expression. Blood 126(16):1930–1939. https://doi.org/10.1182/blood-2015-06-649087
7. Kubicek S, O'Sullivan RJ, August EM, Hickey ER, Zhang Q, Teodoro ML, Rea S, Mechtler K, Kowalski JA, Homon CA, Kelly TA, Jenuwein T (2007) Reversal of H3K9me2 by a small-molecule inhibitor for the G9a histone methyltransferase. Mol Cell 25(3):473–481. https://doi.org/10.1016/j.molcel.2007.01.017
8. Liu F, Chen X, Allali-Hassani A, Quinn AM, Wigle TJ, Wasney GA, Dong A, Senisterra G, Chau I, Siarheyeva A, Norris JL, Kireev DB, Jadhav A, Herold JM, Janzen WP, Arrowsmith CH, Frye SV, Brown PJ, Simeonov A, Vedadi M, Jin J (2010) Protein lysine methyltransferase G9a inhibitors: design, synthesis, and structure activity relationships of 2,4-diamino-7-aminoalkoxy-quinazolines. J Med Chem 53(15):5844–5857. https://doi.org/10.1021/jm100478y
9. Liu F, Barsyte-Lovejoy D, Allali-Hassani A, He Y, Herold JM, Chen X, Yates CM, Frye SV, Brown PJ, Huang J, Vedadi M, Arrowsmith CH, Jin J (2011) Optimization of cellular activity of G9a inhibitors 7-aminoalkoxy-quinazolines. J Med Chem 54(17):6139–6150. https://doi.org/10.1021/jm200903z
10. Vedadi M, Barsyte-Lovejoy D, Liu F, Rival-Gervier S, Allali-Hassani A, Labrie V, Wigle TJ, Dimaggio PA, Wasney GA, Siarheyeva A, Dong A, Tempel W, Wang SC, Chen X, Chau I, Mangano TJ, Huang XP, Simpson CD, Pattenden SG, Norris JL, Kireev DB, Tripathy A, Edwards A, Roth BL, Janzen WP, Garcia BA, Petronis A, Ellis J, Brown PJ, Frye SV, Arrowsmith CH, Jin J (2011) A chemical probe selectively inhibits G9a and GLP methyltransferase activity in cells. Nat Chem Biol 7(8):566–574. https://doi.org/10.1038/nchembio.599
11. Lehnertz B, Pabst C, Su L, Miller M, Liu F, Yi L, Zhang R, Krosl J, Yung E, Kirschner J, Rosten P, Underhill TM, Jin J, Hebert J, Sauvageau G, Humphries RK, Rossi FM (2014) The methyltransferase G9a regulates HoxA9-dependent transcription in AML. Genes Dev 28(4):317–327. https://doi.org/10.1101/gad.236794.113
12. Liu F, Barsyte-Lovejoy D, Li F, Xiong Y, Korboukh V, Huang XP, Allali-Hassani A, Janzen WP, Roth BL, Frye SV, Arrowsmith CH, Brown PJ, Vedadi M, Jin J (2013) Discovery of an in vivo chemical probe of the lysine methyltransferases G9a and GLP. J Med Chem 56(21):8931–8942. https://doi.org/10.1021/jm401480r
13. Kim Y, Lee HM, Xiong Y, Sciaky N, Hulbert SW, Cao X, Everitt JI, Jin J, Roth BL, Jiang YH (2017) Targeting the histone methyltransferase G9a activates imprinted genes and improves survival of a mouse model of Prader-Willi syndrome. Nat Med 23(2):213–222. https://doi.org/10.1038/nm.4257
14. Xiong Y, Li F, Babault N, Dong A, Zeng H, Wu H, Chen X, Arrowsmith CH, Brown PJ, Liu J, Vedadi M, Jin J (2017) Discovery of potent and selective inhibitors for G9a-like protein (GLP) lysine methyltransferase. J Med Chem 60(5):1876–1891. https://doi.org/10.1021/acs.jmedchem.6b01645
15. Sweis RF, Pliushchev M, Brown PJ, Guo J, Li F, Maag D, Petros AM, Soni NB, Tse C, Vedadi M, Michaelides MR, Chiang GG, Pappano WN (2014) Discovery and development of potent and selective inhibitors of histone methyltransferase g9a. ACS Med Chem Lett 5(2):205–209. https://doi.org/10.1021/ml400496h
16. Pappano WN, Guo J, He Y, Ferguson D, Jagadeeswaran S, Osterling DJ, Gao W, Spence JK, Pliushchev M, Sweis RF, Buchanan FG, Michaelides MR, Shoemaker AR, Tse C, Chiang GG (2015) The histone methyltransferase inhibitor A-366 uncovers a role for G9a/GLP in the epigenetics of leukemia. PLoS One 10(7):e0131716. https://doi.org/10.1371/journal.pone.0131716

17. Yuan Y, Wang Q, Paulk J, Kubicek S, Kemp MM, Adams DJ, Shamji AF, Wagner BK, Schreiber SL (2012) A small-molecule probe of the histone methyltransferase G9a induces cellular senescence in pancreatic adenocarcinoma. ACS Chem Biol 7(7):1152–1157. https://doi.org/10.1021/cb300139y
18. Cherblanc FL, Chapman KL, Brown R, Fuchter MJ (2013) Chaetocin is a nonspecific inhibitor of histone lysine methyltransferases. Nat Chem Biol 9(3):136–137. https://doi.org/10.1038/nchembio.1187
19. Devkota K, Lohse B, Liu Q, Wang MW, Staerk D, Berthelsen J, Clausen RP (2014) Analogues of the natural product sinefungin as inhibitors of EHMT1 and EHMT2. ACS Med Chem Lett 5(4):293–297. https://doi.org/10.1021/ml4002503
20. Kondengaden SM, Luo LF, Huang K, Zhu M, Zang L, Bataba E, Wang R, Luo C, Wang B, Li KK, Wang PG (2016) Discovery of novel small molecule inhibitors of lysine methyltransferase G9a and their mechanism in leukemia cell lines. Eur J Med Chem 122:382–393. https://doi.org/10.1016/j.ejmech.2016.06.028
21. Herz HM, Garruss A, Shilatifard A (2013) SET for life: biochemical activities and biological functions of SET domain-containing proteins. Trends Biochem Sci 38(12):621–639. https://doi.org/10.1016/j.tibs.2013.09.004
22. Lawson KA, Teteak CJ, Gao J, Li N, Hacquebord J, Ghatan A, Zielinska-Kwiatkowska A, Song G, Chansky HA, Yang L (2013) ESET histone methyltransferase regulates osteoblastic differentiation of mesenchymal stem cells during postnatal bone development. FEBS Lett 587 (24):3961–3967. https://doi.org/10.1016/j.febslet.2013.10.028
23. Karanth AV, Maniswami RR, Prashanth S, Govindaraj H, Padmavathy R, Jegatheesan SK, Mullangi R, Rajagopal S (2017) Emerging role of SETDB1 as a therapeutic target. Expert Opin Ther Targets 21(3):319–331. https://doi.org/10.1080/14728222.2017.1279604
24. Rice JC, Briggs SD, Ueberheide B, Barber CM, Shabanowitz J, Hunt DF, Shinkai Y, Allis CD (2003) Histone methyltransferases direct different degrees of methylation to define distinct chromatin domains. Mol Cell 12(6):1591–1598
25. Loyola A, Bonaldi T, Roche D, Imhof A, Almouzni G (2006) PTMs on H3 variants before chromatin assembly potentiate their final epigenetic state. Mol Cell 24(2):309–316. https://doi.org/10.1016/j.molcel.2006.08.019
26. Peters AH, O'Carroll D, Scherthan H, Mechtler K, Sauer S, Schofer C, Weipoltshammer K, Pagani M, Lachner M, Kohlmaier A, Opravil S, Doyle M, Sibilia M, Jenuwein T (2001) Loss of the Suv39h histone methyltransferases impairs mammalian heterochromatin and genome stability. Cell 107(3):323–337
27. Muramatsu D, Singh PB, Kimura H, Tachibana M, Shinkai Y (2013) Pericentric heterochromatin generated by HP1 protein interaction-defective histone methyltransferase Suv39h1. J Biol Chem 288(35):25285–25296. https://doi.org/10.1074/jbc.M113.470724
28. Sone K, Piao L, Nakakido M, Ueda K, Jenuwein T, Nakamura Y, Hamamoto R (2014) Critical role of lysine 134 methylation on histone H2AX for gamma-H2AX production and DNA repair. Nat Commun 5:5691. https://doi.org/10.1038/ncomms6691
29. Friedman J, Cho WK, Chu CK, Keedy KS, Archin NM, Margolis DM, Karn J (2011) Epigenetic silencing of HIV-1 by the histone H3 lysine 27 methyltransferase enhancer of Zeste 2. J Virol 85(17):9078–9089. https://doi.org/10.1128/JVI.00836-11
30. Greiner D, Bonaldi T, Eskeland R, Roemer E, Imhof A (2005) Identification of a specific inhibitor of the histone methyltransferase SU(VAR)3-9. Nat Chem Biol 1(3):143–145. https://doi.org/10.1038/nchembio721
31. Greiner D, Bonaldi T, Eskeland R, Roemer E, Imhof A (2013) Reply to "Chaetocin is a nonspecific inhibitor of histone lysine methyltransferases". Nat Chem Biol 9(3):137. https://doi.org/10.1038/nchembio.1188
32. Chaib H, Nebbioso A, Prebet T, Castellano R, Garbit S, Restouin A, Vey N, Altucci L, Collette Y (2012) Anti-leukemia activity of chaetocin via death receptor-dependent apoptosis and dual modulation of the histone methyl-transferase SUV39H1. Leukemia 26(4):662–674. https://doi.org/10.1038/leu.2011.271

33. Margueron R, Li G, Sarma K, Blais A, Zavadil J, Woodcock CL, Dynlacht BD, Reinberg D (2008) Ezh1 and Ezh2 maintain repressive chromatin through different mechanisms. Mol Cell 32(4):503–518. https://doi.org/10.1016/j.molcel.2008.11.004
34. Ketel CS, Andersen EF, Vargas ML, Suh J, Strome S, Simon JA (2005) Subunit contributions to histone methyltransferase activities of fly and worm polycomb group complexes. Mol Cell Biol 25(16):6857–6868. https://doi.org/10.1128/MCB.25.16.6857-6868.2005
35. Vasanthakumar A, Xu D, Lun AT, Kueh AJ, van Gisbergen KP, Iannarella N, Li X, Yu L, Wang D, Williams BR, Lee SC, Majewski IJ, Godfrey DI, Smyth GK, Alexander WS, Herold MJ, Kallies A, Nutt SL, Allan RS (2017) A non-canonical function of Ezh2 preserves immune homeostasis. EMBO Rep 18(4):619–631. https://doi.org/10.15252/embr.201643237
36. Kim KH, Roberts CW (2016) Targeting EZH2 in cancer. Nat Med 22(2):128–134. https://doi.org/10.1038/nm.4036
37. Margueron R, Reinberg D (2011) The Polycomb complex PRC2 and its mark in life. Nature 469(7330):343–349. https://doi.org/10.1038/nature09784
38. Lee TI, Jenner RG, Boyer LA, Guenther MG, Levine SS, Kumar RM, Chevalier B, Johnstone SE, Cole MF, Isono K, Koseki H, Fuchikami T, Abe K, Murray HL, Zucker JP, Yuan B, Bell GW, Herbolsheimer E, Hannett NM, Sun K, Odom DT, Otte AP, Volkert TL, Bartel DP, Melton DA, Gifford DK, Jaenisch R, Young RA (2006) Control of developmental regulators by Polycomb in human embryonic stem cells. Cell 125(2):301–313. https://doi.org/10.1016/j.cell.2006.02.043
39. Nichol JN, Dupere-Richer D, Ezponda T, Licht JD, Miller WHJ (2016) H3K27 methylation: a focal point of epigenetic deregulation in cancer. Adv Cancer Res 131:59–95. https://doi.org/10.1016/bs.acr.2016.05.001
40. Glazer RI, Knode MC, Tseng CK, Haines DR, Marquez VE (1986) 3-Deazaneplanocin A: a new inhibitor of S-adenosylhomocysteine synthesis and its effects in human colon carcinoma cells. Biochem Pharmacol 35(24):4523–4527
41. Miranda TB, Cortez CC, Yoo CB, Liang G, Abe M, Kelly TK, Marquez VE, Jones PA (2009) DZNep is a global histone methylation inhibitor that reactivates developmental genes not silenced by DNA methylation. Mol Cancer Ther 8(6):1579–1588. https://doi.org/10.1158/1535-7163.MCT-09-0013
42. Sun F, Lee L, Zhang Z, Wang X, Yu Q, Duan X, Chan E (2015) Preclinical pharmacokinetic studies of 3-deazaneplanocin A, a potent epigenetic anticancer agent, and its human pharmacokinetic prediction using GastroPlus. Eur J Pharm Sci 77:290–302. https://doi.org/10.1016/j.ejps.2015.06.021
43. Knutson SK, Wigle TJ, Warholic NM, Sneeringer CJ, Allain CJ, Klaus CR, Sacks JD, Raimondi A, Majer CR, Song J, Scott MP, Jin L, Smith JJ, Olhava EJ, Chesworth R, Moyer MP, Richon VM, Copeland RA, Keilhack H, Pollock RM, Kuntz KW (2012) A selective inhibitor of EZH2 blocks H3K27 methylation and kills mutant lymphoma cells. Nat Chem Biol 8(11):890–896. https://doi.org/10.1038/nchembio.1084
44. McCabe MT, Graves AP, Ganji G, Diaz E, Halsey WS, Jiang Y, Smitheman KN, Ott HM, Pappalardi MB, Allen KE, Chen SB, Della Pietra 3rd A, Dul E, Hughes AM, Gilbert SA, Thrall SH, Tummino PJ, Kruger RG, Brandt M, Schwartz B, Creasy CL (2012) Mutation of A677 in histone methyltransferase EZH2 in human B-cell lymphoma promotes hyper-trimethylation of histone H3 on lysine 27 (H3K27). Proc Natl Acad Sci U S A 109(8):2989–2994. https://doi.org/10.1073/pnas.1116418109
45. Yap TA, Winter JN, Leonard JP, Ribrag V, Constantinidou A, Giulino-Roth L, Michot J-M, Khan TA, Horner T, Carver J, Pene Dumetrescu T, He Z, McCabe MT, Creasy CL, Dhar A, Carpenter C, Johnson PM (2016) A phase I study of GSK2816126, an enhancer of zeste homolog 2(EZH2) inhibitor, in patients (pts) with relapsed/refractory diffuse large B-cell lymphoma (DLBCL), other non-Hodgkin lymphomas (NHL), transformed follicular lymphoma (tFL), solid tumors and multiple myeloma (MM). Blood 128(22):4203
46. Verma SK, Tian X, LaFrance LV, Duquenne C, Suarez DP, Newlander KA, Romeril SP, Burgess JL, Grant SW, Brackley JA, Graves AP, Scherzer DA, Shu A, Thompson C, Ott HM,

Aller GS, Machutta CA, Diaz E, Jiang Y, Johnson NW, Knight SD, Kruger RG, McCabe MT, Dhanak D, Tummino PJ, Creasy CL, Miller WH (2012) Identification of potent, selective, cell-active inhibitors of the histone lysine methyltransferase EZH2. ACS Med Chem Lett 3(12): 1091–1096. https://doi.org/10.1021/ml3003346

47. Qi W, Chan H, Teng L, Li L, Chuai S, Zhang R, Zeng J, Li M, Fan H, Lin Y, Gu J, Ardayfio O, Zhang JH, Yan X, Fang J, Mi Y, Zhang M, Zhou T, Feng G, Chen Z, Li G, Yang T, Zhao K, Liu X, Yu Z, Lu CX, Atadja P, Li E (2012) Selective inhibition of Ezh2 by a small molecule inhibitor blocks tumor cells proliferation. Proc Natl Acad Sci U S A 109(52):21360–21365. https://doi.org/10.1073/pnas.1210371110
48. Konze KD, Ma A, Li F, Barsyte-Lovejoy D, Parton T, Macnevin CJ, Liu F, Gao C, Huang XP, Kuznetsova E, Rougie M, Jiang A, Pattenden SG, Norris JL, James LI, Roth BL, Brown PJ, Frye SV, Arrowsmith CH, Hahn KM, Wang GG, Vedadi M, Jin J (2013) An orally bioavailable chemical probe of the lysine methyltransferases EZH2 and EZH1. ACS Chem Biol 8(6): 1324–1334. https://doi.org/10.1021/cb400133j
49. Xu B, On DM, Ma A, Parton T, Konze KD, Pattenden SG, Allison DF, Cai L, Rockowitz S, Liu S, Liu Y, Li F, Vedadi M, Frye SV, Garcia BA, Zheng D, Jin J, Wang GG (2015) Selective inhibition of EZH2 and EZH1 enzymatic activity by a small molecule suppresses MLL-rearranged leukemia. Blood 125(2):346–357. https://doi.org/10.1182/blood-2014-06-581082
50. Nasveschuk CG, Gagnon A, Garapaty-Rao S, Balasubramanian S, Campbell R, Lee C, Zhao F, Bergeron L, Cummings R, Trojer P, Audia JE, Albrecht BK, Harmange JC (2014) Discovery and optimization of tetramethylpiperidinyl benzamides as inhibitors of EZH2. ACS Med Chem Lett 5(4):378–383. https://doi.org/10.1021/ml400494b
51. Gehling VS, Vaswani RG, Nasveschuk CG, Duplessis M, Iyer P, Balasubramanian S, Zhao F, Good AC, Campbell R, Lee C, Dakin LA, Cook AS, Gagnon A, Harmange JC, Audia JE, Cummings RT, Normant E, Trojer P, Albrecht BK (2015) Discovery, design, and synthesis of indole-based EZH2 inhibitors. Bioorg Med Chem Lett 25(17):3644–3649. https://doi.org/10.1016/j.bmcl.2015.06.056
52. Vaswani RG, Gehling VS, Dakin LA, Cook AS, Nasveschuk CG, Duplessis M, Iyer P, Balasubramanian S, Zhao F, Good AC, Campbell R, Lee C, Cantone N, Cummings RT, Normant E, Bellon SF, Albrecht BK, Harmange JC, Trojer P, Audia JE, Zhang Y, Justin N, Chen S, Wilson JR, Gamblin SJ (2016) Identification of (R)-N-((4-methoxy-6-methyl-2-oxo-1,2-dihydropyridin-3-yl)methyl)-2-methyl-1-(1-(1-(2,2,2-trifluoroethyl)piperidin-4-yl)ethyl)-1H-indole-3-carboxamide (CPI-1205), a potent and selective inhibitor of histone methyltransferase EZH2, suitable for phase I clinical trials for B-cell lymphomas. J Med Chem 59(21):9928–9941. https://doi.org/10.1021/acs.jmedchem.6b01315
53. Knutson SK, Warholic NM, Wigle TJ, Klaus CR, Allain CJ, Raimondi A, Porter Scott M, Chesworth R, Moyer MP, Copeland RA, Richon VM, Pollock RM, Kuntz KW, Keilhack H (2013) Durable tumor regression in genetically altered malignant rhabdoid tumors by inhibition of methyltransferase EZH2. Proc Natl Acad Sci U S A 110(19):7922–7927. https://doi.org/10.1073/pnas.1303800110
54. Morera L, Lubbert M, Jung M (2016) Targeting histone methyltransferases and demethylases in clinical trials for cancer therapy. Clin Epigenetics 8:57. https://doi.org/10.1186/s13148-016-0223-4
55. Ribrag V, Soria J-C, Michot J-M, Schmitt A, Postel-Vinay S, Bijou F, Thomson B, Keilhack H, Blakemore SJ, Reyderman L (2015) Phase 1 study of tazemetostat (EPZ-6438), an inhibitor of enhancer of zeste-homolog 2 (EZH2): preliminary safety and activity in relapsed or refractory non-Hodgkin lymphoma (NHL) patients. Blood 126(23):473–473
56. Clinical Trials. www.clinicaltrials.gov. Accessed 26 Jan 2018
57. Lu B, Shen X, Zhang L, Liu D, Zhang C, Cao J, Shen R, Zhang J, Wang D, Wan H, Xu Z, Ho M-H, Zhang M, Zhang L, He F, Tao W (2018) Discovery of EBI-2511: a highly potent and orally active EZH2 inhibitor for the treatment of non-Hodgkin's lymphoma. ACS Med Chem Lett 9(2):98–102. https://doi.org/10.1021/acsmedchemlett.7b00437

58. Honma D, Kanno O, Watanabe J, Kinoshita J, Hirasawa M, Nosaka E, Shiroishi M, Takizawa T, Yasumatsu I, Horiuchi T, Nakao A, Suzuki K, Yamasaki T, Nakajima K, Hayakawa M, Yamazaki T, Yadav AS, Adachi N (2017) Novel orally bioavailable EZH1/2 dual inhibitors with greater antitumor efficacy than an EZH2 selective inhibitor. Cancer Sci 108(10):2069–2078. https://doi.org/10.1111/cas.13326
59. McCabe MT, Mohammad HP, Barbash O, Kruger RG (2017) Targeting histone methylation in cancer. Cancer J 23(5):292–301. https://doi.org/10.1097/PPO.0000000000000283
60. Kim W, Bird GH, Neff T, Guo G, Kerenyi MA, Walensky LD, Orkin SH (2013) Targeted disruption of the EZH2-EED complex inhibits EZH2-dependent cancer. Nat Chem Biol 9 (10):643–650. https://doi.org/10.1038/nchembio.1331
61. Kong X, Chen L, Jiao L, Jiang X, Lian F, Lu J, Zhu K, Du D, Liu J, Ding H, Zhang N, Shen J, Zheng M, Chen K, Liu X, Jiang H, Luo C (2014) Astemizole arrests the proliferation of cancer cells by disrupting the EZH2-EED interaction of polycomb repressive complex 2. J Med Chem 57(22):9512–9521. https://doi.org/10.1021/jm501230c
62. Chen H, Gao S, Li J, Liu D, Sheng C, Yao C, Jiang W, Wu J, Chen S, Huang W (2015) Wedelolactone disrupts the interaction of EZH2-EED complex and inhibits PRC2-dependent cancer. Oncotarget 6(15):13049–13059. https://doi.org/10.18632/oncotarget.3790
63. Li L, Zhang H, Zhang M, Zhao M, Feng L, Luo X, Gao Z, Huang Y, Ardayfio O, Zhang JH, Lin Y, Fan H, Mi Y, Li G, Liu L, Feng L, Luo F, Teng L, Qi W, Ottl J, Lingel A, Bussiere DE, Yu Z, Atadja P, Lu C, Li E, Gu J, Zhao K (2017) Discovery and molecular basis of a diverse set of polycomb repressive complex 2 inhibitors recognition by EED. PLoS One 12(1): e0169855. https://doi.org/10.1371/journal.pone.0169855
64. Lingel A, Sendzik M, Huang Y, Shultz MD, Cantwell J, Dillon MP, Fu X, Fuller J, Gabriel T, Gu J, Jiang X, Li L, Liang F, McKenna M, Qi W, Rao W, Sheng X, Shu W, Sutton J, Taft B, Wang L, Zeng J, Zhang H, Zhang M, Zhao K, Lindvall M, Bussiere DE (2017) Structure-guided design of EED binders allosterically inhibiting the epigenetic polycomb repressive complex 2 (PRC2) methyltransferase. J Med Chem 60(1):415–427. https://doi.org/10.1021/acs.jmedchem.6b01473
65. Huang Y, Zhang J, Yu Z, Zhang H, Wang Y, Lingel A, Qi W, Gu XJ, Zhao K, Shultz MD, Wang L, Fu X, Sun Y, Zhang Q, Jiang X, Zhang JW, Zhang C, Li L, Zeng J, Feng L, Zhang C, Liu Y, Zhang M, Zhang L, Zhao M, Gao Z, Liu X, Fang D, Guo H, Mi Y, Gabriel T, Dillon MP, Atadja P, Oyang C (2017) Discovery of first-in-class, potent and orally bioavailable EED inhibitor with robust anti-cancer efficacy. J Med Chem 60(6):2215–2226. https://doi.org/10.1021/acs.jmedchem.6b01576
66. He Y, Selvaraju S, Curtin ML, Jakob CG, Zhu H, Comess KM, Shaw B, The J, Lima-Fernandes E, Szewczyk MM, Cheng D, Klinge KL, Li HQ, Pliushchev M, Algire MA, Maag D, Guo J, Dietrich J, Panchal SC, Petros AM, Sweis RF, Torrent M, Bigelow LJ, Senisterra G, Li F, Kennedy S, Wu Q, Osterling DJ, Lindley DJ, Gao W, Galasinski S, Barsyte-Lovejoy D, Vedadi M, Buchanan FG, Arrowsmith CH, Chiang GG, Sun C, Pappano WN (2017) The EED protein-protein interaction inhibitor A-395 inactivates the PRC2 complex. Nat Chem Biol 13(4):389–395. https://doi.org/10.1038/nchembio.2306
67. Qi W, Zhao K, Gu J, Huang Y, Wang Y, Zhang H, Zhang M, Zhang J, Yu Z, Li L, Teng L, Chuai S, Zhang C, Zhao M, Chan H, Chen Z, Fang D, Fei Q, Feng L, Feng L, Gao Y, Ge H, Ge X, Li G, Lingel A, Lin Y, Liu Y, Luo F, Shi M, Wang L, Wang Z, Yu Y, Zeng J, Zeng C, Zhang L, Zhang Q, Zhou S, Oyang C, Atadja P, Li E (2017) An allosteric PRC2 inhibitor targeting the H3K27me3 binding pocket of EED. Nat Chem Biol 13(4):381–388. https://doi.org/10.1038/nchembio.2304
68. Barnash KD, The J, Norris-Drouin JL, Cholensky SH, Worley BM, Li F, Stuckey JI, Brown PJ, Vedadi M, Arrowsmith CH, Frye SV, James LI (2017) Discovery of peptidomimetic ligands of EED as allosteric inhibitors of PRC2. ACS Comb Sci 19(3):161–172. https://doi.org/10.1021/acscombsci.6b00174

69. Lai WKM, Pugh BF (2017) Understanding nucleosome dynamics and their links to gene expression and DNA replication. Nat Rev Mol Cell Biol 18(9):548–562. https://doi.org/10.1038/nrm.2017.47
70. Chan HM, Gu X-J, Huang Y, Li L, Mi Y, Qi W, Sendzik M, Sun Y, Wang L, Yu Z (2017) Triazolopyrimidine compounds and uses thereof. United States of America Patent US9580437 (B2)
71. Stazi G, Zwergel C, Mai A, Valente S (2017) EZH2 inhibitors: a patent review (2014–2016). Expert Opin Ther Pat 27(7):797–813. https://doi.org/10.1080/13543776.2017.1316976
72. Grinshtein N, Rioseco CC, Marcellus R, Uehling D, Aman A, Lun X, Muto O, Podmore L, Lever J, Shen Y, Blough MD, Cairncross GJ, Robbins SM, Jones SJ, Marra MA, Al-Awar R, Senger DL, Kaplan DR (2016) Small molecule epigenetic screen identifies novel EZH2 and HDAC inhibitors that target glioblastoma brain tumor-initiating cells. Oncotarget 7(37): 59360–59376. https://doi.org/10.18632/oncotarget.10661
73. Frederiks F, Tzouros M, Oudgenoeg G, van Welsem T, Fornerod M, Krijgsveld J, van Leeuwen F (2008) Nonprocessive methylation by Dot1 leads to functional redundancy of histone H3K79 methylation states. Nat Struct Mol Biol 15(6):550–557. https://doi.org/10.1038/nsmb.1432
74. Anglin JL, Song Y (2013) A medicinal chemistry perspective for targeting histone H3 lysine-79 methyltransferase DOT1L. J Med Chem 56(22):8972–8983. https://doi.org/10.1021/jm4007752
75. Woo Park J, Kim KB, Kim JY, Chae YC, Jeong OS, Seo SB (2015) RE-IIBP methylates H3K79 and induces MEIS1-mediated apoptosis via H2BK120 ubiquitination by RNF20. Sci Rep 5:12485. https://doi.org/10.1038/srep12485
76. Feng Q, Wang H, Ng HH, Erdjument-Bromage H, Tempst P, Struhl K, Zhang Y (2002) Methylation of H3-lysine 79 is mediated by a new family of HMTases without a SET domain. Curr Biol 12(12):1052–1058
77. Min J, Feng Q, Li Z, Zhang Y, Xu RM (2003) Structure of the catalytic domain of human DOT1L, a non-SET domain nucleosomal histone methyltransferase. Cell 112(5):711–723
78. Schubert HL, Blumenthal RM, Cheng X (2003) Many paths to methyltransfer: a chronicle of convergence. Trends Biochem Sci 28(6):329–335. https://doi.org/10.1016/S0968-0004(03)00090-2
79. Zee BM, Levin RS, Xu B, LeRoy G, Wingreen NS, Garcia BA (2010) In vivo residue-specific histone methylation dynamics. J Biol Chem 285(5):3341–3350. https://doi.org/10.1074/jbc.M109.063784
80. Jo SY, Granowicz EM, Maillard I, Thomas D, Hess JL (2011) Requirement for Dot1l in murine postnatal hematopoiesis and leukemogenesis by MLL translocation. Blood 117(18): 4759–4768. https://doi.org/10.1182/blood-2010-12-327668
81. Zhang L, Deng L, Chen F, Yao Y, Wu B, Wei L, Mo Q, Song Y (2014) Inhibition of histone H3K79 methylation selectively inhibits proliferation, self-renewal and metastatic potential of breast cancer. Oncotarget 5(21):10665–10677. https://doi.org/10.18632/oncotarget.2496
82. Daigle SR, Olhava EJ, Therkelsen CA, Majer CR, Sneeringer CJ, Song J, Johnston LD, Scott MP, Smith JJ, Xiao Y, Jin L, Kuntz KW, Chesworth R, Moyer MP, Bernt KM, Tseng JC, Kung AL, Armstrong SA, Copeland RA, Richon VM, Pollock RM (2011) Selective killing of mixed lineage leukemia cells by a potent small-molecule DOT1L inhibitor. Cancer Cell 20(1): 53–65. https://doi.org/10.1016/j.ccr.2011.06.009
83. Yu W, Chory EJ, Wernimont AK, Tempel W, Scopton A, Federation A, Marineau JJ, Qi J, Barsyte-Lovejoy D, Yi J, Marcellus R, Iacob RE, Engen JR, Griffin C, Aman A, Wienholds E, Li F, Pineda J, Estiu G, Shatseva T, Hajian T, Al-Awar R, Dick JE, Vedadi M, Brown PJ, Arrowsmith CH, Bradner JE, Schapira M (2012) Catalytic site remodelling of the DOT1L methyltransferase by selective inhibitors. Nat Commun 3:1288. https://doi.org/10.1038/ncomms2304
84. Chen L, Deshpande AJ, Banka D, Bernt KM, Dias S, Buske C, Olhava EJ, Daigle SR, Richon VM, Pollock RM, Armstrong SA (2013) Abrogation of MLL-AF10 and CALM-AF10-

mediated transformation through genetic inactivation or pharmacological inhibition of the H3K79 methyltransferase Dot1l. Leukemia 27(4):813–822. https://doi.org/10.1038/leu.2012.327

85. Daigle SR, Olhava EJ, Therkelsen CA, Basavapathruni A, Jin L, Boriack-Sjodin PA, Allain CJ, Klaus CR, Raimondi A, Scott MP, Waters NJ, Chesworth R, Moyer MP, Copeland RA, Richon VM, Pollock RM (2013) Potent inhibition of DOT1L as treatment of MLL-fusion leukemia. Blood 122(6):1017–1025. https://doi.org/10.1182/blood-2013-04-497644
86. Waters NJ, Smith SA, Olhava EJ, Duncan KW, Burton RD, O'Neill J, Rodrigue ME, Pollock RM, Moyer MP, Chesworth R (2016) Metabolism and disposition of the DOT1L inhibitor, pinometostat (EPZ-5676), in rat, dog and human. Cancer Chemother Pharmacol 77(1):43–62. https://doi.org/10.1007/s00280-015-2929-y
87. Basavapathruni A, Olhava EJ, Daigle SR, Therkelsen CA, Jin L, Boriack-Sjodin PA, Allain CJ, Klaus CR, Raimondi A, Scott MP, Dovletoglou A, Richon VM, Pollock RM, Copeland RA, Moyer MP, Chesworth R, Pearson PG, Waters NJ (2014) Nonclinical pharmacokinetics and metabolism of EPZ-5676, a novel DOT1L histone methyltransferase inhibitor. Biopharm Drug Dispos 35(4):237–252. https://doi.org/10.1002/bdd.1889
88. Yao Y, Chen P, Diao J, Cheng G, Deng L, Anglin JL, Prasad BV, Song Y (2011) Selective inhibitors of histone methyltransferase DOT1L: design, synthesis, and crystallographic studies. J Am Chem Soc 133(42):16746–16749. https://doi.org/10.1021/ja206312b
89. Scheufler C, Mobitz H, Gaul C, Ragot C, Be C, Fernandez C, Beyer KS, Tiedt R, Stauffer F (2016) Optimization of a fragment-based screening hit toward potent DOT1L inhibitors interacting in an induced binding pocket. ACS Med Chem Lett 7(8):730–734. https://doi.org/10.1021/acsmedchemlett.6b00168
90. Chen C, Zhu H, Stauffer F, Caravatti G, Vollmer S, Machauer R, Holzer P, Mobitz H, Scheufler C, Klumpp M, Tiedt R, Beyer KS, Calkins K, Guthy D, Kiffe M, Zhang J, Gaul C (2016) Discovery of novel Dot1L inhibitors through a structure-based fragmentation approach. ACS Med Chem Lett 7(8):735–740. https://doi.org/10.1021/acsmedchemlett.6b00167
91. Mobitz H, Machauer R, Holzer P, Vaupel A, Stauffer F, Ragot C, Caravatti G, Scheufler C, Fernandez C, Hommel U, Tiedt R, Beyer KS, Chen C, Zhu H, Gaul C (2017) Discovery of potent, selective, and structurally novel Dot1L inhibitors by a fragment linking approach. ACS Med Chem Lett 8(3):338–343. https://doi.org/10.1021/acsmedchemlett.6b00519
92. Rao RC, Dou Y (2015) Hijacked in cancer: the KMT2 (MLL) family of methyltransferases. Nat Rev Cancer 15(6):334–346. https://doi.org/10.1038/nrc3929
93. Li Y, Han J, Zhang Y, Cao F, Liu Z, Li S, Wu J, Hu C, Wang Y, Shuai J, Chen J, Cao L, Li D, Shi P, Tian C, Zhang J, Dou Y, Li G, Chen Y, Lei M (2016) Structural basis for activity regulation of MLL family methyltransferases. Nature 530(7591):447–452. https://doi.org/10.1038/nature16952
94. Dou Y, Milne TA, Ruthenburg AJ, Lee S, Lee JW, Verdine GL, Allis CD, Roeder RG (2006) Regulation of MLL1 H3K4 methyltransferase activity by its core components. Nat Struct Mol Biol 13(8):713–719. https://doi.org/10.1038/nsmb1128
95. Yokoyama A, Somervaille TC, Smith KS, Rozenblatt-Rosen O, Meyerson M, Cleary ML (2005) The menin tumor suppressor protein is an essential oncogenic cofactor for MLL-associated leukemogenesis. Cell 123(2):207–218. https://doi.org/10.1016/j.cell.2005.09.025
96. Cao F, Townsend EC, Karatas H, Xu J, Li L, Lee S, Liu L, Chen Y, Ouillette P, Zhu J, Hess JL, Atadja P, Lei M, Qin ZS, Malek S, Wang S, Dou Y (2014) Targeting MLL1 H3K4 methyltransferase activity in mixed-lineage leukemia. Mol Cell 53(2):247–261. https://doi.org/10.1016/j.molcel.2013.12.001
97. Karatas H, Li Y, Liu L, Ji J, Lee S, Chen Y, Yang J, Huang L, Bernard D, Xu J, Townsend EC, Cao F, Ran X, Li X, Wen B, Sun D, Stuckey JA, Lei M, Dou Y, Wang S (2017) Discovery of a highly potent, cell-permeable macrocyclic peptidomimetic (MM-589) targeting the WD repeat

domain 5 protein (WDR5)-mixed lineage leukemia (MLL) protein-protein interaction. J Med Chem 60(12):4818–4839. https://doi.org/10.1021/acs.jmedchem.6b01796
98. Grebien F, Vedadi M, Getlik M, Giambruno R, Grover A, Avellino R, Skucha A, Vittori S, Kuznetsova E, Smil D, Barsyte-Lovejoy D, Li F, Poda G, Schapira M, Wu H, Dong A, Senisterra G, Stukalov A, Huber KVM, Schonegger A, Marcellus R, Bilban M, Bock C, Brown PJ, Zuber J, Bennett KL, Al-Awar R, Delwel R, Nerlov C, Arrowsmith CH, Superti-Furga G (2015) Pharmacological targeting of the Wdr5-MLL interaction in C/EBPalpha N-terminal leukemia. Nat Chem Biol 11(8):571–578. https://doi.org/10.1038/nchembio.1859
99. Zhu J, Sammons MA, Donahue G, Dou Z, Vedadi M, Getlik M, Barsyte-Lovejoy D, Al-awar R, Katona BW, Shilatifard A, Huang J, Hua X, Arrowsmith CH, Berger SL (2015) Gain-of-function p53 mutants co-opt chromatin pathways to drive cancer growth. Nature 525(7568):206–211. https://doi.org/10.1038/nature15251
100. Borkin D, He S, Miao H, Kempinska K, Pollock J, Chase J, Purohit T, Malik B, Zhao T, Wang J, Wen B, Zong H, Jones M, Danet-Desnoyers G, Guzman ML, Talpaz M, Bixby DL, Sun D, Hess JL, Muntean AG, Maillard I, Cierpicki T, Grembecka J (2015) Pharmacologic inhibition of the Menin-MLL interaction blocks progression of MLL leukemia in vivo. Cancer Cell 27(4):589–602. https://doi.org/10.1016/j.ccell.2015.02.016
101. Zhou JX, Dhawan S, Fu H, Snyder E, Bottino R, Kundu S, Kim SK, Bhushan A (2013) Combined modulation of polycomb and trithorax genes rejuvenates beta cell replication. J Clin Invest 123(11):4849–4858. https://doi.org/10.1172/JCI69468
102. He S, Senter TJ, Pollock J, Han C, Upadhyay SK, Purohit T, Gogliotti RD, Lindsley CW, Cierpicki T, Stauffer SR, Grembecka J (2014) High-affinity small-molecule inhibitors of the menin-mixed lineage leukemia (MLL) interaction closely mimic a natural protein-protein interaction. J Med Chem 57(4):1543–1556. https://doi.org/10.1021/jm401868d
103. Keating ST, El-Osta A (2013) Transcriptional regulation by the Set7 lysine methyltransferase. Epigenetics 8(4):361–372. https://doi.org/10.4161/epi.24234
104. Barsyte-Lovejoy D, Li F, Oudhoff MJ, Tatlock JH, Dong A, Zeng H, Wu H, Freeman SA, Schapira M, Senisterra GA, Kuznetsova E, Marcellus R, Allali-Hassani A, Kennedy S, Lambert JP, Couzens AL, Aman A, Gingras AC, Al-Awar R, Fish PV, Gerstenberger BS, Roberts L, Benn CL, Grimley RL, Braam MJ, Rossi FM, Sudol M, Brown PJ, Bunnage ME, Owen DR, Zaph C, Vedadi M, Arrowsmith CH (2014) (R)-PFI-2 is a potent and selective inhibitor of SETD7 methyltransferase activity in cells. Proc Natl Acad Sci U S A 111(35): 12853–12858. https://doi.org/10.1073/pnas.1407358111
105. Niu Y, Shi D, Li L, Guo J, Liu H, Yao X (2017) Revealing inhibition difference between PFI-2 enantiomers against SETD7 by molecular dynamics simulations, binding free energy calculations and unbinding pathway analysis. Sci Rep 7:46547. https://doi.org/10.1038/srep46547
106. Francis NJ, Rowlands M, Workman P, Jones K, Aherne W (2012) Small-molecule inhibitors of the protein methyltransferase SET7/9 identified in a high-throughput screen. J Biomol Screen 17(8):1102–1109. https://doi.org/10.1177/1087057112452137
107. Mori S, Iwase K, Iwanami N, Tanaka Y, Kagechika H, Hirano T (2010) Development of novel bisubstrate-type inhibitors of histone methyltransferase SET7/9. Bioorg Med Chem 18(23): 8158–8166. https://doi.org/10.1016/j.bmc.2010.10.022
108. Meng F, Cheng S, Ding H, Liu S, Liu Y, Zhu K, Chen S, Lu J, Xie Y, Li L, Liu R, Shi Z, Zhou Y, Liu YC, Zheng M, Jiang H, Lu W, Liu H, Luo C (2015) Discovery and optimization of novel, selective histone methyltransferase SET7 inhibitors by pharmacophore- and docking-based virtual screening. J Med Chem 58(20):8166–8181. https://doi.org/10.1021/acs.jmedchem.5b01154
109. Takemoto Y, Ito A, Niwa H, Okamura M, Fujiwara T, Hirano T, Handa N, Umehara T, Sonoda T, Ogawa K, Tariq M, Nishino N, Dan S, Kagechika H, Yamori T, Yokoyama S, Yoshida M (2016) Identification of cyproheptadine as an inhibitor of SET domain containing lysine methyltransferase 7/9 (Set7/9) that regulates estrogen-dependent transcription. J Med Chem 59(8):3650–3660. https://doi.org/10.1021/acs.jmedchem.5b01732

110. Spellmon N, Holcomb J, Trescott L, Sirinupong N, Yang Z (2015) Structure and function of SET and MYND domain-containing proteins. Int J Mol Sci 16(1):1406–1428. https://doi.org/10.3390/ijms16011406
111. Reynoird N, Mazur PK, Stellfeld T, Flores NM, Lofgren SM, Carlson SM, Brambilla E, Hainaut P, Kaznowska EB, Arrowsmith CH, Khatri P, Stresemann C, Gozani O, Sage J (2016) Coordination of stress signals by the lysine methyltransferase SMYD2 promotes pancreatic cancer. Genes Dev 30(7):772–785. https://doi.org/10.1101/gad.275529.115
112. Ferguson AD, Larsen NA, Howard T, Pollard H, Green I, Grande C, Cheung T, Garcia-Arenas R, Cowen S, Wu J, Godin R, Chen H, Keen N (2011) Structural basis of substrate methylation and inhibition of SMYD2. Structure 19(9):1262–1273. https://doi.org/10.1016/j.str.2011.06.011
113. Sweis RF, Wang Z, Algire M, Arrowsmith CH, Brown PJ, Chiang GG, Guo J, Jakob CG, Kennedy S, Li F, Maag D, Shaw B, Soni NB, Vedadi M, Pappano WN (2015) Discovery of A-893, a new cell-active benzoxazinone inhibitor of lysine methyltransferase SMYD2. ACS Med Chem Lett 6(6):695–700. https://doi.org/10.1021/acsmedchemlett.5b00124
114. Nguyen H, Allali-Hassani A, Antonysamy S, Chang S, Chen LH, Curtis C, Emtage S, Fan L, Gheyi T, Li F, Liu S, Martin JR, Mendel D, Olsen JB, Pelletier L, Shatseva T, Wu S, Zhang FF, Arrowsmith CH, Brown PJ, Campbell RM, Garcia BA, Barsyte-Lovejoy D, Mader M, Vedadi M (2015) LLY-507, a cell-active, potent, and selective inhibitor of protein-lysine methyltransferase SMYD2. J Biol Chem 290(22):13641–13653. https://doi.org/10.1074/jbc.M114.626861
115. Eggert E, Hillig RC, Koehr S, Stockigt D, Weiske J, Barak N, Mowat J, Brumby T, Christ CD, Ter Laak A, Lang T, Fernandez-Montalvan AE, Badock V, Weinmann H, Hartung IV, Barsyte-Lovejoy D, Szewczyk M, Kennedy S, Li F, Vedadi M, Brown PJ, Santhakumar V, Arrowsmith CH, Stellfeld T, Stresemann C (2016) Discovery and characterization of a highly potent and selective aminopyrazoline-based in vivo probe (BAY-598) for the protein lysine methyltransferase SMYD2. J Med Chem 59(10):4578–4600. https://doi.org/10.1021/acs.jmedchem.5b01890
116. Huang L, Xu AM (2017) SET and MYND domain containing protein 3 in cancer. Am J Transl Res 9(1):1–14
117. Mitchell LH, Boriack-Sjodin PA, Smith S, Thomenius M, Rioux N, Munchhof M, Mills JE, Klaus C, Totman J, Riera TV, Raimondi A, Jacques SL, West K, Foley M, Waters NJ, Kuntz KW, Wigle TJ, Scott MP, Copeland RA, Smith JJ, Chesworth R (2016) Novel oxindole sulfonamides and sulfamides: EPZ031686, the first orally bioavailable small molecule SMYD3 inhibitor. ACS Med Chem Lett 7(2):134–138. https://doi.org/10.1021/acsmedchemlett.5b00272
118. Van Aller GS, Graves AP, Elkins PA, Bonnette WG, McDevitt PJ, Zappacosta F, Annan RS, Dean TW, Su DS, Carpenter CL, Mohammad HP, Kruger RG (2016) Structure-based design of a novel SMYD3 inhibitor that bridges the SAM-and MEKK2-binding pockets. Structure 24(5):774–781. https://doi.org/10.1016/j.str.2016.03.010
119. McDaniel SL, Strahl BD (2017) Shaping the cellular landscape with Set2/SETD2 methylation. Cell Mol Life Sci 74(18):3317–3334. https://doi.org/10.1007/s00018-017-2517-x
120. Zhu Y, Zhu L, Lu L, Zhang L, Zhang G, Wang Q, Yang P (2014) Role and mechanism of the alkylglycerone phosphate synthase in suppressing the invasion potential of human glioma and hepatic carcinoma cells in vitro. Oncol Rep 32(1):431–436. https://doi.org/10.3892/or.2014.3189
121. Zheng W, Ibanez G, Wu H, Blum G, Zeng H, Dong A, Li F, Hajian T, Allali-Hassani A, Amaya MF, Siarheyeva A, Yu W, Brown PJ, Schapira M, Vedadi M, Min J, Luo M (2012) Sinefungin derivatives as inhibitors and structure probes of protein lysine methyltransferase SETD2. J Am Chem Soc 134(43):18004–18014. https://doi.org/10.1021/ja307060p
122. Beck DB, Oda H, Shen SS, Reinberg D (2012) PR-Set7 and H4K20me1: at the crossroads of genome integrity, cell cycle, chromosome condensation, and transcription. Genes Dev 26(4):325–337. https://doi.org/10.1101/gad.177444.111

123. Milite C, Feoli A, Viviano M, Rescigno D, Cianciulli A, Balzano AL, Mai A, Castellano S, Sbardella G (2016) The emerging role of lysine methyltransferase SETD8 in human diseases. Clin Epigenetics 8:102. https://doi.org/10.1186/s13148-016-0268-4
124. Reinberg D, Trojer P, Sbardella G (2007) Selective inhibitors for transferases. Google Patents
125. Valente S, Lepore I, Dell'Aversana C, Tardugno M, Castellano S, Sbardella G, Tomassi S, Di Maro S, Novellino E, Di Santo R, Costi R, Altucci L, Mai A (2012) Identification of PR-SET7 and EZH2 selective inhibitors inducing cell death in human leukemia U937 cells. Biochimie 94(11):2308–2313. https://doi.org/10.1016/j.biochi.2012.06.003
126. Williams DE, Dalisay DS, Li F, Amphlett J, Maneerat W, Chavez MA, Wang YA, Matainaho T, Yu W, Brown PJ, Arrowsmith CH, Vedadi M, Andersen RJ (2013) Nahuoic acid A produced by a Streptomyces sp. isolated from a marine sediment is a selective SAM-competitive inhibitor of the histone methyltransferase SETD8. Org Lett 15(2):414–417. https://doi.org/10.1021/ol303416k
127. Williams DE, Izard F, Arnould S, Dalisay DS, Tantapakul C, Maneerat W, Matainaho T, Julien E, Andersen RJ (2016) Structures of nahuoic acids B-E produced in culture by a Streptomyces sp. isolated from a marine sediment and evidence for the inhibition of the histone methyl transferase SETD8 in human cancer cells by nahuoic acid A. J Org Chem 81(4):1324–1332. https://doi.org/10.1021/acs.joc.5b02569
128. Ma A, Yu W, Li F, Bleich RM, Herold JM, Butler KV, Norris JL, Korboukh V, Tripathy A, Janzen WP, Arrowsmith CH, Frye SV, Vedadi M, Brown PJ, Jin J (2014) Discovery of a selective, substrate-competitive inhibitor of the lysine methyltransferase SETD8. J Med Chem 57(15):6822–6833. https://doi.org/10.1021/jm500871s
129. Ma A, Yu W, Xiong Y, Butler KV, Brown PJ, Jin J (2014) Structure-activity relationship studies of SETD8 inhibitors. MedChemComm 5(12):1892–1898. https://doi.org/10.1039/C4MD00317A
130. Veschi V, Liu Z, Voss TC, Ozbun L, Gryder B, Yan C, Hu Y, Ma A, Jin J, Mazur SJ, Lam N, Souza BK, Giannini G, Hager GL, Arrowsmith CH, Khan J, Appella E, Thiele CJ (2017) Epigenetic siRNA and chemical screens identify SETD8 inhibition as a therapeutic strategy for p53 activation in high-risk neuroblastoma. Cancer Cell 31(1):50–63. https://doi.org/10.1016/j.ccell.2016.12.002
131. Butler KV, Ma A, Yu W, Li F, Tempel W, Babault N, Pittella-Silva F, Shao J, Wang J, Luo M, Vedadi M, Brown PJ, Arrowsmith CH, Jin J (2016) Structure-based design of a covalent inhibitor of the SET domain-containing protein 8 (SETD8) lysine methyltransferase. J Med Chem 59(21):9881–9889. https://doi.org/10.1021/acs.jmedchem.6b01244
132. Blum G, Ibanez G, Rao X, Shum D, Radu C, Djaballah H, Rice JC, Luo M (2014) Small-molecule inhibitors of SETD8 with cellular activity. ACS Chem Biol 9(11):2471–2478. https://doi.org/10.1021/cb500515r
133. Judge RA, Zhu H, Upadhyay AK, Bodelle PM, Hutchins CW, Torrent M, Marin VL, Yu W, Vedadi M, Li F, Brown PJ, Pappano WN, Sun C, Petros AM (2016) Turning a substrate peptide into a potent inhibitor for the histone methyltransferase SETD8. ACS Med Chem Lett 7(12):1102–1106. https://doi.org/10.1021/acsmedchemlett.6b00303
134. Bromberg KD, Mitchell TR, Upadhyay AK, Jakob CG, Jhala MA, Comess KM, Lasko LM, Li C, Tuzon CT, Dai Y, Li F, Eram MS, Nuber A, Soni NB, Manaves V, Algire MA, Sweis RF, Torrent M, Schotta G, Sun C, Michaelides MR, Shoemaker AR, Arrowsmith CH, Brown PJ, Santhakumar V, Martin A, Rice JC, Chiang GG, Vedadi M, Barsyte-Lovejoy D, Pappano WN (2017) The SUV4-20 inhibitor A-196 verifies a role for epigenetics in genomic integrity. Nat Chem Biol 13(3):317–324. https://doi.org/10.1038/nchembio.2282
135. Fraga MF, Ballestar E, Villar-Garea A, Boix-Chornet M, Espada J, Schotta G, Bonaldi T, Haydon C, Ropero S, Petrie K, Iyer NG, Perez-Rosado A, Calvo E, Lopez JA, Cano A, Calasanz MJ, Colomer D, Piris MA, Ahn N, Imhof A, Caldas C, Jenuwein T, Esteller M (2005) Loss of acetylation at Lys16 and trimethylation at Lys20 of histone H4 is a common hallmark of human cancer. Nat Genet 37(4):391–400. https://doi.org/10.1038/ng1531

Top Med Chem (2020) 33: 159–196
DOI: 10.1007/7355_2019_73

Published online: 30 October 2019

PRMT Inhibitors

Matthijs J. van Haren and Nathaniel I. Martin

Contents

1 Introduction 161
2 Nonspecific Protein Methyltransferase Inhibitors 163
2.1 Background 163
2.2 Inhibitors: In Vitro and Cell-Based Activities 164
3 PRMT1 165
3.1 Background 165
3.2 Inhibitors: In Vitro and Cell-Based Activities 166
3.3 Biological Relevance of Inhibitors and Current Outlook 170
4 PRMT2 170
4.1 Background 170
4.2 Inhibitors: In Vitro and Cell-Based Activities 170
4.3 Biological Relevance of Inhibitors and Current Outlook 171
5 PRMT3 172
5.1 Background 172
5.2 Inhibitors: In Vitro and Cell-Based Activities 172
5.3 Biological Relevance of Inhibitors and Current Outlook 173
6 CARM1 (PRMT4) 173
6.1 Background 173
6.2 Inhibitors: In Vitro and Cell-Based Activities 173
6.3 Biological Relevance of Inhibitors and Current Outlook 177
7 PRMT5 178
7.1 Background 178
7.2 Inhibitors: In Vitro and Cell-Based Activities 178
7.3 Biological Relevance of Inhibitors and Current Outlook 180
8 PRMT6 181
8.1 Background 181
8.2 Inhibitors: In Vitro and Cell-Based Activities 181
8.3 Biological Relevance of Inhibitors and Current Outlook 182
9 PRMT7 183
9.1 Background 183
9.2 Inhibitors: In Vitro and Cell-Based Activities 183

M. J. van Haren and N. I. Martin (✉)
Biological Chemistry Group, Institute of Biology Leiden, Leiden University, Leiden, The Netherlands
e-mail: n.i.martin@biology.leidenuniv.nl

9.3 Biological Relevance of Inhibitors and Current Outlook 184
10 PRMT8 184
10.1 Background 184
10.2 Inhibitors: In Vitro and Cell-Based Activities 184
10.3 Biological Relevance of Inhibitors and Current Outlook 185
11 PRMT9 185
12 Conclusions 185
References 188

Abstract The methylation of arginine residues in numerous protein targets is a post-translational modification that has gained increased interest in the scientific community over the past two decades. Arginine methylation is performed by the dedicated family of protein arginine methyltransferases and is known to be involved in a plethora of cellular pathways and biochemical mechanisms in both healthy and disease states. The development of inhibitors for these enzymes for use as biological tools can lead to a more detailed understanding of the functions of the different members of the PRMT family. In addition, a number of recent studies point towards PRMTs as therapeutic targets for a number of diseases and the first clinical trials with compounds inhibiting PRMTs are now underway. We here provide a broad overview of the current status of the inhibitors that have been developed against PRMTs using both high-throughput screening and rational design approaches.

Keywords Activity, Inhibition, Methylation, Protein arginine *N*-methyltransferase, Therapeutics

Abbreviations

aDMA	Asymmetrically dimethylated arginine
AdoHcy	*S*-adenosyl-L-homocysteine
AdoMet	*S*-adenosyl-L-methionine
Adox	Adenosine dialdehyde
AMI	Arginine methyltransferase inhibitor
AML	Acute myeloid leukaemia
CARM1	Coactivator-associated arginine methyltransferase
DNA	Deoxyribonucleic acid
EBV	Epstein-Barr virus
EC50	Half maximal effective concentration
GAR	Glycine-arginine rich
HEK293T	Human embryonic kidney cell line
HepG2	Hepatocellular carcinoma cell line
HIV	Human immunodeficiency virus
IC_{50}	Half maximal inhibitory concentration
Ki	Inhibition constant

LNCaP	Lymph node carcinoma of the prostate, prostate cancer cell line
MCF7	Michigan Cancer Foundation-7, breast cancer cell line
MCL	Mantle cell lymphoma
MEP50	Methylosome protein 50
MLL	Mixed lineage leukaemia
MMA	Monomethylated arginine
MTA	Methylthioadenosine
MTAP	5-Methylthioadenosine phosphorylase
PABP1	Poly(A)-binding protein-1
PAD	Protein arginine deiminase
PGM	Proline, glycine, methionine-rich
PK/PD	Pharmacokinetic/pharmacodynamic
PRMT	Protein arginine *N*-methyltransferase
RNA	Ribonucleic acid
RSF1	Repressor splicing factor
SAH	*S*-adenosyl-L-homocysteine
SAHH	*S*-adenosyl-L-homocysteine hydrolase
SAM	*S*-adenosyl-L-methionine
SAR	Structure-activity relationship
sDMA	Symmetrically dimethylated arginine
SET7	SET domain containing protein 7
SGC	Structural genomics consortium
Tat	Trans-activator of transcription

1 Introduction

The methylation of arginine residues in proteins is an important post-translational modification, performed by the family of protein arginine *N*-methyltransferases (PRMTs). The enzymes use nature's ubiquitous methyl donor *S*-adenosyl-L-methionine (AdoMet, also known as SAM) as a cofactor to form the methylated protein product with concomitant release of *S*-adenosyl-L-homocysteine (AdoHcy, also known as SAH) as a byproduct. Mechanistically, target arginine residues are first monomethylated by all types of PRMTs and subsequently dimethylated asymmetrically by type I PRMTs and symmetrically by type II PRMTs. Type III PRMTs produce only monomethylated arginine (MMA). The PRMTs share highly conserved active site sequences, including a number of residues involved in AdoMet cofactor recognition as well as two glutamate residues that hydrogen bond with the guanidine moiety of the substrate peptide. These glutamate residues are part of the so-called double E-loop which ensures that the guanidine group is positioned in close proximity to the AdoMet cofactor to facilitate the methyl transfer via an S_N2-like substitution reaction (Fig. 1).

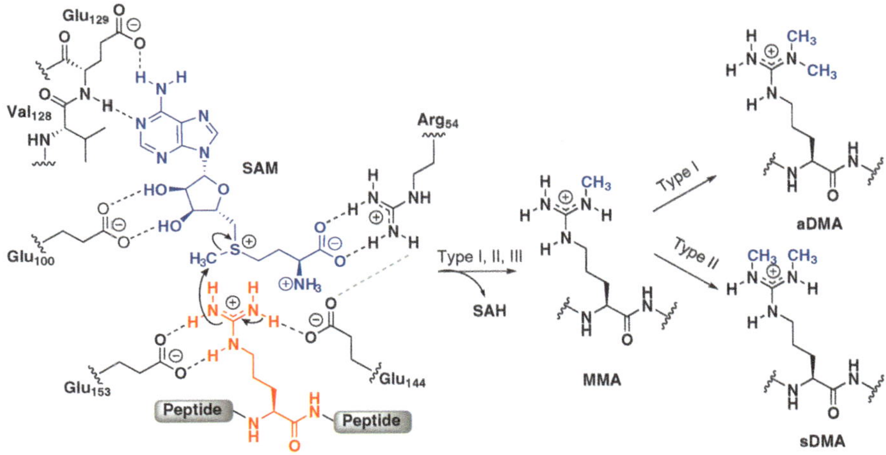

Fig. 1 Schematic overview of the PRMT-binding site (residue numbering PRMT1) showing the hydrogen bonding interactions of the double E-loop (residues Glu144 and Glu153) with the guanidine moiety of the arginine sidechain. Cofactor *S*-adenosyl-L-methionine (SAM) is depicted in blue and the protein arginine sidechain in red. First monomethylarginine (MMA) is formed releasing *S*-adenosyl-L-homocysteine (SAH). MMA is directly converted to asymmetrically dimethylated arginine (aDMA) by type I PRMTs and to symmetrically dimethylated arginine (sDMA) by type II PRMTs

The presence of (di-)methylated arginines in proteins and an enzyme responsible for this process were discovered 50 years ago [1, 2]. A decade later it was found that the majority of proteins with methylated arginines are nuclear proteins that contain asymmetrically dimethylated arginine (aDMA) sidechains [3]. Subsequently, it became clear that not one enzyme but a distinct family of enzymes was responsible for this post-translational modification [4–8]. The seminal study of Herschman and co-workers published in 1996 [9] clearly identified and characterized the first member of the family, PRMT1. In the following decade, new PRMT family members were identified on a near yearly basis resulting in the now known total of nine PRMTs [10–18].

Whether or not arginine methylation is a dynamic process remains to be proven unambiguously. Protein arginine deiminases (PADs) are known to hydrolyse the guanidine moiety of arginine side chains in proteins into citrulline. However, PADs are not able to convert dimethylated arginines to citrulline [19]. In addition, while in vivo data suggested the possibility that PADs can convert monomethylated arginines to citrulline, in vitro studies have not been able to reproduce this activity [19]. In addition, citrulline can only be converted back into arginine after proteolysis, but not in the context of an intact protein. Given that it is unlikely that PADs are the active player in removing methyl groups from methylated arginines, much effort has been spent at trying to identify a true arginine demethylase. In 2007, the lysine demethylating enzyme JMJD6 was reported to also demethylate arginines in histones making it the first example of an arginine demethylase [20]. These findings have, however, become the source of some debate as the results could not be reproduced by other groups [21–26]. Recently, other members of the Jumonji-domain containing lysine demethylases (KDM4E and KDM5C) were also shown to demethylate arginines in histones in vitro, suggesting the process is indeed dynamic [27]. However, at this point in time, in vivo proof for the presence of arginine-specific demethylases is still lacking. In this regard, the only certain way to control arginine methylation remains via inhibition of the PRMTs.

A great number of protein substrates have been identified for the different PRMTs, ranging from general substrates to others that are only acted upon by one specific PRMT [28–30]. Arginine methylation by PRMTs is involved in many cellular processes, including RNA processing, gene transcription, signal transduction and DNA repair [31, 32]. In terms of their biological roles, the PRMTs have little redundancy as indicated by the dramatic phenotypes observed in knockout mice [31]. A growing body of evidence implicates dysregulated arginine methylation in a variety of diseases, including numerous cancers [32], cardiovascular [33], pulmonary [34–36] and viral diseases [37, 38]. In an attempt to address the connection between aberrant PRMT activity and human disease, the Bedford group published the first report describing PRMT inhibitors in 2004 [39]. Since this time, the number of publications describing new, more potent and selective PRMT inhibitors has steadily increased.

The recent interest in the field of PRMTs has led to growing amounts of data along with reviews describing the progress made in different areas of research [40–46]. We here provide an overview on the current status of the development of compounds aimed at inhibiting PRMTs. Starting with a brief summary of general methyltransferase inhibitors, we then address the development of inhibitors specific for each of the nine PRMTs, followed by a brief discussion on their biological relevance. In cases where a compound has inhibitory activity against more than one PRMT, it will be discussed in the context of the enzyme that is inhibited with the highest activity and selectivity.

2 Nonspecific Protein Methyltransferase Inhibitors

2.1 Background

The enzymatic reaction performed by PRMTs and other methyltransferases using AdoMet as the methyl donor is self-regulating due to the inhibitory properties of cofactor byproduct, AdoHcy (**1**). Structurally similar compounds include Aza-AdoMet (**2**), where the sulphur of AdoMet is replaced by a nitrogen atom, and the bacterially produced natural product sinefungin (**3**, Fig. 2). These compounds are known to inhibit all AdoMet-dependent methyltransferases by competition with AdoMet. It is due to this mechanism of action that the AdoMet analogues lack specificity of inhibition. Therefore, the primary use of these analogues is as reference inhibitors in both biochemical and cellular assays. Adenosine dialdehyde

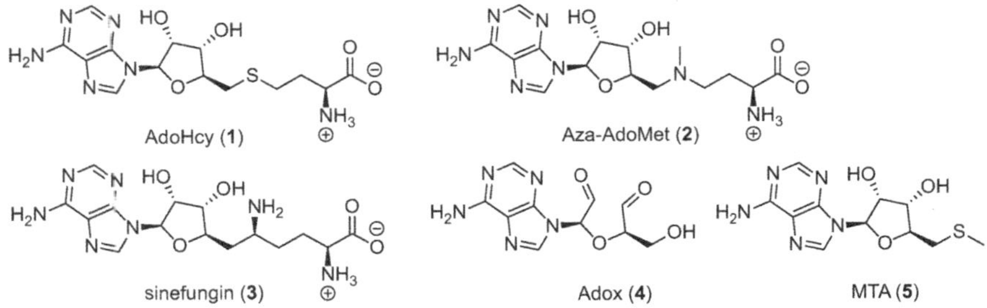

Fig. 2 General methyltransferase inhibitors *S*-adenosyl-L-homocysteine (AdoHcy, **1**), nitrogen analogue Aza-AdoMet (**2**), bacterial compound sinefungin (**3**), indirect inhibitor adenosine dialdehyde (Adox, **4**) and methylthioadenosine (MTA, **5**)

(Adox, **4**) is a known inhibitor of *S*-adenosyl-L-homocysteine hydrolase (SAHH) which leads to a build-up of AdoHcy. Via this mechanism Adox is an indirect methyltransferase inhibitor that can be used in cellular assays. Methylthioadenosine (MTA, **5**) is sometimes mentioned as a general methyltransferase inhibitor, but recent publications show MTA is rather specific in inhibiting PRMTs (see subchapter on PRMT5) [47–49].

2.2 Inhibitors: In Vitro and Cell-Based Activities

Recently, a more targeted general PRMT inhibitor was reported by Jin and co-workers [50]. The compound, designated as MS023 (**6**, Fig. 3), inhibits all type I PRMTs. Its design was based on the structures of other PRMT inhibitors specific for PRMT4 and PRMT6 also containing an ethylenediamino group which was found to function as an arginine mimetic. Building upon structure-activity relationship (SAR) studies, MS023 was developed and, along with the negative control compound MS094 (lacking the ethylenediamino moiety), tested against a panel of methyltransferases. The results showed potent inhibition of all type I PRMTs with activities from 4 to 119 nM with no inhibition of type II and III PRMTs, lysine methyltransferases, DNA methyltransferases, histone demethylases and methyl-readers [50]. A co-crystal structure of MS023 bound to PRMT6 in the presence of AdoHcy (Fig. 3) shows the interactions of the ethylenediamino group in the active site where the arginine moiety would be expected to bind. Replacement of the terminal amine with a hydroxyl group as in MS094 (**7**, Fig. 3) leads to complete loss of inhibitory activity. The cellular assays performed with MS023 showed a decrease in asymmetric dimethylation of histone H4R3 by PRMT1 with an IC_{50} value of 9 nM and of H3R2 by PRMT6 with an IC_{50} value of 56 nM. Also of note, the decrease in cellular aDMA levels was accompanied by a measurable increase in MMA and symmetrically dimethylated arginine (sDMA) levels. Interestingly, no explanation has yet been provided for the striking selectivity observed for MS023 in

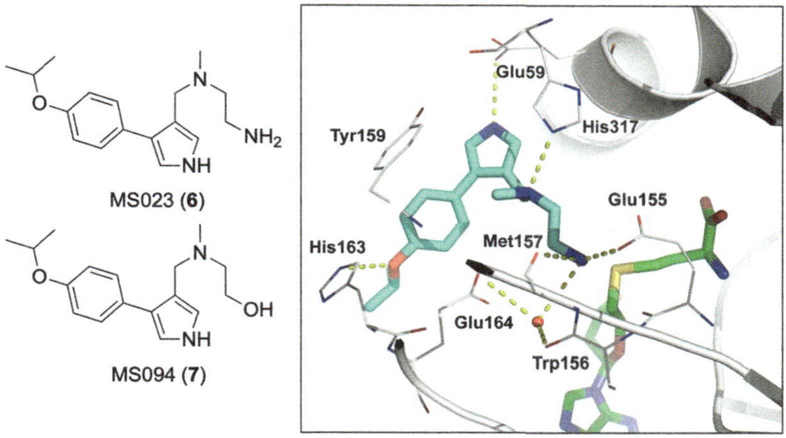

Fig. 3 Left, structures of general type I PRMT inhibitor MS023 (**6**) and its negative control MS094 (**7**). Right, co-crystal structure of MS023 (blue) bound to PRMT6 in the presence of AdoHcy (green). MS023 binds in the substrate-binding site with the ethylenediamino moiety interacting with His317 of the "THW-loop" and Glu155 and Glu164 of the "double E-loop" (PDB ID: 5E8R) [50]

the inhibition of type I over type II/III PRMTs. While such type I-specific PRMT inhibitor can be of value when looking into the role of aDMA in complex systems, their therapeutic potential remains to be demonstrated. Nevertheless, in order to investigate the specific roles of the individual PRMTs in detail, a comprehensive toolbox of inhibitory compounds is essential.

3 PRMT1

3.1 Background

PRMT1 was the first identified member of the protein arginine methyltransferase family and the most abundant [9, 51, 52]. Alternative splicing results in the possibility of at least seven variants (v1–v7) with differences in the length of the N-terminal tail [53]. These variants can differ in localization and substrate specificity. PRMT1 produces asymmetrically dimethylated arginine (aDMA) and is responsible for more than 85% of all arginine methylations [52]. As expected, knockout of PRMT1 in cells results in a significant decrease in aDMA and a corresponding increase in MMA and sDMA [54].

In its target substrates, PRMT1 preferentially methylates the RGG motif, although this is not a strict prerequisite [6, 8, 55]. Known PRMT1 substrates are diverse and include histone H4R3, Sam68 (sarcoma associated in mitosis of 68 kDa), MRE11 (meiotic recombination 11), 53BP1 (p53-binding protein 1) and hnRNP A1 (heterogeneous nuclear ribonucleoprotein A1), among many others which are involved in

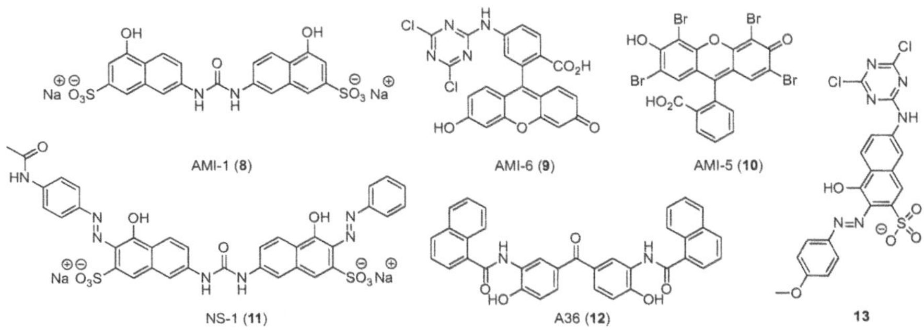

Fig. 4 PRMT1 inhibitors **8–13** interacting with the histone H4 and glycine-arginine-rich (GAR) substrates

gene transcription, nuclear transport, DNA repair and RNA processing, respectively [32, 56]. PRMT1 is overexpressed in breast [53, 57], prostate [58], lung [59], colon [60, 61], head and neck [62] and bladder cancer [63] and in leukaemia [32, 64, 65]. It is further involved in a number of other serious conditions ranging from pulmonary disease [35, 66], cardiovascular disease [33] and diabetes [67] to cocaine addiction [68].

3.2 *Inhibitors: In Vitro and Cell-Based Activities*

The first report on the development of PRMT inhibitors was published in 2004, indicating the relative infancy of the field. A screening campaign resulted in the identification of *a*rginine *m*ethyltransferase *i*nhibitors AMI-1 (**8**) and AMI-6 (**9**, Fig. 4) with IC_{50} values against PRMT1 of 8.8 and 5.1 μM, respectively [39]. Small molecules based on AMI scaffolds, including AMI-5 (**10**, Fig. 4), were studied via docking and binding studies [69], and a follow-up study with simplified AMI-5 analogues showed inhibitory effects on different methyltransferases, including PRMT1 and PRMT4 and lysine methyltransferase SET7 (SET domain containing protein) [70]. For several active compounds, cellular activity in human leukaemia U937 cell lines was found with specific effects on cell cycle arrest, apoptosis and granulocyte differentiation [70].

Follow-up studies revealed analogues of the AMI series, including naphthyl-sulfo derivatives, such as compound NS-1 (**11**, Fig. 4), [71] and pharmacophore-based small molecule inhibitors, like analogue A36 (**12**, Fig. 4), both of which exhibit IC_{50} values in the low μM range [72]. Interestingly, it was later discovered that the mechanism of action of these inhibitors is not by interaction with the PRMT but rather with the histone H4 substrate and other glycine- and arginine-rich (GAR) substrates, explaining – at least in part – the observed (lack of) specificities for these compounds. Inspired by the findings of the AMI compounds, Mowen and co-workers combined structural features of the different AMIs to generate new

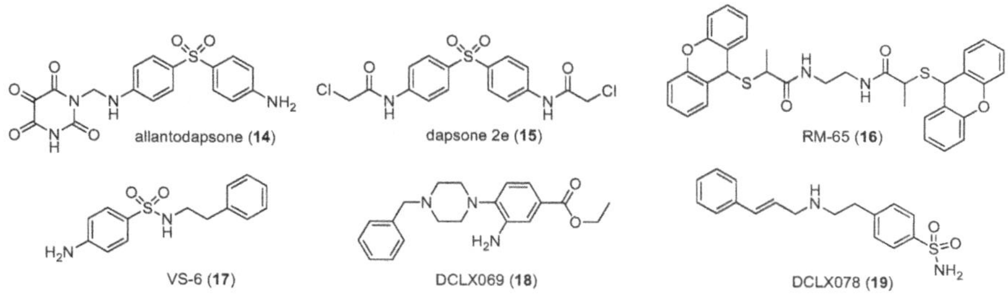

Fig. 5 Virtual screening hits **14–19** for PRMT1

PRMT inhibitors [73]. Compound **13** of this series showed an IC_{50} value of 4.2 μM against PRMT1 and 2.7 μM against PRMT4 and also demonstrated activity against PRMT3, PRMT5, PRMT6 and PRMT8 as shown by Western blot analysis. The same inhibitor was only slightly active against lysine methyltransferase SET7 and was found to be cell-permeable and decrease levels of H3R17 methylation and lowers the secretion of interferon IFN-gamma and interleukin IL-4 from T helper cells [73].

As another approach to discovering PRMT inhibitors, virtual screening methods have also been employed. Such strategies have revealed allantodapsone (**14**) [74]; dapsone analogues, including compound **15** [75]; thioglycolic amides, including RM-65 (**16**) [76]; and other virtual screening hits such as VS-6 (**17**) [77] (Fig. 5). These compounds were evaluated for their inhibitory activity against PRMT1 using a target-based virtual screening approach. Compound **15** was found to be cytotoxic in MCF7 breast cancer cells and LNCaP (lymph node carcinoma of the prostate) prostate cancer cells, while RM-65 was demonstrated to result in hypomethylation of proteins in HepG2 hepatocellular carcinoma cancer cells. Similar virtual screening approaches by other research groups led to the discovery of compounds DCLX069 (**18**) and DCLX078 (**19**) [78] both of which exhibit micromolar inhibition of PRMT1 (IC_{50} 17.9–26.2 μM) with some selectivity over PRMT4 and PRMT6. The compounds also showed antiproliferative action in three different cancer cell lines (HepG2, MCF7 and the monocytic leukaemia cell line THP1). These compounds represent promising starting points for hit-to-lead optimizations in pursuit of selective PRMT inhibitors.

Applying a bisubstrate approach, Dowden and co-workers designed PRMT inhibitors linking structural features of the AdoMet cofactor to a guanidine or amine moiety (**20–22**, Fig. 6) [79, 80]. This approach resulted in inhibitors that are active against PRMT1 with low micromolar potency (IC_{50} values ranged from 2.9 to 6.2 μM) and inactive against SET7 and PRMT4. No cellular assays were performed with these bisubstrates.

Working together with Frankel and co-workers, our group has also investigated N^{η}-substituted arginine-containing peptides as PRMT inhibitors (**23–25**, Fig. 6) [81–84]. Using an established PRMT substrate derived from a fibrillarin peptide sequence, incorporation of fluorinated ethyl groups on the side chain of the target

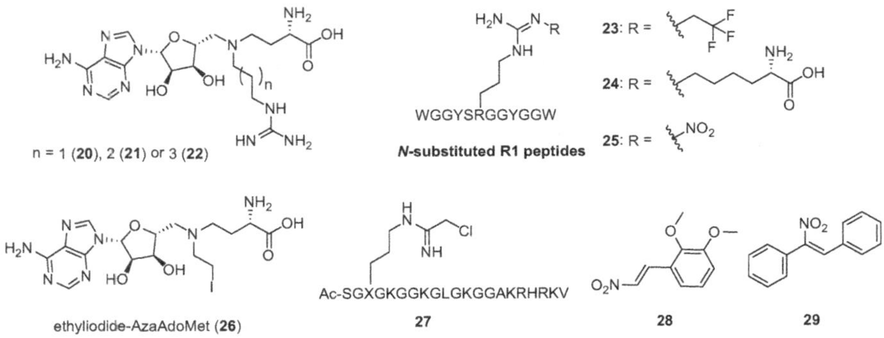

Fig. 6 Bisubstrate inhibitors **20–25** and covalent inhibitors **26–29** of PRMT1

arginine as in compound **23** (Fig. 6) resulted in micromolar IC_{50} values for PRMT1 (27.5 μM) and PRMT6 (9.4 μM) with no significant inhibition of PRMT4 (168 μM) [81]. By attaching an amino acid moiety to the side chain of the target arginine, peptidic partial bisubstrate **24** (Fig. 6) was obtained that showed inhibition against PRMT1 (13.9 μM), PRMT4 (35.7 μM) and PRMT6 (29.0 μM) [82]. A similar effect was observed for N^{η}-nitro-substituted arginine **25** (IC_{50} 26–47 μM for PRMT1, 4 and 6) when studying the effects of substitution on the methylation kinetics of PRMTs [84]. Small modifications of the side chain of an arginine residue in an HIV Tat peptide sequence resulted in micromolar K_i values against PRMT1 (2.7–7.6 μM) and PRMT6 (19.9–100 μM) with no significant inhibition of PRMT4 [83].

An in situ bisubstrate approach was also applied by Thompson and co-workers using ethyliodide-Aza-AdoMet (**26**, Fig. 6) [85]. This *N*-mustard-containing AdoMet analogue rearranges to form an aziridinium ion which will react with the substrate but only in the presence of the methyltransferase enzyme. In this way the histone H4 (1–21) substrate was enzymatically linked to the AdoMet cofactor. The conjugate was found to inhibit PRMT1 with an IC_{50} of 11.9 μM and a 4.4-fold specificity over PRMT4. In another covalent inhibitor approach, applied to inhibiting PRMT1, a chloroacetamidine warhead was incorporated in a H4 peptide substrate (**27**, Fig. 6). This approach was inspired by the success obtained for inhibiting PAD4, an arginine deiminase, that bears an active site cysteine [86]. Compound **27** inhibited PRMT1 and PRMT6 with IC_{50} values of 1.8 μM and 8.8 μM, respectively, and was >250-fold selective over PRMT3 and PRMT4 [87, 88]. Although a specific target cysteine was not discussed in these papers, other studies found a reactive cysteine (C101) in the active site of PRMT1 involved in binding AdoMet [89, 90]. The reactivity of this cysteine was confirmed for compounds **28** and **29** (Fig. 6) when testing against both a wild-type PRMT1 and the corresponding C101A mutant [91]. These covalent inhibitors were also found to be active against PRMT8 and inactive against PRMT4 and SET7 (tested at 10 and 100 μM only).

Recently, a small library of diamidine compounds, structurally similar to stilbamidine (**30**, Fig. 7), was screened for activity against PRMT1 [92]. The results revealed furamidine (**31**), a known antiparasitic agent, to be a fairly active (9.4 μM)

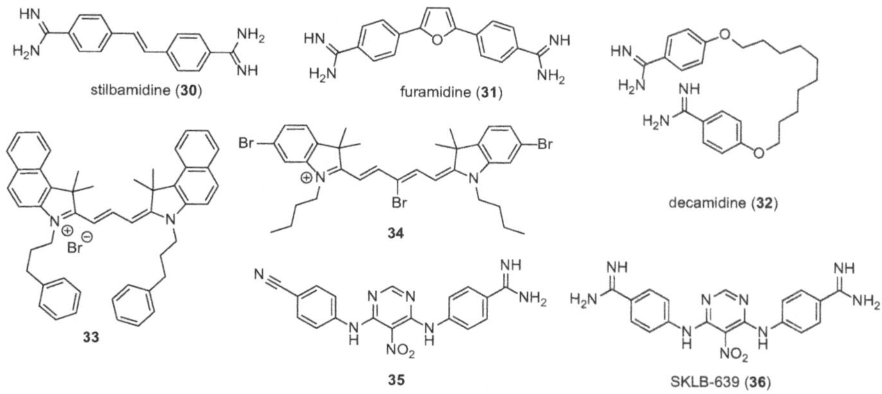

Fig. 7 Diamidines **30–32**, carbocyanine dyes **33–34** and nitropyrimidines **35–36** as inhibitors of PRMT1

and selective (18-, 30- and >42-fold over PRMT5, PRMT6 and PRMT4, respectively) inhibitor of PRMT1. Interestingly, furamidine was found to inhibit proliferation in leukaemia cell lines with higher sensitivity in cell lines derived from Down's syndrome patients and individuals diagnosed with mixed lineage leukaemia (MLL) [92]. Subsequent investigation of spacer length in diamidines revealed decamidine (**32**), bearing a 10-carbon spacer, to be a slightly more active (13 μM IC_{50} compared to 22 μM found for furamidine) but less specific PRMT1 inhibitor compared to furamidine. No additional analogues containing longer spacers were reported so the optimal spacing has yet to be confirmed [93]. The same group also investigated cyanine dyes as PRMT inhibitors and found compound **33** (Fig. 7) as an active hit with IC_{50} values of 0.61–1.74 μM against PRMT1, PRMT3, PRMT4, PRMT5, PRMT6 and PRMT8. This study also revealed compound **34** as a partially selective hit with an IC_{50} of 3.38 μM against PRMT1 with 6-, 10- and 25-fold selectivity over PRMT4, PRMT5 and PRMT8, respectively, but with no selectivity for PRMT3 or PRMT6 [94, 95]. Compound **34** was also tested in three different leukaemia cell lines and showed significant cell growth inhibition at 100–200 nM.

Also recently, an SAR study was reported by Yang and co-workers based upon a nitropyrimidine hit from a screening campaign [96]. This work showed that the amidine moiety was important for PRMT1 inhibition with modelling studies suggesting that it binds with the double E-loop of the enzyme. The optimized nitropyrimidine compound **35** (Fig. 7) showed an IC_{50} of 2.0 μM for PRMT1 with fivefold selectivity over PRMT4 and no activity against PRMT5 and PRMT6. Cellular assays revealed low micromolar IC_{50} activity against colon cancer (4.4 μM), bladder cancer (13.1 μM) and neuroblastoma (11.4 μM) tumour cell lines [96]. In a parallel study, the same group reported nitropyrimidine-diamidine compound SKLB-639 (**36**, Fig. 7) with similar activity against PRMT1 (2.4 μM) and selectivities of 15-fold over PRMT3, 30-fold over PRMT4 and no activity against

PRMT5, PRMT6 and PRMT8 [68]. In vivo studies show a decrease of H4R3me2a, but not H3R2me2a, H3R17me2a or sDMA in cocaine treated mice.

3.3 Biological Relevance of Inhibitors and Current Outlook

As the predominant member of the arginine methyltransferase family, PRMT1 has been the subject of thorough investigation and aims at identifying potent and selective inhibitors. Thus far, limited success has been achieved. To date, the most potent inhibitors reported show activity in the micromolar range, and the compounds are often active against at least one more PRMT. At present, no potent and selective inhibitors have been described for PRMT1. Studies investigating the role of PRMT1 in healthy and disease states show its upregulation in malignant cell lines. Cellular assays performed with a number of the inhibitors described above show their efficacy in killing tumour cell lines. Such promising results underscore the importance of developing potent and selective PRMT1 inhibitors.

4 PRMT2

4.1 Background

The second member identified as part of the PRMT family [10], PRMT2, is among the least studied PRMTs. PRMT2 is a type I PRMT, producing both MMA and aDMA. While it exhibits weak activity for histone H4 methylation [97], only few unique substrates for PRMT2 have been identified to date [98]. PRMT2 resides mainly in the nucleus, interacts with splicing factors and is a coactivator of nuclear hormone receptors, including the androgen and oestrogen α receptors [99–101]. PRMT2 regulates leptin signalling by methylation of STAT3 (signal transducer and activator of transcription 3) [102] and is downregulated under high glucose conditions leading to increased atherosclerosis through reduced cholesterol efflux [103]. PRMT2 is also associated with survival outcome and tumour grade in breast cancer via transcriptional activation of oestrogen receptor α [43, 104, 105].

4.2 Inhibitors: In Vitro and Cell-Based Activities

In addition to its weak methylating activity towards histones and the limited knowledge on nonhistone substrates, very few PRMT2 inhibitors have been reported. In a recent paper describing the crystal structure of PRMT2 [98], repressor splicing factor 1 (RSF1) was identified as a new nonhistone substrate. PRMT2 was found to methylate RSF1 much more efficiently than histone H3 or H4. Crystal structures

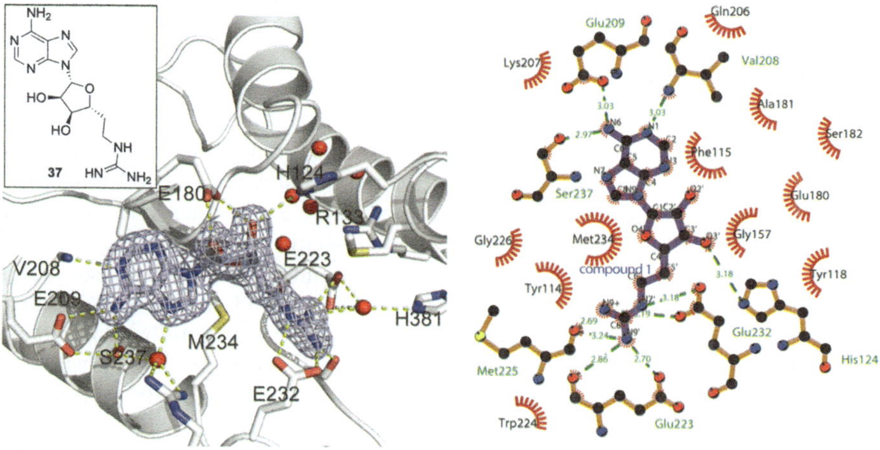

Fig. 8 Co-crystal structure of compound **37** bound to mouse PRMT2 (PDB ID: 5FWA), showing the interactions in the active site [98]. The "double E-loop" residues E223 and E232 and "THW-loop" residue H381 interact with the guanidine moiety of **37**. In addition, the adenosine moiety of **37** interacts with the conserved active site residues E180, E209, V208 and S237 in the cofactor-binding site

of PRMT2 from zebrafish and from mice were solved in complex with AdoHcy, sinefungin or a small molecule bisubstrate inhibitor known as Cp1 [106] (**37**, Fig. 8). Thermal shift assays of PRMT2 demonstrated an increased stability for the enzyme when bound to compound **37** compared to sinefungin or AdoHcy, indicating a stronger binding affinity. Furthermore, when using RSF1 as a substrate, the IC_{50} values for AdoHcy and inhibitor compound **37** could be determined. Compound **37** was found to inhibit PRMT2 with an IC_{50} of 16.3 μM, similar to AdoHcy (18.3 μM). In this regard, **37** is the first synthetic compound described as an inhibitor of PRMT2. However, compound **37** is not selective for PRMT2 as it displays much more potent inhibition of PRMT4.

4.3 Biological Relevance of Inhibitors and Current Outlook

PRMT2 is an elusive target with potential roles in a wide range of diseases, including obesity [102], diabetes [103] and cancer [104, 105]. The structural information now available combined with the discovery of an efficient nonhistone substrate for PRMT2 should provide the tools needed for future inhibitor development.

5 PRMT3

5.1 Background

PRMT3 was identified by Herschman and co-workers in 1998 based on sequence similarity with PRMT1 [11]. The crystal structure of the core of PRMT3, resolved shortly thereafter, revealed the highly conserved active site residues and its dimeric nature [107]. PRMT3 is a type I PRMT, predominantly present in the cytoplasm. PRMT3 contains a zinc finger for substrate recognition with known substrates including the 40S ribosomal protein S2, tumour protein p53 and (at least in vitro) histone H4R3 [108]. A recent screening via bio-orthogonal profiling of protein methylation using engineered methyltransferases resulted in the identification of over 80 substrates of PRMT3, 70% of which were cytoplasmic [109]. However, their functions in a biological setting remain to be elucidated. The human tumour suppressor protein DAL-1/4.1B (differentially expressed in adenocarcinoma of the lung) has also been shown to interact with PRMT3, inhibiting its methylating activity and leading to apoptosis in breast cancer cells [110].

5.2 Inhibitors: In Vitro and Cell-Based Activities

In 2012, the groups of Vedadi and Schapira published the crystal structure of PRMT3 bound to an allosteric inhibitor, identified through virtual screening [111]. The initial hit compound (**38**, Fig. 9) was reported to inhibit PRMT3 with an IC_{50} of 2.5 μM. As revealed in the co-crystal structure in Fig. 9, it binds close to the dimerization arm, preventing the enzyme to form a catalytically active state. The inhibitor was optimized through extensive SAR studies to yield compound SGC707

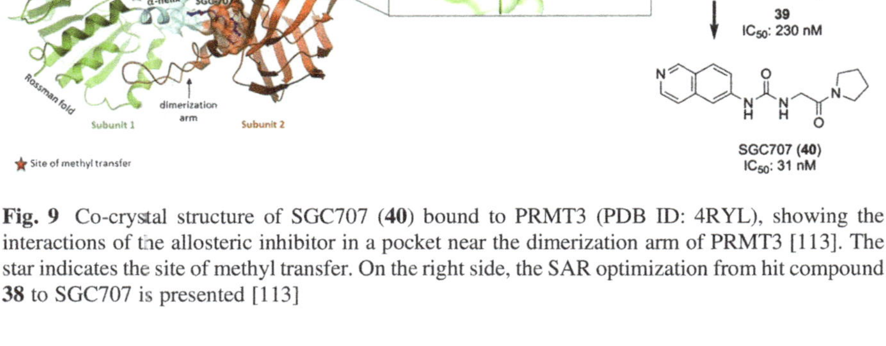

Fig. 9 Co-crystal structure of SGC707 (**40**) bound to PRMT3 (PDB ID: 4RYL), showing the interactions of the allosteric inhibitor in a pocket near the dimerization arm of PRMT3 [113]. The star indicates the site of methyl transfer. On the right side, the SAR optimization from hit compound **38** to SGC707 is presented [113]

(**40**, Fig. 9), a potent allosteric inhibitor of PRMT3 with an IC_{50} of 31 nM and high selectivity for PRMT3 over almost 300 other methyltransferase enzymes, including PRMT1 and PRMT4–PRMT8 [112, 113]. SGC707 is cell-active and non-toxic, except for high concentrations and long exposures. Initial PK data in mice show it can be used in animal models. No disease relevant studies have yet been reported.

5.3 Biological Relevance of Inhibitors and Current Outlook

The nanomolar activity of SGC707 combined with its high selectivity provides a strong tool for probing the biological role(s) of PRMT3. While links with human disease are currently limited, its interaction with tumour suppressor DAL-1/4.1B suggests PRMT3 as a potential target in breast cancer.

6 CARM1 (PRMT4)

6.1 Background

First identified as *c*oactivator-*a*ssociated *a*rginine *m*ethyltransferase [12], CARM1 – or PRMT4 – was the fourth PRMT to be found through its sequence homology with PRMT1, PRMT2 and PRMT3. Unlike the other PRMTs, CARM1 doesn't recognize the GAR motif but prefers proline-, glycine- and methionine-rich (PGM) motifs [114, 115]. CARM1 asymmetrically dimethylates histone residues H3R17, H3R26 and H3R42 [116–118] and a large number of nonhistone substrates, including splicing factors, RNA-binding proteins, transcription factors, coactivators and itself [114, 119, 120].

In a well-characterized example of crosstalk between post-translational modifications, the methylation of Arg17 in the histone H3-tail peptide is regulated by the acetylation state of the neighbouring Lys18. Specifically, acetylation of Lys18 makes the H3 tail a better substrate for CARM1 [121]. Automethylation and O-GlcNAcylation of CARM1 is also known to regulate substrate specificity [122]. Upregulation of CARM1 is associated with a variety of diseases, including breast [43, 123], colon [124, 125], prostate [124, 126] and liver [127] cancers. Therefore, CARM1 has received increasing attention as a therapeutic target.

6.2 Inhibitors: In Vitro and Cell-Based Activities

Through a series of high-throughput screens and hit-to-lead SAR optimizations, different pyrazole (**40–42**, **47** and CMPD-2 (**43**)), benzo[d]imidazole (**48**) and indole-type (CMPD-1 (**50**) and **53**) inhibitors were developed with IC_{50} values as

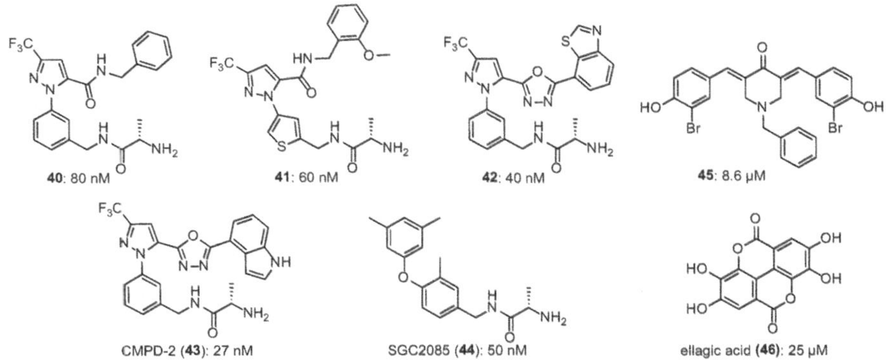

Fig. 10 CARM1 inhibitors **40–44** containing the alanine-amide moiety as a mimic of the guanidine of the arginine substrate. Compound **45** is a curcumin analogue, and ellagic acid (**46**) is a natural product extracted from pomegranates

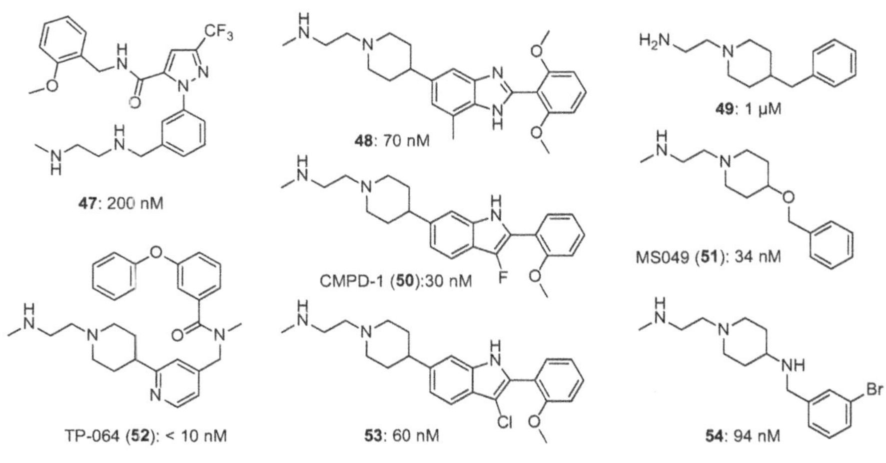

Fig. 11 CARM1 inhibitors **47–54** containing the ethylenediamino moiety mimicking the guanidine of the arginine substrate

low as 27 nM against CARM1 (Figs. 10 and 11) [128–132]. These inhibitors were found to be selective for CARM1 over PRMT1 and PRMT3 (other PRMTs were not tested). The most active compounds (CMPD-1 and CMPD-2) were co-crystallized with CARM1 in the presence of AdoHcy or sinefungin showing that they compete for the substrate-binding site (Fig. 12) [133].

Based upon the structural features present in the potent CARM1 inhibitors described above, it was recognized that the ethylenediamino (Fig. 10) and alanine-amide moieties (Fig. 11) are good mimics of the guanidine moiety which interacts with the double E-loop in the active site of PRMTs. Using a fragment-based approach, initially aimed at developing inhibitors of PRMT6, the ethylenediamino-containing

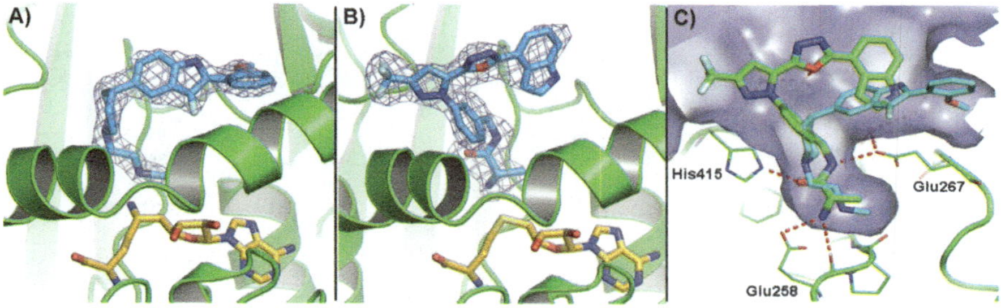

Fig. 12 Co-crystal structures of CARM1 bound to CMPD-1 (**50**) and CMPD-2 (**43**) (PDB ID: 2Y1W and 2Y1X). (**a**) CMPD-1 bound to CARM1 in the presence of sinefungin, (**b**) CMPD-2 bound to CARM1 in the presence of AdoHcy, (**c**) overlay of CMPD-1 and CMPD-2 showing their interactions with the glutamate residues of the double E-loop (Glu258 and Glu267) and the histidine residue of the THW-loop (His415). In CMPD-1, the ethylenediamino group is mimicking the guanidine, whereas in CMPD-2 the guanidine is mimicked by the alanine-amide moiety [133]

fragment **49** (Fig. 11) was identified [134]. This fragment exhibited the low micromolar IC_{50} values against type I PRMTs, PRMT1 (12 μM), PRMT3 (19 μM), CARM1 (1 μM), PRMT6 (0.3 μM) and PRMT8 (2.1 μM), and was not active against PRMT5 and PRMT7. In the same study, compound **53** (Fig. 11), a chlorine analogue of the previously described CMPD-1, was also found to be equally active against CARM1 (60 nM) and PRMT6 (70 nM), while IC_{50} values against PRMT1, PRMT3 and PRMT8 ranged from 1.7 to 4.1 μM. Further optimization of fragment **49** led to compound MS049 (**51**) [135], a dual inhibitor of CARM1 and PRMT6 with activities similar to compound **53**, CARM1; $IC_5$34 nM, PRMT6; IC_{50} 43 nM, PRMT8; and IC_{50} 1.6 μM, while no activity was detected against PRMT1, PRMT3, PRMT5 and PRMT7. Cellular activity of MS049 was detected in HEK293T human embryonic kidney cells where the compound exhibited an IC_{50} value of 0.97 μM for the PRMT6-mediated methylation of H3R2 and 1.4 μM for the methylation of Med12 by CARM1 [135].

Additional SAR studies, starting from fragment **49**, revealed that substitutions on the aromatic ring were generally well tolerated [136]. This led to the identification of the structurally similar but more specific CARM1 inhibitor compound **54** (Fig. 11) with an IC_{50} of 94 nM and 23-fold selectivity over PRMT6 (2.2 μM). However, no cellular assays were reported with this compound. A more complex analogue of MS049 is TP-064 (**52**, Fig. 11) discovered in a collaboration between Takeda Pharmaceutical and the Structural Genomics Consortium (SGC) [137]. TP-064 has an IC_{50} of <10 nM, a 100-fold selectivity over other histone methyltransferases and potent cellular activity with an IC_{50} of 43 nM for CARM1/Med12-Rme2a and growth inhibition observed in multiple myeloma cell lines. A similar hit-to-lead approach was undertaken by the group of Shapira and Vedadi focusing on the alanine-amide moiety as a guanidine mimic [138]. This SAR study yielded compound SGC2085 (**44**) with an IC_{50} of 50 nM against CARM1 and 100-fold selectivity over PRMT6 (5.2 μM). However, the compound did not show any appreciable cellular activity, presumably due to poor cell permeability.

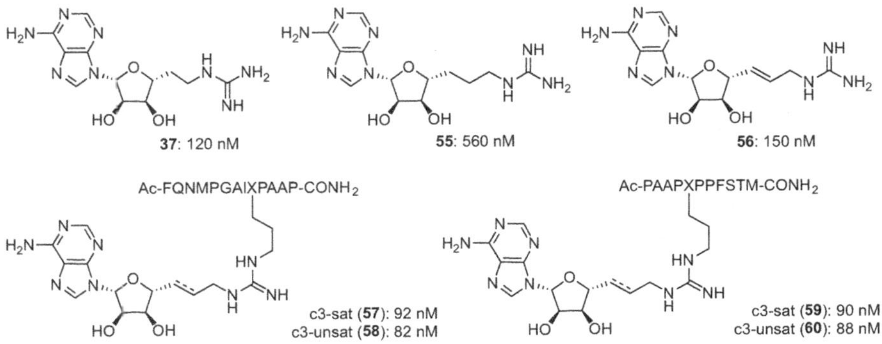

Fig. 13 Bisubstrate inhibitors **37** and **55–60** mimicking parts of the transition state of the methylation reaction

Curcumin-like structures (**45**, Fig. 10) have also been found to act as moderate and somewhat selective CARM1 inhibitors with IC_{50} values in the low micromolar range and no inhibition of PRMT1 or the lysine methyltransferase SET7 [139]. PRMT3 and PRMT6 were inhibited to some extent when **45** was tested at 100 μM. In human LNCaP cells, the prostate-specific antigen promoter was decreased in a dose-dependent manner by compound **45**. Pomegranate-derived compound ellagic acid (**46**, Fig. 10) was found to be a site-specific inhibitor of CARM1, inhibiting methylation of H3R17 but not of H3R26 [140]. Modelling suggests ellagic acid binds the KAPRK motif present around H3R17. Treatment of DNA-damaged HEK293T cells with ellagic acid showed a significant decrease in H3R17 methylation and p21 expression.

Recently, we described the development of inhibitors designed to mimic the transition state of the CARM1 methylation reaction [106]. To do so, bisubstrate-based inhibitors **37**, **55** and **56** (Fig. 13) were prepared by linking an adenosine moiety (mimicking that of AdoMet) to a guanidine moiety via different spacers. Although these compounds were initially expected to be nonspecific PRMT inhibitors, surprising selectivity was found. While three carbon-spaced inhibitor **55** was found to be equally active towards PRMT1, PRMT4 and PRMT6 (IC_{50} 0.56–1.30 μM), the two carbon-spaced inhibitor **37** and the unsaturated three carbon-spaced inhibitor **56** showed 34- to 169-fold selectivity for CARM1. All compounds were inactive against lysine methyltransferase G9a.

In order to increase the specificity of the bisubstrate compounds, a peptidic fragment of the poly(A)-binding protein-1 (PABP1), a well-known substrate of CARM1 [114], was appended to the guanidine group [141]. These peptidic transition state mimics (**57–60**, Fig. 13) showed potent inhibition of CARM1 (IC_{50} 82–92 nM) with high selectivity over PRMT1 (no other PRMTs were tested). Importantly, during co-crystallization studies, it was found that these transition state mimics stabilize the enzyme-substrate complex, thereby greatly facilitating crystallization. Figure 14 shows the complex formed between transition state mimics

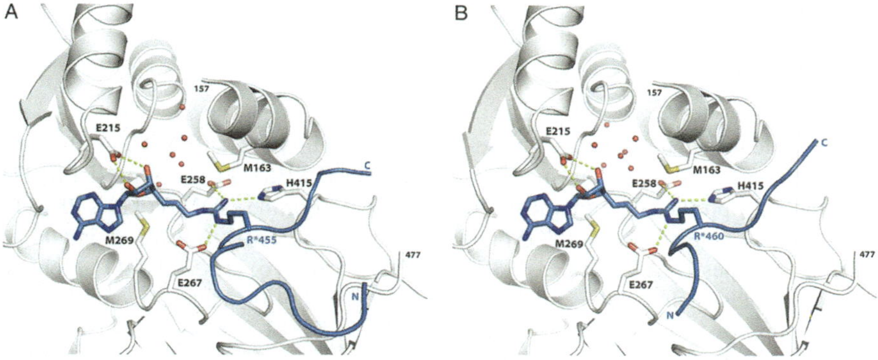

Fig. 14 Co-crystal structures of transition state mimics **57** and **59** bound to CARM1 (PDB ID: 5LGP and 5LGQ). (**a**) Compound **57** consists of residues 447–459 of poly(A)-binding protein-1 (PABP1) linked to adenosine via a fully saturated three-carbon linker; (**b**) compound **59** consists of residues 456–466 of PABP1 linked to adenosine via a fully saturated three-carbon linker. The most important interactions with active site residues have been indicated, including E215 interacting with the hydroxyls of the ribose and E258/E267 of the double E-loop and His415 of the THW-loop interacting with the guanidine [141]

57 and **59** and CARM1, clearly revealing the expected active site interactions with the adenosine and guanidine groups.

6.3 Biological Relevance of Inhibitors and Current Outlook

The involvement of CARM1 in a wide range of human cancers has led to increasing interest in the discovery of a potent and selective CARM1 inhibitor. The first inhibitors described for CARM1 came from high-throughput screening campaigns and showed selectivity for CARM1 over PRMT1 and PRMT3 but were not tested against other PRMTs. Results of later studies showed that structurally similar compounds were equally active against CARM1 and PRMT6. Both CARM1 and PRMT6 are involved in a variety of cancers, and synergy between the two has been described in stimulating oestrogen receptor α-dependent transcription. Dual active compounds, like MS049, are great tool compounds to investigate this synergy in detail. TP-064 is the most potent and selective CARM1 inhibitor described to date, showing activity in enzymatic assays and cell proliferation assays. In addition, the potent and selective peptidic transition state mimics developed by our group are valuable tools for structural studies to investigate PRMT-substrate interactions. The structural information made possible by this approach may yield new insights for the discovery of new PRMT inhibitors.

7 PRMT5

7.1 Background

Initially identified as a Janus kinase-binding protein [142], PRMT5 is the first identified and most abundant member of the type II PRMTs. The type II PRMTs preferentially produce symmetrically dimethylated arginine (sDMA) [13]. PRMT5 recognizes both the GAR and PGM motifs in a wide variety of substrates, including histone arginines H2AR3, H4R3, H3R2 and H3R8 (in vivo) [13, 143, 144]. Nonhistone substrates include ribosomal proteins (RPS10) [145], nuclear factor NF-κB, tumour suppressor protein p53, transcription factor E2F-1 [143, 146, 147], tumour suppressor PDCD4 (programmed cell death protein 4) [148] and the MAPK/ERK pathway (mitogen-activated protein kinase/extracellular signal-regulated kinase) [149], among many others [56]. PRMT5 has multiple associations with binding partners of which MEP50 (methylosome protein 50) is known to be necessary for regulating its specificity in methylating H2A and H4 [150, 151]. Other binding partners regulate the activity and substrate specificity of PRMT5 [152].

PRMT5 is upregulated in wide variety of human cancers, including breast [148], colorectal [153], lung [154, 155] and epithelial ovarian cancer [156], lymphomas [157–159] and melanoma [160]. In addition, recent studies have shown PRMT5 to be a unique anticancer target [47–49]. These recent findings suggest that methylthioadenosine (MTA) plays a role in regulating the activity of PRMT5. This finding originated with the discovery that 5-methylthioadenosine phosphorylase (MTAP) is often co-deleted with a commonly deleted tumour suppressor gene, CDKN2A (cyclin-dependent kinase inhibitor 2A), through close chromosomal proximity [161]. This leads to the accumulation of MTA, which in turn inhibits PRMT5 in a surprisingly specific manner. Through this pathway the MTAP/CDKN2A-deleted tumours have a hypomorphic PRMT5 state, making them sensitive towards further inhibition of PRMT5.

7.2 Inhibitors: In Vitro and Cell-Based Activities

The majority of inhibitors developed against PRMT5 have only been described in recent years. A hit-to-lead optimization study was recently published by Epizyme Inc. describing in detail the optimization of compounds identified through HTS [162]. Extensive SAR studies were performed with compound EPZ007345 (**61**, IC_{50} of 326 nM), yielding compound EPZ015666 (**62**, Fig. 15a) [163]. EPZ015666 was found to inhibit PRMT5:MEP50 with an IC_{50} of 22 nM with no activity detected against a panel of 20 other methyltransferases [162]. Furthermore, the compound was found to be substrate-competitive and AdoMet-uncompetitive. The co-crystal structure confirms EPZ015666 binds in the substrate-binding pocket of PRMT5 (see Fig. 15c). Interestingly, binding to the PRMT5:MEP50 complex was observed only

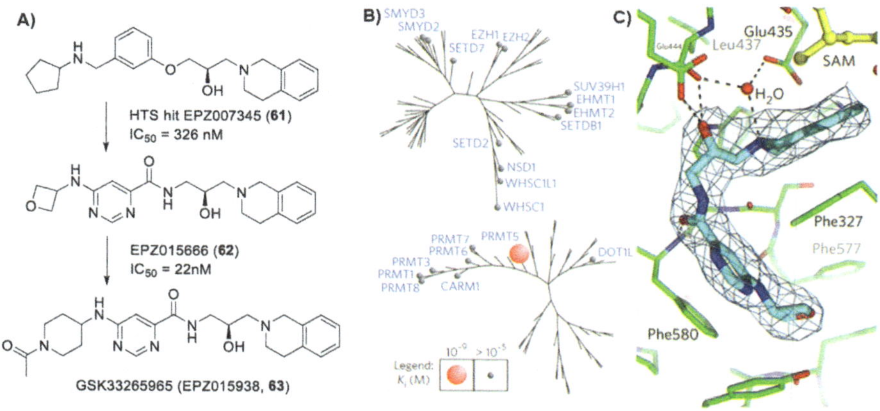

Fig. 15 Overview of the results of the studies performed for compound EPZ015666 (**62**) [163]. (**a**) Hit-to-lead optimization from EPZ007345 (**61**) to GSK3326595 (**63**). GSK3326595 is currently in phase I clinical trials, (**b**) arginine and lysine methyltransferase family trees showing the selectivity of EPZ015666, (**c**) co-crystal structure of EPZ015666 with PRMT5 (PDB ID: 4X61) showing the interactions with the glutamate residues of the "double E-loop" and the substrate-competitive nature of the compound [162]

in the presence of AdoMet or a cofactor analogue like AdoHcy or sinefungin. When testing the compound in mantle cell lymphoma (MCL) cells, EPZ015666 showed concentration-dependent antiproliferative effects, with IC_{50} values of 96 nM and 450 nM against Z-138 and Maver-1 MCL cells, respectively. In addition, upon oral dosing in mice, the compound showed dose-dependent antitumour activity in MCL xenograft models. The correlating decrease in sDMA strongly suggests a direct link with PRMT5 inhibition. Compound EPZ015666 has subsequently been further improved to compound GSK3326595 (previously EPZ015938, **63**, Fig. 15a) in collaboration with GlaxoSmithKline (GSK) and has entered phase I clinical trials with patients that have advanced or recurrent solid tumours and non-Hodgkin's lymphoma.

PRMT5 overexpression was also found in Epstein-Barr virus (EBV)-induced B-cell transformation by the group of Baiocchi and co-workers [38]. The PRMT5 expression was limited to EBV-transformed cells and not found in resting or activated B lymphocytes. From a virtual screening approach, compound **64** (Fig. 16) was identified as a selective inhibitor of PRMT5 over PRMT1, PRMT4 and PRMT7 (tested at a fixed concentration only). Compound **64** was capable of blocking EBV-induced B-cell transformation and survival without affecting the viability of normal B cells. In addition, chromatin immunoprecipitation assays show a decrease of PRMT5 overexpression and its histone marks at H3R8 and H4R3 upon treatment with **64**. No effect on asymmetrically dimethylated arginine at H4R3 was found. Compound **64** was optimized to compound **65** (Fig. 16), by replacing the pyridine ring with an ortho-methoxyphenyl group [164]. Its activity was shown at 10 μM against PRMT5 with no inhibition of PRMT1, PRMT4 and PRMT7. PRMT5 was shown to be upregulated in

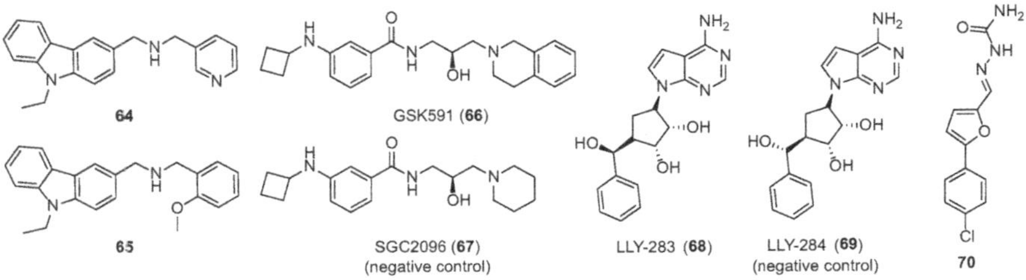

Fig. 16 Chemical structures of PRMT5 inhibitors **64–70**

acute myeloid leukaemia (AML) [164]. When tested against AML samples, compound **65** inhibited H4R3me2s and H3R8me2s methylation while decreasing cell viability in a dose-dependent manner with IC_{50} values of 7.2–21.5 μM for AML cell lines and 4.0–8.7 μM for AML patient blasts.

The SGC has also recently reported two chemical probes for PRMT5. One of the probes (GSK591 [165], **66**, Fig. 16) derives from a compound series explored by Epizyme and GSK, and the other was developed in collaboration with Eli Lilly (LLY-283 [166], **68**, Fig. 16). Probe compound GSK591 inhibits PRMT5:MEP50 methylation of histone H4 in vitro with an IC_{50} of 11 nM and in Z-138 lymphoma cells; it also inhibits the methylation of PRMT5 substrate Sm protein D3 with EC_{50} of 56 nM [167]. Compound LLY-283 has an IC_{50} of 20 nM in vitro (against H4R3 methylation), and in cellular assays, LLY-283 inhibited the methylation of RNA-associated Sm proteins B/B′ with an IC_{50} of 25 nM in MCF7 cells and also affected MDM4 (mouse double minute 4 protein) splicing with a relative IC_{50} of 40 nM in A375 cells [166].

A virtual screening campaign and subsequent SAR studies performed by Ji et al. [168] led to the discovery of a new PRMT5 inhibitor (**70**, Fig. 16), which showed an IC_{50} value of 0.57 μM with selectivity for PRMT5 over all other PRMTs tested in biochemical assays (all but PRMT2 and PRMT9). In DLD-1 colorectal cancer cells, a time- and dose-dependent growth inhibition was demonstrated using a cell viability assay. In addition, a decrease in sDMA marks on H4R3 and H3R8 was observed, but no change in aDMA on H4R3. No cellular IC_{50} or EC_{50} values were calculated.

7.3 Biological Relevance of Inhibitors and Current Outlook

Recently, there has been a significant increase in the number of published reports aimed at identifying new roles of PRMT5 in different disease states as well as the development of inhibitors against PRMT5. With the first clinical trial for a PRMT5 inhibitor against non-Hodgkin's lymphoma currently underway, the biological relevance and therapeutic potential of PRMT5 inhibition will become clearer. With a growing body of knowledge highlighting the involvement of PRMT5 in different cancers and the steady increase in the development of novel PRMT5 inhibitors, PRMT5 is likely the most interesting therapeutic target among the PRMTs at this time.

8 PRMT6

8.1 Background

PRMT6 is a nuclear type I PRMT that methylates histone residues H2AR3, H2AR29, H3R2, H3R42 and H4R3 [14, 169, 170]. PRMT6-mediated aDMA methylation of H3R2 blocks the mixed lineage leukaemia (MLL) complex-mediated di- and tri-methylation of H3K4 and vice versa [171]. Other substrates of PRMT6 include HMGA1a, involved in chromatin structure organization [172] and DNA polymerase β, involved in DNA base excision repair [173]. Furthermore, automethylation increases the stability and anti-HIV-1 activity of PRMT6, and methylation of HIV-Tat protein reduces HIV-1 production and viral replication [37, 174, 175]. PRMT6 has also been found to be overexpressed in a variety of cancers, including bladder and lung cancer [63] and prostate cancer [176], but is downregulated in melanoma [177]. In addition, PRMT6 dysregulation was also recently found to be associated in pulmonary disorders [35].

8.2 Inhibitors: In Vitro and Cell-Based Activities

The ethylenediamino compounds discussed in the section above on CARM1/PRMT4 generally demonstrated similar potency towards both CARM1 and PRMT6. This includes the potent dual CARM1-PRMT6 inhibitor MS049 (**51**, Fig. 11) [135]. The same moiety is also present in compound **71** (Fig. 17), recently developed in our group. Compound **71** showed selective inhibitory activity

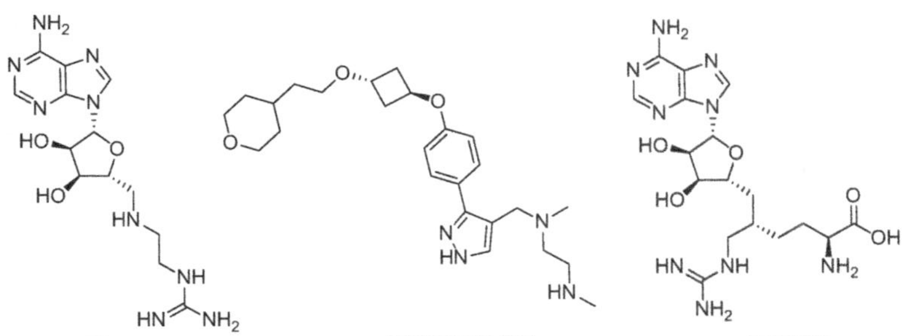

71 EPZ020411 (**72**) GMS (**73**)

Fig. 17 Structures of PRMT6 inhibitors **71–73**

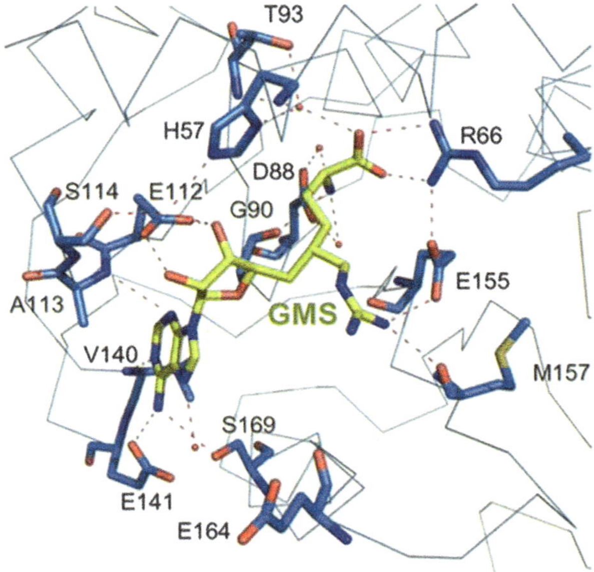

Fig. 18 Co-crystal structure of PRMT6 with bisubstrate compound GMS (**73**), showing the interactions with the active site residues. The adenosine binds with conserved active site residues, such as Arg66, Glu112, Val140 and Glu141 in the AdoMet cofactor-binding pocket, and the guanidine moiety interacts with Glu155 of the "double E-loop" [179]

(IC_{50} 3.2 μM) towards PRMT6 over PRMT1, CARM1 and lysine methyltransferase G9a [106].

Epizyme developed compound EPZ020411 (**72**, Fig. 17), which exhibited an IC_{50} of 10 nM against PRMT6 [178]. It also showed IC_{50} values of 119 nM against PRMT1 and 223 nM against PRMT8 but was more than 100-fold selective over PRMT3, PRMT4, PRMT5 and PRMT7. In fact, most of the ethylenediamino aryl pyrazole compounds tested showed high affinity for PRMT1, PRMT6 and PRMT8. The crystal structure obtained with compound EPZ020411 showed the interactions in the active site of PRMT6, although the structure of the inhibitor was not fully resolved. Treatment of A375 cells with EPZ020411 resulted in a dose-dependent decrease in H3R2 methylation (IC_{50} 0.64 μM).

In a recent study investigating the structural basis of PRMT6-mediated asymmetric dimethylation [179], a bisubstrate guanidine-sinefungin analogue (GMS, **73**, Fig. 17) was synthesized. GMS showed an IC_{50} value of 90 nM for PRMT6 but, not surprisingly, was also active against most other PRMTs. In the co-crystal structure of GMS with PRMT6, the compound binds in the cofactor-binding site with the guanidine moiety interacting with residues in the substrate-binding pocket as depicted in Fig. 18.

8.3 Biological Relevance of Inhibitors and Current Outlook

PRMT6 has been shown to be a valid therapeutic target for a range of diseases. Many compounds developed for PRMT6 (or other PRMTs) show low nanomolar inhibition but often lack PRMT6 selectivity. To date, only limited cellular assays have been performed with the aim of establishing PRMT6 inhibition and its potential role in specific disease relevance. In this regard, more work is necessary in the development of potent, selective, and biologically relevant inhibitors of PRMT6.

9 PRMT7

9.1 Background

In 2004, the first type III PRMT, PRMT7, was identified by Clarke and co-workers [16]. As a type III arginine methyltransferase, PRMT7 produces only monomethylarginine (MMA) and preferentially methylates RxR motifs in lysine- and arginine-rich regions of target proteins [180]. Known substrates for PRMT7 include histones H2AR3, H2BR29, H2BR31 H2BR33, H4R3, H4R17 and H4R19 [180]. In addition, in conjunction with PRMT5, PRMT7 aids in the sDMA methylation of Sm proteins non-redundantly [181]. Interestingly, mutation of the Glu181 residue to Asp in the double E-loop in the PRMT7 active site switched the type III PRMT activity into type I, producing aDMA [182]. Furthermore, the additional mutation of Gln329 to Ala in the canonical THW-loop converted PRMT7 into a type II PRMT, capable of producing sDMA [183].

As related to human disease, PRMT7 has been linked to breast cancer metastasis [184, 185], DNA damage [186] and parasite infection [187]. Interestingly, several studies have also demonstrated the involvement of PRMT7 in increasing the sensitivity of tumour cells to chemotherapeutic agents [186, 188–190].

9.2 Inhibitors: In Vitro and Cell-Based Activities

The group of Vedadi and co-workers developed compound DS-437 (**74**, Fig. 19) [191], an AdoMet analogue, which showed dual inhibition of PRMT5 and PRMT7 (IC_{50} values for both at 6 μM) with no activity against a panel of 29 other methyltransferases, including PRMTs.

Fig. 19 Structure of dual PRMT5-PRMT7-inhibitor DS-437 (**74**)

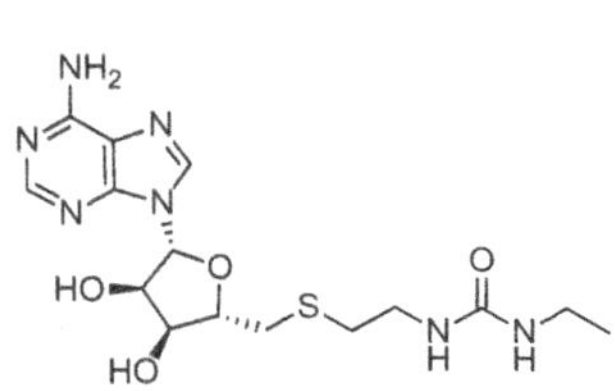

DS-437 (**74**)

9.3 Biological Relevance of Inhibitors and Current Outlook

Very limited work has been done in the development of PRMT7 inhibitors. Currently, no potent and specific PRMT7 inhibitors are available. However, interesting recent findings suggest PRMT7 is a viable therapeutic target, e.g. for (re-)sensitising tumour cells to chemotherapeutic agents or for the treatment of breast cancer.

10 PRMT8

10.1 Background

The eighth member of the PRMT family is a type I PRMT that is primarily expressed in the brain and is myristoylated at the N-terminal glycine, both unique features within the PRMT family [17]. The myristoylation of PRMT8 results in its association with the plasma membrane. Cleavage of the N-terminal domain results in an increase in methylation activity [192], and a variant (PRMT8v2) lacking the N-terminal glycine was found to be located primarily in the nucleus rather than the cell membrane [193, 194]. PRMT8 has high sequence similarity with PRMT1 (about 80%) and methylates the GAR motif.

Recent studies point to an involvement of PRMT8 in amyotrophic lateral sclerosis (ALS) [195] and high expression levels of PRMT8 are linked to a variety of cancers [196]. Furthermore, knockdown of PRMT8 halted cell proliferation and caused cell death in both healthy human dermal fibroblasts and U87MG glioblastoma cells [194].

10.2 Inhibitors: In Vitro and Cell-Based Activities

No specific inhibitors have been developed against PRMT8. Interestingly, ethylenediamino-containing compounds (fragment **49** [134], MS049 (**51**) [135], Fig. 11) that were active against PRMT4 and PRMT6 were also generally active against PRMT8. In addition, the covalent inhibitors **28** and **29** [91] (Fig. 6) designed to interact with the active site cysteine found in PRMT1 were also found to be active against PRMT8. This is not surprising given the high sequence similarity, including the presence of an active site cysteine, between these two PRMTs.

10.3 Biological Relevance of Inhibitors and Current Outlook

Although relatively little is known about the roles of PRMT8 in both healthy and disease states, recent publications suggest its potential as a therapeutic target. In order to study these in greater detail, the development of specific inhibitors against PRMT8 will be necessary.

11 PRMT9

The most recently added member of the PRMT family, PRMT9, took longer to identify due to its low sequence similarity to other PRMTs [17, 18]. Initially described in the literature as Fbox Only Protein 11 (FBXO11), PRMT9 is the second type II PRMT after PRMT5 and is found to localize mainly in the cytoplasm [18, 197]. Recently, spliceosome-associated protein SAP145 was identified as a substrate for PRMT9 [197]. sDMA methylation of SAP145 results in a Tudor domain-binding site for the survival of motor neuron (SMN) protein, thereby regulating alternative splicing.

Recently, the preferred substrate recognition motif of PRMT9 was described as R-F-(K/R/F)-(R/W)-**R**-(M/F)-P-X-P [198], which may facilitate the identification of new PRMT9 substrates. Interestingly, however, the only substrate identified to date (spliceosome-associated protein 145 (SAP145)) does not bear this motif (CFK**R**KYL). Currently, a very limited number of studies focused on PRMT9 have been reported, and much work is needed to elucidate the roles played by PRMT9 in both healthy and disease states.

12 Conclusions

The protein arginine *N*-methyltransferases play crucial roles in a wide range of biological processes associated with both human health and disease. Their functions concentrate on gene regulation, RNA transcription, splicing processes and DNA repair. Not surprisingly, the dysregulation of PRMTs has been shown to contribute to a variety of diseases, including many different cancers. As a result, the development of PRMT inhibitors has also received increasing attention over the past decade.

Comparing different studies focused on inhibiting the same PRMT can prove challenging as different assay methods can result in different IC_{50} values for the same compound. Furthermore, many factors are involved in obtaining a reliable measure of an inhibitor's potency, including substrate and enzyme concentrations, the reference compounds used, signal readout and many other variable assay-specific conditions. In addition, PRMT inhibitor specificity is rarely tested in the context of a large panel of methyltransferases.

Despite these limitations, the overview of the PRMT inhibitors here presented revealed that potent and highly selective compounds have already been developed against PRMT3 (SGC707 (**40**) [113]), CARM1/PRMT4 (TP-064 (**52**) [137]) and PRMT5 (GSK3326595 (**63**) [199], GSK591 (**66**) [167] and LLY-283 (**68**) [166]). In addition, the dual PRMT4/PRMT6 (**51**) [135] and PRMT5/PRMT7 (**74**) [191] inhibitors that have been developed, as well as the more general type I PRMT inhibitor (**6**) [50], serve to strengthen the toolbox of small molecules needed for detailed investigations of PRMTs. Furthermore, the technology recently developed for the co-crystallization of PRMTs with their respective substrates using transition state mimics [141] is expected to yield key new structural insights of value in the discovery of new PRMT inhibitors.

To date, PRMT5 is the first PRMT for which an inhibitor has been taken into phase I clinical trials, illustrating the rapid progress being made in this field. The increasing number of reports describing the involvements of PRMTs in cellular processes and the development of new PRMT inhibitors can be expected to result in a clearer picture of the role(s) played by this important family of enzymes (Table 1).

Table 1 PRMT substrate(s), function(s) in normal and disease-related processes, and effect(s) of inhibition

PRMT	Type	Substrates	Function/disease relation	Effect by inhibition
PRMT1	Type I	GAR substrates Histone H2A Arg3 Histone H4 Arg3 Splicing factors DNA damage proteins RNA-binding proteins Transcription factors Viral proteins Signalling proteins	Involved in gene transcription, nuclear transport, DNA repair and RNA processing Overexpressed in breast, colon, prostate, lung, head and neck, bladder cancer and leukaemia Involved in pulmonary disease, cardiovascular disease, diabetes and cocaine addiction	Inhibition increases cellular MMA and sDMA, and inhibitors show growth inhibition in a variety of cancer cell lines
PRMT2	Type I, III	Histone 4 Splicing factors Transcription factors	Regulates leptin signalling Interacts with splicing factors and nuclear hormone receptors Involved in atherosclerosis and breast cancer	Unknown
PRMT3	Type I	Histone H4 Arg3 Ribosomal proteins Tumour suppressors	Involved in breast cancer	Unknown
CARM1 (PRMT4)	Type I	PGM motif Histone H3 Arg17 Histone H3 Arg26 Histone H3 Arg42 PRMT4	Involved in gene transcription, nuclear transport, DNA repair and RNA processing Overexpressed in breast, colon, prostate and liver cancer	Inhibition results in a decrease in prostate-specific antigen reporter

(continued)

Table 1 (continued)

PRMT	Type	Substrates	Function/disease relation	Effect by inhibition
		(auto) Splicing factors RNA-binding proteins Transcription factors Coactivators		
PRMT5	Type II	Histone H2A Arg3 Histone H3 Arg8 Histone H4 Arg3 Sm proteins Ribosomal proteins Nuclear factors Transcription factors Tumour suppressors	Involved in embryogenesis and gene transcription Binding partners regulate activity and substrate specificity Upregulated in breast, colorectal, lung, epithelial ovarian cancer, lymphomas and melanoma	Antiproliferative in MCL cells and MCL mice. Phase I clinical trials on solid tumours and non-Hodgkin's lymphoma
PRMT6	Type I	Histone H2A Arg3 Histone H2A Arg29 Histone H3 Arg2 Histone H3 Arg8 Histone H3 Arg42 Histone H4 Arg3 Chromatin proteins DNA-binding proteins PRMT6 (auto) Viral proteins	Overexpressed in bladder, lung and prostate cancer Downregulated in melanoma Associated with pulmonary disorders Reduces HIV-1 production and viral replication	Dose-dependent decrease in H3R2 methylation in A375 cells
PRMT7	Type III	RxR motif Histone H2A Arg3 Histone H2B Arg29 Histone H2B Arg31 Histone H2B Arg33 Histone H4 Arg3 Histone H4 Arg17 Histone H4 Arg19 Sm proteins	Linked to breast cancer metastasis, DNA damage and parasite infection Increases tumour sensitivity to chemotherapeutics	Unknown

(continued)

Table 1 (continued)

PRMT	Type	Substrates	Function/disease relation	Effect by inhibition
PRMT8	Type I	GAR substrates Histone H2A Histone 4 Arg3 Tumour proteins RNA-binding proteins PRMT8 (auto)	Essential for cell proliferation Highly expressed in breast, head and neck, glandular, cervical, prostate and thyroid cancer Involved in ALS	Knockdown halted cell proliferation and caused cell death in fibroblasts and glioblastoma cells
PRMT9	Type II	SAP145	Regulates splicing	Unknown

Compliance with Ethical Standards

Funding: The support of Leiden University is kindly acknowledged.

Conflict of Interest: Matthijs van Haren declares that he has no conflict of interest. Nathaniel I. Martin declares that he has no conflict of interest.

Ethical Approval: This article does not contain any studies with human participants or animals performed by any of the authors.

References

1. Paik WK, Kim S (1967) Enzymatic methylation of protein fractions from calf thymus nuclei. Biochem Biophys Res Commun 29:14–20
2. Paik WK, Kim S (1968) Protein methylase I. Purification and properties of the enzyme. J Biol Chem 243:2108–2114
3. Boffa LC et al (1977) Distribution of NG, NG,-dimethylarginine in nuclear protein fractions. Biochem Biophys Res Commun 74:969–976
4. Lee HW et al (1977) S-adenosylmethionine: protein-arginine methyltransferase. Purification and mechanism of the enzyme. Biochemistry 16:78–85
5. Ghosh SK et al (1988) Purification and molecular identification of two protein methylases I from calf brain. Myelin basic protein- and histone-specific enzyme. J Biol Chem 263:19024–19033
6. Najbauer J et al (1993) Peptides with sequences similar to glycine, arginine-rich motifs in proteins interacting with RNA are efficiently recognized by methyltransferase(s) modifying arginine in numerous proteins. J Biol Chem 268:10501–10509
7. Rajpurohit R et al (1994) Enzymatic methylation of recombinant heterogeneous nuclear RNP protein A1. Dual substrate specificity for S-adenosylmethionine:histone-arginine N-methyltransferase. J Biol Chem 269:1075–1082
8. Liu Q, Dreyfuss G (1995) In vivo and in vitro arginine methylation of RNA-binding proteins. Mol Cell Biol 15:2800–2808
9. Lin WJ et al (1996) The mammalian immediate-early TIS21 protein and the leukemia-associated BTG1 protein interact with a protein-arginine N-methyltransferase. J Biol Chem 271:15034–15044
10. Scott HS et al (1998) Identification and characterization of two putative human arginine methyltransferases (HRMT1L1 and HRMT1L2). Genomics 48:330–340

11. Tang J et al (1998) PRMT 3, a type I protein arginine N-methyltransferase that differs from PRMT1 in its oligomerization, subcellular localization, substrate specificity, and regulation. J Biol Chem 273:16935–16945
12. Chen D (1999) Regulation of transcription by a protein methyltransferase. Science 284:2174–2177
13. Branscombe TL et al (2001) PRMT5 (Janus kinase-binding protein 1) catalyzes the formation of symmetric dimethylarginine residues in proteins. J Biol Chem 276:32971–32976
14. Frankel A et al (2002) The novel human protein arginine N-methyltransferase PRMT6 is a nuclear enzyme displaying unique substrate specificity. J Biol Chem 277:3537–3543
15. Lee JH et al (2005) PRMT7, a new protein arginine methyltransferase that synthesizes symmetric dimethylarginine. J Biol Chem 280:3656–3664
16. Miranda TB et al (2004) PRMT7 is a member of the protein arginine methyltransferase family with a distinct substrate specificity. J Biol Chem 279:22902–22907
17. Lee J et al (2005) PRMT8, a new membrane-bound tissue-specific member of the protein arginine methyltransferase family. J Biol Chem 280:32890–32896
18. Cook JR et al (2006) FBXO11/PRMT9, a new protein arginine methyltransferase, symmetrically dimethylates arginine residues. Biochem Biophys Res Commun 342:472–481
19. Thompson PR, Fast W (2006) Histone citrullination by protein arginine deiminase: is arginine methylation a green light or a roadblock? ACS Chem Biol 1:433–441
20. Chang B et al (2007) JMJD6 is a histone arginine demethylase. Science 318:444–447
21. Webby CJ et al (2009) Jmjd6 catalyses lysyl-hydroxylation of U2AF65, a protein associated with RNA splicing. Science 325:90–93
22. Unoki M et al (2013) Lysyl 5-hydroxylation, a novel histone modification, by jumonji domain containing 6 (JMJD6). J Biol Chem 288:6053–6062
23. Wang F et al (2014) JMJD6 promotes Colon carcinogenesis through negative regulation of p53 by hydroxylation. PLoS Biol 12:e1001819
24. Boeckel J-N et al (2011) Jumonji domain-containing protein 6 (Jmjd6) is required for angiogenic sprouting and regulates splicing of VEGF-receptor 1. Proc Natl Acad Sci U S A 108:3276–3281
25. Han G et al (2012) The hydroxylation activity of Jmjd6 is required for its homo-oligomerization. J Cell Biochem 113:1663–1670
26. Böttger A et al (2015) The oxygenase Jmjd6–a case study in conflicting assignments. Biochem J 468:191–202
27. Walport LJ et al (2016) Arginine demethylation is catalysed by a subset of JmjC histone lysine demethylases. Nat Commun 7:11974
28. Uhlmann T et al (2012) A method for large-scale identification of protein arginine methylation. Mol Cell Proteomics 11:1489–1499
29. Sylvestersen KB et al (2014) Proteomic analysis of arginine methylation sites in human cells reveals dynamic regulation during transcriptional arrest. Mol Cell Proteomics 13:2072–2088
30. Larsen SC et al (2016) Proteome-wide analysis of arginine monomethylation reveals widespread occurrence in human cells. Sci Signal 9:rs9
31. Bedford MT, Clarke SG (2009) Protein arginine methylation in mammals: who, what, and why. Mol Cell 33:1–13
32. Yang Y, Bedford MT (2013) Protein arginine methyltransferases and cancer. Nat Rev Cancer 13:37–50
33. Franceschelli S et al (2013) Biological functional relevance of asymmetric dimethylarginine (ADMA) in cardiovascular disease. Int J Mol Sci 14:24412–24421
34. Zakrzewicz D, Eickelberg O (2009) From arginine methylation to ADMA: a novel mechanism with therapeutic potential in chronic lung diseases. BMC Pulm Med 9:5
35. Zakrzewicz D et al (2012) Protein arginine methyltransferases (PRMTs): promising targets for the treatment of pulmonary disorders. Int J Mol Sci 13:12383–12400

36. Jeong S-J et al (2006) Coactivator-associated arginine methyltransferase 1 enhances transcriptional activity of the human T-cell Lymphotropic virus type 1 long terminal repeat through direct interaction with tax. J Virol 80:10036–10044
37. Xie B et al (2007) Arginine methylation of the human immunodeficiency virus type 1 tat protein by PRMT6 negatively affects tat interactions with both cyclin T1 and the tat transactivation region. J Virol 81:4226–4234
38. Alinari L et al (2015) Selective inhibition of protein arginine methyltransferase 5 blocks initiation and maintenance of B-cell transformation. Blood 125:2530–2543
39. Cheng D et al (2004) Small molecule regulators of protein arginine methyltransferases. J Biol Chem 279:23892–23899
40. Peng C, Wong CC (2017) The story of protein arginine methylation: characterization, regulation, and function. Expert Rev Proteomics 14:157–170
41. Kaniskan HÜ et al (2017) Inhibitors of protein methyltransferases and demethylases. Chem Rev 118:989–1068
42. Blanc RS, Richard S (2017) Arginine methylation: the coming of age. Mol Cell 65:8–24
43. Morettin A et al (2015) Arginine methyltransferases as novel therapeutic targets for breast cancer. Mutagenesis 30:177–189
44. Boriack-Sjodin PA, Swinger KK (2016) Protein methyltransferases: a distinct, diverse, and dynamic family of enzymes. Biochemistry 55:1557–1569
45. Schapira M, Ferreira de Freitas R (2014) Structural biology and chemistry of protein arginine methyltransferases. Med Chem Commun 5:1779–1788
46. Scheer S et al (2019) A chemical biology toolbox to study protein methyltransferases and epigenetic signaling. Nat Commun 10:19
47. Kryukov GV et al (2016) MTAP deletion confers enhanced dependency on the PRMT5 arginine methyltransferase in cancer cells. Science 351:1214–1218
48. Marjon K et al (2016) MTAP deletions in cancer create vulnerability to targeting of the MAT2A/PRMT5/RIOK1 axis. Cell Rep 15:574–587
49. Mavrakis KJ et al (2016) Disordered methionine metabolism in MTAP/CDKN2A-deleted cancers leads to dependence on PRMT5. Science 351:1208–1213
50. Eram MS et al (2016) A potent, selective, and cell-active inhibitor of human type I protein arginine methyltransferases. ACS Chem Biol 11:772–781
51. Tang J et al (2000) Protein-arginine methyltransferase I, the predominant protein-arginine methyltransferase in cells, interacts with and is regulated by interleukin enhancer-binding factor 3. J Biol Chem 275:19866–19876
52. Tang J et al (2000) PRMT1 is the predominant type I protein arginine methyltransferase in mammalian cells. J Biol Chem 275:7723–7730
53. Goulet I et al (2007) Alternative splicing yields protein arginine methyltransferase 1 isoforms with distinct activity, substrate specificity, and subcellular localization. J Biol Chem 282:33009–33021
54. Dhar S et al (2013) Loss of the major type I arginine methyltransferase PRMT1 causes substrate scavenging by other PRMTs. Sci Rep 3:1311
55. Wooderchak WL et al (2008) Substrate profiling of PRMT1 reveals amino acid sequences that extend beyond the "RGG" paradigm. Biochemistry 47:9456–9466
56. Wei H et al (2014) Protein arginine methylation of non-histone proteins and its role in diseases. Cell Cycle 13:32–41
57. Baldwin RM et al (2012) Alternatively spliced protein arginine methyltransferase 1 isoform PRMT1v2 promotes the survival and invasiveness of breast cancer cells. Cell Cycle 11:4597–4612
58. Seligson DB et al (2005) Global histone modification patterns predict risk of prostate cancer recurrence. Nature 435:1262–1266
59. Avasarala S et al (2015) PRMT1 is a novel regulator of epithelial-mesenchymal-transition in non-small cell lung cancer. J Biol Chem 290:13479–13489

60. Mathioudaki K et al (2008) The PRMT1 gene expression pattern in colon cancer. Br J Cancer 99:2094–2099
61. Papadokostopoulou A et al (2009) Colon cancer and protein arginine methyltransferase 1 gene expression. Anticancer Res 29:1361–1366
62. Chuang C et al (2017) PRMT1 expression is elevated in head and neck cancer and inhibition of protein arginine methylation by adenosine dialdehyde or PRMT1 knockdown downregulates proliferation and migration of oral cancer cells. Oncol Rep 38:1115–1123
63. Yoshimatsu M et al (2011) Dysregulation of PRMT1 and PRMT6, type I arginine methyltransferases, is involved in various types of human cancers. Int J Cancer 128:562–573
64. Cheung N et al (2007) Protein arginine-methyltransferase-dependent oncogenesis. Nat Cell Biol 9:1208–1215
65. Shia W-J et al (2012) PRMT1 interacts with AML1-ETO to promote its transcriptional activation and progenitor cell proliferative potential. Blood 119:4953–4962
66. Sun Q et al (2015) PRMT1 upregulated by epithelial Proinflammatory cytokines participates in COX2 expression in fibroblasts and chronic antigen-induced pulmonary inflammation. J Immunol 195:298–306
67. Iwasaki H (2009) Impaired PRMT1 activity in the liver and pancreas of type 2 diabetic Goto-Kakizaki rats. Life Sci 85:161–166
68. Li Y et al (2015) Arginine methyltransferase 1 in the nucleus Accumbens regulates behavioral effects of cocaine. J Neurosci 35:12890–12902
69. Ragno R et al (2007) Small molecule inhibitors of histone arginine methyltransferases: homology modeling, molecular docking, binding mode analysis, and biological evaluations. J Med Chem 50:1241–1253
70. Mai A et al (2008) Epigenetic multiple ligands: mixed histone/protein methyltransferase, acetyltransferase, and class III deacetylase (Sirtuin) inhibitors. J Med Chem 51:2279–2290
71. Feng Y et al (2010) Discovery and mechanistic study of a class of protein arginine methylation inhibitors. J Med Chem 53:6028–6039
72. Wang J et al (2012) Pharmacophore-based virtual screening and biological evaluation of small molecule inhibitors for protein arginine methylation. J Med Chem 55:7978–7987
73. Bonham K et al (2010) Effects of a novel arginine methyltransferase inhibitor on T-helper cell cytokine production. FEBS J 277:2096–2108
74. Spannhoff A et al (2007) Target-based approach to inhibitors of histone arginine methyltransferases. J Med Chem 50:2319–2325
75. Bissinger E-M et al (2011) Acyl derivatives of p-aminosulfonamides and dapsone as new inhibitors of the arginine methyltransferase hPRMT1. Bioorg Med Chem 19:3717–3731
76. Spannhoff A et al (2007) A novel arginine methyltransferase inhibitor with cellular activity. Bioorg Med Chem Lett 17:4150–4153
77. Heinke R et al (2009) Virtual screening and biological characterization of novel histone arginine methyltransferase PRMT1 inhibitors. ChemMedChem 4:69–77
78. Xie Y et al (2014) Virtual screening and biological evaluation of novel small molecular inhibitors against protein arginine methyltransferase 1 (PRMT1). Org Biomol Chem 12:9665–9673
79. Dowden J et al (2010) Toward the development of potent and selective bisubstrate inhibitors of protein arginine methyltransferases. Bioorg Med Chem Lett 20:2103–2105
80. Dowden J et al (2011) Small molecule inhibitors that discriminate between protein arginine N-methyltransferases PRMT1 and CARM1. Org Biomol Chem 9:7814
81. Lakowski TM et al (2010) Nη-substituted Arginyl peptide inhibitors of protein arginine N-methyltransferases. ACS Chem Biol 5:1053–1063
82. 't Hart P et al (2011) Peptidic partial bisubstrates as inhibitors of the protein arginine N-methyltransferases. Chembiochem 12:1427–1432
83. 't Hart P et al (2012) Analogues of the HIV-Tat peptide containing Nη-modified arginines as potent inhibitors of protein arginine N-methyltransferases. Med Chem Commun 3:1235–1244

84. Thomas D et al (2014) Protein arginine N-methyltransferase substrate preferences for different Nη-substituted Arginyl peptides. Chembiochem 15:1607–1613
85. Osborne T et al (2008) In situ generation of a Bisubstrate analogue for protein arginine methyltransferase 1. J Am Chem Soc 130:4574–4575
86. Luo Y et al (2006) Inhibitors and inactivators of protein arginine deiminase 4: functional and structural characterization. Biochemistry 45:11727–11736
87. Obianyo O et al (2010) A chloroacetamidine-based inactivator of protein arginine methyltransferase 1: design, synthesis, and in vitro and in vivo evaluation. Chembiochem 11:1219–1223
88. Obianyo O et al (2011) Activity-based protein profiling of protein arginine methyltransferase 1. ACS ChemBio 6:1127–1135
89. Weerapana E et al (2010) Quantitative reactivity profiling predicts functional cysteines in proteomes. Nature 468:790–795
90. Zhang X, Cheng X (2003) Structure of the predominant protein arginine methyltransferase PRMT1 and analysis of its binding to substrate peptides. Structure 11:509–520
91. Dillon MBC et al (2012) Novel inhibitors for PRMT1 discovered by high-throughput screening using activity-based fluorescence polarization. ACS Chem Biol 7:1198–1204
92. Yan L et al (2014) Diamidine compounds for selective inhibition of protein arginine methyltransferase 1. J Med Chem 57:2611–2622
93. Zhang J et al (2017) Discovery of decamidine as a new and potent PRMT1 inhibitor. Med Chem Commun 8:440–444
94. Sinha SH et al (2012) Synthesis and evaluation of carbocyanine dyes as PRMT inhibitors and imaging agents. Eur J Med Chem 54:647–659
95. Hu H et al (2015) Exploration of cyanine compounds as selective inhibitors of protein arginine methyltransferases: synthesis and biological evaluation. J Med Chem 58:1228–1243
96. Yu XR et al (2015) Discovery and structure-activity analysis of 4-((5-nitropyrimidin-4-yl) amino)benzimidamide derivatives as novel protein arginine methyltransferase 1 (PRMT1) inhibitors. Bioorg Med Chem Lett 25:5449–5453
97. Lakowski TM, Frankel A (2009) Kinetic analysis of human protein arginine N-methyltransferase 2: formation of monomethyl- and asymmetric dimethyl-arginine residues on histone H4. Biochem J 421:253–261
98. Cura V et al (2017) Structural studies of protein arginine methyltransferase 2 reveal its interactions with potential substrates and inhibitors. FEBS J 284:77–96
99. Qi C (2002) Identification of protein arginine methyltransferase 2 as a coactivator for estrogen receptor alpha. J Biol Chem 277:28624–28630
100. Meyer R et al (2007) PRMT2, a member of the protein arginine methyltransferase family, is a coactivator of the androgen receptor. J Steroid Biochem Mol Biol 107:1–14
101. Vhuiyan MI et al (2017) PRMT2 interacts with splicing factors and regulates the alternative splicing of BCL-X. J Biochem 162:17–25
102. Iwasaki H et al (2010) Disruption of protein arginine N-methyltransferase 2 regulates leptin signaling and produces leanness in vivo through loss of STAT3 methylation. Circ Res 107:992–1001
103. Hussein MA et al (2015) LXR-mediated ABCA1 expression and function are modulated by high glucose and PRMT2. PLoS One 10:6–8
104. Zhong J et al (2014) Nuclear loss of protein arginine N-methyltransferase 2 in breast carcinoma is associated with tumor grade and overexpression of cyclin D1 protein. Oncogene 33:5546–5558
105. Oh TG et al (2014) PRMT2 and RORγ expression are associated with breast cancer survival outcomes. Mol Endocrinol 28:1166–1185
106. van Haren M et al (2015) Synthesis and evaluation of protein arginine N-methyltransferase inhibitors designed to simultaneously occupy both substrate binding sites. Org Biomol Chem 13:549–560

107. Zhang X et al (2000) Crystal structure of the conserved core of protein arginine methyltransferase PRMT3. EMBO J 19:3509–3519
108. Frankel A, Clarke S (2000) PRMT3 is a distinct member of the protein arginine N-methyltransferase family: conferral of substrate specificity by a zinc-finger domain. J Biol Chem 275:32974–32982
109. Guo H et al (2014) Profiling substrates of protein arginine N-methyltransferase 3 with S-adenosyl-L-methionine analogues. ACS Chem Biol 9:476–484
110. Singh V et al (2004) DAL-1/4.1B tumor suppressor interacts with protein arginine N-methyltransferase 3 (PRMT3) and inhibits its ability to methylate substrates in vitro and in vivo. Oncogene 23:7761–7771
111. Siarheyeva A et al (2012) An allosteric inhibitor of protein arginine methyltransferase 3. Structure 20:1425–1435
112. Liu F et al (2013) Exploiting an allosteric binding site of PRMT3 yields potent and selective inhibitors. J Med Chem 56:2110–2124
113. Kaniskan HÜ et al (2015) A potent, selective and cell-active allosteric inhibitor of protein arginine methyltransferase 3 (PRMT3). Angew Chem Int Ed 54:5166–5170
114. Lee J (2002) PABP1 identified as an arginine methyltransferase substrate using high-density protein arrays. EMBO Rep 3:268–273
115. Cheng D et al (2007) The arginine methyltransferase CARM1 regulates the coupling of transcription and mRNA processing. Mol Cell 25:71–83
116. Schurter BT et al (2001) Methylation of histone H3 by coactivator-associated arginine methyltransferase 1. Biochemistry 40:5747–5756
117. Jacques SL et al (2016) CARM1 preferentially methylates H3R17 over H3R26 through a random kinetic mechanism. Biochemistry 55:1635–1644
118. Casadio F et al (2013) H3R42me2a is a histone modification with positive transcriptional effects. Proc Natl Acad Sci 110:14894–14899
119. Feng Q et al (2006) Signaling within a coactivator complex: methylation of SRC-3/AIB1 is a molecular switch for complex disassembly. Mol Cell Biol 26:7846–7857
120. Kuhn P et al (2011) Automethylation of CARM1 allows coupling of transcription and mRNA splicing. Nucleic Acids Res 39:2717–2726
121. Daujat S et al (2002) Crosstalk between CARM1 methylation and CBP acetylation on histone H3. Curr Biol 12:2090–2097
122. Charoensuksai P et al (2015) O-GlcNAcylation of co-activator-associated arginine methyltransferase 1 regulates its protein substrate specificity. Biochem J 466:587–599
123. Cheng H et al (2013) Overexpression of CARM1 in breast cancer is correlated with poorly characterized clinicopathologic parameters and molecular subtypes. Diagn Pathol 8:129
124. Kim Y-RR et al (2010) Differential CARM1 expression in prostate and colorectal cancers. BMC Cancer 10:197
125. C-YY O et al (2011) A coactivator role of CARM1 in the dysregulation of -catenin activity in colorectal cancer cell growth and gene expression. Mol Cancer Res 9:660–670
126. Hong H et al (2004) Aberrant expression of CARM1, a transcriptional coactivator of androgen receptor, in the development of prostate carcinoma and androgen-independent status. Cancer 101:83–89
127. Osada S et al (2013) Elevated expression of coactivator-associated arginine methyltransferase 1 is associated with early hepatocarcinogenesis. Oncol Rep 30:1669–1674
128. Purandare AV et al (2008) Pyrazole inhibitors of coactivator associated arginine methyltransferase 1 (CARM1). Bioorg Med Chem Lett 18:4438–4441
129. Allan M et al (2009) N-Benzyl-1-heteroaryl-3-(trifluoromethyl)-1H-pyrazole-5-carboxamides as inhibitors of co-activator associated arginine methyltransferase 1 (CARM1). Bioorg Med Chem Lett 19:1218–1223
130. Huynh T et al (2009) Optimization of pyrazole inhibitors of coactivator associated arginine methyltransferase 1 (CARM1). Bioorg Med Chem Lett 19:2924–2927

131. Wan H et al (2009) Benzo[d]imidazole inhibitors of coactivator associated arginine methyltransferase 1 (CARM1)-hit to Lead studies. Bioorg Med Chem Lett 19:5063–5066
132. Therrien E et al (2009) 1,2-diamines as inhibitors of co-activator associated arginine methyltransferase 1 (CARM1). Bioorg Med Chem Lett 19:6725–6732
133. Sack JS et al (2011) Structural basis for CARM1 inhibition by indole and pyrazole inhibitors. Biochem J 436:331–339
134. Ferreira De Freitas R et al (2016) Discovery of a potent class i protein arginine methyltransferase fragment inhibitor. J Med Chem 59:1176–1183
135. Shen Y et al (2016) Discovery of a potent, selective, and cell-active dual inhibitor of protein arginine methyltransferase 4 and protein arginine methyltransferase 6. J Med Chem 59:9124–9139
136. Kaniskan HÜ et al (2016) Design and synthesis of selective, small molecule inhibitors of coactivator-associated arginine methyltransferase 1 (CARM1). Med Chem Commun 7:1793–1796
137. Nakayama K et al (2018) TP-064, a potent and selective small molecule inhibitor of PRMT4 for multiple myeloma. Oncotarget 9:18480–18493
138. Ferreira de Freitas R et al (2016) Discovery of a potent and selective coactivator associated arginine methyltransferase 1 (CARM1) inhibitor by virtual screening. J Med Chem 59:6838–6847
139. Cheng DH et al (2011) Novel 3,5-bis(bromohydroxybenzylidene)piperidin-4-ones as coactivator-associated arginine methyltransferase 1 inhibitors: enzyme selectivity and cellular activity. J Med Chem 54:4928–4932
140. Selvi BR et al (2010) Identification of a novel inhibitor of coactivator-associated arginine methyltransferase 1 (CARM1)-mediated methylation of histone H3 Arg-17. J Biol Chem 285:7143–7152
141. van Haren MJ et al (2017) Transition state mimics are valuable mechanistic probes for structural studies with the arginine methyltransferase CARM1. Proc Natl Acad Sci U S A 114:3625–3630
142. Pollack BP et al (1999) The human homologue of the yeast proteins Skb1 and Hsl7p interacts with Jak kinases and contains protein methyltransferase activity. J Biol Chem 274:31531–31542
143. Wei H et al (2013) PRMT5 dimethylates R30 of the p65 subunit to activate NF- B. Proc Natl Acad Sci U S A 110:13516–13521
144. Migliori V et al (2012) Symmetric dimethylation of H3R2 is a newly identified histone mark that supports euchromatin maintenance. Nat Struct Mol Biol 19:136–144
145. Ren J et al (2010) Methylation of ribosomal protein S10 by protein-arginine methyltransferase 5 regulates ribosome biogenesis. J Biol Chem 285:12695–12705
146. Jansson M et al (2008) Arginine methylation regulates the p53 response. Nat Cell Biol 10:1431–1439
147. Zheng S et al (2013) Arginine methylation-dependent reader-writer interplay governs growth control by E2F-1. Mol Cell 52:37–51
148. Powers MA et al (2011) Protein arginine methyltransferase 5 accelerates tumor growth by arginine methylation of the tumor suppressor programmed cell death 4. Cancer Res 71:5579–5587
149. Andreu-Perez P et al (2011) Protein arginine methyltransferase 5 regulates ERK1/2 signal transduction amplitude and cell fate through CRAF. Sci Signal 4:ra58
150. Antonysamy S et al (2012) Crystal structure of the human PRMT5:MEP50 complex. Proc Natl Acad Sci U S A 109:17960–17965
151. Ho M-C et al (2013) Structure of the arginine methyltransferase PRMT5-MEP50 reveals a mechanism for substrate specificity. PLoS One 8:e57008
152. Morales Y et al (2016) Biochemistry and regulation of the protein arginine methyltransferases (PRMTs). Arch Biochem Biophys 590:138–152

153. Cho E-C et al (2012) Arginine methylation controls growth regulation by E2F-1. EMBO J 31:1785–1797
154. Wei T-YW et al (2012) Protein arginine methyltransferase 5 is a potential oncoprotein that upregulates G1 cyclins/cyclin-dependent kinases and the phosphoinositide 3-kinase/AKT signaling cascade. Cancer Sci 103:1640–1650
155. Győrffy B et al (2013) Online survival analysis software to assess the prognostic value of biomarkers using transcriptomic data in non-small-cell lung cancer. PLoS One 8:e82241
156. Bao X et al (2013) Overexpression of PRMT5 promotes tumor cell growth and is associated with poor disease prognosis in epithelial ovarian cancer. J Histochem Cytochem 61:206–217
157. Pal S et al (2007) Low levels of miR-92b/96 induce PRMT5 translation and H3R8/H4R3 methylation in mantle cell lymphoma. EMBO J 26:3558–3569
158. Wang L et al (2008) Protein arginine methyltransferase 5 suppresses the transcription of the RB family of tumor suppressors in leukemia and lymphoma cells. Mol Cell Biol 28:6262–6277
159. Chung J et al (2013) Protein arginine methyltransferase 5 (PRMT5) inhibition induces lymphoma cell death through reactivation of the retinoblastoma tumor suppressor pathway and Polycomb repressor complex 2 (PRC2) silencing. J Biol Chem 288:35534–35547
160. Nicholas C et al (2013) PRMT5 is upregulated in malignant and metastatic melanoma and regulates expression of MITF and p27Kip1. PLoS One 8:e74710
161. Zhang H et al (1996) Codeletion of the genes for p16INK4, methylthioadenosine phosphorylase, interferon-α1, interferon-β1, and other 9p21 markers in human malignant cell lines. Cancer Genet Cytogenet 86:22–28
162. Chan-Penebre E et al (2015) A selective inhibitor of PRMT5 with in vivo and in vitro potency in MCL models. Nat Chem Biol 11:432–437
163. Duncan KW et al (2016) Structure and property guided design in the identification of PRMT5 tool compound EPZ015666. ACS Med Chem Lett 7:162–166
164. Tarighat SS et al (2016) The dual epigenetic role of PRMT5 in acute myeloid leukemia: gene activation and repression via histone arginine methylation. Leukemia 30:789–799
165. Gerhart SV et al (2018) Activation of the p53-MDM4 regulatory axis defines the anti-tumour response to PRMT5 inhibition through its role in regulating cellular splicing. Sci Rep 8:1–15
166. Bonday ZQ et al (2018) LLY-283, a potent and selective inhibitor of arginine methyltransferase 5, PRMT5, with antitumor activity. ACS Med Chem Lett 9:612–617
167. SGC website: GSK591, a chemical probe for PRMT5. http://www.thesgc.org/chemical-probes/GSK591. Accessed Jan 2019
168. Ji S et al (2016) Discovery of selective protein arginine methyltransferase 5 inhibitors and biological evaluations. Chem Biol Drug Des:585–598
169. Waldmann T et al (2011) Methylation of H2AR29 is a novel repressive PRMT6 target. Epigenetics Chromatin 4:11
170. Hyllus D et al (2007) PRMT6-mediated methylation of R2 in histone H3 antagonizes H3 K4 trimethylation. Genes Dev 21:3369–3380
171. Guccione E et al (2007) Methylation of histone H3R2 by PRMT6 and H3K4 by an MLL complex are mutually exclusive. Nature 449:933–937
172. Sgarra R et al (2006) The AT-hook of the chromatin architectural transcription factor high mobility group A1a is arginine-methylated by protein arginine methyltransferase 6. J Biol Chem 281:3764–3772
173. El-Andaloussi N et al (2006) Arginine methylation regulates DNA polymerase β. Mol Cell 22:51–62
174. Boulanger M-C et al (2005) Methylation of tat by PRMT6 regulates human immunodeficiency virus type 1 gene expression. J Virol 79:124–131
175. Singhroy DN et al (2013) Automethylation of protein arginine methyltransferase 6 (PRMT6) regulates its stability and its anti-HIV-1 activity. Retrovirology 10:73
176. Vieira FQ et al (2014) Deregulated expression of selected histone methylases and demethylases in prostate carcinoma. Endocr Relat Cancer 21:51–61

177. Limm K et al (2013) Deregulation of protein methylation in melanoma. Eur J Cancer 49:1305–1313
178. Mitchell LH et al (2015) Aryl Pyrazoles as potent inhibitors of arginine methyltransferases: identification of the first PRMT6 tool compound. ACS Med Chem Lett 6:655–659
179. Wu H et al (2016) Structural basis of arginine asymmetrical dimethylation by PRMT6. Biochem J 473:3049–3063
180. Feng Y et al (2013) Mammalian protein arginine methyltransferase 7 (PRMT7) specifically targets RXR sites in lysine- and arginine-rich regions. J Biol Chem 288:37010–37025
181. Gonsalvez GB et al (2007) Two distinct arginine methyltransferases are required for biogenesis of Sm-class ribonucleoproteins. J Cell Biol 178:733–740
182. Debler EW et al (2016) A glutamate/aspartate switch controls product specificity in a protein arginine methyltransferase. Proc Natl Acad Sci U S A 113:2068–2073
183. Jain K et al (2016) Protein arginine methyltransferase product specificity is mediated by distinct active-site architectures. J Biol Chem 291:18299–18308
184. Yao R et al (2014) PRMT7 induces epithelial-to-mesenchymal transition and promotes metastasis in breast cancer. Cancer Res 74:5656–5667
185. Baldwin RM et al (2015) Protein arginine methyltransferase 7 promotes breast cancer cell invasion through the induction of MMP9 expression. Oncotarget 6:3013–3032
186. Karkhanis V et al (2012) Protein arginine methyltransferase 7 regulates cellular response to DNA damage by methylating promoter histones H2A and H4 of the polymerase δ catalytic subunit gene, POLD1. J Biol Chem 287:29801–29814
187. Ferreira TR et al (2014) Altered expression of an RBP-associated arginine methyltransferase 7 in Leishmania major affects parasite infection. Mol Microbiol 94:1085–1102
188. Gros L et al (2003) Identification of new drug sensitivity genes using genetic suppressor elements: protein arginine N-methyltransferase mediates cell sensitivity to DNA-damaging agents. Cancer Res 63:164–171
189. Gros L et al (2006) Characterization of prmt7α and β isozymes from Chinese hamster cells sensitive and resistant to topoisomerase II inhibitors. Biochim Biophys Acta 1760:1646–1656
190. Verbiest V et al (2008) Protein arginine (N)-methyl transferase 7 (PRMT7) as a potential target for the sensitization of tumor cells to camptothecins. FEBS Lett 582:1483–1489
191. Smil D et al (2015) Discovery of a dual PRMT5-PRMT7 inhibitor. ACS Med Chem Lett 6:408–412
192. Sayegh J et al (2007) Regulation of protein arginine methyltransferase 8 (PRMT8) activity by its N-terminal domain. J Biol Chem 282:36444–36453
193. Kousaka A et al (2009) The distribution and characterization of endogenous protein arginine N-methyltransferase 8 in mouse CNS. Neuroscience 163:1146–1157
194. Hernandez S, Dominko T (2016) Novel protein arginine methyltransferase 8 isoform is essential for cell proliferation. J Cell Biochem 117:2056–2066
195. Scaramuzzino C et al (2013) Protein arginine methyltransferase 1 and 8 interact with FUS to modify its sub-cellular distribution and toxicity in vitro and in vivo. PLoS One 8:e61576
196. Hernandez SJ et al (2017) PRMT8 demonstrates variant-specific expression in cancer cells and correlates with patient survival in breast, ovarian and gastric cancer. Oncol Lett 13:1983–1989
197. Yang Y et al (2015) PRMT9 is a type II methyltransferase that methylates the splicing factor SAP145. Nat Commun 6:6428
198. Gayatri S et al (2016) Using oriented peptide array libraries to evaluate methylarginine-specific antibodies and arginine methyltransferase substrate motifs. Sci Rep 6:28718
199. ClinicalTrials.gov ID: NCT02783300. A phase I, open-label, dose escalation study to investigate the safety, pharmacokinetics, pharmacodynamics and clinical activity of GSK3326595 in subjects with solid tumors and non-Hodgkin's Lymp. Accessed Jan 2019

Top Med Chem (2020) 33: 197–220
DOI: 10.1007/7355_2019_74

Published online: 15 January 2020

Lysine-Specific Histone Demethylases 1/2 (LSD1/2) and Their Inhibitors

Takayoshi Suzuki

Contents

1 Introduction ... 198
2 Biology of Lysine-Specific Histone Demethylases 1/2 (LSD1/2) ... 199
3 Structural Studies and Catalytic Mechanism of LSD1/2 ... 200
4 Link of LSD1/2 to Diseases ... 200
4.1 Link of LSD1/2 to Cancer ... 201
4.2 Link of LSD1 to Viral Infection ... 201
4.3 Link of LSD1 to Globin Disorders ... 202
4.4 Link of LSD1 to Metabolic Diseases ... 202
4.5 Link of LSD1 to Central Nervous System (CNS) Disorders ... 203
5 LSD1 Inhibitors and Their Biological/Therapeutic Applications ... 203
6 Summary ... 214
References ... 214

Abstract Histone lysine methylation, one of the epigenetic mechanisms, plays a pivotal role in various biological events, including cell cycle progression, immune response, and signal transduction. Histone methylation is closely associated with the oncogenesis and proliferation of cancer cells, and its alteration has been identified in many cancer cells. In addition, histone methylation is involved in such non-cancerous diseases as globin disorders and neurological disorders. Several enzymes that control histone methylation have been identified, including lysine-specific histone demethylases 1/2 (LSD1/2). As LSD1/2 are involved in various diseases, their inhibitors are considered useful not only as a chemical tool for probing the biology of LSD1/2 but also as therapeutic agents. In this chapter, the biology, pharmacology, and inhibitors of LSD1/2 are presented, and the potential of LSD inhibitors as therapeutic agents is discussed.

T. Suzuki (✉)
The Institute of Scientific and Industrial Research, Osaka University, Ibaraki, Osaka, Japan
e-mail: tkyssuzuki@sanken.osaka-u.ac.jp

Keywords Demethylase, Disease, Epigenetics, Histone, Inhibitor, Lysine, Methylation

Abbreviations

ADC	Antibody-drug conjugates
AML	Acute myelogenous leukemia
AR	Androgen receptor
ATRA	All-*trans*-retinoic acid
CNS	Central nervous system
DDS	Drug delivery systems
ERα	Estrogen receptor α
FAD	Flavin adenine dinucleotide
GSC	Glioma stem cells
HCF-1	Host cell factor-1
HSV	Herpes simplex virus
KDM	Lysine demethylase
KMT	Lysine methyltransferase
LSD	Lysine-specific histone demethylase
MAO	Monoamine oxidase
MDS	Myelodysplastic syndrome
NSCLC	Non-small cell lung cancer
PCPA	Phenylcyclopropylamine
PDC	PCPA-drug conjugate
SCLC	Small cell lung carcinoma
siRNA	Small interfering RNA
VZV	Varicella zoster virus

1 Introduction

Histone lysine methylation is one of the epigenetic mechanisms that regulate the expression of genes independently of the changes in DNA sequence. The methylation of histone (H) lysine (K) residues occurs at H1K26, H3K4, H3K9, H3K27, H3K36, H3K79, and H4K20 and is responsible for transcriptional activation as well as silencing [1, 2]. In addition, the ε-amino group of the lysine residues can undergo mono-, di-, or trimethylation, and this differential methylation gives functional diversity to each lysine methylation site. For example, the dimethylation of H3K4 occurs in both inactive and active genes, whereas the trimethylation is exclusive to active genes [3]. Similarly, the monomethylation of H3K9 is seen in active genes, whereas the trimethylation of H3K9 is associated with gene repression [4].

Histone lysine methylation is reversibly controlled by two kinds of enzymes, lysine methyltransferases (KMTs) and lysine demethylases (KDMs) [5]. KMTs add a methyl group to histone lysine residues, whereas KDMs remove the methyl group from methylated histone lysine residues, discriminating the methylated positions and states.

Lysine-specific histone demethylases 1/2 (LSD1/2) (KDM1A/B) are KDMs and are associated with several diseases, such as cancer and neurological disorders. Therefore, small-molecule inhibitors of LSD are of interest as potential therapeutic agents. In this chapter, the biology and pharmacology of LSD and hitherto reported LSD inhibitors are presented, and their potential as therapeutic agents is discussed.

2 Biology of Lysine-Specific Histone Demethylases 1/2 (LSD1/2)

Histone methylation had been regarded as an irreversible modification because of the high thermodynamic stability of the N–C bond. Indeed, whereas a number of KMTs had been identified by 2003 [1], no KDMs had been identified. However, in 2004, LSD1 was the first histone demethylase to be identified [6].

LSD1 removes the methyl groups from mono- and dimethylated Lys4 of histone H3 (H3K4me1/2) through flavin adenine dinucleotide (FAD)-dependent enzymatic oxidation (Fig. 1) [6]. In prostate cell lines, LSD1 also demethylates H3K9me1/2 and regulates androgen receptor (AR)-mediated transcription [7]. The targets of LSD1 regulatory demethylation are not limited to histone H3; LSD1 also demethylates nonhistone proteins, such as p53 [8], DNA methyltransferase 1 [9], STAT3 [10], E2F1 [11], and MYPT1 [12], and regulates their cellular functions.

LSD2 (KDM1B), the other flavin-dependent lysine demethylase, was found in 2009 and exhibits the same H3K4 demethylase activity as LSD1 [13]. However, the function and role of LSD2 are likely to be different from those of LSD1, although they remain unclear so far. It has been reported that LSD2 establishes maternal genomic

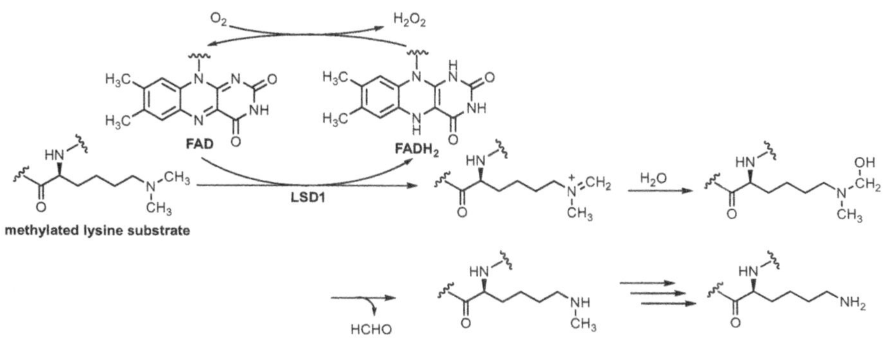

Fig. 1 Catalytic mechanism for the demethylation of methylated lysine substrate by LSD1

imprints, activates gene expression via H3K4 methylation [14], and possesses a demethylase-activity-independent repression function [15]. Recently, it has been reported that LSD2 possesses unexpected E3 ubiquitin ligase activity and inhibits lung cancer cell growth by promoting the ubiquitylation and degradation of *O*-linked *N*-acetylglucosamine transferase [16].

3 Structural Studies and Catalytic Mechanism of LSD1/2

The X-ray crystal structure of LSD1 complexed with CoREST and a histone H3 peptide was determined by Yang et al. [17]. This crystal structure was able to shed light on how histone H3 is recognized. The structural data revealed that histone H3 adopts three consecutive γ-turns, establishing a side chain spacing that places its N terminus in an anionic pocket comprised of Asn, Trp, and two Asp residues. The structural data also confirmed the positioning of the lysine methyl groups in sufficient proximity to FAD for FAD-mediated catalysis.

The crystal structures of LSD1 and the detailed analysis of the catalytic mechanism have led to a solid understanding of the catalytic mechanism for the demethylation of methylated lysine substrates by LSD1 (Fig. 1) [6, 17–19]. First, the methylated lysine substrate is converted into an iminium cation, presumably through a two single-electron oxidation reactions of the amine by FAD. Next, the addition of a water molecule to the iminium cation and the subsequent deformylation afford demethylated lysine. The $FADH_2$ generated in the first step is oxidized by molecular oxygen to FAD, which is utilized again for lysine demethylation. As would be expected from the mechanism, the demethylation by LSD1 is limited to mono- or dimethylated lysine; LSD1 cannot demethylate trimethylated lysine. This proposed catalytic mechanism for the demethylation of methylated lysine substrates provides a basis for the design of selective LSD1 inhibitors.

For the structural study of LSD2, Fang et al. characterized NPAC protein (also known as GLYR1) as an LSD2-specific cofactor that facilitates LSD2-mediated H3K4me1 and H3K4me2 demethylation [20]. They also determined the crystal structures of LSD2 alone and LSD2 in complex with the NPAC protein in the absence and presence of a histone H3 peptide. The structures revealed that the NPAC protein stabilizes the interaction between LSD2 and the histone H3 peptide, thus enhancing the enzymatic activity of LSD2 [21].

4 Link of LSD1/2 to Diseases

Whereas LSD1 is involved in many normal biological events, such as organogenesis [22, 23] and adipocyte differentiation [24], it is associated with several disease states as well, including cancer, viral infection, globin disorders, metabolic syndromes, and neurological disorders. In this section, the links of LSD1/2 to diseases are presented.

4.1 Link of LSD1/2 to Cancer

LSD1 is overexpressed in various cancer cells and tissues, such as neuroblastoma [25], prostate cancer [7, 26], breast cancer [27–29], lung cancer [30], and bladder cancer cells [30]. Furthermore, the results of RNAi-mediated knockdown or LSD1 inhibition suggest that this enzyme is associated with cancer cell growth by modulating pro-survival gene expression and p53 transcriptional activity [25, 27, 31].

Schulte et al. reported that LSD1 expression is correlated with adverse outcome in neuroblastic tumors [25]. The RNAi-mediated knockdown of LSD1 suppresses cell growth, and LSD1 inhibition results in growth inhibition of neuroblastoma cells in both in vitro and in vivo assays. In prostate cancer cells, AR binds an enhancer in the AR second intron and represses AR gene expression through LSD1 recruitment and H3K4me1/2 demethylation [32]. It was also reported that LSD1 is involved in breast cancer proliferation in estrogen receptor α (ERα)-dependent and ERα-independent manners [33]. Whereas LSD1 interacts with ERα in ERα-positive breast cancer cells [34], it is highly expressed also in ER-negative breast cancer cells, and the pharmacological LSD1 inhibition or knockdown of LSD1 using small interfering RNA (siRNA) results in the growth inhibition of ER-negative breast cancer cells and induces the regulation of several proliferation-associated genes, such as p21, ERBB2, and CCNA2 [28]. LSD1 expression is higher also in lung cancer tissue than normal lung tissue [35]. The overexpression of LSD1 protein is associated with shorter overall survival in non-small cell lung cancer (NSCLC) patients, and the interruption of LSD1 using siRNA or chemical LSD1 inhibition suppresses the proliferation, migration, and invasion of NSCLC A549, H460, and 293T cells. It has been reported that LSD1 is overexpressed in leukemia cells and is involved in leukemia cell proliferation and differentiation [36]. Importantly, LSD1 is closely associated with the growth of cancer cells with pluripotent stem cell properties expressing Oct4 and SOX2 [37–39].

It has also been reported recently that LSD2 is involved in breast cancer progression [40]. LSD2 protein level is significantly elevated in malignant breast cancer cell lines compared with normal breast epithelial cell lines. Whereas the overexpression of LSD2 in MDA-MB-231 cells significantly promotes cell proliferation, the siRNA-mediated knockdown of endogenous LSD2 inhibits the growth of multiple breast cancer cell lines, suggesting the critical role of LSD2 in the regulation of breast cancer progression.

4.2 Link of LSD1 to Viral Infection

LSD1 regulates viral gene transcription [41]. In herpes simplex virus (HSV) and varicella zoster virus (VZV), an increase in H3K4 methylation and a decrease in

H3K9 methylation are needed for viral gene transcription in a host cell [42]. To increase methylation, the virus recruits host cell factor-1 (HCF-1) and an HKMT complex. Kristie and co-workers showed that LSD1 interacts with the HCF-1 component of the HKMT complex and demethylates H3K9 [41]. They also showed that blocking LSD1 activity leads to the inhibition of viral gene transcription, suggesting that LSD1 inhibitors could work as anti-HSV and anti-VZV agents. In addition to HSV and VZV, LSD1 has also been reported to be involved in latent HIV infection [43] and hepatitis B virus-induced liver carcinogenesis [44].

4.3 *Link of LSD1 to Globin Disorders*

The human β-globin locus consists of embryonic, fetal, and adult globin genes that are expressed during development. Mutations in the globin locus result in β-globin disorders, such as β-globinopathies, β-thalassemia, and sickle cell disease. Although the fetal globin genes are autonomously silenced in adult-stage erythroid cells, mutations lying both within and outside the locus lead to natural variations in the level of fetal globin gene expression, and some of the mutations ameliorate the clinical symptoms of β-globin disorders. LSD1 is associated with fetal globin gene repression in adult-stage erythroid cells. LSD1 has been shown to interact with the transcription factor BCL11A through a complex containing CoREST [45] and to mediate part of BCL11A's strong γ-globin gene silencing activity. LSD1 also has been shown to interact with the TR2-TR4-DNMT1-LSD1 complex, along with several other corepressor complexes [46]. LSD1 inhibition results in increased γ-globin gene expression in β-globin locus-bearing transgenic mice and cultured primary human erythroid cells [45, 47], suggesting the effectiveness of LSD1 inhibitors as therapeutic agents for β-globin disorders.

4.4 *Link of LSD1 to Metabolic Diseases*

It has been suggested that LSD1 is involved in metabolic diseases [48]. LSD1 regulates energy-expenditure genes in adipocytes, and the loss of LSD1 function in adipocytes induces a number of regulators of energy expenditure and mitochondrial metabolism, resulting in the activation of mitochondrial respiration. The expression of LSD1-target genes is downregulated in the adipose tissues of mice on a high-fat diet as compared with that in tissues of mice on a normal diet. This downregulation is reverted by suppressing the function of LSD1, indicating the involvement of LSD1 in metabolic diseases.

4.5 Link of LSD1 to Central Nervous System (CNS) Disorders

LSD1 has been reported to be involved in central nervous system (CNS) disorders, such as depression and Alzheimer's disease. LSD1 regulates the expression of genes associated with cognitive function, neuroplasticity, and memory in senescence-accelerated SAMP8 mice [49]. LSD1 also controls the expression of genes related to immune reaction and inflammation, including S100A9, which is emerging as an important contributor to inflammation-related neurodegeneration.

It was reported that neuroLSD1, a dominant-negative splicing isoform of LSD1, is responsible for emotional behavior [50]. The knockout of neuroLSD1 in mutant mice reduces the expression of psychosocial-stress-induced genes, resulting in low anxiety-like behavior.

5 LSD1 Inhibitors and Their Biological/Therapeutic Applications

As mentioned above, LSD1 represents an interesting target for epigenetic drugs as supported by data related to its link to several diseases, including cancer, viral infection, globin disorders, metabolic diseases, and CNS disorders. Thus, expectations are high regarding the use of LSD1 inhibitors as therapeutic agents for cancer and non-cancer diseases. In this section, some of the previously reported LSD1 inhibitors (Fig. 2) and their potential as therapeutic agents are presented.

LSD1 is an amine oxidase that catalyzes the demethylation of mono- or dimethylated histone lysine residues and shows homology with monoamine oxidases (MAOs) A and B [51]. Indeed, *trans*-2-phenylcyclopropylamine (PCPA) (Fig. 2), a MAO inhibitor used as an antidepressant, was found to be also able to inhibit LSD1 and LSD2 [13, 51]. It was shown that PCPA is a mechanism-based irreversible inhibitor of LSD1. Kinetics, MS, and X-ray analysis data suggested that PCPA inhibits LSD1 through the formation of a covalent adduct with the flavin ring following one-electron oxidation and cyclopropyl ring opening (Fig. 3) [51, 52]. PCPA at high concentrations induces an increase of global H3K4 methylation and growth inhibition of neuroblastoma cells and bladder cancer cells [25, 53]. In addition, the combination of PCPA and all-*trans*-retinoic acid (ATRA) is an effective therapy for acute myelogenous leukemia (AML) [54]. In addition to cancer, PCPA has been reported to show pharmacological effects in α-herpes virus latent infection [41], globin disorders [47], metabolic disorders [48], and neurodegenerative disorders [55], suggesting that LSD1 inhibitors are useful as therapeutic agents for not only cancer but also non-cancerous diseases.

Ueda et al. designed LSD1-selective inhibitors on the basis of the structures of the methylated lysine substrate and PCPA (Fig. 4) [56]. PCPA-lysine analog hybrid compounds are expected to be potent LSD1-selective inhibitors because they can

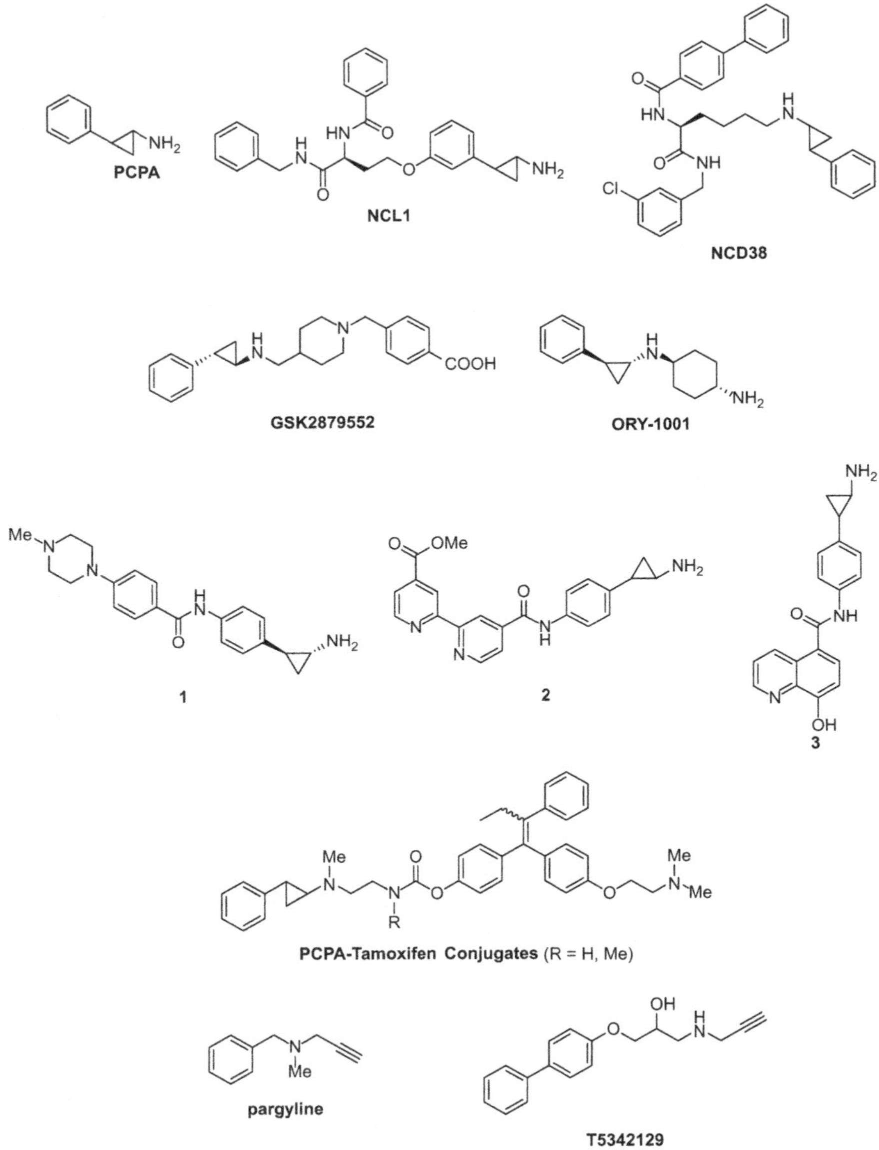

Fig. 2 Examples of LSD1 inhibitors

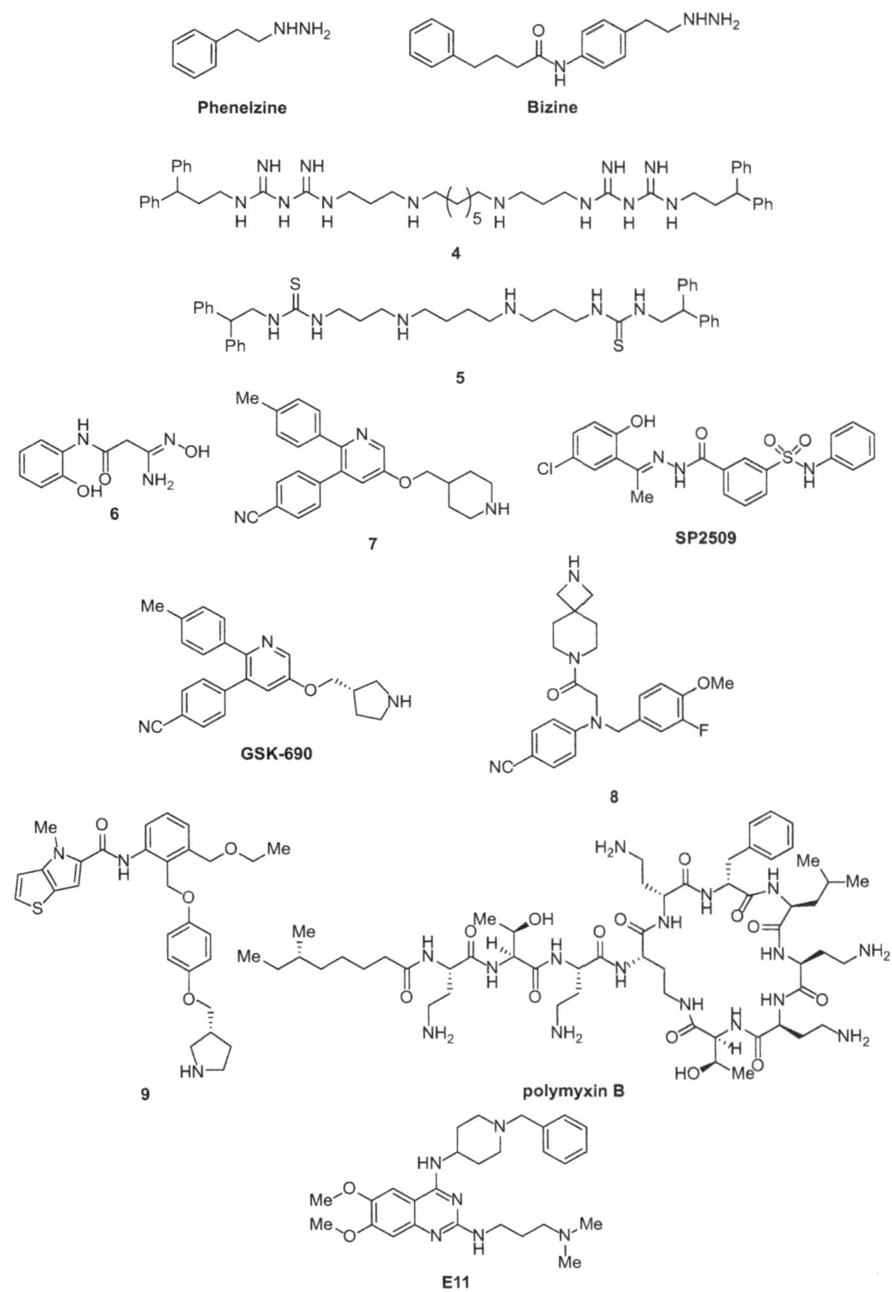

Fig. 2 (continued)

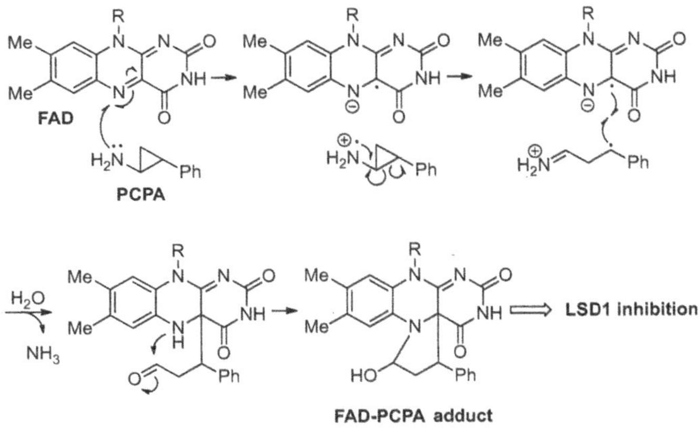

Fig. 3 Mechanism of LSD1 inhibition by PCPA

bind to the methylated lysine substrate-binding site of LSD1 and they possess the PCPA structure that reacts with FAD in the active site of LSD1. Small-molecule PCPA-lysine analog hybrid compounds were designed on the basis of the X-ray crystal structures of the FAD-PCPA adduct and the FAD-*N*-propargyl lysine peptide adduct in the active site of LSD1 [17, 52]. The superimposition of the two structures is shown in Fig. 4a. The FAD parts of the two adducts are well superimposed, and the phenyl ring of the FAD-PCPA adduct overlaps with ε-N and δ-C of the FAD-*N*-propargyl lysine peptide adduct. Based on these superimposed structures, PCPA-lysine analog hybrid compound NCL1 was designed (Figs. 2 and 4a), in which the side chain of the amino acid is linked to the phenyl ring of PCPA through an ether bond at the meta and para positions, respectively. Benzylamino and benzoyl groups were chosen as the substituents of the carbonyl and amino groups of the amino acid, respectively, because they were expected to be recognized by hydrophobic amino acid residues (Val 333, Ile 356, Phe 382, Leu 386, Leu 536, Ala 539, Thr 566, and Leu 677) at the entrance to the *N*-methylated lysine binding channel of LSD1 (Fig. 4b). In addition, the attachment of these small, hydrophobic groups could enhance membrane permeability. Furthermore, NCL1 was expected to selectively inhibit LSD1 over MAO-A and MAO-B, as the X-ray crystal structures of MAO-A and MAO-B indicated that their active-site cavities are not sufficiently capacious to accommodate the large group attached to the phenyl ring of PCPA in NCL1. NCL1 was prepared and its inhibitory activities toward human LSD1 and MAO-A and MAO-B were evaluated. Kinetic analysis and MS analysis suggested that the inhibitory activity of NCL1 occurs via the LSD1-directed synthesis of the FAD-PCPA adduct in the active site of LSD1 in a similar manner to PCPA (Fig. 3). As shown in Fig. 4c, NCL1 is a highly selective LSD1 inhibitor. Furthermore, NCL1 inhibits cancer cell growth at μM concentration, consistent with its effect on the methylation of H3K4, a substrate of LSD1. In addition, antiestrogen and NCL1 combination therapy suppresses the growth of drug-resistant breast cancer cells [57, 58]. NCL1 also reduces tumor volume in mice injected

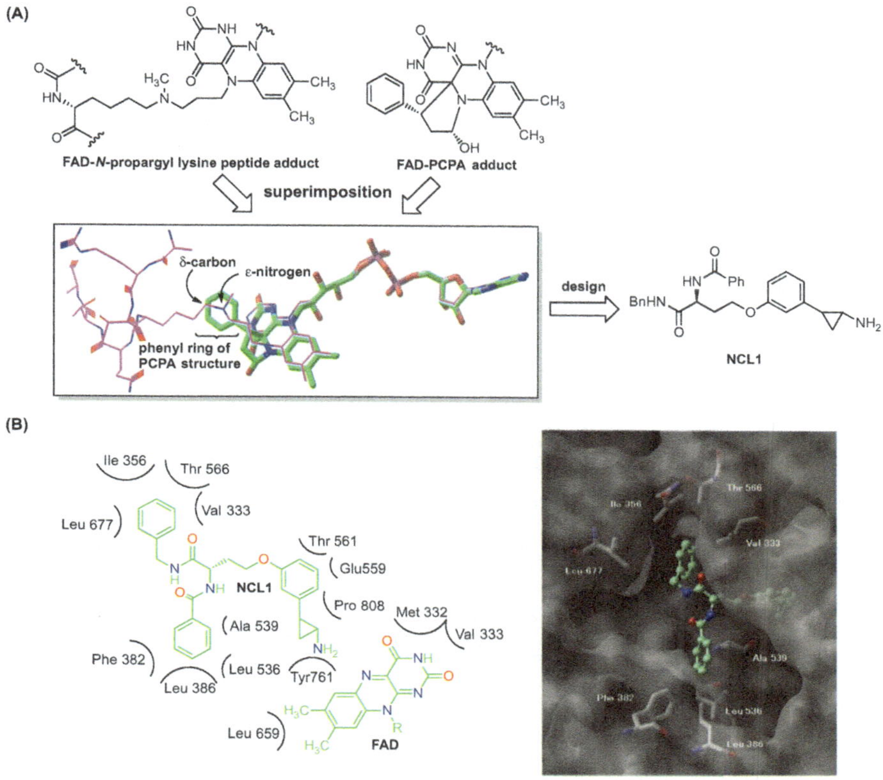

(C)

	IC_{50} (μM)			selectivity	
compound	LSD1	MAO-A	MAO-B	MAO-A/LSD1	MAO-B/LSD1
PCPA	32	7.3	4.3	0.23	0.13
NCL1	2.5	230	500	92	200

Fig. 4 (**a**) Design of an LSD1-selective inhibitor NCL1 based on superimposition of the FAD-PCPA adduct (PDB code: 2UXX) (tube) and the reduced FAD-*N*-propargyl lysine peptide adduct (PDB code: 2UXN) (wire) in the active site of LSD1. Amino acid residues in the active site are not shown for the sake of clarity. (**b**) View of the conformation of NCL1 (ball and stick) docked in the LSD1 catalytic core. Residues within 5 Å from NCL1 are displayed in the tube graphic. (**c**) In vitro LSD1, MAO-A, and MAO-B inhibitory activities of NCL1

subcutaneously with hormone-resistant prostate cancer PCai1 cells without adverse effects, suggesting the potential of LSD1 inhibitors as therapeutic agents for hormone-resistant prostate cancer [59]. These results point to the possibility of NCL1 as an anticancer agent.

Although NCL1 is a potent and selective LSD1 inhibitor, its activity in cell-based assays is insufficient. Ogasawara et al. had aimed to find novel LSD1 inactivators on the basis of the new concept "protein-targeted drug delivery" [60].

As mentioned above, PCPA inhibits LSD1 by forming a covalent bond with FAD (Figs. 3 and 5a). In the course of LSD1 inactivation by PCPA, the nitrogen atom of

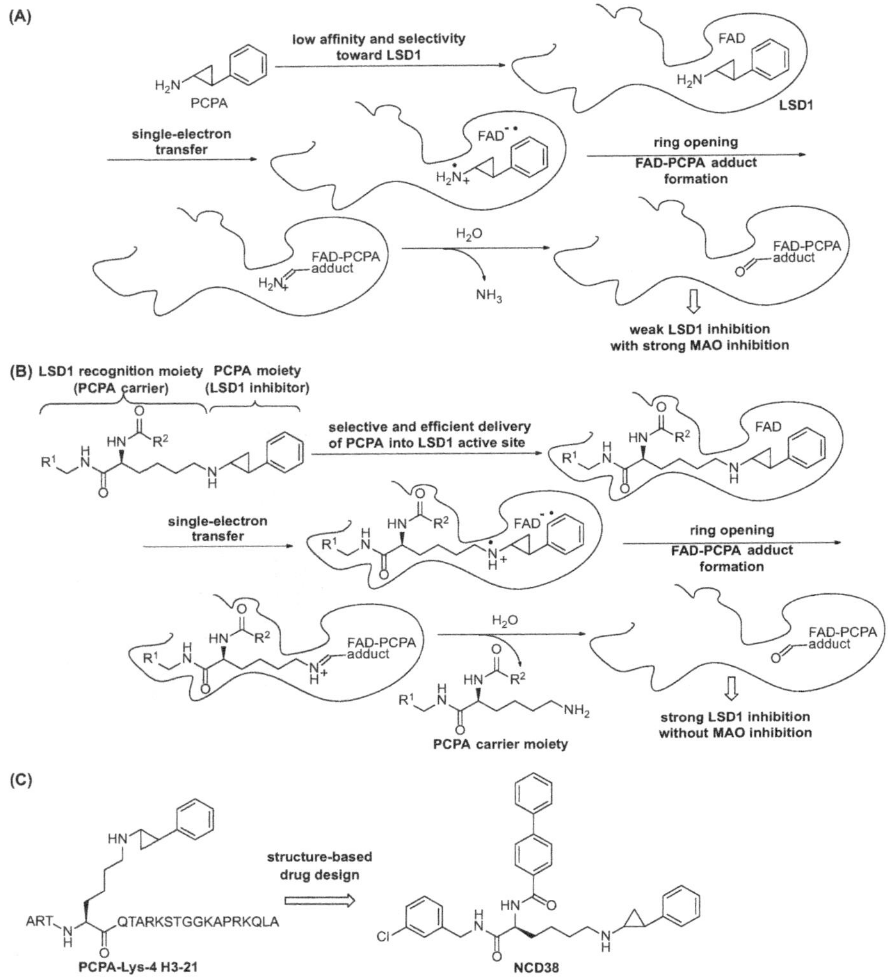

(D)

compound	IC$_{50}$ (μM)			LSD1 inhibition		
	LSD1	MAO-A	MAO-B	k_{inact} (s^{-1})	K_i (μM)	k_{inact}/K_i ($M^{-1}s^{-1}$)
PCPA	32	7.3	4.3	0.0070	180	38
NCD38	0.41	>100	>100	0.0052	2.3	2200

Fig. 5 (**a**) Mechanism of LSD1 inhibition by PCPA. (**b**) Mechanism of LSD1-targeted delivery of PCPA. (**c**) Structures of PCPA-Lys-4 H3-21 and NCD38. (**d**) LSD1 and MAO inhibitory activities and kinetic parameters of PCPA and NCD38

PCPA is released as an ammonia molecule through the hydrolysis of the imine intermediate. The nitrogen atom of the ammonia molecule corresponds to the ε-nitrogen atom of the lysine substrate in the proposed demethylation mechanism by LSD1 (Fig. 1). Taking these mechanisms into account, together with the idea of delivering PCPA directly to the active site of LSD1, candidate LSD1 inactivators were designed (Fig. 5b), in which PCPA is coupled to a lysine carrier moiety at the nitrogen atom. Because methylated lysine is the substrate of LSD1, it is expected that the lysine moiety of the candidate inactivator would be efficiently recognized by LSD1, which would lead to high selectivity over MAO-A and MAO-B. After the PCPA moiety of the candidate inactivator is carried to the active site of LSD1, it is expected that the PCPA moiety would inactivate LSD1 in a similar manner to PCPA itself, namely, through single-electron transfer, radical opening of the cyclopropyl ring, and covalent bond formation with FAD (Fig. 5b). Then, the lysine moiety is expected to be released through the hydrolysis of the imine intermediate (Fig. 5b). Thus, the lysine moiety of the candidate inactivator serves as a carrier that delivers PCPA into the active site of LSD1 selectively and efficiently.

Initially, as a proof of concept study, PCPA-Lys-4 H3-21 (Fig. 5c), which bears a PCPA moiety at Lys-4 of a 21-amino-acid LSD1 substrate peptide (H3-21), was designed and prepared. As expected, PCPA-Lys-4 H3-21 strongly inhibited LSD1 with an IC_{50} of 0.16 μM in a time- and concentration-dependent manner, but did not inhibit MAO-A or MAO-B ($IC_{50} > 100$ μM). However, PCPA-Lys-4 H3-21 showed only weak antiproliferative activity in cancer cells where LSD1 was overexpressed. It was speculated that PCPA-Lys-4 H3-21 has poor membrane permeability, likely as a result of the high polarity of its peptide structure. Thus, based on this proof of concept of the LSD1-targeted PCPA delivery strategy, this strategy was applied to the design of nonpeptide, small-molecule LSD1 inactivators that show activity in cell-based assays. A number of candidate small-molecule, drug-delivery-type LSD1 inactivators were designed and synthesized guided by the X-ray crystal structure of LSD1, and NCD38 was eventually identified as a potent and selective LSD1 inactivator (Figs. 2 and 5c). NCD38 also showed potent antiproliferative activity in solid cancer cells.

In addition, the LSD1 inactivation mechanism was investigated to confirm that NCD38 indeed inhibits LSD1 by delivering PCPA to the LSD1 active site (Fig. 5b). Kinetic analysis revealed that NCD38 is a time-dependent LSD1 inactivator, in accordance with the irreversible mechanism we proposed (Fig. 5b). The kinetic parameters of NCD38 are shown in Fig. 5d. The k_{inact}/K_i value of NCD38 is much larger than that of PCPA, thus confirming that NCD38 is a much more potent LSD1 inactivator than PCPA. MALDI MS analysis of the inactivated mixture of LSD1 with NCD38 was also performed. Peaks with m/z 918 and 900, corresponding to the FAD-PCPA adduct and the dehydrated adduct, respectively, were observed in the mixture of LSD1/NCD38. The lysine moiety released from LSD1/NCD38 was also detected. These mechanistic data strongly support the idea that NCD38 inhibits LSD1 through the efficient and selective delivery of PCPA to the active site of LSD1 with the assistance of its lysine moiety (Fig. 5b).

Interestingly, a recent report has shown that NCD38 derivatives inhibit LSD1 in preference to LSD2 [61].

Recently, it has been reported that NCD38 inhibits the growth of MLL-AF9 leukemia as well as erythroleukemia, megakaryoblastic leukemia, and myelodysplastic syndrome (MDS) overt leukemia cells in the concentration range in which normal hematopoiesis is spared [62]. A single administration of NCD38 causes the in vivo eradication of primary MDS-related leukemia cells with a complex karyotype. Mechanistic studies showed that NCD38 elevates H3K27ac level on enhancers of the LSD1 signature genes and derepresses the super-enhancers of hematopoietic regulators that are silenced abnormally by LSD1.

In addition to the antiproliferative activity of NCD38 in differentiated cancer cells, NCD38 is also able to inhibit cancer stem cell formation and the maintenance of human metastatic breast cancer cells, thus reverting them to epithelial form [63]. The pharmacological inhibition of LSD1 using NCD38 significantly reduces cell viability and neurosphere formation and induces apoptosis of glioma stem cells (GSCs) with little effect on differentiated cells [64]. In preclinical studies using orthotopic models, NCD38 significantly reduces GSC-driven tumor progression and improves mouse survival. Mechanistic studies showed that NCD38 causes apoptosis of GSCs by inducing the activation of the unfolded protein response pathway.

Thus, NCD38 and its derivatives are considered candidates for anticancer agents as well as tools for probing the biology of LSD1. Currently, IMG-7289, an NCD38 mimetic, is being evaluated in a phase 1/2 clinical trial for the treatment of AML and MDS (http://www.imagobio.com/imago-biosciences-doses-first-patients-in-phase-12-study-of-img-7289-in-myelofibrosis/).

GlaxoSmithKline and Oryzon Genomics discovered potent and selective LSD1 inhibitors GSK2879552 and ORY-1001, respectively (Fig. 2) [65, 66]. GSK2879552 and ORY-1001 exhibit antileukemia activity and are currently undergoing clinical trials for AML treatment. In addition, screening for cancer cell lines revealed that small cell lung carcinoma (SCLC) is sensitive to LSD1 inhibition by PCPA analogs, including GSK2879552 [65]. GSK2879552 exhibits DNA hypomethylation in SCLC lines, suggesting that DNA hypomethylation can be used as a predictive biomarker of LSD1 inhibitory activity.

In addition, Oryzon Genomics developed LSD1/MAO-B inhibitor ORY-2001. ORY-2001 prevents the development of memory deficit in SAMP8 mice through the induction of neuronal plasticity and the reduction of neuroinflammation [49]. Currently, ORY-2001 is being evaluated in a phase 1 clinical trial for the treatment of Alzheimer's disease.

Vianello et al. reported a novel PCPA derivative **1** (Fig. 2) as a potent inhibitor of LSD1 [67]. PCPA derivative **1** strongly inhibits the clonogenic potential of acute leukemia cell lines. Furthermore, compound **1** exhibits in vivo efficacy after oral administration, inducing a 62% increase in survival in a mouse leukemia model.

Mai and co-workers identified hybrid LSD1/JmjC-domain containing histone lysine demethylase inhibitors by coupling the skeleton of PCPA, an LSD1 inhibitor, with 4-carboxy-4-carbomethoxy-2,2-bipyridine or 5-carboxy-8-hydroxyquinoline, two 2-oxoglutarate competitive templates developed for JmjC inhibition [68]. The hybrid compounds were validated as potential antitumor agents in cells. Compounds **2** and **3** (Fig. 2) caused growth arrest and substantial apoptosis in LNCaP prostate and HCT116 colon cancer cells with an increase in H3K4 and H3K9 methylation levels in the cells.

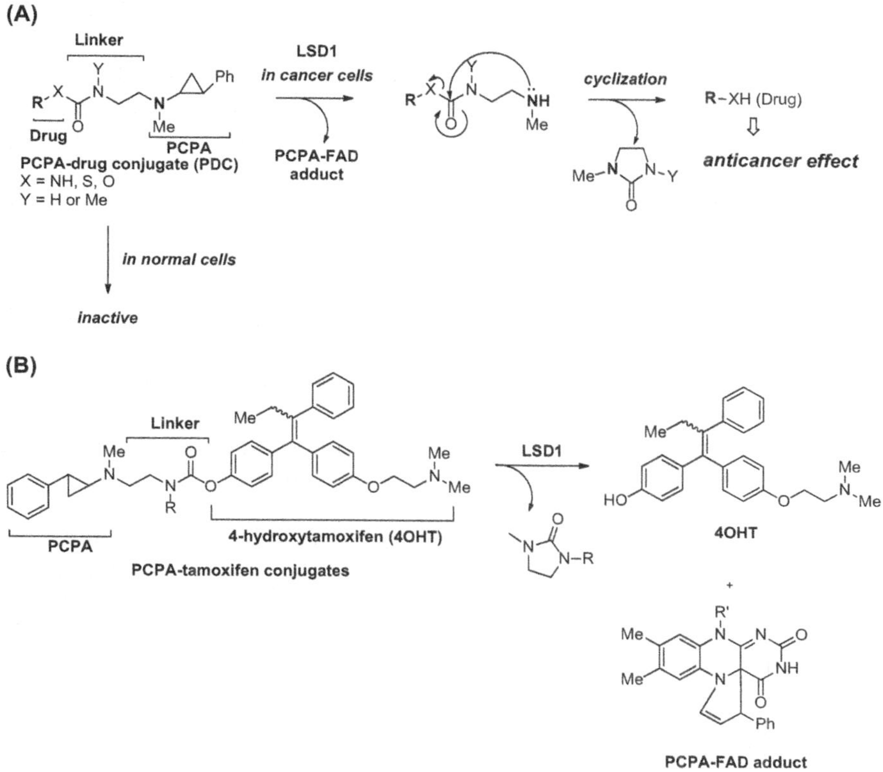

Fig. 6 (**a**) Concept of small-molecule-based drug delivery system for cancer therapy using PCPA-drug conjugates (PDCs). (**b**) Release of 4-hydroxytamoxifen (4OHT) from PCPA-tamoxifen conjugates in the presence of LSD1

The chemotherapy of cancer, including targeted therapy using anticancer drugs, usually provides a certain level of beneficial therapeutic effect while simultaneously causing serious adverse effects on account of the cytotoxicity of the employed drugs toward normal cells. In order to reduce the adverse effects of anticancer drugs, several drug delivery systems (DDSs) for anticancer drugs have been developed. One example is antibody-drug conjugates (ADCs) that show both potent and selective cytotoxicity toward cancer cells that express a specific antigen. Some of these ADCs, such as brentuximab vedotin and trastuzumab emtansine, are currently used in clinical practice. However, because of their macromolecular structure, ADCs have several limitations, including poor tissue penetration, immunogenicity, low bioavailability, and high cost. To overcome the problems of macromolecule-based DDSs, such as ADCs, Ota et al. focused on LSD1 to trigger the controlled release of anticancer drugs in cancer cells where LSD1 is highly expressed [69]. For that purpose, conjugates of the LSD1 inhibitor PCPA are used as novel anticancer drug delivery molecules to selectively release anticancer drugs through the inhibition of LSD1 (Fig. 6a). PCPA-drug conjugates (PDCs) are expected to be recognized by LSD1 and to inactivate it in a similar manner to PCPA itself, i.e., via a single-electron transfer mechanism (Fig. 6b). Subsequently, the drug should be released together with the linker moiety of the PDCs through hydrolysis of the imine intermediate, and an ensuing intramolecular cyclization should eventually separate the linker from the drug. Thus, PDCs could serve as delivery molecules that selectively release an anticancer drug upon binding to LSD1. This method would significantly reduce adverse effects as such molecules are inactive toward normal cells where the expression of LSD1 is low. As a PDC prototype, we have designed and synthesized PCPA-tamoxifen conjugates (Fig. 2) targeting breast cancer cells (Fig. 6b), which release 4-hydroxytamoxifen in the presence of LSD1 in in vitro assays. Furthermore, PCPA-tamoxifen conjugates inhibit the growth of breast cancer cells through the simultaneous inhibition of LSD1 and ERα without exhibiting cytotoxicity toward normal cells. These results demonstrated that PDCs are a useful anticancer drug delivery tool that may facilitate the selective release of drugs in cancer cells.

Previously, Schüle and co-workers tested whether pargyline (Fig. 2), a well-known MAO inhibitor, inhibits LSD1 and found that it blocks the demethylation of H3K9 by LSD1 and consequently blocks AR-dependent transcription [7]. Based on this report, Jung and co-workers discovered new small-molecule inhibitors of LSD1 containing a propargylamine warhead [70]. Druglike LSD1 inhibitors with a propargylamine moiety, such as T5342129 (Fig. 2), showed histone hypermethylation in breast cancer cells.

Phenelzine (Fig. 2), a MAO inhibitor, has also been reported to inhibit LSD1, although its inhibitory activity and selectivity for LSD1 are very low [51, 52, 71, 72]. Cole and co-workers identified analogs of phenelzine and their LSD1 inhibitory properties [73]. A novel phenelzine analog (bizine) (Fig. 2) containing a phenylbutyrylamide appendage was shown to be a potent LSD1 inhibitor in vitro and was selective versus MAO-A, MAO-B, and LSD2. Bizine was found to be effective in modulating bulk histone methylation in cancer cells, and ChIP-seq experiments revealed a statistically significant overlap in the H3K4 methylation pattern of genes affected by bizine and those altered in LSD1$^{-/-}$ cells. Treatment of two cancer cell lines, LNCaP and H460, with bizine resulted in a reduction of proliferation rate, and bizine showed additive to synergistic effects on cell growth when used in combination with two of the five HDAC inhibitors tested. Moreover, neurons exposed to oxidative stress were protected by the presence of bizine, suggesting potential applications in neurodegenerative diseases.

Not only irreversible LSD1 inhibitors, which are derived from MAO inhibitors, but also reversible LSD1 inhibitors have been identified.

Woster and co-workers reported that (bis)guanidines, (bis)biguanides, and their urea and thiourea isosteres, such as **4** and **5** (Fig. 2), are potent inhibitors of LSD1 and induce the re-expression of aberrantly silenced tumor suppressor genes in tumor cells in vitro [31]. They also reported a series of small-molecule amidoximes, such as **6** (Fig. 2), which are moderate inhibitors of recombinant LSD1 but produce dramatic changes in methylation at the H3K4 chromatin mark in Calu-6 lung carcinoma cells [74]. In addition, these analogs increased cellular levels of LSD1 target genes, such as secreted frizzle-related protein 2, H-cadherin, and the transcription factor GATA4.

Wu et al. reported 3-(piperidin-4-ylmethoxy)pyridine-containing compounds, such as **7** ($K_i = 29$ nM) (Fig. 2), as potent LSD1 inhibitors [75]. These compounds exhibit high selectivity (>160×) against related MAO-A and MAO-B. Enzyme kinetics and docking studies suggested that they are competitive inhibitors against a dimethylated H3K4 substrate and proposed a possible binding mode. The potent LSD1 inhibitors can increase cellular H3K4 methylation and strongly inhibit the proliferation of several leukemia and solid tumor cells with EC_{50} values as low as 280 nM, while they have negligible effects on normal cells.

A structure-based virtual screening of a compound library containing approximately two million small molecular entities has led to the identification of SP2509 (Fig. 2) as a reversible and selective LSD1 inhibitor ($K_i = 31$ nM) [76], although the core *N′*-(2-hydroxybenzylidene)hydrazide motif was previously identified as a pan-assay interference compound [77]. SP2509 inhibits the proliferation and survival of several cancer cell lines, including breast and colorectal cancer. It was also reported that treatment with SP2509 attenuates the binding of LSD1 with the corepressor CoREST, increases the permissive H3K4Me3 mark on the target gene promoters, and increases the levels of p21, p27, and CCAAT/enhancer binding protein α in cultured AML cells [78]. In addition, SP2509 inhibits colony growth of AML cells. SP2509 also induces morphological features of differentiation in cultured and primary AML blasts. Treatment with SP2509 alone significantly improves the survival of immune-depleted mice following tail-vein infusion and engraftment of cultured or primary human AML cells. Co-treatment of panobinostat, a pan-HDAC inhibitor, and SP2509 synergistically improves survival in mice engrafted with the human AML cells without exhibiting any toxicity.

Recently, Mould et al. identified reversible LSD1 inhibitors from a high-throughput screen and subsequent in silico modelling approaches [79]. Based on a hit compound, they carried out scaffold hopping from GSK-690 (Fig. 2) [80, 81], an analog of compound **7**, to find compound **8** (Fig. 2), which has a K_d value of 32 nM and an EC_{50} value of 0.67 μM, in a surrogate cellular biomarker assay. Compound **8** does not display the same level of hERG liability as that observed with GSK-690 and represents a promising lead for the further development of LSD1 inhibitors.

In addition, Sartori, Vianello, and co-workers discovered thieno[3,2-*b*]pyrrole-5-carboxamide analog **9** (Fig. 2) as a potent reversible inhibitor of LSD1 [82, 83]. Compound **9** upregulated the expression of LSD1 target genes such as CD14, CD11b, and CD86 in THP-1 cells, and it showed a remarkable anticlonogenic cell growth effect on MLL-AF9 human leukemia cells.

It was also reported that polymyxins and the antibiotics polymyxins such as polymyxin B and quinazolines such as E11 (Fig. 2) inhibit LSD1 by binding to the entrance of the substrate cleft where their positively charged moiety interacts with anionic amino acid residues of LSD1 [84]. These scaffolds should be useful for further studies on LSD1 inhibitor development.

6 Summary

At present, there are only six approved epigenetic drugs (two DNMT inhibitors and four HDAC inhibitors), and they are utilized only for MDS, cutaneous T-cell lymphoma, peripheral T-cell lymphoma, or multiple myeloma treatment. Additional indications of the current epigenetic drugs are limited. In this regard, researchers need to acquire an integrated understanding of cancer epigenetics in order to discover useful next-generation drugs for cancer therapy.

In this chapter, the biology and pharmacology of LSD and hitherto reported LSD inhibitors have been presented. In particular, small-molecule LSD1 inhibitors have been discussed from the point of view of potential therapeutic agents. It is hoped that the LSD1-selective inhibitors presented here will provide the basis for the development of novel therapeutic agents for both cancer and non-cancerous diseases.

Compliance with Ethical Standards

Funding: This work was partially supported by the JST CREST program (T.S.; JPMJCR14L2).

Conflict of Interest: The author declares no conflict of interest.

Ethical Approval: Not applicable.

Informed Consent: Not applicable.

References

1. Kubicek S, Jenuwein T (2004) A crack in histone lysine methylation. Cell 119:903–906
2. Bannister AJ, Kouzarides T (2005) Reversing histone methylation. Nature 436:1103–1106
3. Santos-Rosa H, Schneider R, Bannister AJ, Sherriff J, Bernstein BE, Emre NC, Schreiber SL, Mellor J, Kouzarides T (2002) Active genes are tri-methylated at K4 of histone H3. Nature 419:407–411
4. Barski A, Cuddapah S, Cui K, Roh TY, Schones DE, Wang Z, Wei G, Chepelev I, Zhao K (2007) High-resolution profiling of histone methylations in the human genome. Cell 129:823–837
5. Itoh Y, Suzuki T, Miyata N (2013) Small-molecular modulators of cancer-associated epigenetic mechanisms. Mol Biosyst 9:873–896
6. Shi Y, Lan F, Matson C, Mulligan P, Whetstine JR, Cole PA, Casero RA, Shi Y (2004) Histone demethylation mediated by the nuclear amine oxidase homolog LSD1. Cell 119:941–953
7. Metzger E, Wissmann M, Yin N, Müller JM, Schneider R, Peters AH, Günther T, Buettner R, Schüle R (2005) LSD1 demethylates repressive histone marks to promote androgen-receptor-dependent transcription. Nature 437:436–439
8. Huang J, Sengupta R, Espejo AB, Lee MG, Dorsey JA, Richter M, Opravil S, Shiekhattar R, Bedford MT, Jenuwein T, Berger SL (2007) p53 is regulated by the lysine demethylase LSD1. Nature 449:105–108

9. Wang J, Hevi S, Kurash JK, Lei H, Gay F, Bajko J, Su H, Sun W, Chang H, Xu G, Gaudet F, Li E, Chen T (2009) The lysine demethylase LSD1 (KDM1) is required for maintenance of global DNA methylation. Nat Genet 41:125–129
10. Yang J, Huang J, Dasgupta M, Sears N, Miyagi M, Wang B, Chance MR, Chen X, Du Y, Wang Y, An L, Wang Q, Lu T, Zhang X, Wang Z, Stark GR (2010) Reversible methylation of promoter-bound STAT3 by histone-modifying enzymes. Proc Natl Acad Sci U S A 107:21499–21504
11. Kontaki H, Talianidis I (2010) Lysine methylation regulates E2F1-induced cell death. Mol Cell 39:152–160
12. Cho HS, Suzuki T, Dohmae N, Hayami S, Unoki M, Yoshimatsu M, Toyokawa G, Takawa M, Chen T, Kurash JK, Field HI, Ponder BA, Nakamura Y, Hamamoto R (2011) Demethylation of RB regulator MYPT1 by histone demethylase LSD1 promotes cell cycle progression in cancer cells. Cancer Res 71:655–660
13. Karytinos A, Forneris F, Profumo A, Ciossani G, Battaglioli E, Binda C, Mattevi A (2009) A novel mammalian flavin-dependent histone demethylase. J Biol Chem 284:17775–17782
14. Fang R, Barbera AJ, Xu Y, Rutenberg M, Leonor T, Bi Q, Lan F, Mei P, Yuan GC, Lian C, Peng J, Cheng D, Sui G, Kaiser UB, Shi Y, Shi YG (2010) Human LSD2/KDM1b/AOF1 regulates gene transcription by modulating intragenic H3K4me2 methylation. Mol Cell 39:222–233
15. Yang Z, Jiang J, Stewart MD, Qi S, Yamane K, Li J, Zhang Y, Wong J (2010) AOF1 is a histone H3K4 demethylase possessing demethylase activity-independent repression function. Cell Res 20:276–287
16. Yang Y, Yin X, Yang H, Xu Y (2015) Histone demethylase LSD2 acts as an E3 ubiquitin ligase and inhibits cancer cell growth through promoting proteasomal degradation of OGT. Mol Cell 58:47–59
17. Yang M, Culhane JC, Szewczuk LM, Gocke CB, Brautigam CA, Tomchick DR, Machius M, Cole PA, Yu H (2007) Structural basis of histone demethylation by LSD1 revealed by suicide inactivation. Nat Struct Mol Biol 14:535–539
18. Baron R, Binda C, Tortorici M, McCammon JA, Mattevi A (2011) Molecular mimicry and ligand recognition in binding and catalysis by the histone demethylase LSD1-CoREST complex. Structure 19:212–220
19. Forneris F, Binda C, Dall'Aglio A, Fraaije MW, Battaglioli E, Mattevi A (2006) A highly specific mechanism of histone H3-K4 recognition by histone demethylase LSD1. J Biol Chem 281:35289–35295
20. Fang R, Chen F, Dong Z, Hu D, Barbera AJ, Clark EA, Fang J, Yang Y, Mei P, Rutenberg M, Li Z, Zhang Y, Xu Y, Yang H, Wang P, Simon MD, Zhou Q, Li J, Marynick MP, Li X, Lu H, Kaiser UB, Kingston RE, Xu Y, Shi YG (2013) LSD2/KDM1B and its cofactor NPAC/GLYR1 endow a structural and molecular model for regulation of H3K4 demethylation. Mol Cell 49:558–570
21. Chen F, Yang H, Dong Z, Fang J, Wang P, Zhu T, Gong W, Fang R, Shi YG, Li Z, Xu Y (2013) Structural insight into substrate recognition by histone demethylase LSD2/KDM1b. Cell Res 23:306–309
22. Zhu X, Wang J, Ju BG, Rosenfeld MG (2007) Signaling and epigenetic regulation of pituitary development. Curr Opin Cell Biol 19:605–611
23. Hirano K, Namihira M (2016) LSD1 mediates neuronal differentiation of human fetal neural stem cells by controlling the expression of a novel target gene, HEYL. Stem Cells 34:1872–1882
24. Chen Y, Kim J, Zhang R, Yang X, Zhang Y, Fang J, Chen Z, Teng L, Chen X, Ge H, Atadja P, Li E, Chen T, Qi W (2016) Histone demethylase LSD1 promotes adipocyte differentiation through repressing Wnt signaling. Cell Chem Biol 23:1228–1240
25. Schulte JH, Lim S, Schramm A, Friedrichs N, Koster J, Versteeg R, Ora I, Pajtler K, Klein-Hitpass L, Kuhfittig-Kulle S, Metzger E, Schüle R, Eggert A, Buettner R, Kirfel J (2009) Lysine-specific demethylase 1 is strongly expressed in poorly differentiated neuroblastoma: implications for therapy. Cancer Res 69:2065–2071

26. Wissmann M, Yin N, Müller JM, Greschik H, Fodor BD, Jenuwein T, Vogler C, Schneider R, Günther T, Buettner R, Metzger E, Schüle R (2007) Cooperative demethylation by JMJD2C and LSD1 promotes androgen receptor-dependent gene expression. Nat Cell Biol 9:347–353
27. Scoumanne A, Chen X (2007) The lysine-specific demethylase 1 is required for cell proliferation in both p53-dependent and -independent manners. J Biol Chem 282:15471–15475
28. Lim S, Janzer A, Becker A, Zimmer A, Schüle R, Buettner R, Kirfel J (2010) Lysine-specific demethylase 1 (LSD1) is highly expressed in ER-negative breast cancers and a biomarker predicting aggressive biology. Carcinogenesis 31:512–520
29. Wang Y, Zhang H, Chen Y, Sun Y, Yang F, Yu W, Liang J, Sun L, Yang X, Shi L, Li R, Li Y, Zhang Y, Li Q, Yi X, Shang Y (2009) LSD1 is a subunit of the NuRD complex and targets the metastasis programs in breast cancer. Cell 138:660–672
30. Hayami S, Kelly JD, Cho H, Yoshimatsu M, Unoki M, Tsunoda T, Field HI, Neal DE, Yamaue H, Ponder BAJ, Nakamura Y, Hamamoto R (2011) Overexpression of LSD1 contributes to human carcinogenesis through chromatin regulation in various cancers. Int J Cancer 128:574–586
31. Huang Y, Stewart TM, Wu Y, Baylin SB, Marton LJ, Perkins B, Jones RJ, Woster PM, Casero RA Jr (2009) Novel oligoamine analogues inhibit lysine-specific demethylase 1 and induce reexpression of epigenetically silenced genes. Clin Cancer Res 15:7217–7228
32. Cai C, He HH, Chen S, Coleman I, Wang H, Fang Z, Chen S, Nelson PS, Liu XS, Brown M, Balk SP (2011) Androgen receptor gene expression in prostate cancer is directly suppressed by the androgen receptor through recruitment of lysine-specific demethylase 1. Cancer Cell 20:457–471
33. Pollock JA, Larrea MD, Jasper JS, McDonnell DP, McCafferty DG (2012) Lysine-specific histone demethylase 1 inhibitors control breast cancer proliferation in ERα-dependent and -independent manners. ACS Chem Biol 7:1221–1231
34. Park UH, Kang MR, Kim EJ, Kwon YS, Hur W, Yoon SK, Song BJ, Park JH, Hwang JT, Jeong JC, Um SJ (2016) ASXL2 promotes proliferation of breast cancer cells by linking ERα to histone methylation. Oncogene 35:3742–3752
35. Lv T, Yuan D, Miao X, Lv Y, Zhan P, Shen X, Song Y (2012) Over-expression of LSD1 promotes proliferation, migration and invasion in non-small cell lung cancer. PLoS One 7: e35065
36. Fu X, Zhang P, Yu B (2017) Advances toward LSD1 inhibitors for cancer therapy. Future Med Chem 9:1227–1242
37. Wang J, Lu F, Ren Q, Sun H, Xu Z, Lan R, Liu Y, Ward D, Quan J, Ye T, Zhang H (2011) Novel histone demethylase LSD1 inhibitors selectively target cancer cells with pluripotent stem cell properties. Cancer Res 71:7238–7249
38. Zhang X, Lu F, Wang J, Yin F, Xu Z, Qi D, Wu X, Cao Y, Liang W, Liu Y, Sun H, Ye T, Zhang H (2013) Pluripotent stem cell protein Sox2 confers sensitivity to LSD1 inhibition in cancer cells. Cell Rep 5:445–457
39. Stewart CA, Byers LA (2015) Altering the course of small cell lung cancer: targeting cancer stem cells via LSD1 inhibition. Cancer Cell 28:4–6
40. Chen L, Vasilatos SN, Qin Y, Katz TA, Cao C, Wu H, Tasdemir N, Levine KM, Oesterreich S, Davidson NE, Huang Y (2017) Functional characterization of lysine-specific demethylase 2 (LSD2/KDM1B) in breast cancer progression. Oncotarget 8:81737–81753
41. Liang Y, Vogel JL, Narayanan A, Peng H, Kristie TM (2009) Inhibition of the histone demethylase LSD1 blocks α-herpesvirus lytic replication and reactivation from latency. Nat Med 15:1312–1317
42. Narayanan A, Ruyechan WT, Kristie TM (2007) The coactivator host cell factor-1 mediates Set1 and MLL1 H3K4 trimethylation at herpesvirus immediate early promoters for initiation of infection. Proc Natl Acad Sci U S A 104:10835–10840
43. Sakane N, Kwon HS, Pagans S, Kaehlcke K, Mizusawa Y, Kamada M, Lassen KG, Chan J, Greene WC, Schnoelzer M, Ott M (2011) Activation of HIV transcription by the viral Tat protein requires a demethylation step mediated by lysine-specific demethylase 1 (LSD1/ KDM1). PLoS Pathog 7:e1002184

44. Andrisani OM (2013) Deregulation of epigenetic mechanisms by the hepatitis B virus X protein in hepatocarcinogenesis. Viruses 5:858–872
45. Xu J, Bauer DE, Kerenyi MA, Vo TD, Hou S, Hsu YJ, Yao H, Trowbridge JJ, Mandel G, Orkin SH (2013) Corepressor-dependent silencing of fetal hemoglobin expression by BCL11A. Proc Natl Acad Sci U S A 110:6518–6523
46. Cui S, Kolodziej KE, Obara N, Amaral-Psarris A, Demmers J, Shi L, Engel JD, Grosveld F, Strouboulis J, Tanabe O (2011) Nuclear receptors TR2 and TR4 recruit multiple epigenetic transcriptional corepressors that associate specifically with the embryonic β-type globin promoters in differentiated adult erythroid cells. Mol Cell Biol 31:3298–3311
47. Shi L, Cui S, Engel JD, Tanabe O (2013) Lysine-specific demethylase 1 is a therapeutic target for fetal hemoglobin induction. Nat Med 19:291–294
48. Hino S, Sakamoto A, Nagaoka K, Anan K, Wang Y, Mimasu S, Umehara T, Yokoyama S, Kosai K, Nakao M (2012) FAD-dependent lysine-specific demethylase-1 regulates cellular energy expenditure. Nat Commun 3:758
49. Buesa C, Mascaró C, Rotllant D, Griñan-Ferré C, Pallàs M, Maes T (2015) The dual LSD1/MAOB inhibitor ORY2001 prevents the development of the memory deficit in SAMP8 mice through induction of neuronal plasticity and reduction of neuroinflammation. Alzheimers Dement 11:P905
50. Rusconi F, Grillo B, Ponzoni L, Bassani S, Toffolo E, Paganini L, Mallei A, Braida D, Passafaro M, Popoli M, Sala M, Battaglioli E (2016) LSD1 modulates stress-evoked transcription of immediate early genes and emotional behavior. Proc Natl Acad Sci U S A 113:3651–3656
51. Schmidt DM, McCafferty DG (2007) *trans*-2-Phenylcyclopropylamine is a mechanism-based inactivator of the histone demethylase LSD1. Biochemistry 46:4408–4416
52. Yang M, Culhane JC, Szewczuk LM, Jalili P, Ball HL, Machius M, Cole PA, Yu H (2007) Structural basis for the inhibition of the LSD1 histone demethylase by the antidepressant *trans*-2-phenylcyclopropylamine. Biochemistry 46:8058–8065
53. Kauffman EC, Robinson BD, Downes MJ, Powell LG, Lee MM, Scherr DS, Gudas LJ, Mongan NP (2011) Role of androgen receptor and associated lysine-demethylase coregulators, LSD1 and JMJD2A, in localized and advanced human bladder cancer. Mol Carcinog 50:931–944
54. Schenk T, Chen WC, Göllner S, Howell L, Jin L, Hebestreit K, Klein HU, Popescu AC, Burnett A, Mills K, Casero RA Jr, Marton L, Woster P, Minden MD, Dugas M, Wang JC, Dick JE, Müller-Tidow C, Petrie K, Zelent A (2012) Inhibition of the LSD1 (KDM1A) demethylase reactivates the all-*trans*-retinoic acid differentiation pathway in acute myeloid leukemia. Nat Med 18:605–611
55. Tsutsumi T, Iwao K, Hayashi H, Kirihara T, Kawaji T, Inoue T, Hino S, Nakao M, Tanihara H (2016) Potential neuroprotective effects of an LSD1 inhibitor in retinal ganglion cells via p38 MAPK activity. Invest Ophthalmol Vis Sci 57:6461–6473
56. Ueda R, Suzuki T, Mino K, Tsumoto H, Nakagawa H, Hasegawa M, Sasaki R, Mizukami T, Miyata N (2009) Identification of cell-active lysine specific demethylase 1-selective inhibitors. J Am Chem Soc 131:17536–17537
57. Son SY, Ma J, Kondou Y, Yoshimura M, Yamashita E, Tsukihara T (2008) Structure of human monoamine oxidase A at 2.2-Å resolution: the control of opening the entry for substrates/inhibitors. Proc Natl Acad Sci U S A 105:5739–5744
58. Binda C, Li M, Hubalek F, Restelli N, Edmondson DE, Mattevi A (2003) Insights into the mode of inhibition of human mitochondrial monoamine oxidase B from high-resolution crystal structures. Proc Natl Acad Sci U S A 100:9750–9755
59. Etani T, Suzuki T, Naiki T, Naiki-Ito A, Ando R, Iida K, Kawai N, Tozawa K, Miyata N, Kohri K, Takahashi S (2015) NCL1, a highly selective lysine-specific demethylase 1 inhibitor, suppresses prostate cancer without adverse effect. Oncotarget 6:2865–2878
60. Ogasawara D, Itoh Y, Tsumoto H, Kakizawa T, Mino K, Fukuhara K, Nakagawa H, Hasegawa M, Sasaki R, Mizukami T, Miyata N, Suzuki T (2013) Lysine-specific demethylase 1-selective inactivators: protein-targeted drug delivery mechanism. Angew Chem Int Ed 52:8620–8624

61. Kakizawa T, Mizukami T, Itoh Y, Hasegawa M, Sasaki R, Suzuki T (2016) Evaluation of phenylcyclopropylamine compounds by enzymatic assay of lysine-specific demethylase 2 in the presence of NPAC peptide. Bioorg Med Chem Lett 26:1193–1195
62. Sugino N, Kawahara M, Tatsumi G, Kanai A, Matsui H, Yamamoto R, Nagai Y, Fujii S, Shimazu Y, Hishizawa M, Inaba T, Andoh A, Suzuki T, Takaori-Kondo A (2017) A novel LSD1 inhibitor NCD38 ameliorates MDS-related leukemia with complex karyotype by attenuating leukemia programs via activating super-enhancers. Leukemia 31:2303–2314
63. Rao S, Zafar A (2014) Methods and compositions comprising lysine-specific demethylase inhibitors (LSD) for inhibiting growth of cancer stem cells. Patent Appl WO2014205511A1
64. Sareddy GR, Viswanadhapalli S, Surapaneni P, Suzuki T, Brenner A, Vadlamudi RK (2017) Novel KDM1A inhibitors induce differentiation and apoptosis of glioma stem cells via unfolded protein response pathway. Oncogene 36:2423–2434
65. Mohammad HP, Smitheman KN, Kamat CD, Soong D, Federowicz KE, van Aller GS, Schneck JL, Carson JD, Liu Y, Butticello M, Bonnette WG, Gorman SA, Degenhardt Y, Bai Y, McCabe MT, Pappalardi MB, Kasparec J, Tian X, McNulty KC, Rouse M, McDevitt P, Ho T, Crouthamel M, Hart TK, Concha NO, McHugh CF, Miller WH, Dhanak D, Tummino PJ, Carpenter CL, Johnson NW, Hann CL, Kruger RG (2015) A DNA hypomethylation signature predicts antitumor activity of LSD1 inhibitors in SCLC. Cancer Cell 28:57–69
66. Maes T, Mascaró C, Ortega A, Lunardi S, Ciceri F, Somervaille TC, Buesa C (2015) KDM1 histone lysine demethylases as targets for treatments of oncological and neurodegenerative disease. Epigenomics 7:609–626
67. Vianello P, Botrugno OA, Cappa A, Dal Zuffo R, Dessanti P, Mai A, Marrocco B, Mattevi A, Meroni G, Minucci S, Stazi G, Thaler F, Trifiró P, Valente S, Villa M, Varasi M, Mercurio C (2016) Discovery of a novel inhibitor of histone lysine-specific demethylase 1A (KDM1A/LSD1) as orally active antitumor agent. J Med Chem 59:1501–1517
68. Rotili D, Tomassi S, Conte M, Benedetti R, Tortorici M, Ciossani G, Valente S, Marrocco B, Labella D, Novellino E, Mattevi A, Altucci L, Tumber A, Yapp C, King ON, Hopkinson RJ, Kawamura A, Schofield CJ, Mai A (2014) Pan-histone demethylase inhibitors simultaneously targeting Jumonji C and lysine-specific demethylases display high anticancer activities. J Med Chem 57:42–55
69. Ota Y, Itoh Y, Kaise A, Ohta K, Endo Y, Masuda M, Sowa Y, Sakai T, Suzuki T (2016) Targeting cancer with PCPA-drug conjugates: LSD1 inhibition-triggered release of 4-hydroxytamoxifen. Angew Chem Int Ed 55:16115–16118
70. Schmitt ML, Hauser AT, Carlino L, Pippel M, Schulz-Fincke J, Metzger E, Willmann D, Yiu T, Barton M, Schüle R, Sippl W, Jung M (2013) Nonpeptidic propargylamines as inhibitors of lysine specific demethylase 1 (LSD1) with cellular activity. J Med Chem 56:7334–7342
71. Lee MG, Wynder C, Schmidt DM, McCafferty DG, Shiekhattar R (2006) Histone H3 lysine 4 demethylation is a target of nonselective antidepressive medications. Chem Biol 13:563–567
72. Culhane JC, Wang D, Yen PM, Cole PA (2010) Comparative analysis of small molecules and histone substrate analogues as LSD1 lysine demethylase inhibitors. J Am Chem Soc 132:3164–3176
73. Prusevich P, Kalin JH, Ming SA, Basso M, Givens J, Li X, Hu J, Taylor MS, Cieniewicz AM, Hsiao PY, Huang R, Roberson H, Adejola N, Avery LB, Casero RA Jr, Taverna SD, Qian J, Tackett AJ, Ratan RR, McDonald OG, Feinberg AP, Cole PA (2014) A selective phenelzine analogue inhibitor of histone demethylase LSD1. ACS Chem Biol 9:1284–1293
74. Hazeldine S, Pachaiyappan B, Steinbergs N, Nowotarski S, Hanson AS, Casero RA Jr, Woster PM (2012) Low molecular weight amidoximes that act as potent inhibitors of lysine-specific demethylase 1. J Med Chem 55:7378–7391
75. Wu F, Zhou C, Yao Y, Wei L, Feng Z, Deng L, Song Y (2016) 3-(Piperidin-4-ylmethoxy) pyridine containing compounds are potent inhibitors of lysine specific demethylase 1. J Med Chem 59:253–263
76. Sorna V, Theisen ER, Stephens B, Warner SL, Bearss DJ, Vankayalapati H, Sharma S (2013) High-throughput virtual screening identifies novel N′-(1-phenylethylidene)-benzohydrazides as potent, specific, and reversible LSD1 inhibitors. J Med Chem 56:9496–9508

77. Baell JB, Holloway GA (2010) New substructure filters for removal of pan assay interference compounds (PAINS) from screening libraries and for their exclusion in bioassays. J Med Chem 53:2719–2740
78. Fiskus W, Sharma S, Shah B, Portier BP, Devaraj SG, Liu K, Iyer SP, Bearss D, Bhalla KN (2014) Highly effective combination of LSD1 (KDM1A) antagonist and pan-histone deacetylase inhibitor against human AML cells. Leukemia 28:2155–2164
79. Mould DP, Alli C, Bremberg U, Cartic S, Jordan AM, Geitmann M, Maiques-Diaz A, McGonagle AE, Somervaille TCP, Spencer GJ, Turlais F, Ogilvie DJ (2017) Development of (4-cyanophenyl)glycine derivatives as reversible inhibitors of lysine specific demethylase 1. J Med Chem 60:7984–7999
80. Mould DP, Bremberg U, Jordan AM, Geitmann M, Maiques-Diaz A, McGonagle AE, Small HF, Somervaille TCP, Ogilvie DJ (2017) Development of 5-hydroxypyrazole derivatives as reversible inhibitors of lysine specific demethylase 1. Bioorg Med Chem Lett 27:3190–3195
81. Mould DP, Bremberg U, Jordan AM, Geitmann M, McGonagle AE, Somervaille TCP, Spencer GJ, Ogilvie DJ (2017) Development and evaluation of 4-(pyrrolidin-3-yl)benzonitrile derivatives as inhibitors of lysine specific demethylase 1. Bioorg Med Chem Lett 27:4755–4759
82. Vianello P, Sartori L, Amigoni F, Cappa A, Fagá G, Fattori R, Legnaghi E, Ciossani G, Mattevi A, Meroni G, Moretti L, Cecatiello V, Pasqualato S, Romussi A, Thaler F, Trifiró P, Villa M, Botrugno OA, Dessanti P, Minucci S, Vultaggio S, Zagarrí E, Varasi M, Mercurio C (2017) Thieno[3,2-b]pyrrole-5-carboxamides as new reversible inhibitors of histone lysine demethylase KDM1A/LSD1. Part 2: structure-based drug design and structure-activity relationship. J Med Chem 60:1693–1715
83. Sartori L, Mercurio C, Amigoni F, Cappa A, Fagá G, Fattori R, Legnaghi E, Ciossani G, Mattevi A, Meroni G, Moretti L, Cecatiello V, Pasqualato S, Romussi A, Thaler F, Trifiró P, Villa M, Vultaggio S, Botrugno OA, Dessanti P, Minucci S, Zagarrí E, Carettoni D, Iuzzolino L, Varasi M, Vianello P (2017) Thieno[3,2-b]pyrrole-5-carboxamides as new reversible inhibitors of histone lysine demethylase KDM1A/LSD1. Part 1: high-throughput screening and preliminary exploration. J Med Chem 60:1673–1692
84. Speranzini V, Rotili D, Ciossani G, Pilotto S, Marrocco B, Forgione M, Lucidi A, Forneris F, Mehdipour P, Velankar S, Mai A, Mattevi A (2016) Polymyxins and quinazolines are LSD1/KDM1A inhibitors with unusual structural features. Sci Adv 2:e1601017

Top Med Chem (2020) 33: 221–254
DOI: 10.1007/7355_2019_75

Published online: 14 January 2020

Inhibitors of JmjC-Containing Histone Demethylases

Miranda Wright, Paul E. Brennan, and Akane Kawamura

Contents

1 Introduction 222
  1.1 JmjC-Domain-Containing Histone Demethylases 222
  1.2 JmjC-KDM Biology 225
2 Mechanism and Structure of JmjC-KDMs 226
  2.1 Structural Alignment of JmjC-KDMs 226
3 Development of JmjC-KDM Inhibitors 234
  3.1 Inhibitors Targeting the 2OG Binding Site 234
  3.2 Substrate Competitive Inhibitors 245
4 Conclusions 248
References 249

Abstract Histone demethylases (KDMs) catalyse the removal of *N*-methyl marks on histones and play important roles in epigenetic regulation. Abnormal histone methylation and dysregulation of KDMs have been linked to multiple diseases, and KDMs are emerging as promising therapeutic targets. This chapter provides an overview of JmjC-domain-containing KDMs (JmjC-KDMs), with a particular focus on recent advances in JmjC-KDM inhibitor development from a structural perspective.

Keywords 2OG oxygenases, JmjC histone demethylases, KDMs

M. Wright and P. E. Brennan
Structural Genomics Consortium (SGC), University of Oxford, Oxford, UK

Nuffield Department of Medicine, Target Discovery Institute, University of Oxford, Oxford, UK

A. Kawamura (✉)
Chemistry Research Laboratory, Department of Chemistry, University of Oxford, Oxford, UK

Radcliffe Department of Medicine, Wellcome Centre for Human Genetics, University of Oxford, Oxford, UK
e-mail: akane.kawamura@chem.ox.ac.uk

Abbreviations

2,4-PDCA	2,4-Pyridine dicarboxylate
2OG	2-Oxoglutarate
8HQ	8-Hydroxyquinoline
Bpy	4′-Carboxy 2,2′-bipyridine
H3	Histone H3
HMT	Histone methyltransferase
JmjC	Jumonji-C
KDM	Lysine demethylase
Kme_n	Methylated lysine ($n = 1$, mono-; $n = 2$, di-; $n = 3$, tri-)
NOG	*N*-oxalylglycine
PHD-finger	Plant homeodomain finger

1 Introduction

Histone demethylases (KDMs) catalyse the removal of *N*-methyl marks on histones and play important roles in epigenetic regulation. Abnormal histone methylation and dysregulation of KDMs have been linked to multiple diseases. The majority of KDMs are JmjC-domain-containing KDMs (JmjC-KDMs) that belong to the superfamily of Fe(II) and 2-oxoglutarate (2OG)-dependent oxygenases. JmjC-KDMs are emerging as promising therapeutic targets, in particular, in cancer. The availability of crystal structures has provided understanding of the mechanism, function and selectivity of JmjC-KDMs at the molecular level and has aided structure-based inhibitor design. This chapter provides an overview of recent advances in JmjC-KDM inhibitor development from a structural perspective.

1.1 JmjC-Domain-Containing Histone Demethylases

Post-translational modifications on histones, including lysine methylation, have key roles in transcriptional control and cellular function. Lysines on histone tails can be mono-(me1), di-(me2) or tri-(me3)methylated and dynamically regulated by histone methyltransferases (HMTs) and lysine demethylases (KDMs) [1].

There are two distinct families of KDMs that catalyse the demethylation of *N*-methyllysines on histones. KDM1/LSDs belong to the flavin adenine dinucleotide (FAD)-dependent monoamine oxidase superfamily, and Jumonji-C (JmjC) domain-containing KDMs (JmjC-KDMs), the larger of the two KDM families, belong to the Fe(II) and 2-oxoglutarate (2OG)-dependent oxygenase superfamily. There are at least six human subfamilies of JmjC-KDMs (KDM2–7), based on homologies of their catalytic JmjC-domains, with each having distinct sequence and *N*-methylation state selectivity (Table 1, see Højfeldt [2] for reviews on function and substrate selectivities of KDMs).

Trimethylated as well as mono- and di-methylated lysines can be demethylated by JmjC-KDMs, unlike KDM1 subfamily which cannot demethylate trimethylated lysines due to the required formation of an imine. The mechanism of demethylation by JmjC-KDMs requires 2OG and dioxygen as co-substrates and Fe(II) as a cofactor. Evidence shows that demethylation is via known 2OG oxygenase catalysis in which 2OG and the substrate, followed by dioxygen, bind to the active site (Fig. 1). A highly reactive Fe(IV)-oxo intermediate, formed through oxidative decarboxylation of 2OG, can subsequently abstract a hydrogen atom from the *N*-methyl group. This results in an unstable hemiaminal which can fragment to give demethylated lysine and formaldehyde.

Despite the conservation in the catalytic mechanism, the preference for the degree of methylation and the lysine position(s) on the histone tails are quite distinct for each JmjC-KDM subfamily (Table 1, see Sect. 2.1.2). KDM2/KDM3/KDM7 subfamilies are di- and mono-*N*ε-methyl lysine demethylases, whereas KDM4/KDM5/KDM6 subfamilies can demethylate tri-, di- and, in some cases, mono-*N*-ε-methylated lysines. KDM5s, KDM6s and KDM2s are highly specific to demethylation at lysine 4 position of histone H3 (H3K4), H3K27 and H3K36 respectively, while KDM3/4/7 can all demethylate at H3K9. KDM4A/B/C can demethylate multiple histone substrates, including at H3K9 and H3K36, whereas KDM4D (and proposed pseudogene KDM4E) demethylate at H3K9 [1, 3]. As the site and degree of methylation on histones have differential effects on chromatin dynamics, recruitment of epigenetic protein complexes and transcription, the KDMs have wide-ranging functions.

There are approximately 30 proteins that have the JmjC-domains in humans, but not all JmjC-containing proteins have demethylase activities (see [4] for review). While the majority (approximately 20, Table 1) are JmjC-KDMs, some JmjC-proteins are hydroxylase enzymes (e.g. FIH, JMJD4).

Functional assignments for some JmjC-proteins remain controversial, with both hydroxylase and demethylase activities being reported (e.g. MINA53 and NO66, JMJD5/KDM8, JMJD6). Recent work has revealed that some JmjC-KDMs can catalyse the demethylation of methylated lysines on non-histone proteins, as well as methylated arginines, thus extending the possibility of a much wider biological role beyond histone lysine modulation [5–7].

In addition to the catalytic JmjC-domain, many JmjC-KDMs have ancillary domains that are involved in recognising histone modifications (e.g. reader domains such as PHD-finger domains or Tudor domains) or DNA-binding domains (e.g. AT-rich interacting domains (ARID), CXXC zinc-finger domains). In some cases, these ancillary domains can allosterically modulate the KDM catalysis and influence substrate specificity [8, 9]. There are also biological functions of KDMs that are independent of catalytic activity, as exemplified by Jarid2, the founding member of JmjC-family. Jarid2 is predicted to be catalytically inactive but associates with polycomb proteins and has essential roles in ES cell differentiation and development [10].

Table 1 Comparison of residues binding to metal centre in JmjC-KDMs

Name	Synonyms	Substrate[a]	PDB (ligand)	Fe(II) metal-binding residues		
KDM2A	FBXL11, JHDM1A	H3K36me2/1	4QX7 (2OG, Ni(II), H3K36me2)[b]	H212	D214	H284
KDM2B[c]	FBXL10, JHDM1B	H3K36me2/1	–	H211	D213	H283
KDM3A[c]	JMJD1A, JHDM2A	H3K9me2/1	–	H1120	D1122	H1175
KDM3B	JMJD1B, JHDM2B	H3K9me2/1	4C8D (2OG, Mn(II))	H1560	D1562	H1689
JMJD1C	KDM3C, JHDM2C		5FZO (Mn(II))	H2336	E2338	H2466
KDM4A	JMJD2A, JHDM3A	H3K9me3/2, H3K36me3/2	5TVR (2OG, Ni(II))	H188	E190	H276
KDM4B	JMJD2B, JHDM3B	H3K9me3/2, H3K36me3/2	4LXL (24PDCA, Ni (II), H3K9me3)	H189	E191	H277
KDM4C	JMJD2C, JHDM3C	H3K9me3/2	4XDO (2OG, Fe(II))	H190	E192	H278
KDM4D	JMJD2D, JHDM3D	H3K9me3/2	4HON (2OG, Ni(II), H3K9me3)	H192	E194	H280
KDM5A	JARID1A, RBP2	H3K4me3/2	5IVB (2OG, Mn(II))	H483	E485	H571
KDM5B	JARID1B, PLU1	H3K4me3/2	5FUP (2OG, Mn(II))	H499	E501	H587
KDM5C	JARID1C, SMCX	H3K4me3/2	5FWJ (**4**, Mn(II))	H514	E516	H602
KDM5D[c]	JARID1D, SMCY	H3K4me3/2		H514	E516	H602
KDM6A	UTX	H3K27me3/ me2	3AVR (2OG, Ni(II), H3K27me3)	H1146	E1148	H1226
KDM6B	JMJD3	H3K27me3/ me2	2XUE (2OG, Fe(II))	H1390	E1392	H1470
KDM6C	UTY		3ZLI (2OG, Fe(II))	H1014	E1016	H1094
KDM7A	KIAA1718, JHDM1D	H3K9me2/1, H3K27me2/1	3KVA (2OG, Fe(II))	H282	D284	H354
KDM7B	PHF8, JHDM1F	H3K9me2/1	3KV4 (2OG, Ni(II), H3K4me3K9me2)	H247	D249	H319
PHF2	KDM7C, JHDM1E		3PU8 (2OG, Fe(II))	H249	D251	Y321

The crystal structures of human JmjC-KDMs reveal three residues that coordinate the active site Fe(II)

[a]Independently verified consensus assignment in vitro and in cells. Other putative substrates have been identified in vitro. Ligands are Ni(II) and NOG unless otherwise stated. All PDBs are from *Homo sapiens* except where indicated

[b]*Mus musculus*

[c]Where the metal-chelating residues are identified from sequence alignments [1]. Note that the KDM nomenclature is not used when histone demethylase activity has not been independently verified. PDB is a representative from the protein when more than one structure is available

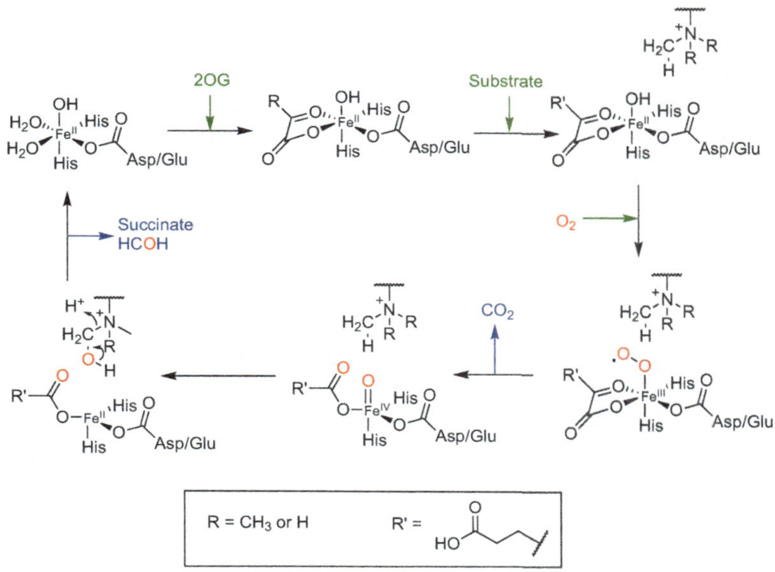

Fig. 1 JmjC-KDM catalytic mechanism. 2OG, the substrate and O_2 bind the enzyme sequentially. Loss of CO_2 from 2OG generates a succinate-bound reactive Fe(IV)-oxo species that radically abstracts a hydrogen from the *N*-methyl group. Dissociation of succinate from the iron, loss of formaldehyde and dissociation of the demethylated product regenerate the active enzyme

1.2 JmjC-KDM Biology

There is now strong evidence that changes in the expression levels of some lysine demethylases can lead to the onset and progression of certain cancers. This occurs either through activating expression of oncogenes, repressing expression of tumour suppressors, disrupting chromosomal stability, altering DNA mismatch repair or interacting with hormonal receptors that control cellular proliferation [11–13]. Overexpression of KDM4s has been linked to various cancers such as lung, breast, oesophageal and prostate cancers and lymphoma [14–18]. KDM5 proteins are also associated with several cancers such as acute myeloid leukaemia [19], hepatocellular carcinoma [20] and lung [21], gastric [22], melanoma [23], breast [24], ovarian [25] and testicular cancer [26] and are essential for the survival of multiple drug-tolerant cancer cells [27, 28]. KDM3 is involved in controlling chemoresistance and tumorigenic growth in human colorectal cancer stem cells through activating Wnt target gene transcription [29].

KDMs also have important roles in development such as KDM5B which blocks differentiation of embryonic and haematopoietic stem cells [30] and KDM5C which has a role in neuronal development [31]. KDM2A (FBXL11) inhibits osteo/dentinogenic differentiation in mesenchymal stem cells [32] and also regulates cell proliferation in human stem cells from apical papilla [33]. It also has a role in the regulation of NF-KB through lysine demethylation of the p65 subunit [34]. KDM6A (UTX) point and deletion mutations have been associated with Kabuki syndrome, a rare congenital disorder with symptoms including growth retardation, intellectual disability and skeletal abnormalities [35, 36]. KDM7B (PHF8) has a key role in midline formation and cognitive development [37], migration of epithelial cells [38], as well as control of inflammation in leukocytes [39].

Due to their interesting biology and their role in development and various diseases, JmjC-KDMs have become popular therapeutic targets for inhibitor design.

2 Mechanism and Structure of JmjC-KDMs

2.1 *Structural Alignment of JmjC-KDMs*

The catalytic core of JmjC-domain in JmjC-KDMs is made up of eight antiparallel β-strands (βI-VIII) arranged in a distorted double-stranded β-helix (DSBH) core fold (Fig. 2). Long β-strands I, VIII, III and VI (major β-sheet) on one face and short β-strands II, VII, IV and V (minor β-sheet) on the other face result in an asymmetric barrel-like DSBH fold, to form an active site. The JmjC-KDMs have a characteristic extended N-terminal loop between DSBH and the N-terminal anti-β strand, as well as an extended insertion between strands IV and V of the DSBH which is involved in substrate binding [4].

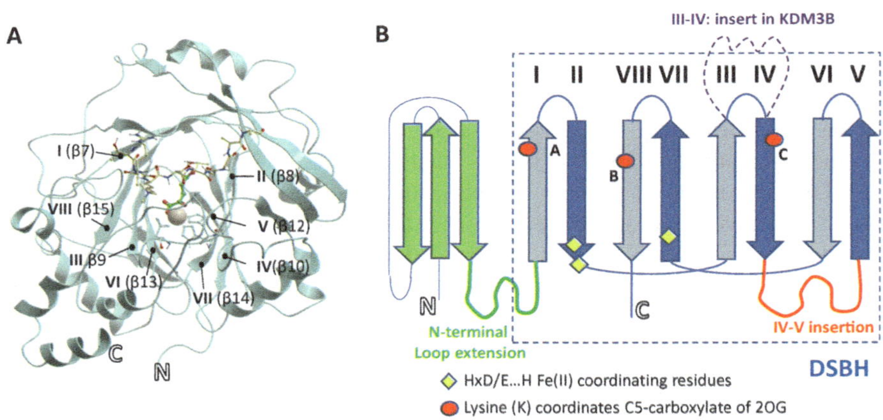

Fig. 2 Structure of JmjC-domain of KDMs. (**a**) Overall structural fold of JmjC-KDMs. Crystal structure of KDM4A in complex with H3K9me3 and *N*-oxalylglycine (NOG, PDB:2OQ6) with the core DSBH labelled. (**b**) Cartoon structure of DSBH fold. Fe(II)-coordinating motif (yellow diamonds) and the basic residue Lys that coordinates the C5 carboxylate of 2OG (red dot, A = KDM6A/B; B = KDM3B; C = KDM2A, KDM4A–E, KDM5A/B, KDM7A–C) are highlighted. Major β-strands are in light blue and minor β-strands are in blue. Adapted from [4]

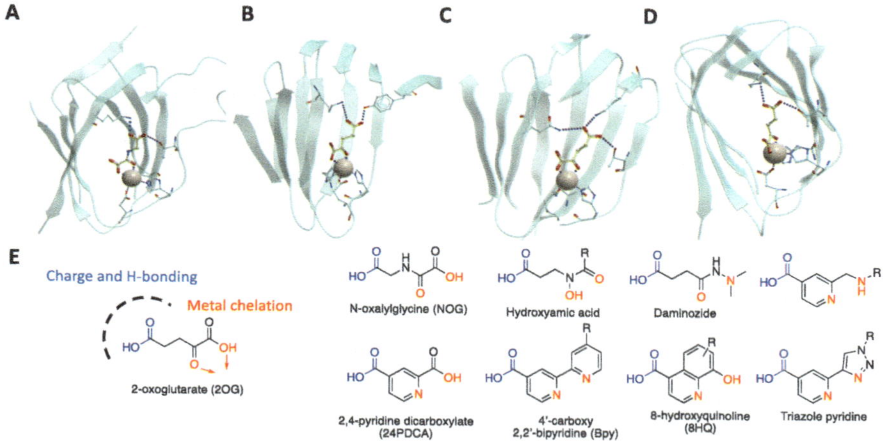

Fig. 3 Polar interactions of JmjC-KDMs with C5-carboxylate of 2OG (or NOG): (**a**) KDM3B/C binding mode; (**b**) KDM4/5 binding mode; (**c**) KDM6A/B/C binding mode; (**d**) KDM2A, KDM7A, B and C binding mode; (**e**) 2OG mimicking JmjC-KDM inhibitor scaffolds

2.1.1 2OG Cofactor Binding Site

The JmjC-KDMs have structurally similar active sites in line with their catalytically similar mechanisms, where 2OG cofactor binds to the catalytic Fe(II) centre through bidentate coordination of the keto acid (Fig. 1). In all KDM structures, the metal ion is positioned through interaction with a triad of residues that forms a HxD/E . . . H motif (Table 1). Note that catalytically inactive 2OG analogue, *N*-oxalylglycine (NOG), and Ni(II) or Mn(II) are often used for JmjC-KDM crystallography. The two His residues are positioned distally on βII, βVII or βVIII, and the Asp or Glu residue is located on βII/βIII loop (Fig. 2). For KDM2A/B, KDM3A/B and KDM7A/B, two histidines and an aspartic acid coordinate to the metal centre. For KDM4 (JMJD2A, B, C and D), KDM5 (JARID1A, B, C and D) and KDM6 (JMJD3, UTX and UTY), the acidic aspartate residue is replaced with a glutamic acid. PHF2 (KDM7C) is anomalous with a histidine, aspartic acid and tyrosine residues coordinating to the metal centre. The three metal coordinating residues are essential for catalysis, and Ala mutants have been shown to abolish the enzyme activity, and often used as catalytically inactive controls [40]. Due to the conserved catalytic mechanism involving Fe(II), the majority of JmjC-KDM inhibitors bind through a monodentate or bidentate coordination to the metal centre and compete with 2OG (Fig. 3).

The other main interaction at the 2OG-binding site is the salt bridge formed between the terminal carboxylate of 2OG and residues in the active site of KDMs. Inhibitors of KDMs therefore commonly contain a carboxylic acid or a bioisosteric replacement to mimic 2OG binding and form an ionic interaction with a lysine in the KDM binding pocket (Fig. 3). Although lysine is found on different β-strands, it is orientated so that the terminal amino group overlaps in different KDM structures.

Table 2 Comparison of residues interacting with carboxylate in KDM structures on different β strands

Strands	N-terminal anti-β strands	DSBH							
		β1	βII	βIII	βIV	βV	βVI	βVII	βVIII
KDM2A			T209		K229				
KDM3B			T1557						K1699
JMJD1C			T2333						Q2476
KDM4A	Y132				K206				
KDM4B	Y133				K207				
KDM4C	Y134				K208				
KDM4D	Y136				K210				
KDM5A	Y409				K501				
KDM5B	Y425				K517				
KDM5C	Y440				K532				
KDM6A		K1137	T1143	N1156					
KDM6B		K1381	T1387	N1400					
KDM6C		K1084	T1090	N1103					
KDM7A			T279	Y292	K299				
KDM7B			T244		K263				
PHF2				Y258	K266			T323	

All JmjC-KDMs have a lysine residue (green) that forms a salt bridge to a carboxylate in 2OG and inhibitors, with the exception of KDM3C which has a glutamine (grey). Further hydrogen bonds to the substrate/inhibitor carboxylate are made by threonine (salmon), tyrosine (yellow) and asparagine (blue)

Unlike all other KDMs, JMJD1C (KDM3C) contains a glutamine rather than a lysine at this position in the sequence which overlaps well with the analogous lysine interaction. Other key residues interacting with the carboxylate include tyrosine and threonine. KDM2A, KDM3C and KDM7A/B/C have a threonine within hydrogen-bonding distance to the carboxylate. KDM6A/B/C have a threonine and asparagine residues in addition to the lysine that can hydrogen bond to the carboxylate (Table 2).

Unlike the other KDMs, KDM4 (JMJD2) and KDM5 (JARID1) contain a tyrosine which interacts with the carboxylate of 2OG. This key structural difference in the binding site can be exploited in inhibitor design to give KDM4/5 inhibitors that are selective over the other KDMs. However, selectivity between KDM4 and KDM5 is difficult to achieve due to identical residues in the primary sphere of the binding pocket. Differences in the second sphere of amino acids can confer different conformations to the binding residues which can result in improved selectivity. Structural information obtained from different co-crystal structures with inhibitors has shown that this key tyrosine residue is more flexible in KDM5 than KDM4. Some selectivity for KDM5 has therefore been observed with cyanopyrazoles, as the tyrosine side chain can be displaced to allow binding in KDM5 [41].

Table 3 Structures of JmjC-KDMs in complex with histone peptides

KDM(JmjC)	Ligands	PDB	References
Kdm2A	H3K36me1, Ni(II), NOG	4QXH	[42]
	H3K36me1, Ni(II), 2OG	4QWN	[42]
	H3K36me2, Ni(II), NOG	4QXC	[42]
	H3K36me2, Ni(II), 2OG	4QX7	[42]
	H3K36me, Ni(II), NOG	4QXB	[42]
	H3K36me3, Ni(II), 2OG	4QX8	[42]
KDM4A	H3K9me3, Ni(II), 2OG	2OQ6	[43]
		2Q8C	[44]
	H3K9me2, Ni(II), 2OG	2OX0	[43]
	H3K9me1, Ni(II), 2OG	2OT7	[43]
	H3K36me2, Ni(II), succinate	2Q8D	[44]
	H3K36me3, Ni(II), 2OG	2OS2	[43]
	H3K36me3, Ni(II), 2OG	2P5B	[44]
	H3K36me3, Ni(II), 2OG	2Q8E	[45]
	H3K36me1, Fe(II), 2OG	2PXJ	[45]
	H4R3me2s, Ni(II), 2OG	5FWE	[7]
	H3K27me3, Ni(II), 2OG	4V2W	[46]
KDM4B	H3K9me3, Ni(II), 24PDCA	4LXL	[47]
KDM4D	H3K9me3, Ni(II), 2OG,	4HON	[48]
KDM6A	H3K27me3, Ni(II), 2OG,	3AVR	[49]
Kdm6B	H3K27me3, Ni(II), NOG	4EZH	[50]
KDM6B	H3A21M, Fe(II), 2OG	5OY3	[51]
KDM7B	H3K4me3K9me2, Ni(II), 2OG	3KV4	[8]

Note all are human JmjC-KDMs except Kdm2A and Kdm6B which are from *Mus musculus*

2.1.2 Histone Substrate Binding

Despite the high conservation in the active site of JmjC-KDMs, the individual subfamilies have distinct substrate specificities (Table 1). Structural information is available for some JmjC-KDMs complexed with histone substrates, which provide some insight into the unique features of these family of proteins (Table 3).

Recognition of Methylated Lysines Structures of KDM4A in complex with H3K9me3/2/1 revealed that the size of the active site region can, to some degree, confer methylation state selectivity in KDM4A [43] (Fig. 4a–c). The tri-Nε-methylated lysine (Kme3) fits tightly in the active site pocket composed of Y177, E190, S288 and N290 and one methyl group projects towards the catalytic metal (position **b**), whereas the other two point towards Y177 (position **c**) and N290 (position **a**). In di-N-ε-methylated lysine (Kme2) structures, the methyl group can either occupy a productive conformation pointing towards the metal or a nonproductive conformation towards Y177 and N290. In the case of mono Nε-methylated lysine, the water molecules are positioned to direct the methyl group away from the metal, stabilising the unproductive

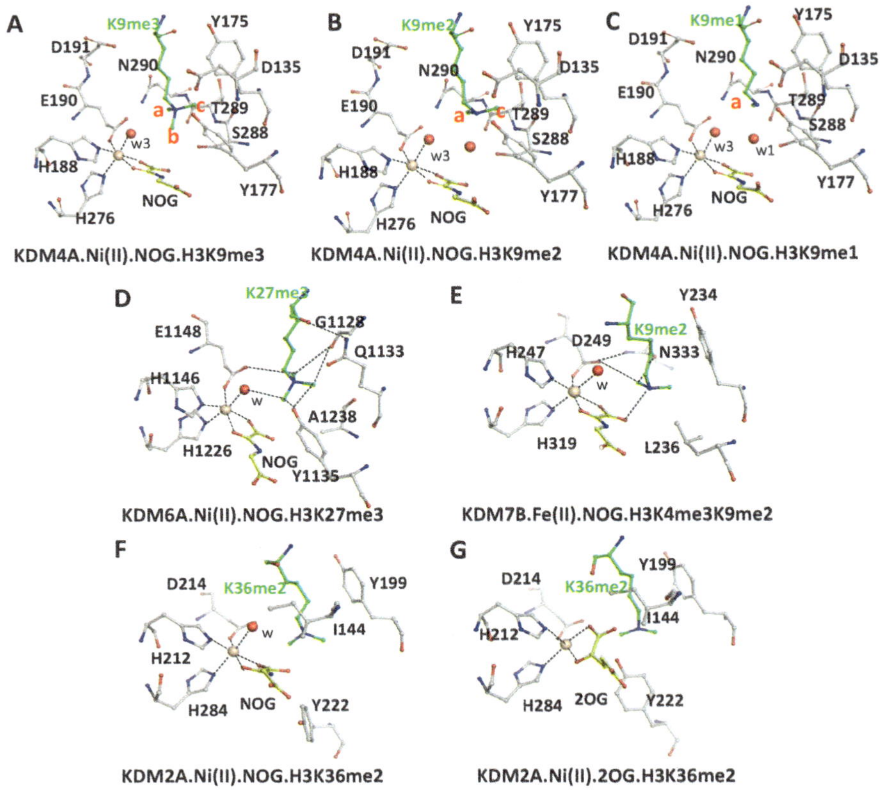

Fig. 4 Crystal structures of JmjC-KDMs in complex with NOG/2OG and histone H3 substrates. (**a–c**) Different orientations of methyl groups are observed for KDM4A complexed with (**a**) H3K9me3 (PDB 2OQ6), (**b**) H3K9me2 (PDB 2OT7) and (**c**) H3K9me1 (PDB 2OTQ). (**d**) KDM6A in complex with H3K27me3 and (**e**) KDM7B in complex with H3K9me2 in the JmjC-domain active sites. (**f, g**) KDM2A in complex with H3K36me2 with NOG (**f**) and 2OG (**g**). NOG adopts an equatorial binding mode to H212, whereas 2OG co-crystal structures can also adopt axial binding mode relative to H212. Two different conformers are observed for Y222 with H3K36me1/2 but only axial for H3K36me3. Histone methyllysines, NOG/2OG, metal and water molecules are in green, yellow, orange and red (sphere), respectively

over productive mode, in agreement with the lack of activity [43]. Similar Kme3 interactions are observed in KDM6A, whereby the H3K27me3 tri-methyl group points towards the active site metal (Fig. 4d). The tri-methyl group is stabilised by Y1135, and G1128 forming H-bonded interactions with methyl groups, and E1148 forms electrostatic interaction with methylammonium group to stabilise the orientation of methyl group for demethylation [49]. From the peptide co-crystal structures available, the Tyr which stabilises the methyl group of Kme3 is conserved across all trimethyllysine demethylases (Y177(KDM4A)/Y1135(KDM6A)/Y1380(KDM6B)), but it is absent in

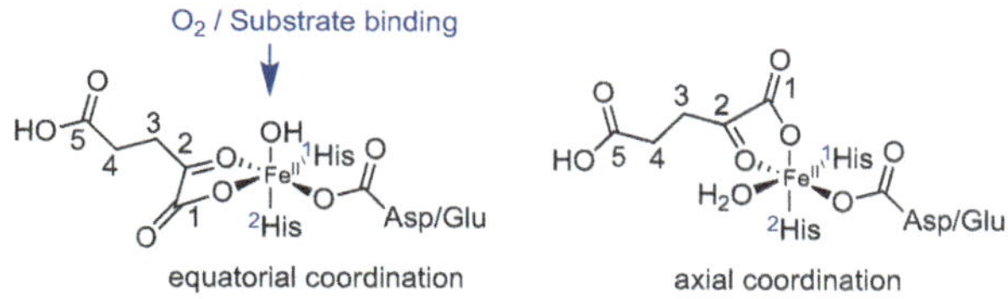

Fig. 5 Equatorial and axial mode of 2OG binding. Two different coordination modes of 2OG. Equatorial coordination of 2OG to His1 is catalytically productive, whereas axial coordination of 2OG cannot bind O_2/substrate. Carbons on 2OG are numbered. NOG has nitrogen instead of C3 position. The first and second histidines in the HxD/E . . . H motif are numbered in blue

KDM2A/KDM7A dimethyllysine demethylases. Thus it is proposed that this Tyr may be important in Kme3 recognition [49].

In KDM7B, a Nε-methyl group of H3K9me2 points towards the aromatic ring of Y234 and the other towards D249 and N333 forming hydrogen bonds. The lone pair from the terminal di-methylated nitrogen forms a hydrogen bond with an oxygen atom of 2OG (NOG in crystal structure); thus a steric clash prevents accommodation of a tri-methyl lysine [8] (Fig. 4e).

In contrast, the crystal structure of mouse Kdm2A in complex with H3K36me3/me2/me1 reveals that the spatial constraint within the methyllysine binding pocket is not the sole determinant of methylation state selectivity in all JmjC-KDMs [42]. Kdm2A is a H3K36me2 di-/mono-demethylase, which is structurally very similar to KDM7s. In H3K36me2 complexed structure, the lysine side-chain binding is stabilised by van der Waals interactions with aromatic (F215, Y199) and hydrophobic interactions (L201, I144) with dimethyl-ammonium group directed towards D214, N298 and NOG (Fig. 4f). However, H3K36me3, which is not a substrate of Kdm2A, can also bind in the same catalytic pocket in a similar manner. Intriguingly, for all NOG-complexed structures, NOG coordinates the metal in an equatorial coordination (also referred to as 'in-line' mode), i.e. C1 carboxylate oxygen is positioned opposite His1 (H284), in plane at equatorial position, whereas all 2OG-complexed structures have axial mode [52], i.e., where C1 carboxylate oxygen coordinates at axial position (also referred to as 'off-line' mode) (Fig. 5). In the 2OG-complexed structures, the side chain of Y222 within the active site can adopt two alternative positions, with major conformer directed towards the metal and minor conformer away from the metal (Fig. 4g). The minor conformer of Y222 reflects the equatorial coordination (as observed in NOG complexes) and is only observed with H3K36me1/2 and not with H3K36me3. It has been proposed that the steric constraints of H3K36me3 binding could prevent the 2OG transitioning from axial to equatorial coordination; thus H3K36me3 cannot be demethylated by Kdm2A despite its binding [42].

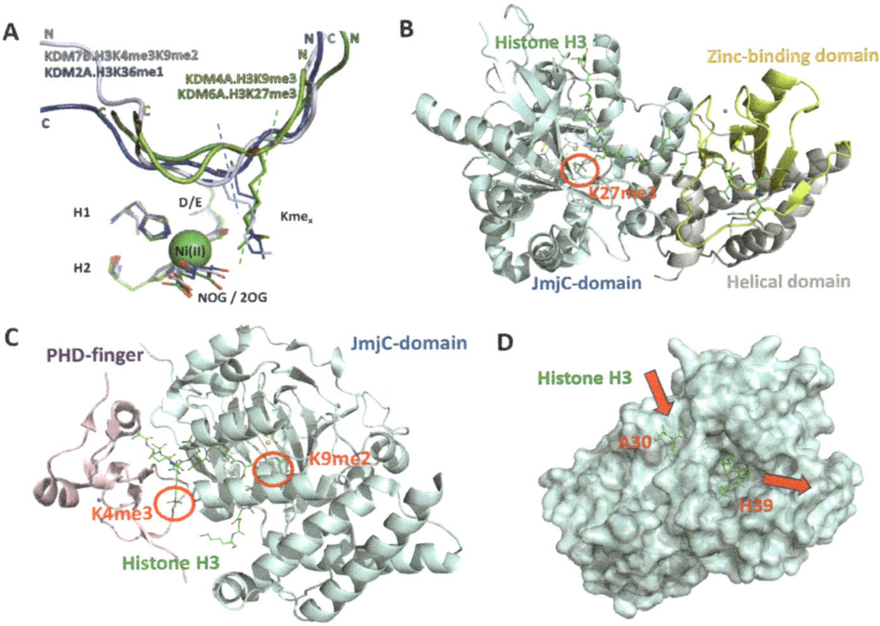

Fig. 6 Histone H3 peptide binding to JmjC-KDMs. (**a**) Overlay of representative crystal structures of JmjC-KDMs in complex with histone H3 peptides. Methyllysines on histone peptides (cartoon) bind in the active site pocket, where it projects towards the catalytic metal coordinated by HxD/E . . . H residues on JmjC-KDMs. The first and the second His are labelled H1 and H2. Methyllysines are projected from different angles for trimethyl-demethylases (KDM4A/6A) relative to dimethyl-demethylases (KDM2A/7A). Overlaid PDB IDs: 4QXH, 2OQ6, 3AVR, 3KV4. (**b**) Histone H3 peptide binds across the JmjC- and zinc-binding domain in KDM6A. H3 peptide binding induces conformational change at the zinc-finger domain upon binding to KDM6A (PDB 3AVR). (**c**) The K4me3 of H3K4me3K9me2 peptide binds to the PHD-finger and places K9me2 at the catalytic site of JmjC-domain in KDM7B (PDB 3KV4). (**d**) 'Molecular threading' of histone H3(30–39) peptide through KDM2A, with K36me3 directing towards the catalytic centre (PDB 4QXB)

Histone Recognition and Binding Pocket The structural overlay of representatives from KDM2/4/6/7 demonstrates that the histone binding orientation (N- to C-terminal orientation of peptides) in the active site pocket is the same for all histone H3 peptide co-crystal structures of JmjC-KDMs except for KDM7B (Fig. 6a). However, in all cases, the methyllysines are positioned in the active site accessed from the same face. Interestingly, Kme3 for KDM4/6s are positioned slightly differently and further away from the catalytic centre compared to the Kme1/2 in KDM2/7s, which may in part account for the inherent differences in the Kme3 and Kme2 demethylases [49].

Histone H3 peptides make multiple interactions with JmjC-KDM proteins, both via their side chains and backbones. In KDM6A, residues surrounding H3K27me3 make interactions with the JmjC-domain, contributing to its substrate specificity and catalytic activity, in addition to the Kme3 binding [49]. Extended interactions of H3 peptide beyond the JmjC-domain are also observed, whereby the zinc-binding domain of KDM6A/B, located C-terminal to JmjC-domain, provides additional affinity and specificity [49, 51] (Fig. 6b). This region undergoes a conformational change upon H3 peptide binding, to expose hydrophobic patches that interact with H3(17–21) region of the peptide. These key interactions have been supported by kinetic studies on mutations to H3 and KDM6A [49], providing evidence that multiple factors contribute towards defining substrate binding and catalysis.

For other binding domains, such as the PHD-fingers, their spatial positioning and distance relative to JmjC-domains can also influence the histone H3 binding sequence specificity. PHD-fingers of KDM7 subfamily recognise the H3K4me3 mark, and in the case of KDM7B, the PHD-finger interacts with H3K4me3 in H3K4me3K9me2 to allow demethylation at K9me2 site by JmjC-domain, while it has negligible binding and catalysis with H3K9me2 alone [53]. The crystal structure of KDM7B in complex with H3K4me3K9me2 reveals that the histone H3 binds across the PHD-finger and JmjC-domain interface in an extended conformation (H3 (1–9)) and forms extensive interactions with both domains or individual domains [53] (Fig. 6c). KDM7A, on the other hand, has a longer linker between the PHD-finger and JmjC-domains and cannot demethylate at the K9 site of H3K4me3K9me2, while it can demethylate at the K27 site of H3K4me3K27me2 [53]. The ancillary reader domain and linker length can thus contribute significantly to the substrate selectivity of the KDM7s. Other JmjC-KDMs, such as the KDM4s and KDM5s, are also allosterically regulated by their ancillary reader domains (PHD-fingers, Tudor domains) [9, 19, 54–56]. It is anticipated that the histone binding complex structures of multidomain KDM5s and KDM4s will provide molecular insight into the substrate binding and cross-talk between the reader and catalytic domains.

The histone H3 binding surfaces can also differ between the JmjC-KDMs. In KDM2A, the H3 peptide 'threads' through a narrow channel that can only be accommodated by Gly residues [42] (Fig. 6d). This steric constrain in the channel provides selectivity towards a unique GG motif at H3K36 region (A29-PATGG-V35). Methylated H3K36 then inserts into a deep cavity ('cavity insertion') and locks the H3K36me2 peptide substrate into the active site pocket [42]. In contrast, in KDM4A, the same H3 peptide sequence binds at the protein surface, and K36me3 inserts into a surface groove ('surface groove' recognition) [42–44]. This binding mode can account for the ability of KDM4s to demethylate multiple histone substrates [7, 46], but KDM2A is highly specific for the H3K36me2.

In summary, the histone substrate recognition and selectivity are achieved through a combination of unique features and mechanisms by JmjC-KDMs, including distinct methyllysine binding pockets, intermolecular side chain and backbone interaction network, different substrate-binding channels and interplay with other domains.

3 Development of JmjC-KDM Inhibitors

There has been significant progress in the development of JmjC-KDM inhibitors since the first inhibitors were reported in 2008 [57]. The overwhelming majority targets the catalytic domain and inhibits the enzyme activity via chelating the active site Fe(II), often competing with 2OG binding. Due to the similarity in the active site pockets of JmjC-KDMs, achieving selectivity has proved challenging, not only between different JmjC-KDM subfamilies but also across a wider 2OG oxygenases superfamily [58]. The recent availability of crystal structures of JmjC-KDM has facilitated medicinal chemistry efforts and enabled the generation of several chemical probes for the JmjC-KDMs (Table 4). In the following sections, we describe the recent progress, focusing on the development of JmjC-KDM inhibitors based on structural rationalisation. For a general overview of JmjC-KDM inhibitors, the readers are referred to recent reviews [59, 60].

3.1 Inhibitors Targeting the 2OG Binding Site

3.1.1 Dual KDM4/5 Inhibitors

Several inhibitors of KDM4/5 have been developed using 2,4-pyridinedicarboxylic acid (2,4-PDCA), a micromolar pan-KDM inhibitor, as a starting point (Fig. 3e).

Table 4 PDB codes for JmjC-KDMs in complex with small molecule inhibitors (top section) or substrate competitive inhibitors (bottom section) described in this chapter

KDM(JmjC)	Ligands	PDB	References
KDM4A	**3**, Ni(II)	5VGI	[62]
	8, Zn(II)	5F3I	[67]
KDM4D	**2**, Co(II)	5FP8	[61]
KDM5A	(*R*)-**6**, Mn(II)	6BH0	[53]
	(*S*)-**6**, Mn(II)	6BH1	[53]
	9, 2OG, Mn(II)	5IW0	[68]
	11, Ni(II)	5CEH	[41]
	13, Ni(II)	5K4L	[70]
	14, Ni(II)	5V9T	[71]
KDM5B	**4**, Mn(II)	5A3T	[63]
	7, Mn(II)	5FUN	[63]
	10, Mn(II)	5A3N	[64]
	15, Mn(II)	5FPU	[63]
KDM6B	**15**, Co(II)	4ASK	[50]
KDM4A	H3K9me3T11C, NOC.Ni(II)	3U4S	[73]
	21, NOG, Ni(II)	5LYI	[76]
	CP2R6Kme3, NOG.Ni(II)	5LY2	[76]
KDM7A	**23**, 2OG.Ni(II)	3U78	[78]

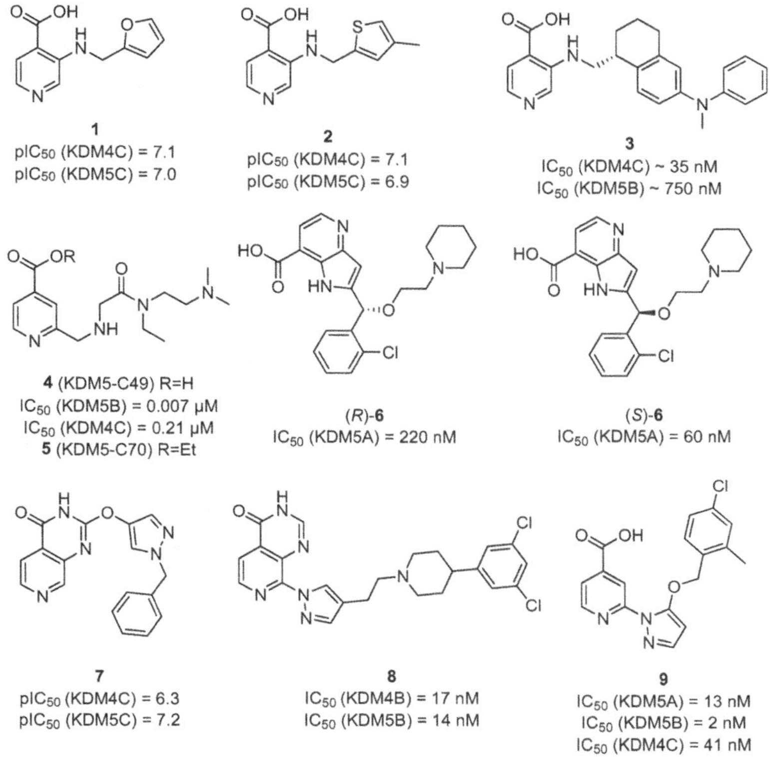

Fig. 7 Structures of dual KDM4/KDM5 inhibitors

3-Amino pyridine carboxylic acid **1** was optimised as a potent inhibitor of KDM4 ($IC_{50} < 100$ nM) and KDM5C ($IC_{50} = 100$–125 nM) in a MS-based biochemical assay [61] (Fig. 7). The compound demonstrated good selectivity (>50-fold) over KDM6B and EGLN3. Some cellular activity against KDM4C was observed ($IC_{50} = 6$–8 μM) with a significant drop in potency from the biochemical assay which was likely due to the poor cell penetration of the carboxylate group. Low cell activity was also observed with overexpressed KDM5C.

A crystal structure of 4-methylthienyl analogue **2** in KDM4D (PDB:5FP8) confirmed binding in the 2OG site through a monodentate coordination of pyridine *N* to the active site metal (Fig. 8a). The carboxylate hydrogen bonded directly to K210 and Y136, while the 4-methylthienyl ring was positioned in a hydrophobic pocket comprised of Y136, A138 and K245.

Using the pan-KDM inhibitor 2,4-PDCA as a starting point, Celgene Quanticel reported 3-aminoisonicotinic acid **3** (QC6352) as a potent inhibitor of KDM4 family [62]. Preliminary SAR showed that (*R*)-tetrahydronaphthalene was the active

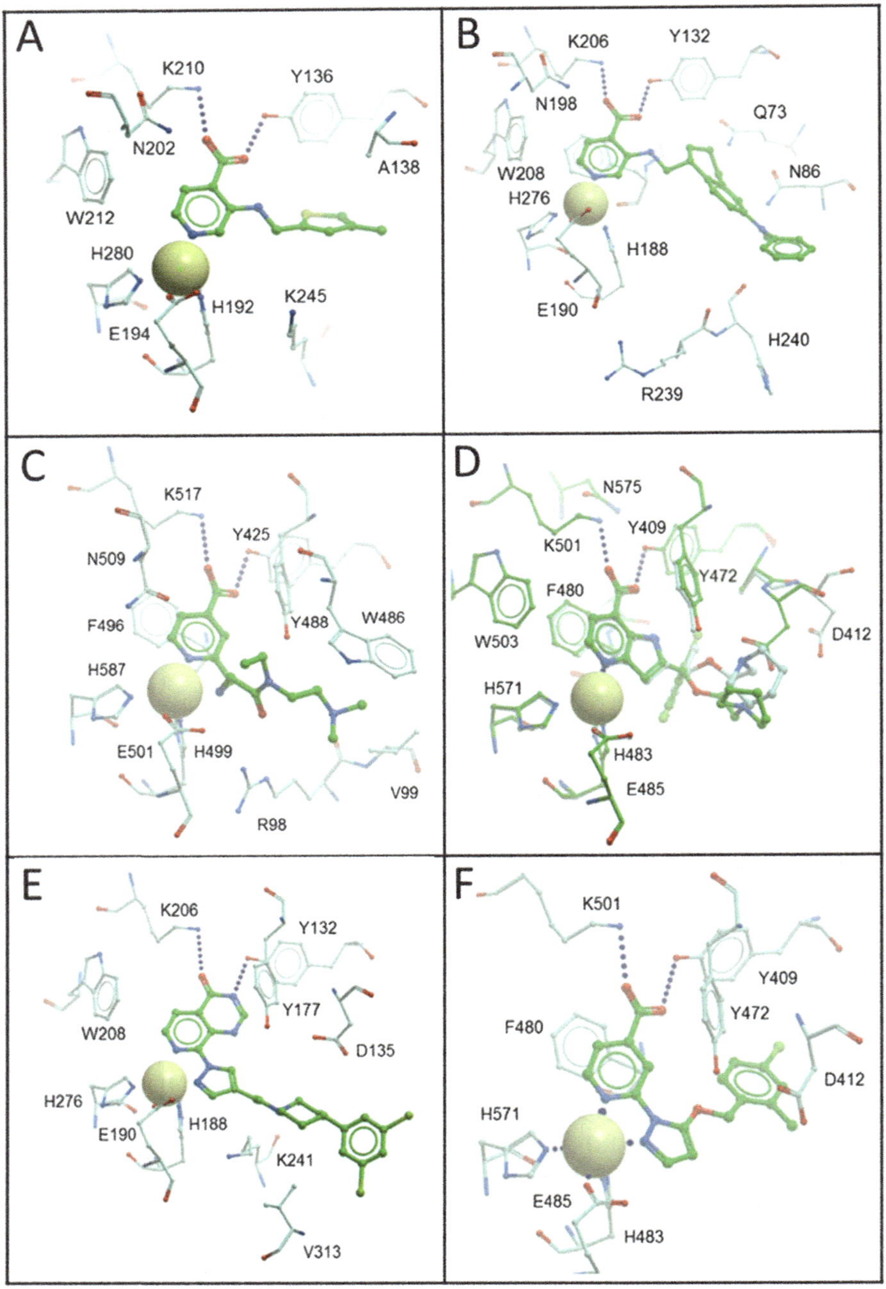

Fig. 8 Crystal structures of KDM4/5 in complex with inhibitors. (**a**) Crystal structure of **2** bound to KDM4D (PDB:5FP8); (**b**) crystal structure of **3** bound to KDM4A (PDB:5VGI); (**c**) crystal structure of **4** bound to KDM5B (PDB:5A3T); (**d**) crystal structure of (R)-**6** (blue) and (S)-**6** (green) bound to KDM5A (PDB:6BH0 and 6BH1); (**e**) crystal structure of **8** bound to KDM4A (PDB:5F3I); (**f**) crystal structure of **9** bound to KDM5A (PDB:5IW0)

enantiomer, while the (*S*)-enantiomer was relatively inactive. Compound **3** was identified as the most potent inhibitor against KDM4C ($IC_{50} = 35$ nM) and showed cell activity against KYSE-150 cells with an EC_{50} of 3.5 nM. Cellular on-target engagement using the same cell line was determined by measuring the increase in H3K9me3 and H3K36me3 through time-resolved fluorescence assay and mass spectroscopy detection to give a H3K36me3 mechanism of action (MOA) EC_{50} of 1.3 nM. The significant improvement in cellular potency from previously reported pyridine carboxylic acids could be attributed to the greater cell permeability from the addition of the lipophilic tetrahydronaphthalene and phenyl rings.

Good selectivity (>1,000-fold) was achieved over other KDM families such as KDM2B, KDM3, KDM6B and KDM7B. However, selectivity over KDM5B was moderate ($IC_{50} = 750$ nM), so some cellular activity may be contributed to KDM5B inhibition. **3** was also efficacious in vivo in breast and colon cancer PDX models showing a reduction in the rate of tumour growth.

A co-crystal structure of **3** with KDM4A (PDB:5VGI) revealed a similar binding mode involving pyridyl *N* coordinating to Ni(II). Key hydrogen bonding interactions of carboxylic acid with Y132 and K206 were observed (Fig. 8b). The methyl group of the methylaniline pointed towards N86 and carbonyl of H240, while the phenyl group is positioned in the R239/H240 loop.

An analogue of 2,4-PDCA, **4** (KDM5-C49), in which the 2-position is substituted with an amino methyl group, was identified as a potent nanomolar inhibitor of KDM5 family (KDM5A/B/C/D IC_{50} 7/4/13/15 nM) with ~50-fold selectivity over closely related KDM4C and >50-fold selectivity over KDM2, 3 and 6 [63, 64]. **5** (KDM5-C70) was designed as an ester pro-drug to permeate the cell and be hydrolysed in the cell to the active carboxylic acid. This compound showed growth inhibition in several breast cancer cell lines and also exhibited increased expression of *MTIF* and *MT1H* genes in comparable levels observed upon KDM5B knockdown [65].

4 binds to KDM5B (PDB:5A3T) in the 2OG binding site with bidentate coordination of pyridine *N* and aminomethyl *N* to the metal (Fig. 8c). As with other pyridine carboxylate derivatives, there are key hydrogen bonding interactions with K517 and Y425. Extensive polar interactions are observed with K517, Y425, N509, E509, R98 and Y488 as well as hydrophobic interactions with W486, V99 and F496. Similar key binding interactions were observed when **4** was crystallised in KDM4 (PDB:5FPV) which resulted in only moderate selectivity.

Chiral enantiomers (*R*)- and (*S*)-2-((2-chlorophenyl)(2-(piperidin-1-yl)ethoxy) methyl)-1*H*-pyrrolo[3,2-*b*]pyridine-7-carboxylic acid, (*R*)-**6** and (*S*)-**6**, were active inhibitors of KDM5A with binding affinities of 220 nM and 60 nM, respectively, and greater than 20-fold selectivity over KDM4A [53]. Cellular activity was not observed, even at high concentrations, with the methyl ester cell-permeable pro-drugs of the racemic compound. The compounds were 2OG competitive with the carboxylic acid ring interacting with K501, N575 and Y409 and pyridine nitrogen coordinating to the metal.

Overlay of the structures of (*R*)-**6** and (*S*)-**6** bound in KDM5A showed differences in their binding and consequently on their potencies ((*R*)-**6** PDB code 6BH0 and (*S*)-**6** PDB code 6BH1) (Fig. 8d). The (*S*)-enantiomer engaged in extensive van der Waals interactions between the chlorine of chlorophenyl with A411, Y409 and Y472. The piperidine nitrogen of (*R*)-**6** hydrogen bonds with Y472, while the corresponding nitrogen in (*S*)-**6** interacts with carboxylate ions of D412. The greater binding affinity of (*S*)-**6** could be attributed to ionic bonding with D412 if the piperidine nitrogen is protonated. A displacement of D412 was observed between the (*R*)-**6** and (*S*)-**6** structures.

A series of pyrido[3,4-d]pyrimidin-4(3H)-one inhibitors were reported by GSK as potent KDM4 and KDM5 inhibitors as a replacement for pyridine-4-carboxylates which suffer from poor cell permeability [66]. Several pyrazolyloxy analogues showed good potency in biochemical assays ($pIC_{50} \sim 6.4$) and some cellular assays ($pIC_{50} \sim 5.3$) against KDM4C. The best compound **7** also showed activity against KDM5C ($pIC_{50} \sim 7.2$) but was selective over KDM6B and EGLN3.

7 was crystallised in KDM5B (PDB:5FUN) which engaged in similar key interactions via a monodentate coordination of pyridine nitrogen with the active site metal and hydrogen bonding interactions of amide with K517 and N591. Substitution at positions 5 and 6 was unfavourable due to good shape complementary at these positions. There were two conformations of Y425 which enabled flexibility for binding to KDM5B depending on the orientation of the phenyl group.

Superimposition of 3-aminoisonicotinic acid **3** and 2-pyridopyrimidinone **7** revealed that their cores were positioned in the same plane (Fig. 9). The carboxylic acid and the amide of pyridopyrimidinone overlap well to interact with the key lysine residue; however, Y425 was not necessary for binding with **7**, unlike with **3**, so it could rotate out of the pocket.

The Institute of Cancer Research and the Structural Genomics Consortium reported 8-(1H-pyrazol-3-yl)pyrido[3,4-d]pyrimidin-4(3H)-one scaffolds as equipotent inhibitors of KDM4 and KDM5 subfamilies [67]. One of the best

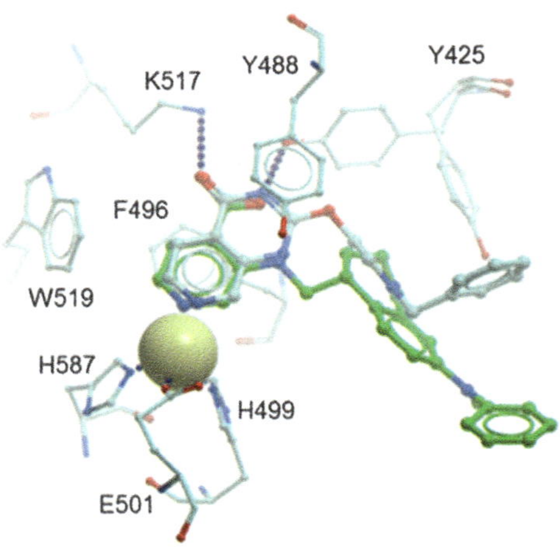

Fig. 9 Overlay of 3-aminoisonicotinic acid 3 in co-crystal of 2-pyridopyrimidinone 7 in KDM5B (PDB:5FUN)

compounds incorporated a dichloro-phenylpiperidine moiety **8** and showed an IC_{50} of <50 nM against KDM4B and KDM5B. Some cellular inhibition of H3K9me3 and H3K4me3 demethylation was observed but only at high compound concentrations.

A crystal structure of dichloro-phenylpiperidine **8** bound in KDM4A was obtained (PDB:5F3I) (Fig. 8e). The compound bound in the 2OG binding site and engaged in a bidentate coordination to the active site metal via the pyridine nitrogen and pyrazole *N*2 nitrogen. The carboxylate engaged in favourable interactions with K206 and Y132, and extending the pyrazole with a phenylpiperidine linker resulted in access to the histone peptide binding pocket for interactions with D135 and Y175. The piperidine ring had an important role in rigidifying the pyrazole substituent from different rotational conformations and ensuring good selectivity over KDM2, KDM3 and KDM6. A hydrophobic interaction was observed between the chlorine and V313 in the *m*-dichlorophenyl analogue which could explain the improved KDM4A/B inhibition compared with the *p*-chlorophenyl analogue in which the chlorine was too distal for an interaction.

Pyrazolylpyridines were designed based on structural information from the known KDM inhibitors [68]. The best compound in this series, **9**, showed nanomolar activity against KDM5 (IC_{50} KDM5A/B = 13/2 nM) as well as KDM4C (IC_{50} = 41 nM) but did exhibit good selectivity across the other KDM families. An increase in H3K4me3 was observed in a breast cancer cell line, ZR-75-1, and in vivo, in a MCF7-breast cancer xenograft PK/PD model upon treatment with **9**.

9 was co-crystallised with KDM5A (PDB:5IW0) confirming bidentate coordination of pyrazolyl and pyridyl nitrogens with the metal centre (Fig. 8f). As previously observed, the carboxylate group interacts with K501 and Y409 while the pyridyl group π-stacks with F480. The benzyl ether also engaged in π-stacking interactions with Y409 resulting in improved selectivity over the other KDMs.

3.1.2 Selective KDM5 Inhibitors

More recently, compounds have been designed that are more potent and selective towards KDM5 using structural information from crystal structures of known KDM5 inhibitors (Fig. 10).

Aminomethylpyridine **10** (KDOAM-25), in which the carboxylic acid of KDM5-C49 was replaced with a primary amide, showed excellent activity on KDM5B (IC_{50} = 19 nM) and also showed improved selectivity (>1,000-fold) over KDM4C as well as other 2OG oxygenases [64]. The compound was partially competitive with 2OG which may explain the low cellular activity (~50 μM) observed in an immunofluorescence assay measuring the increase in H3K4me3 levels upon incubation with **10**.

KDOAM-25 binds to KDM5B (PDB: 5A3N) in a similar pose to KDM5-C49 via metal chelation of the pyridine nitrogen (Fig. 11a). The main difference is the displacement of Y425 to become a hydrogen bond acceptor with the carboxamide rather than a hydrogen bond donor with the carboxylate. A water molecule can

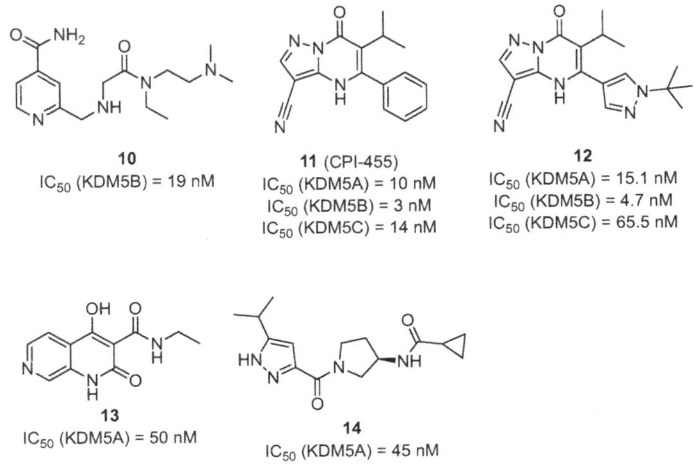

Fig. 10 KDM5 selective inhibitors

occupy the space created by Y425. The flexibility of Y425 in KDM5B explains the selectivity over KDM4 which has a more rigid tyrosine.

11 (CPI-455) was identified by Genentech and Constellation as a potent inhibitor of the KDM5 subfamily with IC_{50} of 10 nM against KDM5A in enzymatic assays. The compounds were 200-fold selective over KDM4C and 700-fold selective over KDM7B and showed no inhibition against KDM2B, KDM3B or KDM6A. A cellular assay measuring global increase in levels of H3K4me3 in PC9 cells after incubation with the compound resulted an EC_{50} of 5.2 μM [41].

Crystallisation of **11** in KDM5A (PDB:5CEH) identified a monodentate interaction between the nitrile and the active site metal and a hydrogen bond between carbonyl oxygen and N575 (Fig. 11b). A π–π stacking interaction was observed between the core and Y472 and F480 aromatic side chains as well as an edge-face interaction with W503. Although K501 and partially charged *N*1 of CPI-455 are positioned in close proximity, hydrogen bonding distance was not observed. As with other KDM inhibitors, an acidic hydrogen was required which was demonstrated by the loss of activity upon alkylation of 4-nitrogen or replacement of pyrimidine with triazene. 5-Phenyl substituent interacted through an edge-face contact with Y409 but otherwise was solvent exposed. Substitution of the isopropyl group was not tolerated due to its tight binding pocket defined by Y409 and S478. CPI-455 binds in the 2OG binding site and demonstrated competitive inhibition with 2OG in biochemical assays.

This compound showed no selectivity over other KDM5 isoforms due to identical amino acids in its binding site. Selectivity over KDM6B could be explained through collisions between the isopropyl and phenyl substituents and K1381, F1328 and T1387 residues. Y409 in KDM5A has conformation freedom to rotate out of the pocket to allow CPI-455 binding, whereas F1328, on a β-sheet next to ARID and PHD1 domains in KDM5A, projects into the pocket in KDM6B. The residues in KDM4C only differ in the second sphere of amino acids that do not directly interact

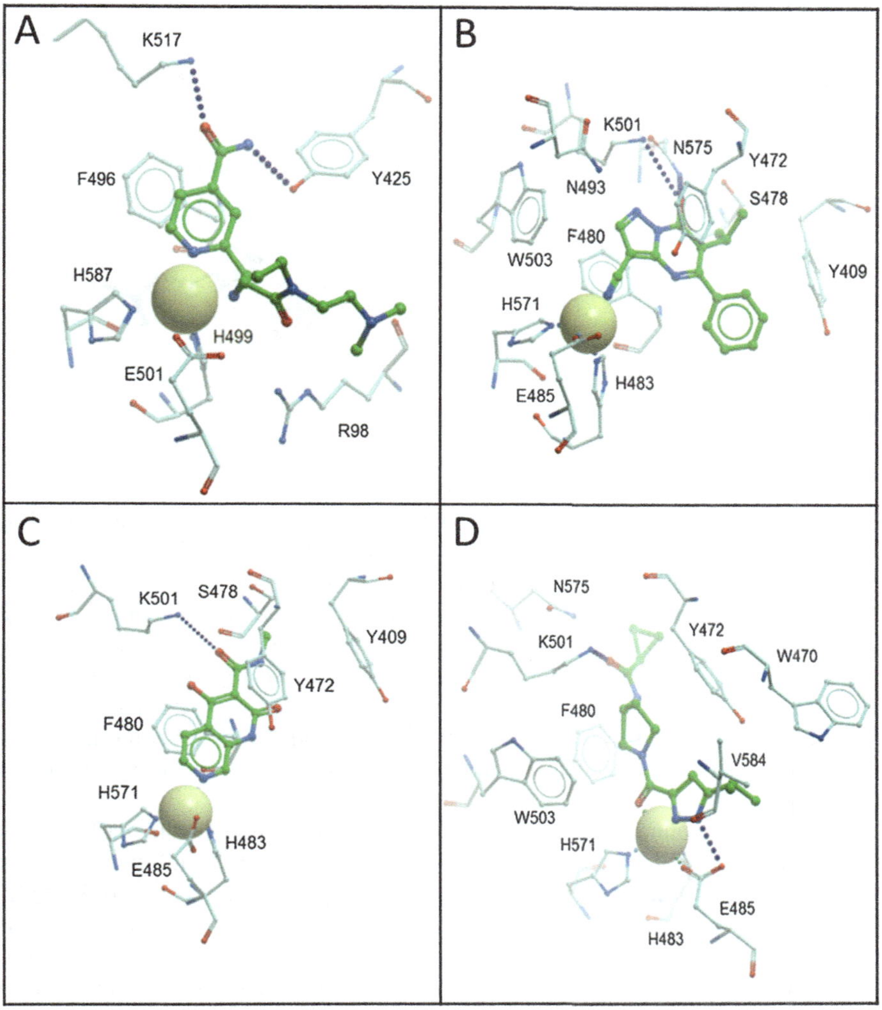

Fig. 11 Crystal structures of KDM5 in complex with selective inhibitors. (**a**) Crystal structure of 10 bound in KDM5B (PDB:5A3N); (**b**) crystal structure of 11 bound in KDM5A (PDB:5CEH); (**c**) crystal structure of 13 bound to KDM5A (PDB:5K4L); (**d**) crystal structure of 14 bound to KDM5A (PDB:5V9T)

with the compound. Selectivity over KDM4C could be explained through replacement of KDM5A A583 with KDM4C S290. The presence of I73 in KDM4C compared to R73 in KDM5A could result in a displacement of Y409.

Further optimisation of **11** led to pyrazolo[1,5-a]pyrimidin-7(4*H*)-one **12**, a potent and selective inhibitor of KDM5 with improved cell potency (PC9 H3K4Me3 $EC_{50} = 0.34\ \mu M$), which has a more balanced human plasma protein binding (hPPB) and cell permeability [69]. **12** showed good selectivity over the other

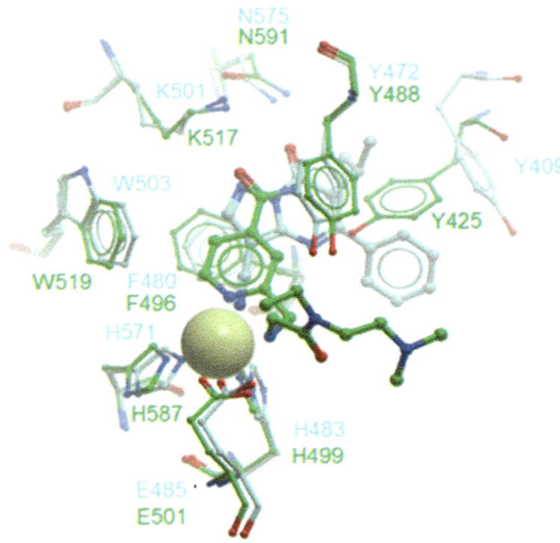

Fig. 12 Overlay of **10** (green) bound in KDM5B (green) and **11** (blue) bound in KDM5A (blue)

KDM families (1A, 2B, 3B, 4C, 6A, 7B) with IC_{50} > 1.9 μM and also against 300 kinase panel with no activity >50% at 10 μM. The compound demonstrated a good pharmacokinetic profile in mice and suitable for further profiling in vivo.

Superimposition of KDOAM-25 **10** and CPI-455 **11** in KDM5A shows that the cyanopyrazole core is tilted from the plane of the pyridine carboxamide (Fig. 12). The core in CPI-455 is offset by the nitrile group coordinating to the metal centre which improves π–π stacking interactions with Y472 and 480. Y409 is displaced away from the active site when CPI-455 is bound which highlights the improved selectivity of CPI-455 over KDM4 in which the rigid tyrosine blocks the isopropyl group from binding.

A hybrid compound of the pyridopyrimidinone and cyanopyrazole scaffolds was developed by Genentech to give a series of 1,7-naphthyridone inhibitors that were potent against KDM5 and showed good selectivity over KDM4C and KDM2B [70]. Various substitutions were trialled at C-3 to extend into the region occupied by isopropyl group of the cyanopyrazoles with KDM5. Introductions of electron withdrawing groups at C-3 resulted in an increase in potency for KDM5A due to improved interactions with K501 and N575. To lower the pKa of the ionisable proton and improve the cell permeability, C-3 carboxamides were explored. Only small alkyl groups on the nitrogen were tolerated due to the tight binding pocket created by S478 and Y409. **13** was identified as the most potent inhibitor against KDM5A (IC_{50} = 50 nM) with excellent MCDK permeability; however, no cellular activity was observed in a PC9 cell-based assay measuring H3K4me3 levels.

A co-crystal structure of **13** in KDM5A (PDB:5K4L) confirmed that 1,7-naphthyridones were 2OG competitive with monodentate metal coordination of *N7* (Fig. 11c). π–π stacking interactions were observed between the ring system and Y472 and F480 and the carbonyl oxygen engaged in hydrogen bonding to *N*ε of K501. Van der Waals interactions were observed between *N*-ethyl of C-3 amide with

S478 and Y409. *N*1 was solvent exposed, and therefore alkylation was not tolerated due to unfavourable hydrophobic interactions. The selectivity over KDM2B was attributed to the loss of aromatic stacking interactions through replacement of Y472 with a leucine. KDM4 does not differ in the direct binding residues, so selectivity over this family was likely due to changes to the second shell of amino acids.

A novel series of KDM5 inhibitors was discovered by Genentech and Constellation [71]. Optimisation of a high-throughput screen hit led to a single enantiomer **14** with a core pyrazole which showed a potent IC_{50} of 45 nM against KDM5A and good selectivity over KDM1, KDM2, KDM3, KDM6 and KDM7 and 91-fold selectivity over most closely related KDM4C. The compound had reasonable cellular potency of 960 nM and demonstrated good in vivo PK properties.

Co-crystallisation of **14** with KDM5A (PDB:5V9T) revealed a bidentate coordination of pyrazole *N*1 and carbonyl oxygen with the active site metal (Fig. 11d). A strong hydrogen bonding interaction was observed between pyrazole *N*2 and E485, while the cyclopropyl carbonyl oxygen engaged in an interaction with K501. As observed with other KDM5 inhibitors, the compound did not require a weakly acidic proton to engage in an ionic interaction with K501, which resulted in improved cell permeability. Most of the compound was positioned in the 2OG binding site, while the isopropyl group resided in the histone peptide binding site. Competition with histone peptide was also confirmed using ^{19}F NMR which represent a novel mechanism for KDM5A inhibition.

3.1.3 KDM6 Inhibitors

A large screening campaign by GSK led to **15** (GSK-J1), a potent ($IC_{50} = 60$ nM) inhibitor of KDM6B (Fig. 13) [50]. The carboxylic acid was necessary for binding; however, its high polarity resulted in poor cell permeability. Ethyl ester analogue, **16** (GSK-J4), was developed as a cell penetrant pro-drug which is rapidly hydrolysed by macrophage esterases to give GSK-J1. **16** inhibited loss of nuclear H3K27me3 by KDM6B in FLAG-KDM6B-transfected HeLa cells and showed increased nuclear H3K27me3 levels in untransfected cells. Although GSK-J1 showed marginal inhibition against KDM4 and other closely related proteins, it was found to be only

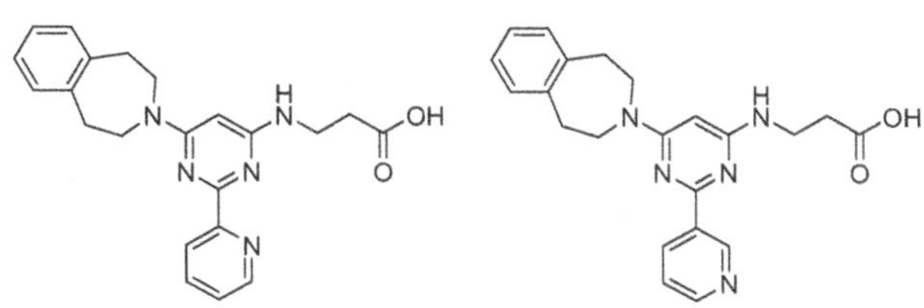

GSK-J1 **15** (X=OH)
IC_{50} (KDM6B) = 60 nM
GSK-J4 **16** (X=OEt)

GSK-J2 **17**

Fig. 13 Structures of GSK-J1, GSK-J2 and GSK-J4

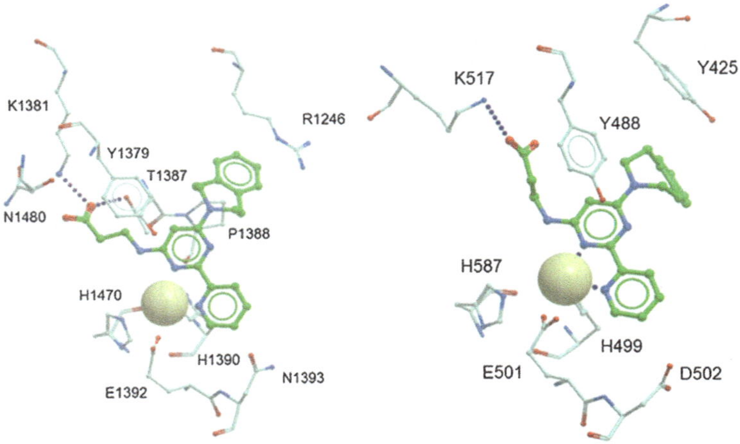

Fig. 14 Co-crystal of GSK-J1 in KDM6B (left-PDB:4ASK) and in KDM5B (right-PDB:5FPU)

fivefold to tenfold selective over KDM5B and KDM5C [72]. Similar IC_{50}s (3–7 μM) were also obtained with GSK-J4 in cells transfected with KDM6B, KDM4C and KDM5B. As a result, inhibition by GSK-J1 is likely a combination of activity towards H3K4me3/me2 and H3K27me3/me2. Despite this, GSK-J1 is the most-selective KDM6B inhibitor and is a useful chemical tool for studying KDM6B biology.

The key binding interactions were determined from a co-crystal structure of GSK-J1 **15** bound to KDM6B (PDB:4ASK) (Fig. 14, left). GSK-J1 bound in the 2OG binding site with the propanoic acid interacting with K1381, T1387 and N1480 and bidentate coordination of the pyridyl-pyrimidine biaryl with catalytic metal. The pyridine regio-isomer **17** (GSK-J2) was a suitable inactive control as it cannot engage in the key bidentate coordination to the metal and was much less active ($IC_{50} > 100$ μM). The catalytic metal in KDM6B was shifted by 2.34 Å away from the HHE triad which was not observed in the other KDM structures. This resulted in the displacement of a water molecule which could engage in a water-mediated interaction between H1470 and the metal. As a result, selectivity could be achieved through designing compounds that can induce this displacement of the metal centre. The tetrahydrobenzazepine ring is positioned in the narrow cleft between R1246 and P1388 which is occupied by P30 in the histone peptide. Despite this, GSK-J1 was found to be only 2OG competitive and not peptide substrate competitive.

15 was also co-crystallised in KDM5B (PDB:5FPU) which showed similar key interactions as in KDM6B with bidentate coordination to the metal and hydrogen bonding of carboxylate to K517 (Fig. 14, Right). The main difference is the lack of a second bonding interaction with the carboxylate in KDM5B as the flexible Y425 is projected towards the tetrahydrobenzazepine ring.

3.2 Substrate Competitive Inhibitors

3.2.1 Histone Peptide-Derived Inhibitors

As discussed in Sect. 2.1.2, histone peptide sequences and methylation states can provide remarkable selectivity towards individual JmjC-KDMs/subfamilies. However, the affinities of histone peptides are generally weaker for the JmjC-domains (as reflected in their K_M values (10^{-5}–10^{-4} M)), compared to 'reader' binding domains, where the K_d values can be in the sub-micromolar range. Thus, efforts have been made to combine the histone peptides with metal-chelating inhibitors to achieve both selectivity and potency. Woon et al. used dynamic chemistry (using multiple thiols) linked to mass spectrometry analysis to identify optimal positions to conjugate a 2OG cofactor mimic, *N*-oxalylcysteine (NOC), to histone peptides [73]. H3K9me3 peptides with Cys at varying positions along the histone peptide were incubated with L-NOC or D-NOC, and various products were tested. H3K9me3T11C was identified to be most stabilising, and a crystal structure (PDB:3U4S) revealed the optimal bidentate coordination of NOC projecting from T11C to Fe(II), as well as H3K9me3 positioning. Replacing the disulphide bond with CH2-S bond using thiol-ene coupling yielded a demethylase inhibitor **18** that was potent for KDM4A/E (IC_{50} = 0.09–0.27 μM) and >100-fold selective over KDM2A/KDM3A/KDM6B/KDM7B and other 2OG oxygenases (Fig. 15). Interestingly, **19** developed using similar approach from H3K36me3, a substrate for KDM4A/B/C but not KDM4D/E [74] conferred selectivity for KDM4A over KDM4E, demonstrating that selective inhibition can be achieved, even within the subfamily, using peptides [73]. In a different approach, Lohse et al. identified minimal histone peptide lengths that maintained demethylase activity for KDM4A/C (H3(7–11)K9me3) and conjugated a bromouracil at the K9 position to chelate the iron (ARK$_{(BRU)}$ST) (**20**) leading to an inhibitory peptide with K_i = 27 μM for

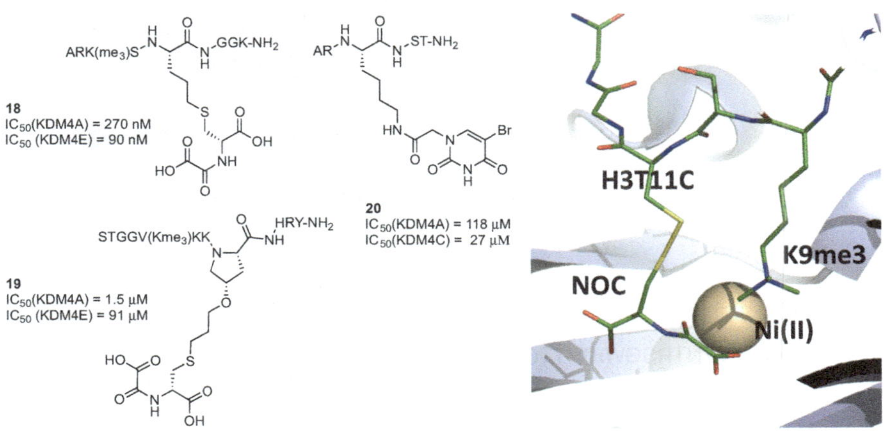

Fig. 15 KDM4 inhibitors derived from histone H3 peptides

KDM4C [75]. While a co-crystal structure was not obtained, the study highlights that a small peptide fragment can confer potency and selectivity for JmjC-KDMs [75].

3.2.2 Substrate-Binding Pocket Targeted Peptide Inhibitors

mRNA-display-based cyclic peptide library screening against KDM4A generated potent and selective inhibitors for the KDM4A/B/C family [76]. Interestingly, the binding cyclic peptides were distinct sequences with no homology to histone H3. A co-crystal structure of KDM4A with CP2 (**21**, PDB:5LY1), a cyclic peptide inhibitor ($IC_{50} = 42$ nM and binding potency $K_d = 30$ nM against KDM4A), revealed that CP2 occupies the substrate-binding pocket (Fig. 16). CP2 forms a β-sheet with two turns; in the β-turn pointing towards the active site pocket, Arg (at position 6 of CP2 (CP2R6)) extends towards the Fe(II) and occupies the same sub-pocket as trimethyllysine of H3K9me3/H3K36me3 in KDM4A. Similar to the lysines, the positively charged guanidino group makes hydrogen bonding interactions to KDM4A (Y177, S288, N290). In addition, Y175 and R309 shift to accommodate CP2 binding, inducing a conformational change in these regions of KDM4A. CP2 demonstrate high selectivity towards KDM4A–C, with >100-fold selectivity over other KDMs and 2OG oxygenases tested and even over KDM4D/E. This selectivity was attributed to the intermolecular interactions that CP2 makes with KDM4A side chains. CP2 interacts with 14 residues in KDM4A, of which five are different between KDM4A/B/C and KDM4D/E. Three (N86, Q88, R309) of these five residues have been identified to be crucial for H3K36me3 binding to KDM4A/B/C

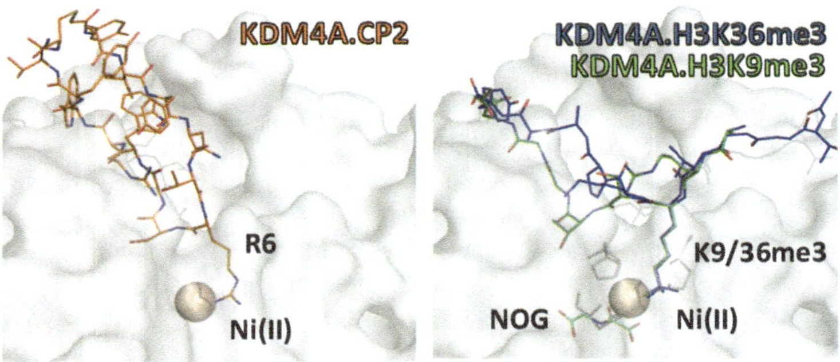

CP2 **21**
IC_{50}(KDM4A) = 42 nM
IC_{50}(KDM4B) = 33 nM
IC_{50}(KDM4C) = 39 nM
IC_{50}(KDM4D) = 6270 nM
IC_{50}(KDM4E) = 9200 nM

Fig. 16 Cyclic peptide CP2 binding to KDM4A

[74]. Thus, subtle differences in the substrate-binding pockets can be utilised for the generation of selective and potent inhibitors against JmjC-KDMs. Structure and MS-guided modifications of CP2 yielded peptides that retained in vitro potency and selectivity and increased cellular permeability and cellular KDM4A target engagement, providing promising cellular tools for the KDM4s.

3.2.3 Small Molecule Targeting Substrate-Binding Pocket

BIX-01294 (diazepin-quinazoline-amines, **22**) and its analogues are potent inhibitors of H3K9 methyltransferase G9a and G9a-linked proteins (GLP) [77]. They inhibit by mimicking the HMT-bound histone H3 (K4-R8) conformation and bind in the substrate-binding groove of HMT. Upadhyay et al. tested BIX-01294 and analogues for inhibition against KDM7A, a H3K9me2/1 demethylase [78]. BIX-01294 was found to be a moderate inhibitor of KDM7A ($IC_{50} = 16.5$ μM), and its analogue E67 (**23**) showed improved potency (Fig. 17). An overlay of the crystal structure of KDM7A.E67.NOG, with KDM7B.H3K4me3K9me2 (PDB:3KV4), shows that E67 occupies the same region where histone H3(K4–K9) lies in KDM7B and that the 5-aminopentyloxyl moiety at the O7-methoxy group extends towards the active site metal. The terminal amino group forms a weak hydrogen bond with carboxyl oxygen of D284, one of the residues that coordinate Fe(II) in the active site. The 3-dimethylamino propyl moiety at 2′-position of E67 extends towards the substrate-binding groove on the protein surface. It is worth noting that the E67 binding leads to a different Y292 conformer to the *apo* structure, inducing NOG to adopt an axial coordination mode (Fig. 17). This suggests Y292 in KDM7B, conserved across KDM2/7 subfamilies, may potentially be attractive residue to target to induce non-productive cofactor binding. Further fine-tuning of selectivity against HMTs generated E67-2 (**24**) which removed HMT inhibition and maintained JmjC-KDM

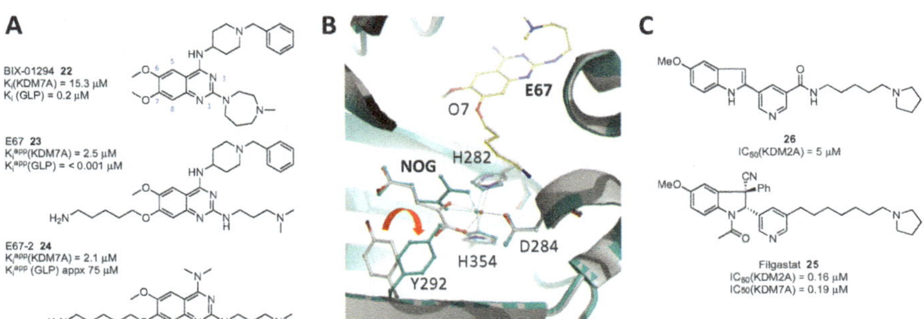

Fig. 17 JmjC-KDM inhibitors inspired from inhibitors targeting other methyllysine-associated epigenetic targets. (**a**) Structures of histone methyltransferase inhibitor BIX-10294 and analogues. (**b**) Co-crystal structure of KDM7A complexed with E6, NOG and Ni(II) (teal, PDB:3U78) overlaid on KDM7A.2OG.Fe(II) (grey, PDB:3KVA). (**c**) Structures of methyllysine binding domain inhibitor and Filgastat

inhibition, providing promising opportunities for 'target-hopping' for histone substrate-mimicking inhibitors. It is worth noting that inhibitor chemotypes targeting other epigenetic proteins have successfully been designed to inhibit the JmjC-KDMs. Filgastat (**25**), derived from methyllysine reader domain inhibitor (**26**), is a potent and selective inhibitor of KDM2/7 [79]. It does not compete with 2OG nor histone H3 peptide, and the mode of inhibition is unclear; however, it demonstrates that selective allosteric inhibition of JmjC-domain is possible.

4 Conclusions

The JmjC-KDMs are a protein family rich in links to disease pathology and are an exciting new hunting ground of epigenetic targets for drug discovery. This family of enzymes has been extensively characterised by structural biology to explain their function and enable inhibitor design. The KMDs all share a similar mechanism of action and conserved features – Fe(II) chelation, salt bridging and hydrogen bonding to 2OG – but differences in their active sites, substrate-binding surface and domain organisation allow selectivity for similar methylated lysine substrates.

Initial efforts at inhibitor discovery yielded nonselective compounds by relying primarily on iron chelation and salt bridging for potency such as NOG, 8HQ and Bpy (Fig. 3). Since those initial inhibitor reports, much effort at inhibitor discovery, aided by thorough characterisation of the family with X-ray crystal structures, has yielded new chemotypes that still rely on iron chelation and ionic interactions with the KDMs but achieve selectivity by exploiting differences in the active sites' first sphere of residues. This led to compounds selective for KDM5 and KDM6 subfamilies such as KDOAM25 (**10**), CPI-455 (**11**) and GSK-J1 (**15**). Selectivity for KDM members with similar active sites was then achieved by taking into account subtle differences in active site residues such as CPI-455 (**11**) which is selective for KDM5 members over KDM4 due to the propensity of an active site Tyr to shift more readily in KDM5 than KDM4. Single KDM selectivity is still elusive, although this may be acceptable and even desirable in drug development as many KDM subfamily members, i.e. KDM5A/B/C/D, have identical substrates and redundant function.

A second class of KDM inhibitors are peptide-competitive and achieve selectivity at the methyllysine binding groove and enzyme surface. CP2 is a cyclic peptide that has no homology to histone H3 yet is a potent and selective inhibitor of KDM4s (Fig. 16). The HMT inhibitors BIX01294 and E67-2 bind at the peptide site and are selective for KDM7, and the related Filgerstat is KDM2A/7A selective although its binding is not well understood.

The future of KDM drug discovery will involve developing a deeper understanding of KDM biology using the inhibitors described herein. To be successful, it will also be necessary to discover new inhibitor chemotypes to continue to improve the potency, selectivity and cellular activity of this important class of epigenetic enzymes.

Acknowledgement and Funding MW is supported by the EPSRC Centre for Doctoral Training in Synthesis for Biology and Medicine (EP/L015838/1). AK gratefully acknowledges the Royal Society for the Dorothy Hodgkin Fellowship, the European Research Council Starting Grant (EPITOOLS-679479) and Cancer Research UK Programme Grant (C8717/A18245). The SGC is a registered charity (number 1097737) that receives funds from AbbVie, Bayer Pharma AG, Boehringer Ingelheim, Canada Foundation for Innovation, Eshelman Institute for Innovation, Genome Canada, Innovative Medicines Initiative (EU/EFPIA) [ULTRA-DD grant no. 115766], Janssen, Merck KGaA Darmstadt Germany, MSD, Novartis Pharma AG, Ontario Ministry of Economic Development and Innovation, Pfizer, São Paulo Research Foundation-FAPESP, Takeda and Wellcome [106169/ZZ14/Z]. The authors acknowledge the EU COST Action CM1406. We apologise for the incomplete citations and for the research that we were not able to cite due to space constraints.

Compliance and Ethical Standards

Conflict of Interest: The authors declare that they have no conflict of interest.

Ethical approval: This article does not contain any studies with human participants or animals performed by any of the authors.

References

1. Klose RJ, Zhang Y (2007) Regulation of histone methylation by demethylimination and demethylation. Nat Rev Mol Cell Biol 8:307–318. https://doi.org/10.1038/nrm2143
2. Højfeldt JW (2013) Histone lysine demethylases as targets for anticancer therapy. Nat Rev Drug Discov 12:917–930. https://doi.org/10.1038/nrd4154
3. Cloos PAC, Christensen J, Agger K, Helin K (2008) Erasing the methyl mark: histone demethylases at the center of cellular differentiation and disease. Genes Dev 22:1115–1140. https://doi.org/10.1101/gad.1652908
4. Markolovic S, Leissing TM, Chowdhury R et al (2016) Structure – function relationships of human JmjC oxygenases – demethylases versus hydroxylases. Curr Opin Struct Biol 41:62–72. https://doi.org/10.1016/j.sbi.2016.05.013
5. Lu T, Jackson MW, Wang B et al (2010) Regulation of NF-kappaB by NSD1/FBXL11-dependent reversible lysine methylation of p65. Proc Natl Acad Sci U S A 107:46–51. https://doi.org/10.1073/pnas.0912493107
6. Solá S, Xavier JM, Santos DM et al (2011) p53 interaction with JMJD3 results in its nuclear distribution during mouse neural stem cell differentiation. PLoS One 6:e18421. https://doi.org/10.1371/journal.pone.0018421
7. Walport L, Hopkinson R, Chowdhury R et al (2016) Arginine demethylation is catalysed by a subset of JmjC histone lysine demethylases. Nat Commun 7:11974. https://doi.org/10.1038/ncomms11974
8. Horton JR, Upadhyay AK, Qi HH et al (2010) Enzymatic and structural insights for substrate specificity of a family of Jumonji histone lysine demethylases. Nat Struct Mol Biol 17:38–43. https://doi.org/10.1038/nsmb.1753
9. Torres IO, Kuchenbecker KM, Nnadi CI et al (2015) Histone demethylase KDM5A is regulated by its reader domain through a positive-feedback mechanism. Nat Commun 6:6204. https://doi.org/10.1038/ncomms7204
10. Landeira D, Fisher AG (2011) Inactive yet indispensable: the tale of Jarid2. Trends Cell Biol 21:74–80. https://doi.org/10.1016/j.tcb.2010.10.004

11. Black JC, Manning AL, Van Rechem C et al (2013) KDM4A lysine demethylase induces site-specific copy gain and rereplication of regions amplified in tumors. Cell 154:541–555. https://doi.org/10.1016/j.cell.2013.06.051
12. Young LC, Hendzel MJ (2012) The oncogenic potential of Jumonji D2 (JMJD2/KDM4) histone demethylase overexpression. Biochem Cell Biol 91:369–377. https://doi.org/10.1139/bcb-2012-0054
13. Young LC, McDonald DW, Hendzel MJ (2013) Kdm4b histone demethylase is a DNA damage response protein and confers a survival advantage following γ-irradiation. J Biol Chem 288:21376–21388. https://doi.org/10.1074/jbc.M113.491514
14. Coffey K, Rogerson L, Ryan-Munden C et al (2013) The lysine demethylase, KDM4B, is a key molecule in androgen receptor signalling and turnover. Nucleic Acids Res 41:4433–4446
15. Italiano A, Attias R, Aurias A et al (2006) Molecular cytogenetic characterization of a metastatic lung sarcomatoid carcinoma: 9p23 neocentromere and 9p23~p24 amplification including JAK2 and JMJD2C. Cancer Genet Cytogenet 167:122–130. https://doi.org/10.1016/j.cancergencyto.2006.01.004
16. Liu G, Bollig-Fischer A, Kreike B et al (2009) Genomic amplification and oncogenic properties of the GASC1 histone demethylase gene in breast cancer. Oncogene 28:4491
17. Vinatzer U, Gollinger M, Müllauer L et al (2008) Mucosa-associated lymphoid tissue lymphoma: novel translocations including rearrangements of ODZ2, JMJD2C, and CNN3. Clin Cancer Res 14:6426–6431
18. Yang Z-Q, Imoto I, Fukuda Y et al (2000) Identification of a novel gene, GASC1, within an amplicon at 9p23–24 frequently detected in esophageal cancer cell lines. Cancer Res 60:4735–4739
19. Wang GG, Song J, Wang Z et al (2009) Haematopoietic malignancies caused by dysregulation of a chromatin-binding PHD finger. Nature 459:847–851. https://doi.org/10.1038/nature08036
20. Liang X, Zeng J, Wang L et al (2013) Histone demethylase retinoblastoma binding protein 2 is overexpressed in hepatocellular carcinoma and negatively regulated by hsa-miR-212. PLoS One 8:e69784
21. Teng Y-C, Lee C-F, Li Y-S et al (2013) Histone demethylase RBP2 promotes lung tumorigenesis and cancer metastasis. Cancer Res 73:4711–4721
22. Zeng J, Ge Z, Wang L et al (2010) The histone demethylase RBP2 is overexpressed in gastric cancer and its inhibition triggers senescence of cancer cells. Gastroenterology 138:981–992. https://doi.org/10.1053/j.gastro.2009.10.004
23. Roesch A, Fukunaga-Kalabis M, Schmidt EC et al (2010) A temporarily distinct subpopulation of slow-cycling melanoma cells is required for continuous tumor growth. Cell 141:583–594. https://doi.org/10.1016/j.cell.2010.04.020
24. Angela B, Bente M, John C et al (2002) PLU-1 nuclear protein, which is upregulated in breast cancer, shows restricted expression in normal human adult tissues: a new cancer/testis antigen? Int J Cancer 101:581–588. https://doi.org/10.1002/ijc.10644
25. Wang L, Mao Y, Du G et al (2015) Overexpression of JARID1B is associated with poor prognosis and chemotherapy resistance in epithelial ovarian cancer. Tumor Biol 36:2465–2472. https://doi.org/10.1007/s13277-014-2859-z
26. Madsen B, Tarsounas M, Burchell JM et al (2003) PLU-1, a transcriptional repressor and putative testis-cancer antigen, has a specific expression and localisation pattern during meiosis. Chromosoma 112:124–132. https://doi.org/10.1007/s00412-003-0252-6
27. Blair LP, Cao J, Zou MR et al (2011) Epigenetic regulation by lysine demethylase 5 (KDM5) enzymes in cancer. Cancers (Basel) 3:1383–1404. https://doi.org/10.3390/cancers3011383
28. Rasmussen PB, Staller P (2014) The KDM5 family of histone demethylases as targets in oncology drug discovery. Epigenomics 6:277–286. https://doi.org/10.2217/epi.14.14
29. Li J, Yu B, Deng P et al (2017) KDM3 epigenetically controls tumorigenic potentials of human colorectal cancer stem cells through Wnt/β-catenin signalling. Nat Commun 8:15146

30. Dey BK, Stalker L, Schnerch A et al (2008) The histone demethylase KDM5b/JARID1b plays a role in cell fate decisions by blocking terminal differentiation. Mol Cell Biol 28:5312–5327. https://doi.org/10.1128/MCB.00128-08
31. Cellot S, Hope KJ, Chagraoui J et al (2013) RNAi screen identifies Jarid1b as a major regulator of mouse HSC activity. Blood 122:1545–1555
32. Juan D, Yushi M, Ping M et al (2012) Demethylation of epiregulin gene by histone demethylase FBXL11 and BCL6 corepressor inhibits osteo/dentinogenic differentiation. Stem Cells 31:126–136. https://doi.org/10.1002/stem.1255
33. Gao R, Dong R, Du J et al (2013) Depletion of histone demethylase KDM2A inhibited cell proliferation of stem cells from apical papilla by de-repression of p15INK4B and p27Kip1. Mol Cell Biochem 379:115–122. https://doi.org/10.1007/s11010-013-1633-7
34. Lu T, Jackson MW, Wang B et al (2009) Regulation of NF-κB by NSD1/FBXL11-dependent reversible lysine methylation of p65. Response 107:1–6. https://doi.org/10.1073/pnas.0912493107
35. Lederer D, Grisart B, Digilio MC et al (2012) Deletion of KDM6A, a histone demethylase interacting with MLL2, in three patients with kabuki syndrome. Am J Hum Genet 90:119–124. https://doi.org/10.1016/j.ajhg.2011.11.021
36. Noriko M, Seiji M, Nobuhiko O et al (2012) KDM6A point mutations cause kabuki syndrome. Hum Mutat 34:108–110. https://doi.org/10.1002/humu.22229
37. Laumonnier F, Holbert S, Ronce N et al (2005) Mutations in PHF8 are associated with X linked mental retardation and cleft lip/cleft palate. J Med Genet 42:780–786. https://doi.org/10.1136/jmg.2004.029439
38. Gu L, Hitzel J, Moll F et al (2016) The histone demethylase PHF8 is essential for endothelial cell migration. PLoS One 11:e0146645
39. Erdoğan Ö, Xie L, Wang L et al (2016) Proteomic dissection of LPS-inducible, PHF8-dependent secretome reveals novel roles of PHF8 in TLR4-induced acute inflammation and T cell proliferation. Sci Rep 6:24833
40. Hatch SB, Yapp C, Montenegro RC et al (2017) Assessing histone demethylase inhibitors in cells: lessons learned. Epigenetics Chromatin 10:1–17. https://doi.org/10.1186/s13072-017-0116-6
41. Vinogradova M, Gehling VS, Gustafson A et al (2016) An inhibitor of KDM5 demethylases reduces survival of drug-tolerant cancer cells. Nat Chem Biol 12:531–538
42. Cheng Z, Cheung P, Kuo AJ et al (2014) A molecular threading mechanism underlies Jumonji lysine demethylase KDM2A regulation of methylated H3K36. Genes Dev 28:1758–1771
43. Ng SS, Kavanagh KL, McDonough MA et al (2007) Crystal structures of histone demethylase JMJD2A reveal basis for substrate specificity. Nature 448:87–91. https://doi.org/10.1038/nature05971
44. Couture J-F, Collazo E, Ortiz-Tello PA et al (2007) Specificity and mechanism of JMJD2A, a trimethyllysine-specific histone demethylase. Nat Struct Mol Biol 14:689–695. https://doi.org/10.1038/nsmb1273
45. Chen Z, Zang J, Kappler J et al (2007) Structural basis of the recognition of a methylated histone tail by JMJD2A. Proc Natl Acad Sci 104:10818–10823. https://doi.org/10.1073/pnas.0704525104
46. Williams ST, Walport LJ, Hopkinson RJ et al (2015) Studies on the catalytic domains of multiple JmjC oxygenases using peptide substrates. Epigenetics 9:1596–1603. https://doi.org/10.4161/15592294.2014.983381
47. Chu C-H, Wang L-Y, Hsu K-C et al (2014) KDM4B as a target for prostate cancer: structural analysis and selective inhibition by a novel inhibitor. J Med Chem 57:5975–5985. https://doi.org/10.1021/jm500249n
48. Krishnan S, Trievel RC (2013) Structural and functional analysis of JMJD2D reveals molecular basis for site-specific demethylation among JMJD2 demethylases. Structure 21:98–108. https://doi.org/10.1016/j.str.2012.10.018

49. Sengoku T, Yokoyama S (2011) Structural basis for histone H3 Lys 27 demethylation by UTX/KDM6A. Genes Dev 25:2266–2277. https://doi.org/10.1101/gad.172296.111
50. Kruidenier L, Chung C, Cheng Z et al (2012) A selective Jumonji H3K27 demethylase inhibitor modulates the proinflammatory macrophage response. Nature 488:404–408. https://doi.org/10.1038/nature11262
51. Jones SE, Olsen L, Gajhede M (2018) Structural basis of histone demethylase KDM6B histone 3 lysine 27 specificity. Biochemistry 57:585–592. https://doi.org/10.1021/acs.biochem.7b01152
52. Hausinger RP (2004) FeII/alpha-ketoglutarate-dependent hydroxylases and related enzymes. Crit Rev Biochem Mol Biol 39:21–68
53. Horton JR, Liu X, Wu L et al (2018) Insights into the action of inhibitor enantiomers against histone lysine demethylase 5A. J Med Chem 61:3193. https://doi.org/10.1021/acs.jmedchem.8b00261
54. Klein BJ, Piao L, Xi Y et al (2014) The histone-H3K4-specific demethylase KDM5B binds to its substrate and product through distinct PHD fingers. Cell Rep 6:325–335. https://doi.org/10.1016/j.celrep.2013.12.021
55. Pack LR, Yamamoto KR, Fujimori DG (2016) Opposing chromatin signals direct and regulate the activity of lysine demethylase 4C (KDM4C). J Biol Chem 291:6060–6070. https://doi.org/10.1074/jbc.M115.696864
56. Zhang Y, Yang H, Guo X et al (2014) The PHD1 finger of KDM5B recognizes unmodified H3K4 during the demethylation of histone H3K4me2/3 by KDM5B. Protein Cell 5:837–850. https://doi.org/10.1007/s13238-014-0078-4
57. Rose NR, Ng SS, Mecinović J et al (2008) Inhibitor scaffolds for 2-oxoglutarate-dependent histone lysine demethylases. J Med Chem 51:7053–7056. https://doi.org/10.1021/jm800936s
58. Joberty G, Boesche M, Brown JA et al (2016) Interrogating the druggability of the 2-oxoglutarate-dependent dioxygenase target class by chemical proteomics. ACS Chem Biol 11:2002–2010. https://doi.org/10.1021/acschembio.6b00080
59. Kaniskan HÜ, Martini ML, Jin J (2017) Inhibitors of protein methyltransferases and demethylases. Chem Rev 118:989–1068. https://doi.org/10.1021/acs.chemrev.6b00801
60. McAllister TE, England KS, Hopkinson RJ et al (2016) Recent progress in histone demethylase inhibitors. J Med Chem 59:1308–1329. https://doi.org/10.1021/acs.jmedchem.5b01758
61. Westaway SM, Preston AGS, Barker MD et al (2015) Cell penetrant inhibitors of the KDM4 and KDM5 families of histone lysine demethylases. 1. 3 – Amino-4-pyridine carboxylate derivatives. J Med Chem 59:1370–1387. https://doi.org/10.1021/acs.jmedchem.5b01537
62. Chen YK, Bonaldi T, Cuomo A et al (2017) Design of KDM4 inhibitors with antiproliferative effects in cancer mModels. ACS Med Chem Lett 8:869–874. https://doi.org/10.1021/acsmedchemlett.7b00220
63. Johansson C, Velupillai S, Tumber A et al (2016) Structural analysis of human KDM5B guides histone demethylase inhibitor development. Nat Chem Biol 12:1–10. https://doi.org/10.1038/nchembio.2087
64. Tumber A, Nuzzi A, Hookway ES et al (2018) Potent and selective KDM5 inhibitor stops cellular demethylation of H3K4me3 at transcription start sites and proliferation of MM1S myeloma cells. Cell Chem Biol 24:371–380. https://doi.org/10.1016/j.chembiol.2017.02.006
65. Horton JR, Liu X, Gale M et al (2016) Structural basis for KDM5A histone lysine demethylase inhibition by diverse compounds. Cell Chem Biol 23:213–221. https://doi.org/10.1016/j.chembiol.2016.06.006
66. Westaway SM, Preston AGS, Barker MD et al (2016) Cell penetrant inhibitors of the KDM4 and KDM5 families of histone lysine demethylases. 2. Pyrido[3,4-d]pyrimidin-4(3H)-one derivatives. J Med Chem 59:1370–1387. https://doi.org/10.1021/acs.jmedchem.5b01538
67. Bavetsias V, Lanigan RM, Ruda GF et al (2016) 8-substituted pyrido[3,4-d]pyrimidin-4(3H)-one derivatives as potent, cell permeable, KDM4 (JMJD2) and KDM5 (JARID1) histone lysine demethylase inhibitors. J Med Chem 59:1388. https://doi.org/10.1021/acs.jmedchem.5b01635

68. Nie Z, Shi L, Lai C et al (2018) Structure-based design and discovery of potent and selective KDM5 inhibitors. Bioorg Med Chem Lett 28:1490–1494. https://doi.org/10.1016/j.bmcl.2018.03.083
69. Liang J, Zhang B, Labadie S et al (2016) Lead optimization of a pyrazolo[1,5-a]pyrimidin-7(4H)-one scaffold to identify potent, selective and orally bioavailable KDM5 inhibitors suitable for in vivo biological studies. Bioorg Med Chem Lett 26:4036–4041. https://doi.org/10.1016/j.bmcl.2016.06.078
70. Labadie SS, Dragovich PS, Cummings RT et al (2016) Design and evaluation of 1,7-naphthyridones as novel KDM5 inhibitors. Bioorg Med Chem Lett 26:4492–4496. https://doi.org/10.1016/j.bmcl.2016.07.070
71. Liang J, Labadie S, Zhang B et al (2017) From a novel HTS hit to potent, selective, and orally bioavailable KDM5 inhibitors. Bioorg Med Chem Lett 27:2974–2981. https://doi.org/10.1016/j.bmcl.2017.05.016
72. Heinemann B, Nielsen JM, Hudlebusch HR et al (2014) Inhibition of demethylases by GSK-J1/J4. Nature 514:E1
73. Woon ECY, Tumber A, Kawamura A et al (2012) Linking of 2-oxoglutarate and substrate binding sites enables potent and highly selective inhibition of JmjC histone demethylases. Angew Chem Int Ed Engl 51:1631–1634. https://doi.org/10.1002/anie.201107833
74. Hillringhaus L, Yue WW, Rose NR et al (2011) Structural and evolutionary basis for the dual substrate selectivity of human KDM4 histone demethylase family. J Biol Chem 286:41616–41625. https://doi.org/10.1074/jbc.M111.283689
75. Lohse B, Nielsen AL, Kristensen JBL et al (2011) Targeting histone lysine demethylases by truncating the histone 3 tail to obtain selective substrate-based inhibitors. Angew Chem Int Ed Engl 50:9100–9103. https://doi.org/10.1002/anie.201101849
76. Kawamura A, Münzel M, Kojima T et al (2017) Highly selective inhibition of histone demethylases by de novo macrocyclic peptides. Nat Commun 8:14773. https://doi.org/10.1038/ncomms14773
77. Chang Y, Zhang X, Horton JR et al (2009) Structural basis for G9a-like protein lysine methyltransferase inhibition by BIX-01294. Nat Struct Mol Biol 16:312–317. https://doi.org/10.1038/nsmb.1560
78. Upadhyay AK, Rotili D, Han JW et al (2012) An analog of BIX-01294 selectively inhibits a family of histone H3 lysine 9 Jumonji demethylases. J Mol Biol 416:319–327. https://doi.org/10.1016/j.jmb.2011.12.036
79. Gerken PA, Wolstenhulme JR, Tumber A et al (2017) Discovery of a highly selective cell-active inhibitor of the histone lysine demethylases KDM2/7. Angew Chem Int Ed 56:15555–15559. https://doi.org/10.1002/anie.201706788

Top Med Chem (2020) 33: 255–286
DOI: 10.1007/7355_2019_76

Published online: 24 October 2019

Chemical Compounds Targeting DNA Methylation and Hydroxymethylation

Roman Belle, Akane Kawamura, and Paola B. Arimondo

Contents

1 Introduction 257
2 DNMT Enzymes 259
  2.1 Structures and Mechanism of DNMTs 261
  2.2 DNMTs in Diseases 261
3 Inhibitors of DNA Methylation 262
  3.1 Cytosine Analogues 262
  3.2 Non-nucleoside Analogues 264
  3.3 Transition State and Bisubstrate Analogues 265
  3.4 New Approaches to DNMT Inhibitor Design 266
4 TET Enzymes 266
  4.1 Discovery and Biological Roles of TETs 266
  4.2 Mechanism of Enzyme Catalysis 268
  4.3 TETs in Development and in Disease 271
5 TET Enzyme Assay and Inhibitor Development 272
  5.1 Kinetic Analysis of Human TETs 273
  5.2 Inhibitors of TET Enzyme Activity 274
6 Conclusions 276
References 278

Abstract DNA methylation and its oxidised forms participate in the interpretation and regulation of the human genome. Many questions arise around the enzymes responsible for these chemical modifications on DNA, and their roles in

---

R. Belle
Department of Chemistry, University of Oxford, Chemistry Research Laboratory, Oxford, UK

A. Kawamura (✉)
Department of Chemistry, University of Oxford, Chemistry Research Laboratory, Oxford, UK

Radcliffe Department of Medicine, University of Oxford, Wellcome Centre for Human Genetics, Oxford, UK
e-mail: akane.kawamura@chem.ox.ac.uk

P. B. Arimondo (✉)
Epigenetic Chemical Biology, Institut Pasteur, CNRS UMR3523, Paris, France
e-mail: paola.arimondo@cnrs.fr

transcriptional regulation. These epigenetic marks are very dynamic and specific in their location and context (tissues, diseases, etc.). We review the major enzymes involved in DNA methylation and oxidation, with a focus on the DNA methyltransferases and TET enzymes. The principal compounds that inhibit these enzymes are presented since they will help address these questions.

Keywords DNA hydroxymethylation, DNA methylation, DNMT, Inhibitors, TET

Abbreviations

1-mA	1-Methyl-adenine
2OG	2-Oxoglutarate
3-mC	3-Methylcytosine
3-mT	3-Methyl-thymine
5-aza-C	5-Aza-cytosine
5-azadC	5-Aza-2′-deoxycytosine
5-caC	5-Carboxycytosine
5-fC	5-Formylcytosine
5-hmC	5-Hydroxymethylcytosine
5-mC	5-Methylcytosine
5-xC	5-Modified cytosine
6-mA	6-Methyl-adenine
AML	Acute myeloid leukaemia
AM-PD	Active modification-passive dilution
BAH1 and BAH2	Bromo-adjacent homology domains 1 and 2
BER	Base excision repair
CFP1	CpG-binding protein, CXXC finger protein 1
CMML	Chronic myelomonocytic leukaemia
CpA	Cytidine pairing adenosine
CpC	Cytidine pairing cytidine
CpG	Cytidine pairing guanosine
CpT	Cytidine pairing thymidine
CXXC	CXXC domain
DMAP domain	DNA methyltransferase-associated protein 1-interacting domain
DNMT	C5-DNA methyltransferase
DSBH	Double-stranded β-ηelix
EGCG	Epigallocatechin gallate
ELISA	Enzyme-linked immunosorbent assay
EMA	European Medicines Agency
FDA	Food and Drug Administration
FH	Fumarate hydratase

FTO	Fat mass and obesity-associated protein
HDAC	Histone deacetylase
IDAX	Inhibition of the Dvl and Axin complex
IDH	Isocitrate dehydrogenase
LCI	Low-complexity insert
LC-MS	Liquid chromatography-mass spectrometry
MALDI-TOF	Matrix-assisted laser desorption/ionisation time-of-flight
MBP	Methyl-binding protein
MDS	Myelodysplastic syndrome
MLL	Mixed lineage leukaemia
mTet1	Murine TET
NgTet1	*Naegleria gruberi* TET
NLS	Nuclear localisation signal
NOG	*N*-Oxalylglycine
PBD	PCNA-binding domain
PHD	Plant homeodomain
PRMT	Protein arginine methyltransferase
PWWP	Proline-tryptophan-tryptophan-proline domain
R/S-2HG	R/S-2-hydroxyglutarate
RFTD	Replication foci targeting sequence (RFTS) domain
ROS1	Repressor of silencing 1
SAH/AdoHys	*S*-Adenosyl-L-homocysteine
SAM/AdoMet	*S*-Adenosyl-L-methionine
SDH	Succinate dehydrogenase
SPR	Surface plasmon resonance
TCA	Tricarboxylic acid
TDG	Thymidine-DNA glycosylase
TET	Ten-eleven translocation
TLC	Thin-layer chromatography
TRDMT1	tRNA aspartic acid methyltransferase

1 Introduction

DNA methylation and its oxidised forms participate in the interpretation and regulation of the genome. Chemical modifications of cytosine in chromosome DNA are (highly) dynamic and are one of the regulatory elements that, together with histone modifications and chromatin remodelling, allow the DNA to interact with the protein machineries that interpret, repair and splice the genetic information. The cytosine modifications identified to date are 5-methylcytosine (5mC), 5-hydroxymethylcytosine (5hmC), 5-formylcytosine (5fC) and 5-carboxycytosine (5caC).

In humans, DNA methylation is the most stable epigenetic mark [1], and it occurs at C5 position of cytosine (5mC), mainly in a CpG dinucleotide context. The CpG dinucleotides are mainly located in the CpG islands (occurring at *ca* 60% of all gene promoters), in repeated sequences and in CpG island shores [2]. If promoter CpG islands are methylated, the corresponding gene is repressed due to a poor recognition by transcription factors and recruitment of proteins involved in chromatin remodelling such as methyl DNA-binding proteins (MBPs) [3].

Failure in maintaining DNA methylation and establishment of new DNA methylation patterns are associated with under- or overexpression of the affected genes, ultimately leading to inflammation, cancer and other diseases. DNA methylation is catalysed by DNA methyltransferases (DNMT) that mediate the transfer of a methyl group from the *S*-adenosyl-L-methionine (SAM or AdoMet, **1**) to position 5 of cytosine in DNA [4] (Figs. 1 and 2a).

The ten-eleven translocator (TET) enzymes catalyse the 2-oxoglutarate (2OG, **3**)-dependent oxidation of 5mC in a cascade of iterative steps to give 5-hydroxymethyl cytosine (5hmC), 5-formylcytosine (5fC) and 5-carboxycytosine (5caC) (Figs. 1 and 2). The discovery of TET oxidation of 5mC to 5hmC in 2009 garnered interest as an important epigenetic regulator [5, 6]. Since the identification of 5hmC in human cells, other DNA modifications (5fC, 5caC as well as 5-hydroxymethyluracil (5hmU)) have been identified and found to be catalysed by TET1–3 [7–11].

Methylation at position 5 of cytosine is a reversible modification, and there are four different demethylation pathways that could lead to the reformation of cytosine identified to date (Fig. 1). First, the 5mC mark is not maintained during DNA replication, thus 5mC is diluted by subsequent rounds of cell division, a process known as passive DNA demethylation. Second, 5mC can be iteratively modified to provide 5hmC, 5fC and 5caC, which then can be diluted in cell division, a combination of active modification followed by passive dilution (AM-PD). The third process is an active pathway via a combination of thymidine-DNA glycosylase (TDG) and base excision repair (BER) enzymes. As part of DNA repair mechanism, TDG recognises a thymidine guanosine pair mismatch and deaminates the thymidine base leaving an a-basic sugar. TDG likewise performs this excision efficiently on 5fC and 5caC [12] followed by DNA repair using the base excision repair (BER) mechanism that introduces C. Finally, recent studies by Iwan et al. demonstrated, by labelling the 5fC sugar and base independently, that 5fC can be directly converted by the human cell to C without the change in base or sugar, suggesting direct deformylation [13]. The protein(s) associated with this process of 5fC, however, have not been identified to date. In plants, a direct 5mC to C pathway is present, following a similar TDG-BER mechanism, where repressor of silencing 1 (ROS1) recognises 5mC [14].

Many questions arise around these enzymes and their roles in regulating cytosine modifications, which are very dynamic but are specific in the location and context (tissues, diseases, etc.). How are they precisely regulated in a concerted manner? Why was this mechanism of DNA demethylation selected in mammals? Their role in diseases, such as cancer, neuronal diseases, inflammation and infection, is well established, but the mechanisms involved are still to be fully understood.

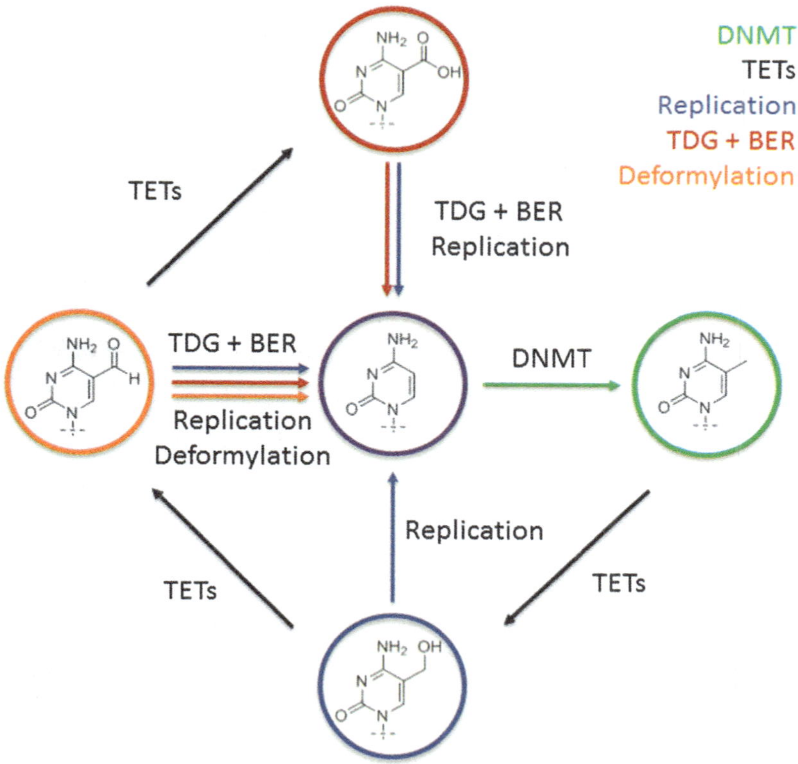

Fig. 1 Methylation of cytosine in DNA and successive oxidation of 5-methylcytosine catalysed by TETs. Cytosine (C), mainly at CpG sites, is methylated by the DNA methyltransferases (DNMT) at position 5 using S-adenosyl-L-methionine (SAM or AdoMet) as methyl donor to give 5-methylcytosine (5mC). Methylated cytosine (5mC) can undergo iterative oxidation by TET enzymes followed by 5-hydroxycytosine (5hmC), 5-formylcytosine (5fC) and 5 carboxycytosine (5caC). Each base has a passive and/or active demethylation pathway leading to cytosine (C). Possible 5mC demethylation pathways include (1) passive dilution; (2) active modification followed by passive dilution; (3) Thymine DNA glycosylase-mediated base excision repair (TDG + BER); and (4) deformylation

Compounds that inhibit the enzymes responsible for DNA methylation and oxidation will help address these questions.

2 DNMT Enzymes

DNA methylation was first described by Holliday and Pugh [15] and Riggs [16], who hypothesised its role in gene regulation during development. Since then it was shown that DNA methylation can induce gene silencing when it occurs on promoters

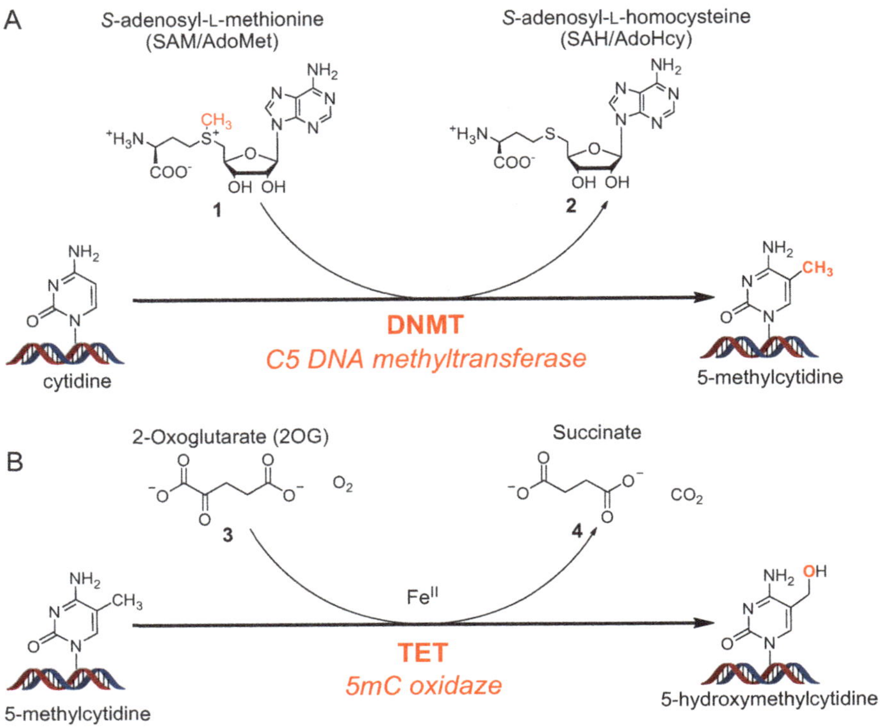

Fig. 2 DNMT- and TET-catalysed cytosine methylation and oxidations, respectively (**a**) and (**b**). The structures of substrates, cofactors and products are indicated

of genes (e.g. imprinted genes, transposon silencing, X-inactivation, tumour suppressor genes in cancer). DNA methyltransferase activities were isolated and observed in several eukaryotic cells [17]. The first mammalian DNMT to be identified was Dnmt1 in mouse cells [18]. Interestingly all C5-DNA methyltransferases share ten motifs that are relatively conserved (the motif IV contains the Pro-Cys dipeptide, the Cys being involved in the catalytic reaction making a covalent bond with the dC). Dnmt1 has a preference for hemi-methylated DNA, and it is thus called maintenance enzyme. Dnmt3A and 3B were discovered later [19] and characterised to be active also on non-methylated DNA and are thus considered de novo DNMTs. Dnmt3L was shown to lack the catalytic motif [20], resulting in a catalytically inactive form. However Dnmt3L plays an important role, especially in development and in imprinting [21]. It enhances the activity of Dnmt3A and 3B [22] and interacts with several proteins involved in chromatin regulation. See [23, 24] for recent reviews on the biology of DNMTs. The highly conserved DNMT2 has been shown to methylate tRNA and has been renamed tRNA aspartic acid (D) methyltransferase 1 (*TRDM1*) [23, 25].

2.1 Structures and Mechanism of DNMTs

DNMTs belong to the C5-DNA methyltransferases, which have a three-dimensionally conserved catalytic pocket. The C-terminal contains the motifs of DNA binding and catalysis of methyl transfer. Upon binding to DNA double helix, cytosine flips and binds in the catalytic pocket. The catalytic cysteine in the Pro-Cys motif (PCQ or PCN) then binds to position 6 of cytosine. Then the methyl group from the SAM is transferred to position 5 of the cytosine, creating a steric clash that releases the 5mC by β-elimination and resolves the DNA-DNMT complex.

In addition to the catalytic domain at the C-terminus, the DNMTs possess regulatory N-terminal domains that include a nuclear localisation signal and domains for protein-protein interactions to chromatin, transcriptional and replication regulators [26]. Several structures of the mammalian DNMTs have been resolved showing the particular features of the different isoforms and some protein partners. In particular, the three DNMT3s have a PHD (plant homeodomain)-like domain, ADD (ATRX-DNMT3-DNMT3L domain) for the interaction with the tail of histone H3 [26] and DNMT3A and 3B have a PWWP domain found in DNA-binding proteins and involved in nucleosome recognition [27], interacting with H3K36me3 [26]. The N-terminus of DNMT1 is rich in protein interaction domains that localise it to the nucleus (NLS), to replication forks (PBD, PCNA-binding domain, DMAP domain, RFTS domain, BAH1 and 2 [28]) and to un-methylated and hemi-methylated DNA (CXXC zinc finger domain; BAH1 and BAH2). These domains are also involved in the enzymatic control of the protein, for example, its inhibition when bound to non-methylated DNA and its switch to catalytically active form when bound to hemi-methylated DNA [26, 29, 30].

2.2 DNMTs in Diseases

It is clearly established that aberrant DNA methylation profile is associated with cancers [31–35]. In parallel to a global hypomethylation, hypermethylation at promoters of specific genes are observed in cancer cells. In particular, genes, such as tumour suppressor genes, are commonly silenced by promoter hypermethylation. These features have been exploited for the development of biomarkers for the detection of many cancers [36, 37], including for colon cancer (ColoVantage® and Epi proColon®) and lung cancer (Epi proLung). Moreover, DNA methylation is an anticancer therapeutic target: 5aza and 5azadC have been approved for the treatment of certain haematological cancers. These drugs are also in clinical trials, mainly in combination, for several solid tumours [38, 39]. However, these aberrant patterns are not limited to cancers [40]. Alteration in DNA methylation, caused by genetic mutations in the DNMTs or DNMTs deregulation, is involved in psychiatric, cognition, neuronal, ageing disorders, cardiovascular diseases, bacterial and viral infections and genetic diseases such as cystic fibrosis [41, 42]. Finally, epigenomics studies are revealing the role of DNA methylation in obesity, allergy, autoimmune diseases, addiction and inflammation [42, 43].

Nevertheless, despite the established roles of DNA methylation and DNMT1 and DNMT3s in diseases [42, 44], which DNMT isoform is best to therapeutically target and in which pathology remains to be determined. Chemical tools can thus be useful to address these important questions. Another unanswered question is in regards to Dnmt2. Mammalian Dnmt2 was discovered by its homology with the most conserved C5-DNA methyltransferases [45], but it was shown not to methylate DNA but rather RNA cytosines, the $tRNA^{Asp}$ [46, 47], and was subsequently renamed TRDMT1. It contains only the catalytic domain and interestingly is present in species that do not have DNMT1 or DNMT3 (such as *Drosophila* and *Schizosaccharomyces pombe*) [25]. Today it is still debated whether in certain species, Dnmt2 is able to methylate DNA or if its action is limited to RNA methylation [48, 49].

Thus, while DNMT enzymes are well-studied, many questions remain to be answered. Chemical tools that specifically inhibit the DNMTs can contribute towards answering these biological questions.

3 Inhibitors of DNA Methylation

Two families of inhibitors have been identified for DNMTs: the *nucleoside analogues*, of which 5-azacytidine (5azaC (**5**)) and 5-aza-deoxycytidine (5azadC(**6**)) are approved anti-leukaemia drugs (known as azacitidine and decitabine), and the *non-nucleoside analogues*, which are composed of very different scaffolds (Fig. 3).

3.1 Cytosine Analogues

5azaC (**5**) was synthetised as an antimetabolite and was described for its anti-leukaemia properties in 1964–1965 by Sorm and Vesely [50, 51]. However, it was its impact on DNA methylation and on the reprogramming of cells by Jones and Taylor [52, 53] that allowed the understanding of its mechanism of action [54–56]. 5azaC and 5azadC incorporate into DNA instead of dC. Once the DNMT is bound to the position 6 of the 5azadC, the β-elimination and restauration of the 5–6 double bond cannot occur and the DNMT is irreversibly trapped on the DNA (suicide complex), inducing its degradation by the proteasome [57]. The trapping of the DNMT1 was elegantly visualised by using fluorescent DNMT1 fusions [56].

5azaC (**5**) was approved by the FDA and then the EMA, together with the deoxy analogue 5azadC (**6**) for the treatment of acute myeloid leukaemia (AML), myelodysplastic syndrome (MDS) and chronic myelomonocytic leukaemia (CMML).

Both 5azaC and 5azadC are chemically unstable (storage and handling is sensitive) and have a very short half-life in patients [58]. Several efforts were made to

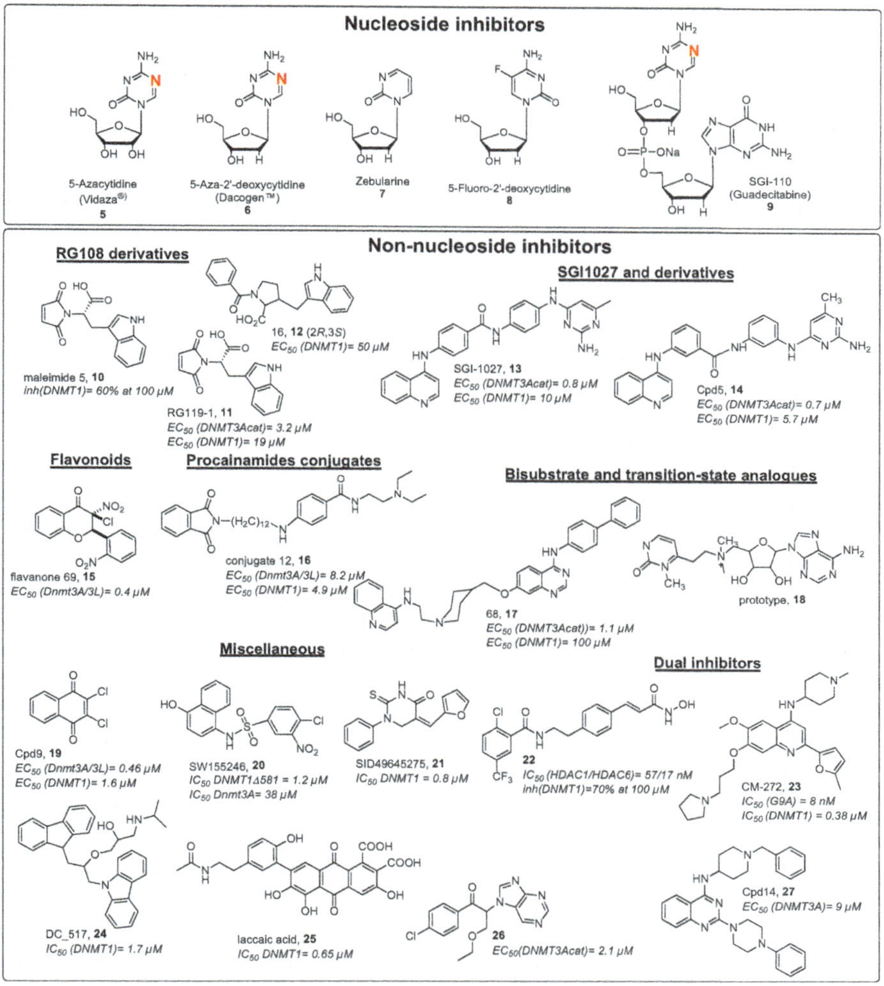

Fig. 3 Collection of inhibitors of DNMTs. Top panel: the nucleoside inhibitors (the substitution on position 5 of cytosine is indicated in bold red). Bottom panel: the non-nucleoside inhibitors. The half maximal inhibitory concentration (IC_{50}) or the half maximal efficacy concentration (EC_{50}) is indicated for DNMT1, DNMT3A and/or the DNMT3A/3L complex

generate stable analogues (such as zebularine (**7**) and 5-fluorocytidine (**8**)) and prodrugs (reviewed in [59–61]). The most promising prodrug is the dinucleotide version of 5azadC (**6**), SGC-110 (**9**, guadecitabine), which is in clinical trials against several haematological and solid cancers. Very interesting results are also being obtained in the clinic by combination studies with other epigenetic inhibitors, conventional chemotherapies or immunotherapies [38, 62].

Nevertheless, despite being the most potent compounds to demethylate DNA identified to date, they induce other effects due to their chemical instability, their incorporation into RNA and DNA, and also by the formation of suicide complexes. Indeed, the nucleoside analogues are incorporated into DNA (and 5azaC also in RNA) instead of C, but not only at CpG sites. They then induce the formation of a suicide substrate with DNMT that is processed by the proteasome and the DNA repair machinery. Because of their mode of action, they are not selective for an isoform of DNMTs and induce effects other than just DNA demethylation. To overcome this, several studies have focused on the identification of DNMT inhibitors that are not nucleoside analogues.

3.2 Non-nucleoside Analogues

The family of non-nucleoside inhibitors has been largely reviewed [24, 42, 59], so this section will focus on a selection of inhibitors that have shown effects on cellular phenotypes and on gene expression or have provided new concepts for the discovery of new inhibitors (Fig. 3). We have chosen to omit most of the compounds that act by binding to DNA, such as intercalators (i.e. acridine derivatives [63]) or minor groove binders (i.e. bisbenzimidazoles [64]), that lack specificity for CpG sites. Natural products have also been excluded since they have been reviewed recently and show little specificity for the DNMT enzymes [65]. Importantly direct comparisons of inhibition properties of these compounds cannot be made, because the assays and conditions used for each study are very different, as well as the choice of the enzyme used (reviewed in [66]), which can have an important impact on the inhibition [67].

The first synthetic non-nucleoside inhibitor, *N*-phthaloyl-l-tryptophan RG108, was identified by in silico screening against a model of DNMT1 and showed reactivation of tumour suppressor genes in different cancer models [68, 69]. It is worth noting that the compound is not active in enzymatic studies against mammalian DNMT1 and DNMT3A [70–72]. However, it is of interest from the perspective of inhibiting DNA methylation since several groups have shown that it leads to demethylation of genes in cellular models [73–77] and in vivo [75, 78]. Modifications on RG108 resulted in compounds with improved activity against the mammalian DNMTs (maleimide **5**, (**10**), (**12**) and RG119-1 (**11**) [70, 71, 79] (Fig. 3)) and highlighted the interactions of the compound within the catalytic pocket. These studies provide new potential for improving this chemotype.

By modulation of DNA minor groove binders, Datta et al. identified SGI-1027 (**13**) [80] which interacts weakly with DNA and inhibits DNMT3A and DNMT1 in the micromolar range [81]. Further modifications by Valente et al. resulted in a more active meta-meta analogue Cpd5 (**14**) ([82] Fig. 3), which shows a stronger interaction with DNA at CG-rich regions [81]. Chemical modifications to better characterise the structure-activity relationships and increase the cellular potency are ongoing.

The screening of a library of flavonoids identified a family of 3-chloro-3-nitroflavanones which was able to inhibit DNMT3A and DNA methylation in zebrafish embryos (flavanone 69 (**15**) Fig. 3; [83]). This family of compounds showed selectivity for the C5-DNA methyltransferases, compared to other non-specific flavones such as epigallocatechin gallate (EGCG).

Procainamide and procaine were described to weakly bind to DNA and inhibit DNMTs [84, 85] and were chemically modified to improve the inhibition [86, 87]. The conjugation to diverse moieties resulted in micromolar inhibitors of DNMT1 and Dnmt3A/3L (conjugate 12 (**16**), Fig. 3, [88]).

High-throughput screening campaigns against different DNMTs and use of diverse biochemical assays resulted in the identification of miscellaneous compounds, the most potent of which are shown in Fig. 3. Among these, the anthraquinone laccaic acid (**25**) was described as a DNA-competitive inhibitor of DNMT1 with a weak activity in breast MCF-7 cancer cells (at 200 μM) [89]. A naphthoquinone, diclone (**19**), a pesticide and fungicide, was specific of DNMT1 and Dnmt3A and inactive against the histone methyltransferase G9A. This observation opened the path to studying the impact on the epigenome of plants, animals and human when addressing the toxicology of pesticides [90]. SW155246 (**20**), an aromatic sulphonamide, showed a weak selectivity against DNMT1 vs DNMT3A and DNMT3B and induced a weak inhibition of methylation and reactivation of TSGs in human lung carcinoma [91]. DC_501, DC_517 (**24**) [92] and SID49645275 (**21**) [93] were identified to inhibit DNMT1 in the low micromolar level and to inhibit cell proliferation. The mechanism of action and the selectivity still need to be explored.

3.3 *Transition State and Bisubstrate Analogues*

Recently, progress has been made in the design of transition state and bisubstrate analogues. We have shown that seven cytosine-adenosine compounds, designed as transition state analogues of the methylation reaction of position 5 of cytosine by DNMT1 and DNMT3, did not result in inhibitors of DNMTs but rather of histone arginine methyltransferases PRMT4 [94].

Interestingly, based on the design of mechanism-based transition analogues, Miletic et al. have described and evaluated a set of adenosyl-1-methyl-pyrimidin-2-one derivatives in silico as leads for the synthesis of mechanism-based suicide inhibitors of DNMT1 ([95], prototype (**18**) Fig. 3).

Another chemical approach we explored is the bisubstrate analogues that have successfully been applied to inhibit adenine DNA methyltransferases. Using this approach we have identified compounds (**17**) as potent inhibitor of DNMT3A able to demethylate promoters of tumour suppressor genes and to reactivate gene expression in cancer cells ([96], Fig. 3).

3.4 New Approaches to DNMT Inhibitor Design

Finally, a very promising approach is the design of dual inhibitors [97]. By modifying BIX-01294, inhibitor of histone H3 lysine 9 methyltransferase G9A, Rotili et al. identified new quinazoline derivatives as inhibitors of DNMT3A with activity in cancer cells (cpd 14 (**27**) Fig. 3 [98]). José-Enériz et al. successfully modulated this family of G9A and DNMT inhibitors to obtain a dual G9A-DNMT1 inhibitor, CMC-272 (**23**) (Fig. 3), with anti-leukaemia effect in an in vivo model of mice engrafted with ALL-derived CEMO-1 cells [99]. This is the first example of DNMT dual inhibitors active in vivo. Other dual inhibitors are currently being explored, such as (**22**) (Fig. 3), resulting from chemical optimisations based on HDAC inhibitors [100].

Taken together, the transition state and bisubstrate analogues, as well as dual inhibitors, are interesting chemical approaches that need to be further explored to obtain new, potent and selective inhibitors of DNMT. Another strategy worth exploring is to develop protein-protein interaction inhibitors (PPI) for DNMTs. PPIs have been successful for other epigenetic targets, and it is worthwhile since the DNMTs are involved in protein complexes that direct DNA methylation [101, 102]. For example, the inhibition of the DNMT1/CFP1 interface with peptides was shown to affect methylation level of cancer cells and to synergise with temozolomide [103]. Most recently, Ye et al. determined the crystal structure of DNMT1 in two different states and suggest the possibility to design inhibitors of the conformational transition necessary for DNMT1 activity [30].

4 TET Enzymes

Methylation at the 5-position of cytosine (5mC) is recognised by the TET enzymes that catalyse the oxidation of 5mC in a cascade of iterative steps (Fig. 1). TETs are part of the oxygenase superfamily that uses Fe(II) for catalysis, using O_2 and 2-oxoglutarate (2OG, **3**) as cofactors, to generate the oxidised substrate, CO_2 and succinate (**4**) [6]. Recent developments reveal that the 5-methyl oxidations at cytosines play an extensive role in epigenetic regulation and key steps in the DNA demethylation pathways. Studies have shown that TETs are fundamental in mammal development and mutations, or overexpression of the protein is linked to various diseases.

4.1 Discovery and Biological Roles of TETs

DNA methylation is a well-established modification, and it has long been known that passive dilution alone cannot fully account for the rapid rate of genome methyl

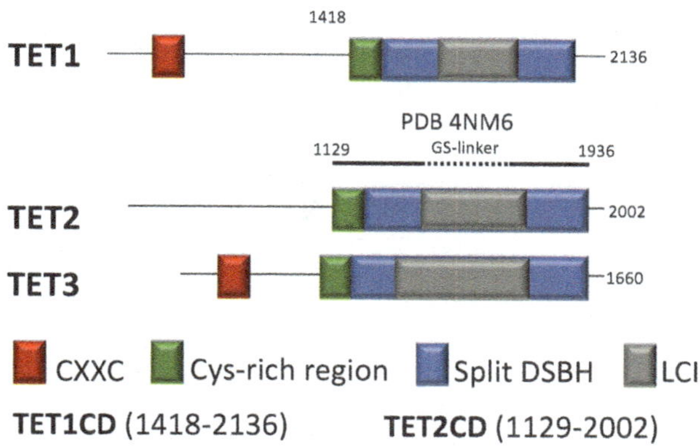

Fig. 4 Domain architecture of the human TET proteins. The 'canonical' sequences for the three human TETs are shown. The minimum regions required for efficient catalysis (often referred to as the 'catalytic domain', CD) have been reported as follows: TET1 (1418–2136), TET2 (1129–2002) and TET3 (689–1596)) [6, 108–110]. The CD includes the Cys-rich region (green), the DSBH domain (blue) and a low-complexity insert (LCI, grey). The DSBH catalytic region is 'split' into two segments by the LCI insertion, with the first two Fe(II) coordinating residues of the HXD . . . H motif in the first segment, and the last residue His and the 2OG binding Arg in the second segment. Additionally, TET1 and TET3 have an N-terminal CXXC zinc finger domain that can bind DNA and act to recruit to target genomic sites

remodelling after mouse fertilisation, indicating the possibility of an additional mechanism that would remove 5mC [104]. The TET proteins were originally linked as being fusion partner of mixed lineage leukaemia (MLL) [105] and later identified as enzymes related to JPB1 that facilitate the oxidation of thymidine (T) in the nucleus to 5-hydroxyuracil (5hmU) [106]. The first discovery of 5hmC modification was made in 2009, where 5hmC was detected in the brain tissue at high abundance [5]. At the same time, it was revealed that TET enzymes are capable of oxidising 5mC to 5hmC [6]. Shortly after, it was discovered that TET enzymes can iteratively oxidise 5mC to 5caC [107].

TET enzymes are widely conserved through evolution [111], including in *Naegleria gruberi* TET (NgTET). Herein, we focus on studies involving human TETs (hTETs), with some references to other model organisms, such as murine TETs (mTET) [110] and NgTETs [112].

To date, three TET enzymes have been found in humans (hTET). The hTET proteins are approximately 180–230 kDa in length, and all human TETs (hTET1–3) can carry out iterative oxidation of 5mC [6]. The catalytic domain (CD) of TETs contains the cysteine-rich domain and a split double-stranded ß-ηelix (DSBH) connected via a flexible region known as the low-complexity insert (LCI) (Fig. 4). The DSBH domain, which contains a jelly roll/cupin fold, is highly conserved among the TETs and a signature motif of the 2OG oxygenase superfamily (see [113] for review). DSBH forms a catalytic core where it positions the 5mC DNA substrate at the active site adjacent to Fe(II) enabling catalytic oxidation, while the cysteine-rich region structurally stabilises the DSBH and DNA interactions.

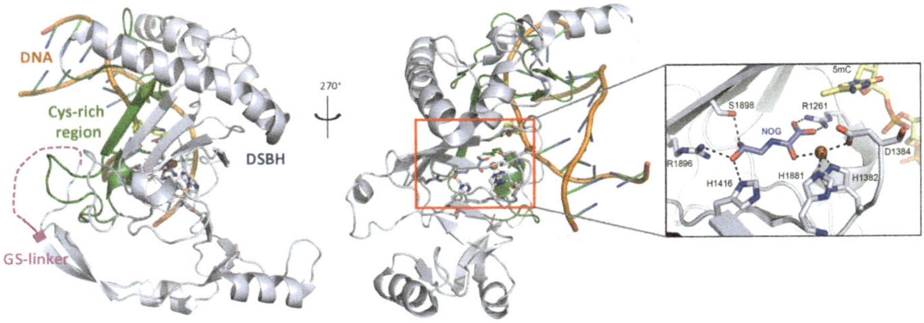

Fig. 5 Crystal structure of human TET2 in complex with DNA. (LHS) Two different views of the human TET2 catalytic domain complexed with dsDNA-5mC and NOG. Each domain is coloured (Cys-rich region, green; DSBH, light blue; GS linker (15 GS repeat replacing LCI, delΔ1099–1936), magenta; dsDNA-5hmC, orange). (RHS) NOG is bound at the active site. PDB:4NM6

TET1 and TET3 have an additional DNA-binding CXXC domain at the N-terminus, which preferentially binds CpG-rich regions. TET2 lacks the CXXC zinc finger domain but instead has been shown to interact with IDAX (CXXC4) which then interacts with DNA [6, 114].

While the cysteine-rich domain is essential for the catalytic activity [114], truncations in LCI retain the activity in all hTET1–3, albeit at reduced catalytic efficiency [108, 114]. The first structural insight came in 2013 from the crystal structure of hTET2 in complex with 5mC-containing double-stranded DNA (dsDNA) (Fig. 5) [114]. To enable crystallisation of hTET2, Hu et al. replaced the 837 amino acid (aa) flexible loop LCI by a GS (15 aa) linker. The structure revealed that the phosphate backbone of DNA interacts with multiple arginine and lysine residues of hTET2. The bound helical DNA structure is distorted, where the G-5mC hydrogen bond interaction is disrupted and 5mC flips into the catalytic core of the DSBH scaffold. The methyl group of 5mC is placed in proximity to the catalytic Fe (II) at the active site, which is held in place by H1382, D1384 and H1881 (Fig. 5). N-oxylylglycine (NOG), an inactive 2OG mimic, coordinates to the metal and interacts with R1261, H1416, S1898 and R1896 of hTET2.

4.2 *Mechanism of Enzyme Catalysis*

The human TET enzymes contain a DSBH domain which catalyses the oxidisation of 5mC, using a similar mechanism as other 2OG oxygenases [115–117]. In the first step of the proposed mechanism, 2OG enters the active site and binds to Fe (II) releasing water molecules (Fig. 6). The DNA substrate binds with the 5mC methyl directed towards Fe(II). Molecular oxygen (O_2) then binds and reacts with the Fe(II) to form Fe(III) species. The radical oxygen molecule reacts further with

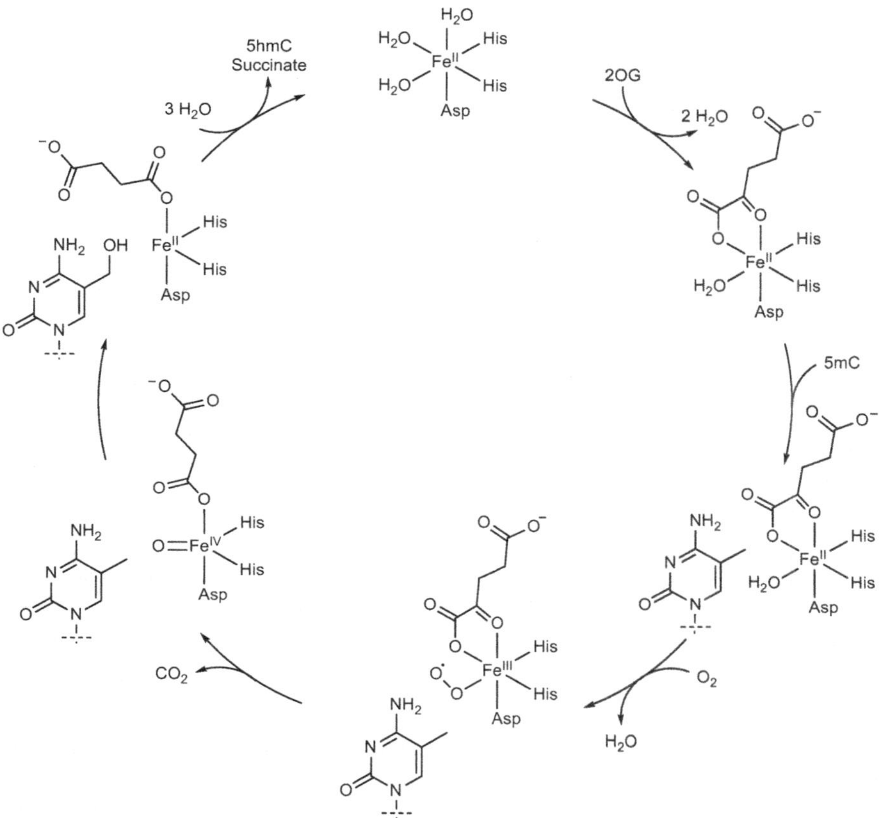

Fig. 6 Proposed mechanism of the catalytic domain of TET enzymes. TET protein binds to 2OG, molecular oxygen and the 5mC substrate. After generation of highly reactive Fe(IV) species, 5-methyl on cytosine is oxidised. The catalytic cycle is completed by the release of CO_2, succinate and substrate and the coordination of water to Fe(II)

2OG giving a Fe(IV) intermediate and carbon dioxide (CO_2). The highly reactive Fe (IV) metal reacts with a proton on the methyl of 5mC substrate to provide the 5hmC product in two steps, and Fe(II) is regenerated. The newly 5hmC substrate and succinate can then be replaced by water molecules completing the catalytic cycle. Oxidative steps of 5hmC to 5fC and 5fC to 5caC catalysed by TET are thought to follow a similar pathway.

The rate of oxidation by TETs is highly context dependent. The catalytic domain by itself is not believed to bind to specific DNA sequence but instead the catalytic domain 'slides' along the DNA strand until the active site binds the modified cytosine (5xC, 5-position modified cytosine, x = m, hm or f (Fig. 7. highlighted in red)). Enzyme kinetic analysis of hTET2 reveal selectivity for oxidation of 5mC over 5hmC or 5fC on dsDNA substrate, with lower (Kcat: ~3.4- to 4.6-fold) and higher K_M (~1.4- to 2.7-fold) than for 5mC [118]. The C-H proton extraction and inter-intra

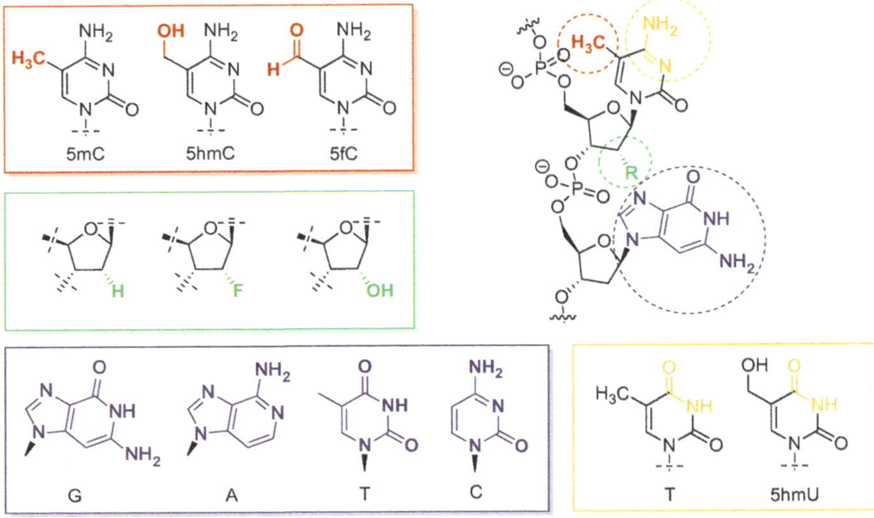

Fig. 7 Double-stranded DNA (dsDNA) representing the variety of different substrates for TET. Red, 5-modified cytosine (5mC, 5hmC, 5fC and 5caC); yellow, cytosine modified to thymidine; green, modified 2′-(H (DNA), OH (RNA), F); purple, phosphate pairing base (CpG, CpA, CpT, CpC); blue, counter strand (single- or double-stranded DNA)

molecular proton stabilisation are understood to be important in determining the differential rates of catalysis. Furthermore, hydroxyl group at the 2′ position on the sugar (i.e. RNA) is tolerated (Fig. 7, highlighted in green) [119].

Comparison of identical main strand sequences of double-stranded DNA (dsDNA), single-stranded DNA (ssDNA) and single-stranded RNA (ssRNA) revealed that both dsDNA and ssDNA can be iteratively oxidised (5mC, 5hmC, 5fC and 5caC) by TETs; however, while 5-methyl RNA cytosine (5mrC) can be oxidised by TETs to form 5hmrC, limited levels of 5frC formation and the absence of 5carC were observed [119]. dsDNA and ssDNA are thus better TET substrates than RNA. It is interesting to note that the 2′-(*R*)-fluorinated derivatives of oxidised 5mC are also substrates of TETs, and although they exhibit decreased catalytic efficiency, they can be used as mechanistic tools [120]. TET activity is thus influenced by the modifications at the 2′ position of the cytosine sugar, suggesting that this site contributes towards substrate specificity of TETs (Fig. 7. Highlighted in green). It has been observed that mTET can oxidise thymidine as an alternative substrate (Fig. 7 highlighted in yellow), both in vitro [121] and in vivo [122], to give 5-hydroxymethyl uracil (5hmU) and 5-formyl uracil (5fU), albeit at significantly reduced activity relative to 5mC oxidation.

In the oligonucleotide sequence context, CpG is the preferred substrate over CpC, CpA and CpT. In the genome, 60–80% of CpG are methylated [123], where symmetry is frequently observed; this gives rise to 21 possible symmetrical (on both strands) CpG combinations (C, 5mC, 5hmC, 5fC and 5caC) which TET can act on. In NgTET, significant drop in catalytic efficiency is observed by altering CpG site to CpA (~1.75-fold), CpT (~3.80-fold) and CpC (~5.80-fold) [112]. The crystal structure of NgTET1 indicates H-bonding interactions between the guanosine in the CpG pair and the NgTET protein which can account for the preference for CpG [112]. Note that 5-C modifications of the CpG on the reverse strand do not influence the catalytic activity.

The alkylation repair protein (AlkB) homologs, subfamily of 2OG oxygenases, are closely related to the TET oxygenases and use Fe(II) catalysis with 2OG (**3**) and O_2 to oxidise DNA or RNA. This subfamily contains ABHs and *f*at mass and *o*besity-associated (FTO) proteins. In particular, ABH2 and ABH3, demethylases involved in repair processes, are structurally similar to hTET2. The preferred substrates for ABHs or FTO are 1*N*′-methyl-adenine (1 mA), 3*N*′-methyl thymidine (3mT), 3*N*′-methyl cytosine (3mC) and 6*N*′-methyl adenosine (6 mA), and their oxidation results in unstable hemiaminal intermediate which decomposes to formaldehyde and demethylated base [114].

4.3 TETs in Development and in Disease

5hmC, 5fC and 5caC are found in many cells, but their concentrations vary depending on the cell types. Interestingly, while the levels of 5mC remain relatively even across different cell types at approximately ~3.5–4.5% of all cytosine in the genomic DNA, this is not observed for 5hmC, 5fC or 5caC. The levels of 5hmC can range from 0.7% in the central nervous system to 0.03% in the spleen [124]. High levels of 5hmC are commonly found in the brain and neurons. Purkinje neurons, for example, contain up to 40% of 5hmC abundance relative to the 5mC levels in the cell [5].

While 5fC and 5caC are stable modifications and believed to be part of signalling pathways, 5fC and 5caC are significantly less abundant than 5mC or 5hmC. Levels of 5fC can range between 0.2 parts per million (ppm) in the lungs and 12 ppm in the brain in mice. Interestingly, there is no direct correlation between 5mC/5hmC and 5fC in cells or among the age of the tissue, suggesting that these marks have independent roles and are actively generated over time [125]. Postnatal mice have undetectable levels of 5caC (<0.1 ppm), but in 12-week-old mice, some tissues, such as the liver, can have elevated levels of 5caC (up to 2.0 ppm), while in others such as the kidney or brain, 5caC levels remain below the detection limit [125].

Aside from their role as intermediates for demethylation, the function of 5hmC, 5fC and 5caC, is not fully understood. Recent studies suggest that each mark has multiple implications. 5hmC formation is detected in active genes and enriched in the promotor regions [126]. While 5fC has been proposed to influence the helical structure of DNA [127], this may be context dependent [128]. Raiber et al. demonstrated that additional anchoring of the nucleosome to the DNA can occur in vitro and in vivo, as a result of histone lysine reacting with the 5fC to form an imine derivative [129]. This would provide enhanced nucleosome organisation within the chromatin remodelling for activation and silencing. Furthermore, 5fC and 5caC can affect the activity of proteins that associate with these marks. For example, with RNA polymerase II, 5fC and 5caC result in a lower rate of incorporation of pairing G [130]. Additionally, modifications at 5mC are distinctly recognised by an array of reader proteins [131] providing additional level of epigenetic regulation mediated by TET oxidation.

TET1 was first identified as a translocation partner of *MLL* gene in patients with acute myeloid leukaemia (AML) [105]. Subsequently, it was demonstrated that MLL fusion protein directly binds to TET1 promoter, causing upregulation of TET1 and global increase in 5hmC levels in MLL-rearranged leukaemia [132]. TET1 plays a critical role in the oncogenesis of MLL-rearranged leukaemia in vitro and in vivo through co-activation of the Hoxa9/Meis1/Pbx3 signalling pathway [132]. In glioblastomas, TET1-mediated 5hmC production plays a critical role in tumorigenicity [133]. These studies highlight the therapeutic potential for targeting TET1 in certain cancers. In contrast, TETs have been shown to also have important tumour suppressor roles in multiple cancers. Mutations in TET2 have been linked to haematopoietic malignancies [134]. In myelodysplastic syndrome (MDS), TET2 is the most highly mutated gene [135], and in AML patients, multiple TET2 mutations (including frameshift, nonsense and missense within the protein creating mutated or truncated proteins) are found with reduced or abolished catalytic activity [136]. Furthermore, abnormalities have been observed in lymphoid malignancies, such as hTET2 B/T-Cell lymphoma [137]. Rare occurrences of mutations are found in hTET1/3 in chronic lymphocytic leukaemia (CLL) [138]. TETs are also linked to various solid tumours including gastric, breast, lung, liver and prostate cancer [139]. Downregulation of *TET* gene expression has been observed in multiple solid tumours, with decrease in 5hmC levels and increasing rate of proliferation [140].

Taken together, the evidence suggests that TETs can have both oncogenic and tumour suppressor roles, depending on the cellular context. Chemical probes for TETs are thus needed to understand their biological functions in development and in diseases.

5 TET Enzyme Assay and Inhibitor Development

The availability of robust and quantitative assays is prerequisite for the biochemical and functional studies of enzymes. For the TET enzymes, there are now a wide range of methodologies available to detect and quantitate oxidised 5mC levels, including global genome-wide mapping at base resolution [141–143]. Analysis techniques of modified nucleosides/oligonucleotides include (1) antibody-based detection of oxidised 5mC, (2) analytical methods using thin-layer chromatography (TLC) [5, 110, 112, 144] or liquid chromatography coupled with mass spectrometry (LC-MS, LC-MS/MS) [107, 124, 145] and (3) chemical conversion or enzymatic labelling of modified cytosines (including glucosylation of 5hmC) [146–153]. These methodologies have enabled studies on tissue and genomic distribution and dynamics of oxidised 5mC in

Table 1 Reported assays used for kinetic analysis and inhibitor development for the TET proteins

Detection method	Analytes; principle	References
Analytical methods		
LC-MS/MS	Nucleosides: enzymatic digestion of oligonucleotides to nucleosides, followed by dephosphorylation by CIP and LC-MS/MS analysis. relative quantification of 5mC and oxidised 5mCs	[114, 118]
MALDI-TOF MS	Oligonucleotides: Relative quantification of intact DNA substrate and products using mass-directed detection	[108, 157]
Antibody-based methods		
Dot blot	Oxidised 5mC (DNA); detection and quantitation of oxidised 5mC containing intact DNA immobilised on membrane, using specific antibodies (5hmC, 5fC, 5caC) (chemiluminescence/fluorescence)	[158]
ELISA	Oxidised 5mC (DNA); immobilised DNA substrate on a plate. Upon incubation with TETs, product formation detected using 5hmC antibody (fluorescence)	Epigentek
AlphaScreen	Oxidised 5mC (DNA): homogenous bead-based assay, with product DNA capture and antibody detection of oxidised 5mC (luminescence)	[158]
Radiolabelling		
TLC	Nucleosides; oligonucleotides digested and dephosphorylated with CIP and labelled with [γ-^{32}P]ATP and T4 polynucleotide kinase. Further treatment with nuclease and analysed on TLC plate. Relative quantification of 5mC and oxidised 5mCs	[159]
$^{14}CO_2$ assay	By-product formation (CO_2); measurement of hydroxylation-coupled $^{14}CO_2$ production using 2-oxo[1-^{14}C]glutarate cofactor	[160]
Other		
Succinate-Glo	Succinate (**5**) by-product of catalysis; Enzyme-coupled assay linking succinate production to ATP production (bioluminescence)	Promega [161]

biological context. The readers are referred to [154] for an overview of sequencing techniques available for mapping oxidised 5mC on the genome and [155, 156] for reviews on some approaches to studying the enzyme activities of TETs. In this section, we highlight key methodologies used for kinetic analysis of TETs in vitro and recent TET assays developed for inhibitor discovery.

5.1 *Kinetic Analysis of Human TETs*

Availability of reagents, such as modified oligonucleotides, oxidised 5mC antibodies and recombinant TET proteins, in recent years have enabled the development of a variety of biochemical assays for the TETs (Table 1).

LC-MS/MS is a direct and reliable method for quantification of multiple cytosine modifications simultaneously. While there are a number of variations on the methods and instruments used, the general procedure is that the DNA is purified, digested to nucleosides by nuclease and phosphodiesterase treatment. The digested nucleosides are separated and analysed by LC in tandem with MS (e.g. triple-quadrupole MS, quadrupole-orbitrap MS) and quantitated against labelled nucleoside internal standards. The large dynamic range (from low femtomole range) allows diverse application of this technique, from gDNA to oligonucleotides. LC-MS/MS analysis has been used to determine the substrate preference for TETs for 5mC oxidation over 5hmC/5fC-DNA [107, 118]. For example, steady-state kinetics analysis of TET2CD (*del*Δ1099–1936) using LC-MS/MS revealed the affinities of modified cytosine to reduce with increasing oxidation (K_M = 0.48, 0.9 and 1.3 μM for 5mC, 5hmC and 5fC, respectively) [118].

While LC-MS/MS methodology provides quantitative accuracy and robustness, the multistep processing of DNA has hampered its use for high-throughput applications. A matrix-assisted laser desorption/ionisation time-of-flight (MALDI-TOF) MS method has provided a middle ground. This assay allows direct measurements of intact DNA without the need for downstream processing steps (e.g. enzymatic and chemical modifications), thus minimising the reagent requirement, processing error and time. Michaelis-Menten kinetic parameters for 2OG (**3**) were determined to be K_M = 15.7 μM (hTET2CD (*del*Δ1099–1936)) and K_M = 24.2 μM (hTET3CD (*del*Δ 689–1596)) using this method. The relatively low 2OG K_M suggests that potency is crucial for 2OG competitive inhibitors.

Assays using antibodies against modified cytosines, such as enzyme-linked immunosorbent assay (ELISA) and AlphaScreen, have also helped improve the sensitivity, the volumes and throughput of the assays [158]. Assays using radiolabelling (e.g. 2-oxo[1-^{14}C]glutarate, [γ-^{32}P]ATP) are also utilised for inhibitor screening.

5.2 *Inhibitors of TET Enzyme Activity*

As for the other 2OG oxygenases, the majority of inhibitors target the catalytic Fe (II) and often mimic or compete with the 2OG binding (see [162] for review of 2OG oxygenase inhibitors (Fig. 8)).

5.2.1 2OG Analogues as TET Inhibitors

The first TET inhibitor to be identified was *R*-2-hydroxyglutarate (*R*-2HG, **28**), an 'oncometabolite' associated with the gain-of-function mutations in isocitrate dehydrogenases (IDH). IDH is a tricarboxylic acid (TCA) cycle enzyme that catalyses the oxidative decarboxylation of isocitrate to 2OG (**3**). IDH with mutations in the active site (Arg100, Arg132 in IDH1, Arg140, Arg172 in IDH2 found in glioma/leukaemia) can further catalyse the reduction of 2OG (**3**) to *R*-2HG (**28**), leading to cellular accumulation of *R*-2HG (**28**) (up to 30–50 mM, compared to <0.1 mM in normal cells) [164–167]. *R*-2HG inhibits human (and mouse) TETs with IC_{50} at 4–5 mM range [159, 160]. Given the inhibitory effect of TETs by *R*-2HG and significantly

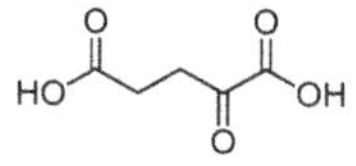

2OG, **3**
2-oxoglutraric acid

K_m (hTET2CDdel)$^{1, A}$ = 15.7 μM
K_m (hTET3CDdel)$^{1, A}$ = 24.2 μM
K_m (mTet1CD)$^{2, B}$ = 55 μM
K_m (mTet2CD)$^{2, B}$ = 60 μM
IC_{50} (NgTET1)$^{3, C}$ = 250 μM

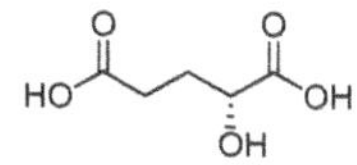

R-2HG (D-2HG), **28**
R-2-hydroxyglutaric acid

IC_{50} (NgTET1)$^{3, C}$ = 5.2 mM
IC_{50} (hTET2CDdel)$^{1, A}$ = 5.3 mM
%I (10 mM, mTet2CD)$^{4, D}$ = 33%
%I (10 mM, mTet1CD)$^{4, D}$ = 28%
IC_{50} (mTet1CD)$^{2, B}$ = 4 mM
IC_{50} (mTet2CD)$^{2, B}$ = 5 mM

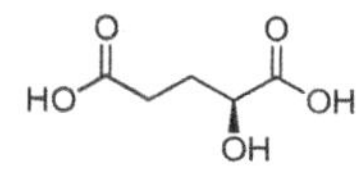

L-2HG (S-2HG), **29**
L-2-hydroxyglutaric acid

IC_{50} (hTET2CDdel)$^{1, A}$ = 12.4 mM
%I (10 mM mTet2CD)$^{4, D}$ = >90%
%I (10 mM mTet1CD)$^{4, D}$ = >90%
IC_{50} (mTet1CD)$^{2, B}$ = 1 mM
IC_{50} (mTet2CD)$^{2, B}$ = 1.6 mM

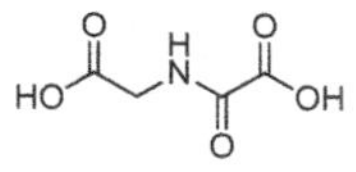

NOG, **30**
N-oxalylglycine

IC_{50} (hTET2CDdel)$^{1, A}$ = 149 μM
IC_{50} (NgTET1, FP)$^{3, C}$ = 49 μM

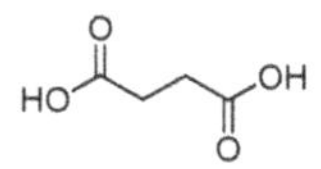

Succinate, **31**

IC_{50} (mTet1CD)$^{2, B}$ = 540 μM
IC_{50} (mTet2CD)$^{2, B}$ = 570 μM

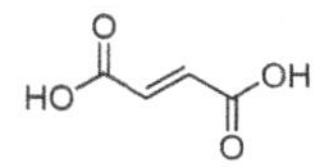

Fumarate, **32**

IC_{50} (mTet1CD)$^{2, B}$ = 390 μM
IC_{50} (mTet2CD)$^{2, B}$ = 400 μM

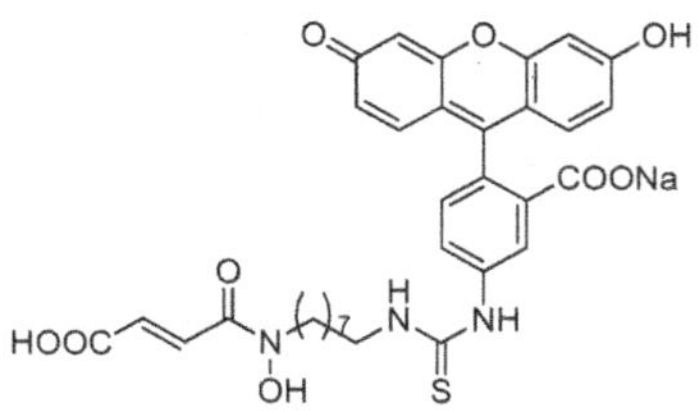

NgTET1 fluorescent probe, **33**
K_d (NgTET1)^{C} = 250 nM

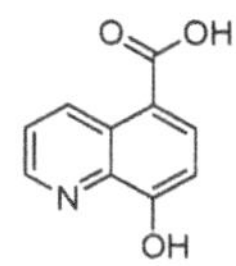

IOX1, **34**
8-hydroxyquinoline
IC_{50} (hTET1CD)$^{5, E}$ = 1 μM
IC_{50} (hTET2CD, AS)$^{5, E}$ = 0.7 μM

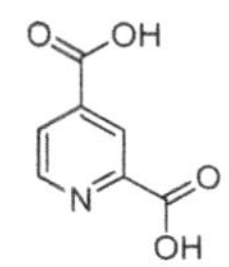

2,4-PDCA, **35**
2,4-pyridinedicarboxylic acid
IC_{50} (NgTET1)$^{3, C}$ = 27 μM

Ac-^{D}YLIYAYYTWWEHSC-NH$_2$ (cyclic, S)
TiP1, **36**

K_d (hTET1CD)$^{6, E}$ = 113 nM
IC_{50} (hTET1CD)$^{5, E}$ = 1.48 μM
IC_{50} (hTET2CD)$^{5, E}$ > 10 μM

Ac-^{D}YWYLPSYRVPWFC-NH$_2$ (cyclic, S)
TiP2, **37**

K_d (hTET1CD)$^{6, E}$ = 116 nM
IC_{50} (hTET1CD)$^{5, E}$ = 1.13 μM
IC_{50} (hTET2CD)$^{5, E}$ = 2.42 μM

Ac-^{D}YYRVKTYYQITYVYLSC-NH$_2$ (cyclic, S)
TiP3M15L, **38**

K_d (hTET1CD)$^{6, E}$ = 216 nM
IC_{50} (hTET1CD)$^{5, E}$ = 1.44 μM
IC_{50} (hTET2CD)$^{5, E}$ = 1.62 μM

Fig. 8 Structures of TET inhibitors. Structures of the cofactor 2OG, TCA cycle intermediates and metabolites and small molecules and cyclic peptides that bind to or inhibit the TET proteins are shown. IC_{50}, K_d or percentage inhibition of activity (%I) are stated at tested concentration. Assay methods used for affinity or inhibition are noted as superscript: (1) MALDI-TOF MS, (2) $^{14}CO_2$ radiolabelling, (3) FP, (4) TLC, (5) AlphaScreen, (6) SPR. (A) Sudhamalla et al. [108], (B) Laukka et al. [160], (C) Marholz et al. [163], (D) Xu et al. [159], (E) Nishio et al. [158]

reduced global levels of 5hmC in *IDH* mutants, TETs are thought to play a contributory role in the tumorigenicity of *IDH* mutant cells. Other TCA cycle intermediates, such as succinate (**31**) and fumarate (**32**), can also accumulate as a result of tumour-associated mutations in the TCA cycle enzymes succinate dehydrogenase (*SDH*) and fumarate hydratase (*FH*) [168]. Biochemical assays using recombinant mouse Tet1 and Tet2 have shown that both succinate and fumarate inhibit at 400–600 μM range, and modest changes in global 5hmC levels are observed in human neuroblastoma cells treated with fumarate and succinate esters [160]. Thus, it appears TET activities are sensitive to concentration flux in some TCA cycle intermediates/oncometabolites in vitro and in cells, as found in other 2OG (**3**) oxygenases [169–171].

In 2015, a fluorescent polarisation assay based on a fluorophore-linked hydroxamic acid probe (**33**) was reported for Tet1 from NgTET, a model protein with 39% similarity to human TETs [163]. The hydroxamic acid motif is thought to chelate the active site Fe(II). Indeed, the probe binds to NgTet1 with $K_d = 250$ nM and competes with 2OG (NgTET $K_d = 250$ μM). *N*-Oxalylglycine (NOG, **30**), a close isostere and an inactive analogue of 2OG (**3**) and a broad-spectrum inhibitor of 2OG oxygenases [172], was also found to bind to NgTet1 ($K_d = 49$ μM), demonstrating the utility of FP assay for the identification of 2OG competitive binders of the TET proteins.

While the human TET2-5hmC-DNA complex had been co-crystallised with NOG [114], it was not until recently that NOG was demonstrated to inhibit the catalytic activity of human TET2CD ($IC_{50} = 149$ μM) using MALDI-TOF MS assay [108]. *R*-2HG was confirmed to weakly inhibit TET2 in a similar range as previously reported and *S*-2HG (**28**) at $IC_{50} > 10$ mM using this assay.

5.2.2 Non-metal-Chelating Inhibitors of TETs

We have recently reported the development of macrocyclic peptide inhibitors for the human TETs [158]. Using mRNA display-based RaPID technology, macrocyclic peptide binders of human TET1CD *del*Δ1099–1936 were selected from a pool of $>10^{12}$ peptides. Three sequences (TiP1 (**36**), TiP2 (**37**), Tip3M15L (**38**)) were confirmed to bind to TET1 ($K_d < 100$–220 nM) using surface plasmon resonance (SPR) and inhibit the catalytic activity at approximately $IC_{50} = 1$ μM using 5hmC antibody-based AlphaScreen™-based assay. Interestingly, TiP1 selectively inhibited TET1 over TET2, demonstrating that some selectivity between different TET proteins is possible. Unlike IOX1 (**34**), a 2OG competitive, metal-chelating broad-spectrum 2OG oxygenase inhibitor control [172], the macrocyclic peptides do not compete directly with 2OG, demonstrating a novel mode of action and a promising approach to developing selective inhibitors.

6 Conclusions

DNA methylation is the most conserved epigenetic modification, and in mammals it plays an important role in gene regulation. It is involved in normal biological process, but it is also aberrant in several human diseases [35]. In particular it is well studied in cancer, and, for example, certain FDA-approved cancer diagnostic kits are based on DNA methylation. Nevertheless, several questions remain to be addressed, and chemical tools can be of help. The most potent inhibitors of DNA methylation are the nucleoside analogues 5azaC and 5azadC that have their limitations as described above. Non-nucleoside inhibitors have been designed or screened to overcome this. Several have shown to demethylate promoters and reactivate tumour suppressor genes, but none have the potency of the nucleoside inhibitors in cells and in vivo. There are ongoing efforts to further improve the design of these compounds, and novel strategies, such as the design of dual inhibitors or bisubstrate analogues, are providing promise. Isoform-selective inhibitors of DNMTs (Dnmt1, Dnmt2 or Dnmt3) will be of great use for the understanding of the roles of each, as well as compounds that specifically demethylate DNA, which are devoid of off-target effects.

In parallel, it is crucial to understand the downstream chemical fate of DNA methylation. While sequential oxidations of 5mC to 5hmC, 5fC and 5caC are key steps in the passive and active demethylation pathways, each oxidised 5mC mark themselves has important regulatory functions, including active recruitment of chromatin reader modules and protein complexes. TETs are thus involved in multiple levels of epigenetic regulation, making them an interesting protein family to study. TETs play key roles in development and in disease. In cancer, TETs have been shown to have both oncogenic and tumour suppressor functions. However, the catalytic and non-catalytically dependent TET functions remain unclear in many biological processes.

Recent progresses in structural and biochemical studies have provided great insight into the functions of TET proteins. The substrate specificity, the allosteric regulation with respect to other domains (e.g. CXXC domain, LCI domain) and the biological impact, however, are not fully understood. Selective domain targeting chemical probes for TETs will be valuable to investigate the regulation of cytosine modifications on chromatin and transcription. Efforts have been made towards developing inhibitors targeting the catalytic domain of TETs; however, many of the small molecule inhibitors reported to date are weak inhibitors and are not selective. Potent cyclic peptide inhibitors show promise, but cell permeability remains a challenge. It is anticipated that the recent advancements in (high-throughput) assays and structural information will aid in the development of chemical probes for the TET proteins.

Acknowledgement RB is supported by the Engineering and Physical Science Research Council and University of Oxford. AK gratefully acknowledges the Royal Society for the Dorothy Hodgkin Fellowship and the European Research Council Starting Grant (EPITOOLS-679479) and the Cancer Research UK Oxford Centre Development Fund (C5255/A18085). We apologise for the incomplete citations of research due to space constraints.

The authors acknowledge the EU COST Action CM1406. PBA is supported by PlanCancer2014-2019 (EPIG-2014-01).

Compliance with Ethical Standards

Funding: RB is supported by the Engineering and Physical Science Research Council and University of Oxford. AK gratefully acknowledges the Royal Society for the Dorothy Hodgkin Fellowship and the European Research Council Starting Grant (EPITOOLS-679479) and the Cancer Research UK Oxford Centre. We apologise for the incomplete citations of research due to space constraints.

The authors acknowledge the EU COST Action CM1406. PBA is supported by PlanCancer 2014–2019 (EPIG-2014-01). PBA was recipient of the French Oversea Fellowship of the French Government and Churchill College Cambridge UK.

Conflict of Interest: Roman Belle declares that he has no conflict of interest. Akane Kawamura declares that she has no conflict of interest and Paola B. Arimondo declares that she has no conflict of interest.

Ethical Approval: This chapter does not contain any studies with human participants or animals performed by any of the authors.

References

1. Reik W (2007) Stability and flexibility of epigenetic gene regulation in mammalian development. Nature 447:425–432. https://doi.org/10.1038/nature05918
2. Gros C, Fahy J, Halby L et al (2012) DNA methylation inhibitors in cancer: recent and future approaches. Biochimie 94:2280–2296. https://doi.org/10.1016/J.BIOCHI.2012.07.025
3. Ludwig AK, Zhang P, Cardoso MC (2016) Modifiers and readers of DNA modifications and their impact on genome structure, expression, and stability in disease. Front Genet 7:115. https://doi.org/10.3389/fgene.2016.00115
4. Jeltsch A, Jurkowska RZ (2014) New concepts in DNA methylation. Trends Biochem Sci 39:310–318. https://doi.org/10.1016/j.tibs.2014.05.002
5. Kriaucionis S, Heintz N (2009) The nuclear DNA Base 5-hydroxymethylcytosine is present in Purkinje neurons and the brain. Science 324:929–930. https://doi.org/10.1126/science.1169786
6. Tahiliani M, Koh KP, Shen Y et al (2009) Conversion of 5-methylcytosine to 5-hydroxymethylcytosine in mammalian DNA by MLL partner TET1. Science 324:930–935. https://doi.org/10.1126/science.1170116
7. Kubik G, Summerer D (2015) Deciphering epigenetic cytosine modifications by direct molecular recognition. ACS Chem Biol 10:1580–1589. https://doi.org/10.1021/acschembio.5b00158
8. Breiling A, Lyko F (2015) Epigenetic regulatory functions of DNA modifications: 5-methylcytosine and beyond. Epigenetics Chromatin 8:24. https://doi.org/10.1186/s13072-015-0016-6
9. Chen H-F, Wu K-J (2016) Epigenetics, TET proteins, and hypoxia in epithelial-mesenchymal transition and tumorigenesis. Biomedicine (Taipei) 6(1). https://doi.org/10.7603/s40681-016-0001-9
10. Spruijt CG, Gnerlich F, Smits AH et al (2013) Dynamic readers for 5-(Hydroxy) methylcytosine and its oxidized derivatives. Cell 152:1146–1159. https://doi.org/10.1016/j.cell.2013.02.004
11. Traube C, Silver G, Reeder RW et al (2017) Delirium in critically ill children. Crit Care Med 45:584–590. https://doi.org/10.1097/CCM.0000000000002250

12. Maiti A, Drohat AC (2011) Thymine DNA glycosylase can rapidly excise 5-formylcytosine and 5-carboxylcytosine. J Biol Chem 286:35334–35338. https://doi.org/10.1074/jbc.C111.284620
13. Iwan K, Rahimoff R, Kirchner A et al (2018) 5-formylcytosine to cytosine conversion by C-C bond cleavage in vivo. Nat Chem Biol 14:72–78. https://doi.org/10.1038/nchembio.2531
14. Zhu J-K (2009) Active DNA demethylation mediated by DNA glycosylases. Annu Rev Genet 43:143–166. https://doi.org/10.1146/annurev-genet-102108-134205
15. Holliday R, Pugh JE (1975) DNA modification mechanisms and gene activity during development. Science 187:226–232
16. Riggs AD (1975) X inactivation, differentiation, and DNA methylation. Cytogenet Genome Res 14:9–25. https://doi.org/10.1159/000130315
17. Drahovsky D, Boehm TLJ (1980) Enzymatic dna methylation in higher eukaryotes. Int J Biochem 12:523–528. https://doi.org/10.1016/0020-711X(80)90002-6
18. Bestor T, Laudano A, Mattaliano R, Ingram V (1988) Cloning and sequencing of a cDNA encoding DNA methyltransferase of mouse cells: the carboxyl-terminal domain of the mammalian enzymes is related to bacterial restriction methyltransferases. J Mol Biol 203:971–983. https://doi.org/10.1016/0022-2836(88)90122-2
19. Okano M, Xie S, Li E (1998) Cloning and characterization of a family of novel mammalian DNA (cytosine-5) methyltransferases. Nat Genet 19:219–220. https://doi.org/10.1038/890
20. Aapola U, Shibuya K, Scott HS et al (2000) Isolation and initial characterization of a novel zinc finger gene, DNMT3L, on 21q22.3, related to the cytosine-5- methyltransferase 3 gene family. Genomics 65:293–298. https://doi.org/10.1006/GENO.2000.6168
21. Bourc'his D, Xu GL, Lin CS et al (2001) Dnmt3L and the establishment of maternal genomic imprints. Science 294:2536–2539. https://doi.org/10.1126/science.1065848
22. Jia D, Jurkowska RZ, Zhang X et al (2007) Structure of Dnmt3a bound to Dnmt3L suggests a model for de novo DNA methylation. Nature 449:248–251. https://doi.org/10.1038/nature06146
23. Lyko F (2018) The DNA methyltransferase family: a versatile toolkit for epigenetic regulation. Nat Rev Genet 19:81–92. https://doi.org/10.1038/nrg.2017.80
24. Rondelet G, Wouters J (2017) Human DNA (cytosine-5)-methyltransferases: a functional and structural perspective for epigenetic cancer therapy. Biochimie 139:137–147. https://doi.org/10.1016/J.BIOCHI.2017.06.003
25. Jurkowski TP, Jeltsch A (2011) On the evolutionary origin of eukaryotic DNA methyltransferases and Dnmt2. PLoS One 6:e28104. https://doi.org/10.1371/journal.pone.0028104
26. Gowher H, Jeltsch A (2018) Mammalian DNA methyltransferases: new discoveries and open questions. Biochem Soc Trans 46:1191–1202. https://doi.org/10.1042/BST20170574
27. Qin S, Min J (2014) Structure and function of the nucleosome-binding PWWP domain. Trends Biochem Sci 39:536–547. https://doi.org/10.1016/j.tibs.2014.09.001
28. Yarychkivska O, Shahabuddin Z, Comfort N et al (2018) BAH domains and a histone-like motif in DNA methyltransferase 1 (DNMT1) regulate de novo and maintenance methylation in vivo. J Biol Chem 293:19466–19475. https://doi.org/10.1074/jbc.RA118.004612
29. Zhang ZM, Liu S, Lin K et al (2015) Crystal structure of human DNA methyltransferase 1. J Mol Biol 427:2520–2531. https://doi.org/10.1016/j.jmb.2015.06.001
30. Ye F, Kong X, Zhang H et al (2018) Biochemical studies and molecular dynamic simulations reveal the molecular basis of conformational changes in DNA methyltransferase-1. ACS Chem Biol 13:772–781. https://doi.org/10.1021/acschembio.7b00890
31. Issa J-PJ, Kantarjian HM (2009) Targeting DNA methylation. Clin Cancer Res 15:3938–3946. https://doi.org/10.1158/1078-0432.CCR-08-2783
32. Jones PA, Baylin SB (2007) The epigenomics of cancer. Cell 128:683–692. https://doi.org/10.1016/j.cell.2007.01.029
33. Baylin SB, Jones PA (2016) Epigenetic determinants of cancer. Cold Spring Harb Perspect Biol 8:a019505. https://doi.org/10.1101/cshperspect.a019505
34. Feinberg AP, Ohlsson R, Henikoff S (2006) The epigenetic progenitor origin of human cancer. Nat Rev Genet 7:21–33

35. Feinberg AP (2018) The key role of epigenetics in human disease prevention and mitigation. N Engl J Med 378:1323–1334. https://doi.org/10.1056/NEJMra1402513
36. Mikeska T, Craig J, Mikeska T, Craig JM (2014) DNA methylation biomarkers: cancer and beyond. Genes (Basel) 5:821–864. https://doi.org/10.3390/genes5030821
37. Leygo C, Williams M, Jin HC et al (2017) DNA methylation as a noninvasive epigenetic biomarker for the detection of cancer. Dis Markers 2017:1–13. https://doi.org/10.1155/2017/3726595
38. Ahuja N, Sharma AR, Baylin SB (2016) Epigenetic therapeutics: a new weapon in the war against cancer. Annu Rev Med 67:73–89. https://doi.org/10.1146/annurev-med-111314-035900
39. Ahuja N, Easwaran H, Baylin SB (2014) Harnessing the potential of epigenetic therapy to target solid tumors. J Clin Invest 124:56–63. https://doi.org/10.1172/JCI69736
40. Feinberg AP (2007) Phenotypic plasticity and the epigenetics of human disease. Nature 447:433–440. https://doi.org/10.1038/nature05919
41. Velasco G, Francastel C (2018) Genetics meets DNA methylation in rare diseases. Clin Genet 95:210–220. https://doi.org/10.1111/cge.13480
42. Lopez M, Halby L, Arimondo PB (2016) DNA methyltransferase inhibitors: development and applications. Adv Exp Med Biol 945:431–473. https://doi.org/10.1007/978-3-319-43624-1_16
43. Andersen GB, Tost J (2018) A summary of the biological processes, disease-associated changes, and clinical applications of DNA methylation. Methods Mol Biol 1708:3–30
44. Jones PA, Issa J-PJ, Baylin S (2016) Targeting the cancer epigenome for therapy. Nat Rev Genet 17:630–641. https://doi.org/10.1038/nrg.2016.93
45. Okano M, Xie S, Li E (1998) Dnmt2 is not required for de novo and maintenance methylation of viral DNA in embryonic stem cells. Nucleic Acids Res 26:2536–2540
46. Tuorto F, Liebers R, Musch T et al (2012) RNA cytosine methylation by Dnmt2 and NSun2 promotes tRNA stability and protein synthesis. Nat Struct Mol Biol 19:900–905. https://doi.org/10.1038/nsmb.2357
47. Goll MG, Kirpekar F, Maggert KA et al (2006) Methylation of tRNAAsp by the DNA methyltransferase homolog Dnmt2. Science 311:395–398. https://doi.org/10.1126/science.1120976
48. Govindaraju G, Jabeena C, Sethumadhavan DV et al (2017) DNA methyltransferase homologue TRDMT1 in *plasmodium falciparum* specifically methylates endogenous aspartic acid tRNA. Biochim Biophys Acta-Gene Regul Mech 1860:1047–1057. https://doi.org/10.1016/j.bbagrm.2017.08.003
49. Capuano F, Mülleder M, Kok R et al (2014) Cytosine DNA methylation is found in *Drosophila melanogaster* but absent in *Saccharomyces cerevisiae, Schizosaccharomyces pombe*, and other yeast species. Anal Chem 86:3697–3702. https://doi.org/10.1021/ac500447w
50. Zadražil S, Fučík V, Bartl P et al (1965) The structure of DNA from *Escherichia coli* cultured in the presence of 5-azacytidine. Biochim Biophys Acta Nucleic Acids Protein Synth 108:701–703. https://doi.org/10.1016/0005-2787(65)90066-3
51. Sorm F, Vesely J (1964) The activity of a new antimetabolite, 5-azacytidine, against lymphoid. Neoplasma 11:123–130
52. Taylor SM, Jones PA (1979) Multiple new phenotypes induced in 10T1/2 and 3T3 cells treated with 5-azacytidine. Cell 17:771–779
53. Jones PA, Taylor SM (1980) Cellular differentiation, cytidine analogs and DNA methylation. Cell 20:85–93
54. Santi DV, Garrett CE, Barr PJ (1983) On the mechanism of inhibition of DNA-cytosine methyltransferases by cytosine analogs. Cell 33:9–10. https://doi.org/10.1016/0092-8674(83)90327-6
55. Santi DV, Norment A, Garrett CE (1984) Covalent bond formation between a DNA-cytosine methyltransferase and DNA containing 5-azacytosine. Proc Natl Acad Sci U S A 81:6993–6997

56. Schermelleh L, Spada F, Easwaran HP et al (2005) Trapped in action: direct visualization of DNA methyltransferase activity in living cells. Nat Methods 2:751–756. https://doi.org/10.1038/nmeth794
57. Egger G, Liang G, Aparicio A, Jones PA (2004) Epigenetics in human disease and prospects for epigenetic therapy. Nature 429:457–463. https://doi.org/10.1038/nature02625
58. Rogstad DK, Herring JL, Theruvathu JA et al (2009) Chemical decomposition of 5-aza-2-′-deoxycytidine (Decitabine): kinetic analyses and identification of products by NMR, HPLC, and mass spectrometry. Chem Res Toxicol 22:1194–1204. https://doi.org/10.1021/tx900131u
59. Erdmann A, Halby L, Fahy J, Arimondo PB (2015) Targeting DNA methylation with small molecules: what's next? J Med Chem 58:2569–2583. https://doi.org/10.1021/jm500843d
60. Fahy J, Jeltsch A, Arimondo PB (2012) DNA methyltransferase inhibitors in cancer: a chemical and therapeutic patent overview and selected clinical studies. Expert Opin Ther Pat 22:1427–1442. https://doi.org/10.1517/13543776.2012.729579
61. Agrawal K, Das V, Vyas P, Hajdúch M (2018) Nucleosidic DNA demethylating epigenetic drugs – a comprehensive review from discovery to clinic. Pharmacol Ther 188:45–79. https://doi.org/10.1016/J.PHARMTHERA.2018.02.006
62. Chiappinelli KB, Zahnow CA, Ahuja N, Bylin SB (2016) Combining epigenetic and immunotherapy to combat cancer. Cancer Res 76:1683–1689. https://doi.org/10.1158/0008-5472.CAN-15-2125
63. Hossain MZ, Healey MA, Lee C et al (2013) DNA-intercalators causing rapid re-expression of methylated and silenced genes in cancer cells. Oncotarget 4:298–309. https://doi.org/10.18632/oncotarget.863
64. Cherepanova NA, Ivanov AA, Maltseva DV et al (2011) Dimeric bisbenzimidazoles inhibit the DNA methylation catalyzed by the murine Dnmt3a catalytic domain. J Enzyme Inhib Med Chem 26:295–300. https://doi.org/10.3109/14756366.2010.499098
65. Zwergel C, Valente S, Mai A (2015) DNA methyltransferases inhibitors from natural sources. Curr Top Med Chem 16:680–696. https://doi.org/10.2174/1568026615666150825141505
66. Lopez M, Leroy M, Etievant C et al (2016) Drug discovery methods. Drug discovery in cancer epigenetics. Elsevier, Amsterdam, pp 63–95
67. Song J, Teplova M, Ishibe-Murakami S, Patel DJ (2012) Structure-based mechanistic insights into DNMT1-mediated maintenance DNA methylation. Science 335:709–712. https://doi.org/10.1126/science.1214453
68. Siedlecki P, Zielenkiewicz P (2006) Mammalian DNA methyltransferases. Acta Biochim Pol 53:245–256
69. Lin X, Asgari K, Putzi MJ et al (2001) Reversal of GSTP1 CpG island hypermethylation and reactivation of pi-class glutathione S-transferase (GSTP1) expression in human prostate cancer cells by treatment with procainamide. Cancer Res 61:8611–8616. https://doi.org/10.1158/0008-5472.can-04-2957
70. Suzuki T, Tanaka R, Hamada S et al (2010) Design, synthesis, inhibitory activity, and binding mode study of novel DNA methyltransferase 1 inhibitors. Bioorg Med Chem Lett 20:1124–1127. https://doi.org/10.1016/J.BMCL.2009.12.016
71. Asgatay S, Champion C, Marloie G et al (2014) Synthesis and evaluation of analogues of *N*-phthaloyl-l-tryptophan (RG108) as inhibitors of DNA methyltransferase 1. J Med Chem 57:421–434. https://doi.org/10.1021/jm401419p
72. Penter L, Maier B, Frede U et al (2015) A rapid screening system evaluates novel inhibitors of DNA methylation and suggests F-box proteins as potential therapeutic targets for high-risk neuroblastoma. Target Oncol 10:523–533. https://doi.org/10.1007/s11523-014-0354-5
73. Stresemann C, Brueckner B, Musch T et al (2006) Functional diversity of DNA methyltransferase inhibitors in human cancer cell lines. Cancer Res 66:2794–2800. https://doi.org/10.1158/0008-5472.CAN-05-2821
74. Graça I, Sousa EJ, Baptista T et al (2014) Anti-tumoral effect of the non-nucleoside DNMT inhibitor RG108 in human prostate cancer cells. Curr Pharm Des 20:1803–1811

75. Machnes ZM, Huang TCT, Chang PKY et al (2013) DNA methylation mediates persistent epileptiform activity in vitro and in vivo. PLoS One 8:e76299. https://doi.org/10.1371/journal.pone.0076299
76. Zhang S, Tang B, Fan C et al (2015) Effect of DNMT inhibitor on bovine parthenogenetic embryo development. Biochem Biophys Res Commun 466:505–511. https://doi.org/10.1016/j.bbrc.2015.09.060
77. Meadows JP, Guzman-Karlsson MC, Phillips S et al (2015) DNA methylation regulates neuronal glutamatergic synaptic scaling. Sci Signal 8:ra61. https://doi.org/10.1126/scisignal.aab0715
78. Chestnut BA, Chang Q, Price A et al (2011) Epigenetic regulation of motor neuron cell death through DNA methylation. J Neurosci 31:16619–16636. https://doi.org/10.1523/JNEUROSCI.1639-11.2011
79. Rondelet G, Fleury L, Faux C et al (2017) Inhibition studies of DNA methyltransferases by maleimide derivatives of RG108 as non-nucleoside inhibitors. Future Med Chem 9:1465–1481. https://doi.org/10.4155/fmc-2017-0074
80. Datta J, Ghoshal K, Denny WA et al (2009) A new class of quinoline-based DNA hypomethylating agents reactivates tumor suppressor genes by blocking DNA methyltransferase 1 activity and inducing its degradation. Cancer Res 69:4277–4285. https://doi.org/10.1158/0008-5472.CAN-08-3669
81. Gros C, Fleury L, Nahoum V et al (2015) New insights on the mechanism of quinoline-based DNA methyltransferase inhibitors. J Biol Chem 290:6293–6302. https://doi.org/10.1074/jbc.M114.594671
82. Valente S, Liu Y, Schnekenburger M et al (2014) Selective non-nucleoside inhibitors of human DNA methyltransferases active in cancer including in cancer stem cells. J Med Chem 57:701–713. https://doi.org/10.1021/jm4012627
83. Ceccaldi A, Rajavelu A, Champion C et al (2011) C5-DNA methyltransferase inhibitors: from screening to effects on zebrafish embryo development. Chembiochem 12:1337–1345. https://doi.org/10.1002/cbic.201100130
84. Villar-Garea A, Fraga MF, Espada J, Esteller M (2003) Procaine is a DNA-demethylating agent with growth-inhibitory effects in human cancer cells. Cancer Res 63:4984–4989
85. Lee BH, Yegnasubramanian S, Lin X, Nelson WG (2005) Procainamide is a specific inhibitor of DNA methyltransferase 1. J Biol Chem 280:40749–40756. https://doi.org/10.1074/jbc.M505593200
86. Castellano S, Kuck D, Sala M et al (2008) Constrained analogues of procaine as novel small molecule inhibitors of DNA methyltransferase-1. J Med Chem 51:2321–2325. https://doi.org/10.1021/jm7015705
87. Castellano S, Kuck D, Viviano M et al (2011) Synthesis and biochemical evaluation of δ(2)-isoxazoline derivatives as DNA methyltransferase 1 inhibitors. J Med Chem 54:7663–7677. https://doi.org/10.1021/jm2010404
88. Halby L, Champion C, Sénamaud-Beaufort C et al (2012) Rapid synthesis of new DNMT inhibitors derivatives of procainamide. Chembiochem 13:157–165. https://doi.org/10.1002/cbic.201100522
89. Fagan RL, Cryderman DE, Kopelovich L et al (2013) Laccaic acid A is a direct, DNA-competitive inhibitor of DNA methyltransferase 1. J Biol Chem 288:23858–23867. https://doi.org/10.1074/jbc.M113.480517
90. Ceccaldi A, Rajavelu A, Ragozin S et al (2013) Identification of novel inhibitors of DNA methylation by screening of a chemical library. ACS Chem Biol 8:543–548. https://doi.org/10.1021/cb300565z
91. Kilgore JA, Du X, Melito L et al (2013) Identification of DNMT1 selective antagonists using a novel scintillation proximity assay. J Biol Chem 288:19673–19684. https://doi.org/10.1074/jbc.M112.443895
92. Chen S, Wang Y, Zhou W et al (2014) Identifying novel selective non-nucleoside DNA methyltransferase 1 inhibitors through docking-based virtual screening. J Med Chem 57:9028–9041. https://doi.org/10.1021/jm501134e

93. Ye Y, Stivers JT (2010) Fluorescence-based high-throughput assay for human DNA (cytosine-5)-methyltransferase 1. Anal Biochem 401:168–172. https://doi.org/10.1016/j.ab.2010.02.032
94. Halby L, Marechal N, Pechalrieu D et al (2018) Hijacking DNA methyltransferase transition state analogues to produce chemical scaffolds for PRMT inhibitors. Philos Trans R Soc Lond B Biol Sci 373:20170072. https://doi.org/10.1098/rstb.2017.0072
95. Miletić V, Odorčić I, Nikolić P, Svedružić ŽM (2017) In silico design of the first DNA-independent mechanism-based inhibitor of mammalian DNA methyltransferase Dnmt 1. PLoS One 12:e0174410. https://doi.org/10.1371/journal.pone.0174410
96. Halby L, Menon Y, Rilova E et al (2017) Rational design of bisubstrate-type analogues as inhibitors of DNA methyltransferases in cancer cells. J Med Chem 60:4665–4679. https://doi.org/10.1021/acs.jmedchem.7b00176
97. Ganesan A (2016) Multitarget drugs: an epigenetic epiphany. ChemMedChem 11:1227–1241. https://doi.org/10.1002/cmdc.201500394
98. Rotili D, Tarantino D, Marrocco B et al (2014) Properly substituted analogues of BIX-01294 lose inhibition of G9a histone methyltransferase and gain selective anti-DNA methyltransferase 3A activity. PLoS One 9:e96941. https://doi.org/10.1371/journal.pone.0096941
99. San José-Enériz E, Agirre X, Rabal O et al (2017) Discovery of first-in-class reversible dual small molecule inhibitors against G9a and DNMTs in hematological malignancies. Nat Commun 8:15424. https://doi.org/10.1038/ncomms15424
100. Yuan Z, Sun Q, Li D et al (2017) Design, synthesis and anticancer potential of NSC-319745 hydroxamic acid derivatives as DNMT and HDAC inhibitors. Eur J Med Chem 134:281–292. https://doi.org/10.1016/J.EJMECH.2017.04.017
101. Erdmann A, Arimondo PB, Guianvarc'h D (2016) Structure-guided optimization of DNA methyltransferase inhibitors. Epi-informatics. Elsevier, Amsterdam, pp 53–73
102. Castillo-Aguilera O, Depreux P, Halby L et al (2017) DNA methylation targeting: the DNMT/HMT crosstalk challenge. Biomol Ther 7:3. https://doi.org/10.3390/biom7010003
103. Mo A, Mukamel EA, Davis FP et al (2015) Epigenomic signatures of neuronal diversity in the mammalian brain. Neuron 86:1369–1384. https://doi.org/10.1016/j.neuron.2015.05.018
104. Mayer W, Niveleau A, Walter J et al (2000) Embryogenesis: demethylation of the zygotic paternal genome. Nature 403:501–502. https://doi.org/10.1038/35000656
105. Lorsbach RB, Moore J, Mathew S et al (2003) TET1, a member of a novel protein family, is fused to MLL in acute myeloid leukemia containing the t(10;11)(q22;q23). Leukemia 17:637–641. https://doi.org/10.1038/sj.leu.2402834
106. Borst P, Sabatini R (2008) Base J: discovery, biosynthesis, and possible functions. Annu Rev Microbiol 62:235–251. https://doi.org/10.1146/annurev.micro.62.081307.162750
107. Ito S, Shen L, Dai Q et al (2011) Tet proteins can convert 5-methylcytosine to 5-formylcytosine and 5-carboxylcytosine. Science 333:1300–1303. https://doi.org/10.1126/science.1210597
108. Sudhamalla B, Dey D, Breski M, Islam K (2017) A rapid mass spectrometric method for the measurement of catalytic activity of ten-eleven translocation enzymes. Anal Biochem 534:28–35. https://doi.org/10.1016/j.ab.2017.06.011
109. He Y-F, Li B-Z, Li Z et al (2011) Tet-mediated formation of 5-carboxylcytosine and its excision by TDG in mammalian DNA. Science 333:1303–1307
110. Ito S, D'alessio AC, Taranova OV et al (2010) Role of Tet proteins in 5mC to 5hmC conversion, ES-cell self-renewal and inner cell mass specification. Nature 466:1129–1133. https://doi.org/10.1038/nature09303
111. Iyer LM, Zhang D, Maxwell Burroughs A, Aravind L (2013) Computational identification of novel biochemical systems involved in oxidation, glycosylation and other complex modifications of bases in DNA. Nucleic Acids Res 41:7635–7655. https://doi.org/10.1093/nar/gkt573
112. Hashimoto H, Pais JE, Zhang X et al (2014) Structure of a Naegleria Tet-like dioxygenase in complex with 5-methylcytosine DNA. Nature 506:391–395. https://doi.org/10.1038/nature12905

113. Aik W, McDonough MA, Thalhammer A et al (2012) Role of the jelly-roll fold in substrate binding by 2-oxoglutarate oxygenases. Curr Opin Struct Biol 22:691–700. https://doi.org/10.1016/j.sbi.2012.10.001
114. Hu L, Li Z, Cheng J et al (2013) Crystal structure of TET2-DNA complex: insight into TET-mediated 5mC oxidation. Cell 155:1545–1555. https://doi.org/10.1016/j.cell.2013.11.020
115. Shen L, Song C-X, He C, Zhang Y (2014) Mechanism and function of oxidative reversal of DNA and RNA methylation. Annu Rev Biochem 83:585–614. https://doi.org/10.1146/annurev-biochem-060713-035513
116. McDonough MA, Loenarz C, Chowdhury R et al (2010) Structural studies on human 2-oxoglutarate dependent oxygenases. Curr Opin Struct Biol 20:659–672. https://doi.org/10.1016/j.sbi.2010.08.006
117. Loenarz C, Schofield CJ (2008) Expanding chemical biology of 2-oxoglutarate oxygenases. Nat Chem Biol 4:152–156. https://doi.org/10.1038/nchembio0308-152
118. Hu L, Lu J, Cheng J et al (2015) Structural insight into substrate preference for TET-mediated oxidation. Nature 527:118–122. https://doi.org/10.1038/nature15713
119. Fu L, Guerrero CR, Zhong N et al (2014) Tet-mediated formation of 5-hydroxymethylcytosine in RNA. J Am Chem Soc 136:11582–11585. https://doi.org/10.1021/ja505305z
120. Schröder AS, Parsa E, Iwan K et al (2016) 2′-(R)-fluorinated mC, hmC, fC and caC triphosphates are substrates for DNA polymerases and TET-enzymes. Chem Commun 52:14361–14364. https://doi.org/10.1039/C6CC07517G
121. Pais JE, Dai N, Tamanaha E et al (2015) Biochemical characterization of a Naegleria TET-like oxygenase and its application in single molecule sequencing of 5-methylcytosine. Proc Natl Acad Sci 112:4316–4321. https://doi.org/10.1073/pnas.1417939112
122. Pfaffeneder T, Spada F, Wagner M et al (2014) Tet oxidizes thymine to 5-hydroxymethyluracil in mouse embryonic stem cell DNA. Nat Chem Biol 10:574–581. https://doi.org/10.1038/nchembio.1532
123. Deaton AM, Bird A (2011) CpG islands and the regulation of transcription. Genes Dev 25:1010–1022. https://doi.org/10.1101/gad.2037511
124. Globisch D, Münzel M, Müller M et al (2010) Tissue distribution of 5-hydroxymethylcytosine and search for active demethylation intermediates. PLoS One 5:e15367. https://doi.org/10.1371/journal.pone.0015367
125. Bachman M, Uribe-Lewis S, Yang X et al (2015) 5-formylcytosine can be a stable DNA modification in mammals. Nat Chem Biol 11:555–557. https://doi.org/10.1038/nchembio.1848
126. Xu Y, Wu F, Tan L et al (2011) Genome-wide regulation of 5hmC, 5mC, and gene expression by Tet1 hydroxylase in mouse embryonic stem cells. Mol Cell 42:451–464. https://doi.org/10.1016/J.MOLCEL.2011.04.005
127. Raiber E-A, Murat P, Chirgadze DY et al (2015) 5-Formylcytosine alters the structure of the DNA double helix. Nat Struct Mol Biol 22:44–49. https://doi.org/10.1038/nsmb.2936
128. Hardwick JS, Ptchelkine D, El-Sagheer AH et al (2017) 5-Formylcytosine does not change the global structure of DNA. Nat Struct Mol Biol 24:544–552. https://doi.org/10.1038/nsmb.3411
129. Raiber E-A, Portella G, Cuesta SM et al (2017) 5-Formylcytosine controls nucleosome positioning through covalent histone-DNA interaction. bioRxiv:224444. https://doi.org/10.1101/224444
130. Kellinger MW, Song C-X, Chong J et al (2012) 5-Formylcytosine and 5-carboxylcytosine reduce the rate and substrate specificity of RNA polymerase II transcription. Nat Struct Mol Biol 19:831–833. https://doi.org/10.1038/nsmb.2346
131. Iurlaro M, Ficz G, Oxley D et al (2013) A screen for hydroxymethylcytosine and formylcytosine binding proteins suggests functions in transcription and chromatin regulation. Genome Biol 14:R119. https://doi.org/10.1186/gb-2013-14-10-r119
132. Huang H, Jiang X, Li Z et al (2013) TET1 plays an essential oncogenic role in MLL-rearranged leukemia. Proc Natl Acad Sci 110:11994–11999. https://doi.org/10.1073/pnas.1310656110

133. Takai H, Masuda K, Sato T et al (2014) 5-Hydroxymethylcytosine plays a critical role in glioblastomagenesis by recruiting the CHTOP-methylosome complex. Cell Rep 9:48–60. https://doi.org/10.1016/j.celrep.2014.08.071
134. Rasmussen KD, Helin K (2016) Role of TET enzymes in DNA methylation, development, and cancer. Genes Dev 30:733–750. https://doi.org/10.1101/gad.276568.115
135. Langemeijer SMC, Kuiper RP, Berends M et al (2009) Acquired mutations in TET2 are common in myelodysplastic syndromes. Nat Genet 41:838–842. https://doi.org/10.1038/ng.391
136. Weissmann S, Alpermann T, Grossmann V et al (2012) Landscape of TET2 mutations in acute myeloid leukemia. Leukemia 26:934–942. https://doi.org/10.1038/leu.2011.326
137. Quivoron C, Couronné L, Della Valle V et al (2011) TET2 inactivation results in pleiotropic hematopoietic abnormalities in mouse and is a recurrent event during human lymphomagenesis. Cancer Cell 20:25–38. https://doi.org/10.1016/J.CCR.2011.06.003
138. Quesada V, Conde L, Villamor N et al (2012) Exome sequencing identifies recurrent mutations of the splicing factor SF3B1 gene in chronic lymphocytic leukemia. Nat Genet 44:47–52. https://doi.org/10.1038/ng.1032
139. Yang H, Liu Y, Bai F et al (2013) Tumor development is associated with decrease of TET gene expression and 5-methylcytosine hydroxylation. Oncogene 32:663–669. https://doi.org/10.1038/onc.2012.67
140. Bachman M, Uribe-Lewis S, Yang X et al (2014) 5-Hydroxymethylcytosine is a predominantly stable DNA modification. Nat Chem 6:1049–1055. https://doi.org/10.1038/nchem.2064
141. Song C-X, Yi C, He C (2012) Mapping recently identified nucleotide variants in the genome and transcriptome. Nat Biotechnol 30:1107–1116. https://doi.org/10.1038/nbt.2398
142. Terragni J, Bitinaite J, Zheng Y, Pradhan S (2012) Biochemical characterization of recombinant β-glucosyltransferase and analysis of global 5-hydroxymethylcytosine in unique genomes. Biochemistry 51:1009–1019. https://doi.org/10.1021/bi2014739
143. Booth MJ, Raiber E-A, Balasubramanian S (2015) Chemical methods for decoding cytosine modifications in DNA. Chem Rev 115:2240–2254. https://doi.org/10.1021/cr5002904
144. Tahiliani M, Koh KP, Shen Y et al (2015) Conversion 5-hydroxymethylcytosine in Mammalian DNA by MuL partner TETi. Science 324:930–936
145. Münzel M, Globisch D, Brückl T et al (2010) Quantification of the sixth DNA base hydroxymethylcytosine in the brain. Angew Chem Int Ed 49:5375–5377. https://doi.org/10.1002/anie.201002033
146. Kinney SM, Chin HG, Vaisvila R et al (2011) Tissue-specific distribution and dynamic changes of 5-hydroxymethylcytosine in mammalian genomes. J Biol Chem 286:24685–24693. https://doi.org/10.1074/jbc.M110.217083
147. Szwagierczak A, Bultmann S, Schmidt CS et al (2010) Sensitive enzymatic quantification of 5-hydroxymethylcytosine in genomic DNA. Nucleic Acids Res 38:e181–e181. https://doi.org/10.1093/nar/gkq684
148. Booth MJ, Branco MR, Ficz G et al (2012) Quantitative sequencing of 5-methylcytosine and 5-hydroxymethylcytosine at single-base resolution. Science 336:934–937. https://doi.org/10.1126/science.1220671
149. Pastor WA, Pape UJ, Huang Y et al (2011) Genome-wide mapping of 5-hydroxymethylcytosine in embryonic stem cells. Nature 473:394–397. https://doi.org/10.1038/nature10102
150. Song CX, Szulwach KE, Fu Y et al (2011) Selective chemical labeling reveals the genome-wide distribution of 5-hydroxymethylcytosine. Nat Biotechnol 29:68–75. https://doi.org/10.1038/nbt.1732
151. Yu M, Hon GC, Szulwach KE et al (2012) Tet-assisted bisulfite sequencing of 5-hydroxymethylcytosine. Nat Protoc 7:2159–2170. https://doi.org/10.1038/nprot.2012.137
152. Flusberg BA, Webster DR, Lee JH et al (2010) Direct detection of DNA methylation during single-molecule, real-time sequencing. Nat Methods 7:461–465. https://doi.org/10.1038/nmeth.1459

153. Song C-X, Clark TA, Lu X-Y et al (2011) Sensitive and specific single-molecule sequencing of 5-hydroxymethylcytosine. Nat Methods 9:75–77. https://doi.org/10.1038/nmeth.1779
154. Wu X, Zhang Y (2017) TET-mediated active DNA demethylation: mechanism, function and beyond. Nat Rev Genet 18:517–534. https://doi.org/10.1038/nrg.2017.33
155. Shen L, Zhang Y (2012) Enzymatic analysis of tet proteins: key enzymes in the metabolism of DNA methylation, 1st edn. Elsevier, Amsterdam
156. Liu MY, Denizio JE, Kohli RM (2016) Quantification of oxidized 5-methylcytosine bases and TET enzyme activity, 1st edn. Elsevier, Amsterdam
157. Song CX, Szulwach KE, Dai Q et al (2013) Genome-wide profiling of 5-formylcytosine reveals its roles in epigenetic priming. Cell 153:678–691. https://doi.org/10.1016/j.cell.2013.04.001
158. Nishio K, Belle R, Katoh T et al (2018) Thioether macrocyclic peptides selected against TET1 compact catalytic domain inhibit TET1 catalytic activity. Chembiochem:1–8. https://doi.org/10.1002/cbic.201800047
159. Xu W, Yang H, Liu Y et al (2011) Oncometabolite 2-hydroxyglutarate is a competitive inhibitor of α-ketoglutarate-dependent dioxygenases. Cancer Cell 19:17–30. https://doi.org/10.1016/j.ccr.2010.12.014
160. Laukka T, Mariani CJ, Ihantola T et al (2016) Fumarate and succinate regulate expression of hypoxia-inducible genes via TET enzymes. J Biol Chem 291:4256–4265. https://doi.org/10.1074/jbc.M115.688762
161. Alves J, Vidugiris G, Goueli SA, Zegzouti H (2018) Bioluminescent high-throughput succinate detection method for monitoring the activity of JMJC histone demethylases and Fe(II)/2-oxoglutarate-dependent dioxygenases. SLAS Discov 23:242–254. https://doi.org/10.1177/2472555217745657
162. Rose NR, Ng SS, Mecinović J et al (2008) Inhibitor scaffolds for 2-oxoglutarate-dependent histone lysine demethylases. J Med Chem 51:7053–7056. https://doi.org/10.1021/jm800936s
163. Marholz LJ, Wang W, Zheng Y, Wang X (2016) A fluorescence polarization biophysical assay for the Naegleria DNA hydroxylase Tet1. ACS Med Chem Lett 7(2):167–171. https://doi.org/10.1021/acsmedchemlett.5b00366
164. Gross S, Cairns RA, Minden MD et al (2010) Cancer-associated metabolite 2-hydroxyglutarate accumulates in acute myelogenous leukemia with isocitrate dehydrogenase 1 and 2 mutations. J Exp Med 207:339–344. https://doi.org/10.1084/jem.20092506
165. Dang L, White DW, Gross S et al (2009) Cancer-associated IDH1 mutations produce 2-hydroxyglutarate. Nature 462:739–744. https://doi.org/10.1038/nature08617
166. Ward PS, Patel J, Wise DR et al (2010) The common feature of leukemia-associated IDH1 and IDH2 mutations is a neomorphic enzyme activity converting α-ketoglutarate to 2-hydroxyglutarate. Cancer Cell 17:225–234. https://doi.org/10.1016/j.ccr.2010.01.020
167. Zhao S, Lin Y, Xu W et al (2009) Glioma-derived mutations in IDH1 dominantly inhibit IDH1 catalytic activity and induce HIF-1α. Science 324:261–265. https://doi.org/10.1126/science.1170944
168. Opocher G, Schiavi F (2011) Functional consequences of succinate dehydrogenase mutations. Endocr Pract 17:64–71. https://doi.org/10.4158/EP11070.RA
169. Rose NR, McDonough MA, King ONF et al (2011) Inhibition of 2-oxoglutarate dependent oxygenases. Chem Soc Rev 40:4364. https://doi.org/10.1039/c0cs00203h
170. Tarhonskaya H, Nowak RP, Johansson C et al (2017) Studies on the interaction of the histone demethylase KDM5B with tricarboxylic acid cycle intermediates. J Mol Biol 429:2895–2906. https://doi.org/10.1016/J.JMB.2017.08.007
171. Koivunen P, Hirsilä M, Remes AM et al (2007) Inhibition of hypoxia-inducible factor (HIF) hydroxylases by citric acid cycle intermediates: possible links between cell metabolism and stabilization of HIF. J Biol Chem 282:4524–4532. https://doi.org/10.1074/jbc.M610415200
172. Hopkinson RJ, Tumber A, Yapp C et al (2013) 5-carboxy-8-hydroxyquinoline is a broad spectrum 2-oxoglutarate oxygenase inhibitor which causes iron translocation. Chem Sci 4:3110. https://doi.org/10.1039/c3sc51122g

Top Med Chem (2020) 33: 287–338
DOI: 10.1007/7355_2019_79

Published online: 28 November 2019

Applied Biophysics for Bromodomain Drug Discovery

William C. K. Pomerantz, Jorden A. Johnson, and Peter D. Ycas

Contents

1 Introduction 289
2 Part I: Timeline of Bromodomain Discovery and Characterization 290
2.1 Discovery of the Bromodomain Structural Motif 290
2.2 Structural Biology Studies of Bromodomains Pave the Way for Drug Discovery Efforts 292
3 Part II: Early Small-Molecule Bromodomain Inhibitor Discovery 298
3.1 PCAF 298
3.2 CBP/p300 299
3.3 BETs 301
3.4 Bromodomain Clinical Trials and Chemical Probes 306
4 Part III: Computational and Experimental Biophysical Methods for Bromodomain Inhibitor Development 306
4.1 High-Throughput Virtual Screens and Molecular Dynamic Simulations for Bromodomain Inhibitor Discovery 306
4.2 Direct-Binding Experiments with Bromodomains 308
4.3 Competition-Based Assays for Inhibitor Discovery 318
4.4 Crystallography Guides Rational Design of Bromodomain Inhibitors 322
5 Future Outlook on Bromodomain Inhibitor Development 324
References 327

Abstract The dynamic regulation of epigenetic processes is dictated by the addition, removal, and recognition of posttranslational modifications on proteins and nucleic acids. These processes further regulate how our genetic information is accessed within chromatin. The recognition of acetylated histones by bromodomain modules is one such process that has been significantly evaluated as a promising interaction to disrupt for developing epigenetic therapies. The discovery of such inhibitors has been aided by the application of a wealth of biophysical and computational tools leading to insights into the structural biology of bromodomains and potent inhibitors that are advancing in the clinic. This chapter will first provide a

W. C. K. Pomerantz (✉), J. A. Johnson, and P. D. Ycas
Department of Chemistry, University of Minnesota, Minneapolis, MN, USA
e-mail: wcp@umn.edu

brief historical overview on the discovery and characterization of bromodomains, followed by several of the seminal discoveries of bromodomain inhibitors. The remainder of the chapter will provide descriptions of the experimental and computational tools that are available to scientists interested in biophysical analysis of bromodomain inhibitor discovery for developing new drugs and chemical probes. The field of chemical epigenetics is rapidly expanding, and the goal of this chapter is to help researchers keep abreast of the new methods being used to study this important epigenetic protein domain.

Keywords Acetyl lysine, Biophysical assays, Bromodomain, Histones, Inhibitor discovery, Structural biology

Abbreviations

3FY	3-Fluorotyrosine
4FF	4-Fluorophenylalanine
5FW	5-Fluorotryptophan
BET	Bromodomain and extraterminal domain
BRET	Bioluminescence resonance energy transfer
Brm	Brahma
CBP	CREB-binding protein
CPMG	Carr-Purcell-Meiboom-Gill
Cryo-TEM	Cryo-transmission electron microscopy
DSF	Differential scanning fluorimetry
FP	Fluorescence polarization
FRAP	Fluorescence recovery after photobleaching
GEM	Group epitope mapping
GST	Glutathione S-transferase
HDAC	Histone deacetylase
HTR-FRET	Homogeneous time-resolved fluorescence resonance energy transfer
IC50	Inhibitory constant
ITC	Isothermal calorimetry
K_d	Dissociation constant
LPS	Lipopolysaccharide
NMC	NUT-midline carcinoma
NOE	Nuclear Overhauser effect
NTA	Nickel nitrilotriacetic acid
PrOF NMR	Protein-observed NMR
SPR	Surface plasmon resonance
STD	Saturation transfer difference
T_2	Transverse relaxation time
T_c	Rotational correlation time
T_m	Thermal melting temperature

1 Introduction

Epigenetic regulation of gene transcription is a highly orchestrated process between proteins engaging chromatin through enzymatic modification and effector domain binding. The result of such enzymatic and binding events is the manipulation of chromatin structure and recruitment of transcriptional complexes leading to either activated or repressed genes. Understanding the mechanisms which govern the transcriptional processes leading to heritable phenotypes that are not simply dictated by the genetic code remains at the heart of the field of epigenetics.

The nucleosome, the fundamental building block of eukaryotic chromatin, is composed of ~147 base pairs of DNA wrapped around an octameric bundle of four conserved histone proteins: H2A, H2B, H3, and H4. It is the exposed unstructured termini of these histones, their core, and the DNA itself, which get posttranslationally modified through a dynamic process to form a complex code for determining the transcriptional program of the cell. The molecular mechanisms of the multidomain proteins which read, write, and erase this code in a dynamic fashion underlie what Allis and Strahl proposed as the "histone code" hypothesis [1].

Histone acetylation of *N*-ε-lysine side chains described in this chapter is one of the many posttranslational chromatin modifications which include serine/threonine/tyrosine phosphorylation, cytosine/lysine/arginine methylation, citrullination, ubiquitination, and ADP-ribosylation [2]. Histone acetylation was first identified by Phillips in 1963 [3] after isolation of calf thymus histones [4] and later described by Allfrey et al. in 1964 to correlate with active transcription states leading to loosely packed chromatin, called euchromatin [5]. Installation of lysine acetyl groups is carried out by lysine acetyl transferases (KATS or HATS in the case of histone acetyltransferases), belonging to a general class of epigenetic enzymes, colloquially termed "writers." Removal of acetylation, which can return chromatin to a condensed state, heterochromatin, is carried out by histone deacetylase (HDACs) or more generally "erasers." Finally, the effector or "reader" domains which bind to the *N*-ε-acetyl groups and more broadly lysine acyl groups are bromodomains [6]. However, since the early discovery of bromodomain histone interactions, an additional effector domain, the YEATS domain [7], has also been shown to bind to lysine acyl groups but will not be discussed further in this chapter.

Whereas epigenetics describes the role of chromatin and chromatin-associated proteins, the field of *chemical* epigenetics seeks to apply chemical tools to manipulate epigenetic processes to further understand the biology or to treat disease [8]. Since the first submicromolar inhibitors of a bromodomain were disclosed in the primary literature in 2010 for the bromodomain and extraterminal (BET) family of bromodomains [9, 10], there have been tremendous interest in the development of highly selective bromodomain chemical probes from the academic community and pharmaceutical interest for developing clinical drug candidates to treat disease. As of 2018, 26 clinical trials have been carried out or are ongoing for bromodomains, but it remains to be validated if bromodomain inhibition will be successful as an effective epigenetic therapy for controlling disease.

Despite drug discovery efforts targeting bromodomains beginning in earnest only in the last 8 years, there has been a sharp rise in the literature associated with chemical epigenetic approaches to study these proteins and manipulate their function as well as several comprehensive reviews [6, 11–15]. To provide a better understanding for researchers interested in drug discovery and chemical probe development for bromodomains, this book chapter has been broken up into three parts. Part I gives a brief timeline on the discovery of bromodomains and the structural biology efforts to characterize human bromodomains. In Part II, we will describe several early inhibitors of bromodomains which have led to the current clinical candidate drugs and chemical probes for bromodomains. Finally, in Part III, we provide brief descriptions of various biophysical and biochemical tools that have been developed to study these proteins alongside case studies demonstrating how some of these methods have led to successful inhibitor development. The chapter will close with a brief outlook on the future of chemical epigenetic approaches for bromodomains.

2 Part I: Timeline of Bromodomain Discovery and Characterization

2.1 Discovery of the Bromodomain Structural Motif

Although early drug discovery efforts targeting bromodomains did not start until 2005 [16], the first seminal studies of a bromodomain-containing gene were carried out 13 years earlier by Tamkun et al. in 1992 when they identified the *brahma* gene (*brm*) in *Drosophila melanogaster* (Fig. 1) [17]. These studies were based on a genetic screen for activators of gene expression. The *brm* gene was shown to remove the suppression of polycomb gene products and activation of *src and antp* genes during embryonic development. *Brm* encodes for a 1638 residue protein, which the authors found to be homologous to the SNF2/SWI2 protein found in yeast, a protein that was associated with DNA binding. Within this large protein, a small structural motif was identified not only in Brm and SNF2 but also in three other proteins fsh, SPT7, and CCG1, demonstrating the conservation of this structural domain from flies to yeast and humans. The motif was coined a bromodomain based on similarity in name to the *brahma* gene. Prior to bromodomains, the chromatin-associated chromodomains, which recognize methylated histones, had been identified. It was later pointed out by Owen et al. that the similarity in nomenclature between chromodomains and bromodomains also played a role in coining this new protein domain name [18].

The prediction of the structure and functional roles of the bromodomains began to slowly emerge after the initial discovery. Based on the original structural prediction algorithms, bromodomains were thought to be approximately 77 residues, consisting of 2 helices, αA and αB [19]. The amphipathic character of the helices led researchers to speculate on their role in protein-protein interactions. Seven conserved

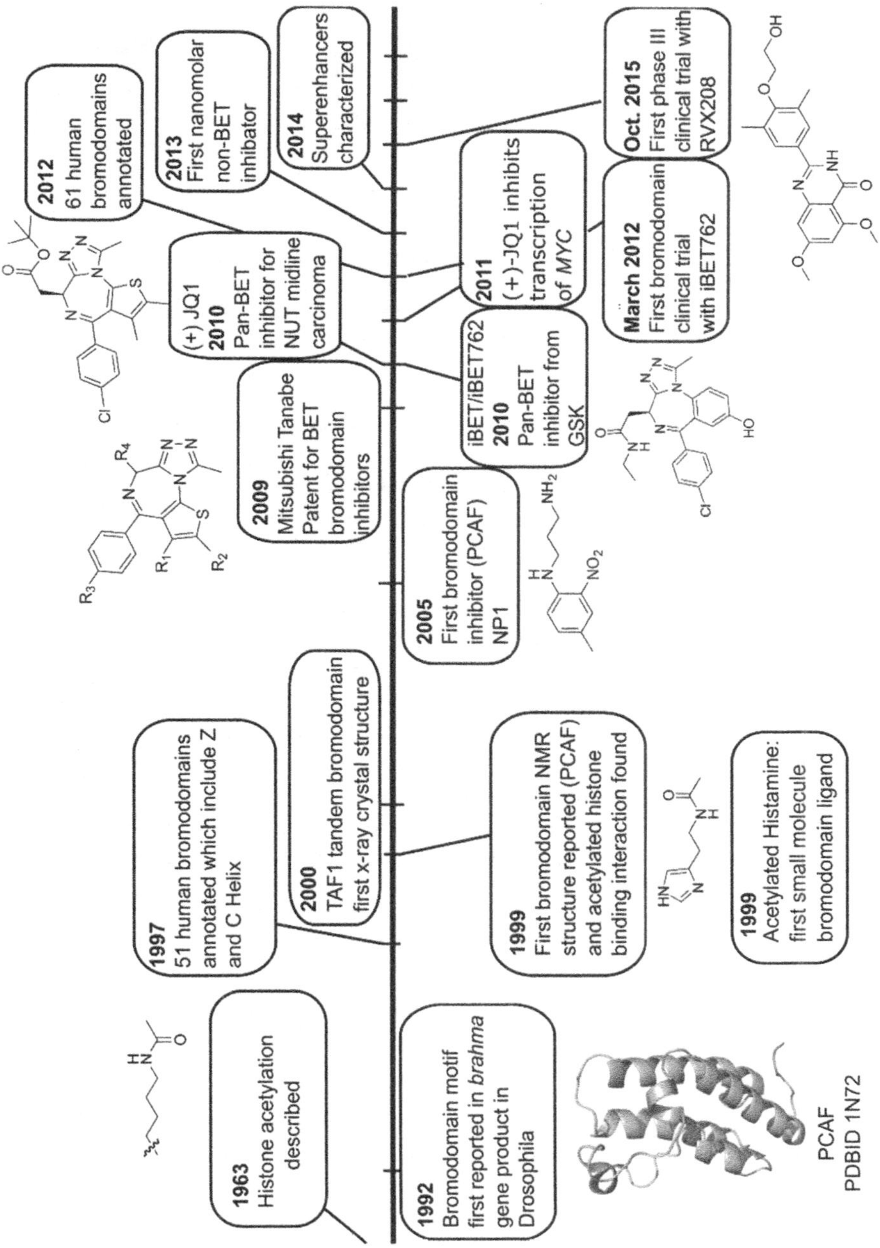

Fig. 1 Timeline of bromodomain discovery, characterization, and early inhibitors

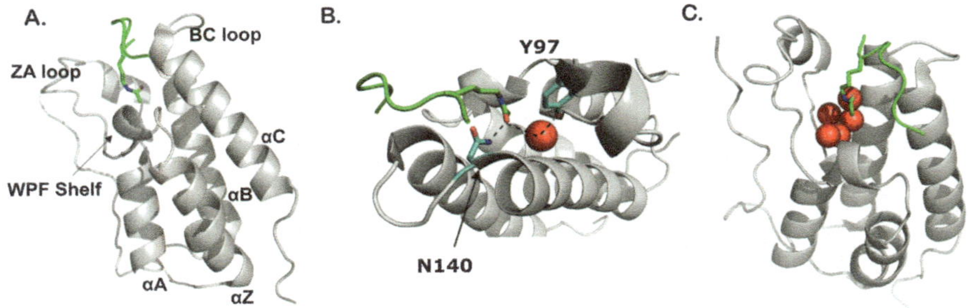

Fig. 2 Bromodomain structural elements. (**a**) The BRD4 bromodomain (gray) is depicted with a bound histone (green), and the four alpha-helices, αZ, αA, αB, and αC, ZA loops, BC loops, and WPF shelf are indicated. (**b**) Conserved binding interactions between a tyrosine and asparagine with the acetylated lysine binding partner. (**c**) Locations of the five strcutural waters in the acetyl lysine binding pocket are indicated in red. Figure adapted using the PDB entry 3UVX

residues were also identified within these helices: three tyrosines, a phenylalanine, an isoleucine, and two asparagines. The conserved phenylalanine was part of what would later be termed the "WPF shelf" motif in bromodomains [20]. One of the conserved asparagine and tyrosine residues would also later be shown to be essential for molecular recognition of the *N*-ε-acetyl of acetylated lysine by bromodomains [18].

By 1997, the family of bromodomains was expanded to include 51 bromodomains within 37 different human bromodomain-containing proteins [21]. Using improved structural algorithms, the domain size was also enlarged to include ~110 amino acids and 2 additional α-helices. The first helix N-terminal to helix αA was termed αZ, and a C-terminal helix was termed αC (Fig. 2). Jeanmougin et al. separated the 51 bromodomains into 6 classes of protein of sequence and functional similarity and 1 unrelated group [21]. This number would increase to the currently established set of 61 human bromodomains, in 8 families across 46 different proteins in 2012 through a large-scale structural analysis from researchers at the Structural Genomics Consortium in Oxford [20].

2.2 *Structural Biology Studies of Bromodomains Pave the Way for Drug Discovery Efforts*

The first structure of a bromodomain was solved by the Zhou group in 1999 using solution phase NMR spectroscopy methods to study the bromodomain of the histone acetyltransferase, p300/CBP-coactivator-associated protein, PCAF (Fig. 1) [22]. This seminal work verified the four-helix structural motif, demonstrating what would become a conserved left-handed four-helix bundle, with two structural loops, connecting the αZ and αA helices and the αB and αC helices, defined as the ZA and BC loops, respectively. Through 2D-HSQC NMR experiments, the authors experimentally showed the first example of an acetylated histone interaction with a bromodomain using a short peptide of H4 with the PCAF bromodomain and determined a 346 μM K_d via NMR titration. Such weak affinities would subsequently be found to be a general characteristic of histone-bromodomain interactions [20].

What would later pave the way for many biophysical and drug discovery studies, the Zhou lab also characterized the first bromodomain small-molecule complex with acetylated histamine [22]. Canonically, a conserved asparagine residue in bromodomains typically forms a hydrogen bond with the acetyl group of histones along with a water-mediated hydrogen bond with tyrosine. In this case, using NOE-derived distance restraints, they localized binding to a hydrophobic cavity between the BC and ZA loops. The mode of binding was distinct, as the *N*-acetyl group on histamine did not engage N803. However, alanine mutagenesis subsequently showed that Y809 was essential for binding interactions.

The following year the first bromodomain crystal structures were solved by Jacobsen et al. of a TAFII250 (also known as TAF1) tandem bromodomain [23] and Owen et al. [18] of GCN5p, co-crystallized with a peptide of histone H4 containing residues 15–29 acetylated at lysine 16. In contrast to the results of Zhou and co-workers, the GCN5p bromodomain structure showed the hydrogen bond-mediated interactions of the acetyl group with the conserved asparagine side chain on the BC loop (N407). A network of conserved water residues helped form a water-mediated hydrogen bond with the phenolic hydroxyl of Y364. In addition to hydrogen bonding, van Der Waals interactions form between the acetyl group and F352 of the WPF shelf at the base of the pocket. Although an unacetylated histone was shown to interact with GCN5p, HSQC NMR results disputed data from Ornaghi et al. Their data favored a secondary low-affinity site with an arginine $i + 3$ residues away [24] over a primary binding interaction with the unacetylated peptide. Wright and co-workers would complement the GCN5p studies with a solution NMR analysis of the protein alone and calculated affinities of acetylated H4 and H2A histone peptides with dissociation constants of ~0.9 mM [25]. Further structural analysis pointed out a negative electrostatic potential near the hydrophobic binding cavity, consistent with binding interactions with the highly cationic peptides.

A network of five structural waters also identified by Owen et al. is conserved in bromodomains. These water molecules need to be considered as targetable groups within the protein complex rather than displaceable for achieving potent inhibition when small molecules are designed to engage this pocket [18, 26]. These waters are shown in Fig. 2 alongside additional bromodomain structural elements. One example demonstrating the importance of structured water interactions was shown by Crawford et al. for tuning selectivity and affinity of a series of new bromodomain inhibitors based on the pyrrolopyridone scaffold (Fig. 3a) [29]. Liu et al. showed a structured water could also be displaced by a polar acetyl lysine mimic, using a hydrazide [27]. Further, in a computational analysis of 24 bromodomains by Vidler et al. [26], they determined the importance of considering 5 structural water molecules present in the acetylated lysine binding site for evaluating druggability.

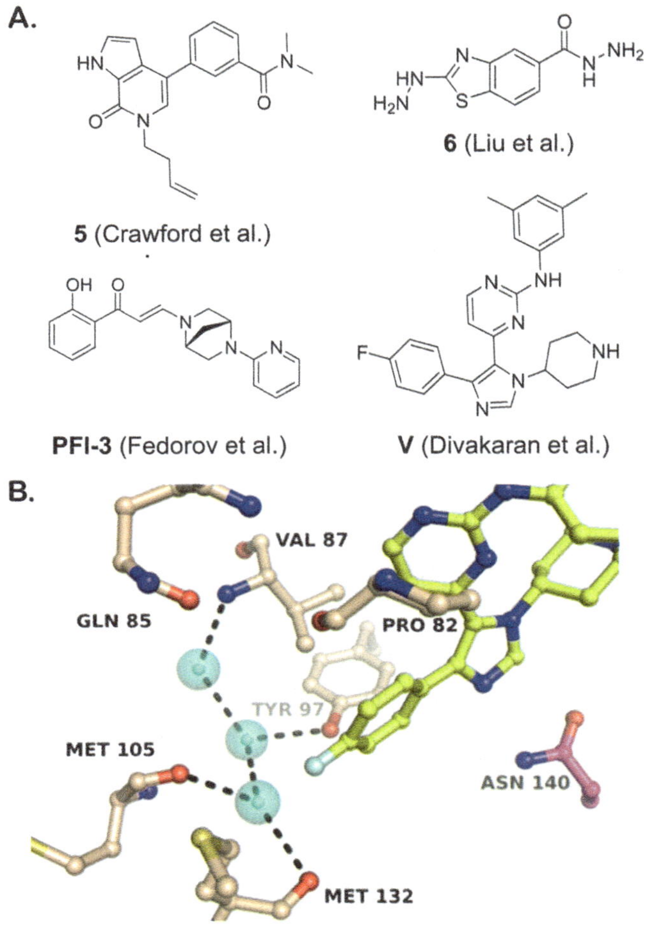

Fig. 3 (**a**) Small-molecule inhibitors **6** [27], **PFI-3** [28], **5** [29], and **V** [30] that perturb the structural waters in the Kac binding site. (**b**) Crystal structure of inhibitor **V** with BRD4 BD1 showing the key binding interactions and displacement of two structured waters. Adapted with permission from Divakaran et al. [30]. Copyright (2018) American Chemical Society

Currently, few ligands are known to displace these structural waters, the presence of which decreases druggability of some bromodomains [1]. However, computational analysis of the structured water network suggests that differential stability of structured waters could be exploited for developing selective inhibitors [31]. For example, structured waters were computationally shown to be more weakly held within the first bromodomains of BRD4 and BRDT relative to the second bromodomain within the same proteins and highly homologous bromodomains within BRD2 and BRD3. In support of this difference in stability, Divakaran et al. showed a 1,4,5 trisubstituted imidazole, **V**, could indeed displace structured waters within the first bromodomain of BRD4 and was identified as one of the origins of high selectivity for this compound [30] (Fig. 3). A similar binding mode and displacement of structured waters has also been reported for a selective BRG1/PB1 bromodomain inhibitor, PFI-3 [28].

Additional chromatin-binding domains are typically found in close proximity to bromodomains as a means to further increase affinity and specificity in both chromatin and transcriptional complexes. These include proximal PHD domains, such as those found in BPTF [32] and BRPF [33]; KIX domains, as seen in p300/CBP [34]; as well as tandem bromodomains [20]. TAFII250/TAF1 was the earliest example where high-resolution structural information was obtained for a multidomain bromodomain-containing protein, as well as being the first x-ray crystal structure of a bromodomain (PDBID: 1EQF) [23]. In this case, Tijan and co-workers crystallized the tandem bromodomains. Isothermal titration calorimetry (ITC) analysis, with both di- and tetraacetylated H4 histones, yielded stoichiometries of either 1:2 or 1:1, supporting engagement of both domains through a single peptide in the hyperacetylated state. In later studies of the bromodomain and extraterminal (BET) family of bromodomains, certain diacetylated histones were shown to engage both the canonical histone binding site, with secondary interactions with the ZA and BC bromodomain loops, whereas in other cases, engagement of two bromodomains was observed such as for histone H4 acetylated at positions 8 and 12 [20]. In addition to histone interactions, Tijan and co-workers speculated on the possibility of weak DNA interactions with TAFII250. These interactions were thought to occur on the basic face of the bromodomain and core histone interactions on its acidic face. Such identifications of weak DNA interactions have since been experimentally supported [35].

By 2012, driven by the growing structural biology data and disease-associated behavior of aberrant bromodomain function, there was considerable interest in the development of small-molecule inhibitors of histone-bromodomain interactions for drug discovery. As a resource for such initiatives, Filippakopolous and co-workers performed a large-scale structural analysis of bromodomains and biophysical determination of histone-bromodomain interactions [20]. These researchers contributed an additional 29 high-resolution bromodomain crystal structures including 25 new protein structures. Due to improved structural algorithms and BLAST searches, they were able to reclassify bromodomains into a cluster of 8 protein families consisting of 61 distinct bromodomains from 46 human proteins. In the class VIII family of bromodomains, a new β-hairpin insert was identified as a new structural motif. These families of bromodomains represent the current state of the field today. As of 2018, there now exist over 1,050 structures of human and nonhuman bromodomains (Table 1) serving as an excellent starting point for characterizing native interactions and design of new inhibitors. A figure of the current bromodomain phylogenetic tree including information on available crystal structures and chemical probes is compiled in Fig. 4.

To address the role of acetylation-dependent bromodomain interactions which underlie the histone code, these researchers further went on to characterize 485 histone-bromodomain interactions using peptide arrays which included both histone

Table 1 Crystal structure information of human and nonhuman bromodomains

Bromodomain[a]	Number of structures	Apo/holo PDB entry	Holo PDB entry
ASH1L	1	3mqm	NA
ATAD2A	43	4tu6	4tte
ATAD2B	1	3lxj	NA
BAZ1A	1	5uiy	NA
BAZ2B	268	5pen	5l8u
BPTF	4	3uv2	NA
BRD1	300	5pqi	5poa
BRD2(1)	19	NA	3yek
BRD2(2)	35	5ibn	5u6v
BRD3(1)	2	3s91	2le5
BRD3(2)	6	2oo1	3s92
BRD4(1)	157	3lyi	3mxf
BRD4(2)	20	NA	3oni
BRD7	2	NA	5mq1
BRD9	30	4yy4	5iy7
BRDT(1)	6	2rfj	4flp
BRDT(2)	1	2wp1	NA
BRPF1B	1	4lc2	NA
CECR2	2	3nxb	5v84
CREBBP	66	4ouf	5i89
EP300	4	NA	5nu5
GCN5L2	2	3d7c	5mlj
PB1(1)	1	3ui5	NA
PB1(5)	14	3g0j	5fh8
PCAF	18	3gg3	5lvq
PHIP(2)	8	3mb3	5enf
SMARCA4	4	2grc	5dkd
SP100	12	4ptb	5pwc
TAF1(2)	8	3uv4	5mg2
TAF1L(2)	2	3hmh	5igl
TRIM24	16	3o33	5h1t
TRIM33A	5	3u5m	5mr8
WDR9(1)	1	3qet	NA
ZMYND11	4	4ns5	NA

[a]Numbering in parenthesis indicates the order of the bromodomain from the N-terminus of the protein. *NA* indicates the structure is not available. Current as of December 2018

acetylation and additional posttranslational modifications [20]. They estimated the affinity limit of detection for this method to be ~0.5 mM, and thus weaker interactions may have been missed. Due to the limited information content from this assay, these interactions should be further verified by additional biophysical methods, such as ITC, which was carried out for a subset of these interactions in that report.

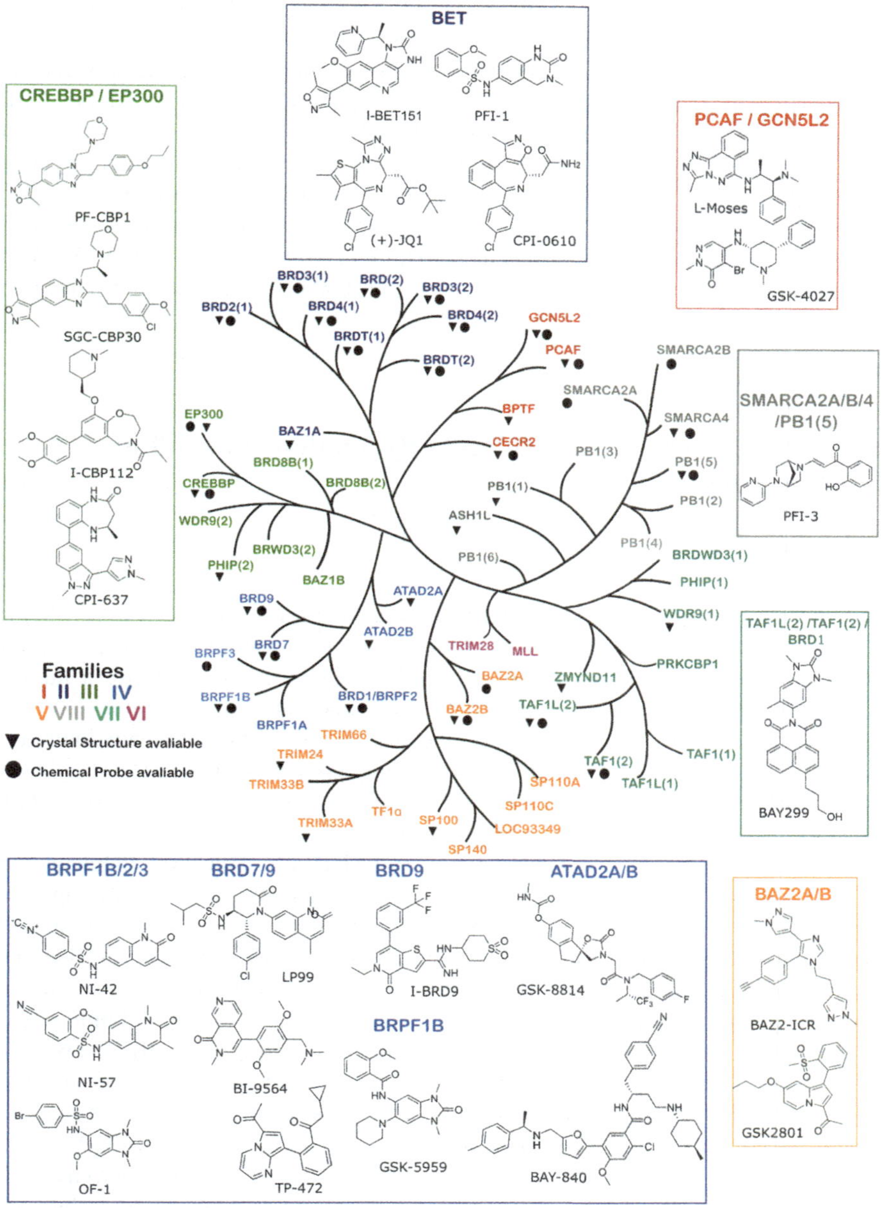

Fig. 4 Bromodomain phylogenetic tree adapted from the Structural Genomics Consortium to estimate the position and similarity of bromodomain family members. Available crystal structures (triangles) and reported chemical probes (circles) according to the Chemical Probes Portal and the Structural Genomics Consortium are indicated when available. Current as of December 2018

As a final commentary on native bromodomain interactions beyond acetylated histones, acetylated transcription factor-bromodomain interactions are relevant for controlling gene expression. Disruption of such interactions can also be targets for drug discovery. Several examples in the literature have included the CBP bromodomain with acetylated p53 [36] and CREB [34], the PCAF bromodomain with acetylated TAT [37], and the BRD4 bromodomain interaction with acetylated NF-κB [38], TWIST [39], and the androgen receptor [40]. In the case of the tandem bromodomains of BRD4, it is thought that the predominant role of the first N-terminal bromodomain is for histone recognition, while the second bromodomain interacts with transcription factors [35]. However, this may not be the mechanism in all cases.

3 Part II: Early Small-Molecule Bromodomain Inhibitor Discovery

3.1 PCAF

The bromodomain of the p300/CBP-associated factor protein, PCAF, holds historical precedent as the first bromodomain for which a small-molecule inhibitor was developed. Until 2005, the functional consequences of inhibiting a human bromodomain were untested due to a lack of small-molecule inhibitors. At this time, Zhou and co-workers described the first reported bromodomain inhibitor, NP1 (Fig. 1), targeting the PCAF bromodomain and demonstrated selectivity over two other bromodomains from CBP and TIF1β [16]. The authors were motivated by the fact that acetylation of the viral transcriptional activator, TAT, at K50 was essential for transcription of essential genes for the HIV life cycle through interaction with the PCAF bromodomain. Blocking such an interaction could then suppress viral replication. Using a 2D ^{1}H-^{15}N HSQC NMR screen of several thousand compounds, the authors identified several molecules which bound outside the histone binding site, targeting the BC and ZA loops. Structure activity studies in an ELISA format using a biotinylated TAT peptide led to NP1 whose binding was further confirmed via an NMR solution structure of the complex (PDBID: 1WUG). Hu et al. would slightly improve on this affinity using a higher-throughput fluorescence polarization assay with a related analog leading to inhibition constants at submicromolar levels ($IC_{50} = 0.93\ \mu M$) versus an IC_{50} of 1.37 μM for NP1 in the same assay [41]. They also demonstrated preliminary anti-HIV cellular activity data.

Since the original work by Zhou and co-workers, several high-affinity inhibitors of PCAF are available with high selectivity against human bromodomains. In addition to HIV infection, aberrant PCAF function has been linked to cancer and neuroinflammation. As such, these inhibitors should be of broad use to the community. One such inhibitor based on a triazolophtalazine scaffold, **L-45** or **L-Moses** (Fig. 4), was reported by Moustakin et al. in 2017 to be cellular active and highly selective [42]. This study used a promiscuous inhibitor scaffold based on the pan-bromodomain inhibitor, bromosporine [43]. Virtual screening efforts of over 12,000 compounds were then applied and a thermal stability assay (differential scanning fluorimetry, DSF) implemented to verify hits. The optimized inhibitor with a K_d of 126 nM for PCAF only had off-target binding to one human bromodomain, the highly homologous bromodomain of GCN5 ($K_d = 600$ nM). Target engagement was carried out by a NanoBRET assay in HEK-293 cells and possessed a 40 min half-life against human liver microsomes demonstrating good metabolic stability for in vivo studies. This compound is available from the Structural Genomics Consortium along with a negative control enantiomer **D-Moses**. During the same time period, Constellation and Genentech patented several PCAF inhibitors, including a pyridazinone scaffold (WO/2016/112298A1), that GSK researchers also reported on [44]. The lead compound GSK4027 (Fig. 3) and negative control GSK4028 are also available from the SGC and have improved binding affinity for PCAF and GCN5 with a K_i of 1.4 nM for both proteins and >70-fold selectivity against other human bromodomains.

3.2 *CBP/p300*

The CREB-binding protein (CBP) and homologous protein p300 are large coactivator proteins possessing histone acetyltransferase activity, histone binding ability via its bromodomain, and additional protein-protein interaction modules for binding transcriptional activators, including the KIX, CH1, and CH2 domains [45]. This protein has been shown to be essential in the development with genetic knockouts resulting in embryonic lethality [46]. Aberrant function of the protein and associated transcription factors are linked to Rubinstein-Taybi syndrome [47], cardiac ischemia [48], leukemia [49], and ovarian, breast, and lung cancers [50].

The bromodomain of CBP was the second target for small-molecule development in 2006 also by the Zhou lab for disrupting an acetylated p53 interaction [48]. The tumor suppressor activity of p53 is often attempted to be preserved as an anticancer therapy; however, in this report Sachchidanand et al. sought to reduce hyperactivity of p53 during myocardial ischemia which leads to cardiomyocyte death. The mechanism for such a response is via acetylation of p53 at K382 leading to p53-mediated recruitment of CBP to chromatin and initiation of gene transcription in response to DNA damage.

Sachchidanand et al. designed a focused library of acetylated lysine mimics, consisting of an *N*-acetyl amine connected to diverse aromatic rings [48]. These 200 compounds were screened by 2D ^{1}H-^{15}N HSQC NMR in mixtures of 8 compounds. Following deconvolution of hits based on chemical shift perturbation, computational docking and an NMR solution structure led to MS7972 (PDBID 2D82). A K_d of 19 μM was subsequently determined via a tryptophan intrinsic fluorescence assay in addition to blocking acetylated p53 interactions in vitro. Preliminary data showed effects at decreasing p21 levels in U2OS cells at high

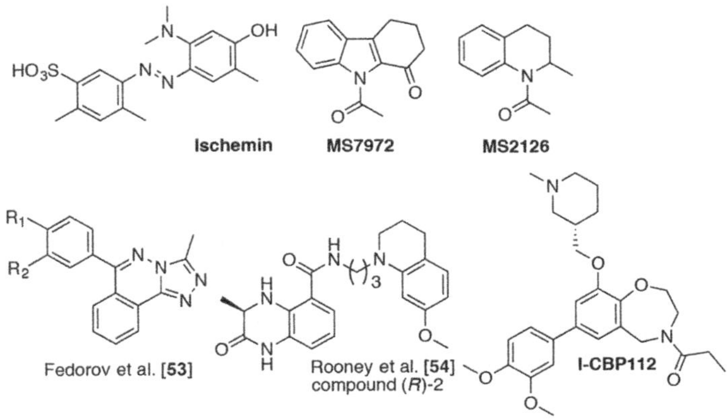

Fig. 5 Select CBP/p300 bromodomain inhibitors, ischemin [52], MS7972 [48], MS2126 [48], triazolophtalazines [53], dihydroquinazolinone [54], and I-CBP112 [55]

micromolar concentrations. A second compound, a tetrahydroquinoline, MS2126, was also shown to display lower levels of activity. The tetrahydroquinoline motif has since been shown to be an effective scaffold against additional bromodomains.

By 2011, Zhou and co-workers would develop improved inhibitors based on cell-permeable peptide macrocycles derived from p53 [51] and a small molecule with a diazobenzene scaffold, termed ischemin [52] (Fig. 5). Ischemin was able to completely inhibit p53 activity at mid-micromolar concentrations in a p53-dependent p21 luciferase assay with an IC_{50} of 5 μM. Moderate selectivity of fivefold was further demonstrated over additional bromodomains, including the first bromodomain of BRD4, PCAF, BAZ1B, and BAZ2B. Doxycycline-induced DNA damage in rat neonatal cardiomyocytes and apoptosis from caspase activity were also blocked when treated with ischemin.

Several submicromolar inhibitors have since been developed for the CBP/p300 bromodomain with increasing levels of selectivity [56]. The first promiscuous but submicromolar binder of the CBP bromodomain was reported in 2013 [53], based on the triazolophtalazines, the same scaffold as the PCAF inhibitor described above. The activity was improved by Conway and co-workers, with their report of a 390 nM inhibitor with increased selectivity for CBP (Fig. 5) [54]. This report is notable, as it was described as the first selective nanomolar inhibitor of a bromodomain outside of the bromodomain and extraterminal (BET) family, for which a variety of inhibitors had already been reported since 2010. Using an AlphaScreen bead-based assay with immobilized histones and the CBP bromodomain, the Conway group developed a dihydroquinazolinone as an acetylated lysine mimic from an original fragment lead based on the solvent molecule *N*-methyl-2-pyrrollidone. X-ray crystallography was used to guide the medicinal chemistry identifying engagement of a ZA channel-structured water, leading to an induced-fit binding mechanism. Notably, a cation-pi interaction with R1173 and the tetrahydroquinoline pendant group on their inhibitor was shown to be essential for the observed affinity. Modest selectivity of 3.6-fold was observed against the first bromodomain of BRD4. When tested in the U2OS cells, target engagement and small-molecule displacement of the CBP bromodomain from chromatin were verified using a fluorescence recovery after photobleaching (FRAP) assay.

An improved inhibitor from the Structural Genomics Consortium in Oxford, I-CBP112 (Figs. 4 and 5), has now been disclosed, with a K_d of 151 and 167 nM for the CBP and p300 bromodomains, respectively, and only low to mid-micromolar affinity for the BRD4 bromodomains [55]. Although active in cells, and competent for displacing an isolated bromodomain from chromatin, I-CBP112 was unable to displace full-length CBP from chromatin in the same FRAP assay, consistent with additional interactions being necessary to stabilize the protein complex at genomic loci. However, this molecule did show anticancer activity in bone marrow cells driven by the MLL-CBP fusion protein impairing clonogenetic growth. Reduced clonogenetic growth from I-CBP112 treatment was also observed in 12 additional human cell lines tested.

3.3 BETs

A seminal breakthrough for bromodomain inhibitor discovery occurred in 2010 with the dual reports of the first submicromolar inhibitors of the BET bromodomains (BRD2, 3, 4, and T) by GSK [10] and a collaborative study between the Structural Genomics Consortium in Oxford and the Bradner laboratory at the Dana-Farber Cancer Institute [9]. These landmark reports which led to invaluable chemical probes for the BET bromodomains have since set the stage for validating the therapeutic potential for inhibiting bromodomain function in human clinical trials.

In the GSK study by Nicodeme et al., the researchers developed the 1,2,4-triazolobenzodiazepine inhibitor termed I-BET based on their investigation of inflammation modulatory compounds for the immune system (Fig. 6) [10]. The lead compound was developed from an initial reporter assay screen for small-molecule activation of ApoA1 followed by target identification of the BET bromodomains using an affinity matrix selection of tethered small molecules. I-BET maintained a K_d of 50.5–61.3 nM for tandem BET bromodomains by ITC and similar affinity in a histone-competitive FRET assay. Selectivity was further verified against 5 other off-target bromodomains in a thermal stability study and 38 additional proteins in biochemical assays. In a cellular context, I-BET-treated bone-derived macrophages stimulated with lipopolysaccharide (LPS) downregulated up to 151 inflammatory genes including expression of cytokines and chemokines, Il6, Ifnb1, Il1b, and cxcl9. Their results with I-BET treatment were similar when compared with siRNA genetic knockdown of individual BETs. However, only the siRNA bromodomain knockdowns affecting TNF levels supported additional mechanisms for BETs outside of bromodomain interactions. In 2012, 20 years after the initial characterization of a bromodomain structural motif, this molecule (also called IBET762 and GSK525762) would become the first

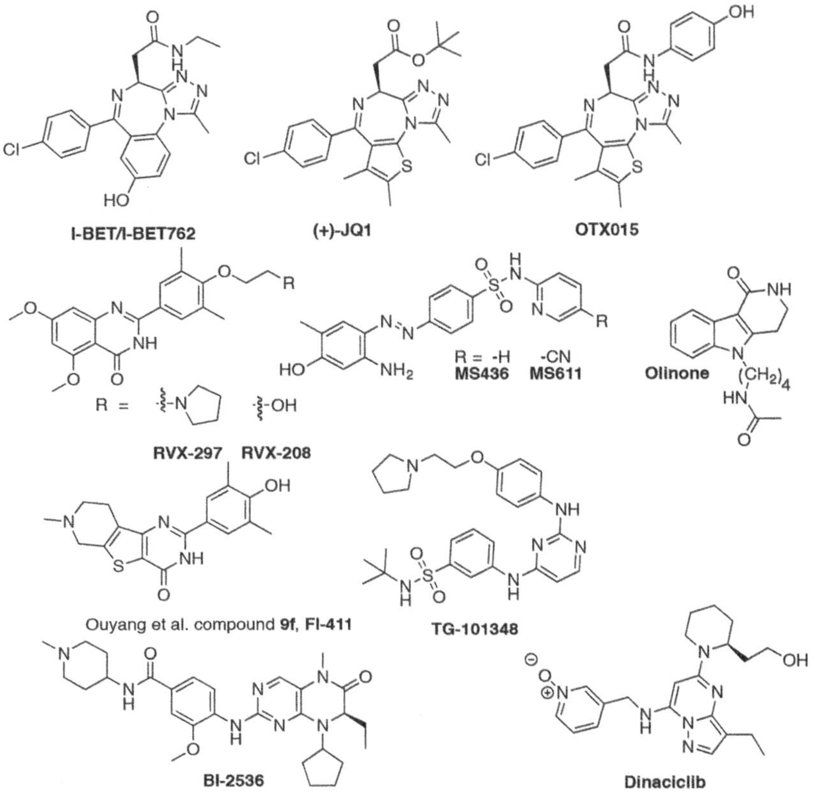

Fig. 6 Select BET bromodomain inhibitors from the literature including Pan-BET inhibitors [9, 10, 57–59], BRD4-selective inhibitor, Fl-411 [60], BET BD1 selective inhibitors [61, 62], or BET BD2 selective inhibitors [63, 64], BD1 selective inhibitor **V**, is shown in Fig. 3

bromodomain inhibitor used in clinical trials for treating NUT-midline carcinoma (NMC, Clinical Trial ID: NCT01587703).

The corresponding study by Filippakopoulos et al. disclosed a similar diazepine, (+)-JQ1, derived from a 1,2,4 triazolothienodiazepine (Fig. 6) [9]. This compound was selected from a prior 2009 patent from Mitsubishi Tanabe Pharma Corporation (WO 2009084693 A1) [65]. Although (+)-JQ1 was not identified as the most efficacious compound in the patent, it was selected based on expectations for reduced off-target receptor binding as well as future derivatization. Interestingly, the more potent compound from the patent contained a 4-hydroxyphenylacetamide group in place of the tert-butyl ester of (+)-JQ1. This molecule has since been licensed by Oncoethix (now OTX-015) and as of 2018 had been tested in phase I and II clinical trials from Merck for treating solid tumors in various cancers.

For chemical probe development, (+)-JQ1 demonstrated high selectivity for BET bromodomains by thermal shift assays against a panel of 36 bromodomains with the highest induced stability for the first bromodomain of BRD4 [9]. X-ray crystallography (PDBID 3mxf) established the binding mechanism, in which the methylated 1,2,4-triazole serving as the acetylated lysine mimetic effectively engaged N140 and Y97 via a water-mediated hydrogen bond.

One advantageous feature of a chemical probe molecule is a suitable negative control with similar physicochemical properties to the active compound. The enantiomer (−)-JQ1 has served this role and has led to robust results describing the biological activity of (+)-JQ1 across a range of disease indications including cancer, cardiac hypertrophy, and inflammation [14]. In the first report, the authors demonstrated the efficacy of (+)-JQ1 for treating patient-derived NMC cells, driven by a BRD4-NUT fusion protein [9]. Treatment led to the cellular differentiation and G1 cell cycle arrest, demonstrating the potential of bromodomain inhibitors for epigenetic reprogramming of cell identity. PET imaging of NMC-derived mouse xenografts also showed tumor reduction, further validating anticancer activity.

(+)-JQ1 is a widely shared chemical probe which has been distributed to researchers across the globe. One of the first collaborative studies to use this probe in combination with an RNAi screen of chromatin regulators was with the Vakoc and Lowe labs demonstrating that BRD4 inhibition with (+)-JQ1 can reduce transcription of the oncogene *c-Myc* in acute myeloid leukemia, leading to differentiation of terminal myeloid cells [66]. At the same time, similar effects were also seen against multiple myeloma, supporting a broader potential for a BET-targeted therapy for hematological diseases [67]. Young and co-workers shed light on how (+)-JQ1 inhibition of a ubiquitous coactivator protein like BRD4, typically involved in transcriptional elongation through association with the positive transcription elongation factor B, P-TEFB, can induce a therapeutic response [68]. In this work, they showed BRD4 associated with Mediator at large closely spaced enhancer regions more than 10 kilobases in total size, termed super-enhancers. Inhibiting transcription of *c-Myc* in blood cancers due to sensitivity to super-enhancer regulation is one of the several mechanisms for which BET bromodomains have now been shown to regulate oncogenic pathways.

Since the disclosure of the triazolodiazepine inhibitor scaffold, numerous inhibitors have since been reported and are actively being investigated in the clinic briefly discussed below. Selectivity within the BET family still remains a significant challenge. Currently the majority of inhibitors in clinical trials are pan-BET inhibitors. Bayer has developed a BRD4 inhibitor BAY123807 with >10fold selectivity over other BETs, but the structure has not been disclosed. In the published literature, Ouyang et al. reported a BRD4-selective inhibitor (FL-411, Fig. 6) with high nanomolar affinity for inducing autophagy-mediated cell death in breast cancer [60]. In lieu of other specific BET family member inhibitors, selectivity between the two terminal bromodomains (BD1 and BD2) within the BET family of proteins has been established in several cases (Fig. 6). RVX-208 [63] and RVX-297 [64], reported by the Structural Genomics Consortium in Oxford and Zenith Epigenetics Corp., now allow for the study of BD2-dependent processes. RVX-208 affected a smaller set of BET-dependent genes when compared to pan-BET inhibitors [63]. However, RVX-208 testing for treating cardiovascular disease is now in phase III clinical trials (NCT02586155).

BET-BD1 inhibition is also being investigated as a potential strategy over inhibiting both bromodomains for reducing potential off-target effects. Early studies in mice showed that deletion of domain 1 of BET, BRDT, which is expressed predominantly in the testis, was sufficient to impair sperm function [69]. The Zhou lab has since developed molecules MS-436 [61], MS-611 [62], and Olinone [62] with BD1 selectivity (Fig. 6). Despite being potent BRD4 BD1 inhibitors, however, MS-436 and MS-611 are not selective for BD1 over BD2 in other BET proteins. Alternatively, Olinone is a weak-affinity ligand for BRD4 BD1 ($K_d = 3.3$ μM) but maintains selectivity against non-BET bromodomains and only exhibits off-target activity against CBP at high micromolar levels. Divakaran has now reported a slightly improved inhibitor BD1-selective inhibitor, **V**, with a 1.2 μM K_d and >55-fold selectivity of BRD4 BD1 over BD2 and the highest affinity for BRD4 BD1 relative to the other BET bromodomains tested [30]. Improved BET BD1 inhibitors, and ultimately single BET bromodomain inhibitors for BRD2, 3, 4, and T, will allow researchers to further dissect BET bromodomain interactions with chromatin, acetylated transcription factors, and their effects on transcription.

As an alternative to improved selectivity, polypharmacology has also been explored for dual kinase and bromodomain inhibition. The BET bromodomain, BRD4, was recently identified as an atypical kinase capable of phosphorylating RNA Pol II [70]. Such behavior led Schönbrunn and co-workers to evaluate a cyclin-dependent kinase inhibitor, dinaciclib, for interactions with BD1 of BRDT (Fig. 7) [59]. Although the binding interaction was weak, $K_d = 37$ μM, selectivity remained high for BETs. A high-resolution x-ray co-crystal structure identified the pyridine oxide of dinaciclib to serve as the acetylated lysine mimic. Moreover, comparison of binding modes with kinase inhibitor co-crystal structures led the researchers to speculate the hinge-binding kinase inhibitor scaffolds could serve as optimal inhibitors for bromodomains.

This observation of kinase hinge binders serving as bromodomain inhibitors was validated the following year with two independent publications from the Schönbrunn lab and the Knapp lab [57, 58]. In these cases, library screening and crystallization studies identified a variety of known kinase inhibitors with affinity for BET bromodomains. Most notably both groups identified PLK1 inhibitor, BI-2536, and JAK2 inhibitor TG-101348 to inhibit BRD4 and BRDT with nanomolar potency (Fig. 6). A subset of the published kinase inhibitor sets I and II from GSK was also used in a tandem bromodomain NMR screen against BPTF and BRD4 BD1 [71], identifying ligands for both proteins including a p38α inhibitor analog of the 1,4,5-trisubstituted imidazole scaffold identified by the Schönbrunn lab for BRD4 BD1.

Although a monotherapy using a dual kinase/bromodomain inhibitor has yet to be investigated in the clinic, promising results of a potent PI3K and BET inhibitor with nanomolar potency have been reported using a thienopyranone scaffold [72]. Supporting a polypharmacology approach, dual inhibition of Myc degradation through PI3K inhibition and Myc transcription via BET bromodomain inhibition, led to less toxic effects in an orthotopic pancreatic cancer model for spontaneous tumor metathesis than when treated with a combination of (+)-JQ1 and PI3K inhibitor BKM120. Of note, their lead thienopyranone, SF2523, also displayed

Fig. 7 Bromodomain inhibitors in clinical trials and their respective targeted disease states

Drug candidate	Condition treated
RVX000222/ RVX208 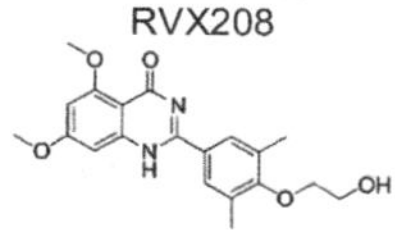	Cardiovascular disease
GSK525762/I-BET-762/iBET 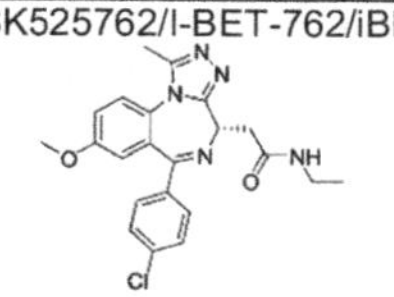	Breast and prostate cancer NUT midline carcinoma
ABBV-075 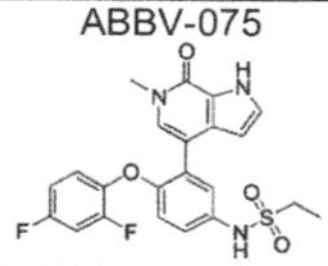	Breast cancer Non-small cell lung cancer
OTX105/OTX015MK8628 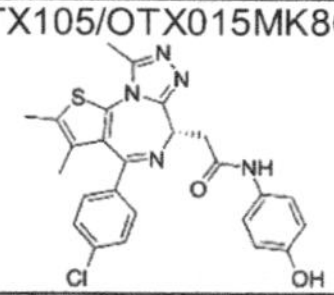	NUT midline carcinoma Triple negative breast cancer Solid tumors
CPI-0610 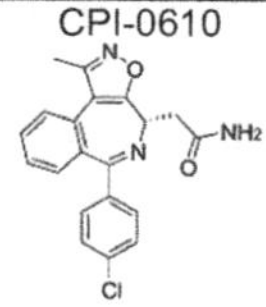	Multiple myeloma Leukemia Myelofibrosis
BMS-986158 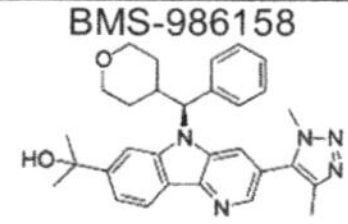	Advanced tumors
RO6870810	Acute myeloid leukemia Ovarian and breast cancers Solid tumors
GSK2820151	Solid tumors
CC-900010	Lymphomas
BAY 1238097	Neoplasms
ZEN003694	Metastatic castration-resistant prostate cancer
INCB054329	Solid tumors and hematologic malignancy
INCB057643	Solid tumors and hematologic malignancy

4.2–10-fold BD1 over BD2 selectivity against BRD2, 3, and 4. Further developments with dual kinase/bromodomain inhibitors have been reviewed elsewhere [73].

3.4 Bromodomain Clinical Trials and Chemical Probes

At this point in time, almost half of the 61 bromodomains now have chemical probes which maintain high selectivity over the majority of other bromodomains to study their mechanisms of action. Following the first phase I clinical trial in 2012 by GSK of BET inhibitor I-BET/I-BET762/GSK525762 for treating patients with NUT-midline carcinoma, 26 additional clinical trials are currently underway or have been either completed or terminated. The most advanced trial, with BET inhibitor, RVX-208, created by Resverlogix, is in phase III clinical trials for cardiovascular disease. In all trials, only BET bromodomain inhibitors are under study. Figure 7 summarizes the current status of both clinical trial candidates, and chemical probes are shown in Fig. 4. These numbers should only grow over time.

4 Part III: Computational and Experimental Biophysical Methods for Bromodomain Inhibitor Development

4.1 High-Throughput Virtual Screens and Molecular Dynamic Simulations for Bromodomain Inhibitor Discovery

Computational methods have been applied to bromodomain ligand discovery using virtual screening of small molecules against virtual protein structures. This is enabled by the wealth of bromodomain structures that have been reported as shown in Table 1. Virtual screening is commonly accomplished using programs such as Maestro's Glide docking software, MOE, or more specialized docking techniques [74–76]. Methods for library selection range from large (millions of commercially available compounds) available in the ZINC database to focused libraries based on previously reported scaffolds or drug repurposing of FDA-approved libraries. Large libraries are useful for discovering previously unreported pharmacophores for bromodomain inhibition. An example of this is the discovery of novel scaffolds for BRD4 reported by Vidler et al. [75]. A large virtual library of compounds was filtered to include pharmacophores similar to the acetylated lysine. Of 2.4 million compounds screened, 240 were taken on for testing in an AlphaScreen assay, and 6 showed activity against BRD4 BD1. Several of these molecules were sufficiently soluble and possessed high enough affinity to be co-crystallized with BRD4 BD1, leading to new substructures for BRD4 inhibitors [75].

An alternative computational approach is to use known small-molecule scaffolds as starting points. An advantage to this is the ability to use known structural biology to rationally design specific interactions to improve affinity and selectivity. This is often challenging due to high structure homology between the 61 human bromodomains. Shadrick et al. used molecular docking of previously reported bromodomain binding epitopes based on a tetrahydroquinoline scaffold [74]. This core is synthetically tractable and was used to develop a selective inhibitor for BET BD2s. They rationally designed analogs to take advantage of the difference in the position of the structural waters in the ZA channel in BRD2 BD1 vs. BD2 and their contacts with tetrahydroquinoline analogs. Using this technique, they developed a modestly selective BET BD2 inhibitor with 7.8-fold selectivity for BD2 of BRD2 as determined by ITC, though the selectivity was less in both SPR and TR-FRET assays. This study demonstrates the continued difficulty of engineering domain or isoform selectivity between bromodomains, despite the wealth of structural biology studies focused on bromodomains.

Continued increases in access to computational power and advances in theory have furthered the use of molecular dynamics in designing selectivity. A study by Aldeghi et al. described a retrospective prediction of the pan-bromodomain inhibitor bromosporine and the BET BD2-selective compound RVX-208. Affinities for a subset of bromodomains using absolute binding free energy calculations, based on molecular dynamics, were calculated. They achieved a good correlation between experimentally measured and predicted affinities; these predictions may lead to designed selectivity in the future [77]. Recent advances in virtual screening use proteochemometric models with small-molecule fingerprints and binding-site descriptors to develop more selective molecules between the bromodomain families [78]. Virtual screening poses do not always recapitulate in co-crystal structures; Allen et al. discovered that molecular dynamics were required to determine accurate poses for their novel class of BRD4 inhibitors [79].

Virtual screening must be followed up by biophysical characterization. A general workflow used in a virtual screen-to-lead study is described in Fig. 8. As an example,

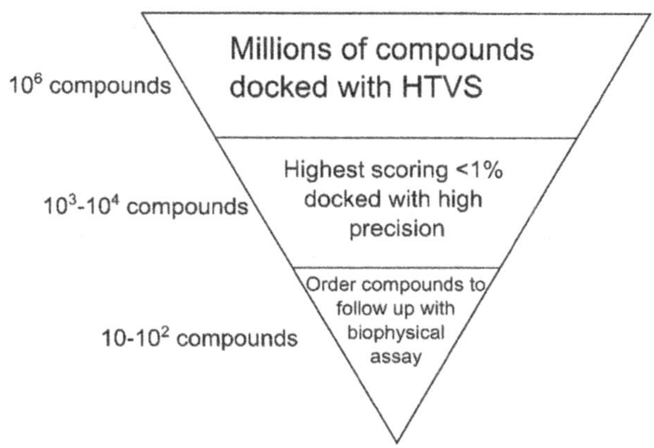

Fig. 8 General virtual screening workflow example: millions of compounds are docked with high-throughput virtual screening; the top 1% are docked with high-precision docking. Commercially available compounds, which scored well in high-precision docking, are followed up in vitro with a biophysical assay to determine affinity

Ayoub et al. conducted a virtual screen of six million compounds against the first bromodomain of BRDT; the top 0.1% were filtered to 200 compounds by precision filtering and removal of potential pan-assay interference compounds. Twenty of these compounds were purchased and tested in a fluorescence polarization (FP) assay, of which nine had activity against BRDT. The lead compound, a dihydropyridine, was further verified as a BET inhibitor using differential scanning fluorimetry, a protein-observed NMR method, and finally the binding pose was elucidated by a co-crystal structure with the BET bromodomain BRD4 BD1. Due to the higher binding affinity for BRD4 BD1, the potency of the molecule was improved leading to in-cell activity in MM.1.S cells by downregulating Myc expression [80].

4.2 Direct-Binding Experiments with Bromodomains

4.2.1 NMR Methods Have Been Instrumental in Bromodomain Inhibitor Discovery

As previously noted, early research by Zhou and co-workers using NMR methods was instrumental in confirming the tertiary structure of bromodomains and understanding binding-site interactions between acetylated histones and bromodomains [17]. Since then, NMR methods have been widely used in screening assays and to characterize bromodomain inhibitors. In a NMR binding assay, the difference in the chemical shift, intensity, or phase of the resonance is analyzed between two experiments: (1) ligand- or protein-only reference and (2) ligand + protein. The two broad classes of NMR binding assays are ligand- and protein-observed, in which the resonances of ligands or the protein are monitored, respectively. Both ligand- and protein-observed NMR methods are sensitive techniques that can detect a wide range of binding affinities (nM to mM) making them applicable assays for any stage of inhibitor development. NMR methods used in drug discovery of bromodomains include, but are not limited to, HSQC, PrOF, STD, and CPMG NMR. These methods are described below.

Both ligand- and protein-observed NMR methods offer distinct advantages and disadvantages. For example, ligand-observed experiments require relatively low concentrations of protein (5–10 μM per experiment), and the behavior of ligands in solution can be monitored by resonance height, line width, or chemical shift. Additionally, mixtures of ligands can be screened in a single NMR tube with no deconvolution needed, because the identity of each ligand resonance is known. However, ligand-observed methods give no information on protein stability or binding-site location. In contrast, protein-observed methods require larger amounts of isotopically labeled protein (typically 50–200 μM of ^{19}F-, ^{13}C-, or ^{15}N-labeled protein per experiment). An advantage of protein-observed methods is the ability to glean protein structural information from the spectrum. Because these methods observe protein resonances, a disadvantage of protein-observed experiments using

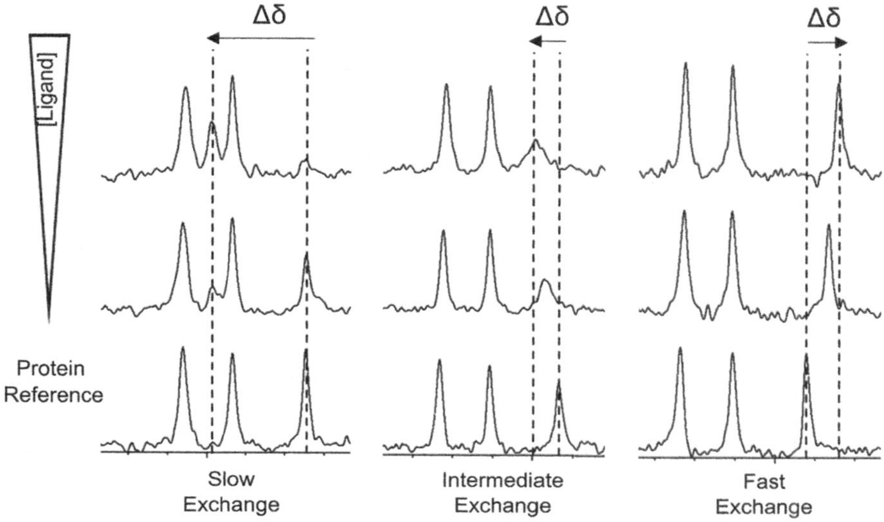

Fig. 9 Examples of resonance behavior indicative of the three chemical exchange regimes. Examples are of protein-observed ^{19}F NMR spectra, but similar behavior is observed using other NMR-active isotopes. Arrows indicate the direction of the free to bound resonance and magnitude of chemical shift change ($\Delta\delta$)

mixtures of ligands is the inability to identify which compound(s) caused the perturbation of resonances as well as the throughput of the experiment.

A key advantage of protein-observed NMR methods is the ability to qualitatively and quantitatively asses binding affinities. For a biomolecular binding interaction, K_d is defined as the ratio of rate constants k_{off}/k_{on} where k_{on} can be estimated to be the rate of diffusion for small molecules (although this may not always be true). Thus, K_d can be estimated from observing k_{off}. On the NMR timescale, compounds sample the protein-bound and unbound states at different rates that can be classified as fast, slow, or intermediate exchange (Fig. 9). A qualitative assessment of the strength of binding can be gleaned by the nature of the protein resonance response. There are three binding regimes on the NMR timescale: (1) fast chemical exchange (change in chemical shift (typically for ligands with low affinity K_d = high μM to mM)), (2) slow chemical exchange (two resonances are present representing the free and bound states of the ligand (typical K_d < 1 μM)), and (3) intermediate chemical exchange (broadening and movement of the resonance (K_d typically ~1–100 μM)). To calculate a K_d of a ligand (or protein) in fast chemical exchange, a titration of various ligand concentrations, ranging from low μM to mM, is performed. The change in chemical shift of the protein resonance is plotted and fitted to Eq. 1:

$$\Delta\delta_{obs} = \Delta\delta_{max} \times \frac{(K_d + [L] + [P]) - \sqrt{(K_d + [L] + [P])^2 - 4[PL]}}{2[PL]} \quad (1)$$

where $\Delta\delta_{max}$ is the maximum change in chemical shift, L is the concentration of ligand, and P is the concentration of protein. Equation 1 is a variation of Eq. 2, the general K_d equation that accounts for ligand or protein depletion.

$$[PL] = \frac{(K_d + [P_{tot}] + [L_{tot}]) - \sqrt{(K_d + [P_{tot}] + [L_{tot}])^2 - 4[P_{tot}][L_{tot}]}}{2} \quad (2)$$

In Eq. 2, P_{tot} is the total concentration of protein, and L_{tot} is the total concentration of the ligand.

4.2.2 Two-Dimensional NMR Methods Provide a Detailed Picture of the Binding Interaction

Two-dimensional protein-observed NMR experiments identify the binding site and give information on the key amino acids involved in binding interactions. Common experiments of this type include ^{1}H-^{15}N- or ^{1}H-^{13}C-HSQC/HMQC experiments. ^{1}H-^{15}N-HSQC/HMQC monitors the protein amide backbone, and ^{1}H-^{13}C-HSQC/HMQC experiments monitor isotopically labeled protein methyl groups resulting in a simpler spectrum than ^{15}N labeling. These experiments can be lengthy, require large amounts of isotopically labeled protein, and give complicated spectra to interpret making them not ideal for screening large libraries of ligands. However, both methods have successfully been used to screen for bromodomain inhibitors. For example, Hasvold et al. used ^{1}H-^{13}C-HSQC to screen 18,000 fragments in mixtures of 30 leading to a novel methyl pyrrole inhibitor of BET bromodomains [81]. Harner et al. screened 13,800 fragments against ADTAD2 using ^{1}H-^{15}N-SOFAST-HMQC (Fig. 10) [82]. More commonly, an assigned ^{1}H-^{15}N-HSQC protein spectrum is used

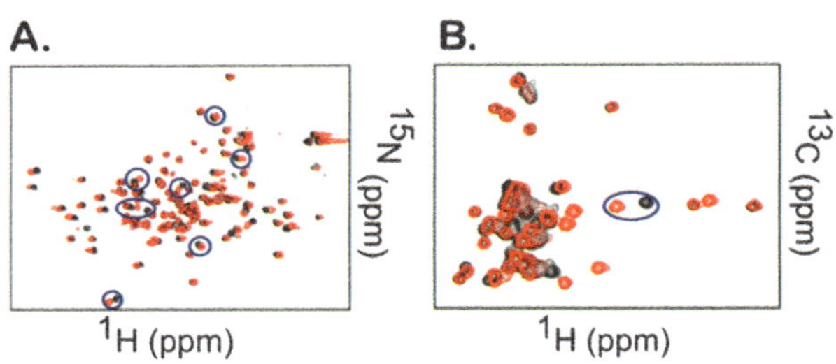

Fig. 10 Example overlays of 2D-NMR spectra. Blue ovals emphasize shifts of protein resonances when a ligand is (red) and is not (black) present in the sample. (**a**) ^{1}H–^{15}N SOFAST-HMQC for ligand discovery ATAD2. Adapted with permission from Harner et al. (**b**) ^{1}H–^{13}C-HSQC for ligand discovery for BRD4 BD2. Adapted with permission from Hasvold et al. permission pending

to characterize binding of known hits or aid in hit development [83]. This method is especially useful to identify the binding site and key amino acids when a co-crystal structure cannot be obtained. Additionally, 2D-NMR methods have been used to study the interaction between bromodomains and acetylated histone peptide mimics [25, 84–87]. As an example, Ferguson et al. used ^{1}H-^{15}N-HSQC chemical shift and CLEAN-chemical exchange experiments to determine hot spot amino acids involved in the native interaction between H3K14ac and BAZ2B [88]. Multidomain proteins have also been studied by ^{1}H-^{15}N-HSQC NMR with chemically defined nucleosomes [35].

4.2.3 PrOF NMR Is a Sensitive Technique That Gives Simple Spectra to Analyze for Binding Interactions

Protein-observed fluorine (PrOF) NMR is an alternate protein-observed method that is fast and gives simple spectra to interpret. For PrOF NMR, bromodomains are fluorine labeled by the incorporation of fluorinated amino acids. Fluorine is an ideal NMR reporting atom, because it is the second most sensitive stable NMR-active nuclei (83% compared to ^{1}H), its hyperresponsiveness to change in its environment evident by its ~400 ppm chemical shift range, ^{19}F is 100% isotopically naturally abundant, and fluorine is not present in biomolecules, resulting in no background signal [89, 90]. Aromatic amino acids are enriched at the bromodomain binding interface but not prevalent throughout the rest of the protein [91, 92]. Thus, nonnatural amino acids such as 5-fluorotryptophan (5FW), 3-fluorotyrosine (3FY), or 4-fluorophenylalanine (4FF) may be incorporated as sensitive NMR reporter side chains. A detailed protocol for fluorinated amino acid incorporation and strategies for fluorine labeling and screening have been described [93]. PrOF NMR experiments with bromodomains are fast (2–5 min with 50 μM protein on a 500 MHz instrument) when cryoprobes are used and the chemical shift perturbations are reproducible making it a good method for screening [94]. Additionally, having multiple fluorine labels allows for monitoring of protein behavior and assessment of where small molecules are binding to the protein (Fig. 11).

A 2015 paper by Urick et al. exemplifies a screening platform that takes advantage of the strengths of PrOF NMR, DSF, FP, and ITC to identify hits, rank order hits, and characterize binding affinity for compounds targeting BRD4 BD1 and BPTF [71]. An advantage of PrOF NMR over ligand-observed NMR methods is the ability to screen two proteins in the same experiment (single NMR tube) because the PrOF NMR spectra are simple, essentially completing two screens at once. Dual-protein PrOF NMR screens give bromodomain selectivity information at the onset of the inhibitor discovery process. The HSQC analog of this experiment, RAMPED-UP NMR, was previously described by Zartler et al. [95].

Urick et al. completed a dual-protein PrOF NMR screen to identify selective hits for both BRD4 and BPTF. BRD4 BD1 and BPTF were sequence selectively labeled with 5-fluorotryptophan (5FW). BRD4 BD1 has three tryptophan residues. W81 is located on the WPF near the binding site and acts as the primary reporting residue.

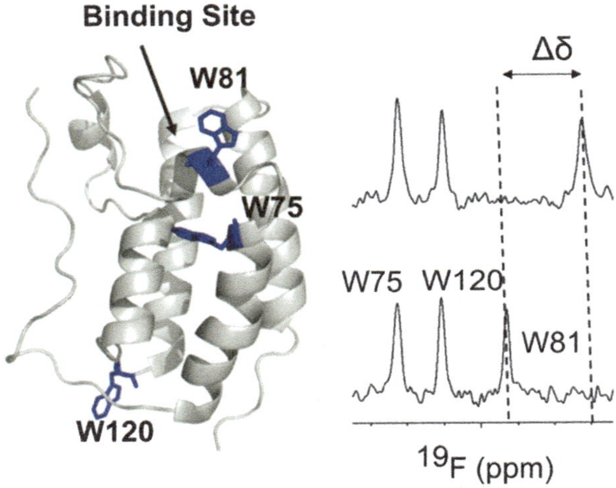

Fig. 11 An example of fast chemical exchange PrOF NMR binding experiment with BRD4 BD1. A change in chemical shift of W81 (the W closest to the binding site) is observed between a sample with just protein (bottom spectrum) and protein + ligand (top spectrum). W120 is located far away from the binding site and is thus not disturbed. PDBID 3UVX

W120 is located on the other side of the protein and acts as a control resonance to assess the stability of the protein. BPTF has a single tryptophan resonance on the WPF shelf. All four 5FW resonances for the proteins are baseline resolved when assayed together (Fig. 12). A total of 229 compounds from the published kinase inhibitor sets I and II (GlaxoSmithKline) were screened against BPTF and BRD4 BD1 simultaneously. Nine hits were found targeting BPTF with three selective over BRD4. Thirty-one hits were found for BRD4 with 15 selective over BPTF. As previously reported, the screen uncovered the 1,4,5- and 2,4,5-trisubstituted imidazole scaffolds as hits for BRD4. Additionally, the screen discovered two new binding classes for both BRD4 and BPTF 1,2,5-oxadiazoles and 2,4-disubstuted pyrimidinyl aryl ureas.

Rank ordering BRD4 BD1 hits by DSF and characterizing affinity via FP revealed SB-284851-BT (**1**, Fig. 12) to be a potent and selective hit. DSF was used to determine ΔT_m for hit compounds against BRD4 BD1. The ΔT_m were plotted on a DSF calibration curve to rank order the hits by predicted affinity (see Sect. 4.2.8 for explanation). A competitive FP assay was used to determine the K_i of select compounds. Compound **1** was the most potent hit with a K_i of 310 nM (another trisubstituted imidazole SB-590885-AAE was the second most potent with K_i of 400 nM). These compounds were previously characterized as p38α and B-Raf kinase inhibitors.

In addition to novel BRD4 inhibitors, the authors discovered and characterized the first known BPTF inhibitor aryl urea 1 (AU1) (GSK1379725A) (Fig. 12). Rank ordering hits with DSF was not possible with BPTF due to a lack of known inhibitors to create a correlation curve. ITC was used to determine a K_d of 2.8 μM for AU1. In a firefly luciferase assay, it was shown that AU1 interferes with the regulatory role of BPTF in cells. As the first known BPTF inhibitor, AU1 provides a starting point for understanding the role of BPTF. AU1 was subsequently used by Frey et al. to evaluate the role of BPTF in breast cancer and shown to affect Myc occupancy on chromatin [96].

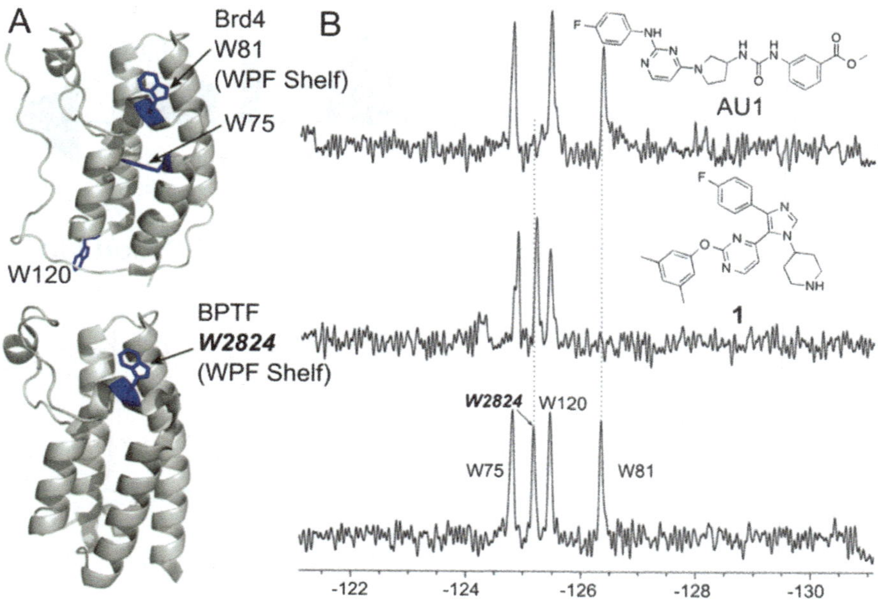

Fig. 12 (**a**) 5FW labels on BRD4 and BPTF. (**b**) PrOF NMR spectra showing the selective binding of **1** to BRD4 and AU1 to BPTF in intermediate chemical exchange. Reprinted (adapted) with permission from Urick et al. [71]. Copyright 2018. American Chemical Society

4.2.4 Based on the NOE, STD NMR Can Provide a Binding Epitope Map

Saturation transfer difference (STD) NMR is a ligand-observed method that has been used to find bromodomain inhibitors. STD NMR is based on the nuclear Overhauser effect (NOE), the transfer of magnetization from protein to ligand. In this experiment, a saturating selective pulse is applied to a region of the spectrum where only protein nuclei will resonate (0 to −2 ppm). The magnetization is transferred through spin diffusion to bound ligands resulting in an increase in their signal intensity described as the *on-resonance spectrum.* The *off-resonance* spectrum (no selective saturation) is subtracted from the *on-resonance* spectrum. Binding ligands will exhibit a peak in the difference spectrum, while non-binding ligand will not have visible resonances (Fig. 13). Because the NOE effect is distance dependent, a binding epitope can be determined by analyzing which proton resonances give the strongest STD signal (group epitope mapping, GEM STD) [97]. Geist et al. conducted BRD4 binding studies using STD GEM with a novel application of waterLOGSY NMR, LOGSY titration, to generate a binding epitope map and identify protein-bound water molecules near the bound ligand [98]. STD NMR has been used as a binding assay to discover bromodomain inhibitors for BAZ2A [59, 99], CBP [100], BRD4 [83], and BRD7 [101]. Despite giving easy to analyze data, STD can be challenging for small proteins (MW $<$ 20 kDa) such as

Fig. 13 (**a**) Standard STD NMR pulse sequence. (**b**) Illustration of the off- and on-resonance spectra as well as the difference spectrum that that shows ligand resonances which is characteristic of a binding event

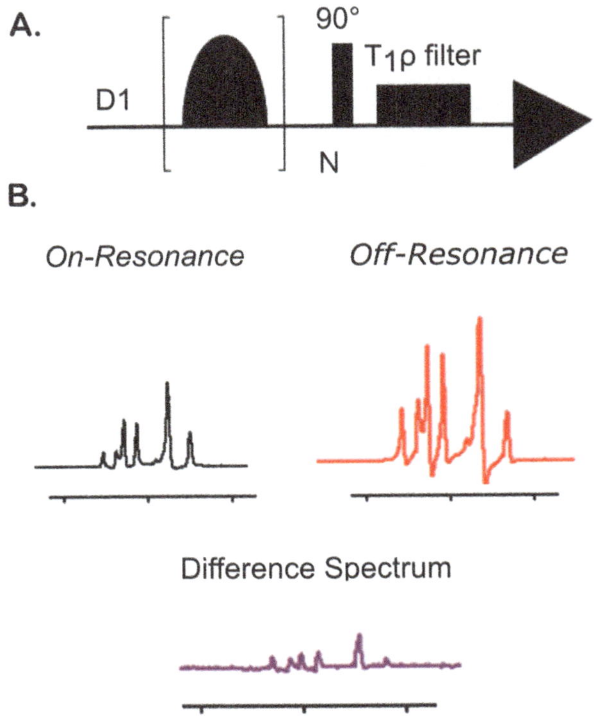

bromodomains (MW ~ 15 kDa). The efficiency of the spin diffusion is aided with increased proton density and longer rotational correlation times, characteristics of larger proteins. The STD effect can be enhanced by irradiating at a more downfield ppm, and protein resonances can be suppressed by using a $T_1\rho$ filter [83, 102].

4.2.5 CPMG NMR Has Been Used to Screen for Bromodomain Inhibitors

The Carr-Purcell-Meiboom-Gill (CPMG) NMR method is a ligand-observed experiment that works well with bromodomain-sized proteins. CPMG NMR is a spin echo train pulse sequence that takes advantage of the difference in rotational correlation time (T_c) of small molecules and proteins to detect a binding event (Fig. 15a). The T_c of a compound is reflected in the transverse relaxation time (T_2) of the compound. T_2 is inversely proportional to T_c. Proteins have shorter T_2 than ligands because they are larger and tumble slower in solution (larger T_c). Ligands that bind to the protein will take on the T_c (and thus the T_2) characteristics of the protein. The pulse sequence allows the difference in T_2 between the bound and free ligand to be represented as a change in resonance intensity. When the ligand-only and ligand + protein spectra are overlaid, a decrease in signal intensity is characteristic of binding (Fig. 14b).

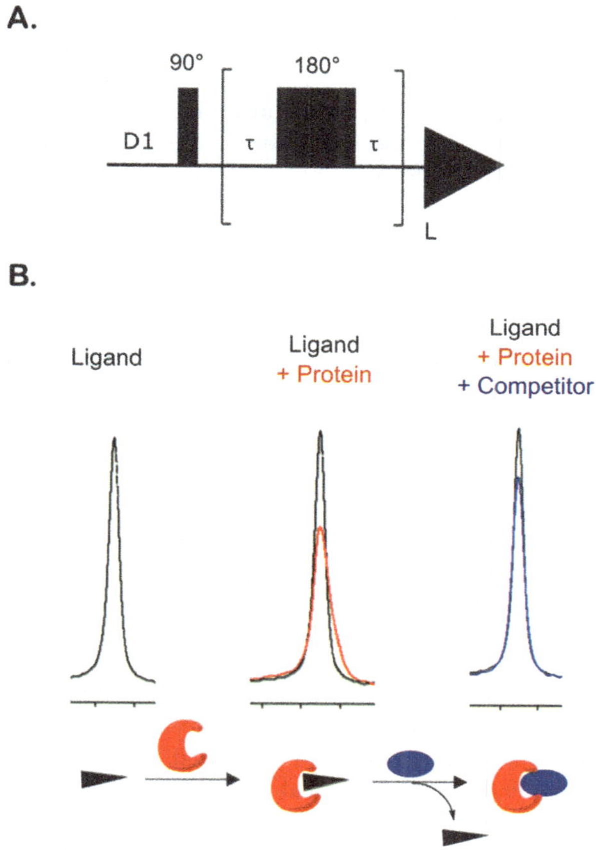

Fig. 14 (**a**) Standard CPMG pulse sequence. (**b**) An example of a three-part CPMG experiment. The black resonance is the ^{1}H proton of a ligand, red is the resonance decreasing in intensity when binding to the protein, and blue is the return in resonance intensity as a competitor is added

Typically, a change of 20% is considered reliable for determining a binding event [102]. If a strong inhibitor that binds in the acetylated lysine binding pocket is known, a competition experiment can be done to determine if the ligand is occupying the same binding site. In a competition experiment, the competitor is added to the ligand + protein tube. An increase in ligand intensity is observed if the competitor displaces the ligand. A 10% return in ligand intensity upon addition of competitor indicates that the ligand of interest is binding in the native binding pocket. Unlike STD, CPMG is not dependent on the proton density of the protein allowing smaller proteins to give CPMG effects. Strong binders (nanomolar dissociation constants) cannot readily be detected by CPMG, and poor shimming can give the illusion of resonance intensity decreasing causing false positives. CPMG has been used to screen compounds against BRD7/BRD9 [101], BAZ2B [59], BRD4 [94], and TRIM24 [27]. Additionally, CPMG has been used to study BPTF recognizing various diacetylated patterns on H2A.Z [103].

When choosing an assay method, discrepancies between different biophysical methods in identifying and characterizing a binding event can be a concern. Urick et al. screened the same 930-compound fragment library against BRD4 using both ^{1}H CPMG and PrOF NMR to investigate the correlation between the ligand- and protein-observed methods [94]. An 85% assay overlap was found for PrOF NMR and ^{1}H CPMG NMR competitive hits. This result indicates that under the assay conditions used, the assays show a good correlation. Additionally, the authors showed rank ordering of hits via DSF and ITC is consistent with rank ordering of hits by PrOF NMR. Finally, K_d values determined by PrOF NMR and ITC were compared. A variation of 1.4-4.7 fold in K_d value indicates a good correlation between the assays for K_d determination.

4.2.6 ITC Can Determine the Thermodynamics and Stoichiometries of Bromodomain Complexes

Isothermal titration calorimetry (ITC) is a gold standard for accurately determining all of the thermodynamic binding properties including the K_d, enthalpy, entropy, and the complex stoichiometry. For example, Filippakopoulos et al. used ITC to determine the affinity and stoichiometry of a variety of bromodomains binding to multi-acetylated peptide histone mimics [20]. The ligand of interest is titrated into a constant concentration of protein (or vice versa). The heat of the binding interaction is measured as the amount of energy used to keep the reaction at a constant temperature. ITC is time-consuming and requires a high concentration of ligand based on a need for a large molar excess above the protein concentration. Although ITC is not ideal for bromodomain screening, ITC is a sensitive technique that does not require modification to the ligand or protein of interest and can accurately measure the K_d of strong and weak binding compounds. This method is often used to characterize hits and fully elaborated chemical inhibitors and probes. Additional label-free methods for determining K_d include bio-layer interferometry and microscale thermophoresis and surface plasmon resonance (SPR).

4.2.7 Kinetic Analysis of Interactions Using SPR

Surface plasmon resonance (SPR) is a direct-binding assay which has been used to screen small-molecule libraries against multiple families of bromodomains [104]. SPR measures the refractive index of a gold-plated surface called a chip. The principle of the assay is that a single wavelength laser is directed at a surface that reflects the light back at a detector. To measure binding affinities, the surface is first immobilized with the protein of interest. If a binding partner is passed over the protein and binds, the effective mass of the protein increases, altering the refractive index of the chip, and causes a change in signal. Data collected via SPR are displayed as sensograms (Fig. 15). Since this assay measures real-time interactions, the full kinetics (k_{on} and k_{off} rates) of the ligand can be determined. If the ligand is measured at several concentrations, the binding sensograms can be plotted as a function of concentration, and a K_d can be determined. This can be done relatively

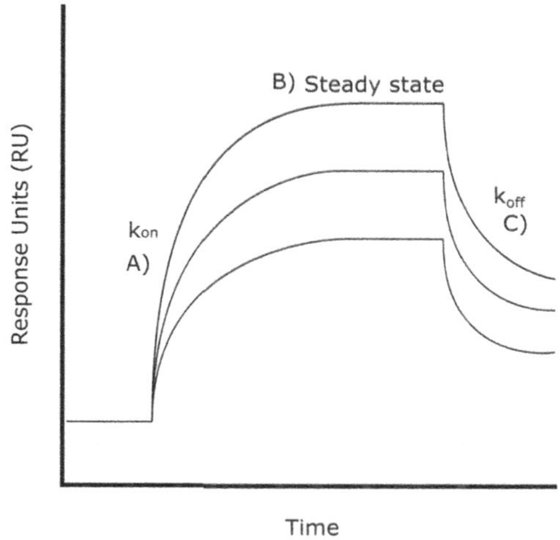

Fig. 15 Simulated dose-response SPR sensograms A) association of small molecule with captured bromodomain on the chip surface, B) steady-state equilibrium response, C) dissociation of molecule from the chip surface

rapidly (20 min to 1 h per compound), and the resulting data gives valuable information about the ligand. Advantages of this assay to ITC include the low reagent amounts required and that once immobilized, a protein chip can be regenerated and reused for typically multiple cycles of SPR. In SPR, several proteins can be tested in parallel, whereas in ITC this can take several hours. A disadvantage of SPR is that the interaction measured is of the protein and ligand, while the protein is immobilized to a chip, which may disrupt or augment the availability of the protein to ligand binding. Due to the speed and sensitivity of SPR, it has been utilized in fragment screening against BRD4 BD1, CREBBP, and PCAF bromodomains [104]. In this study, 656 fragments were screened against BRD4 BD1 and CREBBP, 7 potential hits were measured in a 3-point titration, and 3 were verified as binders to BRD4 BD1 and CREBBP. In a parallel series of experiments, two fragment scaffolds were discovered as binders to the PCAF bromodomain. This study also demonstrated the importance of DMSO as an additive in the SPR running buffer where higher percentages have diminishing effect on observed K_ds. An analogous assay, bio-layer interferometry, has also been used in bromodomain ligand discovery [105].

4.2.8 DSF Has Been Used to Assess Isoform Selectivity of Bromodomain Inhibitors

DSF is a thermal shift assay that has been utilized as a screening technique, to rank order hits, and to access bromodomain isoform selectivity. DSF uses a solvatochromatic fluorescent dye. In polar media, the fluorescence of the dye is quenched, while in hydrophobic environments such as denatured proteins, the fluorescence is increased. In this experiment, the dye is incubated with a protein. As the temperature is gradually increased, the protein denatures exposing the

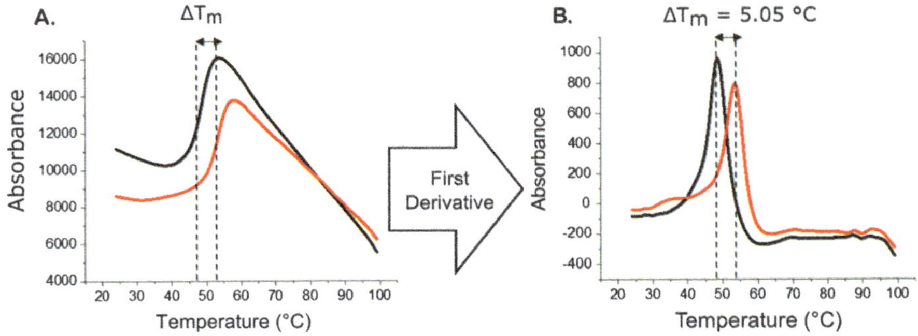

Fig. 16 (**a**) Example of DSF raw data of protein (black) and protein stabilized by ligand (red). (**b**) The first derivative of the raw data is taken to determine the ΔT_m (apex of the curve)

hydrophobic core. The dye binds to the core and increases in fluorescence monitored as a function of temperature. The midpoint of the sigmoidal curve is the thermal melting temperature (T_m) of the protein. Ligand binding will stabilize or destabilize the protein, changing the protein T_m (Fig. 16). Rank ordering bromodomain binders based on ΔT_m has been done by creating a correlation curve of ΔT_m vs K_d of known binders [9, 71, 106]. Although the correlation is only moderate with higher T_m indicative of higher affinity, this has been shown an as effective way to rank order hit compounds for BRD4. This method may not work as well for bromodomains with few known binding ligands. DSF uses small amounts of unlabeled protein, is quick, and thus has been used as a high-throughput screening platform for bromodomains [107–109]. Additionally, DSF has been used as a means to assess inhibitor isoform selectivity of inhibitors [105, 110–112]. An inhibitor that is selective would only perturb the T_m of the protein it targets. For example, the assessment of BET selectivity of (+)-JQ1 was carried out by doing DSF experiments with 36 bromodomains. Only the eight BET family bromodomains had a $\Delta T_m > 7°C$, indicating binding [9]. One challenge with DSF is how to categorize/characterize compounds that destabilize the protein (negative ΔT_m). These compounds may still bind in other assays. There is no consensus on how to effectively describe these compounds. As a new advance, a variant of the DSF assay has been demonstrated for bromodomains to show target engagement in cells called a "cellular engagement thermal shift assay" or CETSA [113, 114].

4.3 Competition-Based Assays for Inhibitor Discovery

4.3.1 AlphaScreen High-Throughput Competition Assays

AlphaScreen is a competition-based assay which is used as an initial screening platform and as a follow-up assay to determine IC_{50} values of compounds

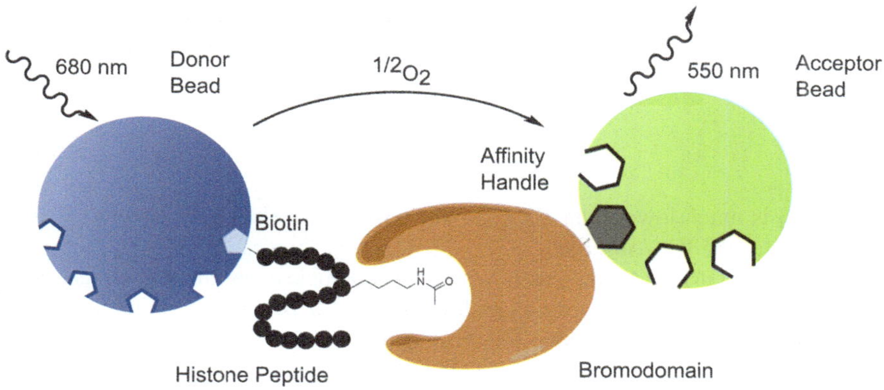

Fig. 17 Illustrated description of an AlphaScreen assay. The acceptor bead (yellow) and donor bead (blue) are brought together by an interaction between the bromodomain (orange) and either a histone peptide (depicted) or a small-molecule partner. When the acceptor and donor bead are proximal and the solution is irradiated with light, a transfer of energy via singlet oxygen occurs, and light is emitted by a chemical reaction with the donor bead and the reactive oxygen species. This interaction may be disrupted by small-molecule inhibitors at a single concentration and dose-dependent screening formats

discovered with other HTS techniques. Typically, the assay consists of an acceptor bead loaded with either a glutathione S-transferase (GST) or nickel nitrilotriacetic acid (NTA) tag to bind to a corresponding recombinant bromodomain and a streptavidin-loaded donor bead to bind to a biotinylated partner known to interact with the bromodomain. Commonly, a polyacetylated histone tail peptide or a small-molecule ligand is biotinylated and used as the binding partner. Due to the multivalency of the beads used in AlphaScreen (multiple proteins are displayed on a single bead), an avidity effect is promoted by the interactions between the beads and their binding partners. This allows for relatively low-affinity partners (such as histone peptides and bromodomains which are generally in the single- to double-digit micromolar affinity range or higher) to be used in very low concentrations (often nanomolar) and still provide robust signal. Signal is produced upon irradiation of the ternary complex (donor bead-protein-ligand-acceptor bead) with 680 nm light which excites the donor beads, leading to the formation of singlet oxygen (Fig. 17). The singlet oxygen may then react with the acceptor bead which emits energy at 550 nm. The observed signal is distance-dependent based on the short lifetime of singlet oxygen in aqueous solutions (200 nm). In conjunction with bromodomains, AlphaScreen is commonly used as a competition-based assay where the IC_{50} of a small molecule is measured by displacing a biotinylated histone tail peptide or small molecule. Due to the avidity effect mentioned above, protein and peptide levels used in the assay are typically much lower than their dissociation constant (K_d). Using low levels of protein and peptide in competition assays allows for estimation of K_i values for small-molecule ligands based on the IC_{50} using the Cheng-Prusoff equation

(Eq. 3) where P is the biotinylated bromodomain binding partner and the $K_{d\ \mathrm{Brd,P}}$ is the binding interaction between the bromodomain and its binding partner.

$$K_i = \frac{\mathrm{IC}_{50}}{1 + \frac{[P]}{K_{d\ \mathrm{Brd},P}}} \tag{3}$$

When $[P]$ is much lower than $K_{d\ \mathrm{Brd},P}$, this term approaches zero, and the K_i then approximates the IC_{50} observed.

Early work using AlphaScreen with bromodomains determined which post-translationally modified histone peptides bound to a panel of bromodomains by taking advantage of the avidity effect to detect the relatively weak interactions between these peptides and proteins [115]. AlphaScreen is an enabling screening technique because it requires only small amounts of material and can be miniaturized for high-throughput library screening in 96- and 384-well formats. Because the assay is based on the release of singlet oxygen, there are less chances of interference due to small-molecule fluorescence. Unlike TR-FRET assays, the beads are light sensitive and are unstable if exposed to light above 100 lux [116]. Since singlet oxygen can travel for about 200 nm in aqueous solutions, a robust signal is possible even when the beads are relatively far away. AlphaScreen is commonly used as a complementary assay to high-throughput virtual or experimental screening [100, 117]. The ability to generate dose-response curves and the knowledge that the native ligand is being displaced provides confidence in the inhibitors mode of action and gives insight into the site of binding.

An AlphaScreen was conducted at the Structural Genomics Consortium which screened a focused library of known kinase inhibitors described above [118]. They began with an initial screen using an AlphaScreen assay using a tetraacetylated H4 biotinylated peptide and His_6-tagged BRD4-BD1 in a 50 μM singleton screen. The compounds were counter-screened against beads without protein. At this concentration, 9 of 628 kinase inhibitors were shown to have 90% inhibition of the bromodomain and less than 50% inhibition of the counter-screen. Hits from the AlphaScreen assay were followed up for bromodomain selectivity assessment with DSF (46 of 61 bromodomains were tested), and binding was confirmed for most of the hits. Dose-response curves of each hit with BRD4-BD1 were generated using the AlphaScreen assay, IC_{50}'s determined ranged from nano- to micromolar affinities. These affinities were confirmed by ITC. The ability of the lead compounds BI-2536 and TG-101348 to displace BRD4 from chromatin in cells was confirmed using FRAP. Downregulation of the oncoprotein c-Myc was determined via western blotting, and cellular activity was confirmed by measuring the inhibition of the proliferation of MM.1S multiple myeloma cells, which are highly sensitive to c-Myc inhibition. At time of press, AlphaScreen services can be purchased from commercial vendors for many His-tagged bromodomains. Assays have been developed for HTS format for BETs [116], CECR2 [119], and BAZ2B [59], though the technology should be applicable to any affinity-tagged bromodomain with the corresponding anti-GST, anti-His, or NTA-linked beads which are commercially available.

4.3.2 TR-FRET: A Complimentary Screening Platform to AlphaScreen

TR-FRET or HTR-FRET (homogeneous time-resolved fluorescence resonance energy transfer) is a bead-based proximity assay which works on a similar principle to AlphaScreen. The assay is based on bringing together a donor and acceptor bead due to a molecular interaction between the protein of interest and a peptide or small-molecule partner. Unlike AlphaScreen, this information is relayed by FRET between the beads. Irradiation of the donor bead by light is followed by relaxation by fluorescence. This fluorescence excites the nearby acceptor bead which again relaxes by fluorescence. The ratio of fluorescence of the acceptor bead to the fluorescence of the donor bead normalizes the signal in each well and can be read as a measurement of the direct-binding interaction between the protein of interest and its binding partner. An advantage of this technique is that the beads are light stable and a prepared plate may be read for as long as the protein and binding partner are stable. The dependence of this assay on fluorescence may cause concern that small molecules which fluoresce could interfere with the assay. To address this issue, rare earth cryptates have been developed as donor dyes such as Eu^{2+} and Tb^{3+} which have long-lived fluorescence decay times, mitigating effects of biological and small-molecule fluorescence. HTR-FRET has been used widely in HTS campaigns and confirmation assays to discover ligands for BAZ2A/B [120], CECR2 [119], CBP [121], BETs [81], PCAF [122], and BRD9 [123]. Variations of this assay are sold commercially for about half of the human bromodomains.

4.3.3 Fluorescence Polarization: A Homogeneous High-Throughput Discovery Assay

An alternative inhibition assay to bead-based assays is fluorescence polarization (FP). FP is a fluorescence assay which requires a fluorescently labeled, high-affinity ligand for the protein of interest. The fluorescent tag must be appended such that binding of the ligand is unhindered to the target protein. Histone peptides can be used as ligands for bromodomains; however, they are often too low affinity to provide sufficient signal. Determination of IC_{50}s begins with measuring the affinity of the fluorescently labeled probe with the protein in a direct-binding experiment. The concentration of protein that is 80% bound is determined with Eq. 4, where Y is anisotropy, a is the fluorescently labeled probe, b is the maximum anisotropy, c is the minimum anisotropy, and P is protein concentration. Lower percent bound protein complex concentrations can also be used, although the dynamic range in obtainable data will be reduced.

$$Y = c + (b - c)\frac{(K_d + a + P) - \sqrt{(K_d + a + P)^2 - 4aP}}{2a} \tag{4}$$

The concentration at 80% bound is determined and used in competition assays where the fluorescent ligand is competed off with potential drug candidates.

IC_{50} values are determined with nonlinear regression fits using software such as GraphPad or Origin. IC_{50} values can be converted into K_i values, which are proximal to dissociation constants, using a variant of the Cheng-Prusoff equation (Eq. 5) as described by Kenakin [124].

$$K_i = \frac{(L_b)(\mathrm{IC}_{50})(K_d)}{(L_0)(R_0) + [L_0(R_0 + L_b - L_0 - K_d)]} \tag{5}$$

Care must be taken that appropriate baseline data is achieved for IC_{50} values to determine relevant K_i values. BET bromodomain assays have taken advantage of co-crystal structures to design fluorescent probes of (+)-JQ1 [125] and BI-2536 [71], both of which are nanomolar inhibitors of BET bromodomains. Advantages of this assay compared to bead-based assays is that there is no need to tag the protein of interest with any kind of affinity tag, as well as freeing the protein from being tethered to a surface or bead. The disadvantages of the assay are in interference of fluorescent compounds, aggregators which commonly cause light scattering, and the need for a fluorescently tagged small molecule, the K_d of which determines the lowest K_i that can be determined by this assay [126].

4.4 Crystallography Guides Rational Design of Bromodomain Inhibitors

Crystallography is a powerful structural biology tool that has aided in the development of bromodomain inhibitors. Crystal structures provide a complete, three-dimensional illustration of the arrangement of all of the atoms of the protein and bound ligand. Co-crystal structures provide information on binding site, binding pose, types, and locations of binding interactions. These co-crystal structures also provide guidance for potential ligand protein interactions with newly designed ligand derivatives to increase potency. Many bromodomain inhibitor campaigns use crystallography to guide analog synthesis and to characterize the binding of their final inhibitor [108–112]. Screening by crystallography has been done to discover bromodomain inhibitors [127, 128]. For example, Ember et al. screened 581 compounds against BRD4 BD1 resulting in 14 co-crystal structures. In addition, both apo and co-crystal structures of bromodomains have enabled improvements for in silico screening and docking methods (see Sect. 4.1) [106]. A total of 953 crystal bromodomain crystal structures are published with 57% of the bromodomains having at least 1 structure (see Table 1).

Although arguably a powerful biophysical method to study bromodomains, x-ray crystallography is not without challenges. First, crystal structures represent a static picture, not capturing the dynamics of proteins in solution. Second, unlike many biophysical methods, the assay conditions are highly dependent on the protein. It can take months to years to develop conditions for crystallization, and these conditions may not be transferable between bromodomains. Furthermore, conditions that give diffracting crystals for apo structures may not be suitable for crystallization with a ligand. Finally, crystallography is material, time, and financially intensive. Large quantities (10–15 mg/mL) of pure and concentrated protein are needed. Protein preparation, scoring crystals, cryo-protecting crystals, data collection, and data processing require a trained researcher. Time at a synchrotron or home x-ray source and crystallography tools are additional infrastructure needs.

A 2012 paper by Fish et al. [129] is an excellent example of the power of crystallography in aiding the process of transforming an unselective, modestly potent acetyl lysine mimic fragment into the potent, BET family-selective, commercially available probe PFI-1 (Fig. 4). Fragments mimicking the endogenous acetyl lysine ligand are attractive starting scaffolds because they anchor binding through a hydrogen bond to asparagine and a water-mediated bond to a tyrosine. To achieve improved potency and selectivity, the starting scaffold must be extended to reach portions of the binding pocket that differ between bromodomains. Inspired by the BRD2 BD1 co-crystal structure by Chung et al. [112] (PDBID 4A9E), Fish and co-workers chose 3,4-dihydro-3-methyl-2(1H)-quinazolinone **2** as acetyl lysine mimics as a starting point. Compound **2** was active in an AlphaScreen assay against BRD4 and CREBBP. While not universally active (no affinity for BPTF or BAZ2B), selectivity for the BET family needed to be improved.

A crystal structure of **2** with BRD4 BD1 was instrumental in guiding decisions on analog synthesis to achieve potency and selectivity (Figs. 18 and 19). Binding of **2** is characterized by a hydrogen bond of the cyclic urea to N140 and lipophilic interactions with V87, L92, L94, Y97, F83, and I146 in the protein. The bromine at the C6 position extends into solvent, a logical place to add functional groups to the molecule to access new binding interactions without disturbing the binding interactions already present. The authors took advantage of the WPF shelf near the binding site that is conserved among the BET family members, but not present in CREBBP, to gain selectivity for the BET family. Sulfonamide linkers with hydrophobic caps were used to introduce a pronounced kink at C6 allowing the proper vector to reach and interact with the WPF shelf.

Two sulfonamide regioisomers were explored. The first generation of sulfonamides had the sulfur alpha to the dihydroquinazolinones. While these analogs showed increased activity compared to the parent, the similar activity of the ethyl and phenyl derivatives (e.g., **3**, Fig. 18) suggests that lipophilicity of the aryl group is not being efficiently delivered to the WPF shelf. This hypothesis was confirmed with the crystal structure of **3**, which has electron density consistent with multiple binding poses (Fig. 19). Aryl groups are ideal for lipophilic and pi-stacking interactions possible at the WPF shelf. The second generation of analogs with the nitrogen atom alpha to the dihydroquinazolinone core showed significant increase in activity compared to the first generation. A co-crystal structure of **PFI-1** with BRD4 BD1 shows not only a binding interaction at the WPF shelf but also an interaction between the sulfonamide oxygens and Q85 and a water-mediated hydrogen bond between the carbonyl of L92 and the methoxy oxygen.

PFI-1 from the second generation of sulfonamides is a potent and BET family-selective chemical probe. A DSF assay was used to assess selectivity of **PFI-1**

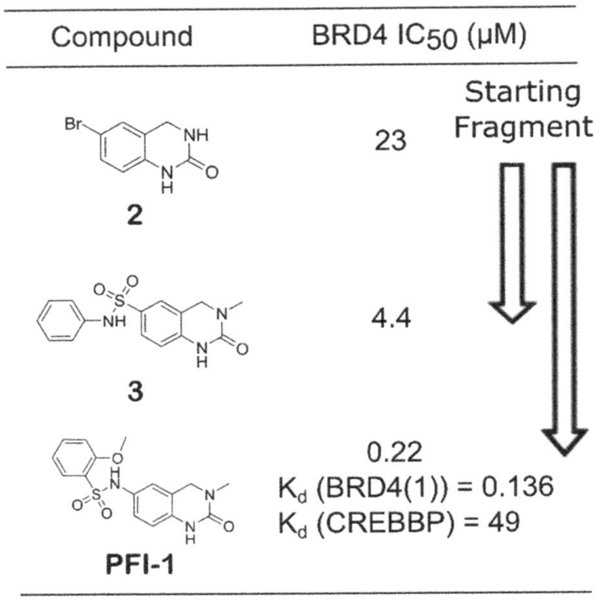

Fig. 18 Fragment development of a BET inhibitor guided by x-ray crystallography [129]

against 11 bromodomains. BET family bromodomains gave ΔT_m values between 2.08 and 6.48°C. CREBBP had a T_m of 1.72°C. Despite showing binding to CREBBP, **PFI-1** is over 100-fold more selective for BRD4 BD1 with K_d of 0.136 μM (SPR, BRD4 BD1) and 50 μM (ITC, CREBBP), respectively. In a subsequent 2013 study [105], **PFI-1** was shown to have antiproliferative effects on leukemic cell lines. It also downregulates MYC expression, induces apoptosis, induces differentiation of leukemic blasts, and downregulates Aurora B kinase.

5 Future Outlook on Bromodomain Inhibitor Development

Beyond traditional non-covalent inhibitor design of single domains, new chemical biology approaches are beginning to emerge to target bromodomains with significant potency. Currently the main focus has been on BET bromodomains. These approaches have included covalently binding small molecules, dimeric compounds for engaging multiple domains, and molecules aimed at degrading bromodomains. Covalently binding small molecules have found use as suicide substrates, permanently inhibiting their protein target. In one report, ethacrynic acid analogs were found in a DNA display screen to covalently attach to bromodomain cysteine residues. Of the 23 bromodomains tested, 22 contained a cysteine, usually found in the acetylated binding pocket. The 22 cysteine-containing bromodomains were all labeled by 1 of 2 ethacrynic acid analogs tested [130]. Sequence and structural

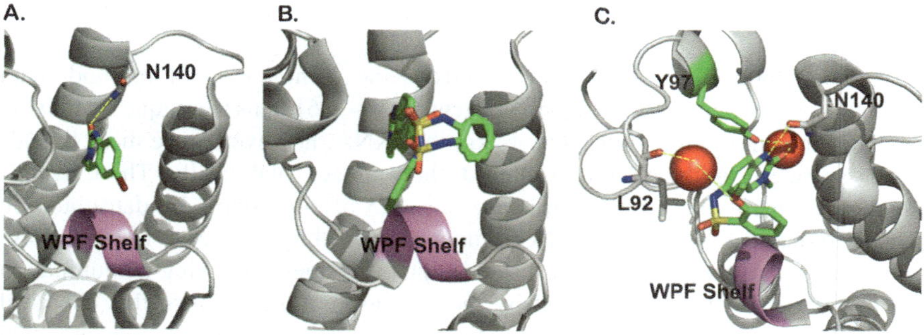

Fig. 19 Crystal structure of BRD4 BD1 alone and in complex with **3**. (**a**) Apo structure of BRD4 BD1. (**b**) Multiple conformations of **3** are visible in the crystal structure of **3** with BRD4 BD1. (**c**) Crystal structure showing the important binding interactions of **PF1-1** with BRD4 BD1 are indicated in red. PDBIDs 4HBV, 4HBY, 4E96

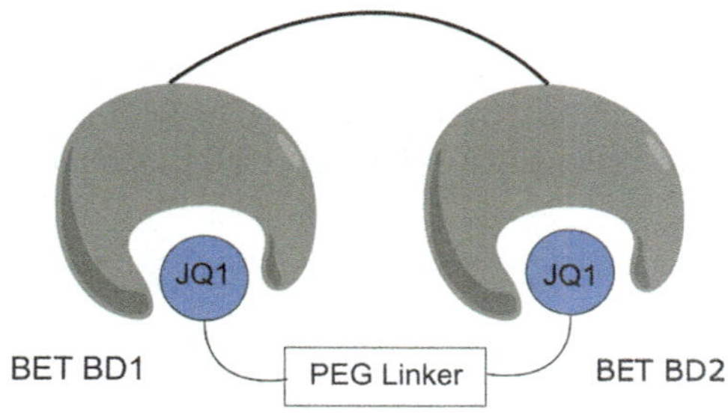

Fig. 20 Depiction of dimeric bromodomain inhibitors [17]. The affinity of the compound is increased by targeting neighboring bromodomains. Since many bromodomain-containing compounds are multidomain proteins, this principle could be applied to non-BET bromodomains, as well

analysis of the bromodomains demonstrated that 40 of the 61 human bromodomains contain cysteines which could be targeted with this strategy.

Bromodomains have been coincidentally and rationally targeted by bivalent small molecules which target multiple bromodomains in the same protein. These molecules have a high affinity due to the potential for an additive free energy of binding from the two individual ligands. A compound was developed by AstraZeneca, BiBET, which binds in a multivalent manner to both bromodomains of BRD4. While no co-crystal structure with both BD1 or BD2 could be solved, a combination of analytical ultracentrifugation and in-cell bioluminescence resonance energy transfer (BRET) experiments demonstrated a conformational change in the dual bromodomain construct tested, as well as an increase in BRET upon addition of the bivalent molecule [131, 132]. In a parallel study, Bradner et al. developed a dimeric construct of (+)-JQ1 (Fig. 20). Through AlphaScreen, ITC, and flow cytometry, they demonstrated their compound engaged multiple bromodomains and dropped the affinity of their compound by nearly 100-fold. This increase in activity was maintained when comparing the bivalent inhibitors with (+)-JQ1 in vivo [133].

Another application of bivalent ligands is the combination of bromodomain inhibitors and ubiquitin ligase targeting agents [134–136]. This technique selectively degrades the entire protein of interest akin to shRNA; however, since these ligands can dissociate from the protein of interest, they can act catalytically. These ligands function by recruiting a ubiquitin ligase (e.g., E3 of von Hippel-Lindau ubiquitin ligase) to the protein of interest. Recruitment of the ubiquitin ligase leads to poly-ubiquitination of the ε-nitrogen of surface-exposed lysines; following ubiquitination the protein of interest is degraded by the proteasome [137]. Since this chemical knockout is induced by dosage, this strategy can be used on protein targets in which a genetic knockout is embryonically lethal. Initial applications of this small-molecule knockout strategy were pioneered by Schneekloth et al. on the androgen receptor using an MDM2 E3 ligase-recruiting molecule. This strategy has since been applied to BET bromodomains [134–136] using both VHL and cereblon targeting molecules (Fig. 21). Exploration of the technology outside of BET bromodomains has so far been limited to BRD9 and TRIM24 [138, 139].

In summary, it has been 26 years since the first description of a bromodomain structural motif. Since this discovery, the sustained effort in structural biology combined with the emerging disease biology of these epigenetic reader proteins

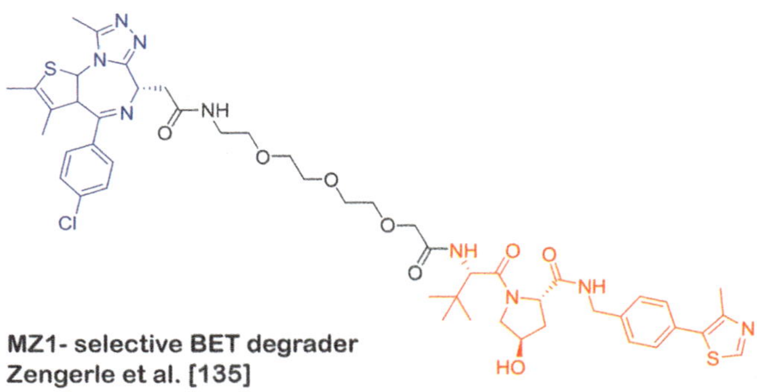

MZ1- selective BET degrader
Zengerle et al. [135]

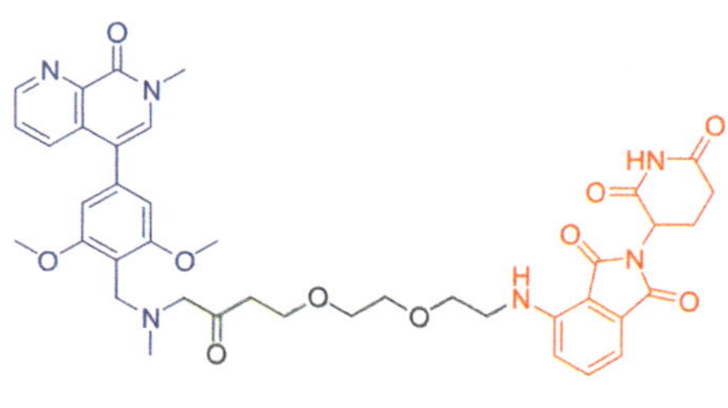

dBRD9- selective BRD9 degrader
Remillard et al. [138]

Fig. 21 Bivalent bromodomain degraders targeting BRD4 [135] and BRD9 [138], respectively. E3 ligase-recruiting molecules are shown in red and bromodomain ligands in blue

has led to many exciting clinical candidates and chemical probes for human bromodomains. Bromodomain motifs relevant for infectious disease are also now being disclosed [42]. Increased structural biology efforts to study the bromodomains as multidomain constructs both by NMR [72] and x-ray crystallography [23] will be important contributions to the fields combined with our growing ability to study these multidomain protein interactions with synthetic chromatin [2, 140]. Improvements with cryo-transmission electron microscopy (cryo-TEM) should also be an enabling tool to study these large complexes. Additionally, as chemical probes continue to be developed for bromodomains, in-cell engagement tools described above, such as gene expression analysis, chromatin-based FRAP experiments, nano-BRET, and CETSA, will need to play significant roles to help validate molecular mechanisms. With the wealth of biophysical methods and cellular tools available to researchers, the chemical epigenetics field of bromodomain inhibitor development should continue to innovate at an accelerated rate.

Compliance with Ethical Standards

Funding: This study was funded by the National Institute of General Medical Sciences of the National Institutes under the award number R01GM121414 (W.C.K.P and P.D.Y). J.A.J. was supported by a National Institutes of Health Biotechnology training grant 5T32GM008347-23.

Conflict of Interest: All authors declare that they have no conflicts of interest.

References

1. Strahl BD, Allis CD (2000) The language of covalent histone modifications. Nature 403(6765):41–45. https://doi.org/10.1038/47412
2. Allis CD, Muir TW (2011) Spreading chromatin into chemical biology. Chembiochem 12(2):264–279. https://doi.org/10.1002/cbic.201000761
3. Phillips DM (1963) Presence of acetyl groups in histones. Biochem J 87(2):258. https://doi.org/10.1042/bj0870258
4. Johns EW, Phillips DM, Simpson P, Butler JAW (1961) The electrophoresis of histones and histone fractions on starch gel. Biochem J 80:189–192
5. Allfrey VG, Faulkner R, Mirsky AE (1964) Acetylation + methylation of histones + their possible role in regulation of RNA synthesis. Proc Natl Acad Sci U S A 51(5):786–794. https://doi.org/10.1073/pnas.51.5.786
6. Filippakopoulos P, Knapp S (2012) The bromodomain interaction module. FEBS Lett 586(17):2692–2704. https://doi.org/10.1016/j.febslet.2012.04.045
7. Li YY, Sabari BR, Panchenko T, Wen H, Zhao D, Guan HP, Wan LL, Huang H, Tang ZY, Zhao YM, Roeder RG, Shi XB, Allis CD, Li HT (2016) Molecular coupling of histone Crotonylation and active transcription by AF9 YEATS domain. Mol Cell 62(2):181–193. https://doi.org/10.1016/j.molcel.2016.03.028
8. Fujimori DG, Conway SJ (2016) Editorial overview: chemical genetics and epigenetics. Curr Opin Chem Biol 33:VI–VII. https://doi.org/10.1016/j.cbpa.2016.08.008
9. Filippakopoulos P, Qi J, Picaud S, Shen Y, Smith WB, Fedorov O, Morse EM, Keates T, Hickman TT, Felletar I, Philpott M, Munro S, McKeown MR, Wang YC, Christie AL, West N, Cameron MJ, Schwartz B, Heightman TD, La Thangue N, French CA, Wiest O, Kung AL, Knapp S, Bradner JE (2010) Selective inhibition of BET bromodomains. Nature 468(7327):1067–1073. https://doi.org/10.1038/nature09504

10. Nicodeme E, Jeffrey KL, Schaefer U, Beinke S, Dewell S, Chung CW, Chandwani R, Marazzi I, Wilson P, Coste H, White J, Kirilovsky J, Rice CM, Lora JM, Prinjha RK, Lee K, Tarakhovsky A (2010) Suppression of inflammation by a synthetic histone mimic. Nature 468(7327):1119–1123. https://doi.org/10.1038/nature09589
11. Arrowsmith CH, Bountra C, Fish PV, Lee K, Schapira M (2012) Epigenetic protein families: a new frontier for drug discovery. Nat Rev Drug Discov 11(5):384–400. https://doi.org/10.1038/nrd3674
12. Belkina AC, Denis GV (2012) BET domain co-regulators in obesity, inflammation and cancer. Nat Rev Cancer 12(7):465–477. https://doi.org/10.1038/nrc3256
13. Brand M, Measures AM, Wilson BG, Cortopassi WA, Alexander R, Hoess M, Hewings DS, Rooney TPC, Paton RS, Conway SJ (2015) Small molecule inhibitors of bromodomain-acetyl-lysine interactions. ACS Chem Biol 10(1):22–39. https://doi.org/10.1021/cb500996u
14. Zhang G, Smith SG, Zhou M-M (2015) Discovery of chemical inhibitors of human bromodomains. Chem Rev 115(21):11625–11668. https://doi.org/10.1021/acs.chemrev.5b00205
15. Chung C-W, Witherington J (2011) Progress in the discovery of small-molecule inhibitors of bromodomain-histone interactions. J Biomol Screen 16(10):1170–1185. https://doi.org/10.1177/1087057111421372
16. Zeng L, Li JM, Muller M, Yan S, Mujtaba S, Pan CF, Wang ZY, Zhou MM (2005) Selective small molecules blocking HIV-1 Tat and coactivator PCAF association. J Am Chem Soc 127(8):2376–2377. https://doi.org/10.1021/ja044885g
17. Tamkun JW, Deuring R, Scott MP, Kissinger M, Pattatucci AM, Kaufman TC, Kennison JA (1992) Brahma - a regulator of drosophila homeotic genes structurally related to the yeast transcriptional activator SNF2 SW12. Cell 68(3):561–572. https://doi.org/10.1016/0092-8674(92)90191-e
18. Owen DJ, Ornaghi P, Yang JC, Lowe N, Evans PR, Ballario P, Neuhaus D, Filetici P, Travers AA (2000) The structural basis for the recognition of acetylated histone H4 by the bromodomain of histone acetyltransferase Gcn5p. EMBO J 19(22):6141–6149. https://doi.org/10.1093/emboj/19.22.6141
19. Randazzo FM, Khavari P, Crabtree G, Tamkun J, Rossant J (1994) BRG1: a putative murine homolog of the drosophila brahma gene, a homeotic gene regulator. Dev Biol 161(1):229–242. https://doi.org/10.1006/dbio.1994.1023
20. Filippakopoulos P, Picaud S, Mangos M, Keates T, Lambert JP, Barsyte-Lovejoy D, Felletar I, Volkmer R, Muller S, Pawson T, Gingras AC, Arrowsmith CH, Knapp S (2012) Histone recognition and large-scale structural analysis of the human bromodomain family. Cell 149(1):214–231. https://doi.org/10.1016/j.cell.2012.02.013
21. Jeanmougin F, Wurtz JM, LeDouarin B, Chambon P, Losson R (1997) The bromodomain revisited. Trends Biochem Sci 22(5):151–153. https://doi.org/10.1016/s0968-0004(97)01042-6
22. Dhalluin C, Carlson JE, Zeng L, He C, Aggarwal AK, Zhou MM (1999) Structure and ligand of a histone acetyltransferase bromodomain. Nature 399(6735):491–496
23. Jacobson RH, Ladurner AG, King DS, Tjian R (2000) Structure and function of a human TAF(II)250 double bromodomain module. Science 288(5470):1422–1425. https://doi.org/10.1126/science.288.5470.1422
24. Ornaghi P, Ballario P, Lena AM, Gonzalez A, Filetici P (1999) The bromodomain of Gcn5p interacts in vitro with specific residues in the N terminus of histone H4. J Mol Biol 287(1):1–7. https://doi.org/10.1006/jmbi.1999.2577
25. Hudson BP, Martinez-Yamout MA, Dyson HJ, Wright PE (2000) Solution structure and acetyl-lysine binding activity of the GCN5 bromodomain. J Mol Biol 304(3):355–370. https://doi.org/10.1006/jmbi.2000.4207
26. Vidler LR, Brown N, Knapp S, Hoelder S (2012) Druggability analysis and structural classification of bromodomain acetyl-lysine binding sites. J Med Chem 55(17):7346–7359. https://doi.org/10.1021/jm300346w
27. Liu J, Li F, Bao H, Jiang Y, Zhang S, Ma R, Gao J, Wu J, Ruan K (2017) The polar warhead of a TRIM24 bromodomain inhibitor rearranges a water-mediated interaction network. FEBS J 284(7):1082–1095. https://doi.org/10.1111/febs.14041

28. Fedorov O, Castex J, Tallant C, Owen DR, Martin S, Aldeghi M, Monteiro O, Filippakopoulos P, Picaud S, Trzupek JD, Gerstenberger BS, Bountra C, Willmann D, Wells C, Philpott M, Rogers C, Biggin PC, Brennan PE, Bunnage ME, Schüle R, Günther T, Knapp S, Müller S (2015) Selective targeting of the BRG/PB1 bromodomains impairs embryonic and trophoblast stem cell maintenance. Sci Adv 1(10):e1500723. https://doi.org/10.1126/sciadv.1500723
29. Crawford TD, Tsui V, Flynn EM, Wang S, Taylor AM, Cote A, Audia JE, Beresini MH, Burdick DJ, Cummings R, Dakin LA, Duplessis M, Good AC, Hewitt MC, Huang H-R, Jayaram H, Kiefer JR, Jiang Y, Murray J, Nasveschuk CG, Pardo E, Poy F, Romero FA, Tang Y, Wang J, Xu Z, Zawadzke LE, Zhu X, Albrecht BK, Magnuson SR, Bellon S, Cochran AG (2016) Diving into the water: inducible binding conformations for BRD4, TAF1(2), BRD9, and CECR2 bromodomains. J Med Chem 59(11):5391–5402. https://doi.org/10.1021/acs.jmedchem.6b00264
30. Divakaran A, Talluri SK, Ayoub AM, Mishra NK, Cui H, Widen JC, Berndt N, Zhu J-Y, Carlson AS, Topczewski JJ, Schonbrunn EK, Harki DA, Pomerantz WCK (2018) Molecular basis for the N-terminal bromodomain-and-extra-terminal-family selectivity of a dual kinase–bromodomain inhibitor. J Med Chem 61(20):9316–9334. https://doi.org/10.1021/acs.jmedchem.8b01248
31. Aldeghi M, Ross GA, Bodkin MJ, Essex JW, Knapp S, Biggin PC (2018) Large-scale analysis of water stability in bromodomain binding pockets with grand canonical Monte Carlo. Commun Chem 1(1):19. https://doi.org/10.1038/s42004-018-0019-x
32. Ruthenburg AJ, Li H, Patel DJ, Allis CD (2007) Multivalent engagement of chromatin modifications by linked binding modules. Nat Rev Mol Cell Biol 8(12):983–994. https://doi.org/10.1038/nrm2298
33. Qin S, Jin L, Zhang J, Liu L, Ji P, Wu M, Wu J, Shi Y (2011) Recognition of unmodified histone H3 by the first PHD finger of bromodomain-PHD finger protein 2 provides insights into the regulation of histone acetyltransferases monocytic leukemic zinc-finger protein (MOZ) and MOZ-related factor (MORF). J Biol Chem 286(42):36944–36955. https://doi.org/10.1074/jbc.M111.244400
34. Paz JC, Park S-H, Phillips N, Matsumura S, Tsai W-W, Kasper L, Brindle PK, Zhang G, Zhou M-M, Wright PE, Montminy M (2014) Combinatorial regulation of a signal-dependent activator by phosphorylation and acetylation. Proc Natl Acad Sci U S A 111 (48):17116–17121. https://doi.org/10.1073/pnas.1420389111
35. Miller TCR, Simon B, Rybin V, Groetsch H, Curtet S, Khochbin S, Carlomagno T, Mueller CW (2016) A bromodomain-DNA interaction facilitates acetylation-dependent bivalent nucleosome recognition by the BET protein BRDT. Nat Commun 7. https://doi.org/10.1038/ncomms13855
36. Mujtaba S, He Y, Zeng L, Yan S, Plotnikova O, Sachchidanand, Sanchez R, Zeleznik-Le NJ, Ronai Z, Zhou MM (2004) Structural mechanism of the bromodomain of the coactivator CBP in p53 transcriptional activation. Mol Cell 13(2):251–263. https://doi.org/10.1016/s1097-2765(03)00528-8
37. Mujtaba S, He Y, Zeng L, Farooq A, Carlson JE, Ott M, Verdin E, Zhou MM (2002) Structural basis of lysine-acetylated HIV-1 Tat recognition by PCAF bromodomain. Mol Cell 9(3):575–586. https://doi.org/10.1016/s1097-2765(02)00483-5
38. Huang B, Yang X-D, Zhou M-M, Ozato K, Chen L-F (2009) Brd4 coactivates transcriptional activation of NF-kappa B via specific binding to acetylated RelA. Mol Cell Biol 29(5):1375–1387. https://doi.org/10.1128/mcb.01365-08
39. Shi J, Wang YF, Zeng L, Wu YD, Deng J, Zhang Q, Lin YW, Li JL, Kang TB, Tao M, Rusinova E, Zhang GT, Wang C, Zhu HN, Yao J, Zeng YX, Evers BM, Zhou MM, Zhou BHP (2014) Disrupting the interaction of BRD4 with diacetylated twist suppresses tumorigenesis in basal-like breast cancer. Cancer Cell 25(2):210–225. https://doi.org/10.1016/j.ccr.2014.01.028
40. Asangani IA, Dommeti VL, Wang XJ, Malik R, Cieslik M, Yang RD, Escara-Wilke J, Wilder-Romans K, Dhanireddy S, Engelke C, Iyer MK, Jing XJ, Wu YM, Cao XH, Qin ZHS, Wang SM, Feng FY, Chinnaiyan AM (2014) Therapeutic targeting of BET

bromodomain proteins in castration-resistant prostate cancer. Nature 510(7504):278. https://doi.org/10.1038/nature13229
41. Hu P, Wang X, Zhang B, Zhang S, Wang Q, Wang Z (2014) Fluorescence polarization for the evaluation of small-molecule inhibitors of PCAF BRD/Tat-AcK50 association. ChemMedChem 9(5):928–931. https://doi.org/10.1002/cmdc.201300499
42. Moustakim M, Clark PGK, Trulli L, de Arriba ALF, Ehebauer MT, Chaikuad A, Murphy EJ, Mendez-Johnson J, Daniels D, Hou CFD, Lin YH, Walker JR, Hui R, Yang HB, Dorrell L, Rogers CM, Monteiro OP, Fedorov O, Huber KVM, Knapp S, Heer J, Dixon DJ, Brennan PE (2017) Discovery of a PCAF bromodomain chemical probe. Angew Chem Int Ed 56(3):827–831. https://doi.org/10.1002/anie.201610816
43. Picaud S, Leonards K, Lambert J-P, Dovey O, Wells C, Fedorov O, Monteiro O, Fujisawa T, Wang C-Y, Lingard H, Tallant C, Nikbin N, Guetzoyan L, Ingham R, Ley SV, Brennan P, Muller S, Samsonova A, Gingras A-C, Schwaller J, Vassiliou G, Knapp S, Filippakopoulos P (2016) Promiscuous targeting of bromodomains by bromosporine identifies BET proteins as master regulators of primary transcription response in leukemia. Sci Adv 2(10):e1600760. https://doi.org/10.1126/sciadv.1600760
44. Humphreys PG, Bamborough P, Chung C-W, Crags PD, Gordon L, Grandi P, Hayhow TG, Hussain J, Jones KL, Lindon M, Michon A-M, Renaux JF, Suckling CJ, Tough DF, Prinjha RK (2017) Discovery of a potent, cell penetrant, and selective p300/CBP-associated factor (PCAF)/general control nonderepressible 5 (GCN5) bromodomain chemical probe. J Med Chem 60(2):695–709. https://doi.org/10.1021/acs.jmedchem.6b01566
45. Goodman RH, Smolik S (2000) CBP/p300 in cell growth, transformation, and development. Genes Dev 14(13):1553–1577
46. Zhang ZW, Hofmann C, Casanova E, Schutz G, Lutz B (2004) Generation of a conditional allele of the CBP gene in mouse. Genesis 40(2):82–89. https://doi.org/10.1002/gene.20068
47. Korzus E (2017) Rubinstein-Taybi syndrome and epigenetic alterations. Adv Exp Med Biol 978:39–62. https://doi.org/10.1007/978-3-319-53889-1_3
48. Sachchidanand, Resnick-Silverman L, Yan S, Mutjaba S, Liu WJ, Zeng L, Manfredi JJ, Zhou MM (2006) Target structure-based discovery of small molecules that block human p53 and CREB binding protein association. Chem Biol 13(1):81–90. https://doi.org/10.1016/j.chembiol.2005.10.014
49. Shankar DB, Cheng JC, Sakamoto KM (2005) Role of cyclic AMP response element binding protein in human leukemias. Cancer 104(9):1819–1824. https://doi.org/10.1002/cncr.21401
50. Iyer NG, Ozdag H, Caldas C (2004) p300/CBP and cancer. Oncogene 23(24):4225–4231. https://doi.org/10.1038/sj.onc.1207118
51. Gerona-Navarro G, Yoel R, Mujtaba S, Frasca A, Patel J, Zeng L, Plotnikov AN, Osman R, Zhou M-M (2011) Rational design of cyclic peptide modulators of the transcriptional coactivator CBP: a new class of p53 inhibitors. J Am Chem Soc 133(7):2040–2043. https://doi.org/10.1021/ja107761h
52. Borah JC, Mujtaba S, Karakikes I, Zeng L, Muller M, Patel J, Moshkina N, Morohashi K, Zhang W, Gerona-Navarro G, Hajjar RJ, Zhou M-M (2011) A small molecule binding to the coactivator CREB-binding protein blocks apoptosis in cardiomyocytes. Chem Biol 18(4):531–541. https://doi.org/10.1016/j.chembiol.2010.12.021
53. Fedorov O, Lingard H, Wells C, Monteiro OP, Picaud S, Keates T, Yapp C, Philpott M, Martin SJ, Felletar I, Marsden BD, Filippakopoulos P, Mueller S, Knapp S, Brennan PE (2014) 1,2,4 Triazolo 4,3-a phthalazines: inhibitors of diverse bromodomains. J Med Chem 57(2):462–476. https://doi.org/10.1021/jm401568s
54. Rooney TPC, Filippakopoulos P, Fedorov O, Picaud S, Cortopassi WA, Hay DA, Martin S, Tumber A, Rogers CM, Philpott M, Wang M, Thompson AL, Heightman TD, Pryde DC, Cook A, Paton RS, Mueller S, Knapp S, Brennan PE, Conway SJ (2014) A series of potent CREBBP bromodomain ligands reveals an induced-fit pocket stabilized by a cation-pi interaction. Angew Chem Int Ed 53(24):6126–6130. https://doi.org/10.1002/anie.201402750
55. Picaud S, Fedorov O, Thanasopoulou A, Leonards K, Jones K, Meier J, Olzscha H, Monteiro O, Martin S, Philpott M, Tumber A, Filippakopoulos P, Yapp C, Wells C, Che KH, Bannister A, Robson S, Kumar U, Parr N, Lee K, Lugo D, Jeffrey P, Taylor S, Vecellio

ML, Bountra C, Brennan PE, O'Mahony A, Velichko S, Mueller S, Hay D, Daniels DL, Urh M, La Thangue NB, Kouzarides T, Prinjha R, Schwaller J, Knapp S (2015) Generation of a selective small molecule inhibitor of the CBP/p300 bromodomain for leukemia therapy. Cancer Res 75(23):5106–5119. https://doi.org/10.1158/0008-5472.Can-15-0236
56. Perez-Salvia M, Esteller M (2017) Bromodomain inhibitors and cancer therapy: from structures to applications. Epigenetics 12(5):323–339. https://doi.org/10.1080/15592294.2016.1265710
57. Ember SWJ, Zhu JY, Olesen SH, Martin MP, Becker A, Berndt N, Georg GI, Schonbrunn E (2014) Acetyl-lysine binding site of bromodomain-containing protein 4 (BRD4) interacts with diverse kinase inhibitors. ACS Chem Biol 9(5):1160–1171. https://doi.org/10.1021/cb500072z
58. Ciceri P, Muller S, O'Mahony A, Fedorov O, Filippakopoulos P, Hunt JP, Lasater EA, Pallares G, Picaud S, Wells C, Martin S, Wodicka LM, Shah NP, Treiber DK, Knapp S (2014) Dual kinase-bromodomain inhibitors for rationally designed polypharmacology. Nat Chem Biol 10(4):305–312. https://doi.org/10.1038/nchembio.1471
59. Ferguson FM, Fedorov O, Chaikuad A, Philpott M, Muniz JRC, Felletar I, von Delft F, Heightman T, Knapp S, Abell C, Ciulli A (2013) Targeting low-druggability bromodomains: fragment based screening and inhibitor design against the BAZ2B bromodomain. J Med Chem 56(24):10183–10187. https://doi.org/10.1021/jm401582c
60. Ouyang L, Zhang L, Liu J, Fu L, Yao D, Zhao Y, Zhang S, Wang G, He G, Liu B (2017) Discovery of a small-molecule bromodomain-containing protein 4 (BRD4) inhibitor that induces AMP-activated protein kinase-modulated autophagy-associated cell death in breast cancer. J Med Chem 60(24):9990–10012. https://doi.org/10.1021/acs.jmedchem.7b00275
61. Zhang GT, Plotnikov AN, Rusinova E, Shen T, Morohashi K, Joshua J, Zeng L, Mujtaba S, Ohlmeyer M, Zhou MM (2013) Structure-guided design of potent diazobenzene inhibitors for the BET bromodomains. J Med Chem 56(22):9251–9264. https://doi.org/10.1021/jm401334s
62. Gacias M, Gerona-Navarro G, Plotnikov AN, Zhang G, Zeng L, Kaur J, Moy G, Rusinova E, Rodriguez Y, Matikainen B, Vincek A, Joshua J, Casaccia P, Zhou M-M (2014) Selective chemical modulation of gene transcription favors oligodendrocyte lineage progression. Chem Biol 21(7):841–854. https://doi.org/10.1016/j.chembiol.2014.05.009
63. Picaud S, Wells C, Felletar I, Brotherton D, Martin S, Savitsky P, Diez-Dacal B, Philpott M, Bountra C, Lingard H, Fedorov O, Mueller S, Brennan PE, Knapp S, Filippakopoulos P (2013) RVX-208, an inhibitor of BET transcriptional regulators with selectivity for the second bromodomain. Proc Natl Acad Sci U S A 110(49):19754–19759. https://doi.org/10.1073/pnas.1310658110
64. Kharenko OA, Gesner EM, Patel RG, Norek K, White A, Fontano E, Suto RK, Young PR, McLure KG, Hansen HC (2016) RVX-297-a novel BD2 selective inhibitor of BET bromodomains. Biochem Biophys Res Commun 477(1):62–67. https://doi.org/10.1016/j.bbrc.2016.06.021
65. Endo J, Hikawa H, Hamada M, Ishibuchi S, Fujie N, Sugiyama N, Tanaka M, Kobayashi H, Sugahara K, Oshita K, Iwata K, Ooike S, Murata M, Sumichika H, Chiba K, Adachi K (2016) A phenotypic drug discovery study on thienodiazepine derivatives as inhibitors of T cell proliferation induced by CD28 co-stimulation leads to the discovery of a first bromodomain inhibitor. Bioorg Med Chem Lett 26(5):1365–1370. https://doi.org/10.1016/j.bmcl.2016.01.084
66. Zuber J, Shi JW, Wang E, Rappaport AR, Herrmann H, Sison EA, Magoon D, Qi J, Blatt K, Wunderlich M, Taylor MJ, Johns C, Chicas A, Mulloy JC, Kogan SC, Brown P, Valent P, Bradner JE, Lowe SW, Vakoc CR (2011) RNAi screen identifies Brd4 as a therapeutic target in acute myeloid leukaemia. Nature 478(7370):524–U124. https://doi.org/10.1038/nature10334
67. Delmore JE, Issa GC, Lemieux ME, Rahl PB, Shi JW, Jacobs HM, Kastritis E, Gilpatrick T, Paranal RM, Qi J, Chesi M, Schinzel AC, McKeown MR, Heffernan TP, Vakoc CR, Bergsagel PL, Ghobrial IM, Richardson PG, Young RA, Hahn WC, Anderson KC, Kung AL, Bradner JE, Mitsiades CS (2011) BET bromodomain inhibition as a therapeutic strategy to target c-Myc. Cell 146(6):903–916. https://doi.org/10.1016/j.cell.2011.08.017

68. Loven J, Hoke HA, Lin CY, Lau A, Orlando DA, Vakoc CR, Bradner JE, Lee TI, Young RA (2013) Selective inhibition of tumor oncogenes by disruption of super-enhancers. Cell 153(2):320–334. https://doi.org/10.1016/j.cell.2013.03.036
69. Berkovits BD, Wolgemuth DJ (2011) The first bromodomain of the testis-specific double bromodomain protein Brdt is required for chromocenter organization that is modulated by genetic background. Dev Biol 360(2):358–368. https://doi.org/10.1016/j.ydbio.2011.10.005
70. Devaiah BN, Lewis BA, Cherman N, Hewitt MC, Albrecht BK, Robey PG, Ozato K, Sims III RJ, Singer DS (2012) BRD4 is an atypical kinase that phosphorylates Serine2 of the RNA polymerase II carboxy-terminal domain. Proc Natl Acad Sci U S A 109(18):6927–6932. https://doi.org/10.1073/pnas.1120422109
71. Urick AK, Hawk LML, Cassel MK, Mishra NK, Liu S, Adhikari N, Zhang W, Dos Santos CO, Hall JL, Pomerantz WCK (2015) Dual screening of BPTF and Brd4 using protein-observed fluorine NMR uncovers new bromodomain probe molecules. ACS Chem Biol 10 (10):2246–2256. https://doi.org/10.1021/acschembio.5b00483
72. Andrews FH, Singh AR, Joshi S, Smith CA, Morales GA, Garlich JR, Durden DL, Kutateladze TG (2017) Dual-activity PI3K-BRD4 inhibitor for the orthogonal inhibition of MYC to block tumor growth and metastasis. Proc Natl Acad Sci U S A 114(7):E1072–E1080. https://doi.org/10.1073/pnas.1613091114
73. Carlino L, Rastelli G (2016) Dual kinase-bromodomain inhibitors in anticancer drug discovery: a structural and pharmacological perspective. J Med Chem 59(20):9305–9320. https://doi.org/10.1021/acs.jmedchem.6b00438
74. Shadrick WR, Slavish PJ, Chai SC, Connelly M, Low JA, Young BM, Bharatham N, Boyd VA, Chen T, Lee RE, Kiplin GR, Potter PM, Waddell B, Tallant C, Knapp S, Morfouace M, Roussel MF, Shelat AA (2018) Exploiting a water network to achieve enthalpy-driven, bromodomain-selective BET inhibitors. Bioorg Med Chem 26:25–36
75. Vidler LR, Filippakopoulos P, Fedorov O, Picaud S, Martin S, Tomsett M, Woodward H, Brown N, Knapp S, Hoelder S (2013) Discovery of novel small-molecule inhibitors of BRD4 using structure-based virtual screening. J Med Chem 56(20):8073–8088. https://doi.org/10.1021/jm4011302
76. Marchand J-R, Vedove AD, Lolli G, Caflisch A (2017) Discovery of inhibitors of four bromodomains by fragment-anchored ligand docking. J Chem Inf Model 57(10):2584–2597. https://doi.org/10.1021/acs.jcim.7b00336
77. Aldeghi M, Heifetz A, Bodkin MJ, Knapp S, Biggin PC (2017) Predictions of ligand selectivity from absolute binding free energy calculations. J Am Chem Soc 139(2):946–957. https://doi.org/10.1021/jacs.6b11467
78. Giblin KA, Hughes SJ, Boyd H, Hansson P, Bender A (2018) Prospectively validated proteochemometric models for the prediction of small-molecule binding to bromodomain proteins. J Chem Inf Model 58(9):1870–1888. https://doi.org/10.1021/acs.jcim.8b00400
79. Allen BK, Mehta S, Ember SWJ, Zhu J-Y, Schönbrunn E, Ayad NG, Schürer SC (2017) Identification of a novel class of BRD4 inhibitors by computational screening and binding simulations. ACS Omega 2(8):4760–4771. https://doi.org/10.1021/acsomega.7b00553
80. Ayoub AM, Hawk LML, Herzig RJ, Jiang J, Wisniewski AJ, Gee CT, Zhao P, Zhu JY, Berndt N, Offei-Addo NK, Scott TG, Qi J, Bradner JE, Ward TR, Schonbrunn E, Georg GI, Pomerantz WCK (2017) BET bromodomain inhibitors with one-step synthesis discovered from virtual screen. J Med Chem 60(12):4805–4817. https://doi.org/10.1021/acs.jmedchem.6b01336
81. Hasvold LA, Sheppard GS, Wang L, Fidanze SD, Liu D, Pratt JK, Mantei RA, Wada CK, Hubbard R, Shen Y, Lin X, Huang X, Warder SE, Wilcox D, Li L, Buchanan FG, Smithee L, Albert DH, Magoc TJ, Park CH, Petros AM, Panchal SC, Sun C, Kovar P, Soni NB, Elmore SW, Kati WM, McDaniel KF (2017) Methylpyrrole inhibitors of BET bromodomains. Bioorg Med Chem Lett 27(10):2225–2233. https://doi.org/10.1016/j.bmcl.2017.02.057
82. Harner MJ, Chauder BA, Phan J, Fesik SW (2014) Fragment-based screening of the bromodomain of ATAD2. J Med Chem 57(22):9687–9692. https://doi.org/10.1021/jm501035j

83. Yu J-L, Chen T-T, Zhou C, Lian F-L, Tang X-L, Wen Y, Shen J-K, Xu Y-C, Xiong B, Zhang N-X (2016) NMR-based platform for fragment-based lead discovery used in screening BRD4-targeted compounds. Acta Pharmacol Sin 37(7):984–993. https://doi.org/10.1038/aps.2016.19
84. Poplawski A, Hu K, Lee W, Natesan S, Peng D, Carlson S, Shi X, Balaz S, Markley JL, Glass KC (2014) Molecular insights into the recognition of N-terminal histone modifications by the BRPF1 bromodomain. J Mol Biol 426(8):1661–1676. https://doi.org/10.1016/j.jmb.2013.12.007
85. Charlop-Powers Z, Zeng L, Zhang Q, Zhou M-M (2010) Structural insights into selective histone H3 recognition by the human polybromo bromodomain 2. Cell Res 20:529. https://doi.org/10.1038/cr.2010.43
86. Liu Y, Wang X, Zhang J, Huang H, Ding B, Wu J, Shi Y (2008) Structural basis and binding properties of the second bromodomain of Brd4 with acetylated histone tails. Biochemistry 47(24):6403–6417. https://doi.org/10.1021/bi8001659
87. Sun H, Liu J, Zhang J, Shen W, Huang H, Xu C, Dai H, Wu J, Shi Y (2007) Solution structure of BRD7 bromodomain and its interaction with acetylated peptides from histone H3 and H4. Biochem Biophys Res Commun 358(2):435–441. https://doi.org/10.1016/j.bbrc.2007.04.139
88. Ferguson FM, Dias DM, Rodrigues JPGLM, Wienk H, Boelens R, Bonvin AMJJ, Abell C, Ciulli A (2014) Binding hotspots of BAZ2B bromodomain: histone interaction revealed by solution NMR driven docking. Biochemistry 53(42):6706–6716. https://doi.org/10.1021/bi500909d
89. Marsh ENG, Suzuki Y (2014) Using 19F NMR to probe biological interactions of proteins and peptides. ACS Chem Biol 9(6):1242–1250. https://doi.org/10.1021/cb500111u
90. Gerig JT. (2001) Fluorine NMR. Online textbook; http://www.biophysics.org/img/jtg2001-2.pdf
91. Moreira IS, Martins JM, Ramos RM, Fernandes PA, Ramos MJ (2013) Understanding the importance of the aromatic amino-acid residues as hot-spots. Biochim Biophys Acta 1834(1):404–414. https://doi.org/10.1016/j.bbapap.2012.07.005
92. Bogan AA, Thorn KS (1998) Anatomy of hot spots in protein interfaces11Edited by J. Wells. J Mol Biol 280(1):1–9. https://doi.org/10.1006/jmbi.1998.1843
93. Arntson KE, Pomerantz WCK (2016) Protein-observed fluorine NMR: a bioorthogonal approach for small molecule discovery. J Med Chem 59(11):5158–5171. https://doi.org/10.1021/acs.jmedchem.5b01447
94. Urick AK, Calle LP, Espinosa JF, Hu H, Pomerantz WCK (2016) Protein-observed fluorine NMR is a complementary ligand discovery method to 1H CPMG ligand-observed NMR. ACS Chem Biol 11(11):3154–3164. https://doi.org/10.1021/acschembio.6b00730
95. Zartler ER, Hanson J, Jones BE, Kline AD, Martin G, Mo H, Shapiro MJ, Wang R, Wu H, Yan J (2003) RAMPED-UP NMR: multiplexed NMR-based screening for drug discovery. J Am Chem Soc 125(36):10941–10946. https://doi.org/10.1021/ja0348593
96. Frey WD, Chaudhry A, Slepicka PF, Ouellette AM, Kirberger SE, Pomerantz WCK, Hannon GJ, dos Santos CO (2017) BPTF maintains chromatin accessibility and the self-renewal capacity of mammary gland stem cells. Stem Cell Rep 9(1):23–31. https://doi.org/10.1016/j.stemcr.2017.04.031
97. Mayer M, Meyer B (2001) Group epitope mapping by saturation transfer difference NMR to identify segments of a ligand in direct contact with a protein receptor. J Am Chem Soc 123(25):6108–6117. https://doi.org/10.1021/ja0100120
98. Geist L, Mayer M, Cockcroft X-L, Wolkerstorfer B, Kessler D, Engelhardt H, McConnell DB, Konrat R (2017) Direct NMR probing of hydration shells of protein ligand interfaces and its application to drug design. J Med Chem 60(21):8708–8715. https://doi.org/10.1021/acs.jmedchem.7b00845
99. Spiliotopoulos D, Wamhoff E-C, Lolli G, Rademacher C, Caflisch A (2017) Discovery of BAZ2A bromodomain ligands. Eur J Med Chem 139(Supplement C):564–572. https://doi.org/10.1016/j.ejmech.2017.08.028

100. Spiliotopoulos D, Zhu J, Wamhoff E-C, Deerain N, Marchand J-R, Aretz J, Rademacher C, Caflisch A (2017) Virtual screen to NMR (VS2NMR): discovery of fragment hits for the CBP bromodomain. Bioorg Med Chem Lett 27(11):2472–2478. https://doi.org/10.1016/j.bmcl.2017.04.001
101. Wang N, Li F, Bao H, Li J, Wu J, Ruan K (2016) NMR fragment screening hit induces plasticity of BRD7/9 bromodomains. Chembiochem 17(15):1456–1463. https://doi.org/10.1002/cbic.201600184
102. Gossert AD, Jahnke W (2016) NMR in drug discovery: a practical guide to identification and validation of ligands interacting with biological macromolecules. Prog Nucl Magn Reson Spectrosc 97(Supplement C):82–125. https://doi.org/10.1016/j.pnmrs.2016.09.001
103. Perell GT, Mishra NK, Sudhamalla B, Ycas PD, Islam K, Pomerantz WCK (2017) Specific acetylation patterns of H2A.Z form transient interactions with the BPTF bromodomain. Biochemistry 56(35):4607–4615. https://doi.org/10.1021/acs.biochem.7b00648
104. Navratilova I, Aristotelous T, Hopkins AL, Picaud S, Chaikuad A, Knapp S, Filappakopoulos P (2016) Discovery of new bromodomain scaffolds by biosensor fragment screening. ACS Med Chem Lett 7(12):1213–1218
105. Picaud S, Da Costa D, Thanasopoulou A, Filippakopoulos P, Fish PV, Philpott M, Fedorov O, Brennan P, Bunnage ME, Owen DR, Bradner JE, Taniere P, O'Sullivan B, Müller S, Schwaller J, Stankovic T, Knapp S (2013) PFI-1, a highly selective protein interaction inhibitor, targeting BET bromodomains. Cancer Res 73(11):3336–3346. https://doi.org/10.1158/0008-5472.can-12-3292
106. Ember SWJ, Zhu J-Y, Olesen SH, Martin MP, Becker A, Berndt N, Georg GI, Schönbrunn E (2014) Acetyl-lysine binding site of bromodomain-containing protein 4 (BRD4) interacts with diverse kinase inhibitors. ACS Chem Biol 9(5):1160–1171. https://doi.org/10.1021/cb500072z
107. Gerstenberger BS, Trzupek JD, Tallant C, Fedorov O, Filippakopoulos P, Brennan PE, Fedele V, Martin S, Picaud S, Rogers C, Parikh M, Taylor A, Samas B, O'Mahony A, Berg E, Pallares G, Torrey AD, Treiber DK, Samardjiev IJ, Nasipak BT, Padilla-Benavides T, Wu Q, Imbalzano AN, Nickerson JA, Bunnage ME, Müller S, Knapp S, Owen DR (2016) Identification of a chemical probe for family VIII bromodomains through optimization of a fragment hit. J Med Chem 59(10):4800–4811. https://doi.org/10.1021/acs.jmedchem.6b00012
108. Chaikuad A, Lang S, Brennan PE, Temperini C, Fedorov O, Hollander J, Nachane R, Abell C, Müller S, Siegal G, Knapp S (2016) Structure-based identification of inhibitory fragments targeting the p300/CBP-associated factor bromodomain. J Med Chem 59(4):1648–1653. https://doi.org/10.1021/acs.jmedchem.5b01719
109. Martin LJ, Koegl M, Bader G, Cockcroft X-L, Fedorov O, Fiegen D, Gerstberger T, Hofmann MH, Hohmann AF, Kessler D, Knapp S, Knesl P, Kornigg S, Müller S, Nar H, Rogers C, Rumpel K, Schaaf O, Steurer S, Tallant C, Vakoc CR, Zeeb M, Zoephel A, Pearson M, Boehmelt G, McConnell D (2016) Structure-based design of an in vivo active selective BRD9 inhibitor. J Med Chem 59(10):4462–4475. https://doi.org/10.1021/acs.jmedchem.5b01865
110. Chen P, Chaikuad A, Bamborough P, Bantscheff M, Bountra C, Chung C-w, Fedorov O, Grandi P, Jung D, Lesniak R, Lindon M, Müller S, Philpott M, Prinjha R, Rogers C, Selenski C, Tallant C, Werner T, Willson TM, Knapp S, Drewry DH (2016) Discovery and characterization of GSK2801, a selective chemical probe for the bromodomains BAZ2A and BAZ2B. J Med Chem 59(4):1410–1424. https://doi.org/10.1021/acs.jmedchem.5b00209
111. Bamborough P, Diallo H, Goodacre JD, Gordon L, Lewis A, Seal JT, Wilson DM, Woodrow MD, Chung C-w (2012) Fragment-based discovery of bromodomain inhibitors part 2: optimization of phenylisoxazole sulfonamides. J Med Chem 55(2):587–596. https://doi.org/10.1021/jm201283q
112. Chung C-w, Dean AW, Woolven JM, Bamborough P (2012) Fragment-based discovery of bromodomain inhibitors part 1: inhibitor binding modes and implications for lead discovery. J Med Chem 55(2):576–586. https://doi.org/10.1021/jm201320w
113. Molina DM, Jafari R, Ignatushchenko M, Seki T, Larsson EA, Dan C, Sreekumar L, Cao Y, Nordlund P (2013) Monitoring drug target engagement in cells and tissues using the cellular thermal shift assay. Science 341(6141):84–87. https://doi.org/10.1126/science.1233606

114. Jafari R, Almqvist H, Axelsson H, Ignatushchenko M, Lundbäck T, Nordlund P, Molina DM (2014) The cellular thermal shift assay for evaluating drug target interactions in cells. Nat Protoc 9:2100. https://doi.org/10.1038/nprot.2014.138. https://www.nature.com/articles/nprot.2014.138#supplementary-information
115. Philpott M, Yang J, Tumber T, Fedorov O, Uttarkar S, Filippakopoulos P, Picaud S, Keates T, Felletar I, Ciulli A, Knapp S, Heightman TD (2011) Bromodomain-peptide displacement assays for interactome mapping and inhibitor discovery. Mol BioSyst 7(10):2899–2908. https://doi.org/10.1039/c1mb05099k
116. Roberts JM, Bradner JE (2015) A bead-based proximity assay for BRD4 ligand discovery. Curr Protoc Chem Biol 7(4):263–278. https://doi.org/10.1002/9780470559277.ch150024
117. Fedorov O, Lingard H, Wells C, Monteiro OP, Picaud S, Keates T, Yapp C, Philpott M, Martin SJ, Felletar I, Marsden BD, Filippakopoulos P, Muller S, Knapp S, Brennan PE (2014) [1,2,4]Triazolo[4,3-a]phthalazines: inhibitors of diverse bromodomains. J Med Chem 57(2):462–476. https://doi.org/10.1021/jm401568s
118. Ciceri P, Mueller S, O'Mahony A, Fedorov O, Filippakopoulos P, Hunt JP, Lasater EA, Pallares G, Picaud S, Wells C, Martin S, Wodicka LM, Shah NP, Treiber DK, Knapp S (2014) Dual kinase-bromodomain inhibitors for rationally designed polypharmacology. Nat Chem Biol 10(4):305–312. https://doi.org/10.1038/nchembio.1471
119. Crawford TD, Audia JE, Bellon S, Burdick DJ, Bommi-Reddy A, Cote A, Cummings RT, Duplessis M, Flynn EM, Hewitt M, Huang H-R, Jayaram H, Jiang Y, Joshi S, Kiefer JR, Murray J, Nasveschuk CG, Neiss A, Pardo E, Romero FA, Sandy P, Sims RJ, Tang Y, Taylor AM, Tsui V, Wang J, Wang S, Wang Y, Xu Z, Zawadzke L, Zhu X, Albrecht BK, Magnuson SR, Cochran AG (2017) GNE-886: a potent and selective inhibitor of the cat eye syndrome chromosome region candidate 2 bromodomain (CECR2). ACS Med Chem Lett 8(7):737–741. https://doi.org/10.1021/acsmedchemlett.7b00132
120. Chen P, Chaikuad A, Bamborough P, Bantscheff M, Bountra C, Chung C-w, Fedorov O, Grandi P, Jung D, Lesniak R, Lindon M, Muller S, Philpott M, Prinjha R, Rogers C, Selenski C, Tallant C, Werner T, Willson TM, Knapp S, Drewry DH (2016) Discovery and characterization of GSK2801, a selective chemical probe for the bromodomains BAZ2A and BAZ2B. J Med Chem 59(4):1410–1424. https://doi.org/10.1021/acs.jmedchem.5b00209
121. Crawford TD, Romero FA, Lai KW, Tsui V, Taylor AM, de Leon Boenig G, Noland CL, Murray J, Ly J, Choo EF, Hunsaker TL, Chan EW, Merchant M, Kharbanda S, Gascoigne KE, Kaufman S, Beresini MH, Liao J, Liu W, Chen KX, Chen Z, Conery AR, Cote A, Jayaram H, Jiang Y, Kiefer JR, Kleinheinz T, Li Y, Maher J, Pardo E, Poy F, Spillane KL, Wang F, Wang J, Wei X, Xu Z, Xu Z, Yen I, Zawadzke L, Zhu X, Bellon S, Cummings R, Cochran AG, Albrecht BK, Magnuson S (2016) Discovery of a potent and selective in vivo probe (GNE-272) for the bromodomains of CBP/EP300. J Med Chem 59(23):10549–10563. https://doi.org/10.1021/acs.jmedchem.6b01022
122. Humphreys PG, Bamborough P, Chung C-w, Craggs PD, Gordon L, Grandi P, Hayhow TG, Hussain J, Jones KL, Lindon M, Michon A-M, Renaux JF, Suckling CJ, Tough DF, Prinjha RK (2017) Discovery of a potent, cell penetrant, and selective p300/CBP-associated factor (PCAF)/general control nonderepressible 5 (GCN5) bromodomain chemical probe. J Med Chem 60(2):695–709. https://doi.org/10.1021/acs.jmedchem.6b01566
123. Theodoulou NH, Bamborough P, Bannister AJ, Becher I, Bit RA, Che KH, Chung C-w, Dittmann A, Drewes G, Drewry DH, Gordon L, Grandi P, Leveridge M, Lindon M, Michon A-M, Molnar J, Robson SC, Tomkinson NCO, Kouzarides T, Prinjha RK, Humphreys PG (2016) Discovery of I-BRD9, a selective cell active chemical probe for bromodomain containing protein 9 inhibition. J Med Chem 59(4):1425–1439. https://doi.org/10.1021/acs.jmedchem.5b00256
124. Kenakin TP (1993) Pharmacological analysis of drug-receptor interaction. Raven, New York
125. Chung C-W, Coste H, White JH, Mirguet O, Wilde J, Gosmini RL, Delves C, Magny SM, Woodward R, Hughes SA, Boursier EV, Flynn H, Bouillot AM, Bamborough P, Brusq J-MG, Gellibert FJ, Jones EJ, Riou AM, Homes P, Martin SL, Uings IJ, Toum J, Clément CA,

Boullay A-B, Grimley RL, Blandel FM, Prinjha RK, Lee K, Kirilovsky J, Nicodeme E (2011) Discovery and characterization of small molecule inhibitors of the BET family bromodomains. J Med Chem 54(11):3827–3838. https://doi.org/10.1021/jm200108t

126. Xinyi H (2003) Fluorescence polarization competition assay: the range of resolvable inhibitor potency is limited by the affinity of the fluorescent ligand. J Biomol Screen 8(1):34–38. https://doi.org/10.1177/1087057102239666
127. Cox OB, Krojer T, Collins P, Monteiro O, Talon R, Bradley A, Fedorov O, Amin J, Marsden BD, Spencer J, von Delft F, Brennan PE (2016) A poised fragment library enables rapid synthetic expansion yielding the first reported inhibitors of PHIP(2), an atypical bromodomain. Chem Sci 7(3):2322–2330. https://doi.org/10.1039/C5SC03115J
128. Zhao L, Cao D, Chen T, Wang Y, Miao Z, Xu Y, Chen W, Wang X, Li Y, Du Z, Xiong B, Li J, Xu C, Zhang N, He J, Shen J (2013) Fragment-based drug discovery of 2-thiazolidinones as inhibitors of the histone reader BRD4 bromodomain. J Med Chem 56(10):3833–3851. https://doi.org/10.1021/jm301793a
129. Fish PV, Filippakopoulos P, Bish G, Brennan PE, Bunnage ME, Cook AS, Federov O, Gerstenberger BS, Jones H, Knapp S, Marsden B, Nocka K, Owen DR, Philpott M, Picaud S, Primiano MJ, Ralph MJ, Sciammetta N, Trzupek JD (2012) Identification of a chemical probe for bromo and extra C-terminal bromodomain inhibition through optimization of a fragment-derived hit. J Med Chem 55(22):9831–9837. https://doi.org/10.1021/jm3010515
130. Daguer JP, Zambaldo C, Abegg D, Barluenga S, Tallant C, Muller S, Adibekian A, Winssinger N (2015) Identification of covalent bromodomain binders through DNA display of small molecules. Angew Chem Int Ed Engl 54(20):6057–6061. https://doi.org/10.1002/anie.201412276
131. Rhyasen GW, Hattersley MM, Yao Y, Dulak A, Wang W, Petteruti P, Dale IL, Boiko S, Cheung T, Zhang J, Wen S, Castriotta L, Lawson D, Collins M, Bao L, Ahdesmaki MJ, Walker G, O'Connor G, Yeh TC, Rabow AA, Dry JR, Reimer C, Lyne P, Mills GB, Fawell SE, Waring MJ, Zinda M, Clark E, Chen H (2016) AZD5153: a novel bivalent BET bromodomain inhibitor highly active against hematologic malignancies. Mol Cancer Ther 15(11):2563–2574. https://doi.org/10.1158/1535-7163.mct-16-0141
132. Waring MJ, Chen H, Rabow AA, Walker G, Bobby R, Boiko S, Bradbury RH, Callis R, Clark E, Dale I, Daniels DL, Dulak A, Flavell L, Holdgate G, Jowitt TA, Kikhney A, McAlister M, Méndez J, Ogg D, Patel J, Petteruti P, Robb GR, Robers MB, Saif S, Stratton N, Svergun DI, Wang W, Whittaker D, Wilson DM, Yao Y (2016) Potent and selective bivalent inhibitors of BET bromodomains. Nat Chem Biol 12:1097. https://doi.org/10.1038/nchembio.2210. https://www.nature.com/articles/nchembio.2210#supplementary-information
133. Tanaka M, Roberts JM, Seo HS, Souza A, Paulk J, Scott TG, DeAngelo SL, Dhe-Paganon S, Bradner JE (2016) Design and characterization of bivalent BET inhibitors. Nat Chem Biol 12(12):1089–1096. https://doi.org/10.1038/nchembio.2209
134. Lu J, Qian Y, Altieri M, Dong H, Wang J, Raina K, Hines J, Winkler JD, Crew AP, Coleman K, Crews CM (2015) Hijacking the E3 ubiquitin ligase cereblon to efficiently target BRD4. Chem Biol 22(6):755–763. https://doi.org/10.1016/j.chembiol.2015.05.009
135. Zengerle M, Chan K-H, Ciulli A (2015) Selective small molecule induced degradation of the BET bromodomain protein BRD4. ACS Chem Biol 10(8):1770–1777. https://doi.org/10.1021/acschembio.5b00216
136. Winter GE, Buckley DL, Paulk J, Roberts JM, Souza A, Dhe-Paganon S, Bradner JE (2015) Drug development. Phthalimide conjugation as a strategy for in vivo target protein degradation. Science 348(6241):1376–1381. https://doi.org/10.1126/science.aab1433
137. Toure M, Crews CM (2016) Small-molecule PROTACS: new approaches to protein degradation. Angew Chem Int Ed 55(6):1966–1973. https://doi.org/10.1002/anie.201507978
138. Remillard D, Buckley DL, Paulk J, Dastjerdi S, Bradner JE, Brien GL, Armstrong SA, Sonnett M, Sonnett M, Wuhr M, Seo H-S, Dhe-Paganon S, Wuhr M (2017) Degradation of

the BAF complex factor BRD9 by heterobifunctional ligands. Angew Chem Int Ed Engl 56(21):5738–5743

139. Gechijian LN, Buckley DL, Lawlor MA, Reyes JM, Paulk J, Ott CJ, Winter GE, Erb MA, Scott TG, Xu M, Seo HS, Dhe-Paganon S, Kwiatkowski NP, Perry JA, Qi J, Gray NS, Bradner JE (2018) Functional TRIM24 degrader via conjugation of ineffectual bromodomain and VHL ligands. Nat Chem Biol. https://doi.org/10.1038/s41589-018-0010-y
140. Tekel SJ, Haynes KA (2017) Molecular structures guide the engineering of chromatin. Nucleic Acids Res 45(13):7555–7570. https://doi.org/10.1093/nar/gkx531

Top Med Chem (2020) 33: 339–400
DOI: 10.1007/7355_2019_78

Published online: 28 November 2019

Methyl-Readers and Inhibitors

Gianluca Sbardella

Contents

1 Introduction 340
2 Methyl-Reader Domains and Families 341
3 PHD Fingers 342
3.1 PHD Structure and Function 342
3.2 Small Molecules Targeting PHD Fingers 345
3.2.1 Targeting JARID1A/KDM5A PHD3 Fingers 345
3.2.2 Targeting Pygo PHD Fingers 346
3.2.3 Targeting BAZ-PHD Fingers 348
4 Protein Reader Domains of the Royal Family 348
4.1 General Characteristics of Royal Family Proteins 348
4.2 The Chromodomain 350
4.2.1 Chromodomain Structure and Function 350
4.2.2 Small Molecules Targeting Chromodomains 350
4.3 The Tudor Domain 356
4.3.1 Single Tudor Domain (PHF1, PHF19, PHF20) 356
4.3.2 Tandem Tudor Domains (53BP1, SGF29, SHH1, UHRF1, Spindlin-1) 357
4.3.3 Hybrid Tudor Domains (JMJ2A/KDM4A) 363
4.3.4 Tudor Domains that Bind Methylarginine Motifs 363
4.4 The MBT Domain 363
4.4.1 Proteins Containing Three MBT Repeats (L3MBTL1 and L3MBTL3) 364
4.4.2 Small Molecules Targeting MBT Domains 365
4.5 The PWWP Domain 371
4.5.1 Structure and Functions 371
5 The WD40 Repeat Domain 372
5.1 Recognition of the Histone H3 Tail by WDR5 373
5.2 Methyllysine-Specific Recognition by EED 374
5.3 Small Molecules Targeting WD40 Domains 375
5.3.1 Targeting WDR5 376
5.3.2 Targeting EED 378
6 Conclusions 382
References 383

G. Sbardella (✉)
Dipartimento di Farmacia, Epigenetic Med Chem Lab, Università di Salerno, Fisciano, SA, Italy
e-mail: gsbardella@unisa.it

Abstract Due to their prevalent role in epigenetic gene regulation, methyllysine and methylarginine domain readers have emerged as potential drug targets for small-molecule intervention. Within this book chapter, the biological role and the associated development of potent small molecules inhibiting the protein-protein interaction of methyllysine readers (Tudor, malignant brain tumor, chromo-, and PHD domain) will be discussed. The druggability of these readers and thus their potential to serve as targets for small-molecule ligands will be evaluated critically. Those domains (PWWP, WD40, ankyrin repeats, and ADD domains) which are not yet targeted will be evaluated for their biological actions and eventual therapeutic implications. To sum up, a comprehensive review of the state of the art for all relevant methyl-readers and their inhibitors if present will be given from a medicinal chemistry standpoint of view.

Keywords Chromatin domain, Epigenetic readers, Inhibitors, Malignant brain tumor domain, Methyl-readers, PHD domain, Royal Family domain, Tudor domain

1 Introduction

Methylation is perhaps the most versatile of all histone posttranslational modifications (PTMs). It can occur on both lysine and arginine residues, and each of them has three possible methylation states. Lysine can be mono (Kme1)-, di (Kme2)-, or trimethylated (Kme3) on its ε-amino group (Fig. 1). The dynamic methylation state of lysine residues is controlled by a balance in the activity of lysine methyltransferases (KMTs) [1], which transfer methyl groups from S-adenosylmethionine (SAM) to the lysine residue, and demethylases (KDMs), which remove methyl groups. Unlike lysine acetylation, methylation does not alter the charge of the amino acid side chain. Instead, there is only a gradual increase in size and lipophilicity from the non-methylated state to mono-, di-, and trimethylation [2–4]. In contrast to lysines, arginines are only mono- (Rme1) or dimethylated, but the dimethylation can be symmetrical (Rme2s) or asymmetrical (Rme2a; Fig. 1) [2, 5]. The three types of arginine methylation are catalyzed by a family of nine SAM-dependent enzymes called the protein arginine methyltransferases (PRMTs). Arginine demethylation activity has been reported for the JmjC domain-containing protein JMJD6 [6], even if this has been challenged by other studies [7]. The methylation of arginine changes its shape, does not alter the charge, but removes potential hydrogen bond donors, which would potentially inhibit certain interactions [8]. On the other hand, the methylation of arginine residues can also increase their affinity to aromatic rings in cation-π interactions, thus promoting other interactions [9].

A significant part of the biological outcome of Lys and/or Arg methylation is quite probably driven by specific reader proteins, which not only recognize the methylation marks but also distinguish between the various methylation states and result in downstream events, including the addition and removal of additional PTMs, chromatin condensation, or recruitment of transcriptional machinery leading

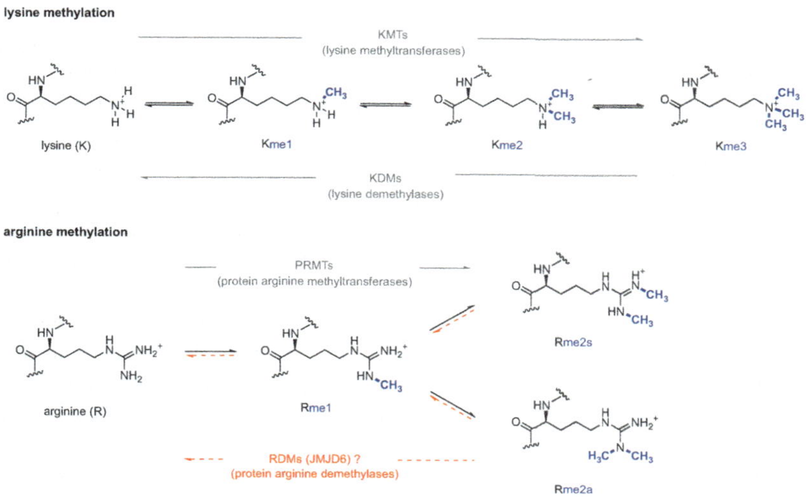

Fig. 1 Regulation of methylation of lysine and arginine residues in proteins. Mono-, di-, or tri-methylation (Kme1, Kme2, or Kme3) of lysine at the ε-amino group is catalyzed by lysine methyltransferases (KMTs) and can be erased by lysine demethylases (KDMs) (top panel). The protein arginine methyltransferases (PRMTs) can modify the guanidinium group of arginine residues and result in either monomethylated (Rme1) or symmetric (Rme2s) or asymmetric (Rme2a) dimethylated arginine (bottom panel). Presently, an arginine demethylation activity in vivo has been proposed for Jumonji domain-containing protein 6 (Jmjd6)

to gene transcription [4]. The crosstalk between methylated Lys and Arg and the resulting control over the dynamic addition and elimination of these PTMs is critical in governing chromatin structure and in the regulation of gene transcription [10]. Dysregulation of Lys and Arg methylation has been linked to breast, prostate, colon, bladder, and lung cancers [11], as well as leukemia, cardiovascular diseases, HIV, multiple sclerosis, and spinal muscular atrophy [11–13]. However, a comprehensive map of these PTMs and their function has not been established. Therefore, a deep understanding of the significance of Lys and Arg methylation and their recognition by reader proteins is of paramount importance to the development of inhibitors, probes, and therapeutics for diseases.

2 Methyl-Reader Domains and Families

After the discovery, made almost 30 years ago, that SRC homology 2 (SH2) domains bind to short protein motifs that are tyrosine phosphorylated [14], it became evident that different modular domains bind distinct types of PTMs [15]. Indeed, even if little is still known about the effect of lysine and arginine methylation on chromatin structure itself, a number of readers of these PTMs have been identified since 2001, when a chromodomain (CD) of HP1 was found to recognize histone H3K9me3 [16–18].

To date, methylated lysines on histone tails appear to be targeted by the largest and most diverse set of readers. This includes plant homeodomains (PHDs), WD-40 domains, chromatin organization modifier domains (chromodomains, CDs), double chromodomains (DCDs), chromo-barrel domains, ADD (ATRX-DNMT3-DNMT3L) domains, ankyrin repeats (ANKs), proline-tryptophan-tryptophan-proline (PWWP) domains, bromo adjacent homology (BAH) domains, chromo-barrel, chromodomain (CD), double chromodomain (DCD), HEAT, malignant brain tumor (MBT) domains, SAWADEE, Tudor domains, tandem Tudor domain (TTD), and zinc finger CW (zf-CW) domains [4, 8, 19–21]. Some methyllysine readers, including PHD and zf-CW, show a high degree of sequence specificity, while others, including MBT and WD40, are more promiscuous though they can select for a certain methylation state of a target lysine (Fig. 2). Despite the wide variety of the readers and histone targets, the majority of these domains have comparable binding affinities, with dissociation constants of the complexes being in the high nanomolar to the low micromolar range [19, 22].

On the other hand, the only protein domain family currently known to bind methylated arginine motifs is the Tudor family (although individual PHD and WD40 domains also harbor this ability) [8, 19].

The main feature of all methyllysine-binding domains identified so far is that they bind this PTM through an aromatic cage, typically formed by two to four aromatic residues. The exact composition and size of the pocket are responsible for the selectivity for mono-, di-, or trimethylated state of lysine [19]. Specificity for a particular methylated lysine is imparted by interaction with surrounding residues. Some histone readers show high degrees of specificity, whereas others are selective for only a certain methylation state and otherwise bind very promiscuously. Beyond caging of the methyllysine, the mechanism of recognition of surrounding residues varies among readers [19].

Like methyllysine, methylated arginine occupies an aromatic cage at the top of the β-barrel structure of a Tudor domain. However, this cage is much narrower than the cage for methyllysine and thus favors the planar guanidinium group [19]. Interestingly, the other structurally related members of the Royal superfamily, namely, chromodomain, chromo-barrel, MBT, PWWP, and TTD modules, which also feature the characteristic β-barrel topology, are not able to bind methylated arginine [8, 19].

3 PHD Fingers

3.1 PHD Structure and Function

The plant homeodomain (PHD) finger was discovered in the *Arabidopsis* protein HAT3.1 in 1993 [23] and has since been found in a variety of proteins implicated in the regulation of chromatin structure and dynamics. The PHD finger is an evolutionarily conserved zinc fingerlike motif which is present either as a single module or in multiple copies in 291 human proteins [24], and, therefore, it is

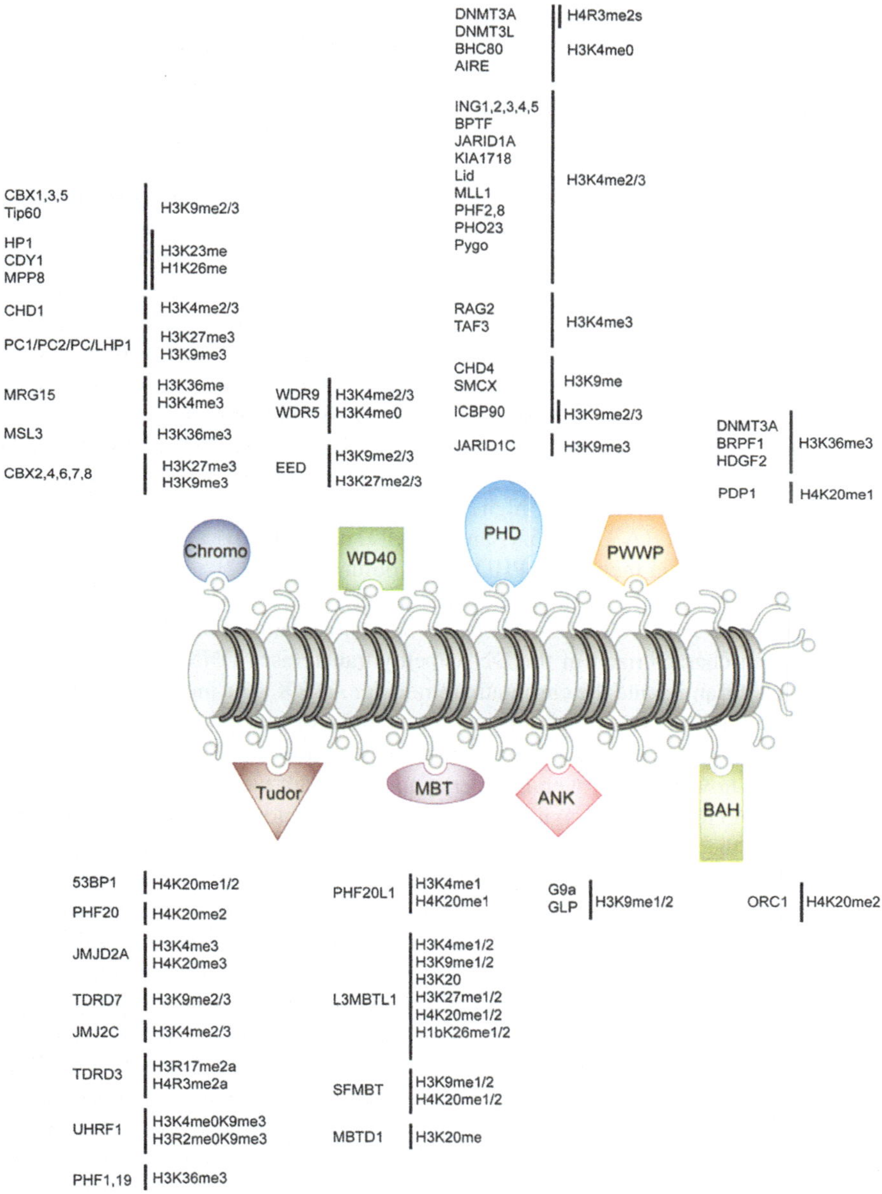

Fig. 2 Different families of methyl-binding proteins and their histone targets

considered one of the largest families among reader domains. The PHD finger is made up of 50–80 amino acid residues including a conserved Cys4-His-Cys3 (C4HC3) motif that binds two zinc ions, necessary for the structural integrity of the domain, in the so-called "cross-braced" manner [25].

A number of studies in the past few years reveal diverse biological roles of the PHD fingers. They recognize unmodified and modified histone H3 tails, interact with nonhistone proteins, and associate with DNA [25, 26], the specificity of binding being dictated by the properties of the particular pocket [4]. Some proteins contain only one canonical PHD finger, but some harbor several PHD fingers that act in concert or have independent functions. The PHD finger can be linked to a zinc knuckle, a zinc finger that coordinates one zinc ion, and is characterized by either C4 or C2HC sequence. The closely coupled tandem PHD fingers create a distinct fold, the double PHD finger (DPF), whereas two PHD fingers connected by a zinc knuckle are assembled into the PZP (PHD-zinc knuckle-PHD) domain. The various combinations of coupled modules greatly influence the overall function and dynamics of the PHD finger. The biological activity of the PHD finger can be further altered through the action of adjacent domains, such as histone readers (bromodomain, chromodomain, Tudor, etc.) or catalytic PTM writers, and erasers.

One of the well-established subsets of PHD fingers, exemplified by the PHD fingers of BPTF and ING2 [27–30], consists of numerous proteins that have been shown to bind H3K4me3 with high specificity and affinity. The lysine-specific demethylases JARID1A (KDM5A) and JARID1B (KDM5B) also has the ability to bind H3K4me3 via their three PHD finger domains [31, 32].

A smaller number of PHD fingers display a preference for the H3K9me3 [33–37]. The ADD (ATRX-DNMT3-DNMT3L) domain is a methyllysine-binding domain that has been characterized in the DNA methyltransferases DNMT3A, DNMT3B, and DNMT3L family and the chromatin remodeler ATRX. The increased prevalence of ATRX loss-of-function mutations in various forms of cancer suggests that it plays an essential role in regard to proper chromatin structure and/or gene regulation in these tissues; however, our overall understanding of both the normal and oncogenic roles of the ATRX ADD domain is still evolving. In addition to cancer, missense mutations in the ATRX ADD domain account for approximately 50% of patients with ATRX syndrome, a congenital disorder that causes intellectual disabilities [38].

In addition to recognizing histone tails, PHD fingers have been implicated in binding to nonhistone proteins and self-association. For example, the third PHD finger (PHD3) of MLL1 is capable of binding both H3K4me3 and the nuclear cyclophilin Cyp33, while the second PHD finger (PHD2) of MLL1 forms a dimer and shows E3 ubiquitin ligase activity in the presence of the E2-conjugating enzyme CDC34 [39]. The mechanistic outcome of histone or nonhistone recognition by PHD fingers is the recruitment or stabilization of their host proteins, i.e., transcription factors, PTM writing and erasing and nucleosome-remodeling enzymes, and other elements of the epigenetic machinery, at chromatin. Yet, the physiological consequence of these interactions is highly context-dependent and is often determined by the overall function of the chromatin-modifying complex in which the PHD finger resides.

Disrupting PHD fingers from properly reading their histone marks has been implicated in a wide variety of human disorders [40]. Mutation, translocation, or deletion of PHD fingers in ING1, JARID1A, PHF23, NSD1–3, MLL, and PHF1 has been identified in multiple forms of leukemia and solid tumors [41]. Germline mutations in PHD finger domains of other proteins have been linked to autoimmune or neurological disorders. To date, the majority of the PHD finger mutations that result in disease prevent its natural function as a chromatin reader; therefore, while small-molecule inhibitors of this domain can serve as essential probes to understand their physiological functions, it is unclear whether inhibition of PHD finger function will be a suitable therapeutic strategy.

3.2 *Small Molecules Targeting PHD Fingers*

3.2.1 Targeting JARID1A/KDM5A PHD3 Fingers

The first attempts toward the discovery of novel small-molecule ligands of PHD finger domains revealed promising results. Using a newly developed HaloTag-based assay, the 446 compounds included in the NIH Clinical Collection 1 library of phase I–III clinical trial drugs were screened against the H3K4me3-recognizing third PHD finger of Jumonji, AT-rich interactive domain 1A (JARID1A, also known as KDM5A), which has been implicated as a potential driver in subsets of acute myeloid leukemia (AML) [42] and is a recurrent genetic abnormality in pediatric acute megakaryoblastic leukemia (AMKL) [43]. The initial screen (200 μM) identified 23 compounds of interest that were further evaluated in a dose-dependent manner using the HaloTag and AlphaScreen™ assay. Following these secondary screens, disulfiram, phenothiazine, amiodarone HCl, and tegaserod maleate (Fig. 3) were identified as weak inhibitors of JARID1A-PHD3/H3K4me3 interactions.

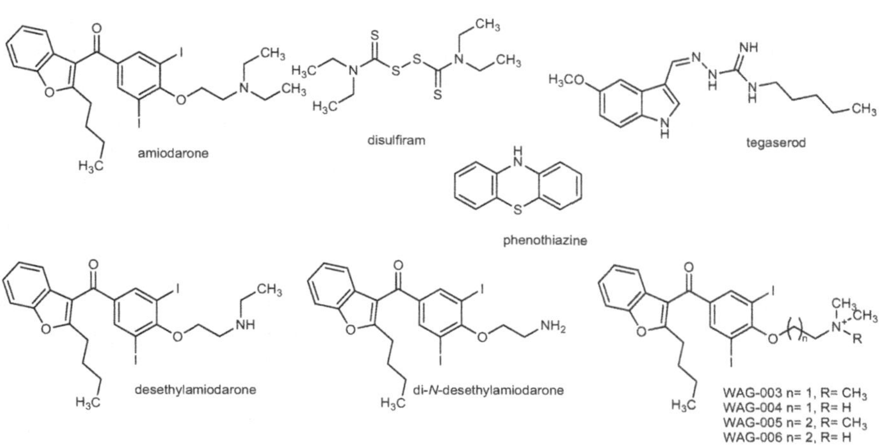

Fig. 3 JARID1A-PHD3 inhibitors

These hits were further validated by affinity pull-down and fluorescence polarization (FP) assays [44].

Disulfiram inhibits acetaldehyde dehydrogenase by covalently modifying cysteine residues and ejecting a zinc ion. Since PHD fingers contain a conserved Cys3-His-Cys4 motif bound to zinc ions, it was hypothesized that inhibition occurred by the release of zinc ions [45]. Indeed, it was found that disulfiram ejects zinc from JARID1A-PHD3 (IC_{50} = 10 μM), thus suggesting that it is a covalent inhibitor for JARID1A-PHD3 as well as other PHD finger domains. Disulfiram's alkylating nature also results in inhibition of other proteins such as dopamine beta-hydroxylase, viral nucleocapsid protein, DNA methyltransferase I, and histone methyltransferases [46]. On the contrary, amiodarone HCl and tegaserod maleate were unable to release zinc. To further probe the structural requirements for amiodarone inhibition of the JARID1A-PHD3, a series of derivatives with varying amine chain lengths and methylation states (compounds WAG-003–WAG-006, Fig. 3) as well as known metabolites, desethylamiodarone, and di-*N*-desethylamiodarone were synthesized and evaluated. The results led to the identification of inhibitors with a tenfold higher potency than amiodarone, including di-*N*-desethylamiodarone (IC_{50}: 26 μM) and trimethylamiodarone (WAG-003, IC_{50}: 30 μM). Docking studies suggest that amiodarone analogs bind to JARID1A-PHD3 primarily in the surface peptide groove and H3R2-binding pocket, but do not interact in the Kme-binding region. Mutational studies supported these docking results; however, they were unable to determine clearly the definitive orientation of the amiodarone scaffold when it interacts with JARID1A-PHD3 [44]. More recently, a few amiodarone derivatives were observed to inhibit catalysis of JARIF1a/KDM5A in a PHD-finger-independent manner, thus suggesting a nonunique mode of action [47].

3.2.2 Targeting Pygo PHD Fingers

A useful starting point for the development of new PHD-finger ligand lead structures was reported in 2014. The potent oncogene β-catenin is a key effector of Wnt signaling and is recruited to target genes through interaction with TCF/LEF transcription factors to activate cell differentiation [48]. Activating mutations in the β-catenin gene have been proven in many types of cancer, and this fact makes β-catenin a potential drug target. Yet, it is rather challenging to target this protein directly with small molecules successfully. In fact, there are no enzymes required for its activity that could be inhibited, and its interface with TCF factors involves most of its structured domain, the armadillo repeat domain (ARD), which is extensive and also constitutes the interface for its negative regulators, including APC and Axin, whose interaction with the ARD overlaps that of TCF [49]. Unsurprisingly, attempts to block specifically the interaction between β-catenin and TCFs have met with little success and failed to uncover any promising leads [48].

Another emerging strategy to target β-catenin signaling is by indirectly affecting its ability to promote H3K4me-mediated transcription. The N-terminus of the ARD

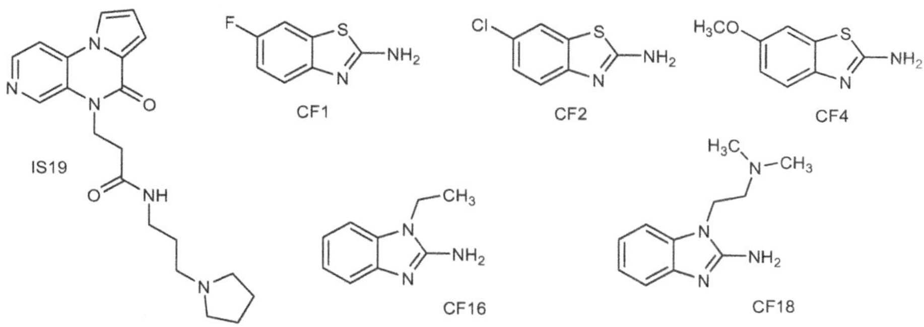

Fig. 4 Pygo-HD1 inhibitors

harbors a separate interaction surface for the BCL9 adaptor proteins, which bind to β-catenin through a short α-helical domain (HD2), simultaneously with TCF. In turn, BCL9 adaptors use a separate domain (HD1) to bind to the rear of the Pygo PHD finger. This interaction induces a slight allosteric modulation of the PHD finger, which facilitates the binding of the protein complex to the histone H3 tail methylated at lysine 4 (H3K4me) through its frontal surface [50]. Thus, oncogenic β-catenin could be inhibited by targeting these three unique and relatively small protein-protein interfaces of the Pygo-BCL9 complex.

The druggability of Pygo PHD fingers to inhibit the Pygo-HD1 complex and deactivate oncogenic β-catenin transcription was explored through a series of consecutive structure-based virtual screens of a library of 225 K commercially available compounds, each screen being increasingly refined and constrained based on the preceding one [51]. Indeed, the potential hits were then further evaluated for direct binding to PHD-HD1 through 2-D NMR spectroscopy. Only a small number of hits was confirmed, all affected by poor solubility. Yet, among them, compound IS19 (Fig. 4) favorably occupied the A1 cavity of the PHD domain, although with a weak affinity (approximately 3.5 mM, as estimated by chemical shift perturbations). Unfortunately, co-crystallization of the PHD-HD1 complex with IS19 was unsuccessful, likely due to the combination of the low solubility and affinity. In an attempt to identify small-molecule scaffolds with improved solubility and ligand efficiency, a fragment-based NMR screen of the Maybridge "rule of three" library of chemical fragments was conducted and has led to the identification of aminobenzothiazole fragments (CF1, CF2, CF4; Fig. 6) which exhibit tight binding to the distal part of the K4me pocket and to a highly conserved narrow cleft at the rear of the PHD neighboring its HD1-binding surface. Continued SAR studies identified benzimidazole analogs CF16 and CF18 (Fig. 4) as improved analogs with two distinct binding sites, the K4me pocket and the benzothiazole cleft, respectively. In particular, CF16 competes with the native ligand (the methylated H3 tail) of the PHD finger and docks into the distal part of the K4me pocket, thus possessing the potential to serve as a template for further chemical development.

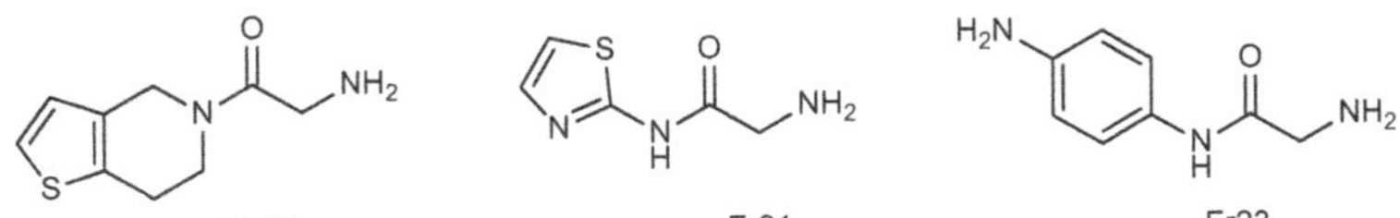

Fig. 5 BAZ2A-/BAZ2B-PHD inhibitors

3.2.3 Targeting BAZ-PHD Fingers

More recently, another fragment-based approach demonstrated the ligandability of the PHD fingers of two proteins of the BAZ (bromodomain adjacent to zinc finger) family, namely, BAZ2A and BAZ2B [52]. First, the ligandability of the histone pocket of BAZ2A was assessed using an in silico analysis that highlighted two potential druggable pockets. Then, targeted virtual screening of a diverse virtual library of a thousand low-molecular-weight compounds identified a set of fragments that were validated experimentally for protein binding using a biophysical screening cascade, using 2-D NMR spectroscopy, thermal shift assay (TSA), AlphaLISA competition assay, and X-ray crystallography. From the set of validated fragments, it was possible to solve the co-crystal structure of one compound (Fr19, Fig. 5) bound to the histone pocket, which guided further in silico optimization of the binding mode, resulting in two more fragments (Fr21 and F3 23, Fig. 5) that were successfully soaked in BAZ2 PHDs to solve the first crystal structures of PHD zinc fingers in complex with fragments bound to an anchoring pocket at the histone-binding site. The best-validated hits were found to displace a histone H3 tail peptide in competition assays. The chemical scaffolds identified in this work are suitable starting points for future ligand optimization using structure-guided approaches and could pave the way for the development of chemical probes to drug PHD reader domains.

4 Protein Reader Domains of the Royal Family

4.1 General Characteristics of Royal Family Proteins

The so-called Royal Family [53] was created for the structural classification of protein modules evolutionarily related to a repeated unit first identified in *Drosophila melanogaster* maternally expressed gene Tudor [54]. Mutations in Tudor result in offspring lethality or infertility [55]. Hence, the reference to English House of Tudor King Henry VIII whose blood group anomaly is thought to be the cause of the numerous stillbirths and miscarriages suffered by his wives. The Royal Family is composed of the chromo-, Tudor, MBT, chromo-barrel, and PWWP domains, which recognize histones and other proteins harboring a methylated lysine and also a methylated arginine for the Tudor domain. The recognition of these posttranslational modifications by Royal Family proteins affects many key cellular processes such as chromatin condensation, DNA transcription, gene silencing, the maintenance of epigenetic expression states, and the DNA damage response. Proteins of the Royal Family usually function in the context of large assemblies typically composed of one or more enzymes and several nonenzymatic subunits. A case in point is that of the evolutionarily conserved SAGA (SPT-ADA-GCN5-acetyltransferase) complex, made of more than 20 proteins, which facilitates DNA transcription [56]. The tandem Tudor domains of a subunit of SAGA, namely, SGF29 (SAGA-associated factor of 29 kDa), recognize H3K4me3, a signal for gene transcription. Binding of SGF29 to H3K4me3 contributes to the recruitment of SAGA to chromatin and thereby allows acetylation of H3 at lysines 9, 14, and 18 by SAGA subunit GCN5 (general control non-derepressible 5, also known as KAT2A), which promotes gene transcription [57].

The members of the Royal Family show structural similarity, and they all have a similar structural fold in common. Four of them (Tudor, MBT, chromo-barrel, and PWWP domains) fold into a roughly 60 amino acid SRC homology 3-like (SH3-like) antiparallel five-stranded twisted β-barrel structure and differ mainly by the number and relative orientations of the β-barrels as well as added secondary structure elements. The first Royal Family structure with a five-stranded β-barrel core motif to be determined was that of the single Tudor domain of human SMN (survival motor neuron) protein which was later shown to recognize a methylarginine [58, 59]. As mentioned above, a defining feature of the Royal Family proteins is an aromatic-binding cage that can accommodate a methyllysine or a methylarginine. A well-characterized example is the complex of 53BP1 tandem Tudor domains with histone H4 methylated at lysine 20 [60].

The chromodomain (chromatin organization modifier domain) differs from other Royal Family members in that in its simplest form it is constituted of a three-stranded antiparallel β-sheet packed against a C-terminal α-helix and does not adopt a barrel structure [61]. A lysine may be mono-, di-, or trimethylated at the ε-amino group (also referred to as the Nζ position). Regardless of the methylation state, the main driving force for methyllysine recognition comes from cation-π interactions between the methylammonium cation and surrounding aromatic residues in the binding cage [62]. For higher methylation states, there is also a significant contribution of the hydrophobic desolvation effect to methyllysine recognition [63]. Some Royal Family domains can selectively recognize a mono- or dimethyllysine and exclude a trimethyllysine. In these domains, a hydrogen bond and an ion-pair interaction between the methylammonium cation and a carboxylate group in the aromatic cage contribute to binding stability and methylation-state specificity.

4.2 The Chromodomain

4.2.1 Chromodomain Structure and Function

Chromodomains are a highly conserved family of methyllysine reader domains that play diverse roles in gene repression/expression in cellular differentiation, cancer progression, and stem cell maintenance [61, 64, 65]. Named after "chromatin organization modifier," the chromodomain was originally identified in *D. melanogaster* proteins HP1 (heterochromatin-associated protein 1) and Polycomb, contributing to genomic imprinting [66]. The chromodomain of HP1, which binds histone H3K9me3, was the first methyllysine reader domain to be discovered [16, 17].

Made up 40–60 amino acid residues, the canonical chromodomain structure is a monomeric three-stranded antiparallel β-sheet flanked by a C-terminal alpha helix [67]. Chromodomains can be generally divided into three structurally distinct classes: the HP1 family, the Polycomb group (PcG) family, and the chromo-ATPase/helicase-DNA-binding (CHD) family [68].

4.2.2 Small Molecules Targeting Chromodomains

4.2.2.1 Targeting CBX7

The members of the Polycomb group of CBX proteins (CBX2, CBX4, CBX6, CBX7, and CBX8) compete with each other for incorporation into multiprotein Polycomb-repressive complexes (PRCs) of which two main complexes are present in mammals (PRC1 and PRC2). Members of the PRC are frequently prone to dysregulation. Representing transcriptional repressors, Polycomb group genes are implemented in development, stem cell maintenance, and cancer progression [69, 70]. Each of the five human CBX proteins recognizes and binds to H3K27me3 (with a different degree of selectivity; see above), a mark that is created by another member of the PRC, the lysine methyltransferase EZH2/KMT6 [71, 72].

CBX7 is among the best-studied representatives of the Polycomb paralogs. It is a master controller that extends cellular lifespan, delays senescence, drives proliferation, and bestows pluripotency to adult and embryonic stem cells [73, 74]. Multiple lines of evidence suggest promise for CBX7 as a therapeutic target in certain forms of cancer. CBX7 expression is strongly proliferative in various stem cell and stemlike cancer cell lines [71, 73, 75, 76]. It has been observed that the depletion or knockdown of CBX7 in prostate cancer cell lines induces a senescent phenotype with reduced cell proliferation [73], whereas CBX7 expression significantly increases in clinical samples of hormone-resistant prostate cancer relative to hormone-dependent prostate cancers [77] and is linked to poor prognosis in ovarian clear cell adenocarcinoma [76]. Additional malignancies, where CBX7 seems to play a central role, are lymphomas [78] and T-cell leukemias [70]. Therefore, even if a few literature reports by the same research group suggested that the role in the regulation of cell growth and tumorigenesis might be tissue- and context-dependent [79, 80], CBX7 constitutes an interesting potential drug target.

A first successful approach to find inhibitors of CBX7 resulted from the truncation of a 25 amino acid long peptide sequence from protein SETDB1, which had demonstrated a twofold increase in binding affinity for CBX7 when compared to H3K27me3-containing sequences [77]. Initial truncation led to a 5-mer compound, Ac-FALKme3S-NH2 (Fig. 6), with promising activity when evaluated in a fluorescence polarization (FP) assay measuring CBX7-H3K27me3 disruption (IC_{50} = 12 μM). From this starting point and with the aid of molecular modeling studies utilizing the crystal structure of CBX7 in complex with H3K27me3 (PDB: 2L1B), a series of small trimethyllysine-containing analogs of the initial peptide were synthesized by the iterative, systematic replacement of one residue on the scaffold at a time (Leu, Ala, Phe, Ser) while keeping the trimethylated lysine residue. Modifications at Leu (−1) revealed a high tolerance for hydrophobic and/or aromatic residues such as Val, Ile, Tyr (analog 11, Fig. 6; IC_{50} = 6.2 μM), or a pentane ring [77]. The co-crystal structures of CBX7 in complex with these analogs confirmed the ligand conformation predicted by modeling studies [77]. It was found that substituting the C-terminal serine by simple H-bond-donating substituents like aminobenzimidazole, propanediols, or a pendent ammonium group increases the binding potency. Finally, a selected set of compounds that comprise all the favorable modifications to the FALKme3S-scaffold previously gained by SAR studies was synthesized, which revealed ligands (e.g., analog 64, Fig. 6) with affinities around 200 nM for CBX7 and tenfold selectivity over CBX8 and even 400-fold selectivity over CBX1, representing the first reported inhibitors of any chromodomain [77]. Starting from Ac-FALKme3S-NH2 as a lead peptide, a series of constrained aza-lysine analogs was also prepared with the aim to study the influence of side chain geometry and ε-amine methylation on affinity [81]. An aza-amino acid scan was performed in which each amino acid residue of the lead peptide was replaced by its corresponding semicarbazide counterpart in order to gain insight into the conformational preferences of the parent peptide. Although the azapeptides maintain all side chain structural components present in parent ligand, a loss of affinity was generally seen across the series. Considering the backbone conformational preferences of azapeptides, as well as the flatter nature of the semicarbazide residue, such constraints on the lead peptide were not tolerated in the protein-binding site, likely because they disturb the preferred β-strand conformation.

Another peptide-based approach led to the identification of the most potent CBX7 inhibitor reported to date, UNC3866 (Fig. 6). The study was based on the amino acid sequence RGFALKme3STHG (Fig. 6), which binds CBX7 with significantly increased affinity compared to H3K27me3 (K_D = 5 μM and 110 μM, respectively) in isothermal calorimetry (ITC) experiments [82]. Several modifications to RGFALKme3STHG designed to increase the hydrophobicity, cell permeability, and stability against lysine demethylases provided the initial lead peptide, UNC3567 (Fig. 6), which bound with comparable affinity to the parent peptide (K_D = 6.7 μM). A series of molecular dynamics simulations investigating the

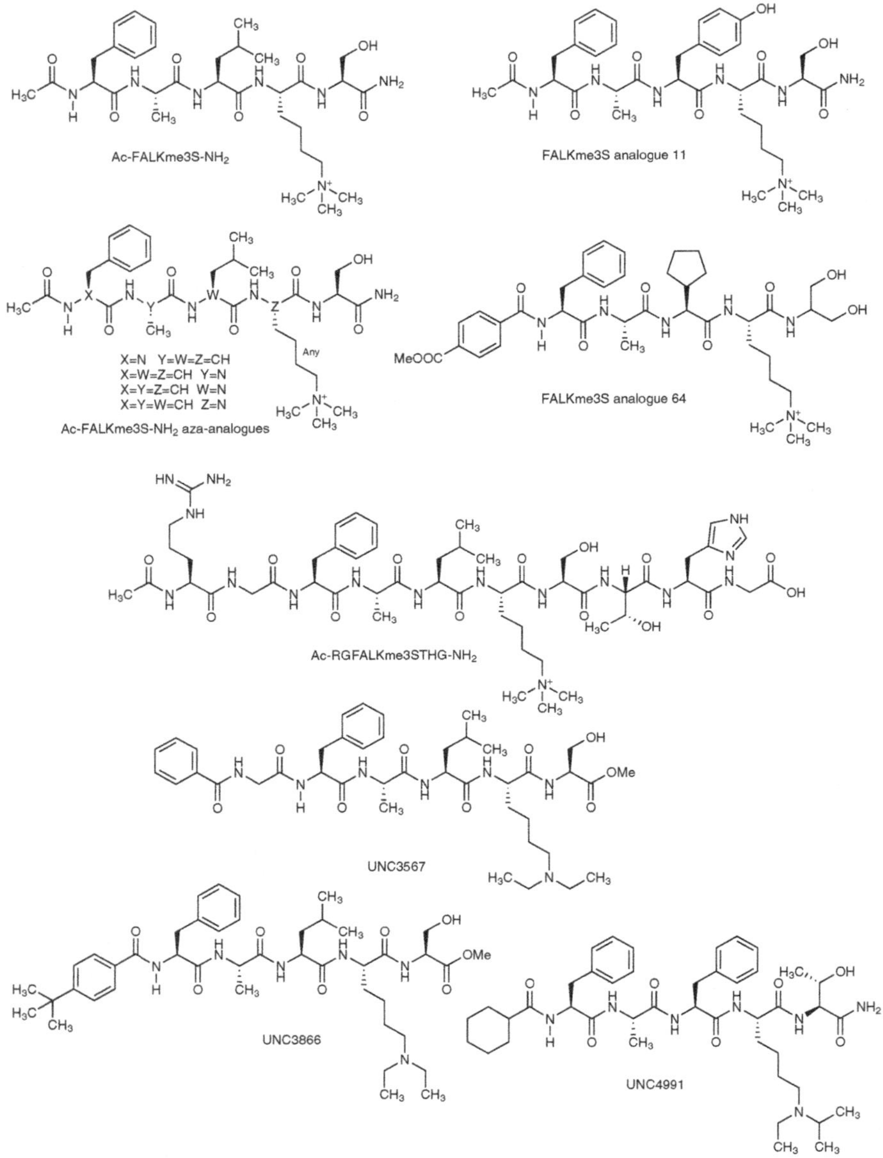

Fig. 6 Peptide-based inhibitors of CBX7

mechanisms through which the CBX7 aromatic cage closes around H3K27me3 suggested that replacement of the N-terminal benzoylglycine of UNC3567 with less flexible structural moieties would enhance hydrogen bonding interactions with CBX7 and facilitate closure of the chromodomain around the peptide. These SAR studies resulted in UNC3866, which binds CBX7 and CBX4 ($K_D = 97$ nM) with equal affinity [82]. UNC3866 bound with modest affinity to chromodomains of the PcG and the lesser explored chromodomain Y chromosome (CDY) [83] protein families but was inactive against HP1 chromodomains and other distinct methyllysine-binding domains. Key insights into UNC3866 selectivity for CBX7 over other CBX proteins were provided by crystallography studies [82, 84]. Consistent with the inhibition of the H3K27me3 recognition by CBX7, UNC3866 demonstrated modest antiproliferative effects in PC3 cells (GI_{50} = 7.6 μM) [82], an advanced stage metastatic prostate cancer model previously shown to have its growth regulated by CBX7 [85].

Due to the off-target affinity of UNC3866 for CDY proteins, specifically CDY1 (K_D = 6.3 μM), CDYL1b (0.91 μM), and CDYL2 (0.85 μM), a series of CDYL inhibitors based on its structure were developed [86]. Roughly 14,000 diverse peptide compounds were synthesized by means of a split-and-pool combinatorial method and then screened through an on-bead strategy utilizing magnetic enrichment [87, 88]. This approach led to the isolation of ligands with the desired selectivity profile and improved affinity for CDYL2. The method successfully identified UNC4991 (Fig. 6) with high affinity for CDYL2. Further evaluation using ITC found UNC4991 to effectively bind both CDYL2 (K_D = 0.64 μM) and CDYL (K_D = 0.49 μM) with ~fivefold to sixfold selectivity over CBX7. Notably, this study further supported the importance of the nature of the N-terminal capping on chromodomain selectivity. In particular, the tert-butyl phenyl moiety of UNC3866 facilitates CBX7 binding, while isobutyl, cyclohexyl, and pyridyl caps favor CDYL2 binding [86].

The first small-molecule inhibitors of the CBX7-H3K27me3 interaction were identified in 2015 by high-throughput screening of a library of over 100 K compounds, including 2,560 FDA-approved drugs and other compounds selected from commercial sources [89]. About 60 compounds were identified as modest inhibitors of CBX7, 6 of which (sennoside A, suramin, aurin tricarboxylic acid, trypan blue, and Evans blue from the FDA database and MS37452 from the commercial compounds library) were confirmed by using 2D ^{1}H-^{15}N HSQC spectra and the fluorescence anisotropy-binding assay. Suramin (IC_{50} = 8.1 μM) and MS37452 (also termed MS452, K_D = 29 μM, IC_{50} = 67 μM; Fig. 7) were chosen for further structural and biological evaluation. The co-crystal structure of suramin in complex with CBX7 demonstrated that two suramin molecules bound to two CBX7 molecules in two distinct orientations to create an interwoven, compact 2:2 complex [89]. Yet, although the trypanocidal suramin has also recently demonstrated anticancer activity [90], it is well-characterized as a promiscuous screening compound in a variety of different screening contexts [91, 92]. Therefore, the study focused on MS37452. The co-crystal structure of the complex between CBX7 and MS37452 showed that the ligand adapts two rotamer conformations in which the methylbenzene moiety is positioned either *cis* or *trans* with respect to the 2,3-dimethoxybenzene [89]. Structural studies demonstrated that the dimethoxybenzene and piperazine interact with the CBX7 aromatic cage residues (Phe11, Trp32, and Trp35) in both orientations, but the methylbenzene moiety in the trans conformation more closely mimics the intermolecular interactions between

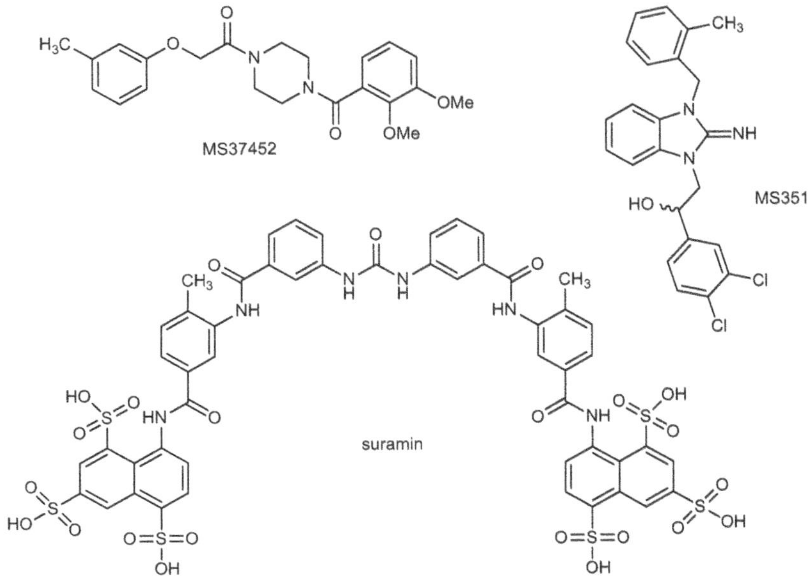

Fig. 7 Small-molecule inhibitors of CBX7

H3K27me3 and CBX7 outside of the methyllysine-binding domain. Selectivity evaluation of MS37452 toward other chromodomain-containing proteins revealed a threefold weaker affinity for CBX4 and even a tenfold lower-binding potency for the other members of the PcG family than for CBX7. Moreover, MS37452 shows almost no binding to the chromodomains of HP1 proteins [89]. Compound MS37452 was further characterized on a cellular level where it induced transcriptional depression of p16/CDKN2A by disrupting CBX7-CHD binding to H3K27me3 at the INK4A/ARF locus in PC3 prostate cancer cells, which was determined by chromatin immunoprecipitation (ChIP) [89].

Later, an in silico screening performed by the same group led to the identification of a structurally distinct aminobenzimidazole compound, MS351 (Fig. 7), endowed with a unique mode of action [93]. The compound binds and forms a ternary complex with the methyllysine-binding pocket of CBX7 and a predicated hairpin RNA derived from long noncoding RNA (lncRNA) ANRIL. Binding of RNA ANRIL is thought to allosterically influence the conformation of the aromatic cage of CBX7, thus "activating" it for methyllysine recognition [93]. The affinity (K_D = 23.8 μM in fluorescence anisotropy assay) is 21-fold higher than for uncomplexed CBX7 protein. MS351 also enhanced RNA ANRIL binding in a dose-dependent manner, while MS37452 did not have a similar effect in the same assay, supporting the differences in their biological actions. Additionally, MS351 effectively induced transcriptional derepression of p16INK4a in PC3 and

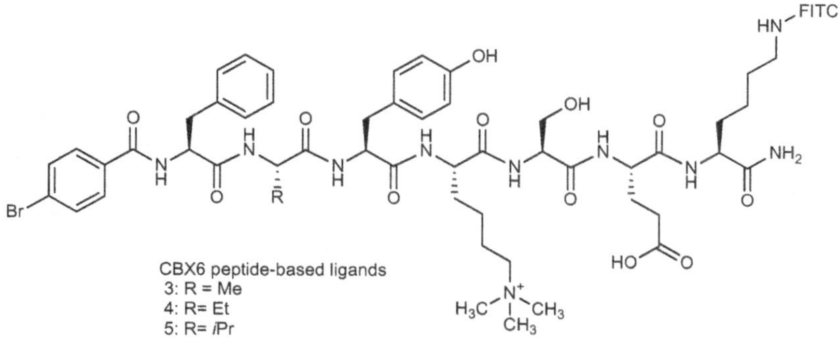

Fig. 8 Peptide-based inhibitors of CBX6

mouse embryonic stem cells at 1–5 μM and 1–2 μM, respectively. The crystal structure of CBX7 complexed with MS351 further supported the allosteric influence RNA binding has on H3K27me3 recognition by CBX7 [93].

4.2.2.2 Targeting CBX6

As described above, the overall similarity of the aromatic cage pockets regarding the chromodomains of the CBX proteins within Polycomb group makes their selective inhibition challenging. Nonetheless, another peptide-based approach, aimed at targeting the more diverse hydrophobic clasp region of CBX chromodomains, successfully led to the identification of CBX6 inhibitors [94]. As mentioned above, CBX4 and CBX7 contain a Val13 residue, whereas CBX2, CBX6, and CBX8 have an Ala residue in the same location at the bottom of the pocket. Starting from the CBX7 inhibitors analogs of peptide Ac-FALKme3S-NH2 (Fig. 6), a series of peptides (Fig. 8) was designed by incorporating slight modifications to the Ala (−2) position. Also, a FITC label was appended to these peptides to aid their evaluation against the CBX proteins. CBX6 ligands 3 (methyl) and 4 (ethyl) were broadly selective toward PcG family proteins over the HP1 protein with moderate-binding affinities ranging from 0.1 to 1.0 μM and 0.4–1.6 μM, respectively. The introduction of an isopropyl group at Ala (−2) (ligand 5; Fig. 8) yielded a potent ($K_D = 0.9$ μM) and highly selective CBX6 inhibitor with 90-, 20-, 18-, 6-, and 7-fold preferential binding over CBX-1, CBX-2, CBX-4, CBX-7, and CBX-8, respectively. Mutational studies revealed that enhanced interactions at the hydrophobic clasp alone did not determine the CBX6 selectivity of the compound. On the other hand, molecular dynamics simulations provided strong evidence that the chromodomain of CBX6, but not that of CBX7, is able to envelop this peptide in the surface groove-binding region completely and that the internal pocket of the CBX6 hydrophobic clasp is less strained when in complex with this ligand [94].

4.3 The Tudor Domain

At around the same time that the chromodomain-containing protein HP1 was discovered as the first "reader" of a methyllysine mark on histone tails [16, 17], the Tudor domain of survival motor neuron protein (SMN1) was characterized as a methylarginine-binding module [59, 95]. Subsequent work revealed that Tudor domains do not only bind methylarginine motifs but also interact with methyllysine motifs [96–98]. Based on the primary amino acid sequence alignment of different Tudor family members, it is not possible to predict whether the Tudor domain belongs to the methyllysine- or methylarginine-binding family or what degree (Kme1, Kme2, or Kme3) or type (Rme1, Rme2s, or Rme2a) of methylation is recognized by a particular domain. As mentioned above, only recently structural insight has been gained into the interaction of Tudor domains with their specific methylarginine motif ligands [99–104] and suggested that the aromatic cage of the methylarginine binders is narrower than that of methyllysine binders, which favors the docking of the planar methyl-guanidinium group of arginine [102].

In mammals, there are at least 41 proteins that harbor Tudor domains [105], and many of these proteins contain multiple copies of this domain. This structurally diverse family of Tudor domain-containing proteins can be divided into three groups: proteins with a single Tudor domain, tandem Tudor domains, and hybrid Tudor domains.

4.3.1 Single Tudor Domain (PHF1, PHF19, PHF20)

4.3.1.1 Structure and Function

The best-characterized single Tudor domains are from the proteins plant homeodomain finger 1 and 19 (PHF1 and PHF19), two components of the Polycomb-like complex (PCL) in mammals. The PCL proteins regulate the Polycomb-repressive complex PRC2 (Polycomb-repressive complex 2) by recruiting PRC2 to its target genes where it triggers transcriptional silencing by trimethylation of H3K27. Although not fully understood, the mechanism involves the recognition by the Tudor domains of PHF1 and PHF19 of histone H3 di- or trimethylated at lysine 36 (H3K36me2/3), which are markers of transcriptionally active genes. This brings PRC1/PHF1 or PRC2/PHF19 complexes and associated H3K36me2/3 demethylases NO66 and KDM2B, and H3K4me3 demethylase KDM2A to target genes. Following H3K36me2/3 and H3K4me3 demethylation, PRC2 trimethylates H3K27, which leads to de novo gene silencing [106–108]. The structures of the complexes of PHF1-Tudor and PHF19-Tudor bound to an H3K36me3 peptide revealed the binding mechanism of a single Tudor domain [106–108]. Another single Tudor domain characterized structurally in complex with a methylated peptide is from human PHF20 (plant homeodomain finger 20), a protein that regulates p53 and NF-κB transcription factors in a lysine methylation-dependent manner [109, 110]. The association of PHF20 with methylated p53 stabilizes p53 by limiting Mdm2-mediated ubiquitylation and subsequent degradation of p53 [109]. PHF20 is also involved in transcriptional regulation and activates p53 through binding the p53 promoter [111, 112]. PHF20 was also recently shown to be necessary for somatic cell reprogramming to generate inducible pluripotent stem cells [113]. In the NF-κB pathway, PHF20 was shown to drive the constitutive activation of this transcription factor in gliomas by preventing the interaction between protein phosphatase PP2A and the p65 subunit of NF-κB. This process strictly depends on the recognition of a methyllysine in p65 (Lys218 or Lys221) by PHF20-Tudor2 [110], suggesting that inhibiting this interaction might be beneficial for cancer treatment.

4.3.2 Tandem Tudor Domains (53BP1, SGF29, SHH1, UHRF1, Spindlin-1)

4.3.2.1 Structure and Function

Tandem Tudor domains were first identified in p53-binding protein 1 (53BP1) [97, 114], a protein that has recently emerged as a major regulator of DNA double-strand break (DSB) repair by nonhomologous end joining (NHEJ) [115] and as an essential protein in the protection against breast cancer due to its ability to contribute significantly to the tumor suppression pathways associated with DNA repair, cell cycle control, apoptosis, and cell senescence [116]. 53BP1 recruits the NHEJ effector protein RIF1 (Rap1-interacting factor 1) to DSBs [117, 118]. 53BP1 is essential for class switch recombination in antibody diversification [119] and for the fusion of deprotected telomeres [120], two processes that depend on NHEJ. Furthermore, 53BP1 inactivates homologous recombination (HR)-mediated DNA repair by inhibiting DNA end resection, the initial step of HR [121]. Remarkably, deletion of the 53bp1 gene restores HR in cells defective in HR protein BRCA1 and alleviates embryonic lethality in Brca1-nullizygous mice [121].

Another TT-containing protein is SAGA-associated factor 29 (SGF29), a subunit of Spt-Ada-Gcn5 acetyltransferase (SAGA) chromatin-modifying complex that regulates gene expression [122]. SAGA is evolutionarily conserved and was initially identified in budding yeast where it was shown to acetylate and deubiquitylate histones [56, 123]. The SGF29 subunit is required for the recruitment of SAGA to gene promoters and for the acetylation of histone H3 by SAGA [57]. In human and *Saccharomyces cerevisiae*, SGF29 recognizes histone marks H3K4me2 and H3K4me3, with a slight preference for H3K4me3, by means of tandem Tudor domains (SGF29-TT) [57, 124]. The structures of budding yeast and human SGF29-TT are virtually identical [57]. Both yeast and human SGF29 are highly specific for the H3 sequence surrounding Lys4. The affinity of SGF29-TT for its target peptide is higher than the affinity of 53BP1-TT for H4K20me2. Unlike 53BP1, SGF29 does not have a stringent methylation-state specificity. SGF29-TT binds tightest to H3K4me3 and H3K4me2, but it also recognizes H3K4me1 with relatively high affinity and even H3K4me0 for the human protein (K_Ds ~0.5 μM, ~1 μM, ~4 μM, and 24 μM, respectively) [57].

Sawadee homeodomain homolog 1 (SHH1) from *Arabidopsis thaliana* is a protein involved in the RNA-directed DNA methylation (RdDM) pathway [125]. This pathway involves plant-specific RNA polymerases such as Pol-IV, which initiates the synthesis of *si*RNAs [126]. SHH1 enables the production of *si*RNAs from a large number of RdDM targets and is required for Pol-IV to occupy the corresponding loci. SHH1 includes tandem Tudor domains (SHH1-TT), also called Sawadee domain. SHH1-TT, which binds histone H3 methylated at Lys9, is essential in vivo for Pol-IV to occupy RdDM targets and for the maintenance of siRNA and DNA methylation levels [127]. SHH1-TT is similar to UHRF1-TT, but with a unique zinc-binding site in Tudor 2. Extensive structural and binding studies have shown that SHH1 binds equally well to H3K9me1, H3K9me2, and H3K9me3 peptides with K_Ds of ~2 μM.

The multidomain ubiquitin ligase UHRF1 (ubiquitin-like, PHD, and ring finger-containing 1) is required for the maintenance of DNA methylation patterns by DNMT1 (DNA methyltransferase 1) at hemimethylated CpG dinucleotides [128, 129]. These sites are recognized by UHRF1 SRA (SET- and RING-associated domain) domain, which also directly binds DNMT1 [130–132]. In addition, UHRF1 associates with H3K9me3 and H3K4me0 or H3K4me1 [133]. UHRF1 ubiquitylates histone H3 at lysine 23, and this mark is required for the recruitment of DNMT1 to DNA replication sites [134]. Therefore, UHRF1 links two layers of epigenetic information: the methylation of DNA associated with transcriptional silencing and histone methylation marks associated with chromatin condensation and inhibition of gene expression [135]. The tandem Tudor domains of UHRF1 (UHRF1-TT) binds H3K4me0K9me3 with a K_D of ~20 μM. Noteworthy, the affinity for H3K4AK9me3 or H3K4me3K9me3 is significantly lower (K_Ds ~210 μM and ~90 μM, respectively) [133]. More recently, it was shown that the PHD finger at the C-terminus of UHRF1 is also involved in recognition of methylated histone H3 [130, 136, 137].

A particular example of TT-containing protein is Spindlin-1, a transcriptional coactivator that has been reported to regulate the expression of rRNA genes, which are regulated by the MAZ transcription factor and Wnt target genes [138, 139]. The protein is named after the meiotic spindle in mice where it was initially found [140]. Later, the gene for the human ortholog (SPIN1) was found as overexpressed in a screen for genes involved in ovarian cancer [141]. Spindlin-1 levels are elevated in a number of different cancers, including non-small cell lung cancers, ovarian tumors, and some hepatic carcinomas [138, 142], and the protein displays a diffuse nuclear localization and is enriched in nucleoli [139]. Spindlin-1-overexpressing cells undergo a complete morphological change and show increased cell growth as well as cell cycle delay in metaphase and chromosome instability [143]. It has been proposed that cancer cell growth promotion by Spindlin-1 occurs via WNT/TCF-4 signaling activation [142].

The structure of human Spindlin-1 revealed a folded globular unit constituted of a tandem of Tudor domains (Tudor 1 and Tudor 2) tightly packed against a third Tudor domain (Tudor 3) via hydrophobic interactions, giving rise to a circular arrangement of three Tudor domains [144]. The first and second Tudor domains have cages made up of four aromatic residues. Spindlin-1 T1 aromatic cage binds an H3K4me3 peptide with a K_D of ~0.2 μM. Interestingly, it was found that the affinity of the binding of Spindlin-1 to H3K4me3 is further increased by an asymmetric methylated arginine in position 8 (H3R8me2a), which is recognized by the second Tudor domain (K_D ~0.05 μM) [145]. Furthermore, in vivo, functional experiments suggested that Spindlin-1 activates the signaling downstream of Wnt/β-catenin from protein arginine methyltransferase 2 (PRMT2), which creates H3R8me2a marks, and the MLL complex, which is responsible for the generation of trimethylated H3K4 residues. Together, this results in a specific "K4me3-R8me2a" pattern which can be recognized by Spindlin-1 [145].

Consistent with the dual recognition of H3K4me3 and H3R8me2a, it was found that Spindlin-1, H3K4me3, and H3R8me2a are all enriched at the promoter region of Wnt target genes, strongly suggesting a direct role of Spindlin-1 in regulating the expression of these genes [145]. Wnt/β-catenin signaling plays a role in a wide range of cellular processes, including proliferation, differentiation, tissue homeostasis, tumorigenesis, apoptosis, and cell survival. Targeting this pathway could open new strategies for the treatment of a variety of diseases [138].

4.3.2.2 Small Molecules Targeting Tandem Tudor Domains

Targeting 53BP1

The first compound (UNC2170, Fig. 9a) that selectively targets the methyllysine-binding domain of 53BP1 was identified from the cross-screening of a series of small molecules developed initially as inhibitors of L3MBTL1/L3MBTL3 (see below) against a panel of methyllysine readers [146]. The binding affinity of UNC2170 for 53BP1 was moderate (K_D = 22 μM), but the compound was completely selective for 53BP1 when compared against a range of other methyllysine readers at concentrations up to 500 μM. Synthetic modifications to the basic amine, linker region, and aromatic moiety of the UNC2170 scaffold together with a structural analysis of the 53BP1–H4K20me2 complex led to the development of SARs.

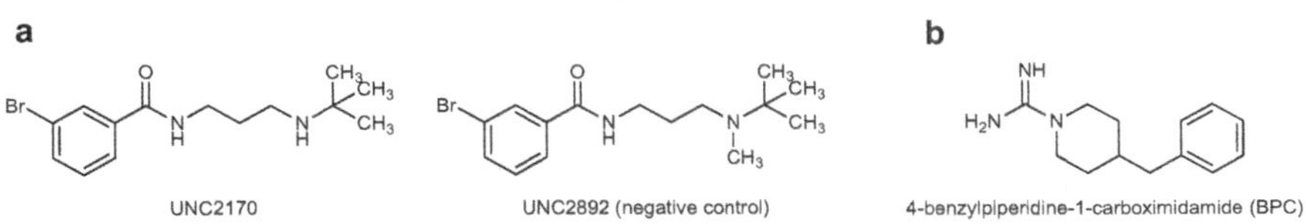

Fig. 9 Small-molecule inhibitors of 53BP1 (**a**) and UHRF1 (**b**)

Interestingly, any changes to the secondary tert-butyl amine of UNC2170 resulted in a complete loss of affinity for 53BP1 [146]. Noteworthy, the tertiary amine derivative UNC2892 (Fig. 9a) was chosen as a negative control for follow-up studies, due to its structural similarity to UNC2170 and its complete lack of binding affinity for 53BP1. In order to confirm that the complete loss of activity for the analogs was due to the lack of the secondary tert-butyl amine, a methyllysine-binding site mutant of 53BP1 (D1521A) was prepared. Removal of the aspartate is known to disrupt the salt bridge and hydrogen bond between 53BP1 and the dimethylammonium of H4K20me2 [96]. UNC2170 did not bind to the 53BP1 mutant, providing strong evidence that the secondary tert-butyl amine was interacting with the Kme2-binding region [146]. The only tolerated substitutions resulted from the replacement of the bromine atom in the C-3 position of the aromatic moiety with bulky, lipophilic substituents, such as iodine, isopropyl, and trifluoromethyl, which all yielded selective and moderate inhibition [146]. Further analysis of UNC2170 confirmed its potential as a chemical probe. Pulldown assays with H4K20me2 and p53K382me2 demonstrated that UNC2170 binds competitively to the TT domain of a His-53BP1 fusion protein. UNC2170 was able to competitively displace p53K382me2 in a concentration-dependent manner ($IC_{50} = 30\ \mu M$). Within a cellular context, UNC2170 was nontoxic and highly permeable with no significant measurable efflux as determined by a bidirectional Caco-2 cell permeability assay. No significant decrease in 53BP1 foci formation was observed in U2OS and γH2AX cells before and after ionizing radiation, possibly due to the bivalent nature of 53BP1 readers also recognizing H2AK15ub at DSB sites [147]. UNC2892 was inactive in each of these cellular assays, providing strong evidence that the cellular effects seen with UNC2170 are due to inhibition of 53BP1 and highlighting the potential of a potent 53BP1 antagonist.

Targeting UHRF1

Histone and DNA binding by UHRF1 are regulated by long-range interdomain and linker-domain interactions within the full-length protein [148–150]. It has been showed that a 15-residue polybasic region (PBR-UHRF1643–657) in the linker between the SRA and RING (really interesting and new gene) domains regulates the transition between PHD- and TTD-mediated histone reader states. This occurs through its reversible binding to the TTD groove or the phospholipid PI5P that in full-length UHRF1 results in the failure of UHRF1 to recognize H3K9me3 caused by a transition from TTD-mediated to PHD-mediated histone binding [148]. The PHD has also been reported to interact with the SRA domain in a UHRF1 state where histone binding is restricted [149, 150]. These studies suggest that disruption of interdomain interactions could be a mechanism to target UHRF1 pharmacologically.

A small-molecule fragment library containing 2,040 compounds was screened against isolated TTD (UHRF1121–UHRF1286) using a fluorescence polarization (FP) assay that tracked the displacement of an N-terminally tagged H3K9me3 peptide (H3K9me3(1–25)). From this screen, eight putative TTD-binding hits were identified. Analysis of amide peak movement in the (^{1}H-^{15}N) HSQC spectra of the TTD in the presence of the fragments indicates that the binding of one of them, 4-benzylpiperidine-1-carboximidamide (BPC, Fig. 9b), occurs in the groove [151]. Further characterization of BPC binding to isolated TTD using ITC and DSF estimated a K_D of 50 μm and a calculated ligand efficiency of 0.38 (which is defined as the binding energy per heavy atom) [151]. Saturating concentrations of this highly soluble compound induced open conformers of the TTD-PHD module and reduced its affinity for H3K9me3 peptides. Therefore, BPC could be a useful tool for in vitro investigations aimed to relate open TTD-PHD conformations with specific UHRF1-binding modes. Due to the small size and high ligand efficiency, the compound can be further optimized for potency, for instance, by linking it to compounds designed for interaction with the aromatic cage, which recognizes the trimethyl lysine of H3K9me3.

Targeting Spindlin-1

The emerging role of Spindlin-1 in transformation and cancer prompted a growing interest in developing inhibitors of the protein-protein interactions that it mediates. Recently, an epigenetic reader assay platform was developed to identify small-molecule inhibitors of the Spindlin-1-H3K4me3 interaction [152]. The methodology included a primary screen using an in vitro AlphaLISA assay, an FP-based verification assay, a fluorescent thermal shift assay (FTSA) to investigate direct binding, a biolayer interferometry (BLI) assay, and finally cellular thermal shift assay (CETSA). The pilot screen of a small library of bioactive compounds identified compound A366 (Fig. 16), previously reported as selective inhibitor of lysine methyltransferase G9a [153], as endowed with good in vitro inhibition of the Spindlin-1-H3K4me3 interaction (IC_{50} ~200 nM) as a result of H3K4me3-competitive direct binding to the Tudor domains of Spindlin-1. However, the compound showed poor cell-based activity (100 μM) [152]. Moreover, similarly to what was previously reported against G9a [153], even minor structural modifications of A366 led to significant reduction of SPIN1-binding affinity. In fact, the structurally related compound YX-11-102 (Fig. 10) did not show any activity in either the AlphaLISA or FP assay [152]. More recently, the same group used an iterative virtual screening approach to generate pharmacophore models based on the crystal structure of Spindlin-1 in complex with H3K4me3 peptide and, subsequently, screen databases of commercially available compounds. After docking the identified potential hits in the crystal structure of Spindlin-1 to verify the occurrence of significant interactions with the residues of the binding site, structural analogs of the validated hits were synthesized and tested employing an AlphaLISA assay. Two molecular scaffolds, 4-aminoquinazoline and quinazolinethione, were identified as promising lead structures (Fig. 10) [154]. However, all the derivatives synthesized displayed only low-to-moderate activity in vitro. Interestingly, the studies highlighted that an appropriate lysine mimetic group is crucial for inhibitory activity [154].

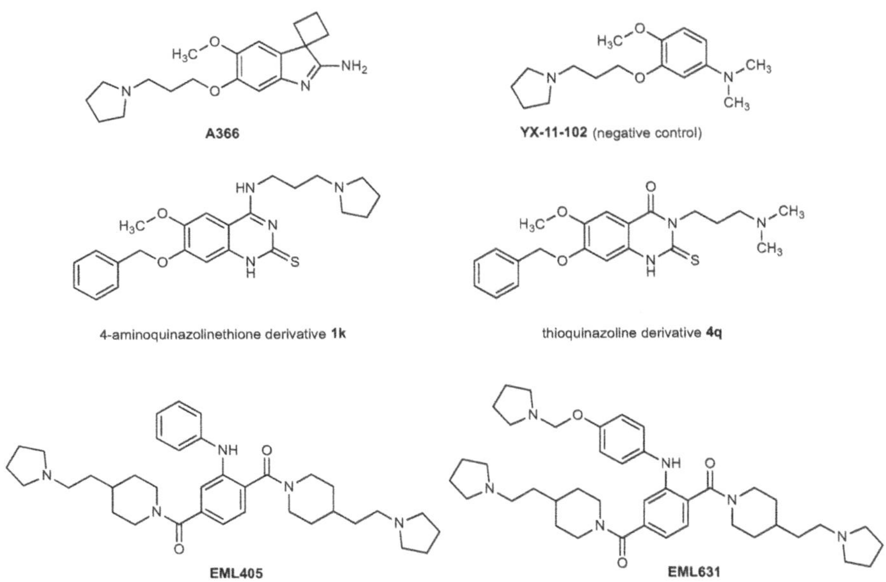

Fig. 10 Small-molecule inhibitors of Spindlin-1

A different approach used a library of biotin-tagged analogs of the small-molecule probe UNC1215, a nonselective inhibitor of MBT domains (see below), to screen a protein domain microarray that harbored about 100 GST fusion proteins, including 41 Tudor domains, 31 chromodomains, and representative PHD, BHA, MBT, PWWP, ANK, AGENET, and HEAT domains. This "library-on-library" approach led to the identification of EML405 (Fig. 10) as a lead compound capable to interact with Spindlin-1 [20]. Structural studies revealed that EML405 engages both the first and second Tudor domains of Spindlin-1 and also revealed a large unoccupied negatively charged pocket [20]. With the aim to engage this pocket, a series of derivatives were designed and synthesized and revealed an improved binding constant with respect to EML405 (3 μM for EML631 vs 14 μM for EML405). Moreover, the novel compounds displayed a significantly weaker binding affinity for the Tudor domain-containing proteins 53BP1 and PHF20 and for the MBT domain-containing proteins L3MBTL1 and L3MBTL3, thus showing a dramatic improvement of selectivity for Spindlin-1. Competition assays, chemiprecipitation experiments, and cellular thermal shift assay (CETSA) experiments demonstrated that the most specific compound identified in this study, EML631 (Fig. 10), was capable to engage Spindlin-1 in cells, block its ability to recognize H3K4me3 marks and inhibit its transcriptional coactivator activity.

Interestingly, the crystal structure of the complex between Spindlin-1 and EML631 revealed an unpredicted interaction of the additional pyrrolidine group with a negatively charged groove that lies between the first two Tudor domains. Recently, EML631 was successfully used as a chemical probe to characterize

Spindlin-1 docking protein (SPIN·DOC), a transcriptional repressor that binds Spindlin-1 and masks its ability to engage the H3K4me3 activation mark [155]. SPIN·DOC associates with TCF4 in a Spindlin-1-dependent manner and dampens Spindlin-1 coactivator activity in a TOPflash luciferase reporter assays. Furthermore, knockdown and overexpression experiments indicated that SPIN·DOC represses the expression of a number of Spindlin-1-regulated genes, including those encoding ribosomal RNA and the cytokine IL1B [155].

4.3.3 Hybrid Tudor Domains (JMJ2A/KDM4A)

The hybrid Tudor domain JMJD2A/KDM4A has been poorly understood, and no inhibitors have been published so far; thus this reader will not be further discussed within this book chapter [156].

4.3.4 Tudor Domains that Bind Methylarginine Motifs

Tudor domains are the only protein domain family that binds methylated arginine motifs (although individual PHD and WD40 domains also harbor this ability) [8, 19]. Several Tudor domain-containing proteins that have been clearly shown to participate in methylarginine-driven protein-protein interactions have been described in the literature. These include proteins that are implicated in the regulation of splicing (SMN, important for the survival motor neuron 1 in spinal muscular atrophy [157] and SPF30, survival of motor neuron-related-splicing factor 30 [102]), in the regulation of gene expression (TDRD3, Tudor domain-containing protein 3 [158], and SND1, staphylococcal nuclease domain-containing protein 1 [159]), and in a gonad-specific small RNA silencing pathway (TDRD1, TDRD6, and TDRD9, and TDRKH [99, 160]). All of these aforementioned Tudor domain proteins are still not fully understood, and no probes or inhibitors can be found in the literature so far.

4.4 The MBT Domain

The MBT (malignant brain tumor) domain was originally discovered as three repeated units in *D. melanogaster* gene *lethal (3) malignant brain tumor* [161]. Mutations in this gene correlate with tumor development in the larval brain. The *Drosophila* L(3)MBT protein functions in transcriptional repression [162]. The corresponding human protein, L3MBTL1, was the first MBT protein to be characterized structurally [163], revealing the general architecture of the MBT repeat conserved in all other MBT structures solved to date [164]. MBT domains always occur in two or more copies, and the arm extension of a given MBT domain packs against the core subunit of another repeat. This generates different spatial organizations of the MBT repeats in different proteins. For example, the three MBT domains of L3MBTL1 adopt a triangular propeller-like architecture where the arm of the first MBT domain contacts the core of the third MBT domain. Structurally, MBT proteins are comprised of two, three, or four MBT domains (MBT1, MBT2, MBT3, MBT4) flanked by other domains such as a zinc finger and/or sterile alpha motif (SAM) [4, 165]. Interestingly, while each MBT-containing protein has multiple MBT domains, only one domain contains the functional aromatic cage that can recognize and bind methylated lysine residues [166, 167].

The cellular functions containing two (such as SCML2) or four (such as L3MBTL2 and Sfbmt) MBT repeats are still poorly understood from a biological point of view. Therefore they will not be discussed further in this more chemical oriented book chapter [168].

4.4.1 Proteins Containing Three MBT Repeats (L3MBTL1 and L3MBTL3)

4.4.1.1 Structure and Function

The first structures of MBT complexes to be reported were those of human L3BMTL1 bound to H4K20me1, H4K20me2, and H1.5K27me2 peptides and single mono- and dimethylated lysines [166]. Only the methylated lysine of H4K20me1, H4K20me2, or H1.5K27me2 contacts extensively L3MBTL1. This is consistent with the relatively low affinity (K_D = ~5 μM for H4K20me1 or H4K20me2) and especially low specificity of this protein. Several other mono- or dimethylated histone peptides such as H1bK26, H3K4, H3K9, and H3K27 bind L3MBTL1 with similar K_Ds [166]. Human L3MBTL1 seems to play multiple roles in the cell although a clear function remains to be determined. It was initially characterized as a transcriptional repressor, and the three MBT domains were shown to be essential for this gene-silencing function [169]. Transcriptional repression requires that L3MBTL1 associates with histones in a mono- or dimethylation-dependent manner at H4K20 or H1bK26, which leads to chromatin condensation at retinoblastoma protein-regulated genes [170].

L3MBTL3, a close homolog of L3MBTL1, is a relatively uncharacterized member of the human MBT family. It also contains three MBT domains but differs from L3MBTL1 in what lysine methylation marks it recognizes. While L3MBTL1 recognizes Kme1 and Kme2 on H2B, H3, and H4, L3MBTL3 exclusively binds to Kme2 marks on these histones [165]. Additionally, L3MBTL1 has been shown to bind nonhistone targets like the tumor suppressor protein p53 (p53K382me1). Unlike other members of the MBT family, both L3MBTL1 and L3MBTL3 are nonselective readers of lysine methylation that do not recognize histone peptides in a sequence-selective manner [171]. Furthermore, L3MBTL1 has been described as a "chromatin lock" with the ability to negatively regulate the expression of E2F regulated genes like c-myc through the binding of the retinoblastoma protein (pRb) to form a repressed E2F/pRb complex [170]. Homozygous deletion of L3MBTL3 was identified in human medulloblastoma [172], and, in general, reduced L3MBTL expression may be relevant in certain subsets of myeloid leukemia [96, 173].

4.4.2 Small Molecules Targeting MBT Domains

The majority of small-molecule probes developed against MBT domains have focused on L3MBTL1 and L3MBTL3 proteins. A crucial first step in targeting methyllysine-binding proteins was the development and optimization of a broadly applicable chemiluminescent AlphaScreen™ bead-based proximity assay capable to identify small molecules that disrupt interactions between methylated histone residues and the reader protein [174]. A preliminary screen identified two compounds, Cefsulodin and I-OME-Tyrphostin AG 538 (Fig. 11; IC50 = 98 nM and

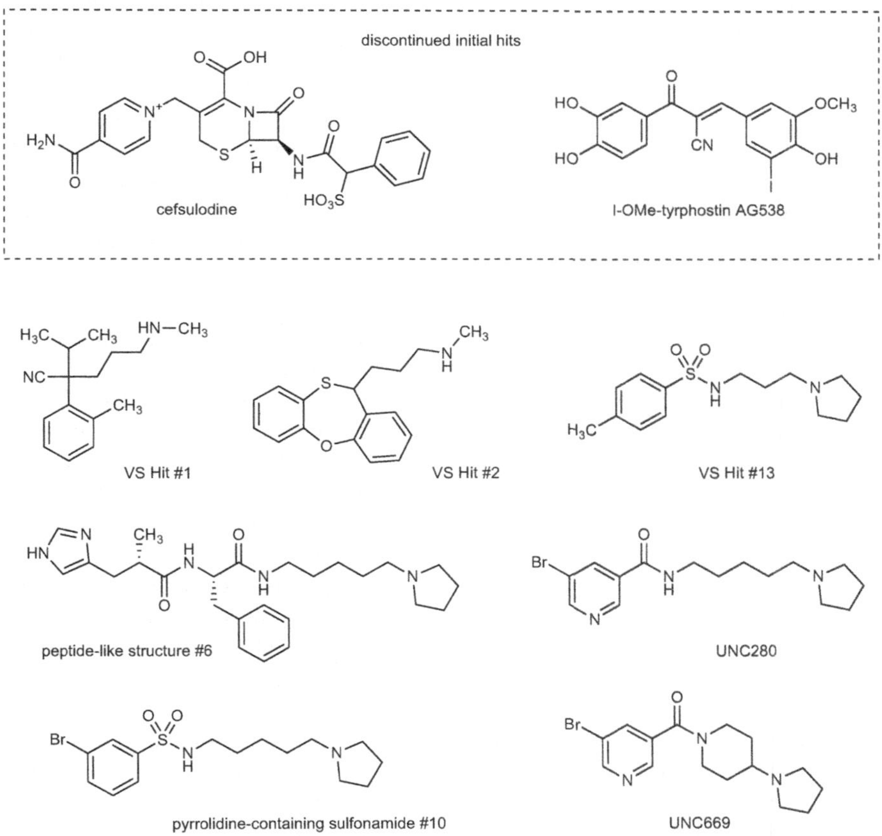

Fig. 11 Initial hits and first generation probes for L3MBTL1 (compound numbering as in the original papers) [175, 176]

282 nM, respectively) as potent inhibitors of L3MBTL1/H3K9me1 interactions. A specific binding of these compounds to the labeled L3MBTL1-His protein was confirmed by follow-up studies. Yet, further reports on these compounds have not been disclosed due to unclearness of the exact mechanism of action.

4.4.2.1 Targeting L3MBTL1

A virtual screening campaign performed on a large library of commercially available compounds containing such moieties led to the identification of several hits, which were predicted to bind with high affinity to the H4K20Me2 pocket on L3MBTL1 [175]. Among them, a few compounds (in particular VS Hits #1, 2, and 13; Fig. 11) that contained a substituted aromatic moiety linked to an alkylated amine through a variety of different tethers were identified as modest inhibitors of L3MBTL1 in the AlphaScreen™ assay (IC_{50} values = 14–17 μM). Docking studies with VS Hit #13 in the MBT domain suggested that the pyrrolidine nitrogen mimics the Kme and forms an ionic bond with Asp355, while the sulfonamide forms hydrogen bonds with the side chains of Tyr386 and Asn358.

A combined ligand- and structure-based approach (using the co-crystal structure of the L3MBTL1-H4K20me2 complex; PDB: 2PQW [166]) led to the design of a series of peptidomimetic inhibitors targeting L3MBTL1 [176]. These compounds included a histidine residue to promote interactions between the small molecule and Thr385 on the protein surface, while interactions with the methyllysine-binding pocket of the MBT2 domain were probed by the addition of various alkylated diamines. Not surprisingly based on the hits from the virtual screening study, the pyrrolidine peptidomimetic structure #6 (Fig. 11) demonstrated significantly improved binding affinity (K_D = 37 μM) compared to several acyclic mono-, di-, or trimethylated amines. Furthermore, the compound occupied the anticipated MBT2 domain site as confirmed by the loss of binding to an L3MBTL1-D355A mutant [176]. To simplify the peptidomimetic structure and determine whether substituted aromatic moieties with improved drug-like properties (cell permeability, ligand efficiency) could enhance the scaffold, a large library of analogs was synthesized and evaluated. While much of the structure-activity relationship (SAR) for this region of the scaffold was initially flat, nicotinamide derivative UNC280 (Fig. 11) did demonstrate moderate-binding affinity (K_D = 26 μM) and a fourfold increase in ligand efficiency when compared to the native peptide, H4K20me1 9-mer (residues 17–25), in ITC studies. UNC280 was not selective for L3MBTL1 but rather demonstrated modest potency against L3MBTL3 and PHF13 domains (IC_{50} values of 28 μM and 48 μM, respectively). The incorporation of the sulfonamide moiety from VS Hit #13 into the UNC280 scaffold led to the pyrrolidine containing sulfonamide derivatives #10 (Fig. 11), which demonstrated a comparable binding affinity to L3MBTL1 (K_D = 25 μM). Both the substitution of the pyrrolidine ring of this derivative and the addition of an oxygen atom to the linker produced significantly reduced or inactive compounds. By contrast, substituting the alkyl linker with a bulky and inflexible piperidine resulted in compound UNC669

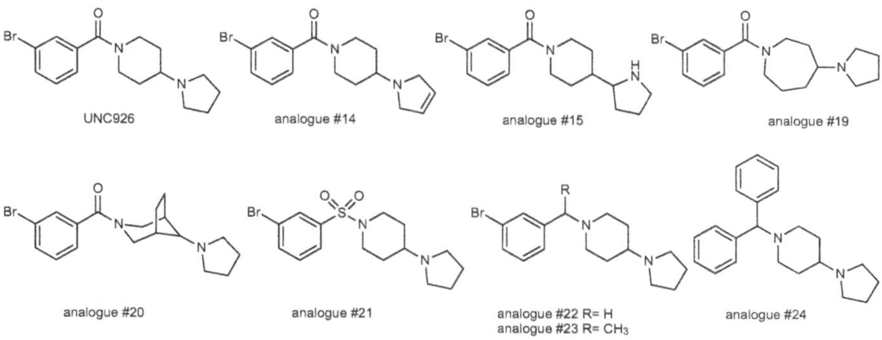

Fig. 12 Second-generation probes for L3MBTL1 (compound numbering as in the original paper) [177]

(K_D values = 5–8.6 μM), which demonstrated approximately a fivefold increase in binding affinity to L3MBTL1 compared to UNC280, but without improved selectivity between L3MBTL1 and L3MBTL3 in the AlphaScreen™ assay (IC_{50} = 6–7.9 μM and 4.8 μM, respectively) [176, 177]. In addition, UNC669 did not bind to the L3MBTL1-D355A mutant suggesting it targets the intended MBT2 domain-binding pocket. ITC and AlphaScreen™ binding results for UNC669 were further confirmed with a competitive fluorescence polarization (FP) inhibition assay using a FAM-labeled H3K9Me1 peptide ($IC_{50} \sim 10$ μM).

Starting from the UNC669 scaffold, a series of second-generation molecules were designed to gain information about SARs for the aromatic ring, the piperidine/pyrrolidine moieties, and the carbonyl linker. The compounds were screened for binding, potency, and selectivity against a panel of methyllysine domains [177]. Unsurprisingly, exchanging the pyridine for a phenyl ring (UNC926, Fig. 12) improved binding twofold for L3MBTL1 (IC_{50} = 3.9 μM), yet it had no effects on selectivity over L3MBTL3 or L3MBTL4 (IC_{50} = 3.2 μM and 15.6 μM, respectively) [177]. Changing the position of the bromine atom as well as removing it from the ring resulted in significant decreases in binding affinity, suggesting favorable interactions between the meta-bromine and the protein surface. Attempts to modify the "right-side" of the UNC926 scaffold in order to obtain SARs regarding the size and disposition of the basic pyrrolidine ring had little success at producing an improved probe of L3MBTL1, with the exception of analog #14 (Fig. 12, compound numbering as in the original paper; IC_{50} = 8.6 μM). However, the compound was not selective with comparable binding to L3MBTL3 (IC_{50} = 8.7 μM) [177]. A follow-up pull-down assay demonstrated that UNC926 inhibits the interaction between the three MBT repeats of L3MBTL1 (L3MBTL13xMBT) and H4K20me1 in a concentration-dependent manner more potently than UNC669.

With the aim to place the basic amine more shallowly in the L3MBTL1 binding pocket, analog #15 was synthesized and evaluated (Fig. 12). Interestingly, while this compound did not interact with L3MBTL1 ($IC_{50} > 30$ μM), it demonstrated modest potency against the MBTD1 domain (IC_{50} = 9.2 μM), indicating amine location may promote reader selectivity. Noteworthy, replacement of the piperidine ring with a (pyrrolidinyl)azepane (analog #19, Fig. 12) or an azabicyclooctane (analog #20, Fig. 12) introduced an increase in hydrophobicity, rigidity, and size to the ligand, resulting in enhanced potency and selectivity for the L3MBTL3 reader (L3MBTL3: IC_{50} values = 6.7 μM and 3.6 μM, respectively; L3MBTL1: IC_{50} values = 28.9 μM and 23.6 μM, respectively) [177]. Modifications to the linker region connecting the phenyl and piperidine rings did not increase potency toward L3MBTL1 when compared to UNC926. Nevertheless, several of these analogs were modest inhibitors of L3MBTL3 and MBTD1 readers. Sulfonamide analog #21 maintained potency against L3MBTL3 (IC_{50} = 3.7 μM) and MBTD1 (IC_{50} = 21.9 μM) but weakly bound L3MBTL1 at 48% inhibition. Similarly, both removal of the carbonyl (analog #22, Fig. 12) or its replacement with a methyl group (analog #23) resulted in compounds that demonstrated modest potency against L3MBTL3 (IC_{50} values = 13.3 μM and 16.0 μM, respectively) and MBTD1 (IC_{50} values = 18.6 μM and 9.0 μM, respectively), but not against L3MBTL1 (IC_{50} > 30 μM). Finally, the addition of a phenyl ring (analog #24) yielded an analog with approximately eightfold selectivity for L3MBTL3 (IC_{50} = 2.4 μM) compared to L3MBTL1 (IC_{50} = 20.3 μM) [177].

4.4.2.2 Targeting L3MBTL3

L3MBTL3 reader antagonists based on the 4-(pyrrolidin-yl)-piperidine scaffold have been recently identified during a campaign aimed to develop small-molecule probes of Tudor domain-containing p53-binding protein 1 (53BP1, see above), which binds Kme2 residues adjacent to lysine or arginine [178, 179]. The primary design strategy for these compounds consisted of appending an additional basic amine to the phenyl ring of the UNC926 scaffold to provide UNC928 (Fig. 13) [179]. Interestingly, while UNC928 was inactive against 53BP1, it demonstrated submicromolar activity against L3MBTL3 (AlphaScreen™ IC_{50} = 0.36 μM).

A LANCE (Lanthanide chelate excite) time-resolved fluorescence resonance energy transfer (TR-FRET) assay was used to confirm the AlphaScreen™ results and found UNC928 had an analogous binding affinity of IC_{50} = 0.29 μM. In addition, UNC928 was moderately active only against L3MBTL1 (IC_{50} = 2.8 μM), suggesting the potential for developing selective probes of the L3MBTL3 reader.

To further probe SAR for this "dimeric" scaffold with regard to both potency and selectivity for the L3MBTL3 reader, an extensive series of small molecules that contained modifications to either the "left-side" amine, the linker, or aromatic core of UNC928 (Fig. 13) were synthesized and evaluated [179]. Repositioning the amine in the para position with respect to the original "right-side" 4-(pyrrolidin-yl)-piperidine resulted in a significant improvement in activity against L3MBTL3 (UNC1021, LANCE IC_{50} = 0.048 μM and AlphaScreen™ IC_{50} = 0.071 μM), without increased affinity for L3MBTL1 (AlphaScreen™ IC_{50} = 2.9 μM). To explore the central aromatic core, a variety of functional groups were appended to the phenyl ring of UNC1021 [179]. While the majority of compounds evaluated did not

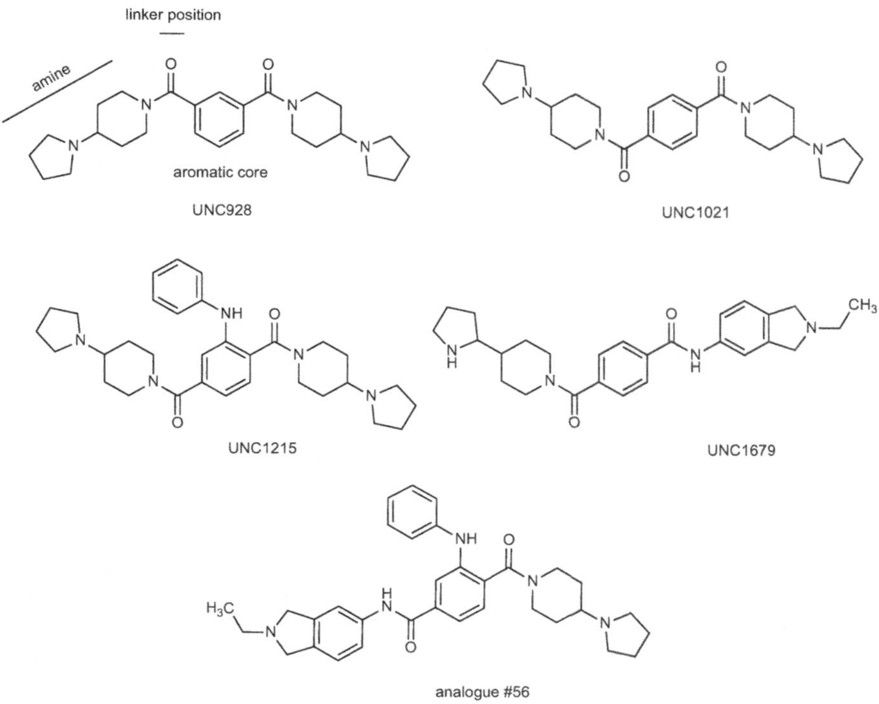

Fig. 13 First-generation probes for L3MBTL3 (compound numbering as in the original paper) [179]

show enhanced activity against L3MBTL3, the addition of an aniline moiety resulted in UNC1215, which demonstrated comparable activity to UNC1021 against L3MBTL3 ($IC_{50} = 0.064\ \mu M$), but with an enhanced binding affinity ($K_D = 0.12$).

Co-crystallographic studies of UNC1215 in complex with L3MBTL3 revealed a unique 2:2 polyvalent binding mode where one UNC1215 molecule interacted with two L3MBTL3 molecules [178]. UNC1215 showed a high degree of selectivity for L3MBTL3 in selectivity studies against a broad panel of methyllysine-binding domains, G protein-coupled receptors, ion channels, and kinases and provided preliminary evidence that cellular effects of this probe could be attributed to its inhibition of this reader protein. In vitro, UNC1215 was nontoxic (100 μM) and capable of competing with cellular factors that bind to MBT domains in fluorescence recovery after photobleaching (FRAP) experiments. Based on its selectivity and cellular activity, UNC1215 was used as a chemical probe to identify cellular proteins that interact with L3MBTL3 [178]. Affinity purification of GFP-fused L3MBTL3 transfected HEK293 cells treated with or without UNC1215 and analysis by mass spectrometry identified BCL2-associated transcription factor 1 (BCLAF1), a transcriptional repressor involved with DNA damage repair, as an antagonized interacting partner of L3MBTL3 [180]. The BCLAF1 interaction with

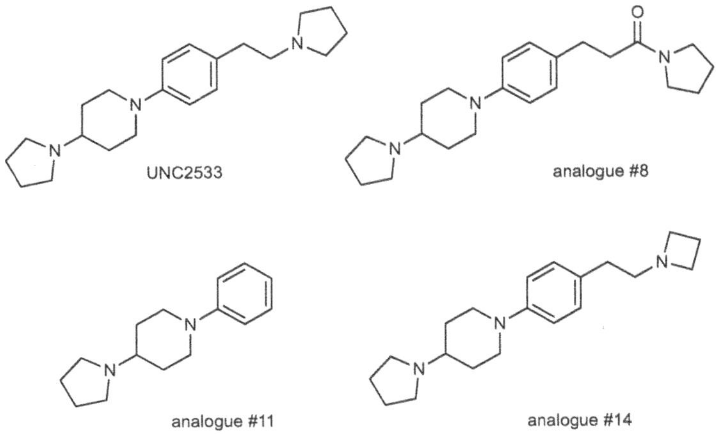

Fig. 14 Second-generation probes for L3MBTL3 (compound numbering as in the original paper) [182]

L3MBTL3 was further confirmed by immunofluorescence, immunoprecipitation, and mutagenesis experiments. Lastly, AlphaScreen™ assay confirmed that K455me, the known methylation mark of BCLAF1, effectively bound to L3MBTL3 with similar affinity as H4K20me histone peptides [178]. The use of UNC1215 to identify this cellular interaction highlights the importance of developing specific probes for each of the methyllysine reader proteins [181].

Parallel attempts to enhance the selectivity of the UNC1215 scaffold for L3MBTL3 focused on substituting a variety of basic amine-containing groups that were predicted to prevent its interaction with the MBT2 domain of L3MBTL1 [179]. These studies resulted in the identification of two compounds, UNC1679 and analog #56 (numbering as in original paper; Fig. 13), that demonstrate comparable potency against L3MBTL3 (IC_{50} values 0.17 μM and 0.13 μM, respectively) and no activity against L3MBTL1 at ~10 μM or greater. A series of molecular modeling and mutational studies provided strong evidence that the enhanced selectivity for L3MBTL3 demonstrated by these compounds is due to favorable stabilizing π-π or CH-π bonds between the 2-ethylisoindoline group and Phe387, interactions that cannot occur in the L3MBTL1 reader domain where the analogous residue is Leu361 [179]. Both UNC1679 and analog #56 were nontoxic in cell culture, and their ability to antagonize L3MBTL3 activity in vitro was confirmed by the eliminating foci formation when HEK293 cells were treated with either compound conjugated to the cell-permeable merocyanine dye mero76 and GFP-FLMBT.

A combination of structural and free energy computations guided the development of second-generation "UNC1215-like" probes targeting L3MBTL3 [182]. These computational studies suggested that small molecules with reduced size might more suitably interact with a compact reader pocket in L3MBTL3. With this in mind, UNC2533 (Fig. 14) was synthesized and identified as a comparable inhibitor of L3MBTL3 (IC_{50} = 0.62 μM and K_D = 0.37 μM) [182]. Similar to

UNC1215, the UNC2533-L3MBTL3 co-crystal structure revealed a 2:2 dimer complex with UNC2533 bridging the L3MBTL3 dimer interface and interacting with the MBT1 domain of one monomer and the MBT2 domain of the second (PDB: 4 L59) [182]. When compared to the UNC1215-L3MBTL3 complex, the two monomers rotated at the dimer interface to accommodate for the smaller size of UNC2533 rather than altering the conformation of the protein backbone in the ligand-binding site [182].

A series of UNC2533-based compounds with modifications to both amine moieties and the two carbon linker between the phenyl and pyrrolidinyl rings were synthesized and evaluated [182]. Increasing the size of the (pyrrolidinyl)piperidine moiety or replacing the piperidine with an aliphatic C-2 chain was not well-tolerated as evidenced by significant decreases in activity against L3MBTL3. Reducing the ethyl linker to a methylene reduced potency approximately sixfold ($IC_{50} = 0.35$ μM), while increasing the tether to three carbons maintained potency against L3MBTL3 ($IC_{50} = 0.07$ μM), but did not demonstrate the same binding affinity in ITC studies ($K_D = 1.8$ μM). The requirement for a basic amine on both sides of the phenyl ring was highlighted by the 40- (analog #8; numbering as in the original paper) and 200-fold (analog #11) loss in potencies seen when these moieties were either modified or removed. Interestingly, without the second basic amine, a decrease in L3MBTL1 activity was also observed, indicating the significance of the pyrrolidine ring in binding to this reader. Replacing the pyrrolidine with an azetidine (analog #14) resulted in a compound with equipotent activity toward L3MBTL3 ($IC_{50} = 0.048$ μM) and increased selectivity with respect to L3MBTL1 ($IC_{50} = 86$ μM) when compared to UNC2533 [182].

From a SAR perspective, it can be concluded that two basic moieties at each end of the molecule are necessary to improve potency and selectivity toward L3MBTL3 due to its unique dimer formation. Furthermore, smaller inhibitors like UNC2533 highlight the rotational mechanism by which L3MBTL3 may use to bind specific ligands or methylated histones without altering Kme-binding pockets. The ability to "tune" molecules for L3MBTL3 inhibition brings promise to the development of potent inhibitors not only targeting L3MBTL3 but also other MBT domain-containing effector proteins.

4.5 The PWWP Domain

4.5.1 Structure and Functions

The PWWP domain was named for the Pro-Trp-Trp-Pro motif initially identified in the Wolf-Hirschhorn syndrome candidate 1 (WHSC1) protein [183]. This motif is loosely conserved in several other proteins, including over 20 human proteins [184]. The first (Pro) and second (Trp) residues can vary, while the third (Trp) and fourth (Pro) are well preserved. As revealed by the first PWWP domain structure to be determined, from mouse DNA methyltransferase DNMT3B, the PWWP domain contains a five-stranded antiparallel β-barrel core followed by a helix bundle of one to six α-helixes following the β-barrel [185]. The PWWP motif (SWWP in DNMT3B) belongs to the β-barrel and is at the interface between the two subdomains. The WWP submotif initiates strand β2. While the PWWP domain of DNMT3B binds DNA, there is evidence that it also interacts with methylated targets, and for several other proteins, it was demonstrated that the PWWP domain is a methyllysine-binding domain [124, 184, 186–188]. General principles for the PWWP fold were derived from the structural characterization of seven human proteins that recognize a methylated lysine [184]. The PWWP domains can be subdivided into three units, namely, a β-barrel, an insertion motif between the second and third strand of the β-barrel which varies in length and secondary structure, and a C-terminal α-helical bundle. WHSC1 is a member of the NSD (nuclear receptor SET domain-containing) family of proteins, large multidomain-containing proteins that, in addition to the PWWP domain, include PHD domains and a catalytic SET (Su(var), E(z), and Trithorax) histone methyltransferase domain [189]. Upregulation of WHSC1 has been linked to several cancers including neuroblastoma [190], gliomas [191], and several others including bladder cancer [190] and has also been linked to increased tumor aggressiveness [192].

Furthermore, a gain-of-function WHSC1 point mutation (E1099K) has also been identified in pediatric acute lymphoblastic leukemia cell lines resulting in a dependency on mutant WHSC1 activity [193]. Amplification of WHSC1L1 has been seen in breast cancer where knockdown modulated the growth and survival of these cells [194]. Interestingly, while no small molecules have been developed to date to target the PWWP domain as a methyllysine reader, very recently a peptide inhibitor targeting the SET domain of WHSC1 has been reported [195].

5 The WD40 Repeat Domain

Proteins that contain WD40 domains constitute a large family, and examples have been identified in roles throughout the cell [196]. In the nucleus, WD40 β-propeller proteins are frequently found as components of chromatin-associated complexes. Here, they mediate inter-complex interactions or facilitate tethering of the complexes to the nucleosome. WD40 repeat proteins are commonly part of large, multiprotein complexes whose protein-protein interactions help orchestrate an array of downstream effects such as vesicle biogenesis [197], cytokinesis [198], chromatin dynamics (i.e., DNA replication and cell cycle progression) [199], and transcriptional regulation [200–203]. More than one WD40 domain protein may be found in a single complex. As yet, a high-resolution structure of a complex including a β-propeller domain, showing its full set of interactions, is not available. Nevertheless, a number of pair-wise interactions between the WD40 proteins and histone proteins, or peptides mimicking histones, have been characterized. These, rather than displaying a common theme, illustrate the versatility of the WD40 β-propeller fold [204].

5.1 *Recognition of the Histone H3 Tail by WDR5*

The plasticity of WD40 β-propeller recognition sites is an important and recurring feature in binding studies, and this is underlined in the investigation of WDR5 and its interaction with histone H3. This β-propeller protein is a member a multiprotein complex that is associated with the KMT2 methyltransferase family [205, 206]. The KMT2s are a family of six histone H3K4-specific methyltransferases, KMT2A-D, KMT2F, and KMT2G, also known as mixed-lineage leukemia protein 1–4 (MLL1–MLL4) and SET domain-containing protein 1A and 1B (SETD1A and SETD1B), respectively. These enzymes are implicated in the positive transcriptional regulation of critical developmental genes [207]. Although the conserved C-terminal catalytic SET domain consists of only 150 amino acids, the KMT2s are large multidomain proteins. It is presumed that the majority of the protein domains are involved in ensuring a highly specific and regulated gene targeting. However, the KMT2 SET domain is not competent for methyl transfer on its own but requires assembly into a multiprotein complex for full biochemical activity [205, 206]. This associated "core complex" consists of four proteins, ASH2L ([absent, small, or homeotic]-like protein), DPY30, and two WD40 β-propeller proteins, WDR5 and RBBP5. Both knockdown experiments in cells and in vitro reconstitution experiments confirm that assembly of the full complex is required for full SET domain activation [205, 208, 209]. Early biochemical studies indicated that WDR5 might function in histone H3 amino tail recognition and recruitment of the complex to the nucleosome and even suggested that WDR5 may sense the H3K4 methylation state [209]. However, subsequently, structural analyses revealed that the β-propeller robustly recognized only a short sequence motif (Ala-Arg-Thr) in the amino tail of histone H3 [210]. The H3-binding site is located in the axial region on the top of the WDR5 propeller [210].

The integral role that the histone H3R2 side chain has in the histone recognition mechanism of the WDR5 β-propeller is particularly significant because this residue is itself a target for posttranslational modifications associated with epigenetic regulation. The R2 arginine side chain can be asymmetrically dimethylated (H3R2me2a) by PRMT6 or symmetrically dimethylated (H3R2me2s) by PRMT5 and PRMT7 [202, 203]. The effect that these two types of arginine methyl modification have on binding to WDR5 is surprisingly divergent. Whereas the H3R2me2a modification effectively ablates binding to WDR5, in contrast, the H3R2me2s modification results in up to tenfold higher binding affinity [202, 203, 210]. WDR5 is rather an arginine reader than a methyllysine reader. This is even more underlined by the observation that WDR5 binds several arginine-containing peptides, with some of them even lacking a lysine residue (e.g., SET1A, SET1B, and MLL1–MLL4). Aside from the arginine side chain itself, the interaction with H3 does not include sequence-specific contacts. In crystallization studies, the plasticity of the site has been highlighted by its occupation by a range of peptide sequences. These included the amino terminus of a neighboring WDR5 molecule in the lattice (sequence ARA), or part of the purification tag (GRE) [211]. The role of WDR5 in the cell is not limited to its assembly into the MLL core complex, but rather it is a ubiquitously expressed protein that has been identified in other chromatin-associated complexes, including the ATAC and NSL histone acetyltransferase complex [212, 213], and associates with the chromodomain helicase DNA-binding protein 8 (CHD8) nucleosome-remodeling enzyme [214]. It is highly likely that the WDR5 histone H3-binding function is integral to its role in these complexes. The epigenetic regulation of this binding activity by arginine methylation may, therefore, have significant downstream consequences through a number of chromatin-mediated pathways. WDR5 was also recently reported to play a role in bladder cancer by mediating the transcription of cyclin B1, cyclin E1, cyclin E2, UHMK1, MCL1, BIRC3, and Nanog by H3K4me3 suggesting that this reader can be a potential biomarker and therapeutic target for the treatment of bladder cancer [215].

5.2 Methyllysine-Specific Recognition by EED

The axial site on the top of the β-propeller is also the site of histone binding by the EED subunit of Polycomb Repressive Complex 2 (PRC2). However, in this case, an aromatic cage senses the methylation state of lysine residues [216]. The PRC2 multiprotein complex is a member of the Polycomb group and functions by copying repressive histone lysine methylation marks from one histone tail to another [217]. This facilitates the spreading of a repressive domain within one cell or copying it into a new daughter cell to control the genetic program [218]. The copying of chromatin domains is needed to ensure the correct development of multicellular organisms but often goes awry in diseases such as cancer [219]. How PRC2 recognizes existing repressive lysine marks and then makes a faithful copy of them to newly formed histones is now beginning to be understood. The PRC2 complex consists of five core subunits: histone-lysine *N*-methyltransferase EZH2 (enhancer of zeste homolog 2; also known as KMT6), which contains the catalytic SET domain, SUZ12 (suppressor of zeste 12 protein homolog), zinc finger protein AEBP2 (adipocyte enhancer-binding protein 2) and two WD40 β-propellers, EED (embryonic ectoderm development protein), and either RBBP4 or RBBP7 (see before) [220]. Similar to the transcriptional activator MLL, the EZH2 SET domain is not very active on its own and requires the assembly with the other PRC2 subunits to methylate histone lysines [217]. The target lysine of PRC2 is histone H3K27, to which EZH2 can add one, two, or three methyl groups. The fully methylated histone H3K27 (H3K27me3) is the epigenetic signature that is classically associated with the repressive chromatin state [221]. However, as the field of epigenetic research has matured, it has become clear that the role of individual marks is more complicated, and for example, H3K27 is known to coexist with H3K4 on bivalent promoters in ESC cells [222]. This more complex view is reflected in how EZH2 activity within the PRC2 complex is regulated through the β-propeller protein, EED. Given that misregulation of PRC2 activity is strongly implicated in the molecular etiology of a range of cancer types, there has been an extensive effort to elucidate the mechanism of its activity [223, 224]. One long-standing experimental observation is that PRC2 activity is markedly higher with native nucleosomes substrates, naturally containing many chemical modifications, than with unmodified recombinant nucleosomes [217]. This implies a role for the pre-existing marks in regulating PRC2 activity, and this model was refined further when it was observed that the complex binds to the H3K27me3 mark in vivo [225]. Thus, pre-existing H3K27me3 marks recruit PRC2 to newly synthesized DNA, and this, in turn, facilitates the H3K27 methylation of the nascent histones. The determination of the EED β-propeller structure in complex with histone peptides has provided an explanation for the role of pre-existing modifications in the recruitment and activation of PRC2 [216].

An initial EED structure, in complex with a helical peptide derived from the amino terminus of EZH2, confirmed that it is a seven-bladed β-propeller, which has a helical insertion of currently unknown function into blade 3 [226]. Later, EED was crystallized in the absence of binding partners, however, serendipitously the conditions needed to produce well-diffracting crystals of the apo-form contained 3-(ethyldimethylammonio)propane-1-sulfonate (NDSB-195), a non-detergent sulfobetaine additive [216]. Fluorescence competition assays were used to assess the relative affinity of EED for histone peptides carrying epigenetically relevant trimethyl marks. These experiments confirmed that EED does bind to histone tails, but in addition, they revealed that the binding was restricted to those associated with repressive chromatin – H1K26, H3K9, H3K27, and H4K20. Conversely, peptides representing marks associated with active chromatin such as H3K4, H3K36, and H3K79 did not bind. The authors of this study went on to determine a series of crystal structures of EED bound to the four repressive trimethylated peptides [216]. The four trimethyl peptides bound to EED β-propeller, in fluorescence and isothermal calorimetry measurements, with dissociation constants in the range 10 μM for H3K36me3 to 45 μM for H3K27me3. H3K27me2 peptide had a K_D of only 114 μM, and H3K27me1 had a K_D of only 434 μM, showing that recognition of H3K27 by EED is methylation state specific. The structure also revealed the mechanism that allows EED to select for only histone marks associated with repressive chromatin domains.

5.3 *Small Molecules Targeting WD40 Domains*

As mentioned above, in addition to being associated with disease gene networks, many WDR domain-containing proteins are target candidates for therapy in cancer, metabolic disorders, neurological diseases, and regenerative medicine. Although these putative targets remain to be fully validated, the WDR domain is potentially a common yet unexploited entry point for drug discovery in many disease areas. Indeed, several compounds have been recently identified as inhibitors of WD40 domain-containing protein complexes.

5.3.1 Targeting WDR5

As a common component of mammalian H3K4 methyltransferase complex, WDR5 is a major H3K4 methylation associated protein [209]. As a bridge between the remainder units of the complex and MLL1, WDR5 bounds to a conserved arginine containing motif of MLL1, called "WIN" or WDR5 interaction motif [205, 227, 228]. The interaction between MLL1 and WDR5 is critical for the integrity of the MLL1 complex and, therefore, its HTM activity [229]. Knockdown of WDR5 affects the levels of global H3K4 methylation, especially decreases the level of H3K4me1/3, and downregulates Hox gene expression in human cells [209]. So disturbing the interaction of WDR5-MLL1 with antagonists to inhibit H3K4 methyltransferase activity can be a potential therapeutic strategy to treat leukemias carrying MLL1 fusion proteins.

The discovery of the first inhibitor of MLL1-WDR5 interaction originated from the identification of peptide Ac-ARA-NH_2 (Fig. 15) as the minimum MLL1-WIN motif able to bind to WDR5 with high affinity [230]. Based on Ac-ARA-NH2, three linear peptidomimetic inhibitors of MLL1-WDR5 PPI were identified (MM-101, MM-102, and MM-103; estimated $K_i < 1$ nM) by modification of the Ala residues [231]. Then cyclic peptidomimetic MM-401 was designed by constraining conformation of the linear peptide MM-102 (Fig. 15) [229]. In a competitive FP experiment, MM-401 showed an IC_{50} value of 0.9 nM in disrupting the interaction of WDR5 with MLL1. MM-401 was further demonstrated to specifically inhibit the enzymatic activity of MLL1 ($IC_{50} = 0.32$ μM) by blocking the MLL1-WDR5 interaction and thus the assembly of the complex assembly. MM-401 is cytotoxic to MLL-associated leukemic cells in an MLL1-mediated H3K4 methylation-dependent manner by inducing cell cycle arrest, apoptosis, and myeloid differentiation, whereas

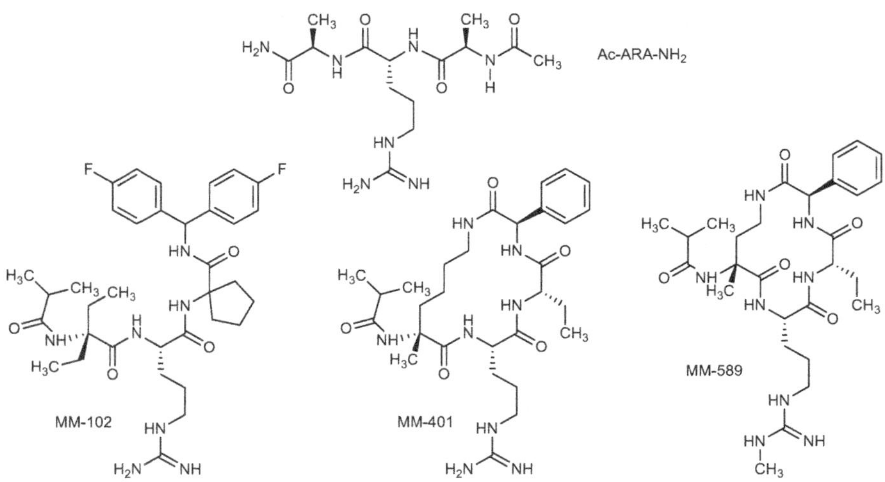

Fig. 15 Peptide-based inhibitors of WDR5

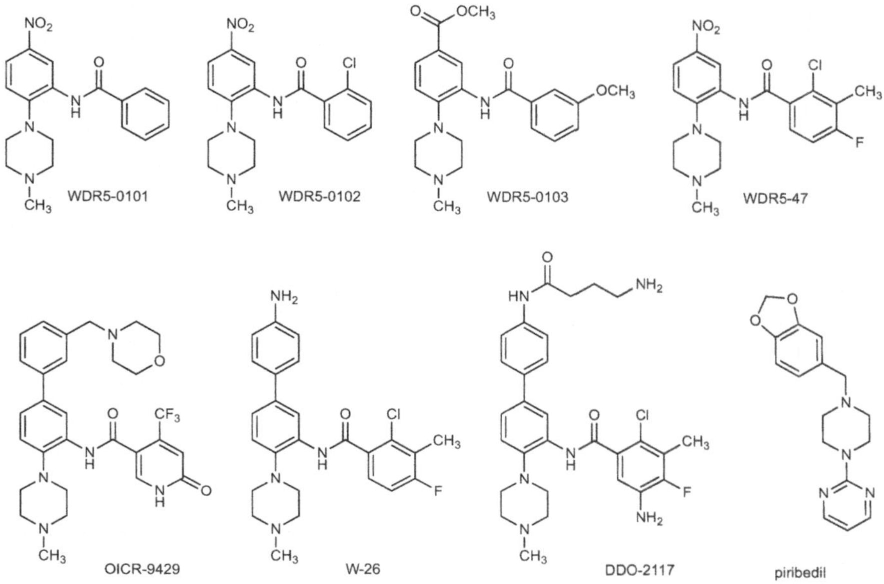

Fig. 16 Small-molecule inhibitors of WDR5

normal cells stay unaffected, making MM-401 an interesting candidate to be further evaluated as a potential therapeutic drug [229]. Based upon the co-crystal structure of MM-401 in complex with WDR5, other macrocyclic peptidomimetics were designed and synthesized to further determine the structure-activity relationships for this class [232]. In particular, this study determined the optimal linker length in these macrocyclic peptidomimetics and discovered a number of potent and promising macrocyclic peptidomimetics with MM-589 (Fig. 15) as the best compound, capable to bind to WDR5 with a K_i value <1 nM and inhibit the MLL HMT activity with an IC_{50} value of 12.7 nM. Significantly, MM-589 resulted >40 times more potent than MM-401 in inhibition of the MLL HMT activity and growth of MOLM-13 and MV4-11 human leukemia cell lines harboring MLL translocation, with >30-fold selectivity over HL-60 leukemia cell line lacking MLL translocation. The compound also displayed excellent metabolic stability in human, mouse, and rat microsomes ($T_{1/2} > 60$ min). Further optimization of MM-589 may ultimately yield a new therapy for the treatment of acute human leukemia carrying MLL translocation and potentially other human diseases and conditions that depend upon the MLL HTM activity, the WDR5-MLL interaction, or WDR5.

The first small-molecule inhibitor of the WDR5-MLL1 interaction was identified in 2013 by screening a library of 16,000 compounds [233]. The most promising hit (WDR5–0101; Fig. 16) revealed a K_D value of 12 μM. Further screening of a library of six million commercially available compounds led to 119 molecules with similarities to WDR5-0101. Among these compounds, WDR5–0103 exhibited the highest binding affinity with a K_D value of 450 nM (Fig. 16) [233]. Based on the co-crystal structure of the inhibitor-WDR5 protein, a more potent antagonist, WDR5-47 (Fig. 16), was obtained from the optimization of WDR5-0102 [234]. A more potent antagonist OICR-9429 (Fig. 16) was reported to explore the mechanism of p30-dependent transformation and establish the essential p30 cofactor WDR5 as a therapeutic target in CEBPA-mutant AML [235]. With the aim to improve the binding affinity of OICR-9429 to WDR5 ($K_D = 50$ nM, as measured by ITC), other groups also worked on this scaffold and designed and synthesized a series of biphenyl inhibitors of MLL1-WDR5 PPI [236–238]. Among them, compounds W-26 [237] and, even better, DDO-2117 (Fig. 16) [236] effectively inhibited MLL1 HMT activity in vitro and in MV4-11 cell line. In particular, DDO-2117 proved to be a high-affinity inhibitor of the MLL1-WDR5 interaction ($IC_{50} = 7.6$ nM, $K_D = 13.6$ nM) and showed the most potent inhibitory activity ($IC_{50} = 0.19$ μM) in HMT assay [236].

Very recently, a "drug repositioning" approach was applied, and a library of 592 FDA-approved drugs was screened for MLL1 inhibitors by measuring alterations in HTRF signal by means of an in vitro histone methyltransferase assay. The dopamine D2/D3 agonist piribedil (Fig. 16), which is used for the treatment of patients with Parkinson's disease and circulatory disorders [239], exhibited a promising antileukemic effect on cells harboring MLL-FPs [240]. Mechanism study showed that piribedil blocked the MLL1-WDR5 interaction and thus selectively reduced MLL1-dependent H3K4 methylation. Importantly, MLL1 depletion-induced gene expression that was similar to that induced by piribedil and rendered the MLL-r cells resistant to piribedil-induced toxicity, revealing piribedil exerted antileukemia effects by targeting MLL1. Furthermore, both the piribedil treatment and MLL1 depletion sensitized the MLL-r cells to doxorubicin-induced apoptosis [240]. Piribedil has been previously shown to inhibit the growth of colorectal cancer DLD1 cells [241], and this study suggested that it could serve as a new drug for the treatment of MLL-r AML and provide new insight for further optimization of targeting MLL1 HMT activity.

5.3.2 Targeting EED

In an effort to identify PRC2 inhibitors, recently a homogeneous time-resolved fluorescence (HTRF) assay was employed to screen approximately 1.4 million compounds using the recombinant 5-member PRC2 complex as an enzyme, and the H3K27me0 peptide (comprising residues 21–44) as a substrate led to the identification of a number of hits with different mechanisms of inhibition. 11,765 compounds at 30% or higher inhibition were identified as primary hits. After chemoinformatic triage of these initial hits and confirmation of the inhibition, 2,911 compounds were confirmed. From the confirmation and counterscreen data, 1,967 compounds were selected and tested in dose-response titration from 15 to 0.1 μM in a 1:2 serial dilution series. Of these, 1,405 compounds produced valid dose-response curves. These hits were further validated in an LC-MS orthogonal

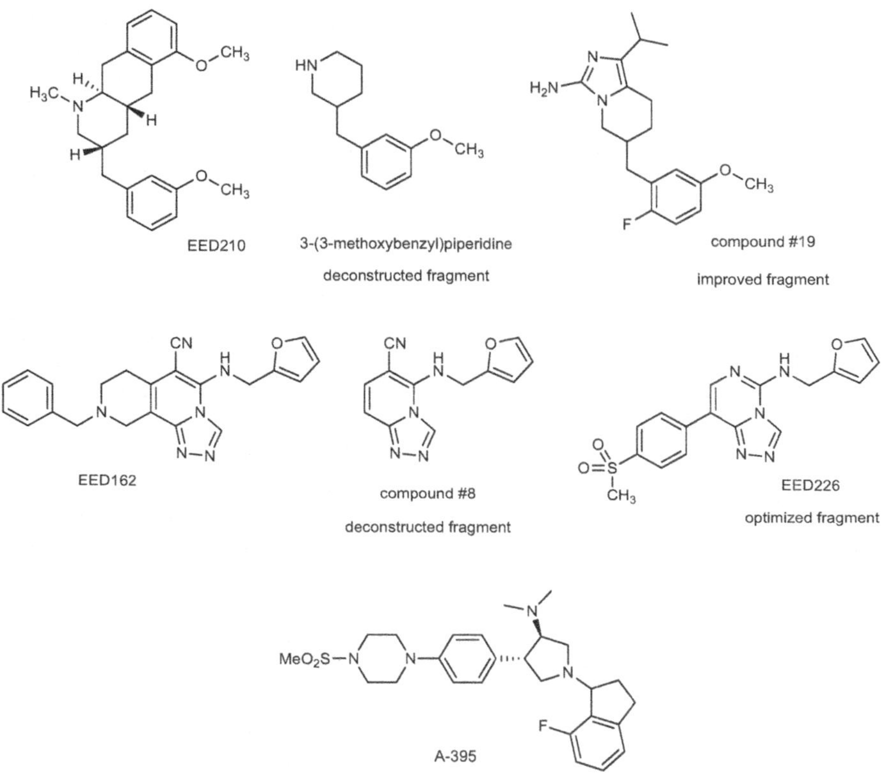

Fig. 17 Small-molecule ligands of the EED WD40 domain (compound numbering as in the original papers [243, 244])

assay, mechanism-of-action assay, cellular H3 trimethylation assay, biophysical assay, and HMT profiling. Data and chemoinformatic clustering analyses allowed to classify some hits as SAM competitive, while others as SAM noncompetitive, and these were further characterized to demonstrate that their mechanism of inhibition (MOI) was indeed through EED binding [242]. An AlphaScreen assay to measure the interaction between His-tagged recombinant EED protein and a biotinylated H3K27me3 peptide showed that these compounds, including derivatives EED162 and EED210 (Fig. 17), reduced the H3K27me3 binding signal to the basal level (with IC_{50} ranges from 0.3 to 11.6 μM), suggesting that they can fully antagonize the binding of the H3K27me3 peptide to EED. Biophysical studies (both ITC and SPR) validated that EED210 indeed directly binds to EED (calculated K_D of 28 μM and 35.1 μM, for ITC and SPR assays, respectively) [242]. Both compounds were found to inhibit not only the H3K27me3-stimulated PRC2 activity but also the basal activity in vitro and showed no inhibitory activity toward other HMTs ($IC_{50} > 100$ μM) in a panel of 22 such enzymes. It is worth noting that they showed similar inhibitory activities to both EZH2 and EZH1 harboring PRC2 complexes.

This is not surprising as EED is the common component in both complexes. The co-crystal structures of EED162 and EED210 bound to EED (as well as those obtained with the other identified hits) demonstrate a common and yet dynamic "induced fit" in the H3K27me3 pocket of EED with a significant conformational change of the aromatic cage residues [242]. These structural studies suggested that optimizing interactions with the key residues of the H3K27me3 pocket can potentially improve the binding affinities of the identified hits, providing an attractive starting point for developing novel EED binders.

Indeed, optimization studies were carried out for EED210. Due to difficulties in synthesizing this scaffold, the structure of EED210 (IC_{50} = 2.5 μM, LE = 0.29, cLogP = 4.9) was deconstructed to its minimal fragment, 3-(3-methoxybenzyl) piperidine (Fig. 17; IC_{50} = 95 μM, LE = 0.36, and cLogP = 2.8). This fragment showed direct binding to EED in the 2D NMR assay with a K_D of 32 μM, and its co-crystal structure with EED showed that the interactions made with the protein are the same as the ones observed for the parent compound EED210, with a total retainment of the binding pose [243]. Aided by the co-crystal structure, the deconstructed ligand was then "regrown," and this approach led to the development of a series of small molecules that allosterically inhibited PRC2 activity by interacting with the methyllysine-binding pocket of EED [243]. All compounds were evaluated for key properties, such as PRC2 inhibition (LC/MS-based assay), ligand efficiency (LE), and cellular permeability (cLogP, Caco-2 cell permeability). The most promising compound resulted derivative #19 (Fig. 17, compound numbering as in the original paper), which demonstrated enhanced potency (IC50 = 1 μM), good ligand efficiency (LE = 0.35), and improved permeability (cLogP = 3.9) [243].

A similar deconstruction-reconstruction approach was applied for the optimization of EED162 [244]. In fact, a closer inspection of the interactions between compound EED162 and EED suggested that the entire piperidine ring connecting C7 to C8 and the benzyl group attached to nitrogen at 9-position did not contribute much to the interaction and likely reduced efficiency of binding due to the nonessential lipophilicity. Compound #8 (Fig. 17, compound numbering as in the original paper), a fragment of EED162, was confirmed to retain most of the key interactions with EED and was found to be as potent as the parent compound but with dramatic improvements of both ligand efficiency (LE; 0.26 and 0.46 for EED162 and compound #8, respectively) and lipophilic efficiency (LipE; 2.6 and 5.0, respectively). Guided by X-ray crystallography, the approach finally led to the discovery of EED226 (Fig. 17) as a potent and selective inhibitor of PRC2 activity (ITC with EED, K_D = 82 nM, and PRC2 complex, K_D = 114 nM) [244, 245]. Like EED162 and EED210, EED226 was found selective against a panel of over 21 other protein methyltransferases, kinases, and other protein classes and was unsurprisingly an inhibitor for the EZH1/2-PRC2 complexes suggesting its potential effectiveness in treating EZH1 and/or EZH2 dependent cancers such as myelodysplastic disorders. EED226 directly binds to the H3K27me3 pocket of EED causing a distinct allosteric effect that inhibits both basal- and H3K27me3-stimulated PRC2 activity. The crystal structure of its complex with EED and EZH2 peptide (40–68, EBD) showed that the binding of EED226 to the aromatic cage of EED (comprised of F97, Y148, W364, and Y365) translocates side chains Y365 and W364 to create a deeper pocket allowing for insertion of EED226 further into the pocket. EED226 did not alter the conformation of the EBD-binding site. In vivo, EED226 induced the regression of tumor xenografts in mouse models. Finally, EED226 was capable of inhibiting cancer cell lines with acquired resistance to S-adenosylmethionine EZH2 inhibitors and also displayed a synergistic effect to inhibit cancer cell growth when combined with EZH2 inhibitors [245].

Another high-throughput screening effort against EED, followed by structure-based medicinal chemistry efforts, led to the identification of A-395 ($K_i = 0.4$ nM) as an allosteric inhibitor of PRC2 (Fig. 17) [246]. The compound selectively and potently binds EED and inhibits the catalytic activity of the trimeric PRC2 complex (EZH2-EED-SUZ12) in an in vitro radioactivity-based assay with an IC_{50} of 18 nM. Crystallographic studies and an AlphaLISA assay demonstrated the capability of A-395 to compete with H3K27me3 for binding to the aromatic cage formed by the WD40 repeat domain of EED ($IC_{50} = 7$ nM). Additionally, A-395 potently reduced H3K27 methylation in rhabdoid tumor cells and inhibited proliferation of human cancer cells, specifically DLBCL Pfeiffer and Karpas422, which are sensitive to EZH2 inhibition. In DLBCL Pfeiffer cell xenograft models, A-395 displayed significant inhibition of tumor growth (84%). Finally, A-395 retained antitumor effects against cell lines resistant to SET domain targeted EZH2 inhibitors [246].

A different approach to discover ligands of EED started from the identification of JARID2 trimethylated at lysine 116 (JARID2-K116me3) as an EED-binding partner and allosteric activator of PRC2 methyltransferase activity at low micromolar potency (K_D = 3 μM), which is tenfold higher than the affinity of EED for H3Kme7 in vitro [247]. The co-crystal structure of EED in complex with JARID2-K116me3 (PDB: 4X3E) provided insight to the structural features necessary for binding, specifically at residues 114–118 [248]. Using a split-and-pool synthesis method, a first-generation library of 1,029 peptides based on a truncated $JARID2_{114-118}$K116me3 scaffold (Fig. 18) was synthesized, and an on-bead magnetic enrichment screening approach for EED was used to identify lead compounds [248]. Compound #1 (Fig. 18, compound numbering as in the original paper) was identified and when evaluated by ITC had improved potency compared to $JARID2_{114-118}$K116me3 (K_D = 4.8 μM and 8.8 μM, respectively). A subsequent second-generation library consisting of 4,410 compounds was designed to enhance ligand affinity and compound #3 (Fig. 18, compound numbering as in the original paper), which demonstrated similar binding values in both ITC and an EED FP assay (K_D = 1.09 μM and IC_{50} = 1.65 μM, respectively) was identified. Crystallographic studies and further modifications to the N- and C-terminus as well as replacement of the Kme3 resulted in the more potent UNC5114 (ITC, K_D = 0.68 μM and FP assay IC_{50} = 1.74 μM) and UNC5115 (ITC, K_D = 1.14 μM and FP Assay, IC_{50} = 3.87 μM). A biotinylated UNC5114 chemiprecipitated PRC2 components, EED, EZH2, and SUZ12, confirming the endogenous interaction of UNC5114 with the Kme reader pocket of EED without disrupting the PRC2 complex. Finally, both UNC5114 and UNC5115 were able to competitively inhibit the PRC2 complex in the presence of H3K27me3 [248].

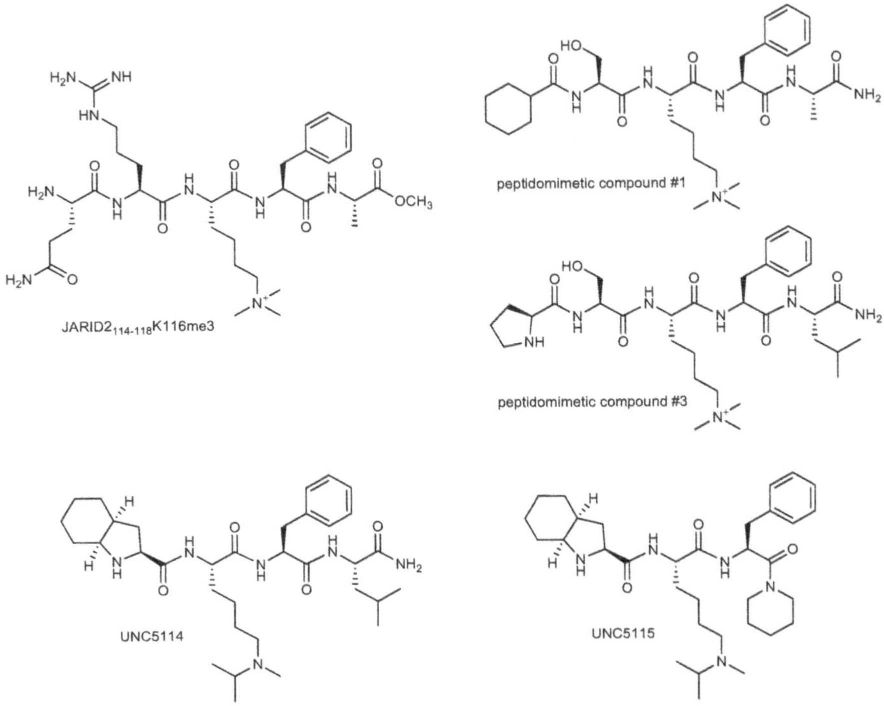

Fig. 18 Peptide-based ligands of the EED WD40 domain (compound numbering as in the original paper [248])

6 Conclusions

Due to their prevalent role in epigenetic gene regulation, methyllysine and methylarginine domain readers have emerged as potential drug targets for small-molecule intervention. The shallow aromatic-binding pockets have initially presented a challenge for targeting such domains with small molecules [249], in contrast with the great success achieved with bromodomain inhibitors. Nonetheless, the development of potent small molecules inhibiting the protein-protein interaction of methyllysine readers (Tudor, MBT, chromo-, and PHD domain) and their modified histones has proven the druggability of these sites, and, even if for many potential targets no small-molecule ligands are known yet or only rather weak inhibitors have been found so far, the increasing knowledge gained about the structure-activity relationships of the identified ligands is paving the way for the development of drug candidates targeting these readers. Future research efforts are needed in order to learn more about the biological actions and therapeutic advantages of these proteins especially those domains (PWWP, WD40, ankyrin repeats, and ADD domains) not yet targeted.

In cases where proteins comprise different properties, like combining writer or eraser functions with reader abilities, targeting the respective reader domain should also be considered as an alternative to the so-far preferred aim of influencing the enzymatic reader or eraser process. It will be interesting to address the question if inhibitors of the reader-histone interaction will phenocopy the inhibition of the enzymatic activity. This strategy would potentially preserve enzymatic nonhistone activities of the respective protein which could help in avoiding possible side effects that might occur upon the treatment of a specific phenotype.

Compliance with Ethical Standards

Funding: G.S. is supported by grants from the Italian Ministero dell'Istruzione, dell'Università e della Ricerca (MIUR), Progetti di Ricerca di Interesse Nazionale (PRIN 20152TE5PK), and from the Università di Salerno (Italy), and by European Cooperation in Science and Technology (COST Action CM1406).

Conflict of Interest: The author declares no competing financial interests.

Ethical Statement: This article does not contain any studies with human participants or animals performed by any of the authors.

References

1. Allis CD, Berger SL, Cote J, Dent S, Jenuwien T, Kouzarides T, Pillus L, Reinberg D, Shi Y, Shiekhattar R, Shilatifard A, Workman J, Zhang Y (2007) New nomenclature for chromatin-modifying enzymes. Cell 131(4):633–636. https://doi.org/10.1016/j.cell.2007.10.039
2. Martin C, Zhang Y (2005) The diverse functions of histone lysine methylation. Nat Rev Mol Cell Biol 6(11):838–849. https://doi.org/10.1038/nrm1761
3. Smith BC, Denu JM (2009) Chemical mechanisms of histone lysine and arginine modifications. Biochim Biophys Acta 1789(1):45–57. https://doi.org/10.1016/j.bbagrm.2008.06.005
4. Taverna SD, Li H, Ruthenburg AJ, Allis CD, Patel DJ (2007) How chromatin-binding modules interpret histone modifications: lessons from professional pocket pickers. Nat Struct Mol Biol 14(11):1025–1040. https://doi.org/10.1038/nsmb1338
5. Bedford MT, Clarke SG (2009) Protein arginine methylation in mammals: who, what, and why. Mol Cell 33(1):1–13. https://doi.org/10.1016/j.molcel.2008.12.013
6. Chang B, Chen Y, Zhao Y, Bruick RK (2007) JMJD6 is a histone arginine demethylase. Science 318(5849):444–447. https://doi.org/10.1126/science.1145801
7. Webby CJ, Wolf A, Gromak N, Dreger M, Kramer H, Kessler B, Nielsen ML, Schmitz C, Butler DS, Yates JR 3rd, Delahunty CM, Hahn P, Lengeling A, Mann M, Proudfoot NJ, Schofield CJ, Bottger A (2009) Jmjd6 catalyses lysyl-hydroxylation of U2AF65, a protein associated with RNA splicing. Science 325(5936):90–93. https://doi.org/10.1126/science.1175865
8. Gayatri S, Bedford MT (2014) Readers of histone methylarginine marks. Biochim Biophys Acta 1839(8):702–710. https://doi.org/10.1016/j.bbagrm.2014.02.015
9. Hughes RM, Waters ML (2006) Arginine methylation in a beta-hairpin peptide: implications for Arg-pi interactions, DeltaCp(o), and the cold denatured state. J Am Chem Soc 128 (39):12735–12742. https://doi.org/10.1021/ja061656g
10. Lee DY, Teyssier C, Strahl BD, Stallcup MR (2005) Role of protein methylation in regulation of transcription. Endocr Rev 26(2):147–170. https://doi.org/10.1210/er.2004-0008

11. Yang Y, Bedford MT (2013) Protein arginine methyltransferases and cancer. Nat Rev Cancer 13(1):37–50. https://doi.org/10.1038/nrc3409
12. Molina-Serrano D, Schiza V, Kirmizis A (2013) Cross-talk among epigenetic modifications: lessons from histone arginine methylation. Biochem Soc Trans 41(3):751–759. https://doi.org/10.1042/BST20130003
13. Dawson MA, Kouzarides T (2012) Cancer epigenetics: from mechanism to therapy. Cell 150(1):12–27. https://doi.org/10.1016/j.cell.2012.06.013
14. Anderson D, Koch CA, Grey L, Ellis C, Moran MF, Pawson T (1990) Binding of SH2 domains of phospholipase C gamma l, GAP, and Src to activated growth factor receptors. Science 250(4983):979–982
15. Deribe YL, Pawson T, Dikic I (2010) Post-translational modifications in signal integration. Nat Struct Mol Biol 17(6):666–672. https://doi.org/10.1038/nsmb.1842
16. Bannister AJ, Zegerman P, Partridge JF, Miska EA, Thomas JO, Allshire RC, Kouzarides T (2001) Selective recognition of methylated lysine 9 on histone H3 by the HP1 chromo domain. Nature 410(6824):120–124. https://doi.org/10.1038/35065138
17. Lachner M, O'Carroll D, Rea S, Mechtler K, Jenuwein T (2001) Methylation of histone H3 lysine 9 creates a binding site for HP1 proteins. Nature 410(6824):116–120. https://doi.org/10.1038/35065132
18. Nakayama J, Rice JC, Strahl BD, Allis CD, Grewal SI (2001) Role of histone H3 lysine 9 methylation in epigenetic control of heterochromatin assembly. Science 292(5514):110–113. https://doi.org/10.1126/science.1060118
19. Musselman CA, Lalonde M-E, Cote J, Kutateladze TG (2012) Perceiving the epigenetic landscape through histone readers. Nat Struct Mol Biol 19(12):1218–1227
20. Bae N, Viviano M, Su X, Lv J, Cheng D, Sagum C, Castellano S, Bai X, Johnson C, Khalil MI, Shen J, Chen K, Li H, Sbardella G, Bedford MT (2017) Developing Spindlin1 small-molecule inhibitors by using protein microarrays. Nat Chem Biol 13(7):750–756. https://doi.org/10.1038/nchembio.2377
21. Musselman CA, Khorasanizadeh S, Kutateladze TG (2014) Towards understanding methyllysine readout. Biochim Biophys Acta 1839(8):686–693. https://doi.org/10.1016/j.bbagrm.2014.04.001
22. Patel DJ, Wang Z (2013) Readout of epigenetic modifications. Annu Rev Biochem 82(1):81–118. https://doi.org/10.1146/annurev-biochem-072711-165700
23. Schindler U, Beckmann H, Cashmore AR (1993) HAT3.1, a novel Arabidopsis homeodomain protein containing a conserved cysteine-rich region. Plant J 4(1):137–150. https://doi.org/10.1046/j.1365-313X.1993.04010137.x
24. Letunic I, Bork P (2018) 20 years of the SMART protein domain annotation resource. Nucleic Acids Res 46(D1):D493–D496. https://doi.org/10.1093/nar/gkx922
25. Sanchez R, Zhou MM (2011) The PHD finger: a versatile epigenome reader. Trends Biochem Sci 36(7):364–372. https://doi.org/10.1016/j.tibs.2011.03.005
26. Oppikofer M, Sagolla M, Haley B, Zhang HM, Kummerfeld SK, Sudhamsu J, Flynn EM, Bai T, Zhang J, Ciferri C, Cochran AG (2017) Non-canonical reader modules of BAZ1A promote recovery from DNA damage. Nat Commun 8(1):862. https://doi.org/10.1038/s41467-017-00866-0
27. Li H, Ilin S, Wang W, Duncan EM, Wysocka J, Allis CD, Patel DJ (2006) Molecular basis for site-specific read-out of histone H3K4me3 by the BPTF PHD finger of NURF. Nature 442(7098):91–95. https://doi.org/10.1038/nature04802
28. Pena PV, Davrazou F, Shi X, Walter KL, Verkhusha VV, Gozani O, Zhao R, Kutateladze TG (2006) Molecular mechanism of histone H3K4me3 recognition by plant homeodomain of ING2. Nature 442(7098):100–103. https://doi.org/10.1038/nature04814
29. Shi X, Hong T, Walter KL, Ewalt M, Michishita E, Hung T, Carney D, Pena P, Lan F, Kaadige MR, Lacoste N, Cayrou C, Davrazou F, Saha A, Cairns BR, Ayer DE, Kutateladze TG, Shi Y, Cote J, Chua KF, Gozani O (2006) ING2 PHD domain links histone H3 lysine 4 methylation to active gene repression. Nature 442(7098):96–99. https://doi.org/10.1038/nature04835

30. Wysocka J, Swigut T, Xiao H, Milne TA, Kwon SY, Landry J, Kauer M, Tackett AJ, Chait BT, Badenhorst P, Wu C, Allis CD (2006) A PHD finger of NURF couples histone H3 lysine 4 trimethylation with chromatin remodelling. Nature 442(7098):86–90. https://doi.org/10.1038/nature04815
31. Wang GG, Song J, Wang Z, Dormann HL, Casadio F, Li H, Luo JL, Patel DJ, Allis CD (2009) Haematopoietic malignancies caused by dysregulation of a chromatin-binding PHD finger. Nature 459(7248):847–851. https://doi.org/10.1038/nature08036
32. Klein BJ, Piao L, Xi Y, Rincon-Arano H, Rothbart SB, Peng D, Wen H, Larson C, Zhang X, Zheng X, Cortazar MA, Pena PV, Mangan A, Bentley DL, Strahl BD, Groudine M, Li W, Shi X, Kutateladze TG (2014) The histone-H3K4-specific demethylase KDM5B binds to its substrate and product through distinct PHD fingers. Cell Rep 6(2):325–335. https://doi.org/10.1016/j.celrep.2013.12.021
33. Dhayalan A, Tamas R, Bock I, Tattermusch A, Dimitrova E, Kudithipudi S, Ragozin S, Jeltsch A (2011) The ATRX-ADD domain binds to H3 tail peptides and reads the combined methylation state of K4 and K9. Hum Mol Genet 20(11):2195–2203. https://doi.org/10.1093/hmg/ddr107
34. Iwase S, Xiang B, Ghosh S, Ren T, Lewis PW, Cochrane JC, Allis CD, Picketts DJ, Patel DJ, Li H, Shi Y (2011) ATRX ADD domain links an atypical histone methylation recognition mechanism to human mental-retardation syndrome. Nat Struct Mol Biol 18(7):769–776. https://doi.org/10.1038/nsmb.2062
35. Mansfield RE, Musselman CA, Kwan AH, Oliver SS, Garske AL, Davrazou F, Denu JM, Kutateladze TG, Mackay JP (2011) Plant homeodomain (PHD) fingers of CHD4 are histone H3-binding modules with preference for unmodified H3K4 and methylated H3K9. J Biol Chem 286(13):11779–11791. https://doi.org/10.1074/jbc.M110.208207
36. Musselman CA, Mansfield RE, Garske AL, Davrazou F, Kwan AH, Oliver SS, O'Leary H, Denu JM, Mackay JP, Kutateladze TG (2009) Binding of the CHD4 PHD2 finger to histone H3 is modulated by covalent modifications. Biochem J 423(2):179–187. https://doi.org/10.1042/BJ20090870
37. Xi Q, Wang Z, Zaromytidou AI, Zhang XH, Chow-Tsang LF, Liu JX, Kim H, Barlas A, Manova-Todorova K, Kaartinen V, Studer L, Mark W, Patel DJ, Massague J (2011) A poised chromatin platform for TGF-beta access to master regulators. Cell 147(7):1511–1524. https://doi.org/10.1016/j.cell.2011.11.032
38. Gibbons RJ, Wada T, Fisher CA, Malik N, Mitson MJ, Steensma DP, Fryer A, Goudie DR, Krantz ID, Traeger-Synodinos J (2008) Mutations in the chromatin-associated protein ATRX. Hum Mutat 29(6):796–802. https://doi.org/10.1002/humu.20734
39. Wang Z, Song J, Milne TA, Wang GG, Li H, Allis CD, Patel DJ (2010) Pro isomerization in MLL1 PHD3-bromo cassette connects H3K4me readout to CyP33 and HDAC-mediated repression. Cell 141(7):1183–1194. https://doi.org/10.1016/j.cell.2010.05.016
40. Musselman CA, Kutateladze TG (2009) PHD fingers: epigenetic effectors and potential drug targets. Mol Interv 9(6):314–323. https://doi.org/10.1124/mi.9.6.7
41. Chi P, Allis CD, Wang GG (2010) Covalent histone modifications--miswritten, misinterpreted and mis-erased in human cancers. Nat Rev Cancer 10(7):457–469. https://doi.org/10.1038/nrc2876
42. van Zutven LJ, Onen E, Velthuizen SC, van Drunen E, von Bergh AR, van den Heuvel-Eibrink MM, Veronese A, Mecucci C, Negrini M, de Greef GE, Beverloo HB (2006) Identification of NUP98 abnormalities in acute leukemia: JARID1A (12p13) as a new partner gene. Genes Chromosomes Cancer 45(5):437–446. https://doi.org/10.1002/gcc.20308
43. de Rooij JD, Hollink IH, Arentsen-Peters ST, van Galen JF, Berna Beverloo H, Baruchel A, Trka J, Reinhardt D, Sonneveld E, Zimmermann M, Alonzo TA, Pieters R, Meshinchi S, van den Heuvel-Eibrink MM, Zwaan CM (2013) NUP98/JARID1A is a novel recurrent abnormality in pediatric acute megakaryoblastic leukemia with a distinct HOX gene expression pattern. Leukemia 27(12):2280–2288. https://doi.org/10.1038/leu.2013.87
44. Wagner EK, Nath N, Flemming R, Feltenberger JB, Denu JM (2012) Identification and characterization of small molecule inhibitors of a plant homeodomain finger. Biochemistry 51(41):8293–8306. https://doi.org/10.1021/bi3009278

45. Bienz M (2006) The PHD finger, a nuclear protein-interaction domain. Trends Biochem Sci 31(1):35–40. https://doi.org/10.1016/j.tibs.2005.11.001
46. Teske KA, Hadden MK (2017) Methyllysine binding domains: structural insight and small molecule probe development. Eur J Med Chem 136:14–35. https://doi.org/10.1016/j.ejmech.2017.04.047
47. Bhushan B, Erdmann A, Zhang Y, Belle R, Johannson C, Oppermann U, Hopkinson RJ, Schofield CJ, Kawamura A (2018) Investigations on small molecule inhibitors targeting the histone H3K4 tri-methyllysine binding PHD-finger of JmjC histone demethylases. Bioorg Med Chem 26(11):2984–2991. https://doi.org/10.1016/j.bmc.2018.03.030
48. Polakis P (2012) Wnt signaling in cancer. Cold Spring Harb Perspect Biol 4(5):a008052. https://doi.org/10.1101/cshperspect.a008052
49. Daniels DL, Eklof Spink K, Weis WI (2001) Beta-catenin: molecular plasticity and drug design. Trends Biochem Sci 26(11):672–678. https://doi.org/10.1016/S0968-0004(01)01952-1
50. Miller TC, Rutherford TJ, Johnson CM, Fiedler M, Bienz M (2010) Allosteric remodelling of the histone H3 binding pocket in the Pygo2 PHD finger triggered by its binding to the B9L/BCL9 co-factor. J Mol Biol 401(5):969–984. https://doi.org/10.1016/j.jmb.2010.07.007
51. Miller TC, Rutherford TJ, Birchall K, Chugh J, Fiedler M, Bienz M (2014) Competitive binding of a benzimidazole to the histone-binding pocket of the Pygo PHD finger. ACS Chem Biol 9(12):2864–2874. https://doi.org/10.1021/cb500585s
52. Amato A, Lucas X, Bortoluzzi A, Wright D, Ciulli A (2018) Targeting ligandable pockets on plant homeodomain (PHD) zinc finger domains by a fragment-based approach. ACS Chem Biol 13(4):915–921. https://doi.org/10.1021/acschembio.7b01093
53. Maurer-Stroh S, Dickens NJ, Hughes-Davies L, Kouzarides T, Eisenhaber F, Ponting CP (2003) The Tudor domain 'Royal Family': tudor, plant Agenet, chromo, PWWP and MBT domains. Trends Biochem Sci 28(2):69–74. https://doi.org/10.1016/S0968-0004(03)00004-5
54. Lu R, Wang GG (2013) Tudor: a versatile family of histone methylation 'readers'. Trends Biochem Sci 38(11):546–555. https://doi.org/10.1016/j.tibs.2013.08.002
55. Golumbeski GS, Bardsley A, Tax F, Boswell RE (1991) Tudor, a posterior-group gene of *Drosophila melanogaster*, encodes a novel protein and an mRNA localized during mid-oogenesis. Genes Dev 5(11):2060–2070. https://doi.org/10.1101/gad.5.11.2060
56. Lee KK, Workman JL (2007) Histone acetyltransferase complexes: one size doesn't fit all. Nat Rev Mol Cell Biol 8(4):284–295. https://doi.org/10.1038/nrm2145
57. Bian C, Xu C, Ruan J, Lee KK, Burke TL, Tempel W, Barsyte D, Li J, Wu M, Zhou BO, Fleharty BE, Paulson A, Allali-Hassani A, Zhou JQ, Mer G, Grant PA, Workman JL, Zang J, Min J (2011) Sgf29 binds histone H3K4me2/3 and is required for SAGA complex recruitment and histone H3 acetylation. EMBO J 30(14):2829–2842. https://doi.org/10.1038/emboj.2011.193
58. Selenko P, Sprangers R, Stier G, Buhler D, Fischer U, Sattler M (2001) SMN tudor domain structure and its interaction with the Sm proteins. Nat Struct Biol 8(1):27–31. https://doi.org/10.1038/83014
59. Friesen WJ, Massenet S, Paushkin S, Wyce A, Dreyfuss G (2001) SMN, the product of the spinal muscular atrophy gene, binds preferentially to dimethylarginine-containing protein targets. Mol Cell 7(5):1111–1117. https://doi.org/10.1016/S1097-2765(01)00244-1
60. Corsini L, Sattler M (2007) Tudor hooks up with DNA repair. Nat Struct Mol Biol 14(2):98–99. https://doi.org/10.1038/nsmb0207-98
61. Eissenberg JC (2012) Structural biology of the chromodomain: form and function. Gene 496(2):69–78. https://doi.org/10.1016/j.gene.2012.01.003
62. Hughes RM, Wiggins KR, Khorasanizadeh S, Waters ML (2007) Recognition of trimethyllysine by a chromodomain is not driven by the hydrophobic effect. Proc Natl Acad Sci U S A 104(27):11184–11188. https://doi.org/10.1073/pnas.0610850104
63. Riemen AJ, Waters ML (2009) Design of highly stabilized beta-hairpin peptides through cation-pi interactions of lysine and n-methyllysine with an aromatic pocket. Biochemistry 48(7):1525–1531. https://doi.org/10.1021/bi801706k

64. Richly H, Aloia L, Di Croce L (2011) Roles of the Polycomb group proteins in stem cells and cancer. Cell Death Dis 2:e204. https://doi.org/10.1038/cddis.2011.84
65. Yap KL, Zhou MM (2011) Structure and mechanisms of lysine methylation recognition by the chromodomain in gene transcription. Biochemistry 50(12):1966–1980. https://doi.org/10.1021/bi101885m
66. Paro R, Hogness DS (1991) The Polycomb protein shares a homologous domain with a heterochromatin-associated protein of Drosophila. Proc Natl Acad Sci U S A 88 (1):263–267. https://doi.org/10.1073/pnas.88.1.263
67. Ball LJ, Murzina NV, Broadhurst RW, Raine AR, Archer SJ, Stott FJ, Murzin AG, Singh PB, Domaille PJ, Laue ED (1997) Structure of the chromatin binding (chromo) domain from mouse modifier protein 1. EMBO J 16(9):2473–2481. https://doi.org/10.1093/emboj/16.9.2473
68. Milosevich N, Hof F (2016) Chemical inhibitors of epigenetic methyllysine reader proteins. Biochemistry 55(11):1570–1583. https://doi.org/10.1021/acs.biochem.5b01073
69. Vandamme J, Volkel P, Rosnoblet C, Le Faou P, Angrand PO (2011) Interaction proteomics analysis of polycomb proteins defines distinct PRC1 complexes in mammalian cells. Mol Cell Proteomics 10(4):M110.002642. https://doi.org/10.1074/mcp.M110.002642
70. Klauke K, Radulovic V, Broekhuis M, Weersing E, Zwart E, Olthof S, Ritsema M, Bruggeman S, Wu X, Helin K, Bystrykh L, de Haan G (2013) Polycomb Cbx family members mediate the balance between haematopoietic stem cell self-renewal and differentiation. Nat Cell Biol 15(4):353–362. https://doi.org/10.1038/ncb2701
71. Morey L, Pascual G, Cozzuto L, Roma G, Wutz A, Benitah SA, Di Croce L (2012) Nonoverlapping functions of the Polycomb group Cbx family of proteins in embryonic stem cells. Cell Stem Cell 10(1):47–62. https://doi.org/10.1016/j.stem.2011.12.006
72. O'Loghlen A, Munoz-Cabello AM, Gaspar-Maia A, Wu HA, Banito A, Kunowska N, Racek T, Pemberton HN, Beolchi P, Lavial F, Masui O, Vermeulen M, Carroll T, Graumann J, Heard E, Dillon N, Azuara V, Snijders AP, Peters G, Bernstein E, Gil J (2012) MicroRNA regulation of Cbx7 mediates a switch of Polycomb orthologs during ESC differentiation. Cell Stem Cell 10(1):33–46. https://doi.org/10.1016/j.stem.2011.12.004
73. Yap KL, Li S, Munoz-Cabello AM, Raguz S, Zeng L, Mujtaba S, Gil J, Walsh MJ, Zhou MM (2010) Molecular interplay of the noncoding RNA ANRIL and methylated histone H3 lysine 27 by polycomb CBX7 in transcriptional silencing of INK4a. Mol Cell 38 (5):662–674. https://doi.org/10.1016/j.molcel.2010.03.021
74. Duan RS, Tang GB, Du HZ, Hu YW, Liu PP, Xu YJ, Zeng YQ, Zhang SF, Wang RY, Teng ZQ, Liu CM (2018) Polycomb protein family member CBX7 regulates intrinsic axon growth and regeneration. Cell Death Differ 25(9):1598–1611. https://doi.org/10.1038/s41418-018-0064-0
75. van den Boom V, Rozenveld-Geugien M, Bonardi F, Malanga D, van Gosliga D, Heijink AM, Viglietto G, Morrone G, Fusetti F, Vellenga E, Schuringa JJ (2013) Nonredundant and locus-specific gene repression functions of PRC1 paralog family members in human hematopoietic stem/progenitor cells. Blood 121(13):2452–2461. https://doi.org/10.1182/blood-2012-08-451666
76. Shinjo K, Yamashita Y, Yamamoto E, Akatsuka S, Uno N, Kamiya A, Niimi K, Sakaguchi Y, Nagasaka T, Takahashi T, Shibata K, Kajiyama H, Kikkawa F, Toyokuni S (2014) Expression of chromobox homolog 7 (CBX7) is associated with poor prognosis in ovarian clear cell adenocarcinoma via TRAIL-induced apoptotic pathway regulation. Int J Cancer 135(2):308–318. https://doi.org/10.1002/ijc.28692
77. Simhadri C, Daze KD, Douglas SF, Quon TT, Dev A, Gignac MC, Peng F, Heller M, Boulanger MJ, Wulff JE, Hof F (2014) Chromodomain antagonists that target the polycomb-group methyllysine reader protein chromobox homolog 7 (CBX7). J Med Chem 57(7):2874–2883. https://doi.org/10.1021/jm401487x
78. Scott CL, Gil J, Hernando E, Teruya-Feldstein J, Narita M, Martinez D, Visakorpi T, Mu D, Cordon-Cardo C, Peters G, Beach D, Lowe SW (2007) Role of the chromobox protein CBX7 in lymphomagenesis. Proc Natl Acad Sci U S A 104(13):5389–5394. https://doi.org/10.1073/pnas.0608721104

79. Pallante P, Federico A, Berlingieri MT, Bianco M, Ferraro A, Forzati F, Iaccarino A, Russo M, Pierantoni GM, Leone V, Sacchetti S, Troncone G, Santoro M, Fusco A (2008) Loss of the CBX7 gene expression correlates with a highly malignant phenotype in thyroid cancer. Cancer Res 68(16):6770–6778. https://doi.org/10.1158/0008-5472.CAN-08-0695
80. Federico A, Pallante P, Bianco M, Ferraro A, Esposito F, Monti M, Cozzolino M, Keller S, Fedele M, Leone V, Troncone G, Chiariotti L, Pucci P, Fusco A (2009) Chromobox protein homologue 7 protein, with decreased expression in human carcinomas, positively regulates E-cadherin expression by interacting with the histone deacetylase 2 protein. Cancer Res 69(17):7079–7087. https://doi.org/10.1158/0008-5472.CAN-09-1542
81. Traore M, Gignac M, Doan ND, Hof F, Lubell WD (2017) Aza-amino acid scanning of chromobox homolog 7 (CBX7) ligands. J Pept Sci 23(4):266–271. https://doi.org/10.1002/psc.2982
82. Stuckey JI, Dickson BM, Cheng N, Liu Y, Norris JL, Cholensky SH, Tempel W, Qin S, Huber KG, Sagum C, Black K, Li F, Huang XP, Roth BL, Baughman BM, Senisterra G, Pattenden SG, Vedadi M, Brown PJ, Bedford MT, Min J, Arrowsmith CH, James LI, Frye SV (2016) A cellular chemical probe targeting the chromodomains of Polycomb repressive complex 1. Nat Chem Biol 12(3):180–187. https://doi.org/10.1038/nchembio.2007
83. Fischle W, Franz H, Jacobs SA, Allis CD, Khorasanizadeh S (2008) Specificity of the chromodomain Y chromosome family of chromodomains for lysine-methylated ARK(S/T) motifs. J Biol Chem 283(28):19626–19635. https://doi.org/10.1074/jbc.M802655200
84. Stuckey JI, Simpson C, Norris-Drouin JL, Cholensky SH, Lee J, Pasca R, Cheng N, Dickson BM, Pearce KH, Frye SV, James LI (2016) Structure-activity relationships and kinetic studies of peptidic antagonists of CBX chromodomains. J Med Chem 59 (19):8913–8923. https://doi.org/10.1021/acs.jmedchem.6b00801
85. Bernard D, Martinez-Leal JF, Rizzo S, Martinez D, Hudson D, Visakorpi T, Peters G, Carnero A, Beach D, Gil J (2005) CBX7 controls the growth of normal and tumor-derived prostate cells by repressing the Ink4a/Arf locus. Oncogene 24(36):5543–5551. https://doi.org/10.1038/sj.onc.1208735
86. Barnash KD, Lamb KN, Stuckey JI, Norris JL, Cholensky SH, Kireev DB, Frye SV, James LI (2016) Chromodomain ligand optimization via target-class directed combinatorial repurposing. ACS Chem Biol 11(9):2475–2483. https://doi.org/10.1021/acschembio.6b00415
87. Astle JM, Simpson LS, Huang Y, Reddy MM, Wilson R, Connell S, Wilson J, Kodadek T (2010) Seamless bead to microarray screening: rapid identification of the highest affinity protein ligands from large combinatorial libraries. Chem Biol 17(1):38–45. https://doi.org/10.1016/j.chembiol.2009.12.015
88. Mendes K, Ndungu JM, Clark LF, Kodadek T (2015) Optimization of the magnetic recovery of hits from one-bead-one-compound library screens. ACS Comb Sci 17(9):506–517. https://doi.org/10.1021/acscombsci.5b00090
89. Ren C, Morohashi K, Plotnikov AN, Jakoncic J, Smith SG, Li J, Zeng L, Rodriguez Y, Stojanoff V, Walsh M, Zhou MM (2015) Small-molecule modulators of methyl-lysine binding for the CBX7 chromodomain. Chem Biol 22(2):161–168. https://doi.org/10.1016/j.chembiol.2014.11.021
90. Li H, Li H, Qu H, Zhao M, Yuan B, Cao M, Cui J (2015) Suramin inhibits cell proliferation in ovarian and cervical cancer by downregulating heparanase expression. Cancer Cell Int 15 (1):52. https://doi.org/10.1186/s12935-015-0196-y
91. Morgan HP, McNae IW, Nowicki MW, Zhong W, Michels PA, Auld DS, Fothergill-Gilmore LA, Walkinshaw MD (2011) The trypanocidal drug suramin and other trypan blue mimetics are inhibitors of pyruvate kinases and bind to the adenosine site. J Biol Chem 286(36):31232–31240. https://doi.org/10.1074/jbc.M110.212613
92. Ragno R, Simeoni S, Castellano S, Vicidomini C, Mai A, Caroli A, Tramontano A, Bonaccini C, Trojer P, Bauer I, Brosch G, Sbardella G (2007) Small molecule inhibitors of histone arginine methyltransferases: homology modeling, molecular docking, binding mode

analysis, and biological evaluations. J Med Chem 50(6):1241–1253. https://doi.org/10.1021/jm061213n
93. Ren C, Smith SG, Yap K, Li S, Li J, Mezei M, Rodriguez Y, Vincek A, Aguilo F, Walsh MJ, Zhou MM (2016) Structure-guided discovery of selective antagonists for the chromodomain of polycomb repressive protein CBX7. ACS Med Chem Lett 7(6):601–605. https://doi.org/10.1021/acsmedchemlett.6b00042
94. Milosevich N, Gignac MC, McFarlane J, Simhadri C, Horvath S, Daze KD, Croft CS, Dheri A, Quon TT, Douglas SF, Wulff JE, Paci I, Hof F (2016) Selective inhibition of CBX6: a methyllysine reader protein in the polycomb family. ACS Med Chem Lett 7(2):139–144. https://doi.org/10.1021/acsmedchemlett.5b00378
95. Brahms H, Meheus L, de Brabandere V, Fischer U, Lührmann R (2001) Symmetrical dimethylation of arginine residues in spliceosomal Sm protein B/B' and the Sm-like protein LSm4, and their interaction with the SMN protein. RNA 7(11):1531–1542
96. Botuyan MV, Lee J, Ward IM, Kim JE, Thompson JR, Chen J, Mer G (2006) Structural basis for the methylation state-specific recognition of histone H4-K20 by 53BP1 and Crb2 in DNA repair. Cell 127(7):1361–1373. https://doi.org/10.1016/j.cell.2006.10.043
97. Huyen Y, Zgheib O, Ditullio RA Jr, Gorgoulis VG, Zacharatos P, Petty TJ, Sheston EA, Mellert HS, Stavridi ES, Halazonetis TD (2004) Methylated lysine 79 of histone H3 targets 53BP1 to DNA double-strand breaks. Nature 432(7015):406–411. https://doi.org/10.1038/nature03114
98. Kim J, Daniel J, Espejo A, Lake A, Krishna M, Xia L, Zhang Y, Bedford MT (2006) Tudor, MBT and chromo domains gauge the degree of lysine methylation. EMBO Rep 7(4):397–403. https://doi.org/10.1038/sj.embor.7400625
99. Chen C, Jin J, James DA, Adams-Cioaba MA, Park JG, Guo Y, Tenaglia E, Xu C, Gish G, Min J, Pawson T (2009) Mouse Piwi interactome identifies binding mechanism of Tdrkh Tudor domain to arginine methylated Miwi. Proc Natl Acad Sci U S A 106(48):20336–20341. https://doi.org/10.1073/pnas.0911640106
100. Liu H, Wang JY, Huang Y, Li Z, Gong W, Lehmann R, Xu RM (2010) Structural basis for methylarginine-dependent recognition of Aubergine by Tudor. Genes Dev 24(17):1876–1881. https://doi.org/10.1101/gad.1956010
101. Liu K, Chen C, Guo Y, Lam R, Bian C, Xu C, Zhao DY, Jin J, MacKenzie F, Pawson T, Min J (2010) Structural basis for recognition of arginine methylated Piwi proteins by the extended Tudor domain. Proc Natl Acad Sci U S A 107(43):18398–18403. https://doi.org/10.1073/pnas.1013106107
102. Liu K, Guo Y, Liu H, Bian C, Lam R, Liu Y, Mackenzie F, Rojas LA, Reinberg D, Bedford MT, Xu RM, Min J (2012) Crystal structure of TDRD3 and methyl-arginine binding characterization of TDRD3, SMN and SPF30. PLoS One 7(2):e30375. https://doi.org/10.1371/journal.pone.0030375
103. Sikorsky T, Hobor F, Krizanova E, Pasulka J, Kubicek K, Stefl R (2012) Recognition of asymmetrically dimethylated arginine by TDRD3. Nucleic Acids Res 40(22):11748–11755. https://doi.org/10.1093/nar/gks929
104. Tripsianes K, Madl T, Machyna M, Fessas D, Englbrecht C, Fischer U, Neugebauer KM, Sattler M (2011) Structural basis for dimethylarginine recognition by the Tudor domains of human SMN and SPF30 proteins. Nat Struct Mol Biol 18(12):1414–1420. https://doi.org/10.1038/nsmb.2185
105. Liu L, Zhen XT, Denton E, Marsden BD, Schapira M (2012) ChromoHub: a data hub for navigators of chromatin-mediated signalling. Bioinformatics 28(16):2205–2206. https://doi.org/10.1093/bioinformatics/bts340
106. Ballare C, Lange M, Lapinaite A, Martin GM, Morey L, Pascual G, Liefke R, Simon B, Shi Y, Gozani O, Carlomagno T, Benitah SA, Di Croce L (2012) Phf19 links methylated Lys36 of histone H3 to regulation of Polycomb activity. Nat Struct Mol Biol 19(12):1257–1265. https://doi.org/10.1038/nsmb.2434
107. Cai L, Rothbart SB, Lu R, Xu B, Chen WY, Tripathy A, Rockowitz S, Zheng D, Patel DJ, Allis CD, Strahl BD, Song J, Wang GG (2013) An H3K36 methylation-engaging Tudor motif

of polycomb-like proteins mediates PRC2 complex targeting. Mol Cell 49(3):571–582. https://doi.org/10.1016/j.molcel.2012.11.026
108. Musselman CA, Avvakumov N, Watanabe R, Abraham CG, Lalonde ME, Hong Z, Allen C, Roy S, Nunez JK, Nickoloff J, Kulesza CA, Yasui A, Cote J, Kutateladze TG (2012) Molecular basis for H3K36me3 recognition by the Tudor domain of PHF1. Nat Struct Mol Biol 19(12):1266–1272. https://doi.org/10.1038/nsmb.2435
109. Cui G, Park S, Badeaux AI, Kim D, Lee J, Thompson JR, Yan F, Kaneko S, Yuan Z, Botuyan MV, Bedford MT, Cheng JQ, Mer G (2012) PHF20 is an effector protein of p53 double lysine methylation that stabilizes and activates p53. Nat Struct Mol Biol 19(9):916–924. https://doi.org/10.1038/nsmb.2353
110. Zhang T, Park KA, Li Y, Byun HS, Jeon J, Lee Y, Hong JH, Kim JM, Huang SM, Choi SW, Kim SH, Sohn KC, Ro H, Lee JH, Lu T, Stark GR, Shen HM, Liu ZG, Park J, Hur GM (2013) PHF20 regulates NF-kappaB signalling by disrupting recruitment of PP2A to p65. Nat Commun 4:2062. https://doi.org/10.1038/ncomms3062
111. Badeaux AI, Yang Y, Cardenas K, Vemulapalli V, Chen K, Kusewitt D, Richie E, Li W, Bedford MT (2012) Loss of the methyl lysine effector protein PHF20 impacts the expression of genes regulated by the lysine acetyltransferase MOF. J Biol Chem 287(1):429–437. https://doi.org/10.1074/jbc.M111.271163
112. Park S, Kim D, Dan HC, Chen H, Testa JR, Cheng JQ (2012) Identification of Akt interaction protein PHF20/TZP that transcriptionally regulates p53. J Biol Chem 287(14):11151–11163. https://doi.org/10.1074/jbc.M111.333922
113. Zhao W, Li Q, Ayers S, Gu Y, Shi Z, Zhu Q, Chen Y, Wang HY, Wang RF (2013) Jmjd3 inhibits reprogramming by upregulating expression of INK4a/Arf and targeting PHF20 for ubiquitination. Cell 152(5):1037–1050. https://doi.org/10.1016/j.cell.2013.02.006
114. Charier G, Couprie J, Alpha-Bazin B, Meyer V, Quemeneur E, Guerois R, Callebaut I, Gilquin B, Zinn-Justin S (2004) The Tudor tandem of 53BP1: a new structural motif involved in DNA and RG-rich peptide binding. Structure 12(9):1551–1562. https://doi.org/10.1016/j.str.2004.06.014
115. Panier S, Boulton SJ (2014) Double-strand break repair: 53BP1 comes into focus. Nat Rev Mol Cell Biol 15(1):7–18. https://doi.org/10.1038/nrm3719
116. Huo Q, Yang Q (2011) P53-binding protein 1: a new player for tumorigenesis and a new target for breast cancer treatment. Med Hypotheses 77(3):359–363. https://doi.org/10.1016/j.mehy.2011.05.015
117. Zimmermann M, Lottersberger F, Buonomo SB, Sfeir A, de Lange T (2013) 53BP1 regulates DSB repair using Rif1 to control 5′ end resection. Science 339(6120):700–704. https://doi.org/10.1126/science.1231573
118. Di Virgilio M, Callen E, Yamane A, Zhang W, Jankovic M, Gitlin AD, Feldhahn N, Resch W, Oliveira TY, Chait BT, Nussenzweig A, Casellas R, Robbiani DF, Nussenzweig MC (2013) Rif1 prevents resection of DNA breaks and promotes immunoglobulin class switching. Science 339(6120):711–715. https://doi.org/10.1126/science.1230624
119. Ward IM, Reina-San-Martin B, Olaru A, Minn K, Tamada K, Lau JS, Cascalho M, Chen L, Nussenzweig A, Livak F, Nussenzweig MC, Chen J (2004) 53BP1 is required for class switch recombination. J Cell Biol 165(4):459–464. https://doi.org/10.1083/jcb.200403021
120. Dimitrova N, Chen YC, Spector DL, de Lange T (2008) 53BP1 promotes non-homologous end joining of telomeres by increasing chromatin mobility. Nature 456(7221):524–528. https://doi.org/10.1038/nature07433
121. Bunting SF, Callen E, Wong N, Chen HT, Polato F, Gunn A, Bothmer A, Feldhahn N, Fernandez-Capetillo O, Cao L, Xu X, Deng CX, Finkel T, Nussenzweig M, Stark JM, Nussenzweig A (2010) 53BP1 inhibits homologous recombination in Brca1-deficient cells by blocking resection of DNA breaks. Cell 141(2):243–254. https://doi.org/10.1016/j.cell.2010.03.012
122. Sanders SL, Jennings J, Canutescu A, Link AJ, Weil PA (2002) Proteomics of the eukaryotic transcription machinery: identification of proteins associated with components of yeast TFIID by multidimensional mass spectrometry. Mol Cell Biol 22(13):4723–4738. https://doi.org/10.1128/mcb.22.13.4723-4738.2002

123. Grant PA, Duggan L, Cote J, Roberts SM, Brownell JE, Candau R, Ohba R, Owen-Hughes T, Allis CD, Winston F, Berger SL, Workman JL (1997) Yeast Gcn5 functions in two multisubunit complexes to acetylate nucleosomal histones: characterization of an Ada complex and the SAGA (Spt/Ada) complex. Genes Dev 11(13):1640–1650. https://doi.org/10.1101/gad.11.13.1640
124. Vermeulen M, Eberl HC, Matarese F, Marks H, Denissov S, Butter F, Lee KK, Olsen JV, Hyman AA, Stunnenberg HG, Mann M (2010) Quantitative interaction proteomics and genome-wide profiling of epigenetic histone marks and their readers. Cell 142(6):967–980. https://doi.org/10.1016/j.cell.2010.08.020
125. Law JA, Vashisht AA, Wohlschlegel JA, Jacobsen SE (2011) SHH1, a homeodomain protein required for DNA methylation, as well as RDR2, RDM4, and chromatin remodeling factors, associate with RNA polymerase IV. PLoS Genet 7(7):e1002195. https://doi.org/10.1371/journal.pgen.1002195
126. Haag JR, Pikaard CS (2011) Multisubunit RNA polymerases IV and V: purveyors of non-coding RNA for plant gene silencing. Nat Rev Mol Cell Biol 12(8):483–492. https://doi.org/10.1038/nrm3152
127. Law JA, Du J, Hale CJ, Feng S, Krajewski K, Palanca AM, Strahl BD, Patel DJ, Jacobsen SE (2013) Polymerase IV occupancy at RNA-directed DNA methylation sites requires SHH1. Nature 498(7454):385–389. https://doi.org/10.1038/nature12178
128. Bostick M, Kim JK, Esteve PO, Clark A, Pradhan S, Jacobsen SE (2007) UHRF1 plays a role in maintaining DNA methylation in mammalian cells. Science 317(5845):1760–1764. https://doi.org/10.1126/science.1147939
129. Sharif J, Muto M, Takebayashi S, Suetake I, Iwamatsu A, Endo TA, Shinga J, Mizutani-Koseki Y, Toyoda T, Okamura K, Tajima S, Mitsuya K, Okano M, Koseki H (2007) The SRA protein Np95 mediates epigenetic inheritance by recruiting Dnmt1 to methylated DNA. Nature 450(7171):908–912. https://doi.org/10.1038/nature06397
130. Arita K, Isogai S, Oda T, Unoki M, Sugita K, Sekiyama N, Kuwata K, Hamamoto R, Tochio H, Sato M, Ariyoshi M, Shirakawa M (2012) Recognition of modification status on a histone H3 tail by linked histone reader modules of the epigenetic regulator UHRF1. Proc Natl Acad Sci U S A 109(32):12950–12955. https://doi.org/10.1073/pnas.1203701109
131. Berkyurek AC, Suetake I, Arita K, Takeshita K, Nakagawa A, Shirakawa M, Tajima S (2014) The DNA methyltransferase Dnmt1 directly interacts with the SET and RING finger-associated (SRA) domain of the multifunctional protein Uhrf1 to facilitate accession of the catalytic center to hemi-methylated DNA. J Biol Chem 289(1):379–386. https://doi.org/10.1074/jbc.M113.523209
132. Bashtrykov P, Jankevicius G, Jurkowska RZ, Ragozin S, Jeltsch A (2014) The UHRF1 protein stimulates the activity and specificity of the maintenance DNA methyltransferase DNMT1 by an allosteric mechanism. J Biol Chem 289(7):4106–4115. https://doi.org/10.1074/jbc.M113.528893
133. Nady N, Lemak A, Walker JR, Avvakumov GV, Kareta MS, Achour M, Xue S, Duan S, Allali-Hassani A, Zuo X, Wang YX, Bronner C, Chedin F, Arrowsmith CH, Dhe-Paganon S (2011) Recognition of multivalent histone states associated with heterochromatin by UHRF1 protein. J Biol Chem 286(27):24300–24311. https://doi.org/10.1074/jbc.M111.234104
134. Nishiyama A, Yamaguchi L, Sharif J, Johmura Y, Kawamura T, Nakanishi K, Shimamura S, Arita K, Kodama T, Ishikawa F, Koseki H, Nakanishi M (2013) Uhrf1-dependent H3K23 ubiquitylation couples maintenance DNA methylation and replication. Nature 502(7470):249–253. https://doi.org/10.1038/nature12488
135. Liu X, Gao Q, Li P, Zhao Q, Zhang J, Li J, Koseki H, Wong J (2013) UHRF1 targets DNMT1 for DNA methylation through cooperative binding of hemi-methylated DNA and methylated H3K9. Nat Commun 4:1563. https://doi.org/10.1038/ncomms2562
136. Lallous N, Legrand P, McEwen AG, Ramon-Maiques S, Samama JP, Birck C (2011) The PHD finger of human UHRF1 reveals a new subgroup of unmethylated histone H3 tail readers. PLoS One 6(11):e27599. https://doi.org/10.1371/journal.pone.0027599

137. Xie S, Jakoncic J, Qian C (2012) UHRF1 double tudor domain and the adjacent PHD finger act together to recognize K9me3-containing histone H3 tail. J Mol Biol 415(2):318–328. https://doi.org/10.1016/j.jmb.2011.11.012
138. Franz H, Greschik H, Willmann D, Ozretic L, Jilg CA, Wardelmann E, Jung M, Buettner R, Schule R (2015) The histone code reader SPIN1 controls RET signaling in liposarcoma. Oncotarget 6(7):4773–4789. https://doi.org/10.18632/oncotarget.3000
139. Wang W, Chen Z, Mao Z, Zhang H, Ding X, Chen S, Zhang X, Xu R, Zhu B (2011) Nucleolar protein Spindlin1 recognizes H3K4 methylation and stimulates the expression of rRNA genes. EMBO Rep 12(11):1160–1166. https://doi.org/10.1038/embor.2011.184
140. Oh B, Hwang SY, Solter D, Knowles BB (1997) Spindlin, a major maternal transcript expressed in the mouse during the transition from oocyte to embryo. Development 124(2):493–503
141. Yue W, Sun LY, Li CH, Zhang LX, Pei XT (2004) Screening and identification of ovarian carcinomas related genes. Ai Zheng 23(2):141–145
142. Wang JX, Zeng Q, Chen L, Du JC, Yan XL, Yuan HF, Zhai C, Zhou JN, Jia YL, Yue W, Pei XT (2012) SPINDLIN1 promotes cancer cell proliferation through activation of WNT/TCF-4 signaling. Mol Cancer Res 10(3):326–335. https://doi.org/10.1158/1541-7786.MCR-11-0440
143. Zhang P, Cong B, Yuan H, Chen L, Lv Y, Bai C, Nan X, Shi S, Yue W, Pei X (2008) Overexpression of spindlin1 induces metaphase arrest and chromosomal instability. J Cell Physiol 217(2):400–408. https://doi.org/10.1002/jcp.21515
144. Zhao Q, Qin L, Jiang F, Wu B, Yue W, Xu F, Rong Z, Yuan H, Xie X, Gao Y, Bai C, Bartlam M, Pei X, Rao Z (2007) Structure of human spindlin1. Tandem tudor-like domains for cell cycle regulation. J Biol Chem 282(1):647–656. https://doi.org/10.1074/jbc.M604029200
145. Su X, Zhu G, Ding X, Lee SY, Dou Y, Zhu B, Wu W, Li H (2014) Molecular basis underlying histone H3 lysine-arginine methylation pattern readout by Spin/Ssty repeats of Spindlin1. Genes Dev 28(6):622–636. https://doi.org/10.1101/gad.233239.113
146. Perfetti MT, Baughman BM, Dickson BM, Mu Y, Cui G, Mader P, Dong A, Norris JL, Rothbart SB, Strahl BD, Brown PJ, Janzen WP, Arrowsmith CH, Mer G, McBride KM, James LI, Frye SV (2015) Identification of a fragment-like small molecule ligand for the methyl-lysine binding protein, 53BP1. ACS Chem Biol 10(4):1072–1081. https://doi.org/10.1021/cb500956g
147. Fradet-Turcotte A, Canny MD, Escribano-Diaz C, Orthwein A, Leung CC, Huang H, Landry MC, Kitevski-LeBlanc J, Noordermeer SM, Sicheri F, Durocher D (2013) 53BP1 is a reader of the DNA-damage-induced H2A Lys 15 ubiquitin mark. Nature 499(7456):50–54. https://doi.org/10.1038/nature12318
148. Gelato KA, Tauber M, Ong MS, Winter S, Hiragami-Hamada K, Sindlinger J, Lemak A, Bultsma Y, Houliston S, Schwarzer D, Divecha N, Arrowsmith CH, Fischle W (2014) Accessibility of different histone H3-binding domains of UHRF1 is allosterically regulated by phosphatidylinositol 5-phosphate. Mol Cell 54(6):905–919. https://doi.org/10.1016/j.molcel.2014.04.004
149. Fang J, Cheng J, Wang J, Zhang Q, Liu M, Gong R, Wang P, Zhang X, Feng Y, Lan W, Gong Z, Tang C, Wong J, Yang H, Cao C, Xu Y (2016) Hemi-methylated DNA opens a closed conformation of UHRF1 to facilitate its histone recognition. Nat Commun 7:11197. https://doi.org/10.1038/ncomms11197
150. Harrison JS, Cornett EM, Goldfarb D, DaRosa PA, Li ZM, Yan F, Dickson BM, Guo AH, Cantu DV, Kaustov L, Brown PJ, Arrowsmith CH, Erie DA, Major MB, Klevit RE, Krajewski K, Kuhlman B, Strahl BD, Rothbart SB (2016) Hemi-methylated DNA regulates DNA methylation inheritance through allosteric activation of H3 ubiquitylation by UHRF1. eLife 5:e17101. https://doi.org/10.7554/eLife.17101
151. Houliston RS, Lemak A, Iqbal A, Ivanochko D, Duan S, Kaustov L, Ong MS, Fan L, Senisterra G, Brown PJ, Wang YX, Arrowsmith CH (2017) Conformational dynamics of the TTD-PHD histone reader module of the UHRF1 epigenetic regulator reveals multiple histone-binding states, allosteric regulation, and druggability. J Biol Chem 292(51):20947–20959. https://doi.org/10.1074/jbc.M117.799700

152. Wagner T, Greschik H, Burgahn T, Schmidtkunz K, Schott AK, McMillan J, Baranauskiene L, Xiong Y, Fedorov O, Jin J, Oppermann U, Matulis D, Schule R, Jung M (2016) Identification of a small-molecule ligand of the epigenetic reader protein Spindlin1 via a versatile screening platform. Nucleic Acids Res 44(9):e88. https://doi.org/10.1093/nar/gkw089
153. Sweis RF, Pliushchev M, Brown PJ, Guo J, Li F, Maag D, Petros AM, Soni NB, Tse C, Vedadi M, Michaelides MR, Chiang GG, Pappano WN (2014) Discovery and development of potent and selective inhibitors of histone methyltransferase g9a. ACS Med Chem Lett 5(2):205–209. https://doi.org/10.1021/ml400496h
154. Robaa D, Wagner T, Luise C, Carlino L, McMillan J, Flaig R, Schule R, Jung M, Sippl W (2016) Identification and structure-activity relationship studies of small-molecule inhibitors of the methyllysine reader protein Spindlin1. ChemMedChem 11(20):2327–2338. https://doi.org/10.1002/cmdc.201600362
155. Bae N, Gao M, Li X, Premkumar T, Sbardella G, Chen J, Bedford MT (2017) A transcriptional coregulator, SPIN.DOC, attenuates the coactivator activity of Spindlin1. J Biol Chem 292(51):20808–20817. https://doi.org/10.1074/jbc.M117.814913
156. Berry WL, Janknecht R (2013) KDM4/JMJD2 histone demethylases: epigenetic regulators in cancer cells. Cancer Res 73(10):2936–2942. https://doi.org/10.1158/0008-5472.CAN-12-4300
157. Singh RN, Howell MD, Ottesen EW, Singh NN (2017) Diverse role of survival motor neuron protein. Biochim Biophys Acta Gene Regul Mech 1860(3):299–315. https://doi.org/10.1016/j.bbagrm.2016.12.008
158. Yang Y, McBride KM, Hensley S, Lu Y, Chedin F, Bedford MT (2014) Arginine methylation facilitates the recruitment of TOP3B to chromatin to prevent R loop accumulation. Mol Cell 53(3):484–497. https://doi.org/10.1016/j.molcel.2014.01.011
159. Zheng S, Moehlenbrink J, Lu YC, Zalmas LP, Sagum CA, Carr S, McGouran JF, Alexander L, Fedorov O, Munro S, Kessler B, Bedford MT, Yu Q, La Thangue NB (2013) Arginine methylation-dependent reader-writer interplay governs growth control by E2F-1. Mol Cell 52(1):37–51. https://doi.org/10.1016/j.molcel.2013.08.039
160. Vagin VV, Wohlschlegel J, Qu J, Jonsson Z, Huang X, Chuma S, Girard A, Sachidanandam R, Hannon GJ, Aravin AA (2009) Proteomic analysis of murine Piwi proteins reveals a role for arginine methylation in specifying interaction with Tudor family members. Genes Dev 23(15):1749–1762. https://doi.org/10.1101/gad.1814809
161. Bonasio R, Lecona E, Reinberg D (2010) MBT domain proteins in development and disease. Semin Cell Dev Biol 21(2):221–230. https://doi.org/10.1016/j.semcdb.2009.09.010
162. Lewis PW, Beall EL, Fleischer TC, Georlette D, Link AJ, Botchan MR (2004) Identification of a Drosophila Myb-E2F2/RBF transcriptional repressor complex. Genes Dev 18(23):2929–2940. https://doi.org/10.1101/gad.1255204
163. Wang WK, Tereshko V, Boccuni P, MacGrogan D, Nimer SD, Patel DJ (2003) Malignant brain tumor repeats: a three-leaved propeller architecture with ligand/peptide binding pockets. Structure 11(7):775–789. https://doi.org/10.1016/S0969-2126(03)00127-8
164. Sathyamurthy A, Allen MD, Murzin AG, Bycroft M (2003) Crystal structure of the malignant brain tumor (MBT) repeats in Sex Comb on Midleg-like 2 (SCML2). J Biol Chem 278 (47):46968–46973. https://doi.org/10.1074/jbc.M306469200
165. Nady N, Krichevsky L, Zhong N, Duan S, Tempel W, Amaya MF, Ravichandran M, Arrowsmith CH (2012) Histone recognition by human malignant brain tumor domains. J Mol Biol 423(5):702–718. https://doi.org/10.1016/j.jmb.2012.08.022
166. Min J, Allali-Hassani A, Nady N, Qi C, Ouyang H, Liu Y, MacKenzie F, Vedadi M, Arrowsmith CH (2007) L3MBTL1 recognition of mono- and dimethylated histones. Nat Struct Mol Biol 14(12):1229–1230. https://doi.org/10.1038/nsmb1340
167. Eryilmaz J, Pan P, Amaya MF, Allali-Hassani A, Dong A, Adams-Cioaba MA, Mackenzie F, Vedadi M, Min J (2009) Structural studies of a four-MBT repeat protein MBTD1. PLoS One 4(10):e7274. https://doi.org/10.1371/journal.pone.0007274

168. Lecona E, Rojas LA, Bonasio R, Johnston A, Fernandez-Capetillo O, Reinberg D (2013) Polycomb protein SCML2 regulates the cell cycle by binding and modulating CDK/CYCLIN/p21 complexes. PLoS Biol 11(12):e1001737. https://doi.org/10.1371/journal.pbio.1001737
169. Boccuni P, MacGrogan D, Scandura JM, Nimer SD (2003) The human L(3)MBT polycomb group protein is a transcriptional repressor and interacts physically and functionally with TEL (ETV6). J Biol Chem 278(17):15412–15420. https://doi.org/10.1074/jbc.M300592200
170. Trojer P, Li G, Sims 3rd RJ, Vaquero A, Kalakonda N, Boccuni P, Lee D, Erdjument-Bromage H, Tempst P, Nimer SD, Wang YH, Reinberg D (2007) L3MBTL1, a histone-methylation-dependent chromatin lock. Cell 129(5):915–928. https://doi.org/10.1016/j.cell.2007.03.048
171. Herold JM, Ingerman LA, Gao C, Frye SV (2011) Drug discovery toward antagonists of methyl-lysine binding proteins. Curr Chem Genomics 5:51–61. https://doi.org/10.2174/1875397301005010051
172. Northcott PA, Nakahara Y, Wu X, Feuk L, Ellison DW, Croul S, Mack S, Kongkham PN, Peacock J, Dubuc A, Ra YS, Zilberberg K, McLeod J, Scherer SW, Sunil Rao J, Eberhart CG, Grajkowska W, Gillespie Y, Lach B, Grundy R, Pollack IF, Hamilton RL, Van Meter T, Carlotti CG, Boop F, Bigner D, Gilbertson RJ, Rutka JT, Taylor MD (2009) Multiple recurrent genetic events converge on control of histone lysine methylation in medulloblastoma. Nat Genet 41(4):465–472. https://doi.org/10.1038/ng.336
173. MacGrogan D, Kalakonda N, Alvarez S, Scandura JM, Boccuni P, Johansson B, Nimer SD (2004) Structural integrity and expression of the L3MBTL gene in normal and malignant hematopoietic cells. Genes Chromosomes Cancer 41(3):203–213. https://doi.org/10.1002/gcc.20087
174. Wigle TJ, Herold JM, Senisterra GA, Vedadi M, Kireev DB, Arrowsmith CH, Frye SV, Janzen WP (2010) Screening for inhibitors of low-affinity epigenetic peptide-protein interactions: an AlphaScreen-based assay for antagonists of methyl-lysine binding proteins. J Biomol Screen 15(1):62–71. https://doi.org/10.1177/1087057109352902
175. Kireev D, Wigle TJ, Norris-Drouin J, Herold JM, Janzen WP, Frye SV (2010) Identification of non-peptide malignant brain tumor (MBT) repeat antagonists by virtual screening of commercially available compounds. J Med Chem 53(21):7625–7631. https://doi.org/10.1021/jm1007374
176. Herold JM, Wigle TJ, Norris JL, Lam R, Korboukh VK, Gao C, Ingerman LA, Kireev DB, Senisterra G, Vedadi M, Tripathy A, Brown PJ, Arrowsmith CH, Jin J, Janzen WP, Frye SV (2011) Small-molecule ligands of methyl-lysine binding proteins. J Med Chem 54(7):2504–2511. https://doi.org/10.1021/jm200045v
177. Herold JM, James LI, Korboukh VK, Gao C, Coil KE, Bua DJ, Norris JL, Kireev DB, Brown PJ, Jin J, Janzen WP, Gozani O, Frye SV (2012) Structure-activity relationships of methyl-lysine reader antagonists. MedChemComm 3(1):45–51. https://doi.org/10.1039/C1MD00195G
178. James LI, Barsyte-Lovejoy D, Zhong N, Krichevsky L, Korboukh VK, Herold JM, MacNevin CJ, Norris JL, Sagum CA, Tempel W, Marcon E, Guo H, Gao C, Huang XP, Duan S, Emili A, Greenblatt JF, Kireev DB, Jin J, Janzen WP, Brown PJ, Bedford MT, Arrowsmith CH, Frye SV (2013) Discovery of a chemical probe for the L3MBTL3 methyllysine reader domain. Nat Chem Biol 9(3):184–191. https://doi.org/10.1038/nchembio.1157
179. James LI, Korboukh VK, Krichevsky L, Baughman BM, Herold JM, Norris JL, Jin J, Kireev DB, Janzen WP, Arrowsmith CH, Frye SV (2013) Small-molecule ligands of methyl-lysine binding proteins: optimization of selectivity for L3MBTL3. J Med Chem 56(18):7358–7371. https://doi.org/10.1021/jm400919p
180. Lee YY, Yu YB, Gunawardena HP, Xie L, Chen X (2012) BCLAF1 is a radiation-induced H2AX-interacting partner involved in γH2AX-mediated regulation of apoptosis and DNA repair. Cell Death Dis 3:e359. https://doi.org/10.1038/cddis.2012.76

181. James LI, Frye SV (2016) Chemical probes for methyl lysine reader domains. Curr Opin Chem Biol 33:135–141. https://doi.org/10.1016/j.cbpa.2016.06.004
182. Camerino MA, Zhong N, Dong A, Dickson BM, James LI, Baughman BM, Norris JL, Kireev DB, Janzen WP, Arrowsmith CH, Frye SV (2013) The structure-activity relationships of L3MBTL3 inhibitors: flexibility of the dimer interface. MedChemComm 4(11):1501–1507. https://doi.org/10.1039/C3MD00197K
183. Stec I, Nagl SB, van Ommen GJ, den Dunnen JT (2000) The PWWP domain: a potential protein-protein interaction domain in nuclear proteins influencing differentiation? FEBS Lett 473(1):1–5. https://doi.org/10.1016/S0014-5793(00)01449-6
184. Wu H, Zeng H, Lam R, Tempel W, Amaya MF, Xu C, Dombrovski L, Qiu W, Wang Y, Min J (2011) Structural and histone binding ability characterizations of human PWWP domains. PLoS One 6(6):e18919. https://doi.org/10.1371/journal.pone.0018919
185. Qin S, Min J (2014) Structure and function of the nucleosome-binding PWWP domain. Trends Biochem Sci 39(11):536–547. https://doi.org/10.1016/j.tibs.2014.09.001
186. Wang Y, Reddy B, Thompson J, Wang H, Noma K, Yates JR 3rd, Jia S (2009) Regulation of Set9-mediated H4K20 methylation by a PWWP domain protein. Mol Cell 33(4):428–437. https://doi.org/10.1016/j.molcel.2009.02.002
187. Dhayalan A, Rajavelu A, Rathert P, Tamas R, Jurkowska RZ, Ragozin S, Jeltsch A (2010) The Dnmt3a PWWP domain reads histone 3 lysine 36 trimethylation and guides DNA methylation. J Biol Chem 285(34):26114–26120. https://doi.org/10.1074/jbc.M109.089433
188. Vezzoli A, Bonadies N, Allen MD, Freund SM, Santiveri CM, Kvinlaug BT, Huntly BJ, Gottgens B, Bycroft M (2010) Molecular basis of histone H3K36me3 recognition by the PWWP domain of Brpf1. Nat Struct Mol Biol 17(5):617–619. https://doi.org/10.1038/nsmb.1797
189. Huang Z, Wu H, Chuai S, Xu F, Yan F, Englund N, Wang Z, Zhang H, Fang M, Wang Y, Gu J, Zhang M, Yang T, Zhao K, Yu Y, Dai J, Yi W, Zhou S, Li Q, Wu J, Liu J, Wu X, Chan H, Lu C, Atadja P, Li E, Wang Y, Hu M (2013) NSD2 is recruited through its PHD domain to oncogenic gene loci to drive multiple myeloma. Cancer Res 73(20):6277–6288. https://doi.org/10.1158/0008-5472.CAN-13-1000
190. Hudlebusch HR, Santoni-Rugiu E, Simon R, Ralfkiaer E, Rossing HH, Johansen JV, Jorgensen M, Sauter G, Helin K (2011) The histone methyltransferase and putative oncoprotein MMSET is overexpressed in a large variety of human tumors. Clin Cancer Res 17(9):2919–2933. https://doi.org/10.1158/1078-0432.CCR-10-1302
191. Li J, Yin C, Okamoto H, Mushlin H, Balgley BM, Lee CS, Yuan K, Ikejiri B, Glasker S, Vortmeyer AO, Oldfield EH, Weil RJ, Zhuang Z (2008) Identification of a novel proliferation-related protein, WHSC1 4a, in human gliomas. Neuro Oncol 10(1):45–51. https://doi.org/10.1215/15228517-2007-036
192. Kassambara A, Klein B, Moreaux J (2009) MMSET is overexpressed in cancers: link with tumor aggressiveness. Biochem Biophys Res Commun 379(4):840–845. https://doi.org/10.1016/j.bbrc.2008.12.093
193. Oyer JA, Huang X, Zheng Y, Shim J, Ezponda T, Carpenter Z, Allegretta M, Okot-Kotber CI, Patel JP, Melnick A, Levine RL, Ferrando A, Mackerell AD Jr, Kelleher NL, Licht JD, Popovic R (2014) Point mutation E1099K in MMSET/NSD2 enhances its methyltransferase activity and leads to altered global chromatin methylation in lymphoid malignancies. Leukemia 28(1):198–201. https://doi.org/10.1038/leu.2013.204
194. Yang ZQ, Liu G, Bollig-Fischer A, Giroux CN, Ethier SP (2010) Transforming properties of 8p11-12 amplified genes in human breast cancer. Cancer Res 70(21):8487–8497. https://doi.org/10.1158/0008-5472.CAN-10-1013
195. Morrison MJ, Boriack-Sjodin PA, Swinger KK, Wigle TJ, Sadalge D, Kuntz KW, Scott MP, Janzen WP, Chesworth R, Duncan KW, Harvey DM, Lampe JW, Mitchell LH, Copeland RA (2018) Identification of a peptide inhibitor for the histone methyltransferase WHSC1. PLoS One 13(5):e0197082. https://doi.org/10.1371/journal.pone.0197082

196. Stirnimann CU, Petsalaki E, Russell RB, Muller CW (2010) WD40 proteins propel cellular networks. Trends Biochem Sci 35(10):565–574. https://doi.org/10.1016/j.tibs.2010.04.003
197. Pryer NK, Salama NR, Schekman R, Kaiser CA (1993) Cytosolic Sec13p complex is required for vesicle formation from the endoplasmic reticulum in vitro. J Cell Biol 120(4):865–875. https://doi.org/10.1083/jcb.120.4.865
198. de Hostos EL, Rehfuess C, Bradtke B, Waddell DR, Albrecht R, Murphy J, Gerisch G (1993) Dictyostelium mutants lacking the cytoskeletal protein coronin are defective in cytokinesis and cell motility. J Cell Biol 120(1):163–173. https://doi.org/10.1083/jcb.120.1.163
199. Shen Z, Sathyan KM, Geng Y, Zheng R, Chakraborty A, Freeman B, Wang F, Prasanth KV, Prasanth SG (2010) A WD-repeat protein stabilizes ORC binding to chromatin. Mol Cell 40(1):99–111. https://doi.org/10.1016/j.molcel.2010.09.021
200. Hoey T, Weinzierl RO, Gill G, Chen JL, Dynlacht BD, Tjian R (1993) Molecular cloning and functional analysis of Drosophila TAF110 reveal properties expected of coactivators. Cell 72(2):247–260. https://doi.org/10.1016/0092-8674(93)90664-C
201. Mersman DP, Du HN, Fingerman IM, South PF, Briggs SD (2012) Charge-based interaction conserved within histone H3 lysine 4 (H3K4) methyltransferase complexes is needed for protein stability, histone methylation, and gene expression. J Biol Chem 287(4):2652–2665. https://doi.org/10.1074/jbc.M111.280867
202. Guccione E, Bassi C, Casadio F, Martinato F, Cesaroni M, Schuchlautz H, Luscher B, Amati B (2007) Methylation of histone H3R2 by PRMT6 and H3K4 by an MLL complex are mutually exclusive. Nature 449(7164):933–937. https://doi.org/10.1038/nature06166
203. Migliori V, Muller J, Phalke S, Low D, Bezzi M, Mok WC, Sahu SK, Gunaratne J, Capasso P, Bassi C, Cecatiello V, De Marco A, Blackstock W, Kuznetsov V, Amati B, Mapelli M, Guccione E (2012) Symmetric dimethylation of H3R2 is a newly identified histone mark that supports euchromatin maintenance. Nat Struct Mol Biol 19(2):136–144. https://doi.org/10.1038/nsmb.2209
204. Schapira M, Tyers M, Torrent M, Arrowsmith CH (2017) WD40 repeat domain proteins: a novel target class? Nat Rev Drug Discov 16(11):773–786. https://doi.org/10.1038/nrd.2017.179
205. Dou Y, Milne TA, Ruthenburg AJ, Lee S, Lee JW, Verdine GL, Allis CD, Roeder RG (2006) Regulation of MLL1 H3K4 methyltransferase activity by its core components. Nat Struct Mol Biol 13(8):713–719. https://doi.org/10.1038/nsmb1128
206. Steward MM, Lee JS, O'Donovan A, Wyatt M, Bernstein BE, Shilatifard A (2006) Molecular regulation of H3K4 trimethylation by ASH2L, a shared subunit of MLL complexes. Nat Struct Mol Biol 13(9):852–854. https://doi.org/10.1038/nsmb1131
207. Ruthenburg AJ, Allis CD, Wysocka J (2007) Methylation of lysine 4 on histone H3: intricacy of writing and reading a single epigenetic mark. Mol Cell 25(1):15–30. https://doi.org/10.1016/j.molcel.2006.12.014
208. Patel A, Dharmarajan V, Vought VE, Cosgrove MS (2009) On the mechanism of multiple lysine methylation by the human mixed lineage leukemia protein-1 (MLL1) core complex. J Biol Chem 284(36):24242–24256. https://doi.org/10.1074/jbc.M109.014498
209. Wysocka J, Swigut T, Milne TA, Dou Y, Zhang X, Burlingame AL, Roeder RG, Brivanlou AH, Allis CD (2005) WDR5 associates with histone H3 methylated at K4 and is essential for H3 K4 methylation and vertebrate development. Cell 121(6):859–872. https://doi.org/10.1016/j.cell.2005.03.036
210. Couture JF, Collazo E, Trievel RC (2006) Molecular recognition of histone H3 by the WD40 protein WDR5. Nat Struct Mol Biol 13(8):698–703. https://doi.org/10.1038/nsmb1116
211. Schuetz A, Allali-Hassani A, Martin F, Loppnau P, Vedadi M, Bochkarev A, Plotnikov AN, Arrowsmith CH, Min J (2006) Structural basis for molecular recognition and presentation of histone H3 by WDR5. EMBO J 25(18):4245–4252. https://doi.org/10.1038/sj.emboj.7601316
212. Guelman S, Kozuka K, Mao Y, Pham V, Solloway MJ, Wang J, Wu J, Lill JR, Zha J (2009) The double-histone-acetyltransferase complex ATAC is essential for mammalian development. Mol Cell Biol 29(5):1176–1188. https://doi.org/10.1128/MCB.01599-08
213. Cai Y, Jin J, Swanson SK, Cole MD, Choi SH, Florens L, Washburn MP, Conaway JW, Conaway RC (2010) Subunit composition and substrate specificity of a MOF-containing

histone acetyltransferase distinct from the male-specific lethal (MSL) complex. J Biol Chem 285(7):4268–4272. https://doi.org/10.1074/jbc.C109.087981

214. Thompson BA, Tremblay V, Lin G, Bochar DA (2008) CHD8 is an ATP-dependent chromatin remodeling factor that regulates beta-catenin target genes. Mol Cell Biol 28(12):3894–3904. https://doi.org/10.1128/MCB.00322-08
215. Chen X, Xie W, Gu P, Cai Q, Wang B, Xie Y, Dong W, He W, Zhong G, Lin T, Huang J (2015) Upregulated WDR5 promotes proliferation, self-renewal and chemoresistance in bladder cancer via mediating H3K4 trimethylation. Sci Rep 5:8293. https://doi.org/10.1038/srep08293
216. Margueron R, Justin N, Ohno K, Sharpe ML, Son J, Drury 3rd WJ, Voigt P, Martin SR, Taylor WR, De Marco V, Pirrotta V, Reinberg D, Gamblin SJ (2009) Role of the polycomb protein EED in the propagation of repressive histone marks. Nature 461(7265):762–767. https://doi.org/10.1038/nature08398
217. Kuzmichev A, Nishioka K, Erdjument-Bromage H, Tempst P, Reinberg D (2002) Histone methyltransferase activity associated with a human multiprotein complex containing the enhancer of Zeste protein. Genes Dev 16(22):2893–2905. https://doi.org/10.1101/gad.1035902
218. Margueron R, Reinberg D (2011) The Polycomb complex PRC2 and its mark in life. Nature 469(7330):343–349. https://doi.org/10.1038/nature09784
219. Suva ML, Riggi N, Bernstein BE (2013) Epigenetic reprogramming in cancer. Science 339(6127):1567–1570. https://doi.org/10.1126/science.1230184
220. Schuettengruber B, Bourbon HM, Di Croce L, Cavalli G (2017) Genome regulation by polycomb and trithorax: 70 years and counting. Cell 171(1):34–57. https://doi.org/10.1016/j.cell.2017.08.002
221. Bannister AJ, Kouzarides T (2011) Regulation of chromatin by histone modifications. Cell Res 21(3):381–395. https://doi.org/10.1038/cr.2011.22
222. Voigt P, Tee WW, Reinberg D (2013) A double take on bivalent promoters. Genes Dev 27(12):1318–1338. https://doi.org/10.1101/gad.219626.113
223. Varambally S, Dhanasekaran SM, Zhou M, Barrette TR, Kumar-Sinha C, Sanda MG, Ghosh D, Pienta KJ, Sewalt RG, Otte AP, Rubin MA, Chinnaiyan AM (2002) The polycomb group protein EZH2 is involved in progression of prostate cancer. Nature 419(6907):624–629. https://doi.org/10.1038/nature01075
224. Kleer CG, Cao Q, Varambally S, Shen R, Ota I, Tomlins SA, Ghosh D, Sewalt RG, Otte AP, Hayes DF, Sabel MS, Livant D, Weiss SJ, Rubin MA, Chinnaiyan AM (2003) EZH2 is a marker of aggressive breast cancer and promotes neoplastic transformation of breast epithelial cells. Proc Natl Acad Sci U S A 100(20):11606–11611. https://doi.org/10.1073/pnas.1933744100
225. Hansen KH, Bracken AP, Pasini D, Dietrich N, Gehani SS, Monrad A, Rappsilber J, Lerdrup M, Helin K (2008) A model for transmission of the H3K27me3 epigenetic mark. Nat Cell Biol 10(11):1291–1300. https://doi.org/10.1038/ncb1787
226. Han Z, Xing X, Hu M, Zhang Y, Liu P, Chai J (2007) Structural basis of EZH2 recognition by EED. Structure 15(10):1306–1315. https://doi.org/10.1016/j.str.2007.08.007
227. Patel A, Dharmarajan V, Cosgrove MS (2008) Structure of WDR5 bound to mixed lineage leukemia protein-1 peptide. J Biol Chem 283(47):32158–32161. https://doi.org/10.1074/jbc.C800164200
228. Patel A, Vought VE, Dharmarajan V, Cosgrove MS (2008) A conserved arginine-containing motif crucial for the assembly and enzymatic activity of the mixed lineage leukemia protein-1 core complex. J Biol Chem 283(47):32162–32175. https://doi.org/10.1074/jbc.M806317200
229. Cao F, Townsend EC, Karatas H, Xu J, Li L, Lee S, Liu L, Chen Y, Ouillette P, Zhu J, Hess JL, Atadja P, Lei M, Qin ZS, Malek S, Wang S, Dou Y (2014) Targeting MLL1 H3K4 methyltransferase activity in mixed-lineage leukemia. Mol Cell 53(2):247–261. https://doi.org/10.1016/j.molcel.2013.12.001

230. Karatas H, Townsend EC, Bernard D, Dou Y, Wang S (2010) Analysis of the binding of mixed lineage leukemia 1 (MLL1) and histone 3 peptides to WD repeat domain 5 (WDR5) for the design of inhibitors of the MLL1-WDR5 interaction. J Med Chem 53(14):5179–5185. https://doi.org/10.1021/jm100139b
231. Karatas H, Townsend EC, Cao F, Chen Y, Bernard D, Liu L, Lei M, Dou Y, Wang S (2013) High-affinity, small-molecule peptidomimetic inhibitors of MLL1/WDR5 protein-protein interaction. J Am Chem Soc 135(2):669–682. https://doi.org/10.1021/ja306028q
232. Karatas H, Li Y, Liu L, Ji J, Lee S, Chen Y, Yang J, Huang L, Bernard D, Xu J, Townsend EC, Cao F, Ran X, Li X, Wen B, Sun D, Stuckey JA, Lei M, Dou Y, Wang S (2017) Discovery of a highly potent, cell-permeable macrocyclic peptidomimetic (MM-589) targeting the WD repeat domain 5 protein (WDR5)-mixed lineage leukemia (MLL) protein-protein interaction. J Med Chem 60(12):4818–4839. https://doi.org/10.1021/acs.jmedchem.6b01796
233. Senisterra G, Wu H, Allali-Hassani A, Wasney GA, Barsyte-Lovejoy D, Dombrovski L, Dong A, Nguyen KT, Smil D, Bolshan Y, Hajian T, He H, Seitova A, Chau I, Li F, Poda G, Couture JF, Brown PJ, Al-Awar R, Schapira M, Arrowsmith CH, Vedadi M (2013) Small-molecule inhibition of MLL activity by disruption of its interaction with WDR5. Biochem J 449(1):151–159. https://doi.org/10.1042/BJ20121280
234. Bolshan Y, Getlik M, Kuznetsova E, Wasney GA, Hajian T, Poda G, Nguyen KT, Wu H, Dombrovski L, Dong A, Senisterra G, Schapira M, Arrowsmith CH, Brown PJ, Al-Awar R, Vedadi M, Smil D (2013) Synthesis, optimization, and evaluation of novel small molecules as antagonists of WDR5-MLL interaction. ACS Med Chem Lett 4(3):353–357. https://doi.org/10.1021/ml300467n
235. Getlik M, Smil D, Zepeda-Velazquez C, Bolshan Y, Poda G, Wu H, Dong A, Kuznetsova E, Marcellus R, Senisterra G, Dombrovski L, Hajian T, Kiyota T, Schapira M, Arrowsmith CH, Brown PJ, Vedadi M, Al-Awar R (2016) Structure-based optimization of a small molecule antagonist of the interaction between WD repeat-containing protein 5 (WDR5) and mixed-lineage leukemia 1 (MLL1). J Med Chem 59(6):2478–2496. https://doi.org/10.1021/acs.jmedchem.5b01630
236. Li DD, Chen WL, Wang ZH, Xie YY, Xu XL, Jiang ZY, Zhang XJ, You QD, Guo XK (2016) High-affinity small molecular blockers of mixed lineage leukemia 1 (MLL1)-WDR5 interaction inhibit MLL1 complex H3K4 methyltransferase activity. Eur J Med Chem 124:480–489. https://doi.org/10.1016/j.ejmech.2016.08.036
237. Li DD, Chen WL, Xu XL, Jiang F, Wang L, Xie YY, Zhang XJ, Guo XK, You QD, Sun HP (2016) Structure-based design and synthesis of small molecular inhibitors disturbing the interaction of MLL1-WDR5. Eur J Med Chem 118:1–8. https://doi.org/10.1016/j.ejmech.2016.04.032
238. Li DD, Wang ZH, Chen WL, Xie YY, You QD, Guo XK (2016) Structure-based design of ester compounds to inhibit MLL complex catalytic activity by targeting mixed lineage leukemia 1 (MLL1)-WDR5 interaction. Bioorg Med Chem 24(22):6109–6118. https://doi.org/10.1016/j.bmc.2016.09.073
239. Lu J, Li X, Wang Q, Pei G (2017) Dopamine D2 receptor and beta-arrestin 2 mediate Amyloid-beta elevation induced by anti-Parkinson's disease drugs, levodopa and piribedil, in neuronal cells. PLoS One 12(3):e0173240. https://doi.org/10.1371/journal.pone.0173240
240. Zhang X, Zheng X, Yang H, Yan J, Fu X, Wei R, Xu X, Zhang Z, Yu A, Zhou K, Ding J, Geng M, Huang X (2018) Piribedil disrupts the MLL1-WDR5 interaction and sensitizes MLL-rearranged acute myeloid leukemia (AML) to doxorubicin-induced apoptosis. Cancer Lett 431:150–160. https://doi.org/10.1016/j.canlet.2018.05.034
241. Xie J, Wang C, Virostko J, Manning HC, Pham W, Bauer J, Gore JC (2013) A novel reporter system for molecular imaging and high-throughput screening of anticancer drugs. Chembiochem 14(12):1494–1503. https://doi.org/10.1002/cbic.201300142
242. Li L, Zhang H, Zhang M, Zhao M, Feng L, Luo X, Gao Z, Huang Y, Ardayfio O, Zhang JH, Lin Y, Fan H, Mi Y, Li G, Liu L, Feng L, Luo F, Teng L, Qi W, Ottl J, Lingel A, Bussiere DE, Yu Z, Atadja P, Lu C, Li E, Gu J, Zhao K (2017) Discovery and molecular basis of a diverse set of polycomb repressive complex 2 inhibitors recognition by EED. PLoS One 12(1):e0169855. https://doi.org/10.1371/journal.pone.0169855

243. Lingel A, Sendzik M, Huang Y, Shultz MD, Cantwell J, Dillon MP, Fu X, Fuller J, Gabriel T, Gu J, Jiang X, Li L, Liang F, McKenna M, Qi W, Rao W, Sheng X, Shu W, Sutton J, Taft B, Wang L, Zeng J, Zhang H, Zhang M, Zhao K, Lindvall M, Bussiere DE (2017) Structure-guided design of EED binders allosterically inhibiting the epigenetic polycomb repressive complex 2 (PRC2) methyltransferase. J Med Chem 60(1):415–427. https://doi.org/10.1021/acs.jmedchem.6b01473
244. Huang Y, Zhang J, Yu Z, Zhang H, Wang Y, Lingel A, Qi W, Gu J, Zhao K, Shultz MD, Wang L, Fu X, Sun Y, Zhang Q, Jiang X, Zhang J, Zhang C, Li L, Zeng J, Feng L, Zhang C, Liu Y, Zhang M, Zhang L, Zhao M, Gao Z, Liu X, Fang D, Guo H, Mi Y, Gabriel T, Dillon MP, Atadja P, Oyang C (2017) Discovery of first-in-class, potent, and orally bioavailable embryonic ectoderm development (EED) inhibitor with robust anticancer efficacy. J Med Chem 60(6):2215–2226. https://doi.org/10.1021/acs.jmedchem.6b01576
245. Qi W, Zhao K, Gu J, Huang Y, Wang Y, Zhang H, Zhang M, Zhang J, Yu Z, Li L, Teng L, Chuai S, Zhang C, Zhao M, Chan H, Chen Z, Fang D, Fei Q, Feng L, Feng L, Gao Y, Ge H, Ge X, Li G, Lingel A, Lin Y, Liu Y, Luo F, Shi M, Wang L, Wang Z, Yu Y, Zeng J, Zeng C, Zhang L, Zhang Q, Zhou S, Oyang C, Atadja P, Li E (2017) An allosteric PRC2 inhibitor targeting the H3K27me3 binding pocket of EED. Nat Chem Biol 13(4):381–388. https://doi.org/10.1038/nchembio.2304
246. He Y, Selvaraju S, Curtin ML, Jakob CG, Zhu H, Comess KM, Shaw B, The J, Lima-Fernandes E, Szewczyk MM, Cheng D, Klinge KL, Li HQ, Pliushchev M, Algire MA, Maag D, Guo J, Dietrich J, Panchal SC, Petros AM, Sweis RF, Torrent M, Bigelow LJ, Senisterra G, Li F, Kennedy S, Wu Q, Osterling DJ, Lindley DJ, Gao W, Galasinski S, Barsyte-Lovejoy D, Vedadi M, Buchanan FG, Arrowsmith CH, Chiang GG, Sun C, Pappano WN (2017) The EED protein-protein interaction inhibitor A-395 inactivates the PRC2 complex. Nat Chem Biol 13(4):389–395. https://doi.org/10.1038/nchembio.2306
247. Sanulli S, Justin N, Teissandier A, Ancelin K, Portoso M, Caron M, Michaud A, Lombard B, da Rocha ST, Offer J, Loew D, Servant N, Wassef M, Burlina F, Gamblin SJ, Heard E, Margueron R (2015) Jarid2 methylation via the PRC2 complex regulates H3K27me3 deposition during cell differentiation. Mol Cell 57(5):769–783. https://doi.org/10.1016/j.molcel.2014.12.020
248. Barnash KD, The J, Norris-Drouin JL, Cholensky SH, Worley BM, Li F, Stuckey JI, Brown PJ, Vedadi M, Arrowsmith CH, Frye SV, James LI (2017) Discovery of peptidomimetic ligands of EED as allosteric inhibitors of PRC2. ACS Comb Sci 19(3):161–172. https://doi.org/10.1021/acscombsci.6b00174
249. Santiago C, Nguyen K, Schapira M (2011) Druggability of methyl-lysine binding sites. J Comput Aided Mol Des 25(12):1171–1178. https://doi.org/10.1007/s10822-011-9505-2

Top Med Chem (2020) 33: 401–428
DOI: 10.1007/7355_2019_83

Published online: 24 October 2019

Altered Long Non-coding RNA Expression in Cancer: Potential Biomarkers and Therapeutic Targets?

David Hanly, Manel Esteller, and María Berdasco

Contents

1 Introduction 403
2 lncRNAs in Epigenetic Processes 404
2.1 Histone Modification and Transcriptional Regulation 404
2.2 LncRNA Epitranscriptomics 407
2.3 Nuclear Architecture and 3D Genome Structure 407
2.4 Alternative Splicing 408
2.5 Crosstalk with miRNAs: Competing Endogenous ncRNAs 408
3 lncRNA and Epigenetic Deregulation in Cancer 409
3.1 Oncogenic lncRNAs: HOTAIR, MALAT1, TUG1 and MIAT 409
3.2 Tumour-Suppressive lncRNAs: lincRNA-p21 and TP53BP1 412
4 Strategies to Target lncRNAs Implicated in Cancer Therapy Resistance 413
4.1 HOTAIR 413
4.2 MALAT1 415
4.3 TUG1 416
5 Conclusion and Future Perspectives 417
References 420

Abstract Long non-coding RNAs (lncRNAs) have been implicated in a wide range of biological processes. One of the most striking functions of lncRNAs is their capacity to act as "backstage" sculptors of the epigenetic landscape by interacting with chromatin-modifying proteins and directing their activities. LncRNAs can

D. Hanly and M. Berdasco (✉)
Cancer Epigenetics Group, Josep Carreras Leukaemia Research Institute (IJC), Barcelona, Spain
e-mail: mberdasco@carrerasresearch.org

M. Esteller
Cancer Epigenetics Group, Josep Carreras Leukaemia Research Institute (IJC), Barcelona, Spain

Department of Physiological Sciences II, School of Medicine, University of Barcelona, Barcelona, Spain

Institució Catalana de Recerca i Estudis Avançats (ICREA), Barcelona, Spain

thereby modulate locus-specific chromatin marks as well as arrange the three-dimensional configurations of entire chromosomes. They may also serve as focal points of microRNA (miRNA) regulatory networks through direct RNA-RNA contacts with miRNAs, shaping miRNA network activities through an "RNA language" yet to be deciphered. In this review, we provide a brief overview of how lncRNAs influence these molecular processes. We also outline a cancer epigenetic perspective on how lncRNAs can help or hinder the development of different types of cancer, how they may be exploited to predict patient prognosis and, most interestingly, how they affect tumour responses to current therapeutic interventions. Finally, we describe recent works that have made substantial progress in bringing to fruition the prospect of therapeutically targeting lncRNAs to correct epigenetic aberrations in human diseases, especially in cancer.

Keywords Cancer, Chromatin modification, Epigenetic therapy, Epigenetics, Long non-coding RNA, RNA therapy

Abbreviations

ALDH1A1	Aldehyde dehydrogenase
ASO	Antisense oligonucleotide
BDNF	Brain-derived neurotrophic factor
BTBB	Blood-tumour-brain barrier
CDDP	Cisplatin
ceRNA	Competing endogenous RNA
CLL	Chronic lymphocytic leukaemia
CSC	Cancer stem cell
DDR	DNA damage response
DDS	Drug delivery system
DNMT	DNA methyltransferase
eRNA	Enhancer RNA
FRDA	Friedreich's ataxia
FXN	Frataxin
GBM	Glioblastoma multiforme
GSC	Glioblastoma stem cells
HAT	Histone acetyltransferase
HDAC	Histone deacetylase
HMT	Histone methyltransferase
hnRNP	Heterogeneous ribonucleoprotein particle
IL	Interleukin
KLF2	Krüppel-like factor 2
lincRNA	Long intergenic non-coding RNA
lncRNA	Long non-coding RNA

m^6A	N^6-adenosine methylation
MALAT1	Metastasis-associated lung adenocarcinoma transcript 1
METTL3	Methyltransferase-Like 3
MIAT	Myocardial infarction-associated transcript
miRNA	microRNA
MMP9	Matrix metalloproteinase 9
MRE	miRNA recognition element
NAT	Natural antisense transcript
ncRNA	Non-coding RNA
NSCLC	Non-small-cell lung cancer
Nt	Nucleotides
OC	Ovarian cancer
pHLIP	pH-low insertion peptide
piRNA	Piwi-interacting RNA
PNA	Peptide nucleic acid
PRC	Polycomb Repressive Complex
RBMP	RNA-binding motif protein
RCC	Renal cell carcinoma
SF1	Splicing factor 1
siRNA	Short interfering RNA
SMA	Spinal muscular atrophy
SMN	Survival motor neuron
SR	Serine/arginine
SRSF	Serine/arginine splicing factor
SSO	Splice-switching oligonucleotide
TAD	Topologically associating domains
TfR	Transferrin receptor
TLS	Translocated in liposarcoma
TMZ	Temozolomide
T-UCR	Transcribed ultraconserved region
TUG1	Taurine Upregulated Gene 1
XCI	X-chromosome inactivation
YTHDC1	YTH domain containing 1

1 Introduction

One of the most striking discoveries of the postgenomic era has been that most of the human genome is transcribed into non-coding RNA (ncRNA), the "dark matter" of the genome, while scarcely 2% consists of protein-coding genes [1]. The non-coding transcriptome spans a vast size range. Some ncRNAs such as miRNAs and Piwi-interacting RNAs (piRNA) are shorter than 40 nucleotides (nt), while long non-coding RNAs (lncRNAs), arbitrarily defined as almost any non-coding transcript of over 200 nt, can exceed 10Knt in length [2]. LncRNAs are further stratified into long intergenic ncRNA (lincRNAs), transcribed ultraconserved regions (T-UCRs), natural antisense transcripts (NATs) and large intronic ncRNAs, among other groups. Although few of the nearly 60,000 human lncRNA "genes" [3] have been well-characterised, they have been implicated in a diverse array of molecular and biological processes, of which we outline herein their roles in the epigenetic regulation of gene expression, nuclear architecture, alternative splicing and RNA-RNA crosstalk [4]. We also provide an overview of key discoveries that reveal how some lncRNAs can misguide epigenetic factors to generate cancer-promoting epigenetic alterations, while others act as barriers to neoplastic growth, and we describe how lncRNAs differentially expressed in cancer can serve as biomarkers for patient prognosis and how they can influence the sensitivity of tumours to anticancer therapy. Finally, we outline recent innovative studies that have brought the prospect of therapeutically targeting oncogenic lncRNAs closer to fruition.

2 lncRNAs in Epigenetic Processes

A central focus of lncRNA research is the hypothesis that lncRNAs directly interact with epigenetic factors to guide them to specific target loci, specifying their sites of reading, writing and erasure of epigenetic marks. In this capacity, they are thought to function as RNA scaffolds of specific and context-dependent chromatin regulatory complexes to fine-tune the chromatin landscape of any given genetic locus or genomic region [5]. This section will outline how the concept of the epigenetic interface with lncRNAs emerged, how this interface may shape the epigenetic landscape and the challenges and discrepancies that confront us in defining these regulatory principals.

2.1 Histone Modification and Transcriptional Regulation

The idea that lncRNAs influence the epigenetic control of gene expression emerged on the back of major breakthroughs in characterising the process of X-chromosome inactivation (XCI) that were made during the 1990s and early 2000s [6]. Recent efforts to comprehensively characterise Xist-interacting proteins using novel biochemical approaches to study protein-RNA interactions have led to the development of a refined model of how Xist enacts XCI. Accordingly, Xist interacts with Xi loci via the DNA-/RNA-binding protein SAFA and recruits histone deacetylase 3 (HDAC3) and the SMRT complex via SHARP to initiate transcriptional silencing. Subsequent recruitment of Polycomb Repressive Complexes 1 and 2 (PRC1/PRC2) by Xist then serves to consolidate gene repression and chromatin compaction [7–9].

The broad notion that other lncRNAs might guide epigenetic factors in regulating gene expression took hold following a study on Homeobox gene regulation [10]. In this study, a lncRNA transcribed from the *HOXC* locus, dubbed HOTAIR, was demonstrated to recruit PRC2 *in trans* to the *HOXD* locus to mediate H3K27me3 deposition and locus silencing [10]. Biochemical confirmation of the reported role of HOTAIR in establishing repressive chromatin at the *HOXD* locus came when HOTAIR was shown to interact not only with PRC2 but also with H3K4 demethylase complex LSD1-CoREST to form an epigenetic silencing scaffold at the *HOXD* locus [11] (Fig. 1a). Two independent global RNA immunoprecipitation studies demonstrated that PRC2 binds to a substantial fraction of lincRNAs and transcripts genome-wide [12, 13]. These findings consolidated the hypothesis of a global regulatory interface between epigenetic machinery and lncRNAs responsible for the dynamic modulation of epigenetic landscapes. However, the centrality of PRC2 in the gene regulatory function of HOTAIR has been disputed. A recent study found that HOTAIR could mediate reporter silencing independently of PRC2 [14]. Perhaps, as seems to be the case for Xist, PRC2 functions more as a sensor of the altered chromatin terrain of the newly HOTAIR-silenced locus and proceeds to enrich it with H3K27me3 to compact the surrounding chromatin and thereby stabilise and maintain its transcriptional repression.

Numerous other lncRNAs have been shown to manipulate Polycomb activity in mediating gene repression. ANRIL, an antisense transcript of the *INK4b(p15)-ARF(p14)-INK4a(p16)* tumour suppressor gene cluster, mediates *cis* recruitment of both PRC1 and PRC2 to the locus, which epigenetically silence it via local enrichment of H3K27me3 and H2AK199ub [15, 16]. KCNQ1OT1, a non-coding transcript of the imprinted *Kcnq1* locus, coordinates lineage-specific transcriptional silencing of the 200 kb proximal region by scaffolding both PRC2 and H3K9 histone methyltransferase (HMT) G9a and enriching the locus with their corresponding repressive marks H3K9me3 and H3K27me3 [17].

LncRNAs can also guide the epigenetic writers of DNA methylation as well as those of histone marks. For example, in the context of reprogramming pre-induced pluripotent stem cells into iPSCs, lincRNA-p21 has been shown to interact not only with H3K9 histone methyltransferase (HMT) but also with DNA methyltransferase 1 (DNMT1), resulting in increasing H3K9me3 deposition and CpG island hypermethylation of the target promoters. LincRNA-p21 thus acts as a barrier to induced pluripotency [18]. It has also been proposed that, in addition to its role in scaffolding PRC2 and G9a, KCNQ1OT1 interacts with DNMT1 to direct CpG island methylation of the paternal *Kcnq1* locus [19].

As well as recruiting and scaffolding epigenetic modification complexes, lncRNAs can also modulate their activity by directly or indirectly inducing structural changes in epigenetic editor proteins. In a classic example, numerous DNA damage-induced ncRNAs transcribed from upstream of the *CCND1* gene recruit and allosterically activate TLS (translocated in liposarcoma) to the locus, which then inhibits locally bound histone acetyltransferase (HAT) CBP/p300, preventing the establishment of the histone acetylation marks necessary for the epigenetic activation of *CCND1* transcription [20]. By contrast, enhancer RNAs (eRNAs) have recently

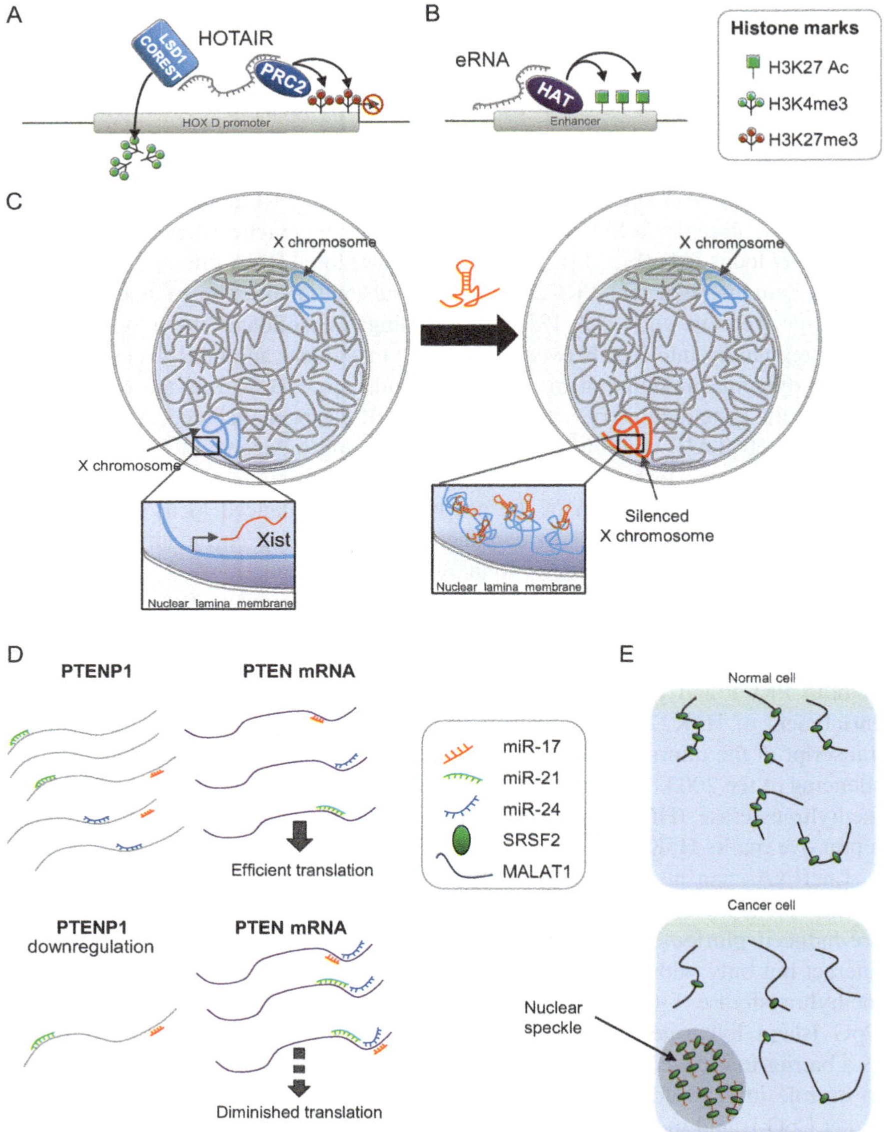

Fig. 1 LncRNAs can scaffold chromatin-modifying complexes to mediate epigenetic silencing of gene transcription: HOTAIR recruits the H3K4 demethylase complex LSD1-COREST and H3K27-specific HMT PRC2 to the *HOXD* locus to silence it (**a**); eRNAs can allosterically activate enhancer-bound HATs (**b**); lncRNAs can rearrange higher-order chromatin conformation: Xist compacts the entire Xi into two silent megadomains and relocates them to the nuclear periphery (**c**); lncRNAs known as ceRNAs can "sponge" miRNAs to upregulate the mRNA transcripts that they target: PTENP1 sponges miRNAs that target PTEN transcripts (**d**); lncRNAs can regulate splicing outcomes by sequestering splicing factors in subnuclear compartments, as MALAT1 can sequester SRSF1 to nuclear speckles to restrict its availability to participate in splicing reactions (**e**)

been found to carry out direct allosteric modulation of enhancer-bound CBP/p300 HAT complexes to the opposite end, that of gene activation. eRNAs can allosterically stimulate CBP HAT activity to increase both enhancer and target promoter H3K27ac levels and epigenetically activate target gene transcription [21] (Fig. 1b).

2.2 LncRNA Epitranscriptomics

Just as the field of epitranscriptomics, the study of post-transcriptional modifications of RNA molecules is beginning to expand, so too has attention turned to chemical modifications of lncRNAs. Although several instances of such lncRNA editing have already been reported, very little is known about their effects on lncRNA function [22]. In the best-characterised lncRNA editing scenario to date, Methyltransferase-Like 3 (METTL3) interacts with Xist RNA via Spen family RNA-binding motif proteins 15 and 18 (RBMP15, RBMP18) to carry out N^6-adenosine methylation (m^6A) of Xist. (m^6A)-Xist is then recognised by YTH domain containing 1 (YTHDC1) [23]. Despite its necessity, it is unclear how Xist-mediated recruitment of YTHDC1, as facilitated by the Xist modification pathway described, contributes to the Xist-directed epigenetic repression of gene transcription during XCI [23].

2.3 Nuclear Architecture and 3D Genome Structure

Recent efforts to characterise the Xist interactome during XCI have also yielded insights into how the higher-order chromatin structure of the Xi is rearranged and compacted. The Xa forms a 3D structural arrangement shaped by various topologically associating domains (TADs). During the process of XCI, Xist first silences and compacts the chromatin surrounding the Xic locus and then spreads to other nearby sites of the designated Xi to form the transcriptionally silent inactive X territory that expands as Xist transcripts spread and pull increasingly distant Xi regions into the inactive X territory as do other chromosomes [24]. This process culminates with the formation of two compacted Xi megadomains that are tethered to the nuclear lamina [25] and several megaloops that escape Xist-mediated compaction to form transcriptionally active foci in the nuclear core [9] (Fig. 1c).

One such XCI escape locus, *Firre*, expresses another lncRNA that sculpts 3D chromosomal conformation. Firre transcripts are anchored by SAFA to both *cis* and *trans* chromatin sites in a multivalent manner to bring specific loci from various chromosomes into proximity with each other. Together they form a distinct Firre-scaffolded nuclear subcompartment that is associated with the co-regulation of genes involved in several pluripotency pathways, chief among which are adipogenesis and electron transport chain-mediated energy metabolism [26].

2.4 Alternative Splicing

Even though the molecular mechanisms of intrinsic pre-mRNA splicing have been well-characterised, it is not well understood how splicing events are directed to determine isoform specificity in selecting between different alternative splicing outcomes [27]. It is appreciated that splicing outcomes are influenced by the dynamic interplay of splicing regulatory factors, histone marks and chromatin-binding proteins and that lncRNAs could also have key roles in selectively directing splicing events towards the production of specific isoforms [28]. Among the lncRNAs that participate in the formation of nuclear subcompartments, some may also regulate splicing factors. MALAT1 (metastasis-associated lung adenocarcinoma transcript 1) forms nuclear speckles in which it preferentially colocalises with serine/arginine (SR) splicing regulatory factors in their phosphorylated forms, most prominently with SRSF1. It is thought that MALAT1 influences alternative splicing by sequestering splicing factors in nuclear speckles [29] (Fig. 1d). A second lncRNA with a well-established influence on alternative splicing regulation is MIAT (myocardial infarction-associated transcript) [30], sometimes alternatively named Gomafu. MIAT directly binds splicing factor 1 (SF1) via a repeat SF1-interacting motif that is homologous to the optimal SF1-binding intronic branch point sequence and thus has a higher affinity for SF1 than endogenous, suboptimal SF1-binding branch points [31]. It was demonstrated that repeat oligos mimicking the MIAT SF1-binding region can delay the onset of splicing of a pre-mRNA substrate containing a suboptimal SF1-interacting branch point, but had no effect on the splicing rate of a substrate containing a strong SF1-binding branch point [31]. MIAT can also bind to other splicing factors including Celf-3 and QKI, further corroborating the notion that MIAT, like MALAT1, can also sequester splicing factors to alter the kinetics of alternative splicing events [32, 33].

2.5 Crosstalk with miRNAs: Competing Endogenous ncRNAs

It is thought that RNA molecules can influence each other's levels by competing for a limited pool of miRNAs, an idea that has given rise to the concept of a "regulatory language" of miRNA-RNA crosstalk [34]. The competing endogenous RNA (ceRNA) hypothesis postulates that protein-coding mRNA transcripts and ncRNAs can act by the reverse logic of miRNA-mediated mRNA suppression to cross-regulate the expression of other RNA molecules with which they share miRNA recognition elements (MREs) by competing for occupancy of mutually recognised miRNAs (Fig. 1e). The elucidation ceRNA interaction networks would thus depend on identifying crosstalking ceRNAs and characterising the precise number and location of MREs that they contain, which would be the "letters" of the RNA cross-regulatory code. Moreover, three conditions that influence the significance of ceRNA regulatory events were outlined: (1) the expression level changes of any given ceRNA must be large enough to relieve or overcome miRNA binding and suppression of competing transcripts; (2) transcripts sharing MREs for multiple miRNAs will form more robust ceRNA networks; and (3) since few transcripts among all those that are bound by any given miRNA are primary targets for suppression, some MREs of any given ceRNA may be more influential than others in its ceRNA network [34].

3 lncRNA and Epigenetic Deregulation in Cancer

Many innovations in non-coding RNA research have been spurred on by the persistent need to better understand the molecular underpinnings of human diseases. Cancer studies in particular have contributed greatly to our overall understanding of the interface between epigenetics and lncRNA. Several lncRNAs have been comprehensively established as playing oncogenic roles in many kinds of tumour, whereas others have emerged as tumour suppressor lncRNAs. In this section, we highlight several interesting insights into the roles of lncRNAs associated with epigenetic regulation in promoting or preventing cancer, how they influence tumour responses to anticancer therapies and how some of these lncRNAs themselves can be exploited as biomarkers or therapeutic targets in cancer (Table 1).

3.1 Oncogenic lncRNAs: HOTAIR, MALAT1, TUG1 and MIAT

HOTAIR was one of the first lncRNAs to be linked to cancer progression, and it remains one of the best-studied oncogenic lncRNAs. HOTAIR overexpression promotes lung cancer metastasis both in vitro and in vivo, and it has also been demonstrated that HOTAIR overexpression can "re-target" PRC2 to repress over 800 genes, including known breast cancer tumour suppressors, and that the oncogenic effect of HOTAIR this context is co-dependent on PRC2 [35]. Furthermore, HOTAIR has been implicated in c-Myc transcriptional activation mechanisms in breast cancer, contributing to breast cancer growth and progression in both in vitro and in vivo models. Therefore, HOTAIR does not exclusively mediate transcriptional silencing but can also coordinate transcriptional activation in breast cancer MALAT1 which earned its name when it was found overexpressed in non-small-cell lung cancer (NSCLC) metastases. It was one of the first long non-coding transcripts found to be relevant in cancer [36]. A landmark MALAT1 knockout study of NSCLC demonstrated that MALAT1 significantly alters the expression levels of metastasis-associated genes in NSCLC cells but does not cause significant mRNA splicing changes [37]. MALAT1 was also shown to be required for lung cancer metastasis of xenograft NSCLC tumours in mice both by *MALAT1* gene knockout

Table 1 Biological effects of lncRNAs associated with epigenetic processes that are up- and downregulated in various types of cancer and their potential clinical applications

lncRNA	Disease	Oncogenic/tumour-suppressive effects	Potential clinical applications	References
Upregulated in cancer				
HOTAIR	Breast cancer	• Cell cycle progression • Proliferation • Metastasis • Chemoresistance	• Prognostic biomarker	[35, 97–100]
	Lung cancer	• Proliferation • Metastasis • Cisplatin resistance	• Prognostic biomarker	[101–104]
	Ovarian cancer (OC)	• Proliferation • Platinum drug resistance	• Prognostic biomarker • Therapeutic target	[52, 53, 105, 106]
	Oral/oesophageal squamous cell carcinoma (OSCC)	• Proliferation • Metastasis • CSC maintenance	• Prognostic biomarker • Therapeutic target	[107–109]
	Hepatocellular carcinoma (HCC)	• Cell viability • Metastasis • CSC maintenance	• Prognostic biomarker	[110–113]
MALAT1	Lung cancer	• Proliferation • Cell motility • Metastasis • Chemoresistance	• Prognostic biomarker • Diagnostic biomarker • Therapeutic target	[36, 37, 114–117]
	Glioblastoma	• Proliferation • Metastasis • Temozolomide resistance	• Prognostic biomarker • Therapeutic target	[64, 65, 118]
	HCC	• Proliferation • Metastasis • Multiple drug resistance	• n/a	[119–122]
	CRC	• Proliferation • Metastasis • Chemoresistance	• Prognostic biomarker • Therapeutic target	[123, 124]
	Osteosarcoma	• Metastasis • Angiogenesis	• Therapeutic target	[40, 125, 126]

(continued)

Table 1 (continued)

lncRNA	Disease	Oncogenic/tumour-suppressive effects	Potential clinical applications	References
MIAT	Chronic lymphocytic leukaemia (CLL)	• Evasion of cell death	• Prognostic biomarker	[47]
	Breast cancer	• Proliferation • Metastasis	• Prognostic biomarker	[127]
	Lung cancer	• Proliferation • Metastasis	• Prognostic biomarker • Therapeutic target	[48, 128, 129]
TUG1	Glioblastoma	• Proliferation • CSC maintenance	• Therapeutic target	[68, 130]
	RCC	• Proliferation • Evasion of cell death • Metastasis	• Prognostic biomarker • Therapeutic target	[42, 131]
	Osteosarcoma	• Proliferation • Evasion of cell death • Metastasis	• Therapeutic target	[44, 132, 133]
ANRIL	Gastric cancer	• Proliferation • Multiple drug resistance	• Prognostic biomarker • Therapeutic target	[134, 135]
	Osteosarcoma	• Proliferation • Metastasis	• Prognostic biomarker • Therapeutic target	[136, 137]
Downregulated in cancer				
lincRNA-p21	CRC	• Growth suppression • Self-renewal suppression • Radiosensitivity	• Prognostic biomarker • Therapeutic target	[138–140]
	Prostate cancer (PCa)	• Proliferation suppression • Apoptosis	• Therapeutic target	[141, 142]
TB53TG1	CRC	• Proliferation suppression • Metastasis inhibition	• Prognostic biomarker	[50]
MEG3	Pancreatic cancer	• Growth inhibition • Apoptosis • Metastasis inhibition • Gemcitabine sensitivity	• Prognostic biomarker	[143, 144]
	Lung cancer	• Proliferation suppression • Apoptosis • Chemosensitivity	• Therapeutic target	[145–148]

and by MALAT1 RNA depletion using antisense oligonucleotides (ASOs), providing an early demonstration of lncRNA targeting as a potentially useful therapeutic strategy in cancer treatment [37].

Subsequent studies ranging across several types of solid tumour such as bladder cancer have corroborated the notion that MALAT1 overexpression mainly contributes to tumour metastasis through the epigenetic dysregulation of gene expression, rather than through its molecular activities of nuclear body formation and pre-mRNA splicing regulation [38]. MALAT1 upregulation has also been linked to epigenetic dysregulation in renal cell carcinoma (RCC), as well as reduced overall survival among RCC patients [39]. In addition, the MALAT1-EZH2 interaction has been linked to metastasis-promoting E-Cadherin repression in osteosarcoma tumours associated with reduced survival [40]. Another lncRNA implicated in driving cancer is Taurine Upregulated Gene 1 (TUG1) [41]. Overexpression of TUG1 is generally associated with poor prognosis across several tumour types including RCC [42] and bladder cancer [43]. In osteosarcoma, TUG1 has been implicated as a positive regulator of EZH2 expression by directly binding and inhibiting anti-EZH2 miRNA miR-144-3p, which enhances the metastatic potential of osteosarcoma cells [44]. TUG1 is also upregulated in HCC, in which it promotes cell proliferation and apoptosis resistance by directing PRC2, via EZH2, to epigenetically silence Krüppel-like factor 2 (KLF2) [45]. Thus, in many solid tumours, TUG1 appears to act as an oncogenic lncRNA [46].

MIAT is rapidly emerging as a novel cancer-promoting long non-coding RNA. The first indication of its relevance in cancer came when it was found to be upregulated in aggressive forms of chronic lymphocytic leukaemia (CLL) [47]. In CLL, MIAT participates in a positive feedback loop with pluripotency-associated transcription factor Oct4 that facilitates evasion of apoptosis in malignant B cells. Recently, several studies have implicated MIAT in several different solid tumours, among which MIAT has been found to be significantly upregulated in advanced-stage NSCLC tumours. Upregulated MIAT has been shown both in vitro and in vivo to be responsible for increased NSCLC tumour cell proliferation as well as metastasis, in which it promotes the epigenetic activation of metastasis-associated gene matrix metalloproteinase 9 (*MMP9*) [48].

3.2 Tumour-Suppressive lncRNAs: lincRNA-p21 and TP53BP1

LincRNA-p21 is a p53-directed lncRNA that is required to transcriptionally regulate a substantial proportion of genes regulated by p53 and is itself a tumour suppressor, as its knockdown significantly reduces apoptosis in cancer cells [49]. It is thought that lincRNA-p21 may perform its tumour-suppressive function by guiding heterogeneous ribonucleoprotein particle K (hnRNP-K), both a coactivator and corepressor of p53, to its target genes to coordinate its branch of the p53-directed DNA damage response (DDR) gene regulatory network [49].

Another lncRNA of the p53 network, TP53TG1, undergoes CpG hypermethylation and transcriptional repression of its genomic locus in CRC cell line HCT-116 and acts as a tumour suppressor when overexpressed in p53-positive CRC cells [50]. TP53TG1 also directly interacts with YBX1, a dual DNA-/RNA-binding protein that plays regulatory roles in gene transcription, pre-mRNA splicing and translation [51]. TP53TG1 recovery in HCT116 cells reduces YBX1 occupancy of the PI3K promoter, which in turn lowers PI3K expression and stabilises p53 levels. In addition, TP53TG1 promoter hypermethylation and increased nuclear YBX1 are both associated with worse progression-free survival of CRC patients [50].

4 Strategies to Target lncRNAs Implicated in Cancer Therapy Resistance

4.1 HOTAIR

Of all tumour types in which the therapeutic relevance of HOTAIR is being investigated, some of the most remarkable insights have been made in ovarian cancer (OC) studies. Two recent studies have done much to deepen our understanding of the influence of HOTAIR on OC response to chemotherapy [52, 53]. A functional overlap had previously been established between PRC2 and NF-κB in the response of breast cancer cells to genotoxic agents [54], and since HOTAIR negatively regulates the expression of NF-κB pathway inhibitor Iκ-Bα [55], it may in turn influence DDR via modulation of NF-κB activity in OC cells that display HOTAIR upregulation, NF-κB hyperactivation and platinum drug resistance. HOTAIR overexpression in OC cells sensitive to cisplatin increased their clonogenic potential, while the reverse was the case following HOTAIR knockdown in cisplatin-resistant cells. When HOTAIR-overexpressing cisplatin-sensitive cells were injected into mice, the tumours formed showed enhanced growth and cisplatin resistance [52]. Correspondingly, there were significant differences in the expression levels of NF-κB pathway genes. Increased secretion of interleukin 6 (IL-6), along with several other cytokines, was also observed following HOTAIR vector transfection [52]. Furthermore, treatment of non-transfected cells with "HOTAIR-conditioned" media was sufficient to increase their resistance to cisplatin in the presence of functional IL-6, implying that HOTAIR upregulation may contribute to the previously reported role of IL-6 in shaping tumour microenvironments into "chemoresistant niches" [52, 56].

To therapeutically exploit the HOTAIR-NFκB axis in chemoresistant OC, a strategy to target HOTAIR using peptide nucleic acids (PNAs) was developed (Fig. 2a). In essence, PNAs are ASOs in which the sugar-phosphate backbone is replaced with a synthetic peptide backbone, rendering them more stable than conventional short interfering RNA (siRNA)s due to nuclease resistance and increased thermal stability [57, 58]. A PNA was designed to target a minimum

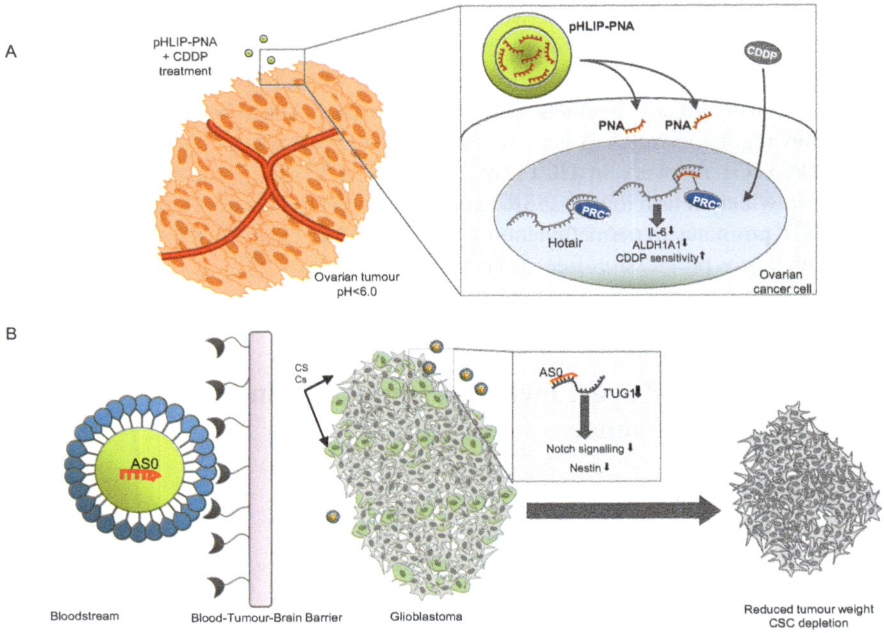

Fig. 2 Recent strategies to therapeutically target lncRNAs in cancer: delivery of HOTAIR-targeting PNA to the acidic tumour microenvironments of ovarian tumours, sensitising them to cisplatin (CDDP) (**a**); delivery of TUG1-targeting ASOs to glioblastoma tumours across the BTBB via integrin-specific peptide delivery nanoparticles, reducing tumour weight by depleting the glioblastoma CSC population (**b**)

89mer within the EZH2-interacting region of HOTAIR RNA to disrupt its interaction with PRC2. This PNA, named PNA3 in this study, inhibits the interaction by 80% [53]. OC and breast cancer cell lines ectopically overexpressing HOTAIR were then treated with PNA3 in combination with either cisplatin or etoposide. PNA3 co-treatment reduced clonogenic survival and migrational capacity of cisplatin-resistant cells, but PNA3 had no effect on either cell proliferation or on HOTAIR levels when administered as a single agent [53], indicating that HOTAIR inhibition could resensitise chemoresistant OC cells to these agents. Co-treatment of OC cells with cisplatin and PNA3 also led to the downregulation of ovarian cancer stem cell (CSC)-associated aldehyde dehydrogenase (ALDH1A1) [59], suggesting that the combined treatment could confer chemosensitivity by depleting chemoresistant CSC subpopulations of OC tumours [53].

Next, to specifically target PNA3 to the acidic tumour microenvironment, PNA3 was conjugated with a pH-low insertion peptide (pHLIP) [60], akin to a strategy previously developed to deliver an anti-miR155 PNA to lymphomas [61]. Cisplatin-resistant OC cells were co-treated with cisplatin and pHLIP-PNA3 under both neutral and acidic ($pH < 6$) conditions. No survival differences were observed under the pH-neutral condition, whereas reduced survival was exhibited by co-treated cells under the pH-low condition, indicating that pHLIP-PNA3 specifically resensitised cisplatin-resistant cells in an acidic environment mimicking that of the tumour microenvironment [53]. In mice, the combined treatment decreased tumour volume and increased survival and led to reduced blood levels of secreted IL-6 and reduced tumour expression of IL-6 and ALDH1A1 [53].

4.2 *MALAT1*

MALAT1 overexpression has been linked with poor prognosis in glioblastoma [62], a form of childhood neurological cancer that remains notoriously difficult to successfully treat because of the frequent emergence of resistance to front-line treatment with the chemotherapeutic agent temozolomide (TMZ). This typically manifests as a multiple drug resistance phenotype in which drug efflux channels and pumps of the MDR and MRP families are upregulated [63]. There is an urgent need to find strategies to increase the effectiveness of TMZ treatment and substantially improve survival outcomes for glioblastoma patients. Recent experimental evidence indicates that MALAT1 is a key factor in conferring TMZ resistance to glioblastoma multiforme (GBM) tumours. It is upregulated in TMZ-resistant GBM cell lines that also exhibit upregulation of multiple drug resistance genes [64]. siRNA knockdown of MALAT1 resensitised TMZ-resistant GBM cells to TMZ, and gene expression profiling showed downregulation of all three of the active multiple drug resistance-associated genes [64].

Another factor involved in the TMZ resistance problem in GBM treatment is the blood-tumour-brain barrier (BTBB). It has an intrinsically low permeability to free anticancer compounds in circulation and thus severely hinders their access to GBM tumours. To overcome this barrier, an immunoliposome-based system has been developed to deliver drugs across the BTBB to GBM tumours [65]. This comprises a positively charged liposome conjugated with a single-chain variable region fragment of a human anti-transferrin receptor (TfR) monoclonal antibody (TfRscFv) [66], named scL. It exploits the endogenous activity of TfR in translocating its ligand, diferric transferrin, across the BBB, and also takes advantage of the elevated presentation of TfR on the cell surfaces of tumour cells in order to preferentially deliver the liposome contents across the BTBB and into tumour cells [66]. Following initial success in using scL to enhance GBM tumour uptake of TMZ itself [67], the strategy has recently been adapted to usurp TMZ resistance in GBM by delivering a MALAT1-targeting siRNA to GBM [65]. scL-siMALAT1 treatment is intrinsically effective at inhibiting GBM cell growth and migration capacity in vitro and can also suppress glioblastoma stem cell (GSC) stemness [65]. Crucially, scL-siMALAT1 treatment resensitised TMZ-resistant GSC cells not only to TMZ but also to cisplatin and irinotecan. Finally, even as a single agent, scL-siMALAT1 inhibited tumour growth in mouse xenograft TMZ-resistant GBM tumours, and complete remission was achieved in some mice when scL-siMALAT1 was combined with TMZ treatment. Accordingly, overall survival was exclusively increased among mice administered the combinatorial treatment [65].

Overall, this study has demonstrated a truly effective and feasible approach to smuggle a lncRNA-targeting agent across the BTBB to take down the multiple drug resistance defences of TMZ-resistant GBM and meet the urgent clinical need to improve the effectiveness of TMZ treatment. Indeed, this method of MALAT1 targeting has arguably yielded the best results to this end of any preclinical approach taken to date and should be prioritised in the design and trial of new and improved TMZ treatment regimens that might also be able to effectively use lower and less toxic TMZ doses.

4.3 TUG1

Another recent glioblastoma-focussed study has not only shed light on tumour-promoting epigenetic dysregulation by TUG1 but has also described one of the most promising and feasible lncRNA-targeting therapeutic strategies to date. Katsushima and colleagues were interested in better characterising how GSCs maintain their characteristics [68]. While it had previously been established that Notch signalling is a hallmark of GSCs [69], it was unclear how Notch signalling maintains GSC stemness in glioblastoma (GBM). Based on a previous study indicating that lncRNAs may be involved in Notch signalling regulation in leukaemia [70], expression profiling in GSCs of lncRNAs potentially regulated by Notch signalling revealed that *TUG1* is upregulated in GSCs and contains multiple RBPJκ binding motifs in its promoter and thus is a bona fide transcriptional target of Notch signalling [68]. It was then established that TUG1 promotes GSC cell viability and maintains stem cell characteristics such as neurosphere formation and expression of GSC stemness-associated transcription factors [71] Sox2 and MYC [68]. Two distinct mechanisms by which TUG1 maintains GSC stemness have been suggested. First, TUG1 contains an MRE for miR-145, a repressor of GSC stemness-associated transcription factors *SOX2* and *MYC*, and is proposed to antagonise miR-145 to protect *SOX2* and *MYC* transcripts from miR-145-mediated downregulation. Second, TUG1 binds to both PRC2 and YY1 and is proposed to scaffold them to maintain transcriptional repression of neuronal differentiation genes [68].

To establish the clinical relevance of the findings, not only was it demonstrated that TUG1 is upregulated in a cohort of GBM patient tumours, but it was also confirmed that TUG1 is highly expressed in most *Notch1*-positive cells enriched around perivascular regions [68], in which Notch ligand-positive endothelial cells are thought to form a niche to promote GSC self-renewal via Notch signalling [72]. Having shown that TUG1 overexpression was a physiologically relevant factor in GSC maintenance, a strategy to target TUG1 in GSCs was developed. An ASO against TUG1 was delivered to xenograft mouse GBMs by using a cyclic Arg-Gly-Asp (cRGD) drug delivery system (DDS). The cRGD DDS has previously been developed to deliver a parent compound of the chemotherapeutic agent oxaliplatin across the blood-tumour-brain barrier (BTBB) to GBM by preferentially targeting tumour-associated neovascular endothelial cells that overexpress $\alpha_v\beta_3/\alpha_v\beta_5$ integrins, for which the Arg-Gly-Asp peptides used are ligands [73]. *TUG1*-DDS treatment effectively repressed TUG1 and achieved more marked tumour weight reduction than γ-secretase inhibitor treatment, the current lead strategy for therapeutically targeting Notch signalling in GBM [74, 75]. No TUG1- nor *Notch1*-positive tumours remained following *TUG1*-DDS treatment, and the GSC populations of the tumours were almost fully eliminated [68] (Fig. 2b).

Given that inhibition of Notch cleavage by γ-secretase is the current focus of GBM-targeted therapy development, these findings imply that targeting TUG1 with anti-TUG1 ASOs specifically delivered to the GSC niche might be a safer, more potent and more specific treatment strategy to target the same signalling pathway. Therefore, taken together, not only do the findings of this study represent a major advance towards the future development of implementable lncRNA-targeting therapeutic agents per se but also make an unprecedentedly strong case for prioritising the development of a lncRNA inhibition approach to signalling pathway shutdown over developing a conventional small molecule approach to inhibit a protein component of the same cancer-driving pathway.

5 Conclusion and Future Perspectives

The realisation that epigenetic traits are reversible and dynamically regulated was swiftly followed by the development of small molecule inhibitors of "epi-enzymes", the first generation of "epidrugs" developed to correct epigenetic deregulation in human disease [76]. Many have been developed into mainstream therapeutic agents, including bromodomain inhibitors against epigenetic readers [77]; inhibitors against the erasers of acetyl marks, HDACs [78]; and inhibitors of the catalytic components of histone writer complexes such as EZH2 in PRC2 [79]. While moderately effective in some cases, epi-enzyme inhibitors are generally limited in their use. This is because they inhibit components of the epigenetic machinery that act globally, so the blockade of their activity affects the distribution of the marks written, erased or acted upon across the whole genome, not only at the specific genomic sites that undergo disease-provoking epigenetic deregulation. Also, because many epigenetic proteins act in complexes, inhibition of one component can have secondary effects on other epigenetic proteins that are not direct targets for inhibition [76]. Moreover, widespread transcriptomic changes corresponding with the sweeping global epigenetic alterations brought about by epi-enzyme inhibitors would not only affect protein-coding genomic loci but would also occur in the non-coding genome. Such transcriptional perturbations may in turn affect the expression levels of lncRNAs that are themselves involved in epigenetic regulation, further convoluting the indirect epigenetic consequences of epi-enzyme inhibition. Therefore, the first generation of epigenetic therapy agents are blunt tools for correcting epigenetic defects, and their effectiveness in treating any given disease inevitably depends on the extent to which the global epigenetic changes they induce preferentially lead to disease-ameliorative changes in the epigenetic landscape over disease-worsening ones.

However, lncRNAs represent a steadily emerging "backstage" layer of epigenetic regulation, in which they are responsible for directing the epigenetic machinery in sculpting and moulding the epigenetic landscape in fine detail. As these "backstage" workings of lncRNAs in health and disease become more clearly illuminated year on year, so too does the potential of therapeutically targeting them in order to precisely correct pathological epigenetic alterations. LncRNA-targeting approaches designed to enact gene- and gene network-specific epigenetic change would in principle minimise the occurrence of non-specific, undesired global perturbations of epigenetic traits that are inevitable with current epidrugs but that do nothing to alleviate the diseases being treated. Therapeutically targeting lncRNAs could also expand the repertoire of epidrug development by opening the previously untapped possibility of reactivating the expression of proteins that are silenced in diseased states. Such an approach would not only be useful for reactivating silenced tumour suppressors, such as KLF2 in certain types of cancer, but also for compensating the inborn errors of the epigenetic machinery that characterise rare epigenetic syndromes. For example, Angelman syndrome is caused by the dominant negative mutational loss-of-function of the maternally expressed allele of the imprinted *Ube3a* gene and the continued silencing of the paternal allele by its NAT *Ube3a-AS* that prevents it from being able to compensate the loss of the maternal allele. However, proof of concept for the effectiveness of reactivating the paternal *Ube3a* allele by targeting the NAT of *Ube3a* with siRNA has been provided by results from a mouse model of Angelman syndrome showing that anti-*Ube3a-NAT* siRNA markedly reduced multiple symptoms of the disease [80]. Early in vitro success has also been gained in applying ASOs to alleviate Friedreich's ataxia (FRDA), the most frequent genetic cause of ataxia that is provoked by expanded GAA motif repeats in the first intron of the Frataxin (FXN)-coding mitochondrial gene *FXN*. It is proposed that when mutated *FXN* is transcribed, the expanded repeat region binds with the complementary genomic DNA sequence to form a transcription-disrupting R-loop, which effectively causes the breakdown of its own transcription and in turn depletes FXN protein expression [81]. Li et al. developed a strategy to recover *FXN* transcription, showing that both duplex RNA and LNA against expanded *FXN* repeats could increase FXN levels in fibroblasts derived from FDRA patients [82].

The potential usefulness of therapeutically targeting lncRNAs in non-cancer diseases has also been demonstrated in the neuromuscular disease that is the most common inherited cause of infant death, spinal muscular atrophy (SMA). It is primarily caused by recessive inactivating mutations in the Survival Motor Neuron 1 (*SMN1*) gene [83]. In addition, the homologous *SMN2* gene is retained in variable copy numbers and harbours a mutation that leads to reduced inclusion of exon 7 during pre-mRNA splicing [84]. Most SMN2 mRNA transcripts thus lack exon 7, and these transcripts in turn produce a truncated and unstable form of SMN protein. The consequent lack of functional SMN protein in spinal motor neurones severely impairs their survival [85]. Various studies have innovated the use of splice switching oligonucleotides (SSOs) to correct this splicing defect in SMN2 pre-mRNA to promote exon 7 inclusion and thereby increase the production of stable, full-length SMN protein [86, 87]. One such SMN2-targeting SSO,

Nusinersen (Spinraza), was effective in a phase III trial in infants with SMA [88] and has received FDA approval. However, this strategy is contingent on the number of *SMN2* copies retained. The most severe cases of SMA are associated with the fewest copies of *SMN2*. Therefore, while SMN2 pre-mRNA splicing correction may be an effective strategy for recovering SMN protein levels in cases with high *SMN2* copy numbers and abundant levels of SMN2 pre-mRNA, it will not suffice in more severe cases with few *SMN2* copies and low amounts of targetable SMN2 pre-mRNA [85]. However, a potential solution involving lncRNA targeting has recently emerged. *SMN2* transcription has recently been shown in two independent studies to be negatively regulated by its NAT, SMN-AS1, through PRC2 recruitment and epigenetic repression [89, 90]. In one of the studies, targeting SMN-AS1 for RNase H-mediated degradation with ASOs was found to be effective at restoring *SMN2* transcription and functional SMN protein production in SMA patient-derived cells, while combined systemic administration of ASOs targeting SMN-AS1 with *SMN2*-targeting SSOs alleviated SMA symptoms in mice [89]. The other study demonstrated the effectiveness of using SMN1-AS-targeting ASOs to sterically inhibit the SMN1-AS:EZH2 interaction to epigenetically derepress and upregulate full-length SMN2 expression in human SMA fibroblasts and SMA neuronal cells [90].

The mechanistic roles of lncRNAs in cardiac physiology and disease that have been discovered to date have been comprehensively reviewed elsewhere [91–93]. However, an intriguing recent example that immediately offers a potential therapeutic target is the novel lncRNA CHAST. It is upregulated in the hearts of aortic stenosis patients, and CHAST silencing with GapmeR treatment effectively prevented cardiac hypertrophy in a mouse model of CHAST-inducible cardiac hypertrophy [94]. CHAST inhibition was also effective in attenuating established cardiac hypertrophy in mice with pre-established cardiac disease. These findings demonstrate that ASO-mediated CHAST inhibition may be an effective strategy to ameliorate or even prevent cardiac hypertrophy in patients with aortic stenosis [94].

There remain many outstanding issues regarding lncRNA and epigenetics. Firstly, owing to their highly tissue-specific expression profiles, it remains necessary to more comprehensively characterise endogenous lncRNA transcriptomes in different human tissues. Second, the methodological hurdles that limit our current knowledge of lncRNAs must be overcome. Recent advances in the biochemical study of RNA-protein interactions have already deepened and refined our mechanistic understanding of some lncRNAs. Sustained progress in this regard, combined with sharper prediction of RNA secondary structure, will enable an increasingly detailed and comprehensive appreciation of how lncRNAs dynamically modulate epigenetic processes in physiological settings and of how these processes go awry in disease. Moreover, the pharmacological development of RNA-targeting agents such as ASOs has been rejuvenated by the emergence of lncRNAs as major players in epigenetics. This is precisely because they offer a novel opportunity to solve the problem of how to upregulate targets that are aberrantly downregulated in disease. As outlined in the various examples described in this review, the use of ASOs to target specific lncRNAs overexpressed in disease contexts of lncRNA overexpression can lead to direct and highly specific transcriptional and post-transcriptional upregulation of the targets whose expression they repress [95].

As in the case of SMA, ASO-mediated lncRNA repression can potentiate the effects of SSOs by specifically derepressing the expression of the very mRNAs whose aberrant splicing is targeted for splicing correction [89]. Aside from the demonstration of ASO use to sterically inhibit the SMN-AS1:EZH2 interaction in SMA cells [90], the use of ASOs to sterically disrupt lncRNA-protein interactions remains to be explored as a general treatment strategy in other disease models. Any future efforts to develop ASOs to this end will rely on more comprehensive characterisation of the structural dynamics of lncRNA-protein interaction. For the time being, it may be more feasible to develop small molecule inhibitors of interactions between lncRNAs and epigenetic modifiers, as a recent small molecule screening study identified compounds from a small panel that could effectively disrupt the interaction of brain-derived neurotrophic factor (BDNF) and EZH2 in cancer cells [96].

As further mechanistic details of lncRNA structural and interactive dynamics emerge, so must anti-lncRNA drug development co-evolve to quickly exploit such insights. As the non-coding side of epigenetics continues to be more clearly illuminated, it is hoped that the front runners of the hotly anticipated novel class of lncRNA-targeting epidrugs will soon reach the clinic.

Compliance with Ethical Standards

Funding: This study was funded by CERCA Programme/Generalitat de Catalunya, Instituto de Salud Carlos III co-funded by European Regional Development Funds (ERDF/FEDER) a way to build Europe (grant number PI10/02267 to M.B.) and COST CM1406 (to M.B.). D.J.H. has received funding from the People Program (MSCA-ITN-2014-ETN) of the European Union's Horizon 2020 Program under Chromatin 3D project (Grant number n° SEP-210147404).

Conflict of Interest: D.J.H. declares that he has no conflict of interest. M.E. declares that he has no conflict of interest. M.B. declares that she has no conflict of interest.

Ethical Approval: This chapter does not contain any studies with human participants or animals performed by any of the authors.

References

1. Djebali S, Davis CA, Merkel A, Gingeras TR (2012) Landscape of transcription in human cells. Nature 489:101–108. https://doi.org/10.1038/nature11233
2. Esteller M (2011) Non-coding RNAs in human disease. Nat Rev Genet 12:861–874. https://doi.org/10.1038/nrg3074
3. Iyer MK, Niknafs YS, Malik R et al (2015) The landscape of long noncoding RNAs in the human transcriptome. Nat Genet 47:199–208. https://doi.org/10.1038/ng.3192
4. Khorkova O, Hsiao J, Wahlestedt C (2015) Basic biology and therapeutic implications of lncRNA. Adv Drug Deliv Rev 87:15–24. https://doi.org/10.1016/j.addr.2015.05.012
5. Rinn JL, Chang HY (2012) Genome regulation by long noncoding RNAs. Annu Rev Biochem 81:145–166. https://doi.org/10.1146/annurev-biochem-051410-092902
6. Brockdorff N (2013) Noncoding RNA and Polycomb recruitment. RNA 19:429–442. https://doi.org/10.1261/rna.037598.112

7. Mchugh CA, Chen C, Chow A et al (2015) The Xist lncRNA directly interacts with SHARP to silence transcription through HDAC3. Nature 521:232–236. https://doi.org/10.1038/nature14443
8. Chu C, Zhang QC, Da Rocha ST et al (2015) Systematic discovery of Xist RNA binding proteins. Cell 161:404–416. https://doi.org/10.1016/j.cell.2015.03.025
9. Engreitz JM, Ollikainen N, Guttman M (2016) Long non-coding RNAs: spatial amplifiers that control nuclear structure and gene expression. Nat Rev Mol Cell Biol 17:756–770. https://doi.org/10.1038/nrm.2016.126
10. Rinn JL, Kertesz M, Wang JK et al (2007) Functional demarcation of active and silent chromatin domains in human HOX loci by noncoding RNAs. Cell 129:1311–1323. https://doi.org/10.1016/j.cell.2007.05.022
11. Tsai M, Manor O, Wan Y et al (2010) Long noncoding RNA as modular scaffold of histone modification complexes. Science 329:689–693. https://doi.org/10.1126/science.1192002
12. Khalil AM, Guttman M, Huarte M et al (2009) Many human large intergenic noncoding RNAs associate with chromatin-modifying complexes and affect gene expression. Proc Natl Acad Sci 106:11667–11672. https://doi.org/10.1073/pnas.0904715106
13. Zhao J, Ohsumi TK, Kung JT et al (2010) Genome-wide identification of Polycomb-associated RNAs by RIP-seq. Mol Cell 40:939–953. https://doi.org/10.1016/j.molcel.2010.12.011
14. Portoso M, Ragazzini R, Brenčič Ž et al (2017) PRC2 is dispensable for HOTAIR-mediated transcriptional repression. EMBO J 36:981–994. https://doi.org/10.15252/embj.201695335
15. Yap KL, Li S, Muñoz-Cabello AM et al (2010) Molecular interplay of the noncoding RNA ANRIL and methylated histone H3 lysine 27 by polycomb CBX7 in transcriptional silencing of INK4a. Mol Cell 38:662–674. https://doi.org/10.1016/j.molcel.2010.03.021
16. Aguilo F, Zhou MM, Walsh MJ (2011) Long noncoding RNA, polycomb, and the ghosts haunting INK4b-ARF-INK4a expression. Cancer Res 71:5365–5369. https://doi.org/10.1158/0008-5472.CAN-10-4379
17. Pandey RR, Mondal T, Mohammad F et al (2008) Kcnq1ot1 antisense noncoding RNA mediates lineage-specific transcriptional silencing through chromatin-level regulation. Mol Cell 32:232–246. https://doi.org/10.1016/j.molcel.2008.08.022
18. Bao X, Wu H, Zhu X et al (2015) The p53-induced lincRNA-p21 derails somatic cell reprogramming by sustaining H3K9me3 and CpG methylation at pluripotency gene promoters. Cell Res 25:80–92. https://doi.org/10.1038/cr.2014.165
19. Mohammad F, Pandey RR, Nagano T et al (2008) Kcnq1ot1/Lit1 noncoding RNA mediates transcriptional silencing by targeting to the perinucleolar region. Mol Cell Biol 28:3713–3728. https://doi.org/10.1128/MCB.02263-07
20. Wang X, Arai S, Song X et al (2010) Induced ncRNAs allosterically modify RNA binding proteins in cis to inhibit transcription. Nature 454:126–130. https://doi.org/10.1038/nature06992.Induced
21. Bose DA, Donahue G, Reinberg D et al (2017) RNA binding to CBP stimulates histone acetylation and transcription. Cell 168:135–149.e22. https://doi.org/10.1016/j.cell.2016.12.020
22. Jacob R, Zander S, Gutschner T (2017) The dark side of the epitranscriptome: chemical modifications in long non-coding rnas. Int J Mol Sci 18:E2387. https://doi.org/10.3390/ijms18112387
23. Patil DP, Chen C-K, Pickering BF et al (2016) m6A RNA methylation promotes XIST-mediated transcriptional repression. Nature 537:369–373. https://doi.org/10.1038/nature19342
24. Engreitz JM, Pandya-jones A, Mcdonel P et al (2013) The Xist lncRNA exploits three-dimensional genome architecture to spread across the X-chromosome. Science 341:1237973. https://doi.org/10.1126/science.1237973

25. Chen C-K, Blanco M, Jackson C et al (2016) Xist recruits the X chromosome to the nuclear lamina to enable chromosome-wide silencing. Science 354:468–472. https://doi.org/10.1126/science.aae0047
26. Hacisuleyman E, Goff LA, Trapnell C et al (2014) Topological Organization of Multichromosomal Regions by Firre. Nat Struct Mol Biol 21:198–206. https://doi.org/10.1038/nsmb.2764
27. Kornblihtt AR, Schor IE, Alló M et al (2013) Alternative splicing: a pivotal step between eukaryotic transcription and translation. Nat Rev Mol Cell Biol 14:153–165. https://doi.org/10.1038/nrm3525
28. Luco RF, Misteli T (2011) More than a splicing code: integrating the role of RNA, chromatin and non-coding RNA in alternative splicing regulation. Curr Opin Genet Dev 21:366–372. https://doi.org/10.1016/j.gde.2011.03.004
29. Tripathi V, Ellis JD, Shen Z et al (2010) The nuclear-retained noncoding RNA MALAT1 regulates alternative splicing by modulating SR splicing factor phosphorylation. Mol Cell 39:925–938. https://doi.org/10.1016/j.molcel.2010.08.011
30. Ishii N, Ozaki K, Sato H et al (2006) Identification of a novel non-coding RNA, MIAT, that confers risk of myocardial infarction. J Hum Genet 51:1087–1099. https://doi.org/10.1007/s10038-006-0070-9
31. Tsuiji H, Yoshimoto R, Hasegawa Y et al (2011) Competition between a noncoding exon and introns: Gomafu contains tandem UACUAAC repeats and associates with splicing factor-1. Genes Cells 16:479–490. https://doi.org/10.1111/j.1365-2443.2011.01502.x
32. Ishizuka A, Hasegawa Y, Ishida K et al (2014) Formation of nuclear bodies by the lncRNA Gomafu-associating proteins Celf3 and SF1. Genes Cells 19:704–721. https://doi.org/10.1111/gtc.12169
33. Barry G, Briggs JA, Vanichkina DP et al (2014) The long non-coding RNA Gomafu is acutely regulated in response to neuronal activation and involved in schizophrenia-associated alternative splicing. Mol Psychiatry 19:486–494. https://doi.org/10.1038/mp.2013.45
34. Salmena L, Poliseno L, Tay Y et al (2011) A ceRNA hypothesis: the Rosetta Stone of a hidden RNA language? Cell 146:353–358. https://doi.org/10.1016/j.cell.2011.07.014
35. Gupta RA, Shah N, Wang KC et al (2010) Long noncoding RNA HOTAIR reprograms chromatin state to promote cancer metastasis. Nature 464:1071–1076. https://doi.org/10.1038/nature08975
36. Ji P, Diederichs S, Wang W et al (2003) MALAT-1, a novel noncoding RNA, and thymosin b4 predict metastasis and survival in early-stage non-small cell lung cancer. Oncogene 22:8031–8041. https://doi.org/10.1038/sj.onc.1206928
37. Gutschner T, Hämmerle M, Eißmann M et al (2013) The noncoding RNA MALAT1 is a critical regulator of the metastasis phenotype of lung cancer cells. Cancer Res 73:1180–1189. https://doi.org/10.1158/0008-5472.CAN-12-2850
38. Fan Y, Shen B, Tan M et al (2014) TGF-b-induced upregulation of malat1 promotes bladder cancer metastasis by associating with suz12. Clin Cancer Res 20:1531–1541. https://doi.org/10.1158/1078-0432.CCR-13-1455
39. Hirata H, Hinoda Y, Shahryari V et al (2015) Long noncoding RNA MALAT1 promotes aggressive renal cell carcinoma through Ezh2 and interacts with miR-205. Cancer Res 75:1322–1331. https://doi.org/10.1158/0008-5472.CAN-14-2931
40. Huo Y, Li Q, Wang X et al (2017) MALAT1 predicts poor survival in osteosarcoma patients and promotes cell metastasis through associating with EZH2. Oncotarget 8:46993–47006. https://doi.org/10.18632/oncotarget.16551
41. Young TL, Matsuda T, Cepko CL (2005) The noncoding RNA taurine upregulated gene 1 is required for differentiation of the murine retina. Curr Biol 15:501–512. https://doi.org/10.1016/j.cub.2005.02.027
42. Zhang M, Lu W, Huang Y et al (2016) Downregulation of the long noncoding RNA TUG1 inhibits the proliferation, migration, invasion and promotes apoptosis of renal cell carcinoma. J Mol Histol 47:421–428. https://doi.org/10.1007/s10735-016-9683-2

43. Iliev R, Kleinova R, Juracek J et al (2016) Overexpression of long non-coding RNA TUG1 predicts poor prognosis and promotes cancer cell proliferation and migration in high-grade muscle-invasive bladder cancer. Tumor Biol 37:13385–13390. https://doi.org/10.1007/s13277-016-5177-9
44. Cao J, Han X, Qi X et al (2017) TUG1 promotes osteosarcoma tumorigenesis by upregulating EZH2 expression via MIR-144-3p. Int J Oncol 51:1115–1123. https://doi.org/10.3892/ijo.2017.4110
45. Huang M, Chen W, Qi F et al (2015) Long non-coding RNA TUG1 is up-regulated in hepatocellular carcinoma and promotes cell growth and apoptosis by epigenetic silencing of KLF2. Mol Cancer 14:165. https://doi.org/10.1186/s12943-015-0431-0
46. Zhou Y, Lu Y, Li R et al (2017) Prognostic role of long non-coding RNA TUG1 expression in various cancers: a meta-analysis. Oncotarget 8:100499–100507. https://doi.org/10.18632/oncotarget.20037
47. Sattari A, Siddiqui H, Moshiri F et al (2016) Upregulation of long noncoding RNA MIAT in aggressive form of chronic lymphocytic leukemias. Oncotarget 7:54174–54182. https://doi.org/10.18632/oncotarget.11099
48. Lai I-L, Yang C-A, Lin P-C et al (2017) Long noncoding RNA MIAT promotes non-small cell lung cancer proliferation and metastasis through MMP9 activation. Oncotarget 8:98148–98162. https://doi.org/10.18632/oncotarget.21465
49. Huarte M, Guttman M, Feldser D et al (2010) A large intergenic noncoding RNA induced by p53 mediates global gene repression in the p53 response. Cell 142:409–419. https://doi.org/10.1016/j.cell.2010.06.040
50. Diaz-Lagares A, Crujeiras AB, Lopez-Serra P et al (2016) Epigenetic inactivation of the p53-induced long noncoding RNA TP53 target 1 in human cancer. Proc Natl Acad Sci 113:E7535–E7544. https://doi.org/10.1073/pnas.1608585113
51. Jürchott K, Kuban RJ, Krech T et al (2010) Identification of Y-box binding protein 1 as a core regulator of MEK/ERK pathway-dependent gene signatures in colorectal cancer cells. PLoS Genet 6:1–19. https://doi.org/10.1371/journal.pgen.1001231
52. Özeş AR, Miller DF, Özeş ON et al (2016) NF-κB-HOTAIR axis links DNA damage response, chemoresistance and cellular senescence in ovarian cancer Ali. Oncogene 35:5350–5361. https://doi.org/10.1038/onc.2016.75
53. Özeş AR, Wang Y, Zong X et al (2017) Therapeutic targeting using tumor specific peptides inhibits long non-coding RNA HOTAIR activity in ovarian and breast cancer. Sci Rep 7:1–11. https://doi.org/10.1038/s41598-017-00966-3
54. Lee ST, Li Z, Wu Z et al (2011) Context-specific regulation of NF-κB target gene expression by EZH2 in breast cancers. Mol Cell 43:798–810. https://doi.org/10.1016/j.molcel.2011.08.011
55. Chu C, Spitale RC, Chang HY (2015) Technologies to probe functions and mechanisms of long noncoding RNAs. Nat Struct Mol Biol 22:29–35. https://doi.org/10.1038/nsmb.2921
56. Gilbert LA, Hemann MT (2010) DNA damage-mediated induction of a chemoresistant niche. Cell 143:355–366. https://doi.org/10.1016/j.cell.2010.09.043
57. Egholm M, Buchardt O, Christensen L et al (1993) PNA hybridizes to complementary oligonucleotides obeying the Watson-Crick hydrogen-bonding rules. Nature 365:566–568. https://doi.org/10.1038/365566a0
58. Bennett CF, Swayze EE (2010) RNA targeting therapeutics: molecular mechanisms of antisense oligonucleotides as a therapeutic platform. Annu Rev Pharmacol Toxicol 50:259–293. https://doi.org/10.1146/annurev.pharmtox.010909.105654
59. Landen CN, Goodman B, Katre AA et al (2010) Targeting aldehyde dehydrogenase cancer stem cells in ovarian cancer. Mol Cancer Ther 9:3186–3199. https://doi.org/10.1158/1535-7163.MCT-10-0563
60. Thévenin D, An M, Engelman DM (2009) pHLIP-mediated translocation of membrane-impermeable molecules into cells. Chem Biol 16:754–762. https://doi.org/10.1016/j.chembiol.2009.06.006

61. Cheng CJ, Bahal R, Babar IA et al (2015) MicroRNA silencing for cancer therapy targeted to the tumor microenvironment. Nature 518:107–110. https://doi.org/10.1038/nature13905
62. Ma KX, Wang HJ, Li XR et al (2015) Long noncoding RNA MALAT1 associates with the malignant status and poor prognosis in glioma. Tumor Biol 36:3355–3359. https://doi.org/10.1007/s13277-014-2969-7
63. Beier D, Schulz JB, Beier CP (2011) Chemoresistance of glioblastoma cancer stem cells – much more complex than expected. Mol Cancer 10:1–11. https://doi.org/10.1186/1476-4598-10-128
64. Li H, Yuan X, Yan D et al (2017) Long non-coding RNA MALAT1 decreases the sensitivity of resistant glioblastoma cell lines to temozolomide. Cell Physiol Biochem 42:1192–1201. https://doi.org/10.1159/000478917
65. Kim S-S, Harford JB, Moghe M et al (2017) Targeted nanocomplex carrying siRNA against MALAT1 sensitizes glioblastoma to temozolomide. Nucleic Acids Res 46:1424–1440. https://doi.org/10.1093/nar/gkx1221
66. Daniels TR, Delgado T, Helguera G, Penichet ML (2006) The transferrin receptor part II: targeted delivery of therapeutic agents into cancer cells. Clin Immunol 121:159–176. https://doi.org/10.1016/j.clim.2006.06.006
67. Kim S-S, Rait A, Kim E et al (2017) Encapsulation of temozolomide in a tumor-targeting nanocomplex enhances anti-cancer efficacy and reduces toxicity in a mouse model of glioblastoma. Cancer Lett 369:250–258. https://doi.org/10.1016/j.canlet.2015.08.022
68. Katsushima K, Natsume A, Ohka F et al (2016) Targeting the Notch-regulated non-coding RNA TUG1 for glioma treatment. Nat Commun 7:1–14. https://doi.org/10.1038/ncomms13616
69. Fan X, Khaki L, Zhu TS et al (2010) NOTCH pathway blockade depletes CD133-positive glioblastoma cells and inhibits growth of tumor neurospheres and xenografts. Stem Cells 28:5–16. https://doi.org/10.1002/stem.254
70. Trimarchi T, Bilal E, Ntziachristos P et al (2014) Genome-wide mapping and characterization of notch-regulated long noncoding RNAs in acute leukemia. Cell 158:593–606. https://doi.org/10.1016/j.cell.2014.05.049
71. Suvà ML, Rheinbay E, Gillespie SM et al (2014) Reconstructing and reprogramming the tumor-propagating potential of glioblastoma stem-like cells. Cell 157:580–594. https://doi.org/10.1016/j.cell.2014.02.030
72. Zhu TS, Costello MA, Talsma CE et al (2011) Endothelial cells create a stem cell niche in glioblastoma by providing NOTCH ligands that nurture self-renewal of cancer stem-like cells. Cancer Res 71:6061–6072. https://doi.org/10.1158/0008-5472.CAN-10-4269
73. Miura Y, Takenaka T, Toh K et al (2013) Cyclic RGD-linked polymeric micelles for targeted delivery of platinum anticancer drugs to glioblastoma through the blood-brain tumor barrier. ACS Nano 7:8583–8592. https://doi.org/10.1021/nn402662d
74. Xu R, Shimizu F, Hovinga K et al (2016) Molecular and clinical effects of notch inhibition in glioma patients: a phase 0/I trial. Clin Cancer Res 22:4786–4796. https://doi.org/10.1158/1078-0432.CCR-16-0048
75. Yahyanejad S, King H, Iglesias VS et al (2016) NOTCH blockade combined with radiation therapy and temozolomide prolongs survival of orthotopic glioblastoma. Oncotarget 7:41251–41264. https://doi.org/10.18632/oncotarget.9275
76. Simó-Riudalbas L, Esteller M (2015) Targeting the histone orthography of cancer: drugs for writers, erasers and readers. Br J Pharmacol 172:2716–2732. https://doi.org/10.1111/bph.12844
77. Pérez-Salvia M, Esteller M (2017) Bromodomain inhibitors and cancer therapy: from structures to applications. Epigenetics 12:323–339. https://doi.org/10.1080/15592294.2016.1265710
78. Bolden JE, Peart MJ, Johnstone RW (2006) Anticancer activities of histone deacetylase inhibitors. Nat Rev Drug Discov 5:769–784. https://doi.org/10.1038/nrd2133
79. Morera L, Lübbert M, Jung M (2016) Targeting histone methyltransferases and demethylases in clinical trials for cancer therapy. Clin Epigenetics 8:16. https://doi.org/10.1186/s13148-016-0223-4

80. Meng L, Ward AJ, Chun S et al (2015) Towards a therapy for Angelman syndrome by reduction of a long non-coding RNA. Nature 518:409–412. https://doi.org/10.1038/nature13975
81. Groh M, Lufino MMP, Wade-Martins R, Gromak N (2014) R-loops associated with triplet repeat expansions promote gene silencing in Friedreich ataxia and fragile X syndrome. PLoS Genet 10:e1004318. https://doi.org/10.1371/journal.pgen.1004318
82. Li L, Matsui M, Corey DR (2016) Activating frataxin expression by repeat-targeted nucleic acids. Nat Commun 7:1–8. https://doi.org/10.1038/ncomms10606
83. Lefebvre S, Buerglen L, Reboullet S et al (1995) Identification and characterization of a spinal muscular atrophy-determining gene. Cell 80:155–165
84. Lorson CL, Hahnen E, Androphy EJ, Wirth B (1999) A single nucleotide in the SMN gene regulates splicing and is responsible for spinal muscular atrophy. Proc Natl Acad Sci 96:6307–6311. https://doi.org/10.1073/pnas.96.11.6307
85. Burghes AHM, Beattie CE (2009) Spinal muscular atrophy: why do low levels of SMN make motor neurons sick? Nat Rev Neurosci 10:597–609. https://doi.org/10.1038/nrn2670
86. Touznik A, Maruyama R, Hosoki K et al (2017) LNA/DNA mixmer-based antisense oligonucleotides correct alternative splicing of the SMN2 gene and restore SMN protein expression in type 1 SMA fibroblasts. Sci Rep 7:1–9. https://doi.org/10.1038/s41598-017-03850-2
87. Robin V, Griffith G, Carter JPL et al (2017) Efficient SMN rescue following subcutaneous tricyclo-DNA antisense oligonucleotide treatment. Mol Ther Nucleic Acids 7:81–89. https://doi.org/10.1016/j.omtn.2017.02.009
88. Finkel RS, Mercuri E, Darras BT et al (2017) Nusinersen versus Sham control in infantile-onset spinal muscular atrophy. N Engl J Med 377:1723–1732. https://doi.org/10.1056/NEJMoa1702752
89. D'Ydewalle C, Ramos DM, Pyles NJ et al (2017) The antisense transcript SMN-AS1 regulates SMN expression and is a novel therapeutic target for spinal muscular atrophy. Neuron 93:66–79. https://doi.org/10.1016/j.neuron.2016.11.033
90. Woo CJ, Maier VK, Davey R et al (2017) Gene activation of SMN by selective disruption of lncRNA-mediated recruitment of PRC2 for the treatment of spinal muscular atrophy. Proc Natl Acad Sci 114:E1509–E1518. https://doi.org/10.1073/pnas.1616521114
91. Gomes CPC, Spencer H, Ford KL et al (2017) The function and therapeutic potential of long non-coding RNAs in cardiovascular development and disease. Mol Ther Nucleic Acids 8:494–507. https://doi.org/10.1016/j.omtn.2017.07.014
92. De Majo F, Calore M (2018) Chromatin remodelling and epigenetic state regulation by non-coding RNAs in the diseased heart. Noncoding RNA Res 3:20–28. https://doi.org/10.1016/j.ncrna.2018.02.003
93. Hung J, Miscianinov V, Sluimer JC et al (2018) Targeting non-coding RNA in vascular biology and disease. Front Physiol 9:1–16. https://doi.org/10.3389/fphys.2018.01655
94. Viereck J, Kumarswamy R, Foinquinos A et al (2016) Long noncoding RNA Chast promotes cardiac remodeling-supplementary materials. Sci Transl Med 8:326ra22. https://doi.org/10.1126/scitranslmed.aaf1475
95. Modarresi F, Faghihi MA, Lopez-Toledano MA et al (2012) Natural antisense inhibition results in transcriptional de-repression and gene upregulation. Nat Biotechnol 30:453–459. https://doi.org/10.1038/nbt.2158
96. Pedram Fatemi R, Salah-Uddin S, Modarresi F et al (2015) Screening for small-molecule modulators of long noncoding RNA-protein interactions using alphascreen. J Biomol Screen 20:1132–1141. https://doi.org/10.1177/1087057115594187
97. Ding W, Ren J, Ren H, Wang D (2017) Long noncoding RNA HOTAIR modulates MiR-206-mediated Bcl-w signaling to facilitate cell proliferation in breast cancer. Sci Rep 7:1–9. https://doi.org/10.1038/s41598-017-17492-x
98. Li Y, Wang Z, Shi H et al (2016) HBXIP and LSD1 scaffolded by lncRNA hotair mediate transcriptional activation by c-Myc. Cancer Res 76:293–304. https://doi.org/10.1158/0008-5472.CAN-14-3607

99. Xue X, Yang YA, Zhang A et al (2016) LncRNA HOTAIR enhances ER signaling and confers tamoxifen resistance in breast cancer. Oncogene 35:2746–2755. https://doi.org/10.1038/onc.2015.340
100. Zhang L, Song X, Wang X et al (2015) Circulating DNA of HOTAIR in serum is a novel biomarker for breast cancer. Breast Cancer Res Treat 152:199–208. https://doi.org/10.1007/s10549-015-3431-2
101. Liu X, Liu Z, Sun M et al (2013) The long non-coding RNA HOTAIR indicates a poor prognosis and promotes metastasis in non-small cell lung cancer. BMC Cancer 13:464. https://doi.org/10.1186/1471-2407-13-464
102. Liu Z, Sun M, Lu K et al (2013) The long noncoding RNA HOTAIR contributes to cisplatin resistance of human lung adenocarcinoma cells via downregulation of p21WAF1/CIP1 expression. PLoS One 8:1–15. https://doi.org/10.1371/journal.pone.0077293
103. Zhai N, Xia Y, Yin R et al (2016) A negative regulation loop of long noncoding RNA HOTAIR and p53 in non-small-cell lung cancer. Onco Targets Ther 9:5713–5720. https://doi.org/10.2147/OTT.S110219
104. Liu M-Y, Li X-Q, Gao T-H et al (2016) Elevated HOTAIR expression associated with cisplatin resistance in non-small cell lung cancer patients. J Thorac Dis 8:3314–3322. https://doi.org/10.21037/jtd.2016.11.75
105. Wang Y, Wang H, Song T et al (2015) HOTAIR is a potential target for the treatment of cisplatin-resistant ovarian cancer. Mol Med Rep 12:2211–2216. https://doi.org/10.3892/mmr.2015.3562
106. Teschendorff AE, Lee S-H, Jones A et al (2015) HOTAIR and its surrogate DNA methylation signature indicate carboplatin resistance in ovarian cancer. Genome Med 7:108. https://doi.org/10.1186/s13073-015-0233-4
107. Wu Y, Zhang L, Zhang L et al (2015) Long non-coding RNA HOTAIR promotes tumor cell invasion and metastasis by recruiting EZH2 and repressing E-cadherin in oral squamous cell carcinoma. Int J Oncol 46:2586–2594. https://doi.org/10.3892/ijo.2015.2976
108. Ge XS, Ma HJ, Zheng XH et al (2013) HOTAIR, a prognostic factor in esophageal squamous cell carcinoma, inhibits WIF-1 expression and activates Wnt pathway. Cancer Sci 104:1675–1682. https://doi.org/10.1111/cas.12296
109. Lu M, Liao Y, Chen P et al (2017) Targeting LncRNA HOTAIR suppresses cancer stemness and metastasis in oral carcinomas stem cells through modulation of. Oncotarget 8:98542–98552. https://doi.org/10.18632/oncotarget.21614
110. Geng YJ, Xie SL, Li Q et al (2011) Large intervening non-coding RNA HOTAIR is associated with hepatocellular carcinoma progression. J Int Med Res 39:2119–2128. https://doi.org/10.1177/147323001103900608
111. Yang Z, Zhou L, Wu L-M et al (2011) Overexpression of long non-coding RNA HOTAIR predicts tumor recurrence in hepatocellular carcinoma patients following liver transplantation. Ann Surg Oncol 18:1243–1250. https://doi.org/10.1245/s10434-011-1581-y
112. Battistelli C, Cicchini C, Santangelo L et al (2017) The Snail repressor recruits EZH2 to specific genomic sites through the enrollment of the lncRNA HOTAIR in epithelial-to-mesenchymal transition. Oncogene 36:942–955. https://doi.org/10.1038/onc.2016.260
113. Li H, An J, Wu M et al (2015) LncRNA HOTAIR promotes human liver cancer stem cell malignant growth through downregulation of SETD2. Oncotarget 6:27847–27864. https://doi.org/10.18632/oncotarget.4443
114. Jen J, Tang Y-A, Lu Y-H et al (2017) Oct4 transcriptionally regulates the expression of long non-coding RNAs NEAT1 and MALAT1 to promote lung cancer progression. Mol Cancer 16:104. https://doi.org/10.1186/s12943-017-0674-z
115. Schmidt LH, Spieker T, Koschmieder S et al (2011) The long noncoding MALAT-1 RNA indicates a poor prognosis in non-small cell lung cancer and induces migration and tumor growth. J Thorac Oncol 6:1984–1992. https://doi.org/10.1097/JTO.0b013e3182307eac
116. Weber DG, Johnen G, Casjens S et al (2013) Evaluation of long noncoding RNA MALAT1 as a candidate blood-based biomarker for the diagnosis of non-small cell lung cancer. BMC Res Notes 6:518. https://doi.org/10.1186/1756-0500-6-518

117. Chen W, Zhao W, Zhang L et al (2017) MALAT1-miR-101-SOX9 feedback loop modulates the chemo-resistance of lung cancer cell to DDP via Wnt signaling pathway. Oncotarget 8:94317–94329
118. Cai T, Liu Y, Xiao J (2017) Long noncoding RNA MALAT1 knockdown reverses chemoresistance to temozolomide via promoting microRNA-101 in glioblastoma. Cancer Med 7:1–12. https://doi.org/10.1002/cam4.1384
119. Malakar P, Shilo A, Mogilevsky A et al (2017) Long noncoding RNA MALAT1 promotes hepatocellular carcinoma development by SRSF1 upregulation and mTOR activation. Cancer Res 77:1155–1167. https://doi.org/10.1158/0008-5472.CAN-16-1508
120. Wang J, Wang H, Zhang Y et al (2014) Mutual inhibition between YAP and SRSF1 maintains long non-coding RNA, Malat1-induced tumourigenesis in liver cancer. Cell Signal 26:1048–1059. https://doi.org/10.1016/j.cellsig.2014.01.022
121. Li C, Miao R, Liu S et al (2017) Down-regulation of miR-146b-5p by long noncoding RNA MALAT1 in hepatocellular carcinoma promotes cancer growth and metastasis. Oncotarget 8:28683–28695. https://doi.org/10.18632/oncotarget.15640
122. Yuan P, Cao W, Zang Q et al (2016) The HIF-2α-MALAT1-miR-216b axis regulates multi-drug resistance of hepatocellular carcinoma cells via modulating autophagy. Biochem Biophys Res Commun 478:1067–1073. https://doi.org/10.1016/j.bbrc.2016.08.065
123. Ji Q, Zhang L, Liu X et al (2014) Long non-coding RNA MALAT1 promotes tumour growth and metastasis in colorectal cancer through binding to SFPQ and releasing oncogene PTBP2 from SFPQ/PTBP2 complex. Br J Cancer 111:736–748. https://doi.org/10.1038/bjc.2014.383
124. Li P, Zhang X, Wang H et al (2017) MALAT1 is associated with poor response to oxaliplatin-based chemotherapy in colorectal cancer patients and promotes chemoresistance through EZH2. Mol Cancer Ther 16:739–751. https://doi.org/10.1158/1535-7163.MCT-16-0591
125. Dong Y, Liang G, Yuan B et al (2014) MALAT1 promotes the proliferation and metastasis of osteosarcoma cells by activating the PI3K/Akt pathway. Tumor Biol 36:1477–1486. https://doi.org/10.1007/s13277-014-2631-4
126. Zhang Z-C, Tang C, Dong Y et al (2017) Targeting the long noncoding RNA MALAT1 blocks the pro-angiogenic effects of osteosarcoma and suppresses tumour growth. Int J Biol Sci 13:1398–1408. https://doi.org/10.7150/ijbs.22249
127. Luan T, Zhang X, Wang S et al (2017) Long non-coding RNA MIAT promotes breast cancer progression and functions as ceRNA to regulate DUSP7 expression by sponging miR-155-5p. Oncotarget 8:1–12. https://doi.org/10.18632/oncotarget.19190
128. Zhang H-Y, Zheng F-S, Yang W, Lu J-B (2017) The long non-coding RNA MIAT regulates zinc finger E-box binding homeobox 1 expression by sponging miR-150 and promoting cell invasion in non-small-cell lung cancer. Gene 633:61–65. https://doi.org/10.1016/j.gene.2017.08.009
129. Fu Y, Li C, Luo Y et al (2018) Silencing of long non-coding RNA MIAT sensitizes lung cancer cells to gefitinib by epigenetically regulating miR-34a. Front Pharmacol 9:1–12. https://doi.org/10.3389/fphar.2018.00082
130. Cai H, Xue Y, Wang P et al (2015) The long noncoding RNA TUG1 regulates blood-tumor barrier permeability by targeting miR-144. Oncotarget 6:19759–19779. https://doi.org/10.18632/oncotarget.4331
131. Wang PQ, Wu YX, Zhong XD et al (2017) Prognostic significance of overexpressed long non-coding RNA TUG1 in patients with clear cell renal cell carcinoma. Eur Rev Med Pharmacol Sci 21:82–86
132. Wang Y, Yang T, Zhang Z et al (2017) Long non-coding RNA TUG1 promotes migration and invasion by acting as a ceRNA of miR-335-5p in osteosarcoma cells. Cancer Sci 108:859–867. https://doi.org/10.1111/cas.13201
133. Zhang Q, Geng P-L, Yin P et al (2013) Down-regulation of long non-coding RNA TUG1 inhibits osteosarcoma cell proliferation and promotes apoptosis. Asian Pac J Cancer Prev 14:2311–2315. https://doi.org/10.7314/APJCP.2013.14.4.2311

134. Zhang E, Kong R, Yin D et al (2014) Long noncoding RNA ANRIL indicates a poor prognosis of gastric cancer and promotes tumor growth by epigenetically silencing of miR-99a/miR-449a. Oncotarget 5:2276–2292. https://doi.org/10.18632/oncotarget.1902
135. Lan W-G, Xu D-H, Xu C et al (2016) Silencing of long non-coding RNA ANRIL inhibits the development of multidrug resistance in gastric cancer cells. Oncol Rep 36:263–270. https://doi.org/10.3892/or.2016.4771
136. Cheng S, Huang T, Li P et al (2017) Long non-coding RNA ANRIL promotes the proliferation, migration and invasion of human osteosarcoma cells. Exp Ther Med 14:5121–5125. https://doi.org/10.3892/etm.2017.5123
137. Yu G, Liu G, Yuan D et al (2018) Long non-coding RNA ANRIL is associated with a poor prognosis of osteosarcoma and promotes tumorigenesis via PI3K/Akt pathway. J Bone Oncol 11:51–55. https://doi.org/10.1016/j.jbo.2018.02.002
138. Zhai H, Fesler A, Schee K et al (2013) Clinical significance of long intergenic noncoding RNA-p21 in colorectal Cancer. Clin Colorectal Cancer 12:261–266. https://doi.org/10.1016/j.clcc.2013.06.003
139. Wang G, Li Z, Zhao Q et al (2014) LincRNA-p21 enhances the sensitivity of radiotherapy for human colorectal cancer by targeting the Wnt/β-catenin signaling pathway. Oncol Rep 31:1839–1845. https://doi.org/10.3892/or.2014.3047
140. Wang J, Lei Z-J, Guo Y et al (2015) miRNA-regulated delivery of lincRNA-p21 suppresses beta-catenin signaling and tumorigenicity of colorectal cancer stem cells. Oncotarget 6:37852–37870. https://doi.org/10.18632/oncotarget.5635
141. Wang X, Ruan Y, Wang X et al (2017) Long intragenic non-coding RNA lincRNA-p21 suppresses development of human prostate cancer. Cell Prolif 50:1–10. https://doi.org/10.1111/cpr.12318
142. Wang X, Xu Y, Wang X et al (2017) LincRNA-p21 suppresses development of human prostate cancer through inhibition of PKM2. Cell Prolif 50:1–10. https://doi.org/10.1111/cpr.12395
143. Zhang Y-Y, Feng H-M (2017) MEG3 suppresses human pancreatic neuroendocrine tumor cells growth and metastasis by down-regulation of Mir-183. Cell Physiol Biochem 44:345–356. https://doi.org/10.1159/000484906
144. Ma L, Wang F, Du C et al (2018) Long non-coding RNA MEG3 functions as a tumour suppressor and has prognostic predictive value in human pancreatic cancer. Oncol Rep 39:1132–1140. https://doi.org/10.3892/or.2018.6178
145. Wang P, Chen D, Ma H, Li Y (2017) LncRNA MEG3 enhances cisplatin sensitivity in non-small cell lung cancer by regulating miR-21-5p/SOX7 axis. Onco Targets Ther 10:5137–5149. https://doi.org/10.2147/OTT.S146423
146. Xia H, Qu X, Liu L et al (2018) LncRNA MEG3 promotes the sensitivity of vincristine by inhibiting autophagy in lung cancer chemotherapy. Eur Rev Med Pharmacol Sci 22:1020–1027. https://doi.org/10.26355/eurrev_201802_14384
147. Xia Y, He Z, Liu B et al (2015) Downregulation of Meg3 enhances cisplatin resistance of lung cancer cells through activation of the WNT/β-catenin signaling pathway. Mol Med Rep 12:4530–4537. https://doi.org/10.3892/mmr.2015.3897
148. Lu K, Li W, Liu X et al (2013) Long non-coding RNA MEG3 inhibits NSCLC cells proliferation and induces apoptosis by affecting p53 expression. BMC Cancer 13:461

Top Med Chem (2020) 33: 429–454
DOI: 10.1007/7355_2019_81

Published online: 10 October 2019

Acetylation and Methylation in Asthma, COPD, and Lung Cancer

Martijn R. H. Zwinderman, Fangyuan Cao, and Frank J. Dekker

Contents

1 Introduction 431
  1.1 Epigenetics 431
  1.2 Epidrugs 431
  1.3 Protein Lysine Acetylation and Deacetylation 432
  1.4 Relevance of Deacetylases in Disease and Their Potential as Drug Targets 432
2 Lysine Deacetylases in Asthma and COPD 433
  2.1 Similarities and Differences Between Asthma and COPD 433
  2.2 Deacetylase and Acetyltransferase Expression in COPD 434
  2.3 Causes and Consequences of a Decrease in HDAC2 Expression in COPD 434
  2.4 Restoring HDAC2 Activity in COPD as a Therapeutic Strategy 436
  2.5 Targeting HDAC Isoenzymes in COPD 436
  2.6 Deacetylase and Acetyltransferase Expression in Asthma 438
  2.7 Which HDAC to Inhibit in Asthma? 439
3 Protein and DNA Methylation in Asthma and COPD 440
  3.1 Cross Talk Between Acetylation and Methylation 441
  3.2 DNA Methylation in Asthma and COPD 442
  3.3 Protein Methylation in Asthma and COPD 443
4 HDACi and DNMTi in Lung Cancer 444
5 Conclusions and Future Directions 445
References 446

Abstract The urgent need for new therapies to treat airway diseases is exemplified by the death of approximately three million patients suffering from chronic obstructive pulmonary disease (COPD) each year. Additionally, lung cancer is the most

M. R. H. Zwinderman, F. Cao, and F. J. Dekker (✉)
Chemical and Pharmaceutical Biology, Groningen Research Institute of Pharmacy (GRIP), University of Groningen, Groningen, The Netherlands
e-mail: r.h.zwinderman@rug.nl; f.cao@rug.nl; f.j.dekker@rug.nl

common cause of cancer death, causing over 1.5 million deaths per year. A promising direction for new therapies for these and other lung diseases, like asthma, comes from the notion that deranged intracellular signaling pathways in asthma, COPD, and lung cancer are critically regulated by protein posttranslational modifications (PTMs). Acetylation and methylation are the quintessential PTMs, found on over a thousand proteins and influencing a diverse range of protein properties. Acetyl- and methyltransferases and deacetyl- and demethylases, the enzymes that control the dynamic process of acetylation and methylation, have consequently been recognized as important drug targets. Hence, inhibitors of these enzymes have been developed, which are currently being evaluated in preclinical models of asthma and COPD and in clinical trials of lung cancer. Significant progress has been made in this area, with many promising results, but several challenges still need to be overcome to provide effective new therapies for these airway diseases.

Keywords Asthma, COPD, Deacetylase inhibitors (HDACi), DNA methyltransferases (DNMTs), Lung cancer, Posttranslational modification (PTM)

Abbreviations

COPD	Chronic obstructive pulmonary disease
DNMT(s)	DNA methyltransferase(s)
DNMTi	DNA methyltransferase inhibitor(s)
GR	Glucocorticoid receptor
H3K4	Lysine 4 of histone H3
HDAC(s)	Histone deacetylase(s)
HDACi	Histone deacetylase inhibitor(s)
hm5C	5-Hydroxymethylcytosine
IL	Interleukin
KDM(s)	Lysine demethylase(s)
m5C	5-Methylcytosine
NF-κB	Nuclear factor-kappa B
NSCLC	Non-small cell lung cancer
PI3K	Phosphoinositide-3-kinase
PRMT(s)	Protein arginine methyltransferase(s)
PTM(s)	Posttranslational modification(s)
SCLC	Small-cell lung cancer
TET	Ten-eleven translocation
TSA	Trichostatin A

1 Introduction

1.1 Epigenetics

Every human cell contains a fascinating 3 m long string of DNA that is tightly coiled around repeating histone octamers, together called chromatin. Chromatin can be packaged lightly, as is the case in actively transcribed DNA regions, or tightly, leading to gene silencing by a decrease in DNA accessibility [1]. Variability in DNA packaging, tightly or lightly, is mostly the result of differences in posttranslational modification (PTM) of histones and is an important requisite for cell-specific gene expression [2]. Numerous histone residues can be modified in a variety of ways, including ubiquitination, phosphorylation, acetylation, and methylation, each having their own specific effect [3]. Histone acetylation only occurs on lysine residues, while both lysine and arginine are methylated. Lysine and arginine are positively charged at physiological pH and are present in the N-terminal DNA-binding tail of histones. Acetylation is thought to facilitate transcriptional activation either by neutralization of the ionic interaction between the tail and DNA or by forming a binding site for chromatin remodeling proteins [4]. Methylation is similarly associated with transcriptionally active genes, although this highly depends on the specific lysine or arginine residues that are methylated and to what extent. Methyl marks are furthermore a binding site for a number of chromatin-modifying proteins [5]. The ability to influence gene expression through the modification of histones raises the question if and how such modifications are being passed down to daughter cells during replication to maintain similar gene expression patterns, which is the main topic of epigenetics. All epigenetic mechanisms together provide an important explanation of how cells with the same DNA are able to differentiate into distinct cell lines.

1.2 Epidrugs

An important area of research centers on the enzymes that install and remove the mentioned PTMs on histones. Such enzymes can be considered to be the principal executors of epigenetic mechanisms. Being highly dynamic in nature, PTMs are not only regulated in a cell-specific but also in a time-specific manner, in turn turning many different genes on or off over the course of a cell cycle. The crucial consequence of this discovery is that gene expression can directly be controlled at any point in time by drugs that target the enzymes that carry out PTMs. To refer to their actions on epigenetic processes, such drugs are often termed "epidrugs" [6]. "Epidrug" is, however, not an all-encompassing name that only refers to an epigenetic component. It becomes increasingly clear that also nonhistone proteins are targeted by enzymes that were initially discovered to act on histones [7]. Therefore, the actions on nonhistone proteins need always be taken into account when evaluating the effects of epidrugs.

1.3 Protein Lysine Acetylation and Deacetylation

Besides histones, principally any protein with a lysine can undergo a dynamic process of acetylation and deacetylation at some point during its lifetime [8]. One study mapped over 3,600 of such acetylation sites on approximately 1,750 proteins [9]. This so-called acetylome is under the control of acetyltransferases and deacetylases, of which many isoforms exist. To date, 18 deacetylases are known, which are grouped into two families based on their dependency on either zinc or nicotinamide adenine dinucleotide as cofactor [10]. Members of the latter family are called sirtuins. Deacetylases that contain zinc are still mostly called histone deacetylases (HDACs), named after their firstly discovered target, and are further divided into four different classes based on sequence similarity. The functionally opposing group of proteins that show acetyltransferase activity is even larger and more diverse [11].

1.4 Relevance of Deacetylases in Disease and Their Potential as Drug Targets

Protein acetylation influences important protein properties. Next to altering DNA-protein interactions, protein-protein interactions, subcellular localization, and even transcriptional activity can all be tuned by the dynamic process of acetylation [12–14]. Acetylation of the transcription factor p53, for example, destabilizes the interaction with its negative regulator Mdm2, thereby activating p53 and ultimately leading to an increase in the expression of apoptosis and growth arrest-inducing genes [15]. In a similar fashion, acetylation of the p65 subunit of nuclear factor-kappa B (NF-κB), an important inflammatory transcription factor [16], influences its subcellular location and transcriptional activity [17]. Since acetylation is a posttranslational modification with fundamental importance for the function of key proteins, it is not surprising that acetylation patterns are distorted in various diseases. In cancer, p53 is acetylated to a lesser extent, allowing cells to grow rapidly [18]. In ulcerative colitis and other inflammatory diseases, the acetylation status of NF-κB is changed, resulting in an increased expression of cytokines [18, 19]. Consequently, small molecule inhibitors of HDACs (HDACi), which increase acetylation by repressing deacetylation, are in over 100 clinical trials for cancer and may also enter clinical trials for inflammatory diseases [20]. So far, four HDACi have been FDA-approved, and these are known to relatively selectively induce apoptosis in a number of tumor cell types [21]. As many of the underlying processes of cancer are also important in inflammation, HDACi may be equally useful in chronic inflammatory diseases [22, 23]. Importantly, their anti-inflammatory effects generally occur at 10–100-fold lower concentrations than needed for their cytotoxic effects on cancer cells [24].

The currently FDA-approved HDACi are pan-HDAC inhibitors, which cause an overall increase in acetylation. Many side effects are expected to originate from this nonselectivity. To improve HDACi in oncology, it is therefore important to develop inhibitors that target individual deacetylase isoforms [25]. Isoenzyme selectivity will be even more crucial in non-oncological applications such as inflammation. For instance, NF-κB is mostly deacetylated by HDAC3 [26], and selective inhibition of this deacetylase enables modulation of the NF-κB pathway. Co-inhibiting other deacetylases might counteract or change these effects. Therefore, the individual contributions of the various deacetylase isoforms in specific disease models need to be understood to enable their exploitation as therapeutic targets.

Additionally, apart from their enzymatic activity, HDACs also have roles in the formation of protein-protein complexes. It is important to be aware of this when comparing conditional knockout studies to studies using HDAC inhibitors [27]. This is exemplified by the HDACs 4, 5, 7, and 9 for which their deacetylase activity might not be crucial for their function [28]. HDAC4, for instance, is unable to efficiently deacetylate proteins on its own [28]. Instead, this deacetylase binds HDAC3, and the thus resulting HDAC4/HDAC3 complex has deacetylase activity [27]. Additionally, these HDACs shuttle between the nucleus and the cytoplasm, and thus their subcellular localization, rather than their expression or activity per se, controls their actions.

In this chapter we focus on the particular roles of zinc-dependent lysine deacetylases in models of asthma, COPD, and lung cancer. Additionally, the role of protein and DNA methylation in these diseases is described.

2 Lysine Deacetylases in Asthma and COPD

2.1 Similarities and Differences Between Asthma and COPD

Asthma and COPD are important examples of chronic inflammatory diseases, which are characterized by an influx of inflammatory cells in the lungs and an associated abundance of secreted cytokines. Each disease constitutes a major health burden, with 328 million people worldwide suffering from moderate to severe COPD and asthma affecting approximately 300 million [26, 27]. Asthma and COPD both involve the small airways and cause airflow obstruction and are usually characterized by mucus secretion and bronchoconstriction, leading to breathlessness, coughing, and eventually severe breathing impairment [29]. The underlying mechanisms of inflammation are, however, different. Asthma is primarily eosinophilic in nature and driven by CD4+ T cells, while COPD is mostly neutrophilic and

CD8-driven [29–31]. This is an important distinction, since it affects drug efficacy. There is now plenty of evidence that inhaled corticosteroids are effective against the eosinophilic inflammation observed in asthma but largely ineffective against the primarily neutrophilic inflammation encountered in COPD [32, 33]. Even more so, currently no pharmaceutical agent effectively stops the progression of COPD, leading to approximately three million deaths per year [30].

2.2 Deacetylase and Acetyltransferase Expression in COPD

In peripheral lung tissue, alveolar macrophages, and bronchial-biopsy specimens from COPD patients, HDAC activity is decreased compared to healthy non-smokers [34]. Notably, there is a marked and selective reduction in the expression of HDAC2 in patients with very severe COPD. They express less than 5% of HDAC2 of non-smokers, whereas smaller reductions were observed for the expression of HDAC5 and HDAC8. Reduced HDAC2 activity is correlated with disease severity and increased expression of inflammatory genes such as IL-8, which is most likely caused by increased histone H4 acetylation at the IL-8 promoter [34]. Reduced HDAC activity is also related to corticosteroid resistance, a characteristic feature of COPD. The expression of HDAC1, HDAC3, HDAC4, HDAC6, and HDAC7 was found unaltered. The expression and activity of HDAC9, HDAC10, and HDAC11 and most of the sirtuins in COPD are unknown, although sirtuin 1 activity is decreased [35]. The expression of acetyltransferases is largely unaffected.

2.3 Causes and Consequences of a Decrease in HDAC2 Expression in COPD

The decrease in HDAC2 activity is believed to be partly caused by nitration of one or two tyrosine residues of HDAC2 through the formation of peroxynitrite, which is generated by a reaction between superoxide anions and nitric oxide upon inhalation of noxious particles [36]. Nitration in turn triggers ubiquitination and subsequent proteosomal degradation [37]. While both HDAC1 and HDAC2 are modified by such reactive nitrogen species, only the level of HDAC2 is decreased, possibly by selective ubiquitin tagging of nitrated HDAC2 [37]. The enzymes responsible for ubiquitination of HDAC1 are unknown but differ from those of HDAC2. Additionally, reactive oxygen species in inflamed lung tissue activate the phosphoinositide-3-kinase (PI3K) pathway, causing phosphorylation of HDAC2's serine residues by either PI3K itself or other downstream kinases, which similarly leads to its ubiquitin-proteosomal breakdown (Fig. 1b) [38].

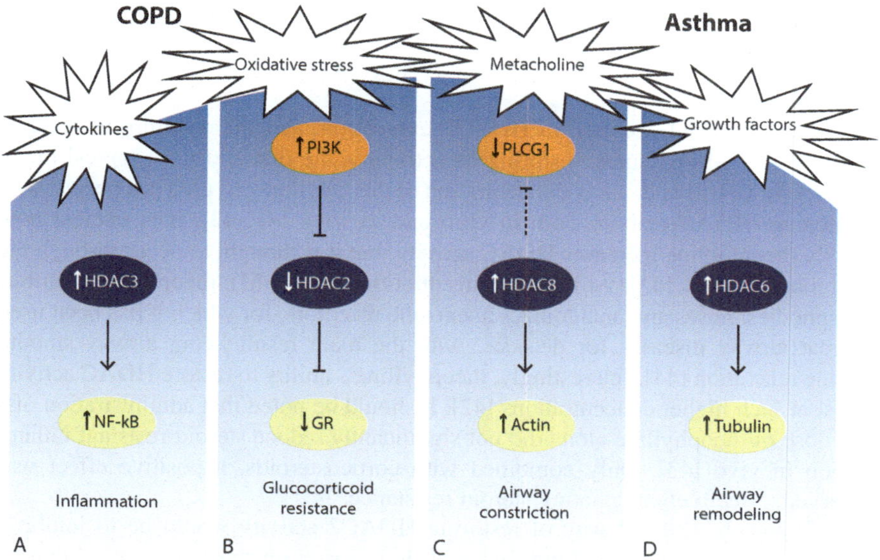

Fig. 1 Roles of HDAC isoforms in asthma and COPD. (**a**) HDAC3 deacetylates NF-κB at several lysines. Depending on the lysine deacetylated, it activates NF-κB to increase the expression of pro-inflammatory genes. (**b**) Oxidative stress following cigarette smoke exposure causes PI3K to phosphorylate HDAC2, which results in its ubiquitin-mediated proteosomal degradation. The glucocorticoid receptor (GR) remains acetylated, rendering it unable to repress NF-κB, thus leading to steroid resistance. (**c**) HDAC8 promotes actin polymerization in smooth muscle cells to facilitate contraction in response to methacholine stimulation. HDAC8 may also repress the activity of PLCG1, potentially preventing sarcoplasmic Ca^{2+} release. (**d**) HDAC6 is implicated in growth factor-activated pathways by decreasing microtubule stability through deacetylation of α-tubulin. In the long term, this may lead to airway remodeling

Besides influencing histone acetylation at the IL-8 promotor, HDAC2 is known to play a crucial role in the glucocorticoid receptor (GR) pathway [38]. When a corticosteroid binds to the GR, the receptor translocates to the nucleus where it is acetylated by an acetyltransferase. Upon acetylation, the receptor binds to the promotor regions of several key anti-inflammatory genes, such as IL-10 and IκB, leading to a subsequent increase in their expression [39]. Even bigger anti-inflammatory effects are, however, seen upon deacetylation of the GR by HDAC2. In its deacetylated form, it forms a complex with an array of other proteins, together able to repress the important NF-κB pro-inflammatory pathway [40]. Altogether, this indicates a clear connection between a decrease in HDAC2 activity and steroid resistance (Fig. 1b).

2.4 Restoring HDAC2 Activity in COPD as a Therapeutic Strategy

Given the observed decrease in HDAC2 expression, and the resulting increase in IL-8 and steroid resistance, restoring its activity is the first and most logical therapeutic aim. To this end, a low concentration of theophylline (10 μM) has been shown to increase HDAC activity both in vitro and ex vivo [41, 42]. It is unclear how exactly theophylline increases HDAC activity, but it is thought to occur through the inhibition of PI3K [43]. At higher concentrations (>10 μM), theophylline inhibits phosphodiesterases and antagonizes adenosine receptors, for which it has been used to treat airway diseases for decades, with the main result being airway smooth muscle relaxation [44]. Interestingly, theophylline's ability to restore HDAC activity is lost at such higher concentrations [42]. It should be noted that administration of a low dose of theophylline alone did not significantly reduce steroid-resistant inflammation in vivo [43]. Only combined with corticosteroids, a positive effect was observed, effectively alleviating steroid resistance.

A hypothetical novel way of restoring HDAC2 activity would be to inhibit a functionally opposing acetyltransferase. This is based on the presumption that an acetyltransferase may have the same substrates as HDAC2, but so far this has not been confirmed. Even more so, the development of clinically useful acetyltransferase inhibitors has fallen behind that of HDACi partly due to the low druggability of the catalytic domain of acetyltransferases. Consequently, only a few acetyltransferase inhibitors exist, and none have been tested in COPD. Other ways to restore HDAC activity are focused on inhibition of PI3K similar to theophylline or on the reduction of oxidative stress and thus peroxynitrite, potentially by the administration of antioxidants [45]. In the latter case, reducing the exposure to harmful particles works in a similar way. The success of such a strategy is exemplified by the fact that in COPD patients who smoke, smoking cessation has been shown to be the single most effective approach for slowing or halting disease progression [46, 47].

2.5 Targeting HDAC Isoenzymes in COPD

Given the decreased activity of HDAC2 in COPD and an observed add-on effect of theophylline as a HDAC activator in corticosteroid therapy, attempts to decrease HDAC activity with HDACi in COPD seem counterintuitive. On the other hand, diverse chemical classes of HDACi demonstrated therapeutic potential in several animal models of inflammation, such as colitis [48], multiple sclerosis [49], and arthritis [50]. The mentioned models are not necessarily predictive of an anti-inflammatory effect in COPD, as the underlying inflammatory dynamics are often very different. However, entinostat, a HDAC1-, HDAC2-, and HDAC3-selective inhibitor (Fig. 2), was recently found to attenuate inflammation in a mouse model of COPD [51]. The observed positive effect was connected to a decrease in neutrophil

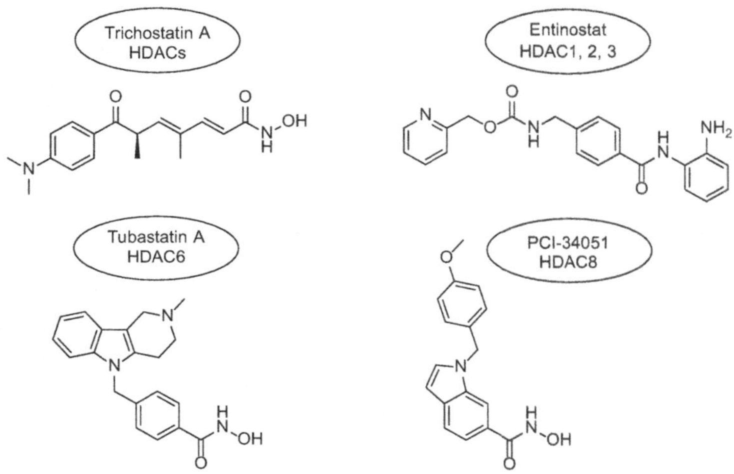

Fig. 2 Deacetylase inhibitors evaluated in asthma and COPD in vivo

influx in the lungs, in turn caused by a decrease in mouse IL-8. The expression of the pro-inflammatory cytokines IL-6 and IL-1ß was also decreased. Surprisingly, the used HDACi restored overall HDAC activity and led to an overall decrease in histone acetylation. This leads to intriguing questions about the exact mechanisms by which HDACs and their inhibitors act and in which cell types they play a leading role.

In the mouse COPD model, at least part of the effects can be explained mechanistically by the indirect action of entinostat on the acetylation status of NF-κB in macrophages [51]. NF-κB contains seven lysine residues that can be acetylated in a site-specific way. Acetylation of lysines 122 and 123 affects DNA binding, thereby acting in an inhibitory manner [52]. In contrast, acetylated lysines 218 and 221 block NF-κB's association with its cytoplasmic partner IκB, thereby enhancing its nuclear translocation and DNA-binding capability [53]. Acetylation at lysine 310 has been shown to boost transcriptional activity without affecting IκB or DNA binding, presumably by generating a binding site for coactivator proteins [54]. Most interestingly, acetylation of lysines 314 and 315 does not affect NF-κB's transcriptional activity but rather increases its promoter selectivity [55]. Overall, this means that acetylation patterns shape NF-κB's transcriptional activity and, importantly, gear it toward certain parts of the chromatin. For all lysine residues, acetylation is under the control of CBP/p300 acetyltransferases [56]. Deacetylation of NF-κB is mostly controlled by HDAC3 (Fig. 1a), making HDAC3 a highly interesting target in inflammation [26]. This is further backed up by the finding that HDAC3-deficient macrophages are unable to activate almost half of LPS-induced inflammatory gene expression [57]. Coming back to entinostat, which inhibits HDAC1, HDAC2, and HDAC3, the acetylation level of NF-κB increases in LPS-/IFNγ-stimulated macrophages upon treatment with the inhibitor, leading to its increased transcriptional activity [51]. Additionally, NF-κB nuclear translocation and binding to the IL-10 promotor increase, resulting in an enhanced expression of the anti-inflammatory cytokine IL-10.

The role described for HDAC3 in the NF-κB pathway in macrophages is an example, highlighting the potential for HDAC3-selective inhibition. Nevertheless, it should be noted that HDACs potentially target thousands of substrates, including many other transcription factors and general transcriptional regulators. However, it is to be expected that only a few potential lysine acetylation and deacetylation events have functional consequences. Therefore, it is important to identify these crucial events and the enzymes involved. This provides chances to find specific effects in specific disease models. Taken together, the current evidence indicates that HDAC3 is an important player in pro- and anti-inflammatory gene transcription and that HDAC inhibitors with isoenzyme selectivity among HDAC1, HDAC2, and HDAC3 have potential as anti-inflammatory agents.

2.6 Deacetylase and Acetyltransferase Expression in Asthma

Asthma is a heterogeneous disease with many different clinically overlapping subtypes. Prototypically, asthma is considered to be an allergic disorder mediated by mast cells, eosinophils, and T-helper cells ($CD4^+$) [58]. The importance of IL-8 in COPD is mirrored by IL-5 in asthma, which induces eosinophil generation in the bone marrow and leads to a subsequent increase in eosinophils in the lung [59]. Additionally, IL-4 is a key cytokine that enhances IgE-mediated immune responses in allergic airway inflammation [60]. Asthma is further clinically characterized by increased sensitivity of the airways to an inhaled constrictor agonist, like methacholine, termed airway hyperresponsiveness [61]. Looking at the balance between acetylases and deacetylases in patients with mild, stable asthma, a much more complicated balance is found compared to COPD. One study showed that histones were acetylated to a significantly higher extent in bronchial biopsies of asthma patients compared to healthy individuals and that this may be caused by a small but significant decrease in the expression of both HDAC1 and HDAC2 [62]. The expression of other HDACs was not evaluated, but it was shown that HDAC1 to HDAC6 were present primarily within the epithelium and to a lesser extent in infiltrating inflammatory cells. Another study analyzed the deacetylase activity in alveolar macrophages from asthmatic patients and found a decrease, related to a decrease in the expression of HDAC1, but not HDAC2 nor HDAC3 [63]. In circulating blood monocytes, no differences in HDAC activity were found, further confirming that differences in HDAC expression only occur locally at or near the site of inflammation. Taken together these results show that the expression levels of HDAC1, HDAC2, and HDAC3 are affected to a much lesser extent in asthma compared to COPD. This may have important consequences for treating this disease with HDACi. Unfortunately, neither the expression and activity of other HDACs nor the role of acetyltransferases has been adequately characterized in asthma.

2.7 Which HDAC to Inhibit in Asthma?

Trichostatin A (TSA, Fig. 2) has originally been reported as a fungistatic antibiotic and is one of the first natural compounds found to inhibit HDACs. This nonselective HDACi has been evaluated in allergen-induced airway inflammation. TSA significantly reduced airway hyperresponsiveness in both acute and chronic asthma models at concentrations of 1, 2, and 5 mg/kg [64–66]. Additionally, the number of eosinophils and the expression of IL-4 and IL-5 were decreased in bronchoalveolar lavage fluid upon treatment with $\geq$1 mg/kg TSA. The anti-inflammatory effects of TSA may partly be the result of enhanced apoptosis in eosinophils and neutrophils through a caspase-activating pathway, not involving PI3K nor NF-κB [67].

At a lower concentration, namely, 0.6 mg/kg, TSA did not reduce airway inflammation, but it did inhibit airway hyperresponsiveness [68]. This could be explained by blockage of airway smooth muscle contraction by TSA, even at lower concentrations, by decreasing the release of Ca^{2+} from the sarcoplasmic reticulum. A specific HDAC might be involved in this pathway, possibly HDAC8. This HDAC is known to deacetylate cortactin, which subsequently promotes actin filament polymerization and smooth muscle contraction and therefore plays an important role in the regulation of cytoskeletal dynamics of airway smooth muscle (Fig. 1c) [68–70]. In contradiction, inhibition of HDAC8 has been shown to increase Ca^{2+} flux and subsequent apoptosis in T cells by activating PLCG1 [71]. In any case, treatment with the HDAC8-selective inhibitor PCI-34051 (Fig. 2) showed that inhibition of HDAC8 alone attenuates airway hyperresponsiveness, airway inflammation, and to some extent even airway remodeling [72]. In view of airway remodeling, HDAC6 should also be considered as a target since its main function is to deacetylate α-tubulin [73, 74]. Reversible acetylation of α-tubulin regulates microtubule stability and function, with acetylation marking stabilized microtubules and an associated decrease in cellular motility (Fig. 1d). In asthma, inhibition of HDAC6 potentially maintains the cellular morphology of airway smooth muscle cells and thus may prevent airway remodeling. Although administration of the HDAC6-selective inhibitor tubastatin A (Fig. 2) in asthmatic mice reduced inflammation and airway hyperresponsiveness to a lesser extent than the HDAC8 inhibitor, it was similarly involved in decreasing α-smooth muscle actin and TGF-β1, both markers of airway remodeling [72]. In summary, inhibition of HDAC6 and HDAC8 might relieve airway constriction and prevent airway remodeling.

From a theoretical standpoint, inhibiting all HDACs, like TSA does, will lead to conflicting effects, but in reality a mostly positive outcome on inflammation and airway hyperresponsiveness is observed in asthma. This is especially fascinating given the fact that it was administered systemically by either subcutaneous or intraperitoneal injection. Although TSA did not seem to be toxic or induce noticeable side effects during the duration of the experiments, respiratory diseases are ideally treated by local administration to reduce the chance of systemic side effects. Since HDACs are ubiquitously expressed, this may offer substantial benefits to treatment with HDACi in airway diseases.

In conclusion, HDAC2 inhibition should be avoided in asthma since it has the potential to shift asthma's eosinophilic character to the more neutrophilic one seen in COPD, potentially by a similar mechanism of derepression of IL-8. Furthermore, the use of HDAC6- and HDAC8-selective inhibitors highlights key modes of action of the respective HDAC isoforms in asthma and shows that their inhibition leads to promising effects on airway hyperresponsiveness, inflammation, and remodeling.

3 Protein and DNA Methylation in Asthma and COPD

The role of other PTMs in lung diseases is also being deciphered. Of these other PTMs, most is known about dynamic methylation, governed by methyltransferases and demethylases. Importantly, while reversible acetylation is only known to occur on proteins, both proteins and DNA are found to be methylated. Similar to post-translational modifications, DNA methylation is a post-replicative modification. However, different enzyme families are at play that control either protein or DNA methylation and demethylation. DNA methyltransferases (DNMTs) methylate the C-5 carbon of cytosines that are next to a guanine, and the presence of 5-methylcytosine (m5C) in a promotor generally leads to a transcriptionally inactive gene [75]. Interestingly, the DNMT inhibitors (DNMTi) azacitidine (Fig. 3) and decitabine are substrates for the DNA replication machinery and incorporated into DNA as cytosine substitutes. Once incorporated, they are recognized by DNMTs as if they were natural cytosine, but upon methylation the enzymes are covalently trapped. This leads to the degradation of DNMTs and so ultimately to DNA hypomethylation [76]. Of note, extensive incorporation of azacitidine and decitabine results in DNA damage and apoptosis [77].

Demethylation of DNA on the other hand is a more complicated process. The DNA demethylation pathway begins with oxidation of the methyl group of m5C by the ten-eleven translocation (TET) family of enzymes to result in 5-hydroxymethylcytosine (hm5C), which is then further oxidized to 5-formyl and 5-carboxylcytosine by the same TET enzymes [78–80]. It is currently unclear if the

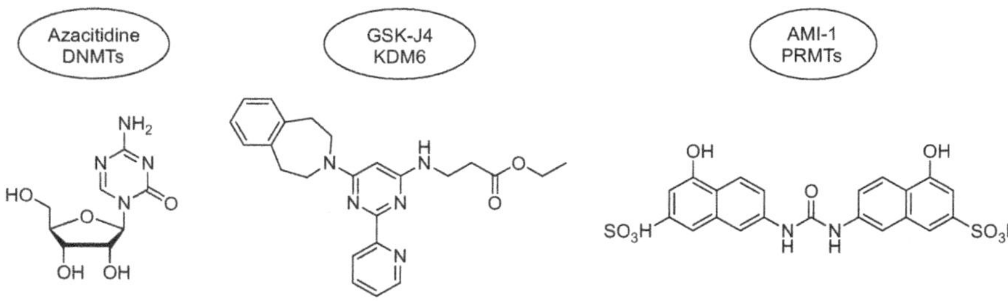

Fig. 3 Selected examples of inhibitors of DNA methyltransferases, lysine demethylases, and protein arginine methyltransferases

oxidized intermediates serve as epigenetic marks in their own right or if they are just short-lived intermediates in the DNA demethylation pathway. Some evidence indicates that hm5C has some signaling properties, also because it is more prevalent than the formyl and carboxyl derivatives [81]. In any case, the final product in the oxidation cascade, 5-carboxylcytosine, could in theory be decarboxylated to reform unmodified cytosine. However, a decarboxylase that is capable of decarboxylating 5-carboxylcytosine is yet to be identified. Known mechanisms to date show that the oxidized versions of 5mC are diluted in a replication-dependent manner to regenerate unmodified cytosine or in the case of 5-formyl and 5-carboxylcytosine are recognized and excised by enzymes involved in base excision repair mechanisms [82, 83]. The first step in this process is to flip the oxidized base out of the DNA double helix followed by cleavage of the *N*-glycosidic bond, which is catalyzed by thymine DNA glycosylase [84]. This leaves a gap in the DNA while leaving the sugar phosphate backbone intact. Other enzymes then refill the gap with an unmodified cytosine.

Protein demethylation is a somewhat simpler process that nonetheless also starts with oxidation [85]. However, in this case oxidation results in an unstable intermediate, a hemiaminal, which degrades to release formaldehyde and the thus demethylated lysine or arginine residue. Similar to HDACs, lysine demethylases (KDMs) come in various isoforms, and KDMs are grouped based on their dependency on either Fe(II) and 2-oxoglutarate or flavin [86–88]. In contrast, so far just one enzyme, the Jumonji domain containing protein 6, is speculated to have arginine demethylation activity in vivo [89]. The functionally opposing group of methyltransferases for both lysines and arginines is even larger and more diverse. Additionally, while acetyltransferases catalyze the addition of just one acetyl group to a lysine residue, methyltransferases mono-, di-, or tri-methylate them and mono- or di-methylate arginines [90]. The extent of methylation and the specific residues involved dictate the functional outcome.

3.1 Cross Talk Between Acetylation and Methylation

An additional layer of complexity is added by the fact that PTMs can be linked mechanistically by processes that are collectively termed "cross talk" [91]. As an example, methylation of histone H3 on lysine 4 (H3K4) recruits acetyltransferases to acetylate several other lysine residues of histone H3 [92]. Furthermore, the degree to which H3K4 is methylated directly influences the extent of H3 acetylation [93]. As a result, trimethylation of H3K4 is generally considered to be a marker of transcriptional activation. Additionally, methylation of H3K4 is known to inhibit DNA methylation, thereby further ensuring transcriptional activation [94, 95]. Conversely, DNA methylation decreases H3K4 methylation and H3 acetylation through a group of proteins that specifically recognize and bind to methylated DNA and in turn recruit a protein complex that contains HDACs and histone demethylases [96, 97]. Cross talk thus not only occurs between histone modifications but also between histone modifications and other epigenetic processes like DNA methylation. Not surprisingly, cross talk between PTMs of many different proteins exists, simply because lysines are found to be both acetylated and methylated, and these modifications affect protein functionality in different ways. It is often not clear how these different modifications compete with each other.

3.2 DNA Methylation in Asthma and COPD

There does not seem to be a clear association between DNA methylation levels and the presence of COPD. While there are studies that suggest that DNA methylation may be a biomarker of COPD [98], a genome-wide analysis of DNA methylation using blood samples of 903 never and 658 current smokers from the general population failed to show any significant association [99]. A systematic review of six articles that assessed the association of COPD with DNA methylation also failed to see consistency in the data [100]. Perhaps due to this lack of a clear link between DNA methylation and COPD, no inhibitors of DNA methyltransferase have been tested so far in in vivo models of COPD. Interestingly, the promotor of HDAC6 is known to be hypomethylated in COPD [101], which may partly explain the elevated HDAC6 expression seen in COPD and further point to treatment with HDAC6-selective inhibitors.

In contrast, a large meta-analysis of DNA methylation in childhood asthma, using data from more than 5,000 children, identified reduced DNA methylation levels in 14 distinct sites to be associated with asthma across childhood from ages 4 to 16 years [102]. Interestingly, whole blood DNA methylation profiles were strongly driven by lower methylation within eosinophils, further highlighting the important role of eosinophils in asthma. Additionally, the association was not found at birth, suggesting that environmental factors, like allergen exposure, could be the main cause of the observed change in the DNA methylation status. This possibility is further supported by results from mouse models of asthma, which showed reduced overall m5C and increased hm5C that correlated with a respective decrease in DNMTs and increase in TET enzymes upon allergen exposure [103, 104]. This indicates that DNMT activity limits asthma severity and thus that increasing DNA methylation may be beneficial in asthma. In stark contrast, treatment with azacitidine, a nonselective DNMTi, has been reported to reduce inflammation and airway hyperreactivity in mice, possibly by increasing the numbers of regulatory T cells [105]. Future research therefore needs to determine whether treatment of asthma with DNMTi or with still to be developed DNA demethylase inhibitors will be most beneficial.

3.3 Protein Methylation in Asthma and COPD

There is limited data on the role of arginine and lysine methylation in asthma and COPD. With respect to arginine methylation, one study reports the expression levels of protein arginine methyltransferase (PRMT) isoforms 1–6 in an animal model of asthma [106]. Except for PRMT4, the expression of all other PRMTs was found to be increased. In asthmatic rats, PRMT1 expression was especially high in epithelial cells [107]. The epithelial cells also expressed and secreted more eosinophil-attracting chemokines and fibroblast-activating cytokines. Consequently, the fibroblast proliferated more and expressed more growth factors, most likely through IL-1β-induced NF-κB activation [108, 109]. Interestingly, PRMT1 expression was also increased in the fibroblasts. Furthermore, treatment of the asthmatic rats with AMI-1 (Fig. 3), a nonselective inhibitor of PRMTs, ameliorated the observed pulmonary inflammation, mucus secretion, and collagen generation. This indicates that inhibition of PRMTs might be a valuable therapeutic strategy in asthma. Another reason for PRMT inhibition in asthma is that methylated arginine in itself, resulting from metabolic turnover of methylated proteins, is known to inhibit nitric oxide synthase [110, 111]. This results in a decrease in nitric oxide and an increase in reactive oxygen species, through which methylated arginine potentiates lung inflammation and airway hyperresponsiveness.

In contrast to asthma, the expression of PRMT6 is downregulated in lung tissue of COPD patients, as well as in mice with emphysema [112]. Furthermore, restoration of PRMT6 expression in mice exposed to cigarette smoke extract resulted in a decrease in inflammation, apoptosis, and oxidative stress. PRMT6 expression was restored by treatment with a lentivirus that was encoded with the PMRT6 gene. The protective effect of PRMT6 expression indicates that COPD treatment should focus on increasing protein methylation. However, the role of the other PRMT isoforms and protein demethylases in COPD is as of yet unknown. Consequently, the outcome of treatment of COPD with either PRMT or demethylase inhibitors is unpredictable. As with HDAC inhibition, the extent of inhibition of specific isoforms of PRMTs or demethylases will dictate the outcome. The field is thereby in need of the discovery and testing of isoform-selective inhibitors of PRMTs and demethylases in models of COPD.

In comparison to arginine methylation, less is known about the functional role of lysine methylation in asthma and COPD. This is in spite of the fact that potent and selective lysine demethylase inhibitors have been developed [113, 114]. One in vitro study found that reduced lysine histone methylation enhances the expression of vascular endothelial growth factor in airway smooth muscle cells from patients with asthma, thereby playing a role in airway remodeling [115]. An in vivo study found that restoration of lysine methylation by treatment with the potent KDM6 inhibitor GSK-J4 (Fig. 3) ameliorated the classical hallmarks of asthma, such as airway hyperresponsiveness, airway inflammation, and remodeling [116]. Treatment with GSK-J4 did however not decrease the expression of the vascular endothelial growth factor. The main explanation for the observed positive effects of GSK-J4 is that the inhibitor decreased the proliferation and migration of airway smooth muscle cells and prevented the upregulation of contractile proteins in these cells. In conclusion, in asthma, an increase in lysine methylation and a decrease in arginine methylation have so far shown to be potentially beneficial treatment strategies. For COPD nothing is known about the role of lysine methylation, but the role of arginine methylation in COPD is starting to be explored. The first studies are showing a beneficial effect of the restoration of arginine methylation in COPD.

4 HDACi and DNMTi in Lung Cancer

Of the different types of inhibitors described in this chapter, primarily HDACi and DNMTi are in clinical trials for lung cancer. Alterations in DNA methylation and protein acetylation are both considered to be major contributors to the development and progression of lung cancer. Lung cancer is generally classified as either small-cell lung cancer (SCLC) or non-small cell lung cancer (NSCLC), with the latter category split into a range of other histological subtypes. NSCLC accounts for approximately 85% of all lung cancer cases, and SCLC takes up the remaining 15% [117]. Most SCLC tumors initially respond to chemotherapy, but, unfortunately, essentially all patients experience relapse within 1 year of receiving first-line treatment. Similar to COPD, smoking is the major risk factor for all forms of lung cancer, particularly for SCLC.

DNMTi and HDACi have both demonstrated anticancer effects in in vitro NSCLC studies. However, single-agent clinical trials with these inhibitors in lung cancer patients mostly failed to show a beneficial effect. The current view is that the combination of DNMTi or HDACi with established therapies might increase efficacy by priming cancer cells to standard chemotherapy, possibly by reactivation of tumor suppressor genes [118]. Consequently, DNMTi and HDACi are in a host of phase I and II clinical trials in combination therapies, for instance, with immune checkpoint inhibitors like the monoclonal anti-PD1 antibody nivolumab. A detailed description of these trials can be found elsewhere [119].

Interestingly, a direct link between HDACs and DNMTs exists, since it has been found that HDACs deacetylate DNMTs [119]. In support of this, knockdown of HDACs and treatment with a HDACi induced DNMT acetylation. Surprisingly, this led to a decrease in DNMT levels. Overall, treatment with HDACi may thus concomitantly decrease DNMT levels. Since increased expression of DNMTs and HDACs crucially occurs in the oncogenic transformation of epithelia, it is therefore speculated that HDAC inhibition may prevent lung cancer. Additionally, combinations of DNMTi and HDACi are under investigation because they may synergize in the re-expression of silenced genes. A phase I/II trial of low-dose azacitidine combined with entinostat in 45 extensively pretreated patients with recurrent metastatic NSCLC showed that the combination was well tolerated and objective responses were observed [120]. This included one complete responder that was free of disease 26 months since enrolment. The combination also compared favorably with erlotinib, the existing treatment option. However, the objective responses to this therapy occurred in only 4% of patients, and the median survival in the entire cohort was nevertheless just 6.4 months. Two other phase II trials using the combination of azacitidine and entinostat in lung cancer are still ongoing [121]. The results of these trials will hopefully clarify the potentially synergistic effect between HDACi and DNMTi.

Finally, an interesting ongoing phase I clinical trial investigates the efficacy of inhaled azacitidine in patients with advanced NSCLC [121]. Generally, azacitidine is administered by subcutaneous injections, leading to systemic exposure. However, localized delivery of the drug to the lungs by means of inhalation was proven highly effective in animal models, because it was associated with longer survival, less toxicity, and less lung cancer burden compared to subcutaneous injection [122]. The results of that study sparked the initiation of the phase I trial. The outcome of this trial will be interesting as it only investigates the effect of inhaled azacitidine, not in combination with other therapies, which as a single-agent therapy has so far been unsuccessful. A caveat for the use of inhaled therapies in lung cancer is that between 30 and 40% of patients will have systemic metastases at the time of clinical diagnosis [123]. Logically, a therapy localized to the lungs will then need to be combined with a more systemic one.

5 Conclusions and Future Directions

Progress has been made in the chemotherapy of lung cancer by employing drugs that influence the reversible posttranslational and post-replicative modifications of proteins and DNA, respectively. However, optimum treatment regimens with such drugs still need to be defined, and it is clear that mostly a combination with another drug is required. In that sense the triple therapy of a HDACi, a DNMTi, and an immune checkpoint inhibitor is an interesting approach that is currently being investigated. Additionally, other influencers of PTMs, like inhibitors of KDMs, are in clinical trials for lung cancer [121]. It is to be expected that in the future an even broader range of compounds that influence the state of PTMs will be studied in combination therapies for lung cancer.

The results of the studies in lung cancer are highly relevant to other lung diseases like asthma and COPD, because one way to view the airway remodeling observed in asthma and COPD is to compare it to benign tumor growth. For instance, in both cases cell proliferation is out of control, and tissue boundaries are weakened. Therefore, as in lung cancer therapy, using drugs that regain control of tissue proliferation and architecture might be promising in asthma and COPD. Yet, as mentioned in this chapter, the study of HDACi and other influencers of PTMs in these diseases is still in the proof-of-principle stage. The initial in vivo preclinical studies are nonetheless promising, especially in the case of HDACi. In mouse models of asthma and COPD, application of several HDAC inhibitors has shown to result in a broad range of beneficial effects. Local administration of HDACi at relatively low concentrations may therefore hold great promise for the treatment of these diseases.

Compliance with Ethical Standards

Conflict of Interest: Martijn R.H. Zwinderman declares that he has no conflict of interest. Fangyuan Cao declares that she has no conflict of interest. Frank J. Dekker declares that he has no conflict of interest.

Ethical approval: This chapter does not contain any studies with human participants or animals performed by any of the authors.

References

1. Li B, Carey M, Workman JL (2007) The role of chromatin during transcription. Cell 128:707–719. https://doi.org/10.1016/j.cell.2007.01.015
2. Venkatesh S, Workman JL (2015) Histone exchange, chromatin structure and the regulation of transcription. Nat Rev Mol Cell Biol 16:178–189. https://doi.org/10.1038/nrm3941
3. Bannister AJ, Kouzarides T (2011) Regulation of chromatin by histone modifications. Cell Res 21:381–395. https://doi.org/10.1038/cr.2011.22
4. Becker PB, Workman JL (2013) Nucleosome remodeling and epigenetics. Cold Spring Harb Perspect Biol 5. https://doi.org/10.1101/cshperspect.a017905
5. Greer EL, Shi Y (2012) Histone methylation: a dynamic mark in health, disease and inheritance. Nat Rev Genet 13:343–357. https://doi.org/10.1038/nrg3173
6. Mai A, Altucci L (2009) Epi-drugs to fight cancer: from chemistry to cancer treatment, the road ahead. Int J Biochem Cell Biol 41:199–213. https://doi.org/10.1016/j.biocel.2008.08.020
7. Altucci L, Rots MG (2016) Epigenetic drugs: from chemistry via biology to medicine and back. Clin Epigenetics 8:56. https://doi.org/10.1186/s13148-016-0222-5
8. Drazic A, Myklebust LM, Ree R, Arnesen T (2016) The world of protein acetylation. Biochim Biophys Acta Proteins Proteomics 1864:1372–1401. https://doi.org/10.1016/j.bbapap.2016.06.007
9. Choudhary C, Kumar C, Gnad F, Nielsen ML, Rehman M, Walther TC, Olsen JV, Mann M (2009) Lysine acetylation targets protein complexes and co-regulates major cellular functions. Science 325:834–840. https://doi.org/10.1126/science.1175371
10. de Ruijter AJM, van Gennip AH, Caron HN, Kemp S, van Kuilenburg ABP (2003) Histone deacetylases (HDACs): characterization of the classical HDAC family. Biochem J 370:737–749. https://doi.org/10.1042/BJ20021321
11. Yang XJ, Seto E (2007) HATs and HDACs: from structure, function and regulation to novel strategies for therapy and prevention. Oncogene 26:5310–5318. https://doi.org/10.1038/sj.onc.1210599
12. Verdin E, Ott M (2015) 50 years of protein acetylation: from gene regulation to epigenetics, metabolism and beyond. Nat Rev Mol Cell Biol 16:258–264. https://doi.org/10.1038/nrm3931
13. Choudhary C, Weinert BT, Nishida Y, Verdin E, Mann M (2014) The growing landscape of lysine acetylation links metabolism and cell signalling. Nat Rev Mol Cell Biol 15:536–550. https://doi.org/10.1038/nrm3841
14. Spange S, Wagner T, Heinzel T, Krämer OH (2009) Acetylation of non-histone proteins modulates cellular signalling at multiple levels. Int J Biochem Cell Biol 41:185–198. https://doi.org/10.1016/j.biocel.2008.08.027
15. Tang Y, Zhao W, Chen Y, Zhao Y, Gu W (2008) Acetylation is indispensable for p53 activation. Cell 133:612–626. https://doi.org/10.1016/j.cell.2008.03.025

16. Lawrence T (2009) The nuclear factor NF-kappa B pathway in inflammation. Cold Spring Harb Perspect Biol 1:1–10. https://doi.org/10.1101/cshperspect.a001651
17. Huang B, Yang X-D, Lamb A, Chen L-F (2010) Posttranslational modifications of NF-kappaB: another layer of regulation for NF-kappaB signaling pathway. Cell Signal 22:1282–1290. https://doi.org/10.1016/j.cellsig.2010.03.017
18. Reed SM, Quelle DE (2014) P53 acetylation: regulation and consequences. Cancers (Basel) 7:30–69. https://doi.org/10.3390/cancers7010030
19. Glauben R, Siegmund B (2011) Inhibition of histone deacetylases in inflammatory bowel diseases. Mol Med 17:1. https://doi.org/10.2119/molmed.2011.00069
20. Mottamal M, Zheng S, Huang TL, Wang G (2015) Histone deacetylase inhibitors in clinical studies as templates for new anticancer agents. Molecules 20:3898–3941. https://doi.org/10.3390/molecules20033898
21. Bolden JE, Shi W, Jankowski K, Kan CY, Cluse L, Martin BP, MacKenzie KL, Smyth GK, Johnstone RW (2013) HDAC inhibitors induce tumor-cell-selective pro-apoptotic transcriptional responses. Cell Death Dis 4:e-519. https://doi.org/10.1038/cddis.2013.9
22. Coussens LM, Werb Z (2002) Inflammation and cancer. Nature 420:860–867. https://doi.org/10.1038/nature01322
23. Shakespear MR, Halili MA, Irvine KM, Fairlie DP, Sweet MJ (2011) Histone deacetylases as regulators of inflammation and immunity. Trends Immunol 32:335–343. https://doi.org/10.1016/j.it.2011.04.001
24. Dinarello CA, Fossati G, Mascagni P (2011) Histone deacetylase inhibitors for treating a spectrum of diseases not related to cancer. Mol Med 17:333–352. https://doi.org/10.2119/molmed.2011.00116
25. Cantley MD, Haynes DR (2013) Epigenetic regulation of inflammation: progressing from broad acting histone deacetylase (HDAC) inhibitors to targeting specific HDACs. Inflammopharmacology 21:301–307. https://doi.org/10.1007/s10787-012-0166-0
26. Leus NGJ, Zwinderman MRH, Dekker FJ (2016) Histone deacetylase 3 (HDAC 3) as emerging drug target in NF-κB-mediated inflammation. Curr Opin Chem Biol 33:160–168. https://doi.org/10.1016/j.cbpa.2016.06.019
27. Fischle W, Dequiedt F, Hendzel MJ, Guenther MG, Lazar MA, Voelter W, Verdin E, Francisco S, Tg A (2002) Enzymatic activity associated with class II HDACs is dependent on a multiprotein complex containing HDAC3 and SMRT/N-CoR. Mol Cell 9:45–57
28. Lahm A, Paolini C, Pallaoro M, Nardi MC, Jones P, Neddermann P, Sambucini S, Bottomley MJ, Lo Surdo P, Carfi A, Koch U, De Francesco R, Steinkuhler C, Gallinari P (2007) Unraveling the hidden catalytic activity of vertebrate class IIa histone deacetylases. Proc Natl Acad Sci 104:17335–17340. https://doi.org/10.1073/pnas.0706487104
29. Buist AS (2003) Similarities and differences between asthma and chronic obstructive pulmonary disease: treatment and early outcomes. Eur Respir J 21:30S–35s. https://doi.org/10.1183/09031936.03.00404903
30. Lõpez-Campos JL, Tan W, Soriano JB (2016) Global burden of COPD. Respirology 21:14–23. https://doi.org/10.1111/resp.12660
31. Croisant S (2014) Epidemiology of asthma: prevalence and burden of disease. Adv Exp Med Biol 795:17–29
32. Barnes PJ (2013) Corticosteroid resistance in patients with asthma and chronic obstructive pulmonary disease. J Allergy Clin Immunol 131:636–645. https://doi.org/10.1016/j.jaci.2012.12.1564
33. Barnes PJ, Ito K, Adcock IM (2004) Corticosteroid resistance in chronic obstructive pulmonary disease: inactivation of histone deacetylase. Lancet 363:731–733. https://doi.org/10.1016/S0140-6736(04)15650-X
34. Ito K, Ito M, Elliott WM, Cosio B, Caramori G, Kon OM, Barczyk A, Hayashi S, Adcock IM, Hogg JC, Barnes PJ (2005) Decreased histone deacetylase activity in chronic obstructive pulmonary disease. N Engl J Med 352:1967–1976. https://doi.org/10.1056/NEJMoa041892

35. Rajendrasozhan S, Yang SR, Kinnula VL, Rahman I (2008) SIRT1, an antiinflammatory and antiaging protein, is decreased in lungs of patients with chronic obstructive pulmonary disease. Am J Respir Crit Care Med 177:861–870. https://doi.org/10.1164/rccm.200708-1269OC
36. Osoata GO, Yamamura S, Ito M, Vuppusetty C, Adcock IM, Barnes PJ, Ito K (2009) Nitration of distinct tyrosine residues causes inactivation of histone deacetylase 2. Biochem Biophys Res Commun 384:366–371. https://doi.org/10.1016/j.bbrc.2009.04.128
37. Adenuga D, Yao H, March TH, Seagrave J, Rahman I (2009) Histone deacetylase 2 is phosphorylated, ubiquitinated, and degraded by cigarette smoke. Am J Respir Cell Mol Biol 40:464–473. https://doi.org/10.1165/rcmb.2008-0255OC
38. Barnes PJ (2009) Targeting the epigenome in the treatment of asthma and chronic obstructive pulmonary disease. Proc Am Thorac Soc 6:693–696. https://doi.org/10.1513/pats.200907-071DP
39. Barnes PJ (2009) Role of HDAC2 in the pathophysiology of COPD. Annu Rev Physiol 71:451–464. https://doi.org/10.1146/annurev.physiol.010908.163257
40. Rao NAS, McCalman MT, Moulos P, Francoijs KJ, Chatziioannou A, Kolisis FN, Alexis MN, Mitsiou DJ, Stunnenberg HG (2011) Coactivation of GR and NFKB alters the repertoire of their binding sites and target genes. Genome Res 21:1404–1416. https://doi.org/10.1101/gr.118042.110
41. Cosio BG, Tsaprouni L, Ito K, Jazrawi E, Adcock IM, Barnes PJ (2004) Theophylline restores histone deacetylase activity and steroid responses in COPD macrophages. J Exp Med 200:689–695. https://doi.org/10.1084/jem.20040416
42. Ito K, Lim S, Caramori G, Cosio B, Chung KF, Adcock IM, Barnes PJ (2002) A molecular mechanism of action of theophylline: induction of histone deacetylase activity to decrease inflammatory gene expression. Proc Natl Acad Sci U S A 99:8921–8926. https://doi.org/10.1073/pnas.132556899
43. To Y, Ito K, Kizawa Y, Failla M, Ito M, Kusama T, Elliott WM, Hogg JC, Adcock IM, Barnes PJ (2010) Targeting phosphoinositide-3-kinase-δ with theophylline reverses corticosteroid insensitivity in chronic obstructive pulmonary disease. Am J Respir Crit Care Med 182:897–904. https://doi.org/10.1164/rccm.200906-0937OC
44. Abramson M, Sim MR (2006) Theophylline for COPD. Thorax 61:741–742. https://doi.org/10.1136/thx.2005.056200
45. Marwick JA, Ito K, Adcock IM, Kirkham PA (2007) Oxidative stress and steroid resistance in asthma and COPD: pharmacological manipulation of HDAC-2 as a therapeutic strategy. Expert Opin Ther Targets 11:745–755. https://doi.org/10.1517/14728222.11.6.745
46. Scanlon PD, Connett JE, Waller LA, Altose MD, Bailey WC, Buist AS, Tashkin DP (2000) Smoking cessation and lung function in mild-to- moderate chronic obstructive pulmonary disease: the lung health study. Am J Respir Crit Care Med 161:381–390
47. Wu J, Sin DD (2011) Improved patient outcome with smoking cessation: when is it too late? Int J COPD 6:259–267. https://doi.org/10.2147/COPD.S10771
48. Lührs H, Gerke T, Müller JG, Melcher R, Schauber J, Boxberger F, Scheppach W, Menzel T (2002) Butyrate inhibits NF-κB activation in lamina propria macrophages of patients with ulcerative colitis. Scand J Gastroenterol 37:458–466. https://doi.org/10.1080/003655202317316105
49. Faraco G, Cavone L (2011) The therapeutic potential of HDAC inhibitors in the treatment of multiple sclerosis. Mol Med 17:1. https://doi.org/10.2119/molmed.2011.00077
50. Grabiec AM, Krausz S, de Jager W, Burakowski T, Groot D, Sanders ME, Prakken BJ, Maslinski W, Eldering E, Tak PP, Reedquist KA (2010) Histone deacetylase inhibitors suppress inflammatory activation of rheumatoid arthritis patient synovial macrophages and tissue. J Immunol 184:2718–2728. https://doi.org/10.4049/jimmunol.0901467
51. Leus NGJ, Van Den Bosch T, Van Der Wouden PE, Krist K, Ourailidou ME, Eleftheriadis N, Kistemaker LEM, Bos S, Gjaltema RAF, Mekonnen SA, Bischoff R, Gosens R, Haisma HJ, Dekker FJ (2017) HDAC1-3 inhibitor MS-275 enhances IL10 expression in RAW264.7 macrophages and reduces cigarette smoke-induced airway inflammation in mice. Sci Rep 7:1–18. https://doi.org/10.1038/srep45047

52. Kiernan R, Brès V, Ng RWM, Coudart MP, El Messaoudi S, Sardet C, Jin DY, Emiliani S, Benkirane M (2003) Post-activation turn-off of NF-κB-dependent transcription is regulated by acetylation of p65. J Biol Chem 278:2758–2766. https://doi.org/10.1074/jbc.M209572200
53. Chen L, Mu Y, Greene WC (2002) Acetylation of RelA at discrete sites regulates distinct nuclear functions of NF-kappaB. EMBO J 21:6539–6548. https://doi.org/10.1093/emboj/cdf660
54. Chen L-F, Greene WC (2004) Shaping the nuclear action of NF-κB. Nat Rev Mol Cell Biol 5:392–401. https://doi.org/10.1038/nrm1368
55. Rothgiesser KM, Fey M, Hottiger MO (2010) Acetylation of p65 at lysine 314 is important for late NF-kappaB-dependent gene expression. BMC Genomics 11:22. https://doi.org/10.1186/1471-2164-11-22
56. Vanden Berghe W, De Bosscher K, Plaisance S, Boone E, Haegeman G (1999) The nuclear factor-kappa B engages CBP/p300 and histone acetyltransferase activity for transcriptional activation of the interleukin-6 gene promoter. J Biol Chem 274:32091–32098
57. Chen X, Barozzi I, Termanini A, Prosperini E, Recchiuti A, Dalli J, Mietton F, Matteoli G, Hiebert S, Natoli G (2012) Requirement for the histone deacetylase Hdac3 for the inflammatory gene expression program in macrophages. Proc Natl Acad Sci U S A 109:E2865–E2874. https://doi.org/10.1073/pnas.1121131109
58. Busse WW, Rosenwasser LJ (2003) Mechanisms of asthma. J Allergy Clin Immunol 111: S799. https://doi.org/10.1067/mai.2003.158
59. Papathanassiou E, Loukides S, Bakakos P (2016) Severe asthma: anti-IgE or anti-IL-5? Eur Clin Respir J 3. https://doi.org/10.3402/ecrj.v3.31813
60. Steinke J, Borish L (2001) Th2 cytokines and asthma Interleukin-4: its role in the pathogenesis of asthma, and targeting it for asthma treatment with interleukin-4 receptor antagonists. Respir Res 2:66–70
61. Brannan JD, Lougheed MD (2012) Airway hyperresponsiveness in asthma: mechanisms, clinical significance, and treatment. Front Physiol 3:1–11. https://doi.org/10.3389/fphys.2012.00460
62. Ito K, Caramori G, Lim S, Oates T, Fan Chung K, Barnes PJ, Adcock IM (2002) Expression and activity of histone deacetylases in human asthmatic airways. Am J Respir Crit Care Med 166:392–396. https://doi.org/10.1164/rccm.2110060
63. Cosío BG, Mann B, Ito K, Jazrawi E, Barnes PJ, Chung KF, Adcock IM (2004) Histone acetylase and deacetylase activity in alveolar macrophages and blood mononocytes in asthma. Am J Respir Crit Care Med 170:141–147. https://doi.org/10.1164/rccm.200305-659OC
64. Choi JH, Oh SW, Kang MS, Kwon HJ, Oh GT, Kim DY (2005) Trichostatin A attenuates airway inflammation in mouse asthma model. Clin Exp Allergy 35:89–96. https://doi.org/10.1111/j.1365-2222.2004.02006.x
65. Royce SG, Dang W, Yuan G, Tran J, El-Osta A, Karagiannis TC, Tang MLK (2012) Effects of the histone deacetylase inhibitor, trichostatin a, in a chronic allergic airways disease model in mice. Arch Immunol Ther Exp 60:295–306. https://doi.org/10.1007/s00005-012-0180-3
66. Toki S, Goleniewska K, Reiss S, Zhou W, Newcomb DC, Bloodworth MH, Stier MT, Boyd KL, Polosukhin VV, Subramaniam S, Peebles RS (2016) The histone deacetylase inhibitor trichostatin A suppresses murine innate allergic inflammation by blocking group 2 innate lymphoid cell (ILC2) activation. Thorax 71:633–645. https://doi.org/10.1136/thoraxjnl-2015-207728
67. Kankaanranta H, Janka-Junttila M, Ilmarinen-Salo P, Ito K, Jalonen U, Ito M, Adcock IM, Moilanen E, Zhang X (2010) Histone deacetylase inhibitors induce apoptosis in human eosinophils and neutrophils. J Inflamm 7:9. https://doi.org/10.1186/1476-9255-7-9
68. Banerjee A, Trivedi CM, Damera G, Jiang M, Jester W, Hoshi T, Epstein JA, Panettieri RA (2012) Trichostatin A abrogates airway constriction, but not inflammation, in murine and human asthma models. Am J Respir Cell Mol Biol 46:132–138. https://doi.org/10.1165/rcmb.2010-0276OC

69. Waltregny D, Glénisson W, Tran SL, North BJ, Verdin E, Colige A, Castronovo V (2005) Histone deacetylase HDAC8 associates with smooth muscle alpha-actin and is essential for smooth muscle cell contractility. FASEB J 19:966–968. https://doi.org/10.1096/fj.04-2303fje
70. Li J, Chen S, Cleary RA, Wang R, Gannon OJ, Seto E, Tang DD (2014) Histone deacetylase 8 regulates cortactin deacetylation and contraction in smooth muscle tissues. Am J Physiol Cell Physiol 307:C288–C295. https://doi.org/10.1152/ajpcell.00102.2014
71. Balasubramanian S, Ramos J, Luo W, Sirisawad M, Verner E, Buggy JJ (2008) A novel histone deacetylase 8 (HDAC8)-specific inhibitor PCI-34051 induces apoptosis in T-cell lymphomas. Leukemia 22:1026–1034. https://doi.org/10.1038/leu.2008.9
72. Ren Y, Su X, Kong L, Li M, Zhao X, Yu N, Kang J (2016) Therapeutic effects of histone deacetylase inhibitors in a murine asthma model. Inflamm Res 65:995–1008. https://doi.org/10.1007/s00011-016-0984-4
73. Hubbert C, Guardiola A, Shao R, Kawaguchi Y, Ito A, Nixon A, Yoshida M, Wang XF, Yao TP (2002) HDAC6 is a microtubule-associated deacetylase. Nature 417:455–458. https://doi.org/10.1038/417455a
74. Zhang Y, Li N, Caron C, Matthias G, Hess D, Khochbin S, Matthias P (2003) HDAC-6 interacts with and deacetylates tubulin and microtubules in vivo. EMBO J 22:1168–1179. https://doi.org/10.1093/emboj/cdg115
75. Razin A, Cedar H (1991) DNA methylation and gene expression. Microbiol Rev 55:451–458
76. Stresemann C, Lyko F (2008) Modes of action of the DNA methyltransferase inhibitors azacytidine and decitabine. Int J Cancer 123:8–13. https://doi.org/10.1002/ijc.23607
77. Tsai HC, Li H, Van Neste L, Cai Y, Robert C, Rassool FV, Shin JJ, Harbom KM, Beaty R, Pappou E, Harris J, Yen RWC, Ahuja N, Brock MV, Stearns V, Feller-Kopman D, Yarmus LB, Lin YC, Welm AL, Issa JP, Minn I, Matsui W, Jang YY, Sharkis SJ, Baylin SB, Zahnow CA (2012) Transient low doses of DNA-demethylating agents exert durable antitumor effects on hematological and epithelial tumor cells. Cancer Cell 21:430–446. https://doi.org/10.1016/j.ccr.2011.12.029
78. Tahiliani M, Koh KP, Shen Y, Pastor WA, Bandukwala H, Brudno Y, Agarwal S, Iyer LM, Liu DR, Aravind L, Rao A (2009) Conversion of 5-Methylcytosine to 5-Hydroxymethylcytosine in mammalian DNA by MLL partner TET1. Science 324:930–935. https://doi.org/10.1126/science.1170116
79. Ito S, Shen L, Dai Q, Wu SC, Collins LB, Swenberg JA, He C, Zhang Y (2011) Tet proteins can convert 5-Methylcytosine to 5-formylcytosine and 5-carboxylcytosine. Science 333:1300–1303. https://doi.org/10.1126/science.1210597
80. Ito S, Dalessio AC, Taranova OV, Hong K, Sowers LC, Zhang Y (2010) Role of Tet proteins in 5mC to 5hmC conversion, ES-cell self-renewal and inner cell mass specification. Nature 466:1129–1133. https://doi.org/10.1038/nature09303
81. Bachman M, Uribe-Lewis S, Yang X, Williams M, Murrell A, Balasubramanian S (2014) 5-Hydroxymethylcytosine is a predominantly stable DNA modification. Nat Chem 6:1049–1055. https://doi.org/10.1038/nchem.2064
82. Yu-Fei H, Bin-Zhong L, Zheng L, Peng L, Yang W, Qingyu T, Jianping D, Yingying J, Zhangcheng C, Lin L, Yan S, Xiuxue L, Qing D, Chun-Xiao S, Kangling Z, Chuan H, Guo-Liang X (2011) Tet-mediated formation of 5-carboxylcytosine and its excision by TDG in mammalian DNA. Science 333:1303–1307. https://doi.org/10.1126/science.1229223
83. Zhang L, Lu X, Lu J, Liang H, Dai Q, Xu GL, Luo C, Jiang H, He C (2012) Thymine DNA glycosylase specifically recognizes 5-carboxylcytosine-modified DNA. Nat Chem Biol 8:328–330. https://doi.org/10.1038/nchembio.914
84. Slupphaug G, Mol CD, Kavli B, Arvai AS, Krokan HE, Tainer JA (1996) A nucleotide-flipping mechanism from the structure of human uracil–DNA glycosylase bound to DNA. Nature 384:87
85. Kooistra SM, Helin K (2012) Molecular mechanisms and potential functions of histone demethylases. Nat Rev Mol Cell Biol 13:297–311. https://doi.org/10.1038/nrm3327

86. Murn J, Shi Y (2017) The winding path of protein methylation research: milestones and new frontiers. Nat Rev Mol Cell Biol 18:517–527. https://doi.org/10.1038/nrm.2017.35
87. De Santa F, Totaro MG, Prosperini E, Notarbartolo S, Testa G, Natoli G (2007) The histone H3 lysine-27 demethylase Jmjd3 links inflammation to inhibition of polycomb-mediated gene silencing. Cell 130:1083–1094. https://doi.org/10.1016/j.cell.2007.08.019
88. Swigut T, Wysocka J (2007) H3K27 demethylases, at long last. Cell 131:29–32. https://doi.org/10.1016/j.cell.2007.09.026
89. Chang B, Chen Y, Zhao Y, Bruick RK (2007) JMJD6 is a histone arginine demethylase. Science 318:444 LP–444447. https://doi.org/10.1126/science.1145801
90. Kakimoto Y, Akazawa S (1970) Isolation and identification of Ng, Ng- and Ng, N'g-dimethylarginine, Nε-mono-, Di-, and trimethyllysine, and glucosylgalactosyl- and galactosyl-δ-hydroxylysine from human urine. J Biol Chem 245:5751–5758
91. Bird A (2001) Methylation talk between histones and DNA. Science 294:2113–2115. https://doi.org/10.1158/1078-0432.CCR-15-1555
92. Wang H, Cao R, Xia L, Erdjument-bromage H, Borchers C, Tempst P, Zhang Y, Hill C, Carolina N (2001) Purification and functional characterization of a histone H3-lysine 4-specific methyltransferase. Mol Cell 8:1207–1217
93. Nightingale KP, Gendreizig S, White DA, Bradbury C, Hollfelder F, Turner BM (2007) Cross-talk between histone modifications in response to histone deacetylase inhibitors: MLL4 links histone H3 acetylation and histone H3K4 methylation. J Biol Chem 282:4408–4416. https://doi.org/10.1074/jbc.M606773200
94. Okitsu CY, Hsieh C-L (2007) DNA methylation dictates histone H3K4 methylation. Mol Cell Biol 27:2746–2757. https://doi.org/10.1128/MCB.02291-06
95. Weber M, Hellmann I, Stadler MB, Ramos L, Pääbo S, Rebhan M, Schübeler D (2007) Distribution, silencing potential and evolutionary impact of promoter DNA methylation in the human genome. Nat Genet 39:457–466. https://doi.org/10.1038/ng1990
96. Nan X, Ng H-H, Johnson CA, Laherty CD, Turner BM, Eisenman RN, Bird A (1998) Transcriptional repression by the methyl-CpG-binding protein MeCP2 involves a histone deacetylase complex. Nature 393:386
97. Hendrich B, Tweedie S (2003) The methyl-CpG binding domain and the evolving role of DNA methylation in animals. Trends Genet 19:269–277. https://doi.org/10.1016/S0168-9525(03)00080-5
98. Qiu W, Baccarelli A, Carey VJ, Boutaoui N, Bacherman H, Klanderman B, Rennard S, Agusti A, Anderson W, Lomas DA, DeMeo DL (2012) Variable DNA methylation is associated with chronic obstructive pulmonary disease and lung function. Am J Respir Crit Care Med 185:373–381. https://doi.org/10.1164/rccm.201108-1382OC
99. De Vries M, Van Der Plaat DA, Vonk JM, Boezen HM (2018) No association between DNA methylation and COPD in never and current smokers. BMJ Open Respir Res 5:5–7. https://doi.org/10.1136/bmjresp-2018-000282
100. Machin M, Amaral AFS, Wielscher M, Rezwan FI, Imboden M, Jarvelin MR, Adcock IM, Probst-Hensch N, Holloway JW, Jarvis DL (2017) Systematic review of lung function and COPD with peripheral blood DNA methylation in population based studies. BMC Pulm Med 17:54. https://doi.org/10.1186/s12890-017-0397-3
101. Lam HC, Cloonan SM, Bhashyam AR, Haspel JA, Singh A, Sathirapongsasuti JF, Cervo M, Yao H, Chung AL, Mizumura K, An CH, Shan B, Franks JM, Haley KJ, Owen CA, Tesfaigzi Y, Washko GR, Quackenbush J, Silverman EK, Rahman I, Kim HP, Mahmood A, Biswal SS, Ryter SW, Choi AMK (2013) Histone deacetylase 6-mediated selective autophagy regulates COPD-associated cilia dysfunction. J Clin Invest 123:5212–5230. https://doi.org/10.1172/JCI69636
102. Xu CJ, Söderhäll C, Bustamante M, Baïz N, Gruzieva O, Gehring U, Mason D, Chatzi L, Basterrechea M, Llop S, Torrent M, Forastiere F, Fantini MP, Carlsen KCL, Haahtela T, Morin A, Kerkhof M, Merid SK, van Rijkom B, Jankipersadsing SA, Bonder MJ, Ballereau S, Vermeulen CJ, Aguirre-Gamboa R, de Jongste JC, Smit HA, Kumar A,

Pershagen G, Guerra S, Garcia-Aymerich J, Greco D, Reinius L, McEachan RRC, Azad R, Hovland V, Mowinckel P, Alenius H, Fyhrquist N, Lemonnier N, Pellet J, Auffray C, van der Vlies P, van Diemen CC, Li Y, Wijmenga C, Netea MG, Moffatt MF, Cookson WOCM, Anto JM, Bousquet J, Laatikainen T, Laprise C, Carlsen KH, Gori D, Porta D, Iñiguez C, Bilbao JR, Kogevinas M, Wright J, Brunekreef B, Kere J, Nawijn MC, Annesi-Maesano I, Sunyer J, Melén E, Koppelman GH (2018) DNA methylation in childhood asthma: an epigenome-wide meta-analysis. Lancet Respir Med 6:379–388. https://doi.org/10.1016/S2213-2600(18)30052-3
103. Cheng RYS, Shang Y, Limjunyawong N, Dao T, Das S, Rabold R, Sham JSK, Mitzner W, Tang W-Y (2014) Alterations of the lung methylome in allergic airway hyper-responsiveness. Environ Mol Mutagen 55:244–255. https://doi.org/10.1002/em.21851
104. Shang Y, Das S, Rabold R, Sham JSK, Mitzner W, Tang WY (2013) Epigenetic alterations by DNA methylation in house dust mite-induced airway hyperresponsiveness. Am J Respir Cell Mol Biol 49:279–287. https://doi.org/10.1165/rcmb.2012-0403OC
105. Wu C-J, Yang C-Y, Chen Y-H, Chen C-M, Chen L-C, Kuo M-L (2013) The DNA methylation inhibitor 5-azacytidine increases regulatory T cells and alleviates airway inflammation in ovalbumin-sensitized mice. Int Arch Allergy Immunol 160:356–364. https://doi.org/10.1159/000343030
106. Sun Q, Yang X, Zhong B, Jiao F, Li C, Li D, Lan X, Sun J, Lu S (2012) Upregulated protein arginine methyltransferase 1 by IL-4 increases Eotaxin-1 expression in airway epithelial cells and participates in antigen-induced pulmonary inflammation in rats. J Immunol 188:3506–3512. https://doi.org/10.4049/jimmunol.1102635
107. Sun Q, Liu L, Wang H, Mandal J, Khan P, Hostettler KE, Stolz D, Tamm M, Molino A, Lardinois D, Lu S, Roth M (2017) Constitutive high expression of protein arginine methyltransferase 1 in asthmatic airway smooth muscle cells is caused by reduced microRNA-19a expression and leads to enhanced remodeling. J Allergy Clin Immunol 140:510–524.e3. https://doi.org/10.1016/j.jaci.2016.11.013
108. Sun Q, Liu L, Roth M, Tian J, He Q, Zhong B, Bao R, Lan X, Jiang C, Sun J, Yang X, Lu S (2015) PRMT1 upregulated by epithelial proinflammatory cytokines participates in COX2 expression in fibroblasts and chronic antigen-induced pulmonary inflammation. J Immunol 195:298–306. https://doi.org/10.4049/jimmunol.1402465
109. Liu L, Sun Q, Bao R, Roth M, Zhong B, Lan X, Tian J, He Q, Li D, Sun J, Yang X, Lu S (2016) Specific regulation of PRMT1 expression by PIAS1 and RKIP in BEAS-2B epithelia cells and HFL-1 fibroblasts in lung inflammation. Sci Rep 6:1–13. https://doi.org/10.1038/srep21810
110. Ahmad T, Mabalirajan U, Ghosh B, Agrawal A (2010) Altered asymmetric dimethyl arginine metabolism in allergically inflamed mouse lungs. Am J Respir Cell Mol Biol 42:3–8. https://doi.org/10.1165/rcmb.2009-0137RC
111. Klein E, Weigel J, Buford MC, Holian A, Wells SM (2010) Asymmetric dimethylarginine potentiates lung inflammation in a mouse model of allergic asthma. Am J Phys 299:L816–L825. https://doi.org/10.1152/ajplung.00188.2010
112. He X, Li T, Kang N, Zeng H, Ren S, Zong D, Li J, Cai S, Chen P, Chen Y (2017) The protective effect of PRMT6 overexpression on cigarette smoke extract-induced murine emphysema model. Int J COPD 12:3245–3254. https://doi.org/10.2147/COPD.S144881
113. McAllister TE, England KS, Hopkinson RJ, Brennan PE, Kawamura A, Schofield CJ (2016) Recent progress in histone demethylase inhibitors. J Med Chem 59:1308–1329. https://doi.org/10.1021/acs.jmedchem.5b01758
114. Heinemann B, Nielsen JM, Hudlebusch HR, Lees MJ, Larsen DV, Boesen T, Labelle M, Gerlach LO, Birk P, Helin K (2014) Inhibition of demethylases by GSK-J1/J4. Nature 514: E1–E2. https://doi.org/10.1038/nature13688
115. Clifford RL, John AE, Brightling CE, Knox AJ (2012) Abnormal histone methylation is responsible for increased vascular endothelial growth factor 165a secretion from airway smooth muscle cells in asthma. J Immunol 189:819–831. https://doi.org/10.4049/jimmunol.1103641

116. Yu Q, Yu X, Zhao W, Zhu M, Wang Z, Zhang J, Huang M, Zeng X (2018) Inhibition of H3K27me3 demethylases attenuates asthma by reversing the shift in airway smooth muscle phenotype. Clin Exp Allergy 48:1439–1452. https://doi.org/10.1111/cea.13244
117. Dela Cruz CS, Tanoue LT, Matthay RA (2011) Lung cancer: epidemiology, etiology, and prevention. Clin Chest Med 32:605–644. https://doi.org/10.1016/j.ccm.2011.09.001
118. Luszczek W, Cheriyath V, Mekhail TM, Borden EC (2010) Combinations of DNA methyltransferase and histone deacetylase inhibitors induce DNA damage in small cell lung cancer cells: correlation of resistance with IFN-stimulated gene expression. Mol Cancer Ther 9:2309–2321. https://doi.org/10.1158/1535-7163.MCT-10-0309
119. Suraweera A, O'Byrne KJ, Richard DJ (2018) Combination therapy with histone deacetylase inhibitors (HDACi) for the treatment of cancer: achieving the full therapeutic potential of HDACi. Front Oncol 8:1–15. https://doi.org/10.3389/fonc.2018.00092
120. Juergens RA, Wrangle J, Vendetti FP, Murphy SC, Zhao M, Coleman B, Sebree R, Rodgers K, Hooker CM, Franco N, Lee B, Tsai S, Delgado IE, Rudek MA, Belinsky SA, Herman JG, Baylin SB, Brock MV, Rudin CM (2011) Combination epigenetic therapy has efficacy in patients with refractory advanced non-small cell lung cancer. Cancer Discov 1:598–607. https://doi.org/10.1158/2159-8290.CD-11-0214
121. Langevin SM, Kelsey KT (2017) Clinical epigenetics of lung cancer. Elsevier, Amsterdam
122. Qiu X, Liang Y, Sellers RS, Perez-Soler R, Zou Y (2016) Toxicity and pharmacokinetic studies of aerosolized clinical grade azacitidine. Clin Lung Cancer 17:214–222e1. https://doi.org/10.1016/j.cllc.2015.09.005
123. Little AG, Gay EG, Gaspar LE, Stewart AK (2007) National survey of non-small cell lung cancer in the United States: epidemiology, pathology and patterns of care. Lung Cancer 57:253–260. https://doi.org/10.1016/j.lungcan.2007.03.012

Top Med Chem (2020) 33: 455–486
DOI: 10.1007/7355_2019_80

Published online: 28 November 2019

Structure-Based Design of Epigenetic Inhibitors

Dina Robaa, Jelena Melesina, Chiara Luise, and Wolfgang Sippl

Contents

1 Introduction 457
2 Computational Methods to Predict the Target Interaction of Small Molecules and Their Binding Affinities 458
3 Histone Deacetylases 459
  3.1 Zinc-Dependent HDACs 459
  3.2 Sirtuins 463
4 Histone Methyltransferases 465
5 Histone Demethylases 469
  5.1 LSD1 (KDM1A) 469
  5.2 Jumonji Histone Demethylases 472
6 Summary 475
References 475

Abstract Computer-aided and structure-based design methods play an important role in the development of inhibitors for epigenetic drug targets. Multiple hits have been discovered over the last years, and several leads have been optimized using molecular modeling methods, such as virtual screening, molecular docking, binding free energy calculations, homology modeling, and others. In this chapter, advances and success stories of computer-assisted development of epigenetic inhibitors are collected. The presented examples give an overview of successfully applied strategies and emphasize the advantage of guidance and rationalization of experimental data with computational means.

D. Robaa, J. Melesina, C. Luise, and W. Sippl (✉)
Department of Pharmaceutical Chemistry, Martin-Luther University of Halle-Wittenberg, Halle/Saale, Germany
e-mail: wolfgang.sippl@pharmazie.uni-halle.de

Keywords Binding free energy calculations, Computational chemistry, Epigenetic inhibitors, Histone deacetylases, Histone demethylases, Histone methyltransferases, Homology modeling, Molecular docking, Molecular modeling, Structure-based drug design, Virtual screening

Abbreviations

3D	Three-dimensional
α-KG	α-Ketoglutarate
ADMET	Absorption, distribution, metabolism, excretion, and toxicity
AML	Acute myeloid leukemia
AOD	Amine oxidase domain
CARM1	Coactivator-associated arginine methyltransferase 1
CoMFA	Comparative molecular field analysis
CoMSIA	Comparative molecular similarity indices analysis
DOT1L	Disruptor of telomeric silencing 1-like
DNA	Deoxyribonucleic acid
DNMT	DNA methyltransferase
EZH	Enhancer of Zeste homologue
FAD	Flavin adenine dinucleotide
FDH	Formaldehyde dehydrogenase
FEP	Free energy perturbation
FIH	Factor inhibiting hypoxia-inducible factor
G9a/EHMT2	Euchromatic histone-lysine N-methyltransferase 2
GLP	G9a-like protein
H3K4	Lysine 4 of histone 3
H3K4me3	Trimethylated lysine 4 of histone 3
H3K9me2	Dimethylated lysine 9 of histone 3
HDAC	Histone deacetylase
HTS	High-throughput screening
IC_{50}	The half maximal inhibitory concentration
ICM	Internal coordinate modeling
JARID	Jumonji and AT-rich interaction domain
Jmj	Jumonji
JMJD	Jumonji domain-containing protein
KDM	Lysine demethylase
KMT	Lysine methyltransferase
LIE	Linear interaction energy
LSD	Lysine-specific demethylase
MAO	Monoamine oxidase
MDS	Myelodysplastic syndromes
MM-GBSA	Molecular mechanics generalized Born surface area

MM-PBSA	Molecular mechanics Poisson-Boltzmann surface area
NAD^+	Nicotinamide adenine dinucleotide
NCI	National Cancer Institute
PAO	Plant amine oxidase
PDB	Protein Data Bank
PHD	Prolyl hydroxylase domain-containing protein
PKMT	Protein lysine methyltransferase
PRMT	Protein arginine methyltransferase
QSAR	Quantitative structure-activity relationship
RMSE	Root-mean-square error
SAH	S-adenosylhomocysteine
SAM	S-adenosyl-L-methionine
SAR	Structure-activity relationship
SET	Su(var)3-9, Enhancer-of-zeste and Trithorax
SETD	SET domain-containing lysine methyltransferase
Sirt	Sirtuin
smHDAC8	*Schistosoma mansoni* HDAC8
SWIRM	SWI3p, Rsc8p, and Moira
TI	Thermodynamic integration
TCF4	Transcription factor 4
VS	Virtual screening

1 Introduction

The two major biochemical pathways of epigenetic regulation are DNA methylation and posttranslational modifications of amino acids in histone proteins. Histone modifications interact with each other and constitute a particular pattern of alterations in the chromatin structure, the so-called histone code [1]. Posttranslational modifications of histones have been shown to participate in a wide array of cellular processes. Regulations of these covalent modifications and their implications are currently of great interest in the scientific community [2]. Although histone proteins are under the control of various posttranslational modifications, such as acetylation, acylation, methylation, phosphorylation, ubiquitination, or glycosylation, the focus of most studies has been mainly on lysine acetylation/deacetylation and methylation/demethylation. Over the last decade, many of the enzymes that regulate these histone modifications have been identified and characterized on a molecular level. With large effort, a wealth of 3D structures of epigenetic targets has been determined, which helped to understand the different epigenetic targets and their regulation on a molecular level. From the early discovery of histone deacetylase inhibitors, such as trichostatin A, to the more recent discovery of novel histone-modifying enzyme inhibitors, structure- and computer-based approaches were applied to analyze target-ligand interactions and to rationalize the development of small molecule modulators [3]. This chapter will illustrate various structure-based approaches which have been successfully implemented to identify and design small molecule inhibitors of histone-modifying enzymes.

2 Computational Methods to Predict the Target Interaction of Small Molecules and Their Binding Affinities

Designing small molecules with desired biological activity has long been a major goal of structure-based drug design. Different approaches have been tested to predict the biological activity of an inhibitor from its calculated interaction with the target. Molecular docking tools are able to predict the interactions in a putative binding site and can easily screen large numbers of molecules and rank them according to their calculated affinities [4, 5]. However, using a single frozen protein structure in most of the cases hampers the true conformational and orientational sampling of an inhibitor in its putative binding pocket [6]. Moreover, currently available scoring functions that estimate the binding affinity of calculated poses for a ligand are still far away from ranking the biologically active molecules in a reasonable agreement with the experimental data [7]. To overcome this challenge, more sophisticated methods that take into account conformational changes, desolvation effects, as well as entropic effects on binding are needed to predict the binding strength of molecules of interest.

In contrast to simple docking methods, free energy calculations were shown to give results that are in better agreement with experimental data [8]. Two different approaches have been widely studied for their applications in estimating the binding affinity of small molecule inhibitors. The first approach includes the so-called pathway methods such as free energy perturbation (FEP) and thermodynamic integration (TI) which consider all small changes that occur between the initial and final states [9]. These methods are more accurate and robust in predicting reliable affinities than docking methods but are expensive and time-consuming, thus, making them inapplicable for large datasets of compounds. Meanwhile, the second group of approaches (endpoint methods) including linear interaction energy (LIE), molecular mechanics Poisson-Boltzmann surface area (MM-PBSA), and molecular mechanics generalized Born surface area (MM-GBSA) methods calculate only the difference between the starting and last step of these two states leading to much faster calculations than FEP and TI [10]. MM-PBSA and MM-GBSA methods have found various applications in drug discovery studies such as pose selection, discrimination of the true actives from decoys or inactive molecules in virtual screening (VS) studies, and prediction of the relative binding free energies in the lead optimization stage [11, 12]. The applicability of such approximations has been tested and shown to be valuable as post-docking filter to rescore docking results and to increase the hit rate in VS [13, 14]. Satisfactory correlations have been obtained between calculated binding affinities and experimental data, especially for ligand series with high structural similarity [15]. Although previous applications of MM-PBSA and MM-GBSA proved to be very useful and efficient in many macromolecular studies, there remains still room for improvement in many aspects of existing methods, including estimation of the solute entropy solvation free energy of polar compounds, sampling of the conformational space of the protein-ligand interaction as well as developing adequate parameters for describing the molecular structures. The above described methods provide a set of computational tools for structure-based design of epigenetic inhibitors.

3 Histone Deacetylases

The family of human histone deacetylases (HDACs) includes 18 isoforms and is classified into two structurally distinct groups and five classes. The first group is called zinc-dependent HDACs. It consists of 11 isoforms subdivided into four classes: class I (HDAC1–HDAC3, HDAC8), class IIa (HDAC4–HDAC5, HDAC7, HDAC9), class IIb (HDAC6 and HDAC10), and class IV (HDAC11). The second group of HDACs is usually referred to as sirtuins. It comprises seven isoforms, which all belong to class III [16]. Structural features of different HDAC groups, classes, and isoforms are often addressed in the structure-based design of HDAC inhibitors. Computational methods allow to visualize those features and to predict protein-ligand interactions in a qualitative and quantitative manner as illustrated further.

3.1 Zinc-Dependent HDACs

Zinc-dependent HDACs share a conserved deacetylase core domain [17]. To date, more than 100 structures of these domains have been solved [18–26]. Their architecture is similar: an α/β fold formed by an ensemble of conserved α-helices, β-sheets, and binding pocket loops as well as less conserved distant loops. The residues of the binding pocket form an approximately 12 Å long substrate binding tunnel with the catalytically important zinc ion at the bottom [27]. This tunnel and adjacent cavities are usually targeted by inhibitors bearing a warhead containing a zinc-binding group (examples are shown in Fig. 1).

The most common zinc-binding group is hydroxamic acid. Interestingly, it is able to interact with the zinc ion either in a bidentate or a monodentate fashion, as seen, for example, in the crystal structures of zebrafish HDAC6 with Panobinostat (**1**) (PDB ID: 5EF8) and Nexturastat A (**2**) (PDB ID: 5G0I), correspondingly (Fig. 2). In both cases, the hydroxamic acid group makes a number of hydrogen bonds to the nearby amino acid residues, the hydrophobic spacer is stabilized by van der Waals

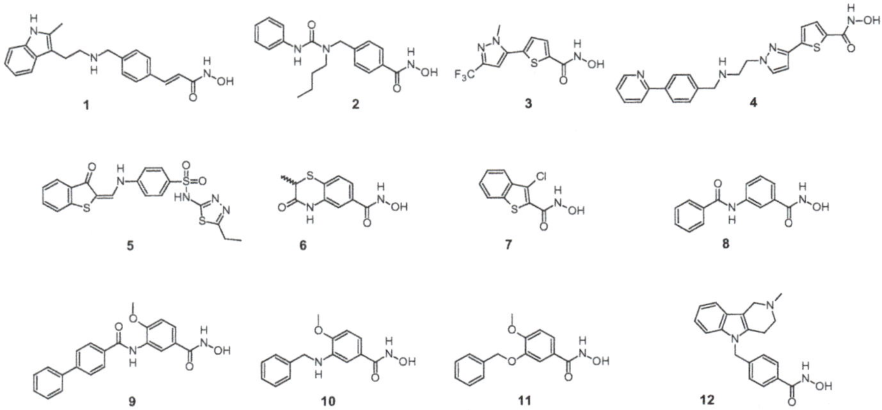

Fig. 1 Molecular structures of HDAC inhibitors mentioned in the text

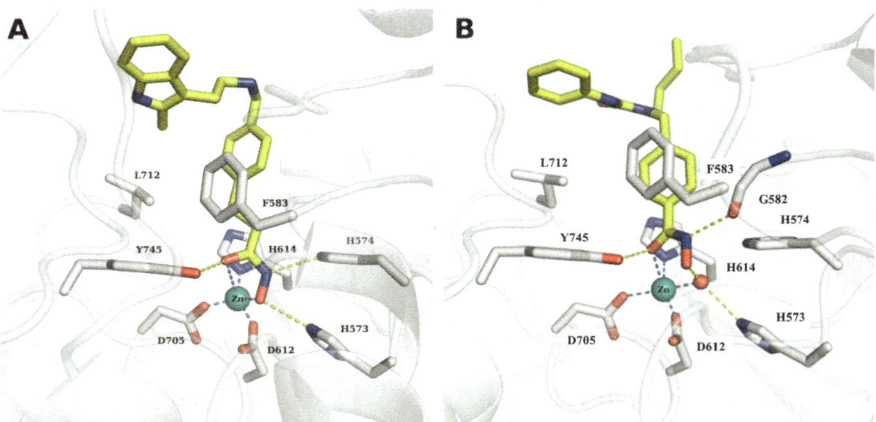

Fig. 2 (**a**, **b**) X-ray structure of DrHDAC6 in complex with inhibitors: (**a**) Panobinostat (**1**) (PDB ID: 5EF8) and (**b**) Nexturastat A (**2**) (PDB ID: 5G0I). Only the residues in the binding pocket are shown. Hydrogen bonds are displayed as yellow dashed lines, metal coordination as blue dashed lines, ligands as yellow sticks, zinc ion as cyan sphere, and water molecules as red spheres

interactions inside the substrate binding tunnel, and the cap group interacts with residues at the rim of the binding pocket [24, 25].

Crystal structures of protein-ligand complexes represent the basic source of structural information for inhibitor optimization and for identification of novel hits. In early studies, the first solved HDAC crystal structure of bacterial HDAC-like protein was readily used for docking and structure-based optimization of inhibitors [28–36]. In the absence of the target protein structure, homology modeling was applied. Homology models of HDAC1 and HDAC6 have been often generated and used to predict the binding modes of inhibitors [37–48]. VS has been a major method to search for novel hits. However, in many cases the identified inhibitors were not optimal due to weak inhibitory activity or unspecific mode of action [49]. Below, several VS campaigns with biologically validated hits suitable for further optimization are described.

Price et al. published a study in which they docked a virtual library of 644 hydroxamic acids, generated from available in-house carboxylic acids to a crystal structure of bacterial HDAC-like protein [50]. Based on the docking poses and scores obtained with FlexX docking program, 75 compounds were selected for biological testing. A promising hit, namely, ADS100380 (**3**, Fig. 1), was identified with an IC_{50} of 0.75 μM in an in vitro HDAC assay and a weak functional activity (IC_{50} of 11.4 μM) in a cell proliferation assay. Structure-based optimization of the hit guided by molecular docking studies resulted in a series of compounds with improved potency, showing up to single-digit nanomolar activity in a HDAC assay and double-digit nanomolar activity in an antiproliferative assay (**4**, Fig. 1) [50].

Tang et al. reported their quantitative structure-activity relationship (QSAR)-based VS implemented to search for human HDAC1 inhibitors. Chemical descriptors were calculated for 59 known HDAC inhibitors to generate QSAR models, which were used to screen a virtual library of 9.5 million compound structures. Out of the 45 retrieved screening hits, 4 structurally diverse compounds were selected for experimental validation. Three of them, which were hydroxamic acids, showed micromolar activity against HDAC1 and HDAC6 [51].

Park et al. identified novel HDAC inhibitors by applying structure-based VS [52]. A virtual library of around 460,000 compounds was pre-filtered by Lipinski's rule of five [53] and structural similarity. The remaining 180,000 compounds were docked to a homology model of HDAC1 using AutoDock program and scored with a modified scoring function which included a novel solvation model. Around 100 top-scored virtual hits were selected for pretesting in a HDAC inhibition assay. The IC_{50} values of the six final hits ranged between 4 and 100 μM. Interestingly, the most active hit (**5**, Fig. 1) was a non-hydroxamate HDAC inhibitor with a putative thiadiazole sulfonamide zinc-binding group [52].

Another successful structure-based VS was conducted by Kannan et al. [54]. The ZINC database of around 15 million compounds was pre-filtered to select only compounds containing one of several known zinc-binding groups (hydroxamic acid, anilinobenzamide, thiazole sulfonamide). About 5,000 retrieved zinc binders were docked and scored with Glide (Glide SP scoring function) and GOLD (GoldScore, ChemScore, ChemPLP, and ASP scoring) docking programs using a homology model of the anti-parasitic target *Schistosoma mansoni* HDAC8 (smHDAC8). This docking protocol was previously validated by re-docking of the cocrystallized human HDAC8 inhibitors. Based on the docking poses and calculated scores, 75 compounds were purchased and tested on smHDAC8. For the six most active hydroxamic acid hits, the IC_{50} values were determined on smHDAC8 (1–6 μM), as well as human HDAC8 (0.6–3 μM), HDAC1 (3–30 μM), and HDAC6 (0.02–4 μM) [54]. Two of the most promising fragment-like compounds T5979345 (**6**) and T6072858 (**7**, Fig. 1) could be cocrystallized with the target protein [55]. Based on the binding mode of **6**, a library of open-ring analogs

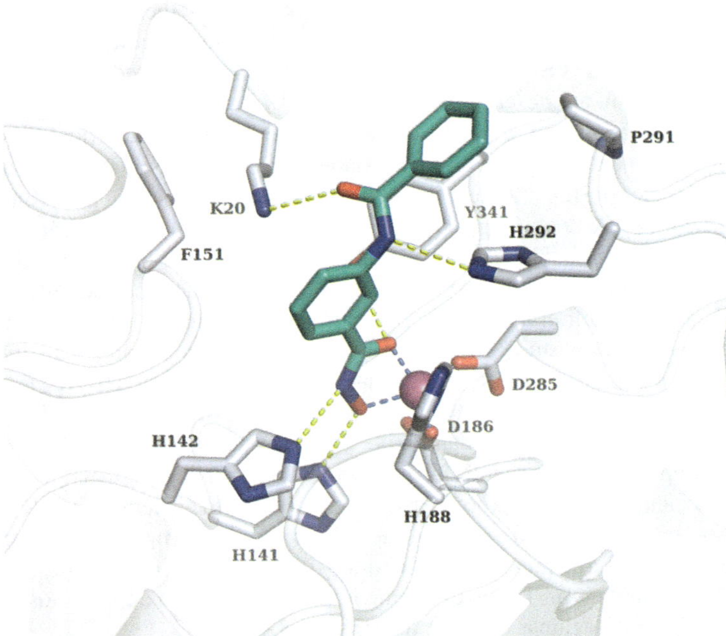

Fig. 3 X-ray structure of smHDAC8 in complex with a benzhydroxamate-based inhibitor (**8**) (PDB ID: 5FUE) showing the interaction with the lysine channel. Only the residues in the binding pocket are shown. Hydrogen bonds are displayed as yellow dashed lines, metal coordination as blue dashed lines, the ligand as cyan sticks, and the zinc ion as a purple sphere

(e.g., compound **8**, Fig. 1) was designed to target smHDAC8 binding pocket features. Docking studies predicted that the benzamide moiety of **8** is embedded into HDAC8-specific side pocket and the amide group undergoes hydrogen bond interactions with the unique smHDAC8 amino acid residue His292. Crystal structure of **8** with the target protein confirmed the predicted binding mode (Fig. 3). Computer-guided optimization of **8** led to a series of benzohydroxamic acids, some of which were potent smHDAC8 inhibitors showing in vitro anti-parasitic activity, for example, compound **9** (Fig. 1) with an IC_{50} of 76 nM [56]. Furthermore, optimization of structural analogs of compound **6** toward human HDAC8 led to potent and selective inhibitors of this enzyme with anti-neuroblastoma activity. Compounds **10** and **11** were among the most potent derivatives with an IC_{50} of 69 nM and 27 nM, respectively (Fig. 1) [57].

Novel HDAC inhibitors have not only been discovered by VS approaches. Structure-based design assisted by molecular modeling methods such as homology modeling, molecular docking, and binding free energy calculations also provided valuable hits. Butler et al. rationally designed the highly potent and selective HDAC6 inhibitor tubastatin A (**12**, Fig. 1). They noticed that the substrate binding tunnel and especially its rim are wider in HDAC6 homology model than in HDAC1.

Ligands with short aromatic linkers and large rigid cap groups were docked into the target binding pocket, and a carbazole cap group was found to fit well. These compounds were synthesized, tested, and found to be indeed highly selective HDAC6 inhibitors [58]. More cases of successful structure-based design of HDAC inhibitors are described in recent reviews [59, 60].

3.2 Sirtuins

Sirtuins, in contrast to the classical zinc-dependent HDACs, use nicotinamide adenine dinucleotide (NAD^+) as cofactor to carry out the deacetylation step. For most of the seven human sirtuin (Sirt) isoforms, crystal structures have already been solved. The reported crystal structures of human Sirt1 [61, 62], Sirt2 [63–66], Sirt3 [65, 67–75], Sirt5 [70, 74, 76–78], and Sirt6 [79], as well as sirtuins derived from other species [80–88] shed light onto the overall structure of sirtuins. Sirt2 has been extensively studied by crystallographers, who solved the 3D structure of this isoform in the apo, cofactor-bound, substrate bound, as well as in inhibited form. So far, no 3D structure of human Sirt4 and Sirt7 is available. A recent review focused on the structural details of sirtuins has been published, and the interested reader is referred to [89]. Sirtuins contain a conserved 275 amino acid catalytic domain with variable N- and C-termini. The structure of the catalytic domain consists of the so-called Rossmann fold and a smaller zinc-binding domain. The sirtuin structure shows several subpockets that accommodate adenine (A), ribose (B), and nicotinamide (C) as well as the acetyllysine substrate peptide (Fig. 4a). Upon binding of the substrate peptide, the zinc-binding domain rotates toward the Rossmann fold domain and induces a "closure" of the active cleft. This closure of both domains helps to correctly orientate the modified lysine group in the hydrophobic tunnel, allowing the formation of a hydrogen bond between the Nε atom of lysine and the backbone carbonyl of a conserved valine residue (Val233, Sirt2 numbering). A further conformation of sirtuins that has been recently reported is the "locked-open conformation" that is observed upon binding of the allosteric and highly selective Sirt2 inhibitor SirReal2 [90] (Fig. 4b). Upon binding, SirReal inhibitors induce a major rearrangement of Sirt2 active site and open up a so far unexploited binding pocket. This unique rearrangement is the basis for the high potency and isotype selectivity of the SirReals. This pocket was therefore termed "selectivity pocket."

Several inhibitors are available for Sirt2 that have been discovered by applying computer-based methods [91–93]. One of the first potent Sirt2 inhibitors, the vinylnitrile derivative AGK2 (**13**, Fig. 5), has been identified by focused-library screening [94], and the interaction of the inhibitor with Sirt2 has been analyzed by docking studies. The docking of the active inhibitors was carried out into different conformations of Sirt2 that were generated by homology modeling. The high flexibility of the active site loop made this strategy necessary [95]. The detailed analysis of the docking results of AGK2 and inactive analogs resulted in the

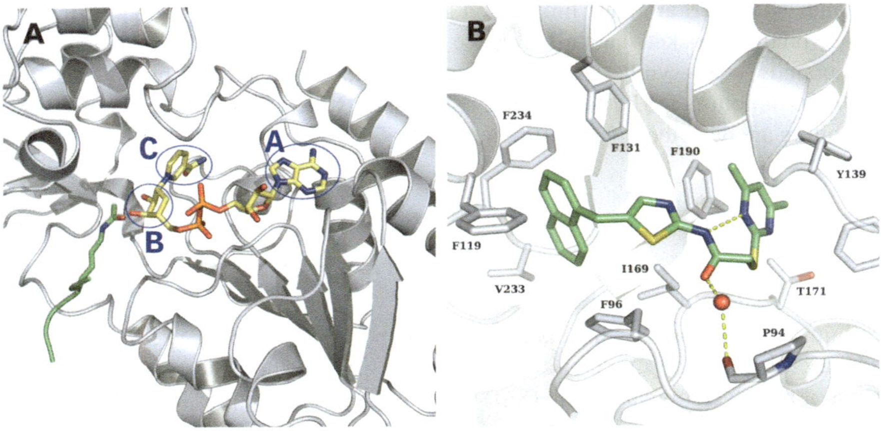

Fig. 4 (**a**) X-ray structure of human Sirt3 (white ribbon) in complex with Carba-NAD^+ (yellow sticks) and acetylated peptide (green) (PDB ID: 4FVT). (**b**) X-ray structure of human Sirt2 in complex with the selective inhibitor SirReal2 (**16**; green sticks) (PDB ID: 4RMH). Only the residues in the binding pocket are shown. Hydrogen bonds are displayed as yellow dashed lines and the water molecule as a red sphere

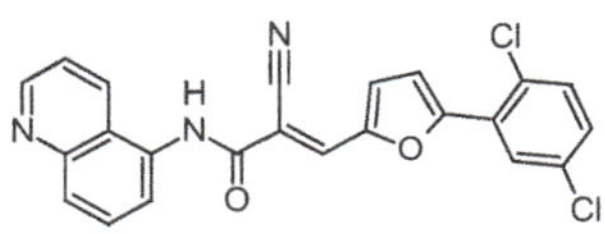

13

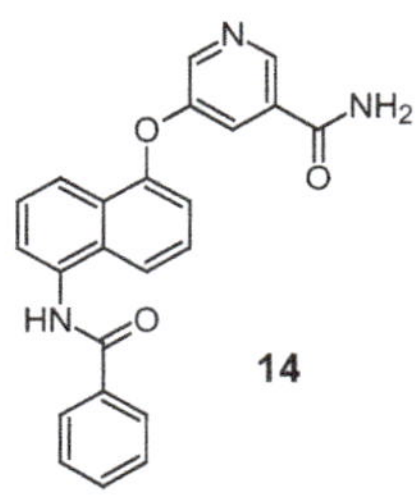

14

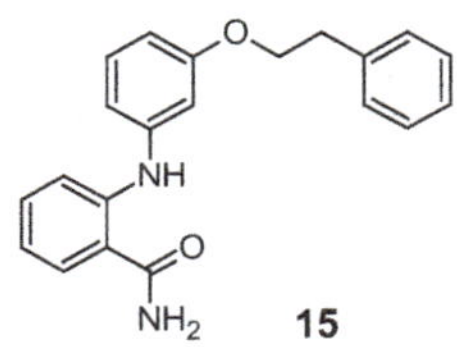

15

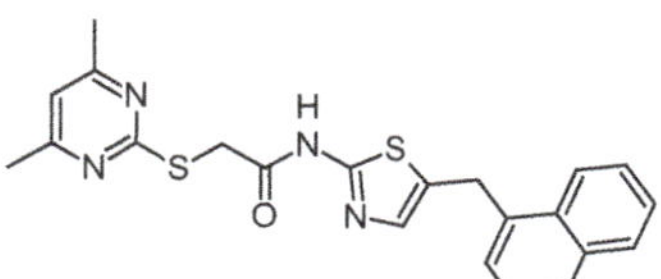

16

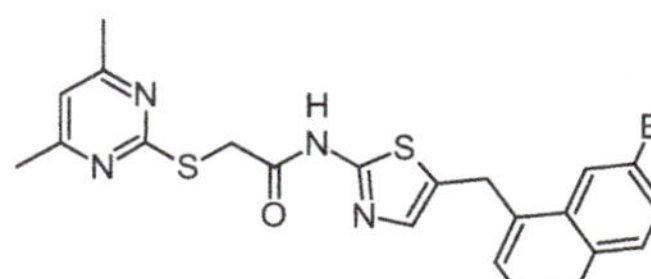

17

Fig. 5 Molecular structures of sirtuin inhibitors mentioned in the text

postulation that the inhibitor is interacting with the nicotinamide-binding pocket of Sirt2. AGK2 inhibits Sirt2 with an IC_{50} value of 3.5 μM and shows more than tenfold selectivity over Sirt1 and Sirt3. AGK2 is a valuable chemical tool since it also displayed the ability to block α-synuclein-mediated toxicity in a Parkinson's disease model, possibly by modulating tubulin acetylation regulated by Sirt2 [94].

Another approach to discover novel sirtuin inhibitors was based on fragments identified by random screening and applying structure-based design to optimize the potency of initial fragments. Cui et al. [96] used chemical fragments derived from suramin and nicotinamide and designed several 5-benzamidonaphthalen-1/2-yloxy-nicotinamide derivatives using the available crystal structure of Sirt2. Among these, compound **14** (Fig. 5) showed nanomolar inhibitory activity against Sirt2 (IC_{50} = 0.0483 μM) and 200–900-fold selectivity over Sirt1 (IC_{50} 12.0 μM) and Sirt3 (IC_{50} 44.2 μM). The docking of inhibitor **14** predicted a binding mode where the nicotinamide moiety binds into the nicotinamide-binding pocket of Sirt2 and forms hydrogen bonds with the conserved residues Ile169 and Asp170, whereas the naphthalene group makes π-π interactions with either Phe96 or Phe119 located in the acetyllysine tunnel. Interestingly, a structurally similar compound (**15**, Fig. 5) was recently crystallized in complex with Sirt2 (PDB ID: 5Y5N) and showed a different binding mode. Unexpectedly, the benzamide moiety of **15** does not occupy the C-pocket as the nicotinamide of NAD^+ (Asp170) but undergoes hydrogen bonds with two cocrystallized water molecules. The hydrophobic aromatic tail of the inhibitor interacts with the hydrophobic selectivity pocket (Tyr139, Phe143, Phe190). This hydrophobic pocket was identified as the binding site of the myristoyl chain, which is also recognized as a substrate modification by Sirt2 [97].

Highly selective Sirt2 inhibitors containing an aminothiazole scaffold were identified by a focused-library screening of putative kinase inhibitors [90]. These inhibitors were named SirReals (sirtuin rearranging ligands) because obtained cocrystal structures of Sirt2 showed that the inhibitors rearrange the Sirt2 binding pocket and behave as allosteric inhibitors. Due to the allosteric inhibition, SirReals were found to be highly selective and represented a promising starting point for structure-based optimization. SirReal2 (**16**, Figs. 4b and 5) binds to a highly hydrophobic pocket adjacent to the zinc-binding domain and does not prevent the binding of NAD^+ in its productive conformation. Guided by the structural insights obtained from the Sirt2-SirReal2 crystal structure, Schiedel et al. aimed to systematically probe the limits of variation within the scaffold of the SirReals [98]. A library of designed aminothiazoles was docked into the Sirt2 crystal structure and was used to guide the structural optimization. About 50 compounds were finally synthesized and tested in vitro resulting in low nanomolar Sirt2 inhibitors. Binding free energies of the Sirt2-inhibitor complexes were calculated using the MM-GBSA approach. The calculated MM-GBSA protein-inhibitor binding energies showed a good correlation with the observed Sirt2 inhibition values ($r^2 = 0.67$, RMSE = 0.60). Moreover, the authors were able to rationalize their results with a further crystal structure of Sirt2 in complex with the most potent inhibitor (**17**, Fig. 5) from this series [98].

4 Histone Methyltransferases

Histone methylation is performed by proteins that belong to two different classes of enzymes called protein arginine methyltransferases (PRMTs) and protein lysine methyltransferases (PKMTs). The latter class is generally subdivided into SET

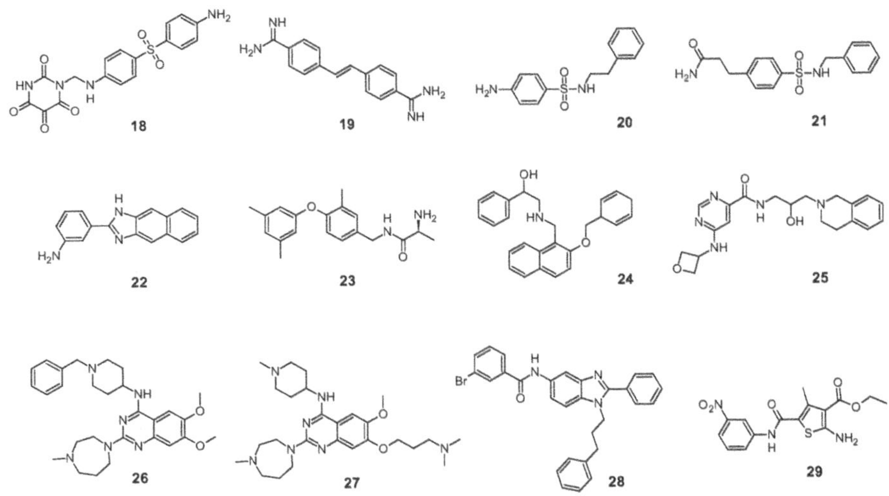

Fig. 6 Molecular structures of histone methyltransferase inhibitors mentioned in the text

domain-containing lysine methyltransferases and non-SET domain lysine methyltransferases. PRMTs and PKMTs share a common mechanism of action, in which S-adenosyl-L-methionine (SAM) acts as cofactor, and hence the methyl donor, and S-adenosylhomocysteine (SAH) is formed as a by-product [99]. Methylation occurs on a nitrogen atom of either an arginine or lysine residue in the histone tails. This type of modification does not alter the charge of the residue, but it affects the basicity, hydrophobicity, and the size of the amino acid side chain which in turn has an impact on the proteins that recognize such modifications. Indeed, despite the subtle changes, many proteins are able to distinguish and bind specific methylation states.

So far, more than 20 PKMTs and 11 PRMTs have been identified [100]. Nevertheless, enzymatic activity could not be demonstrated for all members [101–103]. Over the years, different names and classifications have been attributed to PKMTs, for instance, they have been renamed as lysine methyltransferases (KMTs) and subdivided into eight classes [104]. Many of the methyltransferases have been linked to diverse cellular processes and different types of cancers, and, therefore, the development of selective inhibitors is desirable [105]. Here, we illustrate some computational studies to highlight the utility of ligand- and structure-based methods in the discovery of new inhibitors of histone methyltransferases.

In 2007, the first PRMT1 inhibitors retrieved by applying structure-based VS methods were reported [106]. Through a combination of molecular docking, pharmacophore-based filtering, and biochemical characterization, two drug-like compounds, namely, allantodapsone (**18**) and stilbamidine (**19**, Fig. 6), were identified as substrate-competitive inhibitors of PRMT1. In this study, the authors generated a homology model of PRMT1 which was subsequently used for screening the National Cancer Institute (NCI) diversity set into the substrate binding site.

The analysis of the docking poses could suggest some key interactions between the newly identified inhibitors and PRMT1. For example, a common feature of the active inhibitors is the hydrogen bond of a basic or polar group with the acidic residue Glu152 (human PRMT1 numbering) of the active site. Moreover, van der Waals interactions between the inhibitors and several aromatic residues in the binding pocket (Tyr47, Tyr156, Trp302, human PRMT1 numbering) were present. Interestingly, the inhibitors showed a reduction of estrogen receptor activation by estradiol in a dose-dependent manner. Based on the first positive outcomes and the latest insight into the ligand-protein interactions, further research was conducted. In the new study, Heinke et al. screened the ChemBridge database containing 328,000 molecules using a structure-based pharmacophore generated with LigandScout [107] on the basis of the PRMT1-allantodapsone complex [108]. The hits were then docked using the GOLD program and GoldScore as fitness function into the substrate binding site, and the top-ranked compounds were filtered based on the distances to the amino acids in the binding pocket which were already identified as important binding partners (i.e., the aforementioned Glu152 and Tyr156). Nine compounds among the top-ranked 100 compounds that fulfilled the distance constraints were manually selected and tested in vitro. This approach resulted in the discovery of inhibitors **20**, **21**, and **22** (Fig. 6) with inhibitory effects down to 13 μM [108]. Even though the new hits did not show improvement of the activity in comparison with allantodapsone (IC_{50} 1.7 μM), they could help to prove the importance of specific interactions and to demonstrate the ability of VS to significantly increase the hit rates.

Another successful application of computer-based methods that guided the discovery of several PRMT4 (also named CARM1) inhibitors was reported by de Freitas et al. [109]. Their strategy encompassed a combination of ligand- and structure-based VS approaches followed by rescoring steps. The starting point of the study was the structural analysis of some available crystal structures of PRMT-ligand complexes, which led to the identification of a common feature present in the inhibitors. Indeed, all five inhibitors under analysis show a basic amine tail that is anchored in the PRMT substrate arginine-binding channel. Thus, they constructed a PRMT-focused virtual library by filtering the ZINC database (~22 million compounds) with substructure searches focused on the two basic amine tails featured in known PRMT inhibitors. This resulted in a virtual library of ~132,000 compounds that was consequently subjected to docking studies by means of Glide (Glide SP scoring function) using the tail structures as reference core. The top-ranked 1084 docking poses were then rescored using a MM-PBSA protocol. Out of 51 tested compounds, 11 showed dose-response inhibition of PRMT4. Among these, a valuable hit with an IC_{50} of 1.9 μM was selected for further investigation. As a result of two rounds of structure-based optimization, the nanomolar lead **23** (IC_{50} 50 nM, Fig. 6) was obtained. With the exception of PRMT6 (IC_{50} 5.2 μM), compound **23** showed selectivity against a panel of 21 human protein methyltransferases.

Recently, Ye et al. described a structure-based VS followed by a second round of fingerprint similarity search and a final optimization step for another arginine methyltransferase, PRMT5 [110]. A crystal structure of PRMT5 was employed to dock the filtered SPECS library to the substrate binding pocket with Glide program using Glide SP as scoring function. Afterward, the top-ranked 1,000 compounds were clustered according to their structural similarity and interactions with key residues like Phe327, Phe580, or Glu444 (human PRMT5 numbering). Among the 42 selected and tested hits, DC_P33, a low micromolar hit (IC_{50} 35.6 μM), could be obtained. Based on DC_P33, a chemical fingerprint similarity search was executed which could help in investigating the structure-activity relationship (SAR) of this series. Although the analogs showed lower activity, they could guide the optimization studies that led to discovery of DC_C01 (**24**, Fig. 6) with an IC_{50} of 2.8 μM. The selectivity of DC_C01 was tested and confirmed against PRMT1, EZH2, and DNMT3A. However, the optimized hit is still ~100 times less potent than the reported reference inhibitor for PRMT5, EPZ015666, which shows an IC_{50} of 22 nM (**25**, Fig. 6).

Regarding PKMTs, several studies using computer-based approaches have been reported for G9a [111–115], a SET domain-containing lysine methyltransferase protein also known as euchromatic histone-lysine N-methyltransferase 2 (EHMT2), and some of them are summarized here.

Kubicek et al. combined diverse computer-based methods in order to filter down the Boehringer Ingelheim chemical compound library prior to a HTS assay for G9a [111]. Using nine known small molecule inhibitors of arginine and lysine methyltransferases [116], a similarity searching approach was carried out. Meanwhile for the structure-based approach, a homology model based on the crystal structure of DIM-5 [117], a lysine methyltransferase with 30% identity in the SET domain to G9a, was generated. Thereby, ligand- and structure-based pharmacophore fingerprint and site-point [118] models were developed and used for filtering the library. Other compounds were randomly selected, and a total of ~125,000 compounds were evaluated in HTS which resulted in seven confirmed hits with IC_{50} in the low micromolar range. Of note is BIX-01294 (**26**, Fig. 6), which specifically inhibited G9a at 1.7 μM (IC_{50}), to a lesser extent the closely related GLP (G9a-like protein) with an IC_{50} of 38 μM and showed no activity against SUV39H1 and PRMT1. In addition, the mode of action was investigated and indicated a noncompetitive inhibition with the cofactor SAM. Later, the quinazoline-based scaffold of BIX-01294 was further explored by Liu et al. with the aim to elucidate the SAR and improve potency and selectivity [112]. A highly potent and selective inhibitor, UNC0224 (**27**, Fig. 6) with an IC_{50} of 15 nM, was thus identified, and the first crystal structure of G9a with a small molecule inhibitor was subsequently resolved.

Using the abovementioned G9a crystal structure and the ChemBridge CORE library, Zhang et al. discovered benzoxazole and benzimidazole scaffolds as new G9a inhibitors by means of structure-based VS and optimization steps. The study resulted in compound GA001 (**28**, Fig. 6) with moderate G9a inhibition and cellular activity (G9a IC_{50} of 1.32 μM, MCF7 cells IC_{50} of 5.73 μM) [113].

An integrated procedure combining pharmacophore- and docking-based VS, similarity search, and hit optimization, which led to the discovery of micromolar hits for the lysine methyltransferase SET7 (also named KMT7, SETD7, SET9), has been recently described by Meng et al. [119]. Using the crystal structure of SET7 in complex with SAM, a pharmacophore model was generated and used to screen a filtered SPECS library (182,014 compounds). The pharmacophore hits were subsequently docked into the SAM cofactor binding site using Lys294 as a hydrogen bond constraint. The top-ranked 649 molecules were clustered, and at least one molecule for each cluster was picked to cover a large chemical space. Program Glide was used as docking program due to its better performance in preliminary studies. Indeed, the authors investigated the performance of diverse docking programs with respect to enrichment factors. Moreover, the ability of the selected procedure to reproduce the cocrystallized binding mode of SAM was evaluated by re-docking studies. Using such a protocol, seven compounds, among the 127 tested, were discovered to inhibit SET7 activity. Compound DC-S100 ($IC_{50} = 30.04$ μM) was chosen as a query for the similarity search and structure-guided optimization studies. This approach led to the identification of compound DC-S239 (**29**, Fig. 6) with an IC_{50} of 4.59 μM and selectivity over DNMT1, DOT1L, EZH2, NSD1, SETD8, and G9a.

5 Histone Demethylases

In 2004, lysine-specific demethylase-1 (LSD1, also named KDM1A) was discovered as the first demethylase which can specifically remove methyl groups from histone H3K4 (lysine 4 of histone 3) utilizing flavin adenine dinucleotide (FAD) as a cofactor [120]. Later on, JumonjiC (JmjC) domain-containing proteins were identified as another superfamily of histone demethylases which use Fe^{2+} and α-ketoglutarate (α-KG) as cofactors in an oxygenation reaction to remove the methyl groups from lysine residues [121]. Numerous histone lysine demethylases were found to be overexpressed in primary tumors, which render them as highly promising biological targets [122–124]. Over the past years, huge efforts have been made to explore the structural requirements for modulating these targets. This culminated in the discovery of numerous small molecule inhibitors of diverse histone demethylases; one of them, namely, the irreversible LSD1 inhibitor GSK2879552, is currently undergoing clinical trials for the treatment of acute myeloid leukemia (AML), myelodysplastic syndromes (MDS), and small cell lung carcinoma.

Plenty of crystal structures of LSD1 and various JmjC domains have been solved in the last decade which helped to understand the binding of substrates and some available inhibitors.

5.1 LSD1 (KDM1A)

LSD1 is composed of three domains: the tower domain (aa 419-520), the SWIRM domain (aa 166-260), and the amine oxidase domain (AOD, aa 520-852) [125, 126];

the latter harbors the catalytic site where both the cofactor FAD and the substrate bind [127]. The AOD of LSD1 shares a significant sequence similarity with other FAD-dependent amine oxidases showing 26% homology with PAO (plant amine oxidase) and 20% homology with MAO-A and MAO-B enzymes [128]. The FAD-binding pocket of these FAD-dependent amine oxidases shares a high sequence similarity, whereas the substrate binding pocket, accommodating the natural substrates, shows substantial differences [129]. Importantly, the substrate peptide of LSD1 is embedded in a deep and negatively charged pocket which is much more spacious and open as compared to other FAD-dependent amine oxidases [127, 128]. Moreover, the other amine oxidases characteristically possess an aromatic cage to recognize the positively charged ammonium group through cation-π interactions, which is missing in LSD1 [127, 128]. In LSD1, only one aromatic amino acid residue (Tyr761) is preserved, whereas the second one is replaced by Thr810.

Currently, more than 50 crystal structures of LSD1 are available in the Protein Data Bank (PDB), which have been used for virtual screening and structure-based optimization studies of LSD1 inhibitors. However, docking and VS of LSD1 small molecule inhibitors face two major challenges: the huge size of the substrate binding pocket (more than 1,700 $Å^3$) and the fact that many known LSD1 inhibitors are irreversible and covalently bind to FAD.

Several VS approaches utilizing diverse techniques have been successfully applied to find novel LSD1 inhibitors [130–132]. In 2013, Schmidt et al. reported on a substructure-based screening combined with molecular docking to the LSD1 substrate binding pocket to identify novel propargylamines as irreversible LSD1 inhibitors [130]. The Enamine compound collection (~750,000 compounds) was first screened for compounds bearing an *N*-propargylamine warhead, known from irreversible amine oxidase inhibitors such as pargyline, and the retrieved hits were subsequently docked into LSD1 substrate pocket. This led to the identification of the 3-aryloxy-2-hydroxypropargylamine derivative T5342129 (**30**, Fig. 7) as an LSD1 inhibitor (IC_{50} of 44 μM and 34 μM in a hydrogen peroxide-dependent assay and a formaldehyde dehydrogenase (FDH) assay, respectively). An in vitro dilution assay confirmed that T5342129 is an irreversible LSD1 inhibitor.

Another VS approach was described by Zhou et al. who used a combination of pharmacophore-based VS and docking to identify novel LSD1 inhibitor chemotypes [132]. A common ligand-based pharmacophore model was generated on the basis of 22 known LSD1 inhibitors, although these inhibitors encompassed different chemotypes, including irreversible tranylcypromine and propargylamine inhibitors and FAD-competitive and FAD-noncompetitive inhibitors. The resulting pharmacophore was subsequently screened against the SPECS database (about 172,000 compounds), and the resulting hits (about 7,000 molecules) were further filtered according to their drug-likeness and predicted ADMET properties. Finally, 950 compounds were docked to the LSD1 substrate binding pocket. Among the nine selected compounds, XZ09 (**31**, Fig. 7) was the most potent, showing an IC_{50} value of 2.41 μM and selectivity for LSD1 over MAO-A/MAO-B. However, an

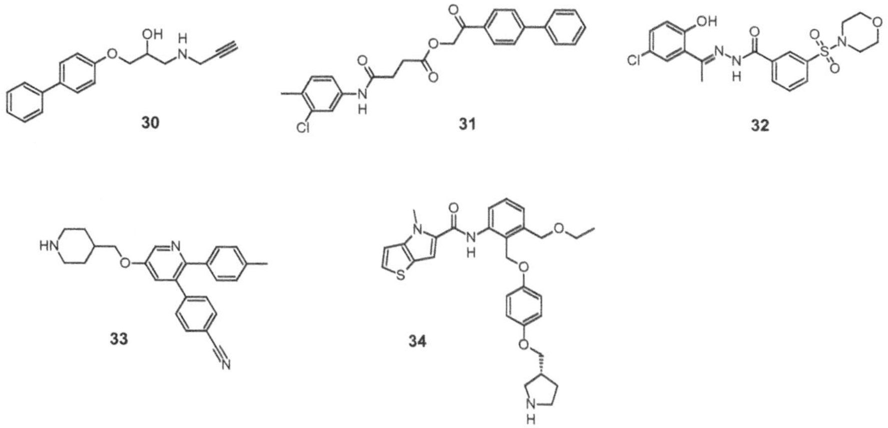

Fig. 7 Molecular structures of LSD1 histone demethylase inhibitors mentioned in the text

experimental validation concerning target engagement and cellular effects has not been reported.

A structure-based VS approach to identify reversible LSD1 inhibitors was described by Sorna et al. [131]. This study involved docking into LSD1 using several docking programs and scoring functions in a stepwise manner. For this purpose, the ZINC database (about 13 million compounds) was first docked using Glide in high-throughput virtual screening mode; the top-ranked 15% compounds were further docked using Glide in Standard Precision mode, followed by docking of the top-ranked 0.5 % using ICM. 121 hits were finally selected for in vitro screening, which led to the identification of a series of six *N*′-(1-phenylethylidene) benzohydrazide derivatives showing submicromolar activity against LSD1. Further chemical optimization and SAR studies resulted in the discovery of compound **32** (Fig. 7), a potent, reversible, and specific inhibitor of LSD1 ($K_i = 31$ nM). Cellular effects such as an increase of the H3K9me2 (dimethylated lysine 9 of histone 3) mark and growth inhibition of several cancer cell lines were also reported.

Molecular docking studies have also been extensively used to predict the binding mode of reversible as well as irreversible LSD1 inhibitors and to rationalize their SAR utilizing various docking software and techniques, such as non-covalent and covalent docking [133–144]. Other molecular modeling approaches aiming to better understand the SAR of LSD1 inhibitors include 3D-QSAR studies as reported by Ding et al. [145]. In this 3D-QSAR study, a series of 25 6-aryl-5-cyano-pyrimidine LSD1 inhibitors was used to generate comparative molecular field analysis (CoMFA) and comparative molecular similarity indices analysis (CoMSIA) models, while a test set of 8 compounds was used to confirm the models' reliability. The best models showed good correlation with experimental data, with the best CoMFA model showing a cross-validation correlation coefficient (q^2) of 0.802 and the best CoMSIA model a q^2 of 0.799. These models could provide a better understanding of the SAR of 6-aryl-5-cyano-pyrimidine LSD1 inhibitors and shed light onto the effect of the different substituents on the inhibitory activity of the compounds.

Moreover, combination of available structural data for LSD1, docking studies and medicinal chemistry approaches were frequently used to guide the optimization of novel LSD1 inhibitors [146–148]. For instance, Wu et al. reported on the structure-based design of a series of 3-(piperidin-4-ylmethoxy)pyridine derivatives, such as compound **33** (IC_{50} 29 nM, Fig. 7), which exhibited high inhibitory activity against LSD1, high selectivity over MAO-A/MAO-B, cellular activity, and inhibition of tumor cells proliferation. Very recently, the structure-based optimization of a novel chemical series of thieno[3,2-*b*]pyrrole-5-carboxamide LSD1 inhibitors, which had been initially identified by an HTS campaign, was also reported. Structural optimization culminated in the discovery of the thieno[3,2-*b*]pyrrole-5-carboxamide **34** (IC_{50} 7.8 nM, Fig. 7) which was also successfully cocrystallized with LSD1 [147].

5.2 *Jumonji Histone Demethylases*

JmjC domain-containing proteins belong to the cupin superfamily of metalloenzymes [149] and comprise about 20 (usually multidomain) proteins which are subdivided into seven subfamilies (KDM2–KDM8) [129, 150]. JmjC domains characteristically show a conserved antiparallel β-barrel fold which harbors the binding site for Fe^{2+} and α-KG cofactors. The catalytic Fe^{2+} is coordinated in an octahedral fashion by three conserved residues in the catalytic pocket (two histidine and one aspartate/glutamate), besides the cofactor α-KG and a water molecule. Meanwhile, the substrate binding pocket of the different JmjC subfamilies shows great structural variations and can thus be addressed for the development of selective inhibitors [151]. Numerous crystal structures of various JmjC domains in complex with inhibitors have appeared in the last few years, which shed light onto the inhibitory mechanism of these compounds. All inhibitors reported so far are metal chelators and bind competitively in the α-KG binding pocket [129, 152]. Hence, many of them lack selectivity and inhibit different JmjC protein subtypes as well as other Fe^{2+}-/α-KG-dependent enzymes, such as prolyl hydroxylase domain-containing proteins (PHDs) and factor inhibiting hypoxia-inducible factor (FIH). While targeting the metal ion is obviously necessary for the activity of the inhibitors, simultaneously addressing the substrate binding pocket represents a promising strategy to achieve selectivity [153].

The first VS for JmjC proteins was reported by Chu et al. [154], who used an available crystal structure of JMJD2A (KDM4A) and docked the NCI dataset (around 240,000 compounds). Among the ten compounds selected for in vitro screening, the 2,4-dinitrobenzene derivative **35** (Fig. 8) showed the highest inhibitory activity against JMJD2A (IC_{50} 6.4 μM) and JMJD2B (IC_{50} 9.3 μM) and selectivity over JMJD2D and JMJD2E. However, the identified inhibitor does not possess a classical metal ion-chelating group, and the binding mode of the molecule is not clear, even if the docking study predicted an interaction between the nitro group and the metal ion.

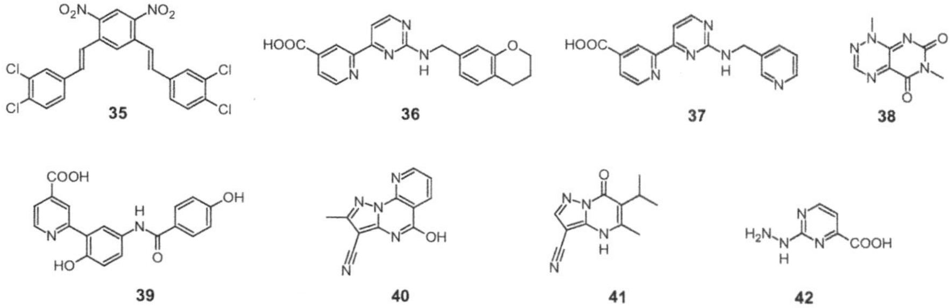

Fig. 8 Molecular structures of JumonjiC histone demethylase inhibitors mentioned in the text

An iterative VS campaign combined with rational structural optimization led to the discovery of another potent class of JMJD2A (KDM4A) inhibitors [155]. Several compound databases were first filtered for molecules containing a metal ion-chelating moiety. The retrieved hits were docked into the α-KG binding pocket of JMJD2A, and the top-ranked compounds were tested in vitro against several KDMs. This led to the identification of two 2-(pyrimidin-4-yl)pyridine-4-carboxylic acid derivatives with promising JMJD2A inhibitory activity (IC_{50} ~1 and 4 μM). A subsequent substructure search for compounds bearing a 2-(pyrimidin-4-yl)pyridine-4-carboxylic acid scaffold resulted in a series of potent JMJD2A inhibitors with IC_{50} between 0.9 and 8.1 μM. Finally, structural optimization led to compound **36** (Fig. 8) which showed a submicromolar inhibition of JMJD2A (IC_{50} 370 nM). Selectivity studies revealed a preference for JMJD2A and JARID1A over JMJD3. As predicted by the docking studies, a cocrystal structure of one of the VS hits (**37**, Fig. 8) showed that this series of compounds act as α-KG competitive inhibitors (Fig. 9), where the pyridyl and pyrimidyl nitrogens chelate the central Fe^{2+} ion.

Another structure-based VS approach to identify JMJD2A inhibitors was recently reported by Franci et al., who performed molecular docking using mcule.com database [156]. They were able to identify PKF118-310 (**38**, Fig. 8), a TCF4/β-catenin signaling antagonist, as JMJD2A inhibitor. The predicted binding mode of this hit compound was, however, not disclosed.

Korczynska et al. reported on a fragment-based docking approach to discover novel JmjC inhibitors [157]. The ZINC fragment library (over 600,000 compounds) was docked into JMJD2A using DOCK3.6. Four of the selected fragments showed an IC_{50} below 100 μM against JMJD2C in the established in vitro assay, three of which possessed a 5-aminosalicylic acid core that was predicted to coordinate the central Fe^{2+} ion via its carboxylate and hydroxyl moieties. Structure-based optimization was performed by an in silico design of a fragment-linked library composed of 12 scaffolds. This resulted in the identification of a series of 2-(2-hydroxyphenyl) pyridine-4-carboxylic acids which showed high potency against JMJD2C and selectivity over related enzymes, as exemplified by compound **39** (K_i 43 nM; Fig. 8). Crystal structures of JMJD2A with **39** and various other analogs showed an

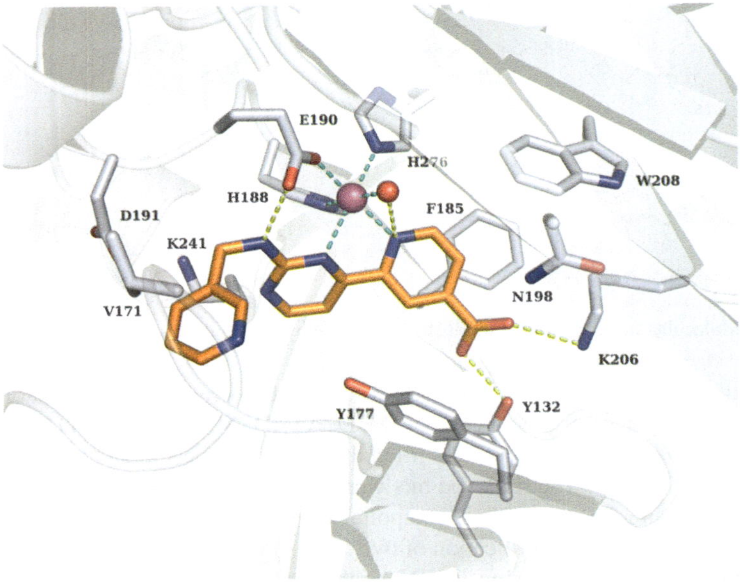

Fig. 9 X-ray structure of the VS hit **37** (orange sticks) with the histone demethylase JMJD2A (PDB ID: 5ANQ). Only the residues in the binding pocket are shown. Hydrogen bonds are displayed as yellow dashed lines, metal coordination as cyan dashed lines, Fe^{2+} as purple sphere, and the water molecule as a red sphere

agreement with the predicted docking poses, where the phenolic group and the pyridyl nitrogen were shown to engage in metal chelation.

Further, a series of pyrazolo[1,5-*a*]pyrimidine-3-carbonitrile was identified as a new class of JMJD2D (KDM4D) inhibitors by another docking-based VS campaign [158]. Various chemical libraries were first docked into the α-KG binding pocket of JMJD2D. Among the 30 compounds selected for primary in vitro screening, two compounds showed considerable inhibition of JMJD2D, with one of them, a pyrazolo[1,5-*a*]pyrimidine-3-carbonitrile derivative, showing selectivity over other tested KDMs. Further chemical modifications resulted in compound **40** (Fig. 8) which showed submicromolar inhibition of JMJD2D (IC_{50} 410 nM). Docking studies predicted that only the nitrile moiety acts as the metal-chelating group, which has been substantiated by the recently reported crystal structure of JARID1A (KDM5A) in complex with an analogous pyrazolopyrimidine-3-carbonitrile **41** (Fig. 8) [159]. However, no cellular activity of compound **40** has been reported so far.

In order to identify novel inhibitors of JARID1A (KDM5A), Wu et al. [160] performed a docking-based VS using a generated homology model of the enzyme. Since no crystal of the JARID1A JmjC domain was available at that time (recently several 3D structures of JARID1A/inhibitor complexes have been resolved), a homology model was built using multiple JMJD crystal structures as templates. Subsequently, several compound libraries were docked into the generated homology model.

Among the top-scored 48 compounds, which were subjected to in vitro screening, 12 compounds showed a micromolar activity against JARIDA. Further structural optimization of one of the identified hits, namely, a 2,4-diaminopyrimidne derivative, resulted in the submicromolar inhibitor **42** (Fig. 8), which bears a pyrimidine-carboxylic acid scaffold, a core structure already known from JMJD2 inhibitors. Cellular testing showed that the compound is able to increase the level of H3K4me3 (trimethylated lysine 4 of histone 3) in a dose-dependent manner.

6 Summary

Despite many technical challenges, structure-based design plays an important role in drug discovery. As a steadily growing number of epigenetic targets are biologically and structurally characterized, structure-based methods are more and more applied to design specific inhibitors in order to elucidate their therapeutic potential. Ligand docking and scoring technologies have steadily improved, and the importance of the adequate validation of pragmatic VS protocols is now well recognized. While both ligand- and structure-based approaches have already demonstrated their worth in the identification of novel lead compounds for epigenetic targets, there is still the challenge to improve the predictive accuracy of scoring functions, particularly to enable scoring-based methods to have a greater impact in guiding lead optimization.

Acknowledgments We thank the European COST actions "Epigenetic Chemical Biology" (CM1406) and "MuTaLig" (CA15135) for support.

Compliance with Ethical Standards

Conflict of Interest: The authors declare that they have no conflict of interest.

Ethical Approval: All procedures performed in studies involving human participants were in accordance with the ethical standards of the institutional and/or national research committee and with the 1964 Helsinki declaration and its later amendments or comparable ethical standards.

Informed Consent: Informed consent was obtained from all individual participants included in the study.

References

1. Wolffe AP, Matzke MA (1999) Epigenetics: regulation through repression. Science 286 (5439):481–486. https://doi.org/10.1126/science.286.5439.481
2. Kouzarides T (2007) Chromatin modifications and their function. Cell 128(4):693–705. https://doi.org/10.1016/j.cell.2007.02.005
3. Finnin MS, Donigian JR, Cohen A, Richon VM, Rifkind RA, Marks PA, Breslow R, Pavletich NP (1999) Structures of a histone deacetylase homologue bound to the TSA and SAHA inhibitors. Nature 401(6749):188–193. https://doi.org/10.1038/43710

4. Warren GL, Andrews CW, Capelli AM, Clarke B, LaLonde J, Lambert MH, Lindvall M, Nevins N, Semus SF, Senger S, Tedesco G, Wall ID, Woolven JM, Peishoff CE, Head MS (2006) A critical assessment of docking programs and scoring functions. J Med Chem 49 (20):5912–5931. https://doi.org/10.1021/jm050362n
5. Gilson MK, Zhou HX (2007) Calculation of protein-ligand binding affinities. Annu Rev Biophys Biomol Struct 36:21–42. https://doi.org/10.1146/annurev.biophys.36.040306.132550
6. Armen RS, Chen J, Brooks 3rd CL (2009) An evaluation of explicit receptor flexibility in molecular docking using molecular dynamics and torsion angle molecular dynamics. J Chem Theory Comput 5(10):2909–2923. https://doi.org/10.1021/ct900262t
7. Durrant JD, McCammon JA (2011) Molecular dynamics simulations and drug discovery. BMC Biol 9:71. https://doi.org/10.1186/1741-7007-9-71
8. Wichapong K, Rohe A, Platzer C, Slynko I, Erdmann F, Schmidt M, Sippl W (2014) Application of docking and QM/MM-GBSA rescoring to screen for novel Myt1 kinase inhibitors. J Chem Inf Model 54(3):881–893. https://doi.org/10.1021/ci4007326
9. Srinivasan J, Cheatham TE, Cieplak P, Kollman PA, Case DA (1998) Continuum solvent studies of the stability of DNA, RNA, and phosphoramidate−DNA helices. J Am Chem Soc 120(37):9401–9409. https://doi.org/10.1021/ja981844+
10. Abel R, Wang L, Mobley DL, Friesner RA (2017) A critical review of validation, blind testing, and real-world use of alchemical protein-ligand binding free energy calculations. Curr Top Med Chem 17(23):2577–2585. https://doi.org/10.2174/1568026617666170414142131
11. Kaus JW, McCammon JA (2015) Enhanced ligand sampling for relative protein-ligand binding free energy calculations. J Phys Chem B 119(20):6190–6197. https://doi.org/10.1021/acs.jpcb.5b02348
12. Klimovich PV, Shirts MR, Mobley DL (2015) Guidelines for the analysis of free energy calculations. J Comput Aided Mol Des 29(5):397–411. https://doi.org/10.1007/s10822-015-9840-9
13. Deng N, Flynn WF, Xia J, Vijayan RSK, Zhang B, He P, Mentes A, Gallicchio E, Levy RM (2016) Large scale free energy calculations for blind predictions of protein–ligand binding: the D3R Grand Challenge 2015. J Comput Aided Mol Des 30(9):743–751. https://doi.org/10.1007/s10822-016-9952-x
14. Lindström A, Edvinsson L, Johansson A, Andersson CD, Andersson IE, Raubacher F, Linusson A (2011) Postprocessing of docked protein−ligand complexes using implicit solvation models. J Chem Inf Model 51(2):267–282. https://doi.org/10.1021/ci100354x
15. Daniel C, Woody S, Thijs B (2017) Calculating water thermodynamics in the binding site of proteins – applications of watermap to drug discovery. Curr Top Med Chem 17 (23):2586–2598. https://doi.org/10.2174/1568026617666170414141452
16. de Ruijter AJ, van Gennip AH, Caron HN, Kemp S, van Kuilenburg AB (2003) Histone deacetylases (HDACs): characterization of the classical HDAC family. Biochem J 370 (Pt 3):737–749. https://doi.org/10.1042/BJ20021321
17. Hodawadekar SC, Marmorstein R (2007) Chemistry of acetyl transfer by histone modifying enzymes: structure, mechanism and implications for effector design. Oncogene 26 (37):5528–5540. https://doi.org/10.1038/sj.onc.1210619
18. Somoza JR, Skene RJ, Katz BA, Mol C, Ho JD, Jennings AJ, Luong C, Arvai A, Buggy JJ, Chi E, Tang J, Sang BC, Verner E, Wynands R, Leahy EM, Dougan DR, Snell G, Navre M, Knuth MW, Swanson RV, McRee DE, Tari LW (2004) Structural snapshots of human HDAC8 provide insights into the class I histone deacetylases. Structure 12(7):1325–1334. https://doi.org/10.1016/j.str.2004.04.012
19. Bottomley MJ, Lo Surdo P, Di Giovine P, Cirillo A, Scarpelli R, Ferrigno F, Jones P, Neddermann P, De Francesco R, Steinkuhler C, Gallinari P, Carfi A (2008) Structural and functional analysis of the human HDAC4 catalytic domain reveals a regulatory structural zinc-binding domain. J Biol Chem 283(39):26694–26704. https://doi.org/10.1074/jbc.M803514200
20. Schuetz A, Min J, Allali-Hassani A, Schapira M, Shuen M, Loppnau P, Mazitschek R, Kwiatkowski NP, Lewis TA, Maglathin RL, McLean TH, Bochkarev A, Plotnikov AN,

Vedadi M, Arrowsmith CH (2008) Human HDAC7 harbors a class IIa histone deacetylase-specific zinc binding motif and cryptic deacetylase activity. J Biol Chem 283 (17):11355–11363. https://doi.org/10.1074/jbc.M707362200
21. Bressi JC, Jennings AJ, Skene R, Wu Y, Melkus R, De Jong R, O'Connell S, Grimshaw CE, Navre M, Gangloff AR (2010) Exploration of the HDAC2 foot pocket: synthesis and SAR of substituted N-(2-aminophenyl)benzamides. Bioorg Med Chem Lett 20(10):3142–3145. https://doi.org/10.1016/j.bmcl.2010.03.091
22. Watson PJ, Fairall L, Santos GM, Schwabe JW (2012) Structure of HDAC3 bound to co-repressor and inositol tetraphosphate. Nature 481(7381):335–340. https://doi.org/10.1038/nature10728
23. Millard CJ, Watson PJ, Celardo I, Gordiyenko Y, Cowley SM, Robinson CV, Fairall L, Schwabe JW (2013) Class I HDACs share a common mechanism of regulation by inositol phosphates. Mol Cell 51(1):57–67. https://doi.org/10.1016/j.molcel.2013.05.020
24. Hai Y, Christianson DW (2016) Histone deacetylase 6 structure and molecular basis of catalysis and inhibition. Nat Chem Biol 12(9):741–747. https://doi.org/10.1038/nchembio.2134
25. Miyake Y, Keusch JJ, Wang L, Saito M, Hess D, Wang X, Melancon BJ, Helquist P, Gut H, Matthias P (2016) Structural insights into HDAC6 tubulin deacetylation and its selective inhibition. Nat Chem Biol 12(9):748–754. https://doi.org/10.1038/nchembio.2140
26. Hai Y, Shinsky SA, Porter NJ, Christianson DW (2017) Histone deacetylase 10 structure and molecular function as a polyamine deacetylase. Nat Commun 8:15368. https://doi.org/10.1038/ncomms15368
27. Zhang Y, Fang H, Jiao J, Xu W (2008) The structure and function of histone deacetylases: the target for anti-cancer therapy. Curr Med Chem 15(27):2840–2849
28. Massa S, Mai A, Sbardella G, Esposito M, Ragno R, Loidl P, Brosch G (2001) 3-(4-aroyl-1H-pyrrol-2-yl)-N-hydroxy-2-propenamides, a new class of synthetic histone deacetylase inhibitors. J Med Chem 44(13):2069–2072
29. Van Ommeslaeghe K, Elaut G, Brecx V, Papeleu P, Iterbeke K, Geerlings P, Tourwe D, Rogiers V (2003) Amide analogues of TSA: synthesis, binding mode analysis and HDAC inhibition. Bioorg Med Chem Lett 13(11):1861–1864
30. Wang DF, Wiest O, Helquist P, Lan-Hargest HY, Wiech NL (2004) On the function of the 14 A long internal cavity of histone deacetylase-like protein: implications for the design of histone deacetylase inhibitors. J Med Chem 47(13):3409–3417. https://doi.org/10.1021/jm0498497
31. Lu Q, Wang DS, Chen CS, Hu YD, Chen CS (2005) Structure-based optimization of phenylbutyrate-derived histone deacetylase inhibitors. J Med Chem 48(17):5530–5535. https://doi.org/10.1021/jm0503749
32. Rodriquez M, Terracciano S, Cini E, Settembrini G, Bruno I, Bifulco G, Taddei M, Gomez-Paloma L (2006) Total synthesis, NMR solution structure, and binding model of the potent histone deacetylase inhibitor FR235222. Angew Chem Int Ed Engl 45(3):423–427. https://doi.org/10.1002/anie.200501995
33. Liu T, Kapustin G, Etzkorn FA (2007) Design and synthesis of a potent histone deacetylase inhibitor. J Med Chem 50(9):2003–2006. https://doi.org/10.1021/jm061082q
34. Maulucci N, Chini MG, Micco SD, Izzo I, Cafaro E, Russo A, Gallinari P, Paolini C, Nardi MC, Casapullo A, Riccio R, Bifulco G, Riccardis FD (2007) Molecular insights into azumamide e histone deacetylases inhibitory activity. J Am Chem Soc 129(10):3007–3012. https://doi.org/10.1021/ja0686256
35. Chen PC, Patil V, Guerrant W, Green P, Oyelere AK (2008) Synthesis and structure-activity relationship of histone deacetylase (HDAC) inhibitors with triazole-linked cap group. Bioorg Med Chem 16(9):4839–4853. https://doi.org/10.1016/j.bmc.2008.03.050
36. Di Micco S, Terracciano S, Bruno I, Rodriquez M, Riccio R, Taddei M, Bifulco G (2008) Molecular modeling studies toward the structural optimization of new cyclopeptide-based

HDAC inhibitors modeled on the natural product FR235222. Bioorg Med Chem 16 (18):8635–8642. https://doi.org/10.1016/j.bmc.2008.08.003
37. Lavoie R, Bouchain G, Frechette S, Woo SH, Abou-Khalil E, Leit S, Fournel M, Yan PT, Trachy-Bourget MC, Beaulieu C, Li Z, Besterman J, Delorme D (2001) Design and synthesis of a novel class of histone deacetylase inhibitors. Bioorg Med Chem Lett 11(21):2847–2850
38. Remiszewski SW, Sambucetti LC, Atadja P, Bair KW, Cornell WD, Green MA, Howell KL, Jung M, Kwon P, Trogani N, Walker H (2002) Inhibitors of human histone deacetylase: synthesis and enzyme and cellular activity of straight chain hydroxamates. J Med Chem 45 (4):753–757
39. Mai A, Massa S, Ragno R, Cerbara I, Jesacher F, Loidl P, Brosch G (2003) 3-(4-aroyl-1-methyl-1H-2-pyrrolyl)-N-hydroxy-2-alkylamides as a new class of synthetic histone deacetylase inhibitors. 1. Design, synthesis, biological evaluation, and binding mode studies performed through three different docking procedures. J Med Chem 46(4):512–524. https://doi.org/10.1021/jm021070e
40. Park H, Lee S (2004) Homology modeling, force field design, and free energy simulation studies to optimize the activities of histone deacetylase inhibitors. J Comput Aided Mol Des 18 (6):375–388
41. Wang DF, Helquist P, Wiech NL, Wiest O (2005) Toward selective histone deacetylase inhibitor design: homology modeling, docking studies, and molecular dynamics simulations of human class I histone deacetylases. J Med Chem 48(22):6936–6947. https://doi.org/10.1021/jm0505011
42. Kim HM, Hong SH, Kim MS, Lee CW, Kang JS, Lee K, Park SK, Han JW, Lee HY, Choi Y, Kwon HJ, Han G (2007) Modification of cap group in delta-lactam-based histone deacetylase (HDAC) inhibitors. Bioorg Med Chem Lett 17(22):6234–6238. https://doi.org/10.1016/j.bmcl.2007.09.034
43. Moradei OM, Mallais TC, Frechette S, Paquin I, Tessier PE, Leit SM, Fournel M, Bonfils C, Trachy-Bourget MC, Liu J, Yan TP, Lu AH, Rahil J, Wang J, Lefebvre S, Li Z, Vaisburg AF, Besterman JM (2007) Novel aminophenyl benzamide-type histone deacetylase inhibitors with enhanced potency and selectivity. J Med Chem 50(23):5543–5546. https://doi.org/10.1021/jm701079h
44. Mukherjee P, Pradhan A, Shah F, Tekwani BL, Avery MA (2008) Structural insights into the Plasmodium falciparum histone deacetylase 1 (PfHDAC-1): a novel target for the development of antimalarial therapy. Bioorg Med Chem 16(9):5254–5265. https://doi.org/10.1016/j.bmc.2008.03.005
45. Schafer S, Saunders L, Eliseeva E, Velena A, Jung M, Schwienhorst A, Strasser A, Dickmanns A, Ficner R, Schlimme S, Sippl W, Verdin E, Jung M (2008) Phenylalanine-containing hydroxamic acids as selective inhibitors of class IIb histone deacetylases (HDACs). Bioorg Med Chem 16(4):2011–2033. https://doi.org/10.1016/j.bmc.2007.10.092
46. Weerasinghe SV, Estiu G, Wiest O, Pflum MK (2008) Residues in the 11 A channel of histone deacetylase 1 promote catalytic activity: implications for designing isoform-selective histone deacetylase inhibitors. J Med Chem 51(18):5542–5551. https://doi.org/10.1021/jm800081j
47. Witter DJ, Harrington P, Wilson KJ, Chenard M, Fleming JC, Haines B, Kral AM, Secrist JP, Miller TA (2008) Optimization of biaryl selective HDAC1&2 inhibitors (SHI-1:2). Bioorg Med Chem Lett 18(2):726–731. https://doi.org/10.1016/j.bmcl.2007.11.047
48. Yan C, Xiu Z, Li X, Li S, Hao C, Teng H (2008) Comparative molecular dynamics simulations of histone deacetylase-like protein: binding modes and free energy analysis to hydroxamic acid inhibitors. Proteins 73(1):134–149. https://doi.org/10.1002/prot.22047
49. Krishna S, Kumar V, Siddiqi MI (2016) Recent advances in computer-assisted structure-based identification and design of histone deacetylases inhibitors. Curr Top Med Chem 16 (9):934–947
50. Price S, Bordogna W, Bull RJ, Clark DE, Crackett PH, Dyke HJ, Gill M, Harris NV, Gorski J, Lloyd J, Lockey PM, Mullett J, Roach AG, Roussel F, White AB (2007) Identification and optimisation of a series of substituted 5-(1H-pyrazol-3-yl)-thiophene-2-hydroxamic acids as

potent histone deacetylase (HDAC) inhibitors. Bioorg Med Chem Lett 17(2):370–375. https://doi.org/10.1016/j.bmcl.2006.10.048
51. Tang H, Wang XS, Huang XP, Roth BL, Butler KV, Kozikowski AP, Jung M, Tropsha A (2009) Novel inhibitors of human histone deacetylase (HDAC) identified by QSAR modeling of known inhibitors, virtual screening, and experimental validation. J Chem Inf Model 49 (2):461–476. https://doi.org/10.1021/ci800366f
52. Park H, Kim S, Kim YE, Lim SJ (2010) A structure-based virtual screening approach toward the discovery of histone deacetylase inhibitors: identification of promising zinc-chelating groups. ChemMedChem 5(4):591–597. https://doi.org/10.1002/cmdc.200900500
53. Lipinski CA, Lombardo F, Dominy BW, Feeney PJ (2001) Experimental and computational approaches to estimate solubility and permeability in drug discovery and development settings (Reprinted from Advanced Drug Delivery Reviews, vol 23, pp 3–25, 1997). Adv Drug Deliver Rev 46(1–3):3–26. https://doi.org/10.1016/S0169-409x(00)00129-0
54. Kannan S, Melesina J, Hauser AT, Chakrabarti A, Heimburg T, Schmidtkunz K, Walter A, Marek M, Pierce RJ, Romier C, Jung M, Sippl W (2014) Discovery of inhibitors of Schistosoma mansoni HDAC8 by combining homology modeling, virtual screening, and in vitro validation. J Chem Inf Model 54(10):3005–3019. https://doi.org/10.1021/ci5004653
55. Marek M, Kannan S, Hauser AT, Moraes Mourao M, Caby S, Cura V, Stolfa DA, Schmidtkunz K, Lancelot J, Andrade L, Renaud JP, Oliveira G, Sippl W, Jung M, Cavarelli J, Pierce RJ, Romier C (2013) Structural basis for the inhibition of histone deacetylase 8 (HDAC8), a key epigenetic player in the blood fluke Schistosoma mansoni. PLoS Pathog 9(9):e1003645. https://doi.org/10.1371/journal.ppat.1003645
56. Heimburg T, Chakrabarti A, Lancelot J, Marek M, Melesina J, Hauser AT, Shaik TB, Duclaud S, Robaa D, Erdmann F, Schmidt M, Romier C, Pierce RJ, Jung M, Sippl W (2016) Structure-based design and synthesis of novel inhibitors targeting HDAC8 from Schistosoma mansoni for the treatment of schistosomiasis. J Med Chem 59(6):2423–2435. https://doi.org/10.1021/acs.jmedchem.5b01478
57. Heimburg T, Kolbinger FR, Zeyen P, Ghazy E, Herp D, Schmidtkunz K, Melesina J, Shaik TB, Erdmann F, Schmidt M, Romier C, Robaa D, Witt O, Oehme I, Jung M, Sippl W (2017) Structure-based design and biological characterization of selective histone deacetylase 8 (HDAC8) inhibitors with anti-neuroblastoma activity. J Med Chem 60(24):10188–10204. https://doi.org/10.1021/acs.jmedchem.7b01447
58. Butler KV, Kalin J, Brochier C, Vistoli G, Langley B, Kozikowski AP (2010) Rational design and simple chemistry yield a superior, neuroprotective HDAC6 inhibitor, tubastatin A. J Am Chem Soc 132(31):10842–10846. https://doi.org/10.1021/ja102758v
59. Micelli C, Rastelli G (2015) Histone deacetylases: structural determinants of inhibitor selectivity. Drug Discov Today 20(6):718–735. https://doi.org/10.1016/j.drudis.2015.01.007
60. De Vreese R, D'Hooghe M (2017) Synthesis and applications of benzohydroxamic acid-based histone deacetylase inhibitors. Eur J Med Chem 135:174–195. https://doi.org/10.1016/j.ejmech.2017.04.013
61. Davenport AM, Huber FM, Hoelz A (2014) Structural and functional analysis of human SIRT1. J Mol Biol 426(3):526–541. https://doi.org/10.1016/j.jmb.2013.10.009
62. Zhao X, Allison D, Condon B, Zhang FY, Gheyi T, Zhang AP, Ashok S, Russell M, MacEwan I, Qian YW, Jamison JA, Luz JG (2013) The 2.5 angstrom crystal structure of the SIRT1 catalytic domain bound to nicotinamide adenine dinucleotide (NAD(+)) and an indole (EX527 analogue) reveals a novel mechanism of histone deacetylase inhibition. J Med Chem 56(3):963–969. https://doi.org/10.1021/jm301431y
63. Finnin MS, Donigian JR, Pavletich NP (2001) Structure of the histone deacetylase SIRT2. Nat Struct Biol 8(7):621–625. https://doi.org/10.1038/89668
64. Moniot S, Schutkowski M, Steegborn C (2013) Crystal structure analysis of human Sirt2 and its ADP-ribose complex. J Struct Biol 182(2):136–143. https://doi.org/10.1016/j.jsb.2013.02.012

65. Rumpf T, Gerhardt S, Einsle O, Jung M (2015) Seeding for sirtuins: microseed matrix seeding to obtain crystals of human Sirt3 and Sirt2 suitable for soaking. Acta Crystallogr F Struct Biol Commun 71:1498–1510. https://doi.org/10.1107/S2053230x15019986
66. Yamagata K, Goto Y, Nishimasu H, Morimoto J, Ishitani R, Dohmae N, Takeda N, Nagai R, Komuro I, Suga H, Nureki O (2014) Structural basis for potent inhibition of SIRT2 deacetylase by a macrocyclic peptide inducing dynamic structural change. Structure 22(2):345–352. https://doi.org/10.1016/j.str.2013.12.001
67. Bao XC, Wang Y, Li X, Li XM, Liu Z, Yang TP, Wong CF, Zhang JW, Hao Q, Li XD (2014) Identification of 'erasers' for lysine crotonylated histone marks using a chemical proteomics approach. Elife 3. https://doi.org/10.7554/eLife.02999
68. Disch JS, Evindar G, Chiu CH, Blum CA, Dai H, Jin L, Schuman E, Lind KE, Belyanskaya SL, Deng J, Coppo F, Aquilani L, Graybill TL, Cuozzo JW, Lavu S, Mao C, Vlasuk GP, Perni RB (2013) Discovery of thieno[3,2-d]pyrimidine-6-carboxamides as potent inhibitors of SIRT1, SIRT2, and SIRT3. J Med Chem 56(9):3666–3679. https://doi.org/10.1021/jm400204k
69. Gertz M, Fischer F, Nguyen GTT, Lakshminarasimhan M, Schutkowski M, Weyand M, Steegborn C (2013) Ex-527 inhibits sirtuins by exploiting their unique NAD(+)-dependent deacetylation mechanism. Proc Natl Acad Sci U S A 110(30):E2772–E2781. https://doi.org/10.1073/pnas.1303628110
70. Gertz M, Giang TTN, Fischer F, Suenkel B, Schlicker C, Franzel B, Tomaschewski J, Aladini F, Becker C, Wolters D, Steegborn C (2012) A molecular mechanism for direct sirtuin activation by resveratrol. Plos One 7(11):e49761. https://doi.org/10.1371/journal.pone.0049761
71. Jin L, Wei WT, Jiang YB, Peng H, Cai JH, Mao C, Dai H, Choy W, Bemis JE, Jirousek MR, Milne JC, Westphal CH, Perni RB (2009) Crystal structures of human SIRT3 displaying substrate-induced conformational changes. J Biol Chem 284(36):24394–24405. https://doi.org/10.1074/jbc.M109.014928
72. Nguyen GTT, Gertz M, Steegborn C (2013) Crystal structures of Sirt3 complexes with 4′-bromo-resveratrol reveal binding sites and inhibition mechanism. Chem Biol 20(11):1375–1385. https://doi.org/10.1016/j.chembiol.2013.09.019
73. Nguyen GTT, Schaefer S, Gertz M, Weyand M, Steegborn C (2013) Structures of human sirtuin 3 complexes with ADP-ribose and with carba-NAD(+) and SRT1720: binding details and inhibition mechanism. Acta Crystallogr D 69:1423–1432. https://doi.org/10.1107/S0907444913015448
74. Szczepankiewicz BG, Dai H, Koppetsch KJ, Qian DM, Jiang F, Mao C, Perni RB (2012) Synthesis of carba-NAD and the structures of its ternary complexes with SIRT3 and SIRT5. J Org Chem 77(17):7319–7329. https://doi.org/10.1021/jo301067e
75. Wu JH, Zhang DY, Chen L, Li JN, Wang JL, Ning CQ, Yu NF, Zhao F, Chen DY, Chen XY, Chen KX, Jiang HL, Liu H, Liu DX (2013) Discovery and mechanism study of SIRT1 activators that promote the deacetylation of fluorophore-labeled substrate. J Med Chem 56(3):761–780. https://doi.org/10.1021/jm301032j
76. Du JT, Zhou YY, Su XY, Yu JJ, Khan S, Jiang H, Kim J, Woo J, Kim JH, Choi BH, He B, Chen W, Zhang S, Cerione RA, Auwerx J, Hao Q, Lin HN (2011) Sirt5 is a NAD-dependent protein lysine demalonylase and desuccinylase. Science 334(6057):806–809. https://doi.org/10.1126/science.1207861
77. Schuetz A, Min JR, Antoshenko T, Wang CL, Allali-Hassani A, Dong AP, Loppnau P, Vedadi M, Bochkarev A, Sternglanz R, Plotnikov AN (2007) Structural basis of inhibition of the human NAD(+)-dependent deacetylase SIRT5 by suramin. Structure 15(3):377–389. https://doi.org/10.1016/j.str.2007.02.002
78. Zhou YY, Zhang HM, He B, Du JT, Lin HN, Cerione RA, Hao Q (2012) The bicyclic intermediate structure provides insights into the desuccinylation mechanism of human sirtuin 5 (SIRT5). J Biol Chem 287(34):28307–28314. https://doi.org/10.1074/jbc.M112.384511

79. Jiang H, Khan S, Wang Y, Charron G, He B, Sebastian C, Du J, Kim R, Ge E, Mostoslavsky R, Hang HC, Hao Q, Lin H (2013) SIRT6 regulates TNF-alpha secretion through hydrolysis of long-chain fatty acyl lysine. Nature 496(7443):110–113. https://doi.org/10.1038/nature12038
80. Avalos JL, Boeke JD, Wolberger C (2004) Structural basis for the mechanism and regulation of Sir2 enzymes. Mol Cell 13(5):639–648
81. Avalos JL, Bever KM, Wolberger C (2005) Mechanism of sirtuin inhibition by nicotinamide: altering the NAD(+) cosubstrate specificity of a Sir2 enzyme. Mol Cell 17(6):855–868. https://doi.org/10.1016/j.molcel.2005.02.022
82. Cosgrove MS, Bever K, Avalos JL, Muhammad S, Zhang X, Wolberger C (2006) The structural basis of sirtuin substrate affinity. Biochemistry 45(24):7511–7521. https://doi.org/10.1021/bi0526332
83. Hoff KG, Avalos JL, Sens K, Wolberger C (2006) Insights into the sirtuin mechanism from ternary complexes containing NAD+ and acetylated peptide. Structure 14(8):1231–1240. https://doi.org/10.1016/j.str.2006.06.006
84. Hawse WF, Hoff KG, Fatkins DG, Daines A, Zubkova OV, Schramm VL, Zheng W, Wolberger C (2008) Structural insights into intermediate steps in the Sir2 deacetylation reaction. Structure 16(9):1368–1377. https://doi.org/10.1016/j.str.2008.05.015
85. Min J, Landry J, Sternglanz R, Xu RM (2001) Crystal structure of a SIR2 homolog-NAD complex. Cell 105(2):269–279
86. Zhao K, Harshaw R, Chai X, Marmorstein R (2004) Structural basis for nicotinamide cleavage and ADP-ribose transfer by NAD(+)-dependent Sir2 histone/protein deacetylases. Proc Natl Acad Sci U S A 101(23):8563–8568. https://doi.org/10.1073/pnas.0401057101
87. Zhao K, Chai X, Marmorstein R (2003) Structure of a Sir2 substrate, Alba, reveals a mechanism for deacetylation-induced enhancement of DNA binding. J Biol Chem 278 (28):26071–26077. https://doi.org/10.1074/jbc.M303666200
88. Pannek M, Simic Z, Fuszard M, Meleshin M, Rotili D, Mai A, Schutkowski M, Steegborn C (2017) Crystal structures of the mitochondrial deacylase Sirtuin 4 reveal isoform-specific acyl recognition and regulation features. Nat Commun 8(1):1513. https://doi.org/10.1038/s41467-017-01701-2
89. Schiedel M, Robaa D, Rumpf T, Sippl W, Jung M (2018) The current state of NAD(+)-dependent histone deacetylases (sirtuins) as novel therapeutic targets. Med Res Rev 38 (1):147–200. https://doi.org/10.1002/med.21436
90. Rumpf T, Schiedel M, Karaman B, Roessler C, North BJ, Lehotzky A, Olah J, Ladwein KI, Schmidtkunz K, Gajer M, Pannek M, Steegborn C, Sinclair DA, Gerhardt S, Ovadi J, Schutkowski M, Sippl W, Einsle O, Jung M (2015) Selective Sirt2 inhibition by ligand-induced rearrangement of the active site. Nat Commun 6. https://doi.org/10.1038/ncomms7263
91. Neugebauer RC, Sippl W, Jung M (2008) Inhibitors of NAD+ dependent histone deacetylases (sirtuins). Curr Pharm Des 14(6):562–573
92. Trapp J, Jochum A, Meier R, Saunders L, Marshall B, Kunick C, Verdin E, Goekjian P, Sippl W, Jung M (2006) Adenosine mimetics as inhibitors of NAD(+)-dependent histone deacetylases, from kinase to sirtuin inhibition. J Med Chem 49(25):7307–7316. https://doi.org/10.1021/jm060118b
93. Karaman B, Jung M, Sippl W (2016) Chapter 11 – structure-based design and computational studies of sirtuin inhibitors. Epi-informatics. Academic Press, Boston, pp 297–325. https://doi.org/10.1016/B978-0-12-802808-7.00011-3
94. Outeiro TF, Kontopoulos E, Altmann SM, Kufareva I, Strathearn KE, Amore AM, Volk CB, Maxwell MM, Rochet JC, McLean PJ, Young AB, Abagyan R, Feany MB, Hyman BT, Kazantsev AG (2007) Sirtuin 2 inhibitors rescue alpha-synuclein-mediated toxicity in models of Parkinson's disease. Science 317(5837):516–519. https://doi.org/10.1126/science.1143780

95. Abagyan R, Totrov M, Kuznetsov D (1994) ICM – a new method for protein modeling and design: applications to docking and structure prediction from the distorted native conformation. J Comput Chem 15(5):488–506. https://doi.org/10.1002/jcc.540150503
96. Cui HQ, Kamal Z, Ai T, Xu YL, More SS, Wilson DJ, Chen LQ (2014) Discovery of potent and selective sirtuin 2 (SIRT2) inhibitors using a fragment-based approach. J Med Chem 57 (20):8340–8357. https://doi.org/10.1021/jm500777s
97. Teng YB, Jing H, Aramsangtienchai P, He B, Khan S, Hu J, Lin H, Hao Q (2015) Efficient demyristoylase activity of SIRT2 revealed by kinetic and structural studies. Sci Rep 5:8529. https://doi.org/10.1038/srep08529
98. Schiedel M, Rumpf T, Karaman B, Lehotzky A, Olah J, Gerhardt S, Ovadi J, Sippl W, Einsle O, Jung M (2016) Aminothiazoles as potent and selective Sirt2 inhibitors: a structure-activity relationship study. J Med Chem 59(4):1599–1612. https://doi.org/10.1021/acs.jmedchem.5b01517
99. Smith BC, Denu JM (2009) Chemical mechanisms of histone lysine and arginine modifications. Biochim Biophys Acta 1789(1):45–57. https://doi.org/10.1016/j.bbagrm.2008.06.005
100. Qian C, Zhou MM (2006) SET domain protein lysine methyltransferases: structure, specificity and catalysis. Cell Mol Life Sci 63(23):2755–2763. https://doi.org/10.1007/s00018-006-6274-5
101. Boisvert FM, Chenard CA, Richard S (2005) Protein interfaces in signaling regulated by arginine methylation. Sci STKE 2005(271):re2. https://doi.org/10.1126/stke.2712005re2
102. Bedford MT (2007) Arginine methylation at a glance. J Cell Sci 120(24):4243–4246. https://doi.org/10.1242/jcs.019885
103. Pal S, Sif S (2007) Interplay between chromatin remodelers and protein arginine methyltransferases. J Cell Physiol 213(2):306–315. https://doi.org/10.1002/jcp.21180
104. Allis CD, Berger SL, Cote J, Dent S, Jenuwien T, Kouzarides T, Pillus L, Reinberg D, Shi Y, Shiekhattar R, Shilatifard A, Workman J, Zhang Y (2007) New nomenclature for chromatin-modifying enzymes. Cell 131(4):633–636. https://doi.org/10.1016/j.cell.2007.10.039
105. Schneider R, Bannister AJ, Kouzarides T (2002) Unsafe SETs: histone lysine methyltransferases and cancer. Trends Biochem Sci 27(8):396–402. PII: S0968-0004(02)02141-2. https://doi.org/10.1016/S0968-0004(02)02141-2
106. Spannhoff A, Heinke R, Bauer I, Trojer P, Metzger E, Gust R, Schule R, Brosch G, Sippl W, Jung M (2007) Target-based approach to inhibitors of histone arginine methyltransferases. J Med Chem 50(10):2319–2325. https://doi.org/10.1021/jm061250e
107. Wolber G, Langer T (2005) LigandScout: 3-d pharmacophores derived from protein-bound ligands and their use as virtual screening filters. J Chem Inf Model 45(1):160–169. https://doi.org/10.1021/ci049885e
108. Heinke R, Spannhoff A, Meier R, Trojer P, Bauer I, Jung M, Sippl W (2009) Virtual screening and biological characterization of novel histone arginine methyltransferase PRMT1 inhibitors. ChemMedChem 4(1):69–77. https://doi.org/10.1002/cmdc.200800301
109. de Freitas RF, Eram MS, Smil D, Szewczyk MM, Kennedy S, Brown PJ, Santhakumar V, Barsyte-Lovejoy D, Arrowsmith CH, Vedadi M, Schapira M (2016) Discovery of a potent and selective coactivator associated arginine methyltransferase 1 (CARM1) inhibitor by virtual screening. J Med Chem 59(14):6838–6847. https://doi.org/10.1021/acs.jmedchem.6b00668
110. Ye Y, Zhang BD, Mao RF, Zhang CH, Wang YL, Xing J, Liu YC, Luo XM, Ding H, Yang YX, Zhou B, Jiang HL, Chen KX, Luo C, Zheng MY (2017) Discovery and optimization of selective inhibitors of protein arginine methyltransferase 5 by docking-based virtual screening. Org Biomol Chem 15(17):3648–3661. https://doi.org/10.1039/c7ob00070g
111. Kubicek S, O'Sullivan RJ, August EM, Hickey ER, Zhang Q, Teodoro ML, Rea S, Mechtler K, Kowalski JA, Homon CA, Kelly TA, Jenuwein T (2007) Reversal of H3K9me2 by a small-molecule inhibitor for the G9a histone methyltransferase. Mol Cell 25 (3):473–481. https://doi.org/10.1016/j.molcel.2007.01.017
112. Liu F, Chen X, Allali-Hassani A, Quinn AM, Wasney GA, Dong AP, Barsyte D, Kozieradzki I, Senisterra G, Chau I, Siarheyeva A, Kireev DB, Jadhav A, Herold JM, Frye

SV, Arrowsmith CH, Brown PJ, Simeonov A, Vedadi M, Jin J (2009) Discovery of a 2,4-diamino-7-aminoalkoxyquinazoline as a potent and selective inhibitor of histone lysine methyltransferase G9a. J Med Chem 52(24):7950–7953. https://doi.org/10.1021/jm901543m

113. Zhang J, Yao D, Jiang Y, Huang J, Yang S, Wang J (2017) Synthesis and biological evaluation of benzimidazole derivatives as the G9a histone methyltransferase inhibitors that induce autophagy and apoptosis of breast cancer cells. Bioorg Chem 72:168–181. https://doi.org/10.1016/j.bioorg.2017.04.005
114. Feng T, Wang H, Zhang X, Sun H, You Q (2014) The discovery of novel histone lysine methyltransferase G9a inhibitors (part 1): molecular design based on a series of substituted 2,4-diamino-7-aminoalkoxyquinazoline by molecular-docking-guided 3D quantitative structure-activity relationship studies. Med Chem 10(4):426–440
115. Chen WL, Wang ZH, Feng TT, Li DD, Wang CH, Xu XL, Zhang XJ, You QD, Guo XK (2016) Discovery, design and synthesis of 6H-anthra[1,9-cd]isoxazol-6-one scaffold as G9a inhibitor through a combination of shape-based virtual screening and structure-based molecular modification. Bioorg Med Chem 24(22):6102–6108. https://doi.org/10.1016/j.bmc.2016.09.071
116. Cheng D, Yadav N, King RW, Swanson MS, Weinstein EJ, Bedford MT (2004) Small molecule regulators of protein arginine methyltransferases. J Biol Chem 279 (23):23892–23899. https://doi.org/10.1074/jbc.M401853200
117. Zhang X, Yang Z, Khan SI, Horton JR, Tamaru H, Selker EU, Cheng X (2003) Structural basis for the product specificity of histone lysine methyltransferases. Mol Cell 12(1):177–185
118. Mason JS, Morize I, Menard PR, Cheney DL, Hulme C, Labaudiniere RF (1999) New 4-point pharmacophore method for molecular similarity and diversity applications: overview of the method and applications, including a novel approach to the design of combinatorial libraries containing privileged substructures. J Med Chem 42(17):3251–3264. https://doi.org/10.1021/jm9806998
119. Meng F, Cheng S, Ding H, Liu S, Liu Y, Zhu K, Chen S, Lu J, Xie Y, Li L, Liu R, Shi Z, Zhou Y, Liu YC, Zheng M, Jiang H, Lu W, Liu H, Luo C (2015) Discovery and optimization of novel, selective histone methyltransferase SET7 inhibitors by pharmacophore- and docking-based virtual screening. J Med Chem 58(20):8166–8181. https://doi.org/10.1021/acs.jmedchem.5b01154
120. Shi YJ, Lan F, Matson C, Mulligan P, Whetstine JR, Cole PA, Casero RA, Shi Y (2004) Histone demethylation mediated by the nuclear amine oxidase homolog LSD1. Cell 119 (7):941–953. https://doi.org/10.1016/j.cell.2004.12.012
121. Tsukada Y, Fang J, Erdjument-Bromage H, Warren ME, Borchers CH, Tempst P, Zhang Y (2006) Histone demethylation by a family of JmjC domain-containing proteins. Nature 439 (7078):811–816. https://doi.org/10.1038/nature04433
122. Wang GG, Allis CD, Chi P (2007) Chromatin remodeling and cancer, part I: covalent histone modifications. Trends Mol Med 13(9):363–372. https://doi.org/10.1016/j.molmed.2007.07.003
123. Lynch JT, Harris WJ, Somervaille TC (2012) LSD1 inhibition: a therapeutic strategy in cancer? Expert Opin Ther Targets 16(12):1239–1249. https://doi.org/10.1517/14728222.2012.722206
124. Johansson C, Tumber A, Che K, Cain P, Nowak R, Gileadi C, Oppermann U (2014) The roles of Jumonji-type oxygenases in human disease. Epigenomics 6(1):89–120. https://doi.org/10.2217/epi.13.79
125. Chen Y, Yang Y, Wang F, Wan K, Yamane K, Zhang Y, Lei M (2006) Crystal structure of human histone lysine-specific demethylase 1 (LSD1). Proc Natl Acad Sci U S A 103 (38):13956–13961. https://doi.org/10.1073/pnas.0606381103

126. Stavropoulos P, Blobel G, Hoelz A (2006) Crystal structure and mechanism of human lysine-specific demethylase-1. Nat Struct Mol Biol 13(7):626–632. https://doi.org/10.1038/nsmb1113
127. Yang M, Culhane JC, Szewczuk LM, Gocke CB, Brautigam CA, Tomchick DR, Machius M, Cole PA, Yu H (2007) Structural basis of histone demethylation by LSD1 revealed by suicide inactivation. Nat Struct Mol Biol 14(6):535–539. https://doi.org/10.1038/nsmb1255
128. Forneris F, Battaglioli E, Mattevi A, Binda C (2009) New roles of flavoproteins in molecular cell biology: histone demethylase LSD1 and chromatin. FEBS J 276(16):4304–4312. https://doi.org/10.1111/j.1742-4658.2009.07142.x
129. Hojfeldt JW, Agger K, Helin K (2013) Histone lysine demethylases as targets for anticancer therapy. Nat Rev Drug Discov 12(12):917–930. https://doi.org/10.1038/nrd4154
130. Schmitt ML, Hauser AT, Carlino L, Pippel M, Schulz-Fincke J, Metzger E, Willmann D, Yiu T, Barton M, Schule R, Sippl W, Jung M (2013) Nonpeptidic propargylamines as inhibitors of lysine specific demethylase 1 (LSD1) with cellular activity. J Med Chem 56(18):7334–7342. https://doi.org/10.1021/jm400792m
131. Sorna V, Theisen ER, Stephens B, Warner SL, Bearss DJ, Vankayalapati H, Sharma S (2013) High-throughput virtual screening identifies novel N′-(1-phenylethylidene)-benzohydrazides as potent, specific, and reversible LSD1 inhibitors. J Med Chem 56(23):9496–9508. https://doi.org/10.1021/jm400870h
132. Zhou C, Kang D, Xu YG, Zhang LY, Zha XM (2015) Identification of novel selective lysine-specific demethylase 1 (LSD1) inhibitors using a pharmacophore-based virtual screening combined with docking. Chem Biol Drug Des 85(6):659–671. https://doi.org/10.1111/cbdd.12461
133. Zhou C, Wu F, Lu L, Wei L, Pai E, Yao Y, Song Y (2017) Structure activity relationship and modeling studies of inhibitors of lysine specific demethylase 1. Plos One 12(2):e0170301. https://doi.org/10.1371/journal.pone.0170301
134. Pieroni M, Annunziato G, Azzali E, Dessanti P, Mercurio C, Meroni G, Trifiró P, Vianello P, Villa M, Beato C, Varasi M, Costantino G (2015) Further insights into the SAR of α-substituted cyclopropylamine derivatives as inhibitors of histone demethylase KDM1A. Eur J Med Chem 92:377–386. https://doi.org/10.1016/j.ejmech.2014.12.032
135. Mould DP, Bremberg U, Jordan AM, Geitmann M, McGonagle AE, Somervaille TCP, Spencer GJ, Ogilvie DJ (2017) Development and evaluation of 4-(pyrrolidin-3-yl)benzonitrile derivatives as inhibitors of lysine specific demethylase 1. Bioorg Med Chem Lett 27(20):4755–4759. https://doi.org/10.1016/j.bmcl.2017.08.052
136. Xi J, Xu S, Wu L, Ma T, Liu R, Liu Y-C, Deng D, Gu Y, Zhou J, Lan F, Zha X (2017) Design, synthesis and biological activity of 3-oxoamino-benzenesulfonamides as selective and reversible LSD1 inhibitors. Bioorg Chem 72:182–189. https://doi.org/10.1016/j.bioorg.2017.04.006
137. Han C, Li Z, Hou J, Wang Z, Xu D, Xue G, Kong L (2018) Bioactivity evaluation of natural product α-mangostin as a novel xanthone-based lysine-specific demethylase 1 inhibitor to against tumor metastasis. Bioorg Chem 76:415–419. https://doi.org/10.1016/j.bioorg.2017.12.004
138. Wang S, Zhao L-J, Zheng Y-C, Shen D-D, Miao E-F, Qiao X-P, Zhao L-J, Liu Y, Huang R, Yu B, Liu H-M (2017) Design, synthesis and biological evaluation of [1,2,4]triazolo[1,5-a]pyrimidines as potent lysine specific demethylase 1 (LSD1/KDM1A) inhibitors. Eur J Med Chem 125:940–951. https://doi.org/10.1016/j.ejmech.2016.10.021
139. Duan Y-C, Guan Y-Y, Zhai X-Y, Ding L-N, Qin W-P, Shen D-D, Liu X-Q, Sun X-D, Zheng Y-C, Liu H-M (2017) Discovery of resveratrol derivatives as novel LSD1 inhibitors: design, synthesis and their biological evaluation. Eur J Med Chem 126:246–258. https://doi.org/10.1016/j.ejmech.2016.11.035
140. Duan Y-C, Ma Y-C, Qin W-P, Ding L-N, Zheng Y-C, Zhu Y-L, Zhai X-Y, Yang J, Ma C-Y, Guan Y-Y (2017) Design and synthesis of tranylcypromine derivatives as novel LSD1/

HDACs dual inhibitors for cancer treatment. Eur J Med Chem 140:392–402. https://doi.org/10.1016/j.ejmech.2017.09.038
141. Schulz-Fincke J, Hau M, Barth J, Robaa D, Willmann D, Kürner A, Haas J, Greve G, Haydn T, Fulda S, Lübbert M, Lüdeke S, Berg T, Sippl W, Schüle R, Jung M (2018) Structure-activity studies on N-substituted tranylcypromine derivatives lead to selective inhibitors of lysine specific demethylase 1 (LSD1) and potent inducers of leukemic cell differentiation. Eur J Med Chem 144:52–67. https://doi.org/10.1016/j.ejmech.2017.12.001
142. Mould DP, Alli C, Bremberg U, Cartic S, Jordan AM, Geitmann M, Maiques-Diaz A, McGonagle AE, Somervaille TCP, Spencer GJ, Turlais F, Ogilvie D (2017) Development of (4-cyanophenyl)glycine derivatives as reversible inhibitors of lysine specific demethylase 1. J Med Chem 60(19):7984–7999. https://doi.org/10.1021/acs.jmedchem.7b00462
143. Li Z-H, Liu X-Q, Geng P-F, Suo F-Z, Ma J-L, Yu B, Zhao T-Q, Zhou Z-Q, Huang C-X, Zheng Y-C, Liu H-M (2017) Discovery of [1,2,3]triazolo[4,5-d]pyrimidine derivatives as novel LSD1 inhibitors. ACS Med Chem Lett 8(4):384–389. https://doi.org/10.1021/acsmedchemlett.6b00423
144. Mould DP, Bremberg U, Jordan AM, Geitmann M, Maiques-Diaz A, McGonagle AE, Small HF, Somervaille TCP, Ogilvie D (2017) Development of 5-hydroxypyrazole derivatives as reversible inhibitors of lysine specific demethylase 1. Bioorg Med Chem Lett 27(14):3190–3195. https://doi.org/10.1016/j.bmcl.2017.05.018
145. Ding L, Wang Z-Z, Sun X-D, Yang J, Ma C-Y, Li W, Liu H-M (2017) 3D-QSAR (CoMFA, CoMSIA), molecular docking and molecular dynamics simulations study of 6-aryl-5-cyano-pyrimidine derivatives to explore the structure requirements of LSD1 inhibitors. Bioorg Med Chem Lett 27(15):3521–3528. https://doi.org/10.1016/j.bmcl.2017.05.065
146. Sartori L, Mercurio C, Amigoni F, Cappa A, Fagá G, Fattori R, Legnaghi E, Ciossani G, Mattevi A, Meroni G, Moretti L, Cecatiello V, Pasqualato S, Romussi A, Thaler F, Trifiró P, Villa M, Vultaggio S, Botrugno OA, Dessanti P, Minucci S, Zagarrí E, Carettoni D, Iuzzolino L, Varasi M, Vianello P (2017) Thieno[3,2-b]pyrrole-5-carboxamides as new reversible inhibitors of histone lysine demethylase KDM1A/LSD1. Part 1: high-throughput screening and preliminary exploration. J Med Chem 60(5):1673–1692. https://doi.org/10.1021/acs.jmedchem.6b01018
147. Vianello P, Sartori L, Amigoni F, Cappa A, Fagá G, Fattori R, Legnaghi E, Ciossani G, Mattevi A, Meroni G, Moretti L, Cecatiello V, Pasqualato S, Romussi A, Thaler F, Trifiró P, Villa M, Botrugno OA, Dessanti P, Minucci S, Vultaggio S, Zagarrí E, Varasi M, Mercurio C (2017) Thieno[3,2-b]pyrrole-5-carboxamides as new reversible inhibitors of histone lysine demethylase KDM1A/LSD1. Part 2: structure-based drug design and structure–activity relationship. J Med Chem 60(5):1693–1715. https://doi.org/10.1021/acs.jmedchem.6b01019
148. Wu F, Zhou C, Yao Y, Wei L, Feng Z, Deng L, Song Y (2016) 3-(Piperidin-4-ylmethoxy)pyridine containing compounds are potent inhibitors of lysine specific demethylase 1. J Med Chem 59(1):253–263. https://doi.org/10.1021/acs.jmedchem.5b01361
149. Clissold PM, Ponting CP (2001) JmjC: cupin metalloenzyme-like domains in jumonji, hairless and phospholipase A2beta. Trends Biochem Sci 26(1):7–9
150. Aik WS, Chowdhury R, Clifton IJ, Hopkinson RJ, Leissing T, McDonough MA, Nowak R, Schofield CJ, Walport LJ (2015) Chapter 2 introduction to structural studies on 2-oxoglutarate-dependent oxygenases and related enzymes. 2-Oxoglutarate-dependent oxygenases. The Royal Society of Chemistry, Cambridge, pp 59–94. https://doi.org/10.1039/9781782621959-00059
151. Hoffmann I, Roatsch M, Schmitt ML, Carlino L, Pippel M, Sippl W, Jung M (2012) The role of histone demethylases in cancer therapy. Mol Oncol 6(6):683–703. https://doi.org/10.1016/j.molonc.2012.07.004
152. Hou H, Yu H (2010) Structural insights into histone lysine demethylation. Curr Opin Struct Biol 20(6):739–748. https://doi.org/10.1016/j.sbi.2010.09.006

153. Hauser A-T, Roatsch M, Schulz-Fincke J, Robaa D, Sippl W, Jung M (2015) Chapter 18 – discovery of histone demethylase inhibitors. Epigenetic technological applications. Academic Press, Boston, pp 397–424. https://doi.org/10.1016/B978-0-12-801080-8.00018-1
154. Chu CH, Wang LY, Hsu KC, Chen CC, Cheng HH, Wang SM, Wu CM, Chen TJ, Li LT, Liu RW, Hung CL, Yang JM, Kung HJ, Wang WC (2014) KDM4B as a target for prostate cancer: structural analysis and selective inhibition by a novel inhibitor. J Med Chem 57 (14):5975–5985. https://doi.org/10.1021/jm500249n
155. Roatsch M, Robaa D, Pippel M, Nettleship JE, Reddivari Y, Bird LE, Hoffmann I, Franz H, Owens RJ, Schule R, Flaig R, Sippl W, Jung M (2016) Substituted 2-(2-aminopyrimidin-4-yl) pyridine-4-carboxylates as potent inhibitors of JumonjiC domain-containing histone demethylases. Future Med Chem 8(13):1553–1571. https://doi.org/10.4155/fmc.15.188
156. Franci G, Sarno F, Nebbioso A, Altucci L (2017) Identification and characterization of PKF118-310 as a KDM4A inhibitor. Epigenetics 12(3):198–205. https://doi.org/10.1080/15592294.2016.1249089
157. Korczynska M, Le DD, Younger N, Gregori-Puigjane E, Tumber A, Krojer T, Velupillai S, Gileadi C, Nowak RP, Iwasa E, Pollock SB, Ortiz Torres I, Oppermann U, Shoichet BK, Fujimori DG (2016) Docking and linking of fragments to discover Jumonji histone demethylase inhibitors. J Med Chem 59(4):1580–1598. https://doi.org/10.1021/acs.jmedchem.5b01527
158. Fang Z, Wang TQ, Li H, Zhang G, Wu XA, Yang L, Peng YL, Zou J, Li LL, Xiang R, Yang SY (2017) Discovery of pyrazolo[1,5-a]pyrimidine-3-carbonitrile derivatives as a new class of histone lysine demethylase 4D (KDM4D) inhibitors. Bioorg Med Chem Lett 27 (14):3201–3204. https://doi.org/10.1016/j.bmcl.2017.05.002
159. Horton JR, Liu X, Gale M, Wu L, Shanks JR, Zhang X, Webber PJ, Bell JS, Kales SC, Mott BT, Rai G, Jansen DJ, Henderson MJ, Urban DJ, Hall MD, Simeonov A, Maloney DJ, Johns MA, Fu H, Jadhav A, Vertino PM, Yan Q, Cheng X (2016) Structural basis for KDM5A histone lysine demethylase inhibition by diverse compounds. Cell Chem Biol 23(7):769–781. https://doi.org/10.1016/j.chembiol.2016.06.006
160. Wu XA, Fang Z, Yang B, Zhong L, Yang QY, Zhang CH, Huang SZ, Xiang R, Suzuki T, Li LL, Yang SY (2016) Discovery of KDM5A inhibitors: homology modeling, virtual screening and structure-activity relationship analysis. Bioorg Med Chem Lett 26(9):2284–2288. https://doi.org/10.1016/j.bmcl.2016.03.048

Top Med Chem (2020) 33: 487–522
DOI: 10.1007/7355_2019_77

Published online: 26 October 2019

Experimental Methodologies for Detection and Mapping of Epigenetic DNA Marks

Christopher L. Seiler, Jenna Fernandez, Qiyuan Han, and Natalia Y. Tretyakova

Contents

1 Introduction 489
2 Methods for Detection and Quantification of Epigenetic Modifications Within the Genome 492
2.1 Initial Detection of 5mC and Its Oxidized Forms via Paper Chromatography and Thin-Layer Chromatography 492
2.2 Antibody-Based Detection of Epigenetic Marks in DNA 494
2.3 High Performance Liquid Chromatography: Mass Spectrometry 496
3 Indirect Methods to Map Epigenetic DNA Modifications Along the Genome 499
3.1 Bisulfite Conversion to Distinguish Between C and 5mC 499
3.2 Microarray Technology for Sequencing 501
3.3 Next-Generation Sequencing 503
3.4 Antibody/Enrichment-Based Sequencing 503
3.5 Whole-Genome Bisulfite Sequencing 505
3.6 Reduced Representation Bisulfite Sequencing (RRBS) 505
3.7 Modified Bisulfite Sequencing Methods for Detection of 5hmC, 5fC, and 5caC ... 506
4 Direct Detection of Epigenetic Modifications 511
4.1 Single-Molecule Real-Time (SMRT) Sequencing 511
4.2 Nanopore Sequencing 512
5 Conclusions 512
References 513

Abstract Epigenetic marks of DNA such as 5-methylcytosine control the levels of gene expression in cells and tissues by altering DNA structure, inhibiting transcription factor binding, and recruiting chromatin-modifying enzymes. DNA methylation marks are introduced in both strands of DNA by DNA methyltransferases, allowing for heritable gene silencing. In addition to 5-methylcytosine, cells contain its oxidized forms 5-hydroxymethylcytosine, 5-formylcytosine, and 5-carboxylcytosine, as well as recently discovered DNA marks, 4-methylcytosine and 6-methyladenosine.

C. L. Seiler, J. Fernandez, Q. Han, and N. Y. Tretyakova (✉)
Department of Medicinal Chemistry and the Masonic Cancer Center, University of Minnesota, Minneapolis, MN, USA
e-mail: trety001@umn.edu

Structural identification, quantitation, and mapping of DNA epigenetic marks are critical for our understanding of epigenetic control in healthy cells and to allow for insight into epigenetic deregulation in human disease and the efficacy of epigenetic therapies. This review focuses on experimental techniques and methodologies that can be used to study epigenetic modifications of DNA, including methods for quantifying global levels of epigenetically modified nucleosides and sequencing techniques for mapping the locations of various epigenetic marks along the genome, leading to better understanding of their biological functions and dynamics.

Keywords Direct detection, Epigenetics, Mass spectrometry, Microarray, Quantitation, Sequencing

Abbreviations

2D-LC	Two-dimensional HPLC
4mC	*N*^{4}-methyl-2′-deoxyadenosine
5caC	5-Carboxylcytosine
5fC	5-Formylcytosine
5hmC	5-Hydroxymethylcytosine
5mC	5-Methylcytosine
6mA-RE-Seq	*N*6MedA restriction enzyme sequencing
anti-CMS-Seq	Anti-cytosine-5-methylenesulfonate sequencing
ARP	Aldehyde reactive probe
caMAB-Seq	Carboxylcytosine modification-assisted bisulfite sequencing
CGIs	CpG islands
CMS	Cytosine-5-methylenesulfonate
DNMT	DNA methyltransferase
ELISA	Enzyme-linked immunosorbent assay
ESI	Electrospray ionization
fCAB-Seq	Formylcytosine chemical-assisted bisulfite sequencing
GLIB-Seq	Glucosylation, periodate oxidation, biotinylation sequencing
hMeSeal-Seq	Hydroxymethyl selective chemical labeling sequencing
HPLC	High-performance liquid chromatography
IP	Immunoprecipitation
LOD	Limit of detection
LOQ	Limit of quantitation
MAB-Seq	Modification-assisted bisulfite sequencing
MeDIP	Methylated DNA immunoprecipitation
MS	Mass spectrometry
MS/MS	Tandem mass spectrometry
MSO	Methylation-specific oligonucleotide
*N*6MedA	*N*^{6}-methyl-2′-deoxyadenosine

oxBS-Seq	Oxidative bisulfite sequencing
redBS-Seq	Reduced bisulfite sequencing
RRBS	Reduced representation bisulfite sequencing
SAM	S-adenosyl-L-methionine
SMRT	Single-molecule real-time
SNPs	Single-nucleotide polymorphisms
TAB	TET-assisted bisulfite
TAB-Seq	TET-assisted bisulfite sequencing
TET	Ten-eleven translocation protein
TLC	Thin-layer chromatography
WGBS	Whole-genome bisulfite sequencing

1 Introduction

The human genome contains three billion base pairs [1]. Of these, nearly 27 million cytosines exist in the CpG sequence context, making them potential targets for DNA methyltransferase (DNMT) activity [2]. 5-Methylcytosine (5mC in Fig. 1) is a key epigenetic modification of DNA with important roles in gene regulation, genetic imprinting, X-chromosome inactivation, and genome stabilization [3]. Methylated CpG sites are recognized by methyl-CpG binding proteins, leading to the recruitment of histone deacetylases, chromatin remodeling, and, in most cases, reduced levels of gene expression [4, 5]. DNMTs introduce 5mC into the genome by catalyzing the transfer of a methyl group from *S*-adenosyl-L-methionine (SAM) to the C-5 position of cytosine in DNA [5, 6]. DNMT3 A/B are responsible for de novo DNA methylation, a process mediated by DNMT3L scaffolding protein [7]. Once established, DNA methylation patterns are faithfully maintained by specialized DNA methyltransferase DNMT1, which recognizes hemi-methylated CpG sites and methylates the newly synthesized strand [5]. This preserves cell identity and maintains tissue-specific gene expression patterns following semiconservative DNA replication.

Recent studies have shown that cytosine methylation marks are reversible. Ten-eleven translocation family of proteins (TET) oxidize 5mC to generate

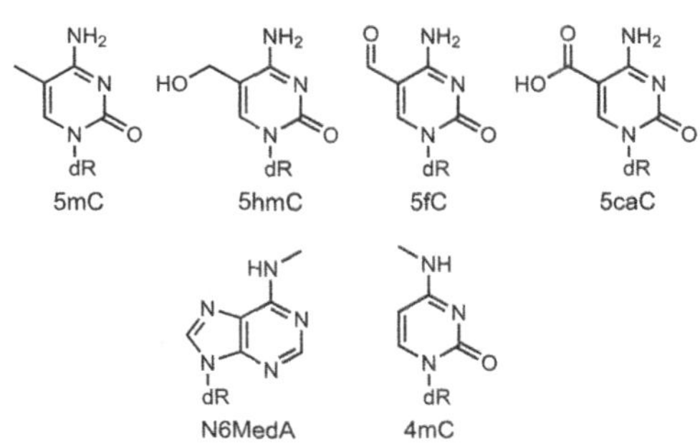

Fig. 1 Chemical structures of epigenetic modifications in DNA

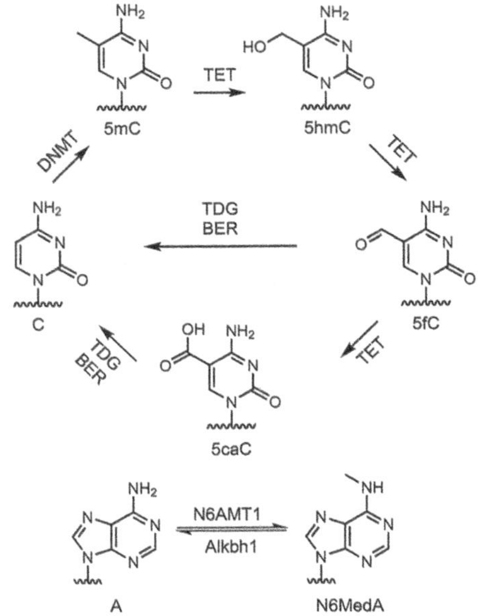

Fig. 2 Enzymatic pathways for the introduction of 5mC, 5hmC, 5fC, 5caC, and *N*6MedA into DNA. Cytosine is methylated by DNA methyltransferase in the presence of S-adenosylmethionine. 5mC can then be oxidized by the TET family of enzymes to produce 5mC, 5hmC, 5fC, and 5caC. *N*6MedA is introduced by *N*6AMT1 and can be demethylated by Alkbh1

5-hydroxymethylcytosine (5hmC), 5-formylcytosine (5fC), and 5-carboxylcytosine (5caC) (Figs. 1 and 2) [8–11]. This leads to passive DNA demethylation since DNMT1 activity is reduced or inhibited at hemi-modified CpG sites [12–14]. Additionally, both 5fC and 5caC can be removed via the base excision repair pathway [11], and all oxidized forms of 5mC can be removed by SRAP1 [15], leading to active DNA demethylation. In addition to their roles in DNA demethylation, oxidized forms of 5mC have been proposed to fulfil their own epigenetic functions by directly participating in epigenetic signaling [16–19]. Mass spectrometry-based proteomics studies have identified a number of proteins that specifically interact with 5mC, 5hmC, 5fC, and 5caC, suggesting that each of these epigenetic marks is recognized by a unique set of "readers" [20, 21]. The distributions of 5mC, 5hmC, 5fC, and 5caC bases along the genome are suggestive of their distinct roles in gene regulation [22]. For instance, 5hmC is preferentially enriched at distal regulatory elements such as enhancers in mouse or human embryonic stem cells, whereas 5fC or 5caC tend to accumulate at poised enhancers and promoters [22].

5mC and its oxidized forms are present in all mammalian tissues examined, but in different amounts. 5mC makes up 4.30 $\pm$ 0.22% of dG [23]. 5hmC levels vary between 0.03 and 0.69% of dG depending on tissue type (Fig. 3) [23]. In most tissues, 5fC and 5caC are present in much lower amounts, with 5fC concentrations approximately three orders of magnitude lower than the levels of 5hmC and the levels of 5caC approximately tenfold less than the amount of 5fC [24].

More recently, several additional epigenetic modifications of DNA were discovered including N^6-methyl-2′-deoxyadenosine (*N*6MedA) and N^4-methyl-2′-deoxycytidine (4mC) (Fig. 1). In prokaryotes, *N*6MedA and 4mC are the most prevalent DNA

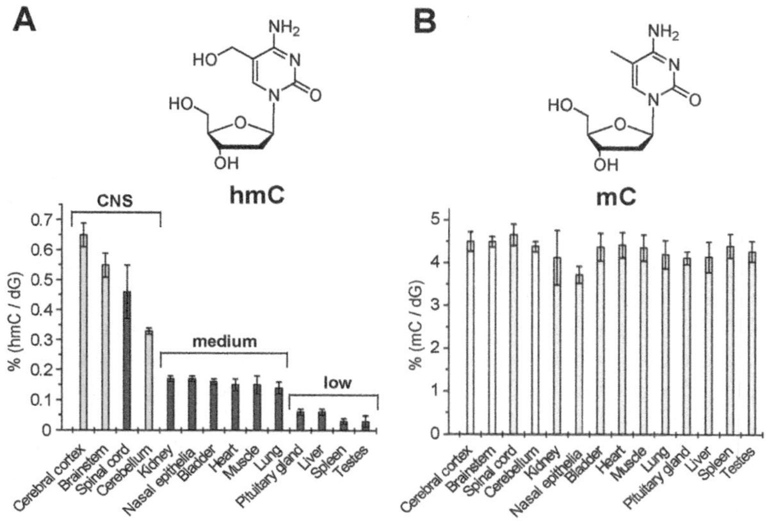

Fig. 3 Tissue-specific amounts of 5mC and 5hmC determined by mass spectrometry. (**a**) Values of 5hmC in percentage of dG. Levels of 5hmC have been shown to be tissue-dependent with the highest levels of 5hmC observed in tissues corresponding to the central nervous system. The lowest levels of 5hmC are observed in the pituitary gland, liver, spleen, and testes. This roughly correlates with proliferation rates, with higher levels of 5hmC found in tissues with the lowest proliferation rates. (**b**) Values of 5mC in percentage of dG. Levels of 5mC appear to stay relatively consistent across all tissue types. Adapted from Globisch D et al. (2010). Tissue distribution of 5-hydroxymethylcytosine and search for active demethylation intermediates. PLoS One. Dec 23;5(12):e15367. doi: https://doi.org/10.1371/journal.pone.0015367. Used in accordance with CC BY license (http://creativecommons.org/licenses/by/4.0/) [23]

modifications that are primarily used for distinguishing host DNA from foreign pathogenic DNA [25, 26]. First indirect evidence for the presence of *N*6MedA in mammals dates to 1983 [27]. Although subsequent studies by Ratel et al. were unable to confirm the presence of *N*6MedA in mammals [28], more recent reports confirmed the presence of *N*6MedA in mouse and human genomes [26, 29] and identified the methylase that creates *N*6MedA, *N*6AMT1 [30]. *N*6MedA was shown to be widely distributed across the genome, but depleted in exonic regions [26]. Furthermore, *N*6MedA is preferentially found at TAGG sites and has been implicated in gene repression [26, 29]. Many features of *N*6MedA in mammals remain to be investigated, including the identities of possible *N*6MedA-interacting proteins [26].

N^4-methyl-2′-deoxycytidine (4mC, Fig. 1) was first detected in the DNA of thermophilic bacteria and bacterial mesophiles [31]. 4mC is generated by the *N*-4 cytosine-specific DNA methyltransferase which specifically methylates the amino group at the *N*-4 position of cytosine [32]. 4mC is a part of a bacterial restriction-modification system used to defend against foreign DNA, with several 4mC methyltransferases and 4mC-sensitive restriction endonucleases identified in various bacterial strains [33]. However, additional biological functions of 4mC have not been fully elucidated, and its relevance in eukaryotes remains to be established. In order to better understand the biological functions of 4mC, it is necessary to map this epigenetic mark along the genome.

Epigenetic deregulation is increasingly recognized as a critical event in the development of cancer and other diseases [34]. Changes in DNA methylation patterns can lead to silencing of tumor suppressor genes and activation of proto-oncogenes, which represents an early critical event in tumor development [34]. Silencing of gene expression is correlated with hypermethylation of promoter regions [3]. Aberrant cytosine methylation is implicated in many other human diseases including asthma, autism, and amyotrophic lateral sclerosis [35–37]. DNMT inhibitors such as 5-azacytidine, 5-aza-2′-deoxycytidine (decitabine), and pyrimidin-2-one ribonucleoside (zebularine) are widely used as single therapeutic agents and in combination with classical chemotherapeutic agents in treatment of various tumors [38–41].

On the other hand, the levels of 5hmC are substantially lowered in human cancers, with a 50–90% reduction of 5hmC in breast cancer, liver cancer, lung cancer, and several other tumor types [42]. In contrast, elevated levels of 5hmC have been observed in patients with systemic lupus erythematosus [43], autism [36], and bronchial asthma [44]. Recently, *N*6MedA concentrations in mouse brain have been shown to be mediated by environmental stress [45], although the biological significance of this finding remains to be established.

2 Methods for Detection and Quantification of Epigenetic Modifications Within the Genome

2.1 Initial Detection of 5mC and Its Oxidized Forms via Paper Chromatography and Thin-Layer Chromatography

Paper chromatography and thin-layer chromatography experiments played a key role in the initial discovery of epigenetically modified DNA bases [46]. In paper chromatography, a compound mixture is spotted on a strip of chromatography paper, which is then hung in a development chamber, and dipped into a solution of volatile alcohols in which to develop the chromatogram. Compounds that are more soluble in the alcohol mobile phase move up the paper strip during development. Paper chromatography was used for the initial detection of 5mC in eukaryotic DNA [46]. Calf thymus DNA was hydrolyzed with acid, and the resulting free bases were separated by paper chromatography using butanol. The resulting chromatographic spots were extracted with alcohol and identified by comparing their UV spectra to those of standard nitrogenous bases. Paper chromatography has since been replaced by newer techniques such as thin-layer chromatography (TLC).

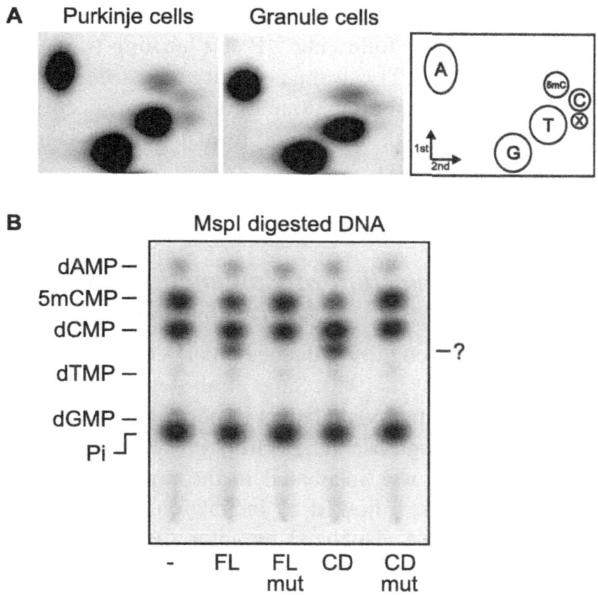

Fig. 4 Identification of 5hmC in brain cells and its formation by TET1 via ^{32}P TLC. (**a**) The 2D-TLC of Purkinje cells and granule cells indicated a new spot marked X, which comigrates with 5hmC monophosphate. From Kriaucionis, S., and Heintz, N. (2009). *The nuclear DNA base 5-hydroxymethylcytosine is present in Purkinje neurons and the brain.* Science 324, 929–930. Reprinted with permission from AAAS [47]. (**b**) DNA was isolated from HEK293 cells after transfection with TET1 plasmid. DNA was digested with MspI, 32P end labeled, and digested to nucleosides. HEK293 DNA revealed an additional nucleotide in the presence of TET1, both full length (FL) and catalytic domain (CD), but not in the mutants of either. From Tahiliani, M. et al. (2009). Conversion of 5-methylcytosine to 5-hydroxymethylcytosine in mammalian DNA by MLL partner TET1. Science 324, 930–935 [8]. Reprinted with permission from AAAS

TLC was instrumental for the initial discovery of 5hmC in Purkinje cells and cerebellum tissue [47] and for the identification of TET family of enzymes responsible for the oxidation of 5mC to 5hmC [8]. To determine the components of DNA by TLC, the DNA was digested to nucleosides and radiolabeled with ^{32}P, followed by 2D-TLC analysis, in which an additional spot was observed on the TLC plate [47]. A new TLC spot, corresponding to 5hmC, was detected in both cell types, but was twice as abundant in Purkinje cells as compared to granule cells [47]. TLC was also used to identify 5hmC as the oxidation product of 5mC in HEK293 cell DNA [8]. Following transfection of HEK293 cells with a plasmid-containing TET1, cellular DNA was isolated and cleaved with MspI. Following radiolabeling with ^{32}P, DNA fragments were enzymatically digested to nucleotides, which were run out on a cellulose TLC plate using a mixture of isobutyric acid, water, and ammonia and imaged using a phosphoimager [8]. The TLC showed a small spot that was only visible when cells had been transfected with TET1 and cleaved with MspI (Fig. 4b) [8].

The strength of TLC in studies aiming to identify new epigenetic modifications of DNA lies in its high sensitivity following ^{32}P end labeling of native and modified nucleosides. A key weakness of TLC is the necessity for a secondary confirmation of the identified spot, which is most often done through mass spectrometry. Additionally, TLC is not scalable to a high-throughput format and thus is not well suited for large studies.

2.2 Antibody-Based Detection of Epigenetic Marks in DNA

Antibodies that specifically recognize epigenetic marks of DNA are commonly used to quantify their global levels. In 1980, Sano et al. developed an antibody against 5mC which could recognize methylated sites in purified DNA [48]. In brief, DNA fragments were separated by agarose gel electrophoresis and transferred to nitrocellulose paper. The membrane was incubated in the presence of purified antibody against 5mC, raised in rabbits, followed by incubation with ^{125}I-labeled goat anti-rabbit IgG. The 5mC was then visualized by autoradiography. This methodology was capable of detecting 20 fmol of 5mC in immobilized restriction fragments of DNA [48]. A more sensitive 5mC antibody, as well as an antibody specific for *N*6MedA, was developed a few years later and utilized for their detection in human and *Drosophila* DNA [27]. This more sensitive 5mC antibody enabled the detection of a minimum of 5 fmol amounts in nanogram quantities of DNA [27].

Today, antibody-based detection of DNA epigenetic marks is a commonly used technique, and antibodies for many epigenetic DNA modifications are commercially available. These antibodies have been used in an enzyme-linked immunosorbent assay (ELISA) [49, 50] (Fig. 5). Commercially available ELISA-based kits for global DNA methylation profiling are available from a number of companies including Zymo and EpiGentek. Generally, these kits capture DNA on an ELISA plate. 5mC is detected through sequential incubations with a primary antibody against 5mC, followed by a secondary antibody linked to luciferase or a similar enzyme, and finally using colorimetric or fluorometric detection reagents [49]. The development of antibodies specific for 5hmC and 5fC allowed for their quantitation using similar commercially available kits (e.g., EpiGentek MethylFlash Global DNA Hydroxymethylation (5-hmC) ELISA Easy Kit; Active Motif Global 5-hmC DNA Quantification Kit; EpiGentek MethylFlash 5-Formylcytosine (5-fC) DNA Quantification Kit). These ELISA-based assays are easy to perform, employ simple equipment that is readily available at most research laboratories, and work well for monitoring relative changes in global DNA methylation. However, it should be noted that ELISA-based methods are prone to high variability, do not allow for absolute quantitation, and are best suited for detecting large changes in global levels of 5hmC or 5fC.

Dot blotting is another simple and rapid antibody-based method to quantify global levels of epigenetic modifications of DNA. In this method, genomic DNA is spotted on a membrane. The membrane is sequentially incubated in blocking

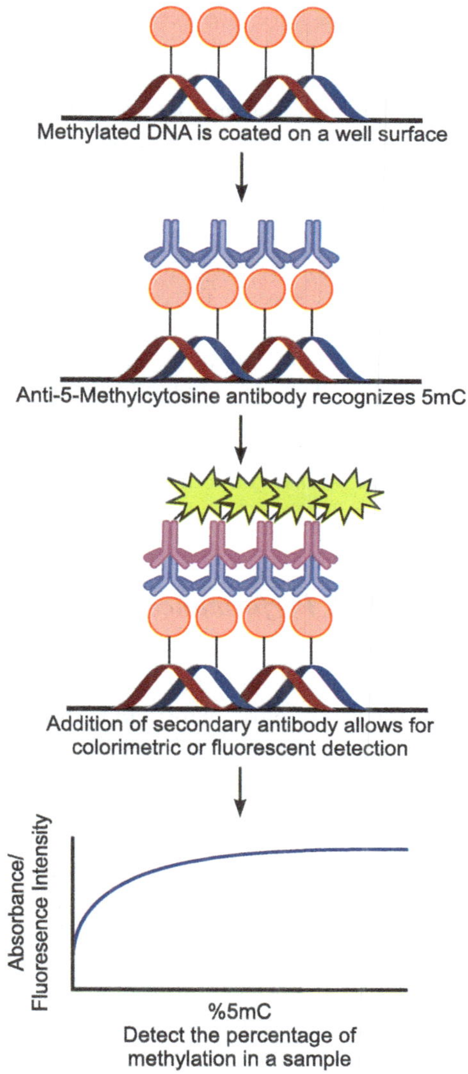

Fig. 5 Scheme of ELISA methodology. In an ELISA the methylated DNA is coated onto the surface of a well plate and incubated with a 5mC-specific antibody. Following binding of the primary antibody, a secondary antibody toward the first is added which is conjugated to a dye or fluorophore. The amount of methylation can then be visualized by the level of absorbance or fluorescence intensity

buffer, antibody solution, and rinsing buffer. Using an antibody specific for epigenetic modifications of interest allows for determination of overall global genomic levels. After chemiluminescence imaging, the intensity of the signal can be used to compare the levels of epigenetic modifications between samples. Generation of antibodies specific for different epigenetic modifications has permitted the quantification of 5mC, 5hmC, 5fC, 5caC, and *N*6MedA in both human and mouse tissues by dot blot [26, 51, 52]. As is the case for ELISA-based detection, dot blots do not allow for absolute quantification and suffer from high variability.

Antibodies have also been used to elucidate the biological functions of DNA epigenetic marks by studying their tissue-specific levels and distribution. Global levels of epigenetic DNA marks can be evaluated by staining cells or tissues with a specific monoclonal antibody [53]. Globisch et al. utilized immunostaining experiments with a commercially available 5hmC-specific antibody to map the distribution of 5hmC in cells and tissues. They found that 5hmC was localized in the cell nuclei and that virtually all cells contained 5hmC. Furthermore, the use of immunostaining allowed these authors to confirm that the highest levels of 5hmC are found in the brain, followed by kidney and nasal epithelia [23].

Antibodies specific for certain epigenetic marks can also be used for affinity enrichment of DNA regions. Antibodies can be used to immunoprecipitate modification-containing DNA fragments, allowing for their enrichment. This technique can then be coupled to methylation-specific oligonucleotide (MSO) microarrays, locus-specific PCR, or whole-genome approaches such as next-generation sequencing or microarrays, to enable for low-resolution mapping of epigenetic marks within the genome (see Sect. 3.2).

2.3 *High Performance Liquid Chromatography: Mass Spectrometry*

Global levels of 5mC, 5hmC, 5fC, 5caC, *N*6MedA, and 4mC in genomic DNA can be accurately quantified by isotope dilution liquid chromatography-mass spectrometry. In this approach, DNA is enzymatically digested using nucleases and phosphatases, and the resulting nucleosides are analyzed by HPLC-ESI-MS/MS. HPLC with UV detection can be used for measuring 5mC in biological samples due to its high content (4–5% of all cytosine bases) [54], while other epigenetic modifications (5hmC, 5fC, 5caC, N6MedA, and 4mC) are significantly less common and thus require more sensitive methods such as HPLC-ESI-MS/MS [54].

HPLC-ESI-MS/MS is widely accepted as the most accurate technique for quantification of global levels of 5mC and its oxidized forms [54]. In this approach, DNA is digested to individual nucleosides. Epigenetically modified DNA nucleosides are detected based on their characteristic molecular weights, HPLC retention times, and MS/MS fragmentation (Fig. 6a). 5mC, 5hmC, 5fC, 5caC, and *N*6MedA can be readily identified by mass spectrometry based on their distinct mass to charge ratios and MS/MS fragmentation patterns (Table 1) [54]. Absolute quantitation is performed using stable isotope-labeled analogues of each nucleoside that are spiked into samples (Fig. 6b) or by using external calibration curves. This methodology makes it possible to accurately quantify global amounts of each epigenetic mark of DNA, although it does not provide their distribution along the genome [55].

Methods that employ HPLC-MS/MS techniques to quantify epigenetic modifications of DNA have been developed for a variety of applications and instruments [56–58]. Due to the large number of methods that have been published for analyzing

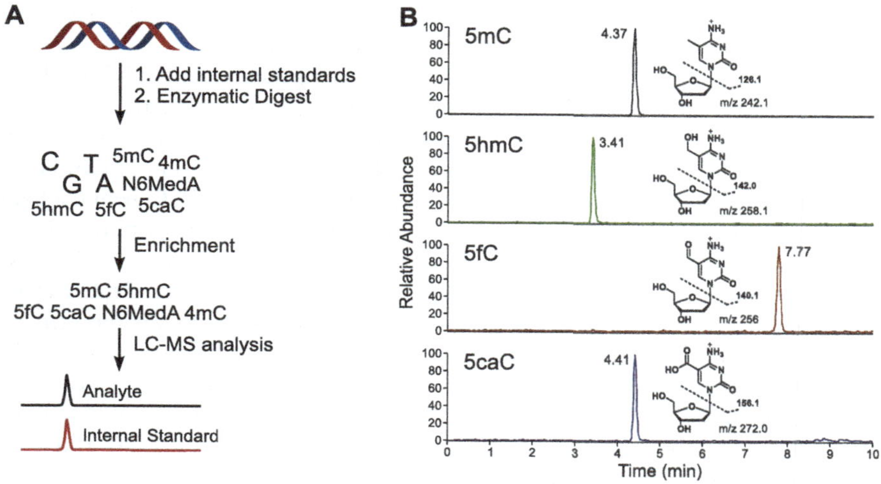

Fig. 6 Quantitation of epigenetic modifications via mass spectrometry. (**a**) Mass spectrometry analyses begin with DNA that is spiked with internal standards and enzymatically digested to nucleosides. The hydrolysate is enriched for the target analytes prior to LC-MS analysis. (**b**) Example of LC-MS/MS trace for the epigenetic modifications of cytosine, 5mC, 5hmC, 5fC, and 5caC. Standards of 5mC, 5hmC, 5fC, and 5caC, 100 fmol each were separated using a Luna Omega C18 column (Phenomenex, Torrance CA) which was eluted with a gradient of 0.1% acetic acid in water and pure acetonitrile. The mass spectrometric analysis was performed using a Thermo Fisher Scientific TSQ Quantiva Mass Spectrometer operating in SRM mode. The fragmentation conditions were optimized to give only the loss of deoxyribose as the primary fragmentation

Table 1 Molecular weights and main MS/MS fragments of standard and epigenetically modified DNA nucleosides

Nucleoside	$[M + H]^+$	$[M + H - \text{deoxyribose}]^+$
dC	228.0979	112.0505
dG	268.1040	152.0567
T	243.0975	127.0502
dA	252.1091	136.0618
5mC	242.1135	126.0662
5hmC	258.1084	142.0611
5fC	256.0928	140.0455
5caC	272.0877	156.0404
N6MedA	266.1248	150.0774
4mC	242.1135	126.0662

a combination of 5mC, 5hmC, 5fC, and 5caC, only a few of them will be discussed here.

Earlier approaches employed offline analyte enrichment prior to HPLC-ESI-MS/MS analysis to minimize chemical noise and improve detection sensitivity. Sample cleanup steps can include offline HPLC [10, 59] or solid-phase extraction [13]. In one example, DNA digests were subjected to offline HPLC cleanup, followed by HPLC-MS/MS/MS method on a LTQ linear ion trap [59]. The methodology used

30–80 μg of genomic DNA, and the limits of detection for 5hmC, 5fC, and 5caC were reported to be 2–3 modifications per 10^5 cytidines [59].

More recent applications of HPLC-MS/MS for epigenetic marks of DNA have employed two-dimensional HPLC (2D-LC) tandem mass spectrometry. A general method for analysis of 5mC, 5hmC, 5fC, and 5caC was reported by Liu et al. [54]. In this approach, DNA samples are enzymatically digested to nucleosides, followed by ultrafiltration to remove proteins. A column switching nano-HPLC methodology using two HPLC columns was used to allow for online sample cleanup prior to high-resolution MS/MS analysis using a Q Exactive hybrid quadrupole-Orbitrap mass spectrometer [54]. Using this methodology, 5mC, 5hmC, 5fC, and 5caC can be quantified at low femtomole levels [54]. A similar methodology was reported by Gackowski et al. [60]. 2D-LC allowed for isolation of peaks of interest (5hmC, 5fC, and 5caC) as they eluted off the first HPLC column and directing them to a second column, which was connected to a Waters Xevo TQ-S mass spectrometer. Using this approach, a high level of sensitivity was achieved for 5hmC (LOD (limit of detection), 3 fmol; LOQ (limit of quantitation), 10 fmol), 5fC (LOD, 0.3 fmol; LOQ, 0.8 fmol), and 5caC (LOD, 0.05 fmol; LOQ, 0.13 fmol), and the amounts of each were quantified in matched cancer and normal human colon tissue samples [60]. By minimizing sample processing and analyte cleanup steps, 2D-LC methodologies significantly simplify sample analyses and minimize the cost, allowing for high-throughput quantification of large numbers of samples.

Although the vast majority of mass spectrometry-based techniques do not allow for mapping epigenetic marks along the genome, a recent report by Lin et al. employed a mass spectrometry-based approach to determine local concentrations of 5mC at specific genes [61]. Following endonuclease cleavage with BbvI, DNA sequences of interest were captured by hybridization to complementary sequences immobilized on streptavidin beads [61]. The captured DNA was heated in formic acid to release 5mC as a free base, which was detected using an LTQ mass spectrometer. Lin et al. used this method to probe methylation status of four promoter regions: GSTP1, BCL2, ESR1, and HIC1 in DNA from clinical prostate tissue samples digested with BbvI [61]. The capture methodology was able to effectively distinguish cancer tissue from normal tissue based on the level of promoter methylation of GSTP1, BCL2, ESR1, and HIC1 [61].

Biological samples from clinical studies are often precious, requiring method minimization to reduce DNA input. Le et al. have developed an LC-MS/MS method for the analyses of 5mC and 5hmC in 50 ng of DNA [62]. This effort employed an Agilent 6,460 QQQ, an Agilent Eclipse C18 (2.1 × 50 mm, 1.8 μm particle size) column eluted at a flow rate of 100 μL/min. External calibration curves were constructed using known concentrations of 5mC and 5hmC. This method was able to quantify 0–10% 5mC and 0–2% 5hmC [62].

The recent discovery of *N*6MedA (Fig. 1) in mammalian embryonic stem cells [29] was accomplished by offline HPLC separation of enzymatic DNA digests, followed by HPLC-ESI-MS/MS analysis of *N*6MedA using a Waters UPLC coupled to a TSQ Quantum Ultra triple-quadrupole mass spectrometer. Quantitation of the *N*6MedA was performed with external calibration curves, and the identity of the novel nucleoside in embryonic stem cells was confirmed using high-resolution mass spectrometry [29].

HPLC-ESI-MS/MS analysis of 4mC is challenging because it is present in relatively low amounts but also has the same molecular mass and similar biophysical properties as 5mC (Fig. 1), making it difficult to detect the two isomers separately. HPLC resolution of 4mC and 5mC was achieved by gradient elution of a C18 column under acidic conditions, allowing for 4mC quantification in *C. kristjanssonii* containing high concentrations of 4mC [33].

3 Indirect Methods to Map Epigenetic DNA Modifications Along the Genome

Mapping epigenetics DNA marks along the genome is essential for our understanding of their biological roles, interrelationships, and dynamics. The majority of techniques used currently employ bisulfite conversion, modified bisulfite conversion, or antibody capture to help locate epigenetic DNA modifications. Although the readout from these techniques is indirect, they have yielded a wealth of important information over the years. Novel methodologies such as nanopore sequencing and single-molecule real-time (SMRT) sequencing allow for a direct sequencing of modified DNA and are discussed in Sect. 4.

3.1 Bisulfite Conversion to Distinguish Between C and 5mC

Early studies of chemical reactivity of native and modified DNA nucleosides have revealed an important difference in chemical reactivity of C and 5mC [63, 64]. Upon treatment with bisulfite, cytosine is readily deaminated to uracil, while much slower reaction is observed for 5mC. The bisulfite reaction with cytosine proceeds through a C-6 sulfonate intermediate, which undergoes hydrolysis in the presence of base to release uracil (Fig. 7a) [63]. Unlike C, 5mC is resistant to deamination under the same conditions [65]. Therefore, bisulfite treatment of genomic DNA can be used to distinguish between methylated and unmethylated cytosines in the genome. Following bisulfite treatment, cytosines are deaminated and subsequently read as T, while 5mC will resist deamination and will be read as C (Fig. 7a). This unique property has been used by various sequencing technologies that employ bisulfite conversion to determine the locations of 5mC bases in the genomes of various species including mammals, plants, and bacteria [66, 67].

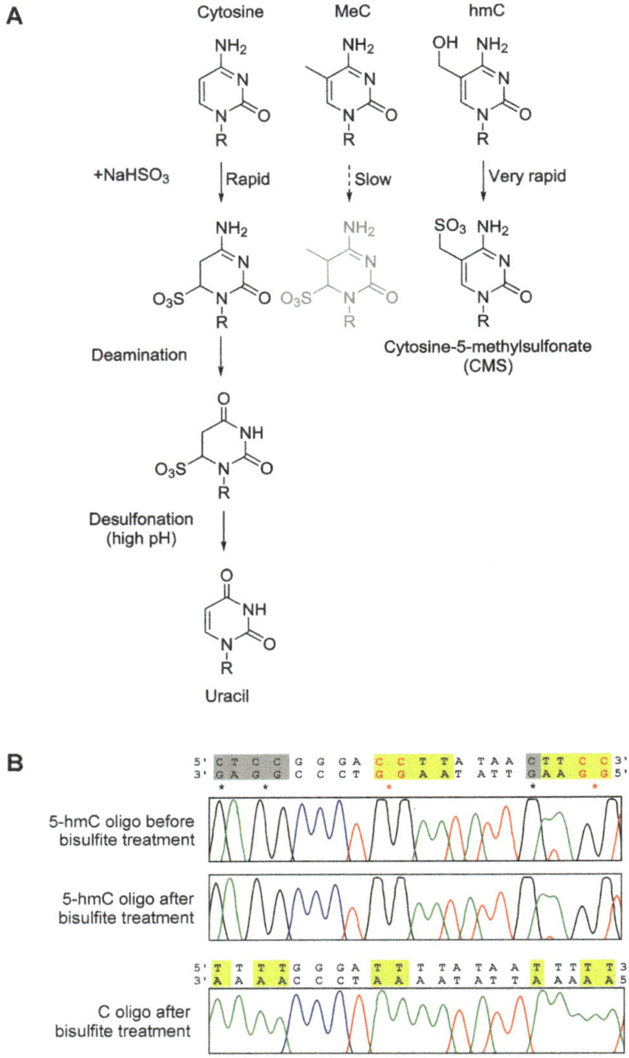

Fig. 7 Effects of bisulfite treatment on 5mC and 5hmC in DNA and the sequencing results. (**a**) Bisulfite-mediated deamination of cytosine. HSO_3^- adds across the 5,6-double bond of cytosine, promoting deamination and conversion to 6-sulfonyluracil. 6-sulfonyluracil is desulfonated to uracil (U) at higher pH. 5mC is also deaminated to thymine by bisulfite conversion, but the rate is approximately two orders of magnitude slower than that of cytosine. Bisulfite quickly converts 5-hydroxymethylcytosine to form cytosine-5-methylenesulfonate (CMS), which does not readily undergo deamination. (**b**) Shown are examples of Sanger sequencing traces of 5hmC-containing DNA before and after bisulfite treatment. 5hmC resists bisulfite-induced deamination and is thus read as cytosine. The control C-containing oligonucleotide, on the other hand, shows complete conversion of all C's in the top strand (highlighted sequences) to T's (*lower panel*). Adapted from Huang Y et al. (2010). The behaviour of 5-hydroxymethylcytosine in bisulfite sequencing. PLoS One. Jan 26;5(1):e8888. doi: https://doi.org/10.1371/journal.pone.0008888. Used in accordance with CC BY license (http://creativecommons.org/licenses/by/4.0/) [78]

3.2 Microarray Technology for Sequencing

A DNA microarray is a collection of DNA segments of known sequence attached in a known pattern to a solid surface. The DNA to be sequenced is fragmented and modified with fluorescent tags, which are incorporated through PCR. These DNA fragments are hybridized to specific sites on the array and can be fluorescently imaged [68]. Differences in hybridization can be quantified using the fluorescence intensity [68]. Microarrays are often coupled to antibody enrichment or used with cDNA to explore RNA abundance under different treatment conditions. The microarray methodologies require small sample input and are easily replicated for multiple samples.

Commercial arrays have been developed to explore methylation status of human samples using bisulfite-treated DNA and offer around 850,000 probes for hybridization [69–73]. In this way, a DNA microarray enables high-throughput screening of many genomic regions [72, 73]. In 2002, Gitan et al. developed the first methylation-specific oligonucleotide (MSO) microarray for DNA methylation analysis [69]. In this technique, DNA samples are treated with bisulfite to convert cytosine to uracil, whereas 5mC remains unchanged (Fig. 7a). Specific CpG-containing regions of the genome are then amplified by PCR and labeled using fluorescent dye-conjugated primers and hybridized to oligonucleotide probes attached to a glass surface. The oligonucleotide probes are specifically designed to discriminate between bisulfite-converted and bisulfite-unconverted sequences containing known CpG sites. The quantitative differences in hybridization, which can be assessed by the fluorescence intensity, indicate the methylation status of a particular genomic locus. This first use of a MSO microarray successfully mapped methylated CpG sites within the CpG island of the estrogen receptor gene in cultured breast cancer cells and clinical samples [69].

Further developments in the MSO microarray methodologies have led to a creation of gene-specific, multiplex systems that can be used to simultaneously examine methylation across multiple CpG sites of numerous genes. This method provides both qualitative and quantitative DNA methylation data [70]. Due to these advantages, the MSO microarray technology was commercialized and is now widely used to study cytosine methylation along the genome.

A variety of commercially available methylation microarrays are available from companies such as Illumina, Affymetrix, Agilent, and Roche NimbleGen. One such commercially available microarray is the Infinium Human Methylation 450 BeadChip (Infinium Methylation 450 K) by Illumina. The Infinium 450 K array makes it possible to assess the methylation status of more than 480,000 cytosines distributed over the entire human genome [71]. This microarray utilizes a similar technique as the MSO microarray developed by Gitan et al. in 2002 and can determine the methylation status of 96% of CpG islands and 99% of known human genes using ~500 ng DNA [49]. The CpG sites being interrogated include promoter regions, 5′-UTRs, gene bodies, 3′-UTRs, and intergenic regions [71]. The Infinium 450 K is also applicable to both cell and tissue samples; however this array lacks coverage of distal regulatory elements [74].

Although the Infinium 450 K had been widely used by the epigenetics research community, the technology has presented some technical challenges [74]. It was reported that many of the probes in the Illumina 450 K array either cross-hybridize to nontargeted genomic regions or target loci that contain known single-nucleotide polymorphisms (SNPs), both of which can interfere with analysis of DNA methylation levels [75]. Because of this, sites of differential DNA methylation identified using the Illumina 450 K array must be confirmed using a secondary independent assay, especially the sites targeted by the cross-reactive probes or that contain known SNPs. This led Illumina to release a new technology, the Infinium MethylationEpic (EPIC) BeadChip. This new array was specifically designed to target enhancer regions and avoid SNPs and cross-reactive probes. It contains over 850,000 probes, which cover more than 90% of the sites on the Infinium 450 K, plus more than 350,000 CpGs at regions identified as potential enhancers by FANTOM5 and the ENCODE project [72, 73]. In this array, probe sequences are designed to be complementary to specific 50 base regions of bisulfite-converted genomic DNA with a CpG site at the 3′ end of the probe. Following hybridization to bisulfite-converted DNA, single-base extension of the probe incorporates a fluorescently labeled ddNTP at the 3′ CpG site to distinguish between the C/T conversion that results from bisulfite treatment [74].

Methylation array methodologies described above represent a cost-effective and high-throughput technology to quantify the methylation level of half a million cytosines all over the genome. Furthermore, these arrays have low input DNA requirements, making them a powerful tool to study DNA methylomes. However, there are potential biases with this technology. As with all bisulfite-based methods, incomplete bisulfite conversion can affect accurate quantitation of methylation differences. In addition, single-nucleotide polymorphisms (SNPs) can interfere with methylation level detection, and commercial arrays are limited to human samples only. No commercial arrays are now available to map methylation of the murine genome, hindering the application of this methodology to the wide range of valuable animal models of disease [76, 77]. Despite these limitations, Needhamsen et al. recently reported the use of the Infinium Human MethylationEPIC BeadChip to study DNA methylation in mouse DNA samples [77]. This study identified over 19,000 EPIC probes which aligned to the bisulfite-converted reference mouse genome [77]. Another limitation of this technology is its inability to differentiate methylation from hydroxymethylation [78].

Traditional array methods do not distinguish between 5mC and 5hmC as both are resistant to deamination. However, recent work by Nazor et al. combined TET-assisted bisulfite (TAB) conversion with Illumina 450 K DNA methylation arrays. This method, termed TAB-array, provides a low-cost, high-throughput approach that distinguishes 5hmC from 5mC signals at base resolution [79]. This technique requires that the genomic DNA from each biological sample is split in two fractions for different downstream processing steps. One fraction undergoes normal bisulfite conversion which identifies cytosines as methylated or not methylated, without distinguishing between 5mC and 5hmC. The other fraction undergoes TAB conversion in which DNA is first glucosylated to protect 5hmC and then oxidized by TET1, followed by bisulfite treatment. This allows 5hmC to be distinguished from 5mC [79].

3.3 *Next-Generation Sequencing*

Next-generation sequencing includes techniques that utilize polymerase synthesis of complementary DNA strand to determine the sequence of input DNA. The technology, developed by Illumina (San Diego, CA), immobilizes DNA fragments onto a chip by hybridizing them to synthetic DNA oligomers that were seeded on the chip during manufacturing. The immobilized DNA is then amplified to create clusters of identical DNA fragments, which will amplify the recorded signals. Following amplification, the resulting DNA clusters are sequenced by synthesis of the complementary DNA using engineered DNA polymerases and specially modified nucleic acids [80]. Each nucleotide has a fluorescent tag and is reversibly modified at the 3′ end to allow for only a single-nucleotide incorporation [80]. Following single-nucleotide addition, a camera images the chip to record a fluorescent signal and records the sequence being made, followed by deprotection of the 3′ end for continued synthesis [80]. These steps for elongation are repeated until the desired read length is achieved. This technology can be combined with antibody/enrichment-based sequencing (Sect. 3.4), whole-genome sequencing (Sect. 3.5), reduced representation sequencing (Sect. 3.6), and modified bisulfite sequencing (Sect. 3.7) as detailed below.

3.4 *Antibody/Enrichment-Based Sequencing*

Low-resolution data for genome-wide distribution of DNA methylation marks can be obtained using antibody-based techniques such as methylated DNA immunoprecipitation (MeDIP). In this assay, a 5mC-targeted antibody is used to specifically immunocapture methylated genomic fragments [81]. The enriched DNA fragments can then be input into high-throughput DNA detection methods such as DNA microarray (MeDIP-Chip) or next-generation sequencing (MeDIP-Seq). Through genome-wide peak calling of the microarray and sequencing results, methylated regions along the whole genome can be identified [81, 82]. The advantages of this method are its low cost and the ability to rapidly identify DNA methylation regions within the genome. However, the enrichment efficiency of 5mC antibody can be affected by the number of methylated CpG sites [81]. Thus, the detection sensitivity is poor at low-CpG-density regions, making it hard to differentiate low methylation differences in these regions.

Similar immunoprecipitation (IP)-based methodologies have been developed for oxidized forms of 5mC (including 5hmC, 5fC, and 5caC), as the corresponding antibodies against them are now readily available [83, 84]. This approach has been used in multiple studies to identify the regions of 5hmC enrichment such as CpG-rich transcription start sites [83] and 5fC and 5caC enrichment at pericentric heterochromatin as well as at distal regulatory elements [84].

Apart from direct antibody enrichment techniques, a number of chemically assisted enrichment techniques have been developed for mapping 5hmC and 5fC. In the anti-cytosine-5-methylenesulfonate (anti-CMS Seq) technique, 5hmC is converted to cytosine-5-methylenesulfonate (CMS) by bisulfite treatment. CMS-containing DNA fragments are enriched via CMS-targeted antibody and then sequenced [85]. This methodology has a lower background and lower reliance on 5hmC density in the fragment as compared to direct antibody pulldown of 5hmC-containing fragments [85, 86].

In GLIB-Seq (glucosylation, periodate oxidation, biotinylation sequencing), 5hmC is glucosylated in the presence of β-glucosyltransferase, and the resulting glucosyl derivatives are oxidized with sodium periodate to yield two aldehyde groups [85]. The latter are derivatized with biotin-containing aldehyde reactive probe (ARP), and the derivatized fragments are enriched via streptavidin beads and sequenced, allowing for mapping of 5hmC marks along the genome [85]. Another chemically assisted enrichment method for 5hmC is hMeSeal-Seq (hydroxymethyl selective chemical labeling) [87]. In this method, an engineered glucose moiety containing an azide group is enzymatically added to the hydroxyl group of 5hmC by T4 bacteriophage β-glucosyltransferase. The azide group can then be chemically modified with biotin for detection, affinity enrichment, and sequencing of 5hmC-containing DNA fragments [88].

Similar mapping methodologies have been developed for 5fC [89, 90]. Since 5fC contains an inherently reactive aldehyde group, it can be directly modified with biotin-containing aldehyde reactive probe (ARP), followed by enrichment of 5fC-containing DNA fragments using streptavidin beads [89]. Another method for mapping 5fC involves enzymatically blocking 5hmC with unmodified UDP-Glc [90]. This is followed by 5fC reduction to 5hmC with $NaBH_4$ and hMeSeal-Seq to tag, enrich, and map the 5fC fragments as newly formed 5hmC sites [90].

The main limitation of all enrichment-based methods is their low resolution (100–300 bp) [91]. This can be a major problem, especially for studies investigating multiple epigenetic marks. Another limitation is that these methods are unable to quantify absolute methylation levels because each enriched fragment may have variable numbers of epigenetic marks. Additional common concerns include antibody cross-reactivity, low capture efficiency, and the limitations of downstream DNA detection methods (microarrays and next-generation sequencing) used to determine the enrichment. However, these techniques are still widely used as they provide a rapid, comparatively less laborious and inexpensive way to study genome-wide methylation.

3.5 Whole-Genome Bisulfite Sequencing

Studies of DNA methylation were initially restricted to localized CpG-rich regions of the genome [92]. However, several methods now exist which can map DNA methylation genome-wide. Whole-genome bisulfite sequencing (WGBS) was developed in 1992 by Frommer et al. to measure CpG methylation genome-wide at a single-base resolution [93]. This method utilizes bisulfite-induced modification of genomic DNA to convert cytosine to uracil; the latter is converted to thymine during PCR amplification. 5mC residues, however, are not converted and remain as cytosines after PCR amplification [93]. The development of high-throughput sequencing has facilitated the generation of genome-wide, single-base resolution DNA methylation maps from bisulfite-converted DNA [94]. The first methylome was generated in 2008 from the *Arabidopsis thaliana* genome [95, 96]. The first human methylome was reported by Lister et al. in 2009 from human embryonic stem cells and fetal fibroblasts [97]. These studies revealed the genome-wide context of DNA methylation. Extensive differences were found between the methylomes of two human cell types, revealing a dynamic nature of this modification. Due to the high genomic coverage of WGBS (95%), this method is preferred for building reference methylomes [98].

WGBS can be combined with nucleosome occupancy and methylome sequencing (NOMe-seq) to measure the relationship between DNA methylation and nucleosome occupancy [99]. Developed by Jones et al. in 2012, this method utilizes the DNA methyltransferase M.CviPI to methylate cytosines that are present in a GpC context. GpC sites that are not protected by nucleosomes or transcription factors are accessible and can be methylated. M.CviPI does not alter endogenous DNA methylation patterns found in a CpG context; thus, NOMe-seq can be paired with WGBS, and the methylation status of both the CpG and GpC sites can be simultaneously detected [100]. This method has enabled footprinting of chromatin architecture at a variety of promoter and non-promoter regions. Work done by Jones et al. has demonstrated that the relationship between nucleosome occupancy and DNA methylation is context-specific, depending on genomic location [99]. Commercially available NOMe-seq kits are available through Active Motif.

3.6 Reduced Representation Bisulfite Sequencing (RRBS)

Although WGBS allows for the greatest coverage of DNA methylation genome-wide, it is time consuming and requires a significant investment to reach the effective depth required for accurate DNA methylation mapping [101]. Although the cost of WGBS has decreased significantly in recent years, it remains around $1,000 per genome according to the National Human Genome Research Institute [102], making this approach non-practical for large studies involving hundreds of samples [103]. Reduced representation bisulfite sequencing (RRBS) is an efficient and high-throughput technique used to analyze methylation profiles at a single-nucleotide level from regions of high CpG content [104, 105]. This technique utilizes a methylation-insensitive restriction enzyme, such as MspI, which cuts at 5′-CCGG-3′ sites, enriching for CpG-rich regions of the genome such as CpG islands by leaving CpG-rich regions at the ends of fragments being sequenced [94]. This portion of the genome is then sequenced using NGS to generate single-base pair resolution DNA methylation maps. RRBS captures 85% of CpG islands and 60% of gene promoters, requiring small sample input (0.01–0.03 μg of DNA) [94, 106]. By combining different restriction enzymes, CpG coverage across the genome can be altered to include or exclude certain regions of interest such as CpG island shores, which are the 2 kb regions adjacent to CpG islands and are known to play an important role in various biological processes including cellular differentiation [107, 108].

One of the main advantages of RRBS over WGBS is the overall reduction of the size of DNA regions to be sequenced, therefore requiring fewer reads for accurate sequencing and thus reducing the cost and allowing for deep sequencing [94]. An important disadvantage of RRBS is that it does not capture all CpG islands and/or promoters. While WGBS can access ~95% of the genome, RRBS only reaches 3.7% genome coverage [94]. However, RRBS enjoys much better coverage depth than WGBS, and the concordance of DNA methylation levels between the two methods is high [101]. Due to these advantages, RRBS has become popular when high-throughput, low-cost methylation analysis is needed such as in clinical applications [109]. It is important to note, however, that RRBS has been mostly applied for DNA methylation profiling of CpG sites within mouse and human genomes, and the ability of MspI coupled with RRBS to cover non-CpG methylation has not been demonstrated [101]. MspI-based RRBS methods can only cover 10% of the genomic regions [106].

More targeted capture-based approaches are available for sequencing specific regions, such as the SeqCap Epi (Roche) and SureSelect (Agilent) systems [110, 111]. These capture methods focus on predetermined genomic regions to reduce sequencing complexity but are limited by design around specific genomic regions which may or may not include the regions of interest. Custom capture libraries can be created if certain regions are not part of the original capture systems [112].

3.7 Modified Bisulfite Sequencing Methods for Detection of 5hmC, 5fC, and 5caC

The discovery of 5hmC, 5fC, and 5caC as oxidation products of 5mC generated by TET enzymes has spurred a flurry of speculation about possible biological roles of these additional epigenetic marks [113]. In addition to their role as demethylation intermediates, these oxidized forms of 5mC may serve as unique epigenetic marks in their own right, interacting with a distinct group of protein readers and eliciting a unique epigenetic response [20, 21]. Therefore, it is of great interest to map the locations of 5hmC, 5fC, and 5caC along the genome.

Standard bisulfite sequencing cannot distinguish between 5mC and its oxidized forms. Similarly to 5mC, 5hmC resists bisulfite-mediated deamination (Fig. 7) [65, 78, 114]. Therefore, previous studies that employed bisulfite conversion to map 5mC detected both hydroxymethylation and methylation together. In contrast, 5fC and 5caC readily undergo deamination upon treatment with bisulfite and thus cannot be distinguished from C [10, 11, 115]. In addition, 5fC also undergoes deformylation which is mediated by sulfonation of the C6 position of the nucleoside to generate the corresponding alcohol. Base-catalyzed deprotonation of the alcohol and β-elimination produce cytosine [11]. Determining the levels and positions of epigenetic modifications in the genome is important to understanding the biological function of these DNA marks. Therefore, novel sequencing methodologies were needed to separately map all four epigenetic marks.

The first method for distinguishing between 5mC and 5hmC during bisulfite sequencing utilized an additional chemical reaction on the genomic DNA. Oxidative bisulfite sequencing (oxBS-Seq) includes DNA treatment with potassium perruthenate ($KRuO_4$) to selectively oxidize the 5hmC to 5fC (Fig. 8) [115]. Following oxidation, the DNA is treated with bisulfite to deaminate cytosine. Booth et al. coupled oxBS-Seq with RRBS to increase the sequencing depth at potentially interesting CG repeats by reducing the amount of DNA that could be sequenced. 5hmC was identified primarily at CG sites located within low-density CpG islands (CGIs) [115].

As an alternative to chemical oxidation to distinguish between 5mC and 5hmC, enzymatic oxidation can be used (TET-assisted bisulfite sequencing, TAB-Seq) [116]. The latter technique utilizes T4 β-glycosyl transferase and UDP-glucose to protect the primary alcohol on 5hmC, followed by oxidation of 5mC to 5caC using recombinant TET1 protein (Figs. 8 and 9) [117]. This is coupled with bisulfite treatment and sequencing of the original sample as well as the TET-treated DNA to provide data for the locations of 5mC and 5hmC. Using the technique on the whole genome, 5hmC was found to be almost exclusively found at CG sites, and nearly half of all 5hmC sites were present in distal regulatory elements [116]. TAB-Seq has been applied to identify changes in 5mC and 5hmC in intergenic regions of kidney tumors [118] and to identify sites of 4mC in bacterial DNA [33].

Mapping 5fC along the genome was accomplished in part by modification of the previous techniques to distinguish between 5mC and 5hmC. The first reported method for 5fC was formylcytosine chemical-assisted bisulfite sequencing (fCAB-Seq) [90]. In this approach, chemical labeling of 5fC is conducted with *O*-ethylhydroxylamine to protect 5fC from conversion during DNA treatment with bisulfite (Figs. 8 and 9) [90, 115]. DNA sequencing is performed on paired samples that have either undergone fCAB protection plus bisulfite treatment or standard bisulfite treatment. Through the mapping of standard bisulfite sequencing results and fCAB-Seq data to the genome, sites specifically modified with 5fC could be

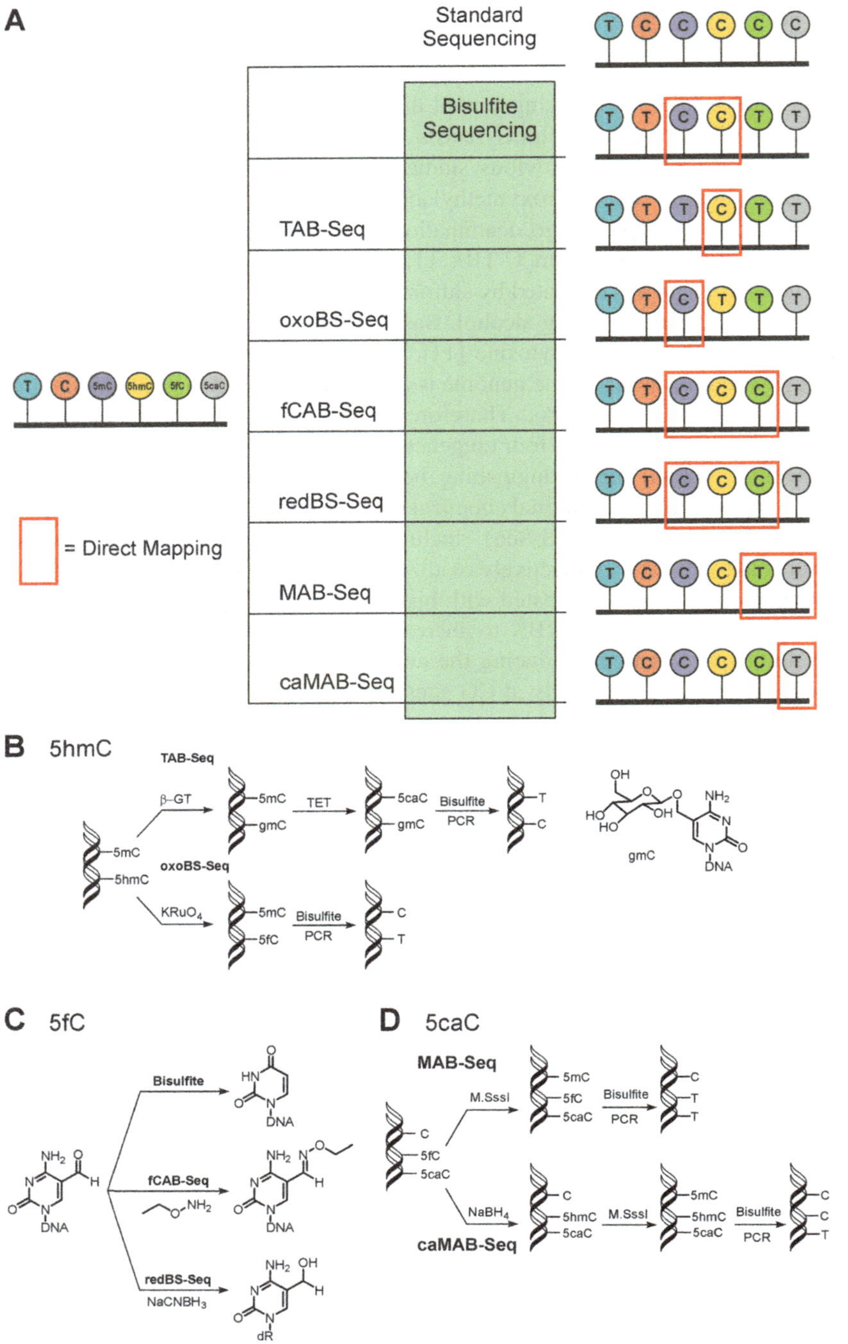

Fig. 8 Methodologies for distinguishing epigenetic modifications in DNA using bisulfite-treated DNA. (**a**) Diagram of the bisulfite-coupled sequencing technologies used for mapping the positions

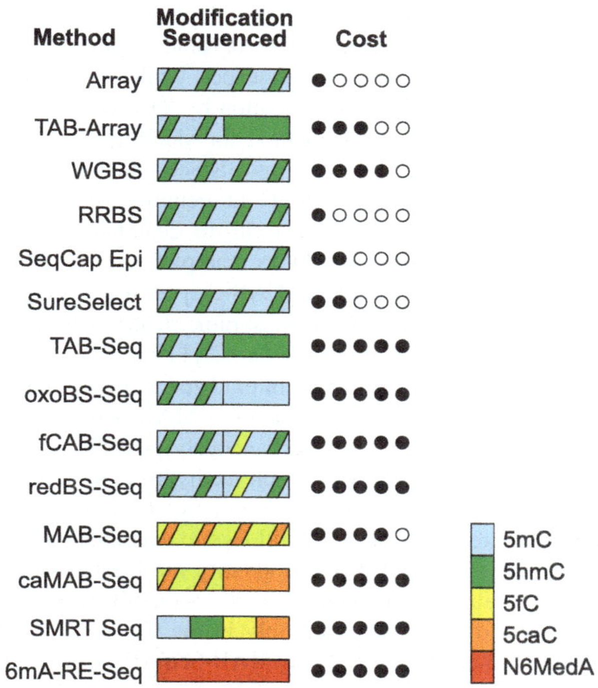

Fig. 9 The effectiveness of various sequencing technologies at discriminating between epigenetic modifications in DNA and the relative cost of their application. The ability of a method to identify 5mC (blue), 5hmC, (green), 5fC (yellow), 5caC (orange), and N6MedA (red). The relative cost is depicted as filled circles, least expensive (●○○○○) to most expensive (●●●●●)

identified. Cytosine formylation amounts at these sites can be estimated by subtracting the percent cytosine methylation determined in the bisulfite only treated sample (which is the sum of 5mC and 5hmC) from the percent of cytosine methylation in the fCAB-Seq sample data (which is the sum of 5mC, 5hmC, and 5fC) [90]. Because 5fC is a minor component of DNA, each region of DNA was

Fig. 8 (continued) of 5mC, 5hmC, 5fC, and 5caC. (**b**) Sample preparation scheme for differentiating 5mC and 5hmC in TAB-Seq [116] and oxBS-Seq [115]. (**c**) 5fC is deaminated and deformylated by treatment with bisulfite. Protection of 5fC by reaction with *O*-ethylhydroxylamine in fCAB-Seq prevents the deamination and deformylation that typically occur when 5fC is treated with bisulfite [90]. In redBS-Seq, 5fC is reduced with sodium cyanoborohydride to 5hmC which will resist deamination [119]. 5fC in the original sample can be determined by comparison of fCAB-Seq or redBS-Seq data with DNA that had undergone standard bisulfite sequencing. (**d**) Because 5fC and 5caC deaminate when treated with bisulfite, MAB-Seq utilizes the methylase M. SssI to methylate unmodified cytosines. Following bisulfite treatment, 5mC and 5hmC will be identified as C, while 5fC and 5caC are deaminated and identified as T [120]. caMAB-Seq first reduces 5fC to 5hmC and then follows MAB-Seq in order to specifically map 5caC [120]

sequenced an average of 1,000 times, which allows estimation of 5fC levels to ±0.1%. Sequencing of DNA from mouse embryonic stem cells showed that 5fC levels were increased at poised enhancer sequences, while 5mC was also reduced at those same enhancers, indicating 5mC oxidation by TET could function in epigenetic priming [90]. Another chemical method for identification of 5fC is reduced bisulfite sequencing (redBS-Seq). The redBS-Seq method employs sodium cyanoborohydride to reduce 5fC to 5hmC in genomic DNA (Figs. 8 and 9) [119]. The DNA is then treated with bisulfite to deaminate cytosine, and both redBS and standard bisulfite-treated DNA samples are sequenced. Cytosine formylation levels are determined as in fCAB-Seq above. Using this technique, 5fC was shown to be present in CpG sites asymmetrically with 5hmC, where one strand contains 5fC and the opposite strand 5hmC [119].

An additional method for determination of oxidized forms of 5mC by sequencing is modification-assisted bisulfite sequencing (MAB-Seq) (Figs. 8 and 9) [120]. Detection of 5caC and 5fC is simultaneously possible by treatment of the DNA with bacterial methylase *M. SssI*, which methylates all unmodified CG dinucleotides to give 5mCG. Following methylation to protect all unmodified cytosine in a CG context, bisulfite treatment of DNA will lead to deamination of 5fC and 5caC. This results in sequencing output that can identify the locations of 5fC and 5caC together. MAB-Seq is reliant on quantitative conversion of C to 5mC by *M. SssI* and is limited to detection of 5fC and 5caC only in the CG context. MAB-Seq was further developed to be specific for 5caC (caMAB-Seq) (Figs. 8 and 9), which takes advantage of the reduction of 5fC by sodium borohydride prior to performing the steps of MAB-Seq [120]. Using MAB-Seq and caMAB-Seq, 5fC and 5caC have been mapped at single-base resolution to regions with higher chromatin accessibility and therefore TET accessibility [121]. The sequencing profiles obtained via MAB-Seq and caMAB-Seq indicate that there might be preferential locations for TET activity, either through chromatin accessibility or sequence-dependent removal of 5fC and 5caC by TDG [120].

The first report for mapping 4mC along the genome employed a modification of the TAB-Seq methodology [33]. This is possible because 4mC undergoes deamination at a rate between C and 5mC [33]. To allow for 4mC detection, the bisulfite conditions were optimized to ensure the retention of 4mC in the genome while deaminating TET-generated 5caC. Using this methodology, 4mC was mapped in *C. kristjanssonii*, a bacterium with similar levels of 5mC and 4mC [33]. This methodology, 4mC-TAB-Seq, is well suited for identification of 4mC in genomes which have high concentrations of 4mC at particular sites, especially in bacteria which utilize 4mC as part of their restriction-modification systems [31]. Application of this methodology to mammalian cells and tissues is limited by the incomplete (40–55%) retention of 4mC after bisulfite conversion, making accurate identification of sites possessing less than complete methylation nearly impossible [33].

Mapping *N*6MedA can also be accomplished using restriction enzyme-based sequencing (6mA-RE-Seq) [122]. This sequencing technique takes advantage of previous information gained from antibody enrichment methods, which found that *N*6MedA was found in CATG and CATC motifs in the *C. reinhardtii* genome, a green algae [122]. Focusing on these two motifs, 6mA-RE-Seq employs digestion of DNA by CviAII or DpnII, sonication to fragment the DNA, end repair, adenylation, and ligation to adapters [122]. The methylated sequences will resist enzymatic digestion, allowing indirect detection of *N*6MedA based on the location of the motif in the sequencing read. The motif at the end of a read indicates a lack of methylation, while a read with the motif in the middle indicates the presence of *N*6MedA. The amount of *N*6MedA can be calculated from the relative number of sequencing reads where the specific adenosine is located at the end of a read (unmethylated) or at an internal position (methylated). Using this technique, *N*6MedA was identified at 3–4% of all available motifs in the *C. reinhardtii* genome [122]. When *N*6MedA-containing sites identified by 6mA-RE-Seq were compared to those from immunoprecipitation (IP), 28% of sites identified by IP were not identified by 6mA-RE-Seq and did not contain the motifs used in 6mA-RE-Seq. Using 6mA-RE-Seq and IP sequencing, Fu et al. found that *N*6MedA was enriched in the regions of DNA linking nucleosomes possibly used for positioning these nucleosomes on DNA [122].

4 Direct Detection of Epigenetic Modifications

The requirement for bisulfite conversion and other chemical reactions to locate DNA epigenetic marks as described above in Sect. 3 introduces an inherent uncertainty because the conversion rate is never 100%. In addition, these methods usually involve several rounds of sequencing under different conditions, which increases the cost, labor, and DNA sample input. Artifactual deamination of cytosine to uracil during sample preparation and the required read alignment pose additional challenges with this approach. Third-generation sequencing methods allow for reading the nucleotide sequences at the single-molecule level, making it possible to directly distinguish between C, 5mC, and its oxidized forms during the sequencing step. These methodologies are currently under active development.

4.1 Single-Molecule Real-Time (SMRT) Sequencing

As DNA sequencing methodologies entered the third generation, single-molecule real-time (SMRT) technology has become a new favorite since it has longer read lengths (10,000 bp up to 60,000 bp), produces higher consensus accuracy, and is characterized by a lower degree of bias as compared to second-generation sequencing [123]. SMRT-Seq can directly read different base modifications via the kinetics of nucleotide incorporation during the normal course of sequencing of an intact DNA [124]. In SMRT-Seq, kinetic signatures of 5mC and 5hmC are very subtle, presenting an accuracy challenge for detection, whereas 5fC and 5caC show very distinct kinetics signatures [125]. More recently, it was shown that glycosylation of

5hmC significantly enhances its kinetics signature, making simultaneous mapping of all three oxidized forms of 5mC possible [126]. Because these methods do not rely on DNA conversion with bisulfite or enzymatic processes and allow for longer reads, read alignment is simplified. Despite these positives, third-generation sequencing still has areas which need improvement, specifically signal intensity for some specific epigenetic marks, the overall accuracy, and larger DNA input requirements. These limitations prevent the application of this technology to whole-genome sequencing without some form of enrichment [29].

4.2 Nanopore Sequencing

Another emerging third-generation sequencing technology is nanopore sequencing, which can directly read different bases out using distinct ionic current signals generated as DNA translocates through a protein nanopore [127]. Generally, a protein nanopore is set in an electrically resistant polymer membrane, and an ionic current is passed through the nanopore by setting a voltage across this membrane. When DNA passes through the pore or near its aperture, this event creates a characteristic disruption in the current. Measurement of that current makes it possible to identify the molecule in question. Previous studies have shown the ability of discriminating between all five cytosine variants by low-throughput nanopore sensors [128]. However, high-throughput nanopore sequencing can only discriminate cytosine, 5mC, and 5hmC [129]. Notably, the overall error rate for nanopore sequencing is 13.4–20.2%, which is higher as compared to SMRT sequencing (1.72–14.2%) [130]. However, both third-generation sequencing techniques remain under active development, which will hopefully bring more robust, cost-effective, and straightforward solutions to the sequencing of epigenetic modifications in DNA.

5 Conclusions

A growing number of DNA epigenetic modifications identified in human cells suggest that epigenetic DNA marks are more common in higher eukaryotes than previously thought. DNA methylation and other epigenetic marks play a central role in controlling the physiological processes of living cells. Epigenetics has been shown to have broad implications for human health, which creates a need for reliable techniques to detect and measure these epigenetic modifications. A variety of methods have been developed and applied to study these epigenetic marks (Table 2). Recent years have seen a marked increase in novel approaches to analyze the expanding number of epigenetic DNA modifications. Information obtained from such studies will be useful both for our understanding of basic cellular regulation and for identifying future targets for drug design.

Table 2 Brief description of techniques for performing base resolution sequencing of oxidized forms of 5-methylcytosine

WGBS	Bisulfite treatment and sequencing are carried out on the whole genome. Reads both 5mC and 5hmC as C after bisulfite conversion
RRBS	CpGs are recognized and cut by specific enzymes. Traditional bisulfite sequencing is carried out on this enriched CpGs region. Reads both 5mC and 5hmC as C after bisulfite conversion
TAB-Seq	5hmC is converted to 5gmC by the β-GT enzyme. This protects 5hmC from TET oxidation, which is used to convert 5mC to 5caC. 5hmC is detected after bisulfite sequencing. 5mC is determined by subtracting 5hmC from standard BS-Seq
oxBS-Seq	5hmC is treated with potassium perruthenate and converted to 5fC. This will cause 5fC to be read as T after bisulfite conversion. 5hmC identified by subtracting the signal for 5mC from traditional BS-Seq
redBS-Seq	5fC is reduced to 5hmC with sodium borohydride. This will cause 5fC to be read as C in BS-Seq. The sites of 5fC can be determined by subtracting BS-Seq methylation from redBS-Seq
fCAB-Seq	5fC is reacted with $EtONH_2$, which protects 5fC from bisulfite-mediated deamination. This allows 5fC to be read as C in BS-Seq. The location of 5fC can be identified by subtracting BS-Seq methylation from fCAB-Seq
MAB-Seq	Unmodified CpGs are methylated by M. SssI. 5fC and 5caC can then be identified in BS-Seq by C to T transition in the sequencing data relative to the reference genome
caMAB-Seq	5fC is reduced to 5hmC with sodium borohydride. Unmodified CpGs are methylated by M. SssI. 5caC can then be identified in BS-Seq by C to T transition in the sequence relative to the reference genome

Compliance with Ethical Standards

Funding: This work was supported by the U.S. National Cancer Institute [2R01 CA-095039].

Conflict of Interest: All authors declare that they do not have a conflict of interest.

Ethical Approval: This chapter does not contain any studies with human participants or animals performed by any of the authors.

References

1. International Human Genome Sequencing C (2004) Finishing the euchromatic sequence of the human genome. Nature 431(7011):931–945. https://doi.org/10.1038/nature03001
2. Bestor TH (2000) The DNA methyltransferases of mammals. Hum Mol Genet 9 (16):2395–2402
3. Warnecke PM, Bestor TH (2000) Cytosine methylation and human cancer. Curr Opin Oncol 12(1):68–73
4. Panning B, Jaenisch R (1998) RNA and the epigenetic regulation of X chromosome inactivation. Cell 93(3):305–308
5. Song J, Teplova M, Ishibe-Murakami S, Patel DJ (2012) Structure-based mechanistic insights into DNMT1-mediated maintenance DNA methylation. Science 335(6069):709–712. https://doi.org/10.1126/science.1214453
6. Klose RJ, Bird AP (2006) Genomic DNA methylation: the mark and its mediators. Trends Biochem Sci 31(2):89–97. https://doi.org/10.1016/j.tibs.2005.12.008

7. Chedin F, Lieber MR, Hsieh CL (2002) The DNA methyltransferase-like protein DNMT3L stimulates de novo methylation by Dnmt3a. Proc Natl Acad Sci U S A 99(26):16916–16921. https://doi.org/10.1073/pnas.262443999
8. Tahiliani M, Koh KP, Shen Y, Pastor WA, Bandukwala H, Brudno Y, Agarwal S, Iyer LM, Liu DR, Aravind L, Rao A (2009) Conversion of 5-methylcytosine to 5-hydroxymethylcytosine in mammalian DNA by MLL partner TET1. Science 324(5929):930–935
9. Pfaffeneder T, Hackner B, Truss M, Munzel M, Muller M, Deiml CA, Hagemeier C, Carell T (2011) The discovery of 5-formylcytosine in embryonic stem cell DNA. Angew Chem Int Ed Engl 50(31):7008–7012
10. Ito S, Shen L, Dai Q, Wu SC, Collins LB, Swenberg JA, He C, Zhang Y (2011) Tet proteins can convert 5-methylcytosine to 5-formylcytosine and 5-carboxylcytosine. Science 333 (6047):1300–1303
11. He YF, Li BZ, Li Z, Liu P, Wang Y, Tang Q, Ding J, Jia Y, Chen Z, Li L, Sun Y, Li X, Dai Q, Song CX, Zhang K, He C, Xu GL (2011) Tet-mediated formation of 5-carboxylcytosine and its excision by TDG in mammalian DNA. Science 333(6047):1303–1307
12. Valinluck V, Sowers LC (2007) Endogenous cytosine damage products alter the site selectivity of human DNA maintenance methyltransferase DNMT1. Cancer Res 67(3):946–950
13. Kotandeniya D, Seiler CL, Fernandez J, Pujari SS, Curwick L, Murphy K, Wickramaratne S, Yan S, Murphy D, Sham YY, Tretyakova NY (2018) Can 5-methylcytosine analogues with extended alkyl side chains guide DNA methylation? Chem Commun (Camb) 54 (9):1061–1064. https://doi.org/10.1039/c7cc06867k
14. Seiler CL, Fernandez J, Koerperich Z, Andersen MP, Kotandeniya D, Nguyen ME, Sham YY, Tretyakova NY (2018) Maintenance DNA methyltransferase activity in the presence of oxidized forms of 5-methylcytosine: structural basis for ten eleven translocation-mediated DNA demethylation. Biochemistry 57(42):6061–6069. https://doi.org/10.1021/acs.biochem.8b00683
15. Kweon SM, Zhu B, Chen Y, Aravind L, Xu SY, Feldman DE (2017) Erasure of Tet-oxidized 5-methylcytosine by a SRAP nuclease. Cell Rep 21(2):482–494. https://doi.org/10.1016/j.celrep.2017.09.055
16. Song CX, He C (2013) Potential functional roles of DNA demethylation intermediates. Trends Biochem Sci 38(10):480–484
17. Bachman M, Uribe-Lewis S, Yang X, Burgess HE, Iurlaro M, Reik W, Murrell A, Balasubramanian S (2015) 5-Formylcytosine can be a stable DNA modification in mammals. Nat Chem Biol 11:555
18. Kriukiene E, Liutkeviciute Z, Klimasauskas S (2012) 5-Hydroxymethylcytosine--the elusive epigenetic mark in mammalian DNA. Chem Soc Rev 41(21):6916–6930
19. Fu Y, He C (2012) Nucleic acid modifications with epigenetic significance. Curr Opin Chem Biol 16(5–6):516–524
20. Spruijt CG, Gnerlich F, Smits AH, Pfaffeneder T, Jansen PW, Bauer C, Munzel M, Wagner M, Muller M, Khan F, Eberl HC, Mensinga A, Brinkman AB, Lephikov K, Muller U, Walter J, Boelens R, van Ingen H, Leonhardt H, Carell T, Vermeulen M (2013) Dynamic readers for 5-(hydroxy)methylcytosine and its oxidized derivatives. Cell 152(5):1146–1159
21. Iurlaro M, Ficz G, Oxley D, Raiber EA, Bachman M, Booth MJ, Andrews S, Balasubramanian S, Reik W (2013) A screen for hydroxymethylcytosine and formylcytosine binding proteins suggests functions in transcription and chromatin regulation. Genome Biol 14 (10):R119. https://doi.org/10.1186/gb-2013-14-10-r119
22. Ngo TT, Yoo J, Dai Q, Zhang Q, He C, Aksimentiev A, Ha T (2016) Effects of cytosine modifications on DNA flexibility and nucleosome mechanical stability. Nat Commun 7:10813. https://doi.org/10.1038/ncomms10813
23. Globisch D, Munzel M, Muller M, Michalakis S, Wagner M, Koch S, Bruckl T, Biel M, Carell T (2010) Tissue distribution of 5-hydroxymethylcytosine and search for active demethylation intermediates. PLoS One 5(12):e15367

24. Gackowski D, Zarakowska E, Starczak M, Modrzejewska M, Olinski R (2015) Tissue-specific differences in DNA modifications (5-hydroxymethylcytosine, 5-formylcytosine, 5-carboxylcytosine and 5-hydroxymethyluracil) and their interrelationships. PLoS One 10 (12):e0144859. https://doi.org/10.1371/journal.pone.0144859
25. Heyn H, Esteller M (2015) An adenine code for DNA: a second life for N6-methyladenine. Cell 161(4):710–713. https://doi.org/10.1016/j.cell.2015.04.021
26. Koziol MJ, Bradshaw CR, Allen GE, Costa ASH, Frezza C, Gurdon JB (2016) Identification of methylated deoxyadenosines in vertebrates reveals diversity in DNA modifications. Nat Struct Mol Biol 23(1):24–30. https://doi.org/10.1038/nsmb.3145
27. Achwal CW, Iyer CA, Chandra HS (1983) Immunochemical evidence for the presence of 5mC, 6mA and 7mG in human, Drosophila and mealybug DNA. FEBS Lett 158(2):353–358
28. Ratel D, Ravanat JL, Charles MP, Platet N, Breuillaud L, Lunardi J, Berger F, Wion D (2006) Undetectable levels of N6-methyl adenine in mouse DNA: cloning and analysis of PRED28, a gene coding for a putative mammalian DNA adenine methyltransferase. FEBS Lett 580 (13):3179–3184. https://doi.org/10.1016/j.febslet.2006.04.074
29. Wu TP, Wang T, Seetin MG, Lai Y, Zhu S, Lin K, Liu Y, Byrum SD, Mackintosh SG, Zhong M, Tackett A, Wang G, Hon LS, Fang G, Swenberg JA, Xiao AZ (2016) DNA methylation on N(6)-adenine in mammalian embryonic stem cells. Nature 532 (7599):329–333. https://doi.org/10.1038/nature17640
30. Xiao CL, Zhu S, He M, Chen D, Zhang Q, Chen Y, Yu G, Liu J, Xie SQ, Luo F, Liang Z, Wang DP, Bo XC, Gu XF, Wang K, Yan GR (2018) N(6)-methyladenine DNA modification in the human genome. Mol Cell 71(2):306–318.e307. https://doi.org/10.1016/j.molcel.2018.06.015
31. Ehrlich M, Wilson GG, Kuo KC, Gehrke CW (1987) N4-methylcytosine as a minor base in bacterial DNA. J Bacteriol 169(3):939–943
32. Timinskas A, Butkus V, Janulaitis A (1995) Sequence motifs characteristic for DNA [cytosine-N4] and DNA [adenine-N6] methyltransferases. Classification of all DNA methyltransferases. Gene 157(1–2):3–11
33. Yu M, Ji L, Neumann DA, Chung DH, Groom J, Westpheling J, He C, Schmitz RJ (2015) Base-resolution detection of N4-methylcytosine in genomic DNA using 4mC-Tet-assisted-bisulfite- sequencing. Nucleic Acids Res 43(21):e148. https://doi.org/10.1093/nar/gkv738
34. Sharma S, Kelly TK, Jones PA (2010) Epigenetics in cancer. Carcinogenesis 31(1):27–36. https://doi.org/10.1093/carcin/bgp220
35. Chapman VL, Terranova R, Moggs JG, Kimber I, Dearman RJ (2016) Evaluation of 5-methylcytosine and 5-hydroxymethylcytosine as potential biomarkers for characterisation of chemical allergens. Toxicology 340:17–26. https://doi.org/10.1016/j.tox.2015.12.003
36. Zhubi A, Chen Y, Dong E, Cook EH, Guidotti A, Grayson DR (2014) Increased binding of MeCP2 to the GAD1 and RELN promoters may be mediated by an enrichment of 5-hmC in autism spectrum disorder (ASD) cerebellum. Transl Psychiatry 4:e349
37. Kadowaki H, Nishitoh H, Urano F, Sadamitsu C, Matsuzawa A, Takeda K, Masutani H, Yodoi J, Urano Y, Nagano T, Ichijo H (2005) Amyloid b induces neuronal cell death through ROS-mediated ASK1 activation. Cell Death Differ 12(1):19–24
38. Gnyszka A, Jastrzebski Z, Flis S (2013) DNA methyltransferase inhibitors and their emerging role in epigenetic therapy of cancer. Anticancer Res 33(8):2989–2996
39. Tsai HC, Li H, van Neste L, Cai Y, Robert C, Rassool FV, Shin JJ, Harbom KM, Beaty R, Pappou E, Harris J, Yen RW, Ahuja N, Brock MV, Stearns V, Feller-Kopman D, Yarmus LB, Lin YC, Welm AL, Issa JP, Minn I, Matsui W, Jang YY, Sharkis SJ, Baylin SB, Zahnow CA (2012) Transient low doses of DNA-demethylating agents exert durable antitumor effects on hematological and epithelial tumor cells. Cancer Cell 21(3):430–446. https://doi.org/10.1016/j.ccr.2011.12.029
40. Flis S, Gnyszka A, Flis K (2014) DNA methyltransferase inhibitors improve the effect of chemotherapeutic agents in SW48 and HT-29 colorectal cancer cells. PLoS One 9(3):e92305. https://doi.org/10.1371/journal.pone.0092305

41. Ruiz-Magana MJ, Rodriguez-Vargas JM, Morales JC, Saldivia MA, Schulze-Osthoff K, Ruiz-Ruiz C (2012) The DNA methyltransferase inhibitors zebularine and decitabine induce mitochondria-mediated apoptosis and DNA damage in p53 mutant leukemic T cells. Int J Cancer 130(5):1195–1207. https://doi.org/10.1002/ijc.26107
42. Pfeifer GP, Xiong W, Hahn MA, Jin SG (2014) The role of 5-hydroxymethylcytosine in human cancer. Cell Tissue Res 356(3):631–641
43. Zhao M, Wang J, Liao W, Li D, Li M, Wu H, Zhang Y, Gershwin ME, Lu Q (2016) Increased 5-hydroxymethylcytosine in CD4(+) T cells in systemic lupus erythematosus. J Autoimmun 69:64–73. https://doi.org/10.1016/j.jaut.2016.03.001
44. Somineni HK, Zhang X, Biagini Myers JM, Kovacic MB, Ulm A, Jurcak N, Ryan PH, Khurana Hershey GK, Ji H (2016) Ten-eleven translocation 1 (TET1) methylation is associated with childhood asthma and traffic-related air pollution. J Allergy Clin Immunol 137 (3):797–805.e5. https://doi.org/10.1016/j.jaci.2015.10.021
45. Yao B, Cheng Y, Wang Z, Li Y, Chen L, Huang L, Zhang W, Chen D, Wu H, Tang B, Jin P (2017) DNA N6-methyladenine is dynamically regulated in the mouse brain following environmental stress. Nat Commun 8(1):1122. https://doi.org/10.1038/s41467-017-01195-y
46. Hotchkiss RD (1948) The quantitative separation of purines, pyrimidines, and nucleosides by paper chromatography. J Biol Chem 175(1):315–332
47. Kriaucionis S, Heintz N (2009) The nuclear DNA base 5-hydroxymethylcytosine is present in Purkinje neurons and the brain. Science 324(5929):929–930
48. Sano H, Royer HD, Sager R (1980) Identification of 5-methylcytosine in DNA fragments immobilized on nitrocellulose paper. Proc Natl Acad Sci U S A 77(6):3581–3585
49. Kurdyukov S, Bullock M (2016) DNA methylation analysis: choosing the right method. Biology (Basel) 5(1). https://doi.org/10.3390/biology5010003
50. Mizugaki M, Itoh K, Yamaguchi T, Ishiwata S, Hishinuma T, Nozaki S, Ishida N (1996) Preparation of a monoclonal antibody specific for 5-methyl-2′-deoxycytidine and its application for the detection of DNA methylation levels in human peripheral blood cells. Biol Pharm Bull 19(12):1537–1540
51. Inoue A, Shen L, Dai Q, He C, Zhang Y (2011) Generation and replication-dependent dilution of 5fC and 5caC during mouse preimplantation development. Cell Res 21(12):1670–1676
52. Valentini E, Zampieri M, Malavolta M, Bacalini MG, Calabrese R, Guastafierro T, Reale A, Franceschi C, Hervonen A, Koller B, Bernhardt J, Slagboom PE, Toussaint O, Sikora E, Gonos ES, Breusing N, Grune T, Jansen E, Dolle ME, Moreno-Villanueva M, Sindlinger T, Burkle A, Ciccarone F, Caiafa P (2016) Analysis of the machinery and intermediates of the 5hmC-mediated DNA demethylation pathway in aging on samples from the MARK-AGE study. Aging (Albany NY) 8(9):1896–1922. https://doi.org/10.18632/aging.101022
53. Habib M, Fares F, Bourgeois CA, Bella C, Bernardino J, Hernandez-Blazquez F, de Capoa A, Niveleau A (1999) DNA global hypomethylation in EBV-transformed interphase nuclei. Exp Cell Res 249(1):46–53. https://doi.org/10.1006/excr.1999.4434
54. Liu MY, DeNizio JE, Kohli RM (2016) Quantification of oxidized 5-methylcytosine bases and TET enzyme activity. Methods Enzymol 573:365–385. https://doi.org/10.1016/bs.mie.2015.12.006
55. Du C, Kurabe N, Matsushima Y, Suzuki M, Kahyo T, Ohnishi I, Tanioka F, Tajima S, Goto M, Yamada H, Tao H, Shinmura K, Konno H, Sugimura H (2015) Robust quantitative assessments of cytosine modifications and changes in the expressions of related enzymes in gastric cancer. Gastric Cancer 18(3):516–525. https://doi.org/10.1007/s10120-014-0409-4
56. Bachman M, Uribe-Lewis S, Yang X, Williams M, Murrell A, Balasubramanian S (2014) 5-Hydroxymethylcytosine is a predominantly stable DNA modification. Nat Chem 6 (12):1049–1055. https://doi.org/10.1038/nchem.2064
57. Pfaffeneder T, Spada F, Wagner M, Brandmayr C, Laube SK, Eisen D, Truss M, Steinbacher J, Hackner B, Kotljarova O, Schuermann D, Michalakis S, Kosmatchev O, Schiesser S, Steigenberger B, Raddaoui N, Kashiwazaki G, Muller U, Spruijt CG, Vermeulen M, Leonhardt H, Schar P, Muller M, Carell T (2014) Tet oxidizes thymine to

5-hydroxymethyluracil in mouse embryonic stem cell DNA. Nat Chem Biol 10(7):574–581. https://doi.org/10.1038/nchembio.1532
58. Tsuji M, Matsunaga H, Jinno D, Tsukamoto H, Suzuki N, Tomioka Y (2014) A validated quantitative liquid chromatography-tandem quadrupole mass spectrometry method for monitoring isotopologues to evaluate global modified cytosine ratios in genomic DNA. J Chromatogr B Analyt Technol Biomed Life Sci 953–954:38–47. https://doi.org/10.1016/j.jchromb.2014.01.050
59. Liu S, Wang J, Su Y, Guerrero C, Zeng Y, Mitra D, Brooks PJ, Fisher DE, Song H, Wang Y (2013) Quantitative assessment of Tet-induced oxidation products of 5-methylcytosine in cellular and tissue DNA. Nucleic Acids Res 41(13):6421–6429
60. Gackowski D, Starczak M, Zarakowska E, Modrzejewska M, Szpila A, Banaszkiewicz Z, Olinski R (2016) Accurate, direct, and high-throughput analyses of a broad Spectrum of endogenously generated DNA base modifications with isotope-dilution two-dimensional ultraperformance liquid chromatography with tandem mass spectrometry: possible clinical implication. Anal Chem 88(24):12128–12136. https://doi.org/10.1021/acs.analchem.6b02900
61. Lin XC, Zhang T, Liu L, Tang H, Yu RQ, Jiang JH (2016) Mass spectrometry based ultrasensitive DNA methylation profiling using target fragmentation assay. Anal Chem 88 (2):1083–1087. https://doi.org/10.1021/acs.analchem.5b04247
62. Le T, Kim KP, Fan G, Faull KF (2011) A sensitive mass spectrometry method for simultaneous quantification of DNA methylation and hydroxymethylation levels in biological samples. Anal Biochem 412(2):203–209. https://doi.org/10.1016/j.ab.2011.01.026
63. Hayatsu H, Wataya Y, Kazushige K (1970) The addition of sodium bisulfite to uracil and to cytosine. J Am Chem Soc 92(3):724–726
64. Hayatsu H (2008) Discovery of bisulfite-mediated cytosine conversion to uracil, the key reaction for DNA methylation analysis--a personal account. Proc Jpn Acad Ser B Phys Biol Sci 84(8):321–330
65. Hayatsu H, Shiragami M (1979) Reaction of bisulfite with the 5-hydroxymethyl group in pyrimidines and in phage DNAs. Biochemistry 18(4):632–637
66. Clark SJ, Statham A, Stirzaker C, Molloy PL, Frommer M (2006) DNA methylation: bisulphite modification and analysis. Nat Protoc 1(5):2353–2364. https://doi.org/10.1038/nprot.2006.324
67. Deng J, Shoemaker R, Xie B, Gore A, LeProust EM, Antosiewicz-Bourget J, Egli D, Maherali N, Park IH, Yu J, Daley GQ, Eggan K, Hochedlinger K, Thomson J, Wang W, Gao Y, Zhang K (2009) Targeted bisulfite sequencing reveals changes in DNA methylation associated with nuclear reprogramming. Nat Biotechnol 27(4):353–360. https://doi.org/10.1038/nbt.1530
68. Taub FE, DeLeo JM, Thompson EB (1983) Sequential comparative hybridizations analyzed by computerized image processing can identify and quantitate regulated RNAs. DNA 2 (4):309–327. https://doi.org/10.1089/dna.1983.2.309
69. Gitan RS, Shi H, Chen CM, Yan PS, Huang TH (2002) Methylation-specific oligonucleotide microarray: a new potential for high-throughput methylation analysis. Genome Res 12 (1):158–164. https://doi.org/10.1101/gr.202801
70. Yan PS, Wei SH, Huang TH (2004) Methylation-specific oligonucleotide microarray. Methods Mol Biol 287:251–260. https://doi.org/10.1385/1-59259-828-5:251
71. Dedeurwaerder S, Defrance M, Calonne E, Denis H, Sotiriou C, Fuks F (2011) Evaluation of the Infinium methylation 450K technology. Epigenomics 3(6):771–784. https://doi.org/10.2217/epi.11.105
72. Lizio M, Harshbarger J, Shimoji H, Severin J, Kasukawa T, Sahin S, Abugessaisa I, Fukuda S, Hori F, Ishikawa-Kato S, Mungall CJ, Arner E, Baillie JK, Bertin N, Bono H, de Hoon M, Diehl AD, Dimont E, Freeman TC, Fujieda K, Hide W, Kaliyaperumal R, Katayama T, Lassmann T, Meehan TF, Nishikata K, Ono H, Rehli M, Sandelin A, Schultes EA, t'Hoen PA, Tatum Z, Thompson M, Toyoda T, Wright DW, Daub CO, Itoh M, Carninci P, Hayashizaki Y, Forrest AR, Kawaji H, Consortium F (2015) Gateways to the FANTOM5

promoter level mammalian expression atlas. Genome Biol 16:22. https://doi.org/10.1186/s13059-014-0560-6

73. Siggens L, Ekwall K (2014) Epigenetics, chromatin and genome organization: recent advances from the ENCODE project. J Intern Med 276(3):201–214. https://doi.org/10.1111/joim.12231
74. Pidsley R, Zotenko E, Peters TJ, Lawrence MG, Risbridger GP, Molloy P, van Djik S, Muhlhausler B, Stirzaker C, Clark SJ (2016) Critical evaluation of the Illumina MethylationEPIC BeadChip microarray for whole-genome DNA methylation profiling. Genome Biol 17(1):208. https://doi.org/10.1186/s13059-016-1066-1
75. Chen YA, Lemire M, Choufani S, Butcher DT, Grafodatskaya D, Zanke BW, Gallinger S, Hudson TJ, Weksberg R (2013) Discovery of cross-reactive probes and polymorphic CpGs in the Illumina infinium HumanMethylation450 microarray. Epigenetics 8(2):203–209. https://doi.org/10.4161/epi.23470
76. Wong NC, Ng J, Hall NE, Lunke S, Salmanidis M, Brumatti G, Ekert PG, Craig JM, Saffery R (2013) Exploring the utility of human DNA methylation arrays for profiling mouse genomic DNA. Genomics 102(1):38–46. https://doi.org/10.1016/j.ygeno.2013.04.014
77. Needhamsen M, Ewing E, Lund H, Gomez-Cabrero D, Harris RA, Kular L, Jagodic M (2017) Usability of human Infinium MethylationEPIC BeadChip for mouse DNA methylation studies. BMC Bioinf 18(1):486. https://doi.org/10.1186/s12859-017-1870-y
78. Huang Y, Pastor WA, Shen Y, Tahiliani M, Liu DR, Rao A (2010) The behaviour of 5-hydroxymethylcytosine in bisulfite sequencing. PLoS One 5(1):e8888
79. Nazor KL, Boland MJ, Bibikova M, Klotzle B, Yu M, Glenn-Pratola VL, Schell JP, Coleman RL, Cabral-da-Silva MC, Schmidt U, Peterson SE, He C, Loring JF, Fan JB (2014) Application of a low cost array-based technique - TAB-array - for quantifying and mapping both 5mC and 5hmC at single base resolution in human pluripotent stem cells. Genomics 104(5):358–367. https://doi.org/10.1016/j.ygeno.2014.08.014
80. Canard B, Sarfati RS (1994) DNA polymerase fluorescent substrates with reversible 3′-tags. Gene 148(1):1–6
81. Weber M, Davies JJ, Wittig D, Oakeley EJ, Haase M, Lam WL, Schubeler D (2005) Chromosome-wide and promoter-specific analyses identify sites of differential DNA methylation in normal and transformed human cells. Nat Genet 37(8):853–862. https://doi.org/10.1038/ng1598
82. Down TA, Rakyan VK, Turner DJ, Flicek P, Li H, Kulesha E, Graf S, Johnson N, Herrero J, Tomazou EM, Thorne NP, Backdahl L, Herberth M, Howe KL, Jackson DK, Miretti MM, Marioni JC, Birney E, Hubbard TJ, Durbin R, Tavare S, Beck S (2008) A Bayesian deconvolution strategy for immunoprecipitation-based DNA methylome analysis. Nat Biotechnol 26(7):779–785. https://doi.org/10.1038/nbt1414
83. Williams K, Christensen J, Pedersen MT, Johansen JV, Cloos PA, Rappsilber J, Helin K (2011) TET1 and hydroxymethylcytosine in transcription and DNA methylation fidelity. Nature 473(7347):343–348. https://doi.org/10.1038/nature10066
84. Shen L, Wu H, Diep D, Yamaguchi S, D'Alessio AC, Fung HL, Zhang K, Zhang Y (2013) Genome-wide analysis reveals TET- and TDG-dependent 5-methylcytosine oxidation dynamics. Cell 153(3):692–706. https://doi.org/10.1016/j.cell.2013.04.002
85. Pastor WA, Pape UJ, Huang Y, Henderson HR, Lister R, Ko M, McLoughlin EM, Brudno Y, Mahapatra S, Kapranov P, Tahiliani M, Daley GQ, Liu XS, Ecker JR, Milos PM, Agarwal S, Rao A (2011) Genome-wide mapping of 5-hydroxymethylcytosine in embryonic stem cells. Nature 473(7347):394–397. https://doi.org/10.1038/nature10102
86. Huang Y, Pastor WA, Zepeda-Martinez JA, Rao A (2012) The anti-CMS technique for genome-wide mapping of 5-hydroxymethylcytosine. Nat Protoc 7(10):1897–1908. https://doi.org/10.1038/nprot.2012.103
87. Song CX, Sun Y, Dai Q, Lu XY, Yu M, Yang CG, He C (2011) Detection of 5-hydroxymethylcytosine in DNA by transferring a keto-glucose by using T4 phage beta-glucosyltransferase. Chembiochem 12(11):1682–1685

88. Song CX, Szulwach KE, Fu Y, Dai Q, Yi C, Li X, Li Y, Chen CH, Zhang W, Jian X, Wang J, Zhang L, Looney TJ, Zhang B, Godley LA, Hicks LM, Lahn BT, Jin P, He C (2011) Selective chemical labeling reveals the genome-wide distribution of 5-hydroxymethylcytosine. Nat Biotechnol 29(1):68–72
89. Raiber EA, Beraldi D, Ficz G, Burgess HE, Branco MR, Murat P, Oxley D, Booth MJ, Reik W, Balasubramanian S (2012) Genome-wide distribution of 5-formylcytosine in embryonic stem cells is associated with transcription and depends on thymine DNA glycosylase. Genome Biol 13(8):R69. https://doi.org/10.1186/gb-2012-13-8-r69
90. Song CX, Szulwach KE, Dai Q, Fu Y, Mao SQ, Lin L, Street C, Li Y, Poidevin M, Wu H, Gao J, Liu P, Li L, Xu GL, Jin P, He C (2013) Genome-wide profiling of 5-formylcytosine reveals its roles in epigenetic priming. Cell 153(3):678–691
91. Yong WS, Hsu FM, Chen PY (2016) Profiling genome-wide DNA methylation. Epigenetics Chromatin 9:26. https://doi.org/10.1186/s13072-016-0075-3
92. Gupta R, Nagarajan A, Wajapeyee N (2010) Advances in genome-wide DNA methylation analysis. Biotechniques 49(4):iii–ixi. https://doi.org/10.2144/000113493
93. Frommer M, McDonald LE, Millar DS, Collis CM, Watt F, Grigg GW, Molloy PL, Paul CL (1992) A genomic sequencing protocol that yields a positive display of 5-methylcytosine residues in individual DNA strands. Proc Natl Acad Sci U S A 89(5):1827–1831
94. Stirzaker C, Taberlay PC, Statham AL, Clark SJ (2014) Mining cancer methylomes: prospects and challenges. Trends Genet 30(2):75–84. https://doi.org/10.1016/j.tig.2013.11.004
95. Lister R, O'Malley RC, Tonti-Filippini J, Gregory BD, Berry CC, Millar AH, Ecker JR (2008) Highly integrated single-base resolution maps of the epigenome in Arabidopsis. Cell 133 (3):523–536. https://doi.org/10.1016/j.cell.2008.03.029
96. Cokus SJ, Feng S, Zhang X, Chen Z, Merriman B, Haudenschild CD, Pradhan S, Nelson SF, Pellegrini M, Jacobsen SE (2008) Shotgun bisulphite sequencing of the Arabidopsis genome reveals DNA methylation patterning. Nature 452(7184):215–219. https://doi.org/10.1038/nature06745
97. Lister R, Pelizzola M, Dowen RH, Hawkins RD, Hon G, Tonti-Filippini J, Nery JR, Lee L, Ye Z, Ngo QM, Edsall L, Antosiewicz-Bourget J, Stewart R, Ruotti V, Millar AH, Thomson JA, Ren B, Ecker JR (2009) Human DNA methylomes at base resolution show widespread epigenomic differences. Nature 462(7271):315–322. https://doi.org/10.1038/nature08514
98. Song Q, Decato B, Hong EE, Zhou M, Fang F, Qu J, Garvin T, Kessler M, Zhou J, Smith AD (2013) A reference methylome database and analysis pipeline to facilitate integrative and comparative epigenomics. PLoS One 8(12):e81148. https://doi.org/10.1371/journal.pone.0081148
99. Kelly TK, Liu Y, Lay FD, Liang G, Berman BP, Jones PA (2012) Genome-wide mapping of nucleosome positioning and DNA methylation within individual DNA molecules. Genome Res 22(12):2497–2506. https://doi.org/10.1101/gr.143008.112
100. Lay FD, Kelly TK, Jones PA (2018) Nucleosome occupancy and methylome sequencing (NOMe-seq). Methods Mol Biol 1708:267–284. https://doi.org/10.1007/978-1-4939-7481-8_14
101. Wang L, Sun J, Wu H, Liu S, Wang J, Wu B, Huang S, Li N, Wang J, Zhang X (2012) Systematic assessment of reduced representation bisulfite sequencing to human blood samples: a promising method for large-sample-scale epigenomic studies. J Biotechnol 157(1):1–6. https://doi.org/10.1016/j.jbiotec.2011.06.034
102. Wetterstrand K DNA sequencing costs: data from the NHGRI genome sequencing program (GSP). Accessed 3 Mar 2018
103. Ziller MJ, Hansen KD, Meissner A, Aryee MJ (2015) Coverage recommendations for methylation analysis by whole-genome bisulfite sequencing. Nat Methods 12(3):230–232. https://doi.org/10.1038/nmeth.3152
104. Meissner A, Mikkelsen TS, Gu H, Wernig M, Hanna J, Sivachenko A, Zhang X, Bernstein BE, Nusbaum C, Jaffe DB, Gnirke A, Jaenisch R, Lander ES (2008) Genome-scale DNA methylation maps of pluripotent and differentiated cells. Nature 454(7205):766–770. https://doi.org/10.1038/nature07107

105. Gu H, Bock C, Mikkelsen TS, Jager N, Smith ZD, Tomazou E, Gnirke A, Lander ES, Meissner A (2010) Genome-scale DNA methylation mapping of clinical samples at single-nucleotide resolution. Nat Methods 7(2):133–136. https://doi.org/10.1038/nmeth.1414
106. Gu H, Smith ZD, Bock C, Boyle P, Gnirke A, Meissner A (2011) Preparation of reduced representation bisulfite sequencing libraries for genome-scale DNA methylation profiling. Nat Protoc 6(4):468–481. https://doi.org/10.1038/nprot.2010.190
107. Doi A, Park IH, Wen B, Murakami P, Aryee MJ, Irizarry R, Herb B, Ladd-Acosta C, Rho J, Loewer S, Miller J, Schlaeger T, Daley GQ, Feinberg AP (2009) Differential methylation of tissue- and cancer-specific CpG island shores distinguishes human induced pluripotent stem cells, embryonic stem cells and fibroblasts. Nat Genet 41(12):1350–1353. https://doi.org/10.1038/ng.471
108. Wang J, Xia Y, Li L, Gong D, Yao Y, Luo H, Lu H, Yi N, Wu H, Zhang X, Tao Q, Gao F (2013) Double restriction-enzyme digestion improves the coverage and accuracy of genome-wide CpG methylation profiling by reduced representation bisulfite sequencing. BMC Genomics 14:11. https://doi.org/10.1186/1471-2164-14-11
109. Lee YK, Jin S, Duan S, Lim YC, Ng DP, Lin XM, Yeo GS, Ding C (2014) Improved reduced representation bisulfite sequencing for epigenomic profiling of clinical samples. Biol Proced Online 16(1):1. https://doi.org/10.1186/1480-9222-16-1
110. Hing B, Ramos E, Braun P, McKane M, Jancic D, Tamashiro KL, Lee RS, Michaelson JJ, Druley TE, Potash JB (2015) Adaptation of the targeted capture methyl-seq platform for the mouse genome identifies novel tissue-specific DNA methylation patterns of genes involved in neurodevelopment. Epigenetics 10(7):581–596. https://doi.org/10.1080/15592294.2015.1045179
111. Fu Y, Springer NM, Ying K, Yeh CT, Iniguez AL, Richmond T, Wu W, Barbazuk B, Nettleton D, Jeddeloh J, Schnable PS (2010) High-resolution genotyping via whole genome hybridizations to microarrays containing long oligonucleotide probes. PLoS One 5(12): e14178. https://doi.org/10.1371/journal.pone.0014178
112. Li Q, Suzuki M, Wendt J, Patterson N, Eichten SR, Hermanson PJ, Green D, Jeddeloh J, Richmond T, Rosenbaum H, Burgess D, Springer NM, Greally JM (2015) Post-conversion targeted capture of modified cytosines in mammalian and plant genomes. Nucleic Acids Res 43:e81
113. Chen K, Zhao BS, He C (2016) Nucleic acid modifications in regulation of gene expression. Cell Chem Biol 23(1):74–85. https://doi.org/10.1016/j.chembiol.2015.11.007
114. Jin SG, Kadam S, Pfeifer GP (2010) Examination of the specificity of DNA methylation profiling techniques towards 5-methylcytosine and 5-hydroxymethylcytosine. Nucleic Acids Res 38(11):e125
115. Booth MJ, Branco MR, Ficz G, Oxley D, Krueger F, Reik W, Balasubramanian S (2012) Quantitative sequencing of 5-methylcytosine and 5-hydroxymethylcytosine at single-base resolution. Science 336(6083):934–937. https://doi.org/10.1126/science.1220671
116. Yu M, Hon GC, Szulwach KE, Song CX, Zhang L, Kim A, Li X, Dai Q, Shen Y, Park B, Min JH, Jin P, Ren B, He C (2012) Base-resolution analysis of 5-hydroxymethylcytosine in the mammalian genome. Cell 149(6):1368–1380
117. Yu M, Hon GC, Szulwach KE, Song CX, Jin P, Ren B, He C (2012) Tet-assisted bisulfite sequencing of 5-hydroxymethylcytosine. Nat Protoc 7(12):2159–2170
118. Chen K, Zhang J, Guo Z, Ma Q, Xu Z, Zhou Y, Xu Z, Li Z, Liu Y, Ye X, Li X, Yuan B, Ke Y, He C, Zhou L, Liu J, Ci W (2016) Loss of 5-hydroxymethylcytosine is linked to gene body hypermethylation in kidney cancer. Cell Res 26(1):103–118. https://doi.org/10.1038/cr.2015.150
119. Booth MJ, Marsico G, Bachman M, Beraldi D, Balasubramanian S (2014) Quantitative sequencing of 5-formylcytosine in DNA at single-base resolution. Nat Chem 6(5):435–440. https://doi.org/10.1038/nchem.1893
120. Wu H, Wu X, Zhang Y (2016) Base-resolution profiling of active DNA demethylation using MAB-seq and caMAB-seq. Nat Protoc 11(6):1081–1100. https://doi.org/10.1038/nprot.2016.069

121. Hu L, Li Z, Cheng J, Rao Q, Gong W, Liu M, Shi YG, Zhu J, Wang P, Xu Y (2013) Crystal structure of TET2-DNA complex: insight into TET-mediated 5mC oxidation. Cell 155 (7):1545–1555
122. Fu Y, Luo GZ, Chen K, Deng X, Yu M, Han D, Hao Z, Liu J, Lu X, Dore LC, Weng X, Ji Q, Mets L, He C (2015) N6-methyldeoxyadenosine marks active transcription start sites in Chlamydomonas. Cell 161(4):879–892. https://doi.org/10.1016/j.cell.2015.04.010
123. Nakano K, Shiroma A, Shimoji M, Tamotsu H, Ashimine N, Ohki S, Shinzato M, Minami M, Nakanishi T, Teruya K, Satou K, Hirano T (2017) Advantages of genome sequencing by long-read sequencer using SMRT technology in medical area. Hum Cell 30(3):149–161. https://doi.org/10.1007/s13577-017-0168-8
124. Rhoads A, Au KF (2015) PacBio sequencing and its applications. Genomics Proteomics Bioinformatics 13(5):278–289. https://doi.org/10.1016/j.gpb.2015.08.002
125. Clark TA, Lu X, Luong K, Dai Q, Boitano M, Turner SW, He C, Korlach J (2013) Enhanced 5-methylcytosine detection in single-molecule, real-time sequencing via Tet1 oxidation. BMC Biol 11:4. https://doi.org/10.1186/1741-7007-11-4
126. Chavez L, Huang Y, Luong K, Agarwal S, Iyer LM, Pastor WA, Hench VK, Frazier-Bowers SA, Korol E, Liu S, Tahiliani M, Wang Y, Clark TA, Korlach J, Pukkila PJ, Aravind L, Rao A (2014) Simultaneous sequencing of oxidized methylcytosines produced by TET/JBP dioxygenases in Coprinopsis cinerea. Proc Natl Acad Sci U S A 111(48):E5149–E5158. https://doi.org/10.1073/pnas.1419513111
127. Niedringhaus TP, Milanova D, Kerby MB, Snyder MP, Barron AE (2011) Landscape of next-generation sequencing technologies. Anal Chem 83(12):4327–4341. https://doi.org/10.1021/ac2010857
128. Wescoe ZL, Schreiber J, Akeson M (2014) Nanopores discriminate among five C5-cytosine variants in DNA. J Am Chem Soc 136(47):16582–16587. https://doi.org/10.1021/ja508527b
129. Rand AC, Jain M, Eizenga JM, Musselman-Brown A, Olsen HE, Akeson M, Paten B (2017) Mapping DNA methylation with high-throughput nanopore sequencing. Nat Methods 14 (4):411–413. https://doi.org/10.1038/nmeth.4189
130. Weirather JL, de Cesare M, Wang Y, Piazza P, Sebastiano V, Wang XJ, Buck D, Au KF (2017) Comprehensive comparison of Pacific Biosciences and Oxford Nanopore Technologies and their applications to transcriptome analysis. F1000Res 6:100. https://doi.org/10.12688/f1000research.10571.2

Top Med Chem (2020) 33: 523–560
DOI: 10.1007/7355_2019_82

Published online: 28 November 2019

Advanced Assays in Epigenetics

Carmela Dell'Aversana, Federica Sarno, Mariarosaria Conte, Cristina Giorgio, and Lucia Altucci

Contents

1 Epi-proteomic Analyses 526
 1.1 Biochemical Studies: In Vitro Assay 526
 1.2 Cell-Based Assays 530
2 Innovative Genome-Wide Technologies 536
 2.1 ChIP: From Standard Procedures to Innovative Applications 536
3 Epigenetic Regulatory Role of miRNA: New Analytical Tools and Technologies 545
 3.1 miRNA Quantification Techniques 546
 3.2 Experimental miRNA Target Identification 551
4 Conclusions 554
References 554

Abstract Epigenetic mechanisms orchestrate the finely tuned regulation of genetic material and play a pivotal role in defining cellular functions and phenotypes. A growing set of tools supports analysis of the epigenome. This chapter will provide an overview of the principle methods of studying complex epigenetic machinery, focusing on recent advancements of tools and techniques in the field of epigenetics. It will also address the advantages, limitations and perspectives of each approach. Increasingly, the high sensitivity, specificity, accuracy, precision and reproducibility of cutting-edge technologies in epigenetics are allowing the identification of new key targets and molecular mechanisms in healthy and pathological states and are becoming methods of choice for clinical investigations.

Keywords Epigenetics, Genome, Histone modification, Methylation, miRNA

C. Dell'Aversana, F. Sarno, C. Giorgio, and L. Altucci (✉)
Department of Precision Medicine, University of Campania "Luigi Vanvitelli", Napoli, Italy
e-mail: lucia.altucci@unicampania.it

M. Conte
IRCCS, SDN, Naples, Italy

Abbreviations

5caC	5-Carboxylcytosine
5fC	5-Formylcytosine
5hmC	5-Hydroxymethylcytosine
5mC	5-Methylcytosine
AlphaScreen	Amplified Luminescent Proximity Homogeneous Assay Screen
BRET	Bioluminescence resonance energy transfer
BS-seq	Bisulfite sequencing
CAB-seq	Chemical modification-assisted bisulfite sequencing
CE-SSCP	Capillary electrophoresis single-strand conformation polymorphism
CETSA	Cellular thermal shift assay
ChIP	Chromatin immunoprecipitation
ChroP	Chromatin proteomics
ddPCR	Droplet digital PCR
EDC	1-Ethyl-3(3-dimethylaminoproyl)-carbodiimide hydrochloride
EnIGMA	Enzyme-assisted identification of genome modification assay
ePL	Enhanced ProLabel
ES	Embryonic stem
EWAS	Epigenome-wide association studies
EXPAR	Exponential amplification reaction
fCAB-seq	5-Formylcytosine chemical modification-assisted bisulfite sequencing
FISH	Fluorescent in situ hybridization
FLIM	Fluorescence lifetime microscopy
FRET	Förster resonance energy transfer
G4	G-quadruplex
HATs	Histone acetyltransferases
HMTs	Histone methyltransferases
HTDR	High-throughput dose-response
HTS	High-throughput screening
HT-seq	High-throughput sequencing
ISH	In situ hybridization
ITC	Isothermal titration colorimetry
LC-MS	Liquid chromatography-mass spectrometry
LNA	Locked nucleic acid
miRNA	microRNA
miR-TRAP	miRNA trapping
MPS	Massive parallel sequencing
MS	Mass spectrometry
MST	Microscale thermophoresis
NGS	Next-generation sequencing
Nluc	NanoLuc luciferase

PAR-CLIP	Photoactivatable ribonucleoside-enhanced cross-linking and immunoprecipitation
QD	Quantum dot
RBPs	RNA binding proteins
RIME	Rapid immunoprecipitation mass spectrometry of endogenous protein
Rluc	Renilla luciferase
RRBS	Reduced representation bisulfite sequencing
scBS-seq	Single-cell bisulfite sequencing
scM&T-seq	Single-cell genome-wide methylome and transcriptome sequencing
scRRBS	Single-cell reduced representation bisulfite sequencing
snmC-seq	Single-nucleus methylcytosine sequencing
SNPs	Single-nucleotide polymorphisms
SPR	Surface plasmon resonance
TAB-seq	Tet-assisted bisulfite sequencing
TCL	Targeted chromatin ligation
Tm	Melting temperature
TR-FRET	Time-resolved fluorescent energy transfer
UV	Ultraviolet
YFP	Yellow fluorescent protein

Epigenetic modifications work with genetic mechanisms to control gene expression and chromatin structure in normal cells [1]. Disruption of epigenetic regulatory processes results in abnormal gene function and/or alterations in cellular signalling pathways, leading to several pathological states such as cancer [2]. Unlike most genetic mutations, epigenetic abnormalities are reversible and potentially good targets for deriving therapeutic strategies. Global changes in the epigenetic landscape can precede genetic alterations at early stages of disease development, making them suitable predictive biomarkers for clinical diagnosis [3]. Furthermore, several studies have identified defects in epigenetic modulators as indicators of both progression and outcome of a disease, as well as biomarkers of patient response to therapy and early detection of recurrence, suggesting a prospective translational application in clinical practice to improve prognosis [4–7]. Advanced technologies able to rapidly analyse epigenetic changes have revealed a challenging scenario and provide an accurate detection of new key epi-targets in pathological states and several multifactorial complex diseases, including cancer. This chapter will describe the most innovative epigenomic technologies and will be divided into three main sections: (1) assays, in vitro and *in cell*, useful to define the activity and binding of epigenetic modulators; (2) genome-wide approaches used to determine the chromatin modification status of cells at nucleotide resolution level and to detect the binding sites of DNA-associated proteins; and (3) methodologies for miRNA expression and functional analysis. The advantages, limitations, cost-effectiveness and therapeutic perspectives of each technique will also be discussed. Further work improving the sensitivity, specificity and reproducibility of epigenetic technologies will not only reveal underlying deregulated epigenetic pathways associated with disease progression and treatment response but will also provide a robust measure of novel biomarkers for risk stratification in terms of disease development and for individualized precision medicine.

1 Epi-proteomic Analyses

Post-translational histone modifications make up a significant portion of epigenetic mechanisms involved in several biological processes. Chromatin remodelling is known to affect up- and downregulation of gene expression, and tumorigenic diseases can be driven by alterations of epigenetic modification patterns. New technologies have therefore been designed to identify histone modification enrichment in specific cellular contexts, and new drugs able to restore the physiological epi-state have been developed. This section of the chapter focuses on the techniques most commonly used to rapidly identify macromolecular complexes and to discover drugs that bind and modify protein activity. The assays have been divided into two main groups: (1) in vitro assays following protein expression and purification and (2) techniques that allow experiments to be performed directly in cells.

1.1 Biochemical Studies: In Vitro Assay

The choice of technique mainly depends on the starting point and objectives. The advantage of biochemical experiments is that the protein of interest can be isolated, purified and overexpressed. Data analysis is straightforward and fast. Indeed, this approach is commonly used to detect protein-protein or protein-drug interactions in high-throughput screening (HTS) strategies. Cell-based assays allow investigation of the target protein, and all the macromolecular complexes of which it is a part, under physiological conditions. In this case it is important to take into account several factors that can complicate studies in cells, such as drug permeability and half-life, cellular enzymatic degradation, small amounts of starting material and long lead times of experiments. The analysis of chemiluminescent protein interactions (AlphaScreen/AlphaLISA), variation in diffraction radius (surface plasmon resonance), or changes in molecular solvation (isothermal titration calorimetry) are the most common techniques used for in vitro macromolecule interaction studies.

1.1.1 AlphaScreen/AlphaLISA

Biomolecular interactions, post-translational modifications or activity of substrates as competitor compounds can be investigated using AlphaScreen and AlphaLISA

(PerkinElmer) technologies (where Alpha stands for Amplified Luminescent Proximity Homogeneous Assay Screen). These are non-enzymatic in vitro assays that detect protein interactions using donor and acceptor beads. When the interaction takes place, a chemical reaction produces an amplified signal. After donor excitation, singlet state oxygen molecules react with acceptor fluorophores, if located in proximity, that emit light at 520–620 nm [8]. Generally, donor beads in AlphaScreen and AlphaLISA contain a phthalocyanine molecule that produces a singlet state oxygen, keeping energy irradiation at 680 nm. Acceptor beads contain thioxene, anthracene and rubene in AlphaScreen and thioxene, europium, terbium or samarium in AlphaLISA [9]. The beads are covered with reactive aldehydes to bind the ligand, antibody or substrate [9]. Compared to other methods such as time-resolved fluorescent energy transfer (TR-FRET), these assays have better sensitivity and can be used for large-scale reactions. Depending on the interaction under investigation, the beads are functionalized in different ways. AlphaLISA epigenetic toolbox (PerkinElmer) is used to detect histone methylation and acetylation marks. Using S-adenosylmethionine or acetyl CoA as a cofactor, epigenetic enzymes such as histone methyltransferases (HMTs) or histone acetyltransferases (HATs) react on the specific biotinylated histone substrate. Antibody AlphaLISA acceptor beads and streptavidin Alpha donor beads are then added to measure methylation and acetylation levels. The donor binds the histone substrate by biotin-streptavidin binding and the acceptor captures the modification. If the enzyme works, the acceptor and donor beads are in proximity, and, after laser irradiation, a chemiluminescent signal at 615 nm is generated [10]. The same methodology was used to develop an HTS platform to identify new G9a [9] and JMJD2A [11] modulator drugs.

AlphaScreen and AlphaLISA offer several advantages:

- The difference between donor and acceptor wavelength reduces interference.
- Interactions are detected from sub-nM to μM range.
- The assay is suitable for use in HTS and adapts to 96-, 384- and 1,536-well formats.
- Many biological interactions can be studied.

1.1.2 Surface Plasmon Resonance (SPR)

To study in real time the association and dissociation constants of protein complexes or drug enzymes, SPR is one of the most widely used in vitro assays. The technique measures refractive index changes when there is a variation in mass on a gold chip. Incident light (Fig. 1) strikes electrons at the surface of the metal sensor and converts them into surface plasmon waves, generating an SPR angle. Interaction between the target immobilized on the chip and the captured analyte induces a perturbation at the gold surface and therefore a change in angle of reflection, which can be measured. To improve the optical quantitative detection method, SPR microscopy or imaging (SPRi) technology is used for high-throughput probing of biomolecular interactions. SPRi Biochips™ and SPRi Slides™ are customized for HORIBA SPR imaging

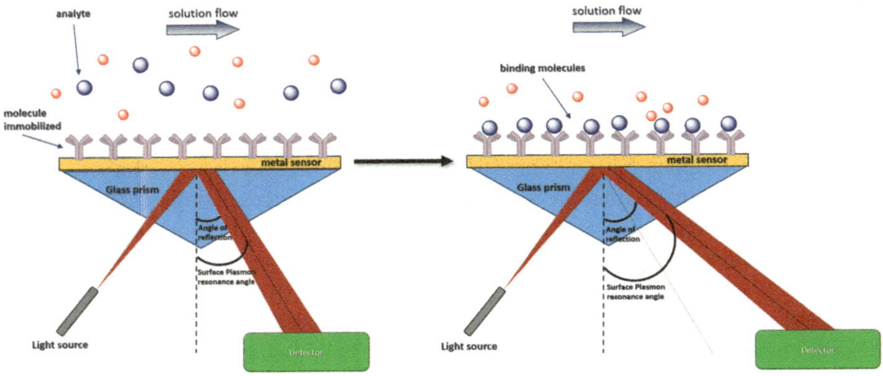

Fig. 1 Scheme of SPR experiment. The molecule is immobilized on metal surface and is injected the analyte solution. When there is the binding between molecules, there is a variation in angle of reflection and of surface plasmon resonance

Table 1 Advantages and disadvantages of SPR

Advantages	Disadvantages
Real-time analysis	Requires a long time for sample preparation
Association (k_a) and dissociation (k_b) constant study	Protein instability
Use of several proteins	It is possible to immobilize only one protein on the gold chip
Does not interfere with physical and chemical molecular properties (e.g. fluorescence of compounds)	
Possible to analyse total protein extraction	
Higher sensibility	

instruments [12]. Although many assays are available to study biomolecular interactions, most require labelled reporter molecules such as fluorophores or antibodies (AlphaScreen, immunoprecipitation [13]) or large quantities of material. The advantage of SPR lies in its reliable instrumentation, automation, disposable sensor chips and versatility of surface chemistries. Table 1 lists the main advantages and disadvantages of the SPR methodology.

1.1.3 Isothermal Titration Colorimetry (ITC)

ITC is a physical technique that allows in vitro study of the thermodynamic parameters of macromolecules or drug interactions in solution. Using temperature variation, it is possible to quantify the interaction by measuring the heat that is released or adsorbed during binding [14, 15]. The microcalorimeter has two cells, one for water (the control) and one for the sample. The sensor detects temperature difference between the cells. The ligand, loaded by a syringe in a cell sample, binds the molecule, and the heat change is measured compared to the control.

This technique has several advantages: (1) the protein can be left in solution and no molecule modification is therefore required, and it measures the affinity of proteins in their native state; (2) the addition of chemofluorescent tags is not necessary; (3) stoichiometry, association constant and binding enthalpy can be characterized in a single experiment; and (4) experiments can be performed in systems such as dispersions, intact organelles or cells. The main disadvantage of ITC is the limited range for measuring binding affinities (association constants from 10^4 to 10^{-1} M). In addition, it is a slow technique for HTS analysis (0.25–2 h/assay). In recent years, new technologies for a more rapid analysis of data have been designed. In 2017, Mitter et al. described the development of a new ITC technique allowing for the acquisition of data in the order of 0.2 s and less.

ITC is widely used for:

- Quantification of binding affinity
- Confirmation of binding targets of small molecules and characterization of mechanisms of action
- Validation of IC50 and EC50
- Measurement of enzyme kinetics

ITC can be performed using MicroCal™ (Malvern Panalytical) ITC instruments such as MicroCal PEAQ-ITC, which gives high sensitivity and quality data with low sample consumption, or MicroCal VP-ITC, an easy-to-use isothermal titration calorimeter. ITC is mainly used for target validation, measuring enzyme kinetics, investigating the binding mechanisms and potency of drugs and studying the design of inhibitors.

1.1.4 Microscale Thermophoresis (MST)

MST, similarly to ITC, quantifies protein binding in a temperature gradient, but the movement of the molecule and conformational state is recorded by the use of fluorophores [16, 17] (Fig. 2). Infrared laser is used for temperature increase. The molecule is placed in a capillary containing a fluorescent solution, and an infrared ray of a wavelength of 1,480 nm is projected against the capillary, causing an increase in temperature. The fluorophores with an increased temperature are excited and emit fluorescence. The ability to emit fluorescence by temperature increase is inversely proportional to the ability to bind the ligand. MST analysis has no limitations in terms of molecule size or molecular weight and is performed in a matter of minutes in free solution.

There are a number of different devices to monitor MST, such as the Monolith NT.Automated or Monolith NT115Pico (NanoTemper) instruments designed to detect even low picomolar concentrations in microlitre volumes of red-emitting fluorophores. With minimal sample consumption, they provide high sensitivity

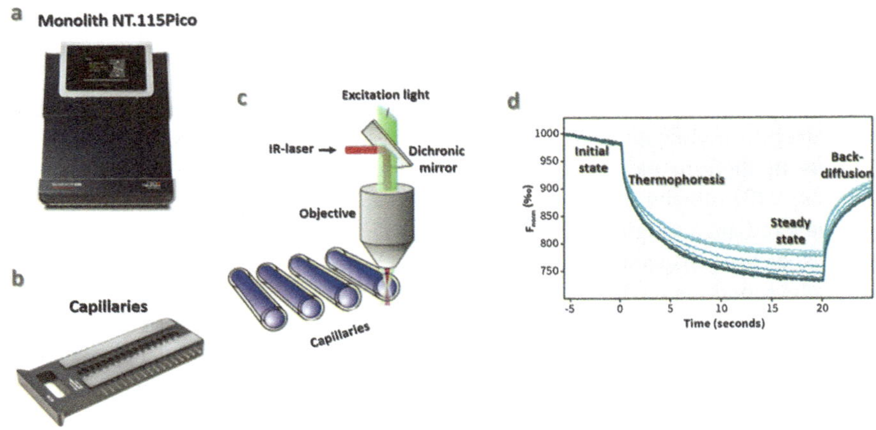

Fig. 2 (**a**) Monolith NT.115Pico. (**b**) Capillary tray. (**c**) Schematic of assay: the temperature is increased inside the capillaries, where the sample is located, by IR-laser. Florescence is detected by the objective. (**d**) Signal obtained in the assay. Molecules in the initial state are homogeneous. When the temperature increases, a T-jump occurs, corresponding to a change in fluorophore properties. The laser is then deactivated and the molecules move by mass diffusion

and affinity interactions. The Monolith NT.Automated is specifically designed with high-throughput capabilities for HTS approaches. In recent years MTS has also been used to study interactions directly in cell lysates using fluorescence fusion proteins. To bypass fluorescence artefacts, MST experiments can be performed with two different cell lysates expressing different fluorescence fusion proteins. These approaches can be used to rapidly test whether and in what manner proteins of interest interact in close-to-native conditions without protein purification. As reported in several studies, MST can be used to characterize protein-histone interactions. Alpatov et al. identified the fragile X mental retardation protein [18] as a chromatin-binding protein that uses its tandem Tudor domain for binding. Josling et al. also demonstrated that the bromodomain protein PfBDP1 binds specifically to H3K9Ac and H3K14Ac peptides with Kd = 110.79 ± 10.31 μM (R2 = 0.995) and Kd = 126.17 ± 14.39 μM (R2 = 0.993), respectively. In conclusion, MST technology is able to perform screening assays, binding assays and competition assays with multiple binding partners [19]. The advantages and disadvantages of these last two techniques are summarized in Table 2.

1.2 Cell-Based Assays

The presence of numerous binding sites, molecular dynamics, interaction complexity and external factors complicates the study of macromolecular complexes in cells. Several new techniques have been developed to study protein-protein interaction

Table 2 Advantages and disadvantages of ITC and MST

Advantages	Disadvantages
• Possibility to measure in one experiment all binding parameters	• Required higher amount of ligand
• It's not required molecular modification to study the interaction	• It's possible to measure only the μM concentrations
• Nothing ligand immobilization	• Kinetics of association and dissociation aren't measured
	• It's a slow technique not used for HTS mode
	• Doesn't read non-covalent complex
• Read small-molecules, vesicles and synthetic compounds	• Required molecular modification with fluorescent tag that can be able to induce a non-specific binding
• Nothing ligand immobilization	• Don't do information on kinetic interaction
• It's possible to measure nM concentrations	
• Faster technique that can be used for HST mode	

complexes, epigenetic modifications and activity of new drugs in living cells. In addition to chromatin immunoprecipitation (ChIP), which has always been one of the most widely used techniques, several methods have been designed to better and more rapidly investigate proteomic complexes, such as cellular thermal shift assay (CETSA) and Förster resonance energy transfer (FRET)/bioluminescence resonance energy transfer (BRET).

1.2.1 ChIP Assay

Immunoprecipitation and ChIP assays are the two main techniques for studying epigenetic modifications and their quantization (ChIP-PCR). The two approaches have also been combined with next-generation sequencing (NGS) to generate profiles of genetic modifications [20] (ChIP-seq), as will be amply described in Sect. 2.1. Over the years, proteomics studies have evolved, coupling conventional ChIP technology that involves cross-linking with formaldehyde, sonication and subsequent quantitative analysis using real-time PCR with mass spectrometry (MS). The combination of these two techniques led to the development of new protocols such as ChIP-MS [21], chromatin proteomics (ChroP) or rapid immunoprecipitation mass spectrometry of endogenous protein (RIME) [22], which allow a faster analysis of all proteins. Standard ChIP assay has several disadvantages, including the possible alteration of chromatin-protein interactions during the experiment, the large amounts of starting material necessary to obtain good results and the long time required to complete the procedure (3 days). A new microfluidic technique was developed to decrease reaction volumes and reduce experiment times [23]. This new microfluidic device integrates sonication and immunoprecipitation, using 1,500–2,000 cells to point (compared to 10^{6-7} in conventional ChIP). Specifically, the cell/DNA sonication step is integrated with the rest of the procedures, and acoustic energy is used to facilitate the mixing and washing of beads. In this way the experiment can be completed in only 40 min. Cross-linked cells are placed on the microfluidic chip to generate sonication of DNA through a magnetic field. The use of crescent structures increases the gas-liquid contact surface, increasing the power of the magnetic field. Low acoustic waves are then used to allow greater interaction between the beads and cell extract. The immunoprecipitation step is completed after 30 min. The beads are simply washed with a buffer to remove impurities and then collected for analysis. The obvious advantages of this technique include small amounts of starting material, ease of experimental procedure, very short experiment times and high resolution.

1.2.2 Cellular Thermal Shift Assay (CETSA)

CETSA allows for the evaluation of protein interactions through macromolecular stabilization or destabilization and therefore variation in the melting temperature (Tm) [24]. Specifically, this technique is based on the biophysical principle that if a protein/enzyme is subjected to temperature increase, it will undergo protein unfolding with different melting curves. CETSA has several advantages: (1) it is not necessary to modify the molecule for immobilization on beads (as in chemo-proteomic approach, AlphaLISA, MST); (2) it is not an expensive technique; and (3) it is very simple to perform. The CETSA protocol includes the following steps: incubation of cells or cell extracts (to ensure permeability of the compound) with the molecule of interest; incubation of cell pellets at different temperatures, usually between 37°C and 70°C (the choice of temperature depends on the protein under investigation); total protein extraction; and detection of stabilization by Western blot.

Over the years, this technique has undergone various modifications, such as the use of different compound concentrations to evaluate dose-dependent enzymatic stability or to study not only the interaction of the molecule with a single protein but with all the cellular proteasomes. A two-dimensional thermal proteomic profile strategy was developed to investigate all the off-target effects of panobinostat (an epigenetic drug used for multiple myeloma) [25]. After treatment of the cells with panobinostat at five different concentrations and incubation at 12 different temperatures, all proteins undergoing a change in Tm, and therefore all proteins that bind the drug, were analysed by liquid chromatography-mass spectrometry (LC-MS)/MS. Using this assay, four new different proteins which interact with the compound were identified. One of the limitations in using CETSA is that it is unsuitable for HTS of protein interactions due to the time and quantity of materials required. In 2017, Holbert and colleagues [26] developed a new method named high-throughput dose-response (HTDR)-CETSA, which allows for HTS using a 384-well plate format of drug-protein interactions by chemiluminescence signal (Fig. 3). In this assay the target protein, SMYD3, is fused and overexpressed in the cell using the BACMam system (ThermoFisher Scientific), with a small enzyme donor fragment of β-galactosidase (β-gal) called enhanced ProLabel® (ePL; DiscoverX).

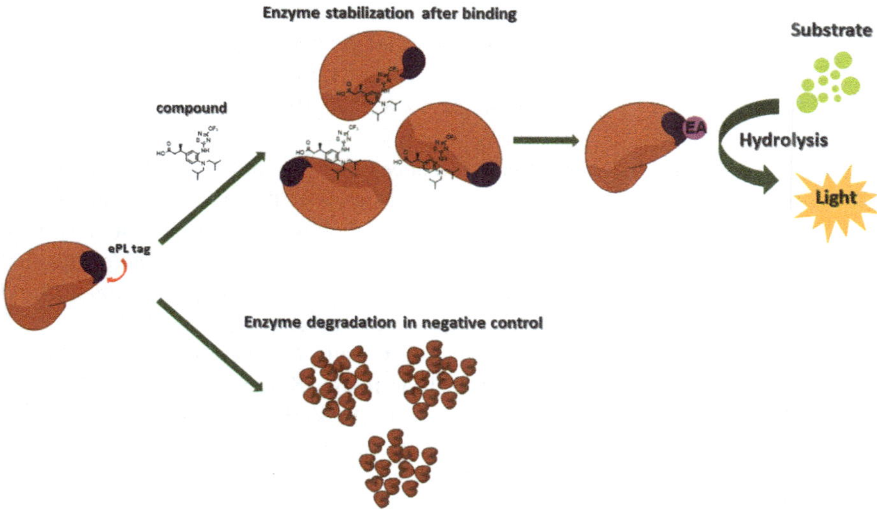

Fig. 3 CETSA in chemiluminescent system. The protein SMYD3 is expressed with ePL tag. After induction and stabilization with GSK5628, the enzyme acceptor binds ePL tag and hydrolyses the substrate

Upon addition of a compound that binds the target, stabilization can be monitored by measuring target protein abundance using chemiluminescent detection. The detection reagents are a chemiluminescent substrate added with a large enzyme acceptor fragment that reacts with the ePL tag to create an active β-gal enzyme, which hydrolyses the substrate to generate a chemiluminescent signal. The signal is directly proportional to protein stabilization (www.discoverx.com/products-applications/target-engagement-assays). The cells were plated in 384-well plates, incubated with the compound for 1 h and treated at different temperatures, as in the standard CETSA protocol. The cell reagent was added and the signal was read after incubation for 3 h at room temperature. Using HTDR-CETSA, (1) it is possible to quantify the protein interaction; (2) centrifugation and washing are not necessary because the results are not affected by the presence of serum or cellular lysate; and (3) data observation is more simple and rapid than in standard CETSA, which uses Western blotting to confirm the protein interaction.

1.2.3 Förster Resonance Energy Transfer (FRET) and Bioluminescence Resonance Energy Transfer (BRET)

FRET and BRET are two effective methods used to identify protein interactions in cells in real time by measuring the interaction between donor and acceptor chromophores. FRET is a physical assay where laser light induces donor excitation and the transfer between excited donor and acceptor chromophores when the molecules are in proximity (100 Å) [27]. Förster proposed that the energy (k_t) was given by:

$$k_t = r^{-6}k^2Jn^{-4}k_F \times 8.71 \times 10^{23}$$

where r is the distance between donor and acceptor, k^2 is the orientation factor, J is the overlap between donor and acceptor spectrum, n is the refractive index of the medium and k_F is the ratio constant emission of donor. FRET is generally read through an imaging-based assay or flow cytometry to measure the molecular distance or the "on" and "off" interaction state. Depending on the experimental aim and the epigenetic protein under investigation, several cell types including U2OS, HEK293, HeLa, COS-7, CHO and HCT-116 cells can be used. A number of instruments are available for imaging such as Image Xpress (Molecular Devices), IN Cell Analyzer (GE Healthcare) and Operetta (PerkinElmer). Alternatively, commercial plate readers can be used to detect the FRET signal, including Infinite M1000 (Tecan), PHERAstarFS (BMG), Synergy 2 and 4 (BioTek) and Envision (PerkinElmer).

Fluorescence lifetime microscopy (FLIM) is a third alternative for measuring FRET in presence and absence of compounds. It only measures changes in donor fluorescence lifetime. Recently, FLIM microplate readers have been produced that are compatible with SBS standard labware and improved to perform 96-well plate reads of tens of minutes instead of hours (see, e.g. http://www.fluorescenceinnovations.com/cells.html) [28, 29]. Using FRET to visualize histone methylation in cells, Tinget al. developed a plasmid-based biosensor which consists of a peptide (substrate, H3 at K9 or K27) that is inserted into a flexible linker with a methyl-lysine binding domain (chromodomain), which binds selectively to lysine-methylated peptides. When methylation occurs at the histone-derived peptide, the methyl-lysine binding domain causes a structural change and induces a FRET signal. Because this mechanism is reversible, demethylation induces the separation of fluorescent units and so lowers the FRET signal back again.

FRET has two main advantages: (1) it does not use additional substrates to generate the signal and (2) it is able to measure the interaction with low protein expression better than other assays, such as BRET. However, the technique also has several disadvantages: (1) the intensity of laser light can cause problems of phototoxicity or photoheating; (2) bleed-through occurs when an acceptor is excited by the donor excitation wavelength and vice versa; and (3) there may be an overlap in donor and acceptor emissions. BRET overcomes some of the problems associated with the FRET approach. BRET is based on the same principles as FRET. The energy is generated by a fluorescent donor, usually Renilla luciferase (Rluc), which matches with yellow fluorescent protein (YFP) [30]. Rluc was originally chosen because its emission spectrum overlaps with YFP emission spectrum. It is now possible to change the Rluc and protein acceptor (Table 3).

In recent years, NanoLuc luciferase (Nluc) assay was developed to increase the distance between the donor and acceptor spectrum and to improve the low quantum

Table 3 Some protein donors and acceptors in FRET

Donor	Donor emission (k, nm)	Acceptor	Acceptor excitation/emission (k, nm)
Rluc	420–530	EYFP	513/527
RLuc	420–530	Topaz	514/527
RLuc	385–420	GFP	405/510
Aequorin	430–500	GFP	489/510

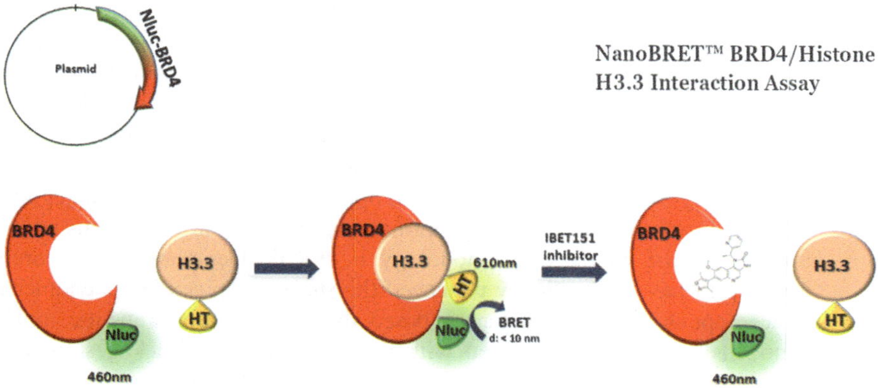

Fig. 4 NanoBRET interaction assay. BRD4 is co-expressed with Nluc with emission at 460 nm. The enzyme interacts with the substrate and the BRET signal is observed. IBET151 binds BRD4 and only the light emission of Nluc is visible

yield and poor luminescent stability of BRET [31]. Nluc uses furimazine, an engineered small protein (19 kDa), as the donor, which produces a higher luminescent signal than Rluc and allows better discrimination with acceptor fluorophores, and a fluorescently labelled HaloTag fusion protein as the acceptor. A recent study demonstrated that chloroalkane derivative of non-chloro TOM dye is the best acceptor substrate [31]. It provides peak light emission at 635 nm, and so the spectra are separated by 175 nm. Nluc is able to visualize subcellular translocation events with sub-second temporal resolution. It is in fact possible to evaluate bromodomain activity and identify compounds that inhibit it using HTS. The interaction between the bromodomain and the acetylated histone substrate can be monitored by expressing Nluc-BRD4 (one of the BET family members [32] in mammalian cells with H3.3-HT protein). When BRD4 binds H3.3 substrate, the distance between Nluc and HT is less than 10 nm and it is possible to see the signal. Treating cells with an inhibitor that competes with the substrate induces the loss of BRD4-H3.3 binding and so a decrease in BRET signal (Fig. 4). Similarly to FRET, with the BRET approach, it is possible to use a cell line that can be transfected (e.g. HEK293, HeLa, HCT-116, NIH3T3, CHO and Jurkat). It is generally better to generate a stable cell line expressing one or both proteins for a higher and clearer signal. Commercial plate readers pre-equipped with the correct filter set-up include GloMax Discover (Promega) and CLARIOstar (BMG). Instruments that can be supplied with optional filters include Varioskan (Thermo Scientific; Edmunds Optics filters: donor 450 nm CWL, 25 mm diameter, 80 nm FWHM, Interference Filter and acceptor 1 in. diameter, RG-610 Long Pass Filter) and EnVision (PerkinElmer; Chroma filters: Emission Filter for EmSlot4).

2 Innovative Genome-Wide Technologies

Numerous accurate and reliable approaches have been gradually introduced to study epigenetic processes both at the level of single genes and genome-wide, collectively called epigenetic technologies [33]. Progress in epigenetics [34] is closely linked to the design of new technologies. Since epigenetic alterations may cause several disorders, studying the (epi)-genome is a valid approach to identify these deregulations. While the genome itself is stable, epi-modifications are in continuous transformation and so the choice of methodology is crucial. Genome-wide research technologies [35] have dramatically enhanced the study of epigenetic phenomena and the development of new epigenetic-based diagnostic and therapeutic tools. Many innovative platforms associated with advanced software packages for the study and interpretation of the epigenome are now available. These platforms are often related to specific diagnostic panels for different diseases. This section will describe the most advanced methodologies for epigenome-wide applications with a particular focus on single-cell epigenomic methodologies [36].

2.1 ChIP: From Standard Procedures to Innovative Applications

ChIP is one of the principal methodologies used to examine the epigenetic state of chromatin, that is to say the regulatory processes including direct methylation of DNA, covalent modifications to histones and protein interactions with genomic regions. The standard ChIP protocol [37] includes a set of specific steps as shown in Fig. 5. Over the years this technique has been extensively revised and improved to provide a better "resolution" of epigenetic markers and profiles. However, ChIP results can still be disappointing for several reasons: high background binding, inappropriate controls, low affinity of antibodies, errors made during DNA amplification and suboptimal quality of reagents used in chromatin preparation procedures. A further critical experimental point involves the starting material. Since ChIP comprises numerous steps, the amount of cells is crucial and pooling immunoprecipitated DNA samples is often a mandatory strategy. Sometimes pooling starting material is not always possible as specific cell types, cancer cells or heterogeneous materials in general are used. The size of DNA fragments is another key issue. DNA fragments are usually obtained by sonication and optimal fragment size

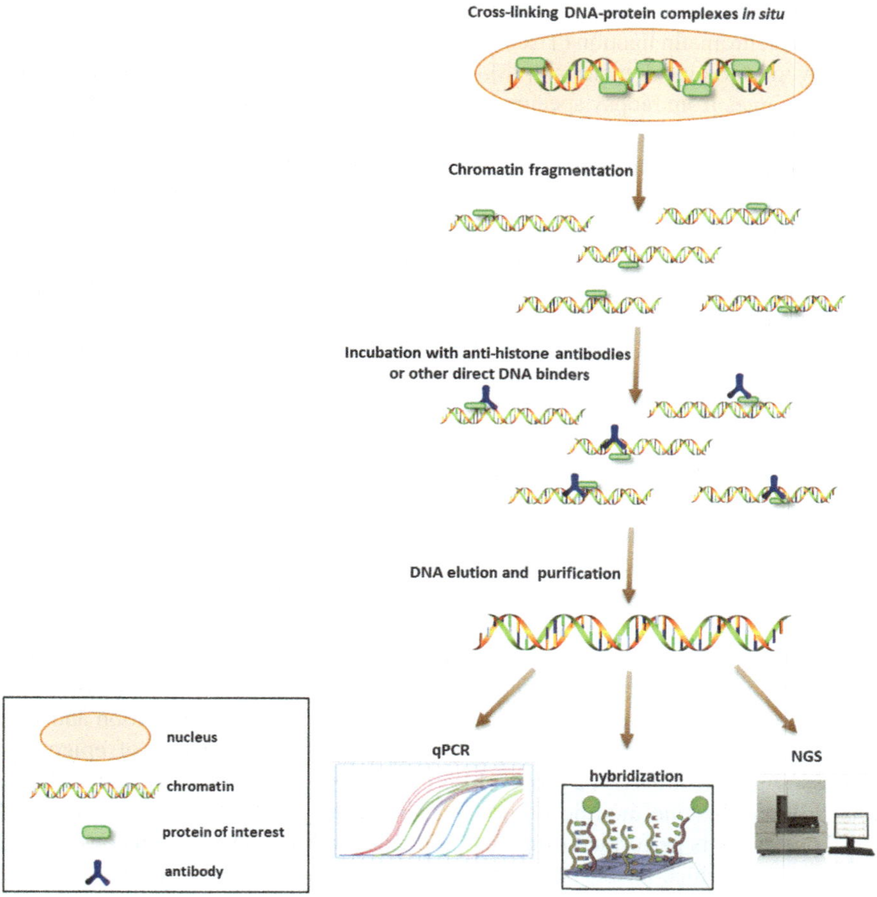

Fig. 5 Schematic representation of standard ChIP procedure. The figure shows the main steps characterizing ChIP basic protocol: DNA-protein complexes are in situ cross-linked, following chromatin fragmentation by sonication and incubation with specific antibodies. DNA is then eluted and purified. Finally the last steps include analysis by quantitative PCR, hybridization platforms or NGS (Chip-seq)

should be in the range of 200–500 bp. This is a very important step which depends on many different variables such as the kind and quantity of starting material (cell line, tissue, etc.), the power and frequency of sonication, the timing for and between each step and the sonicator itself. As previously discussed, conventional ChIP requires a huge amount of starting material, which is often lost during the numerous steps involved in the protocol. This is a crucial drawback, which often does not allow appropriate validation of the method, thus preventing the generation of reliable and precise results. As the use of personalized therapies is increasingly playing a fundamental role in the study and prevention of various diseases, new nonconventional approaches are continuously evolving. One of the latest applications involves the use of targeted chromatin ligation (TLC), by which it is possible to start from a very small quantity of biological material [38]. This is especially important when starting material comes from biopsy specimens of human tissues or rare cell subpopulations. This alternative method preserves the sample without requiring the numerous steps involved in conventional ChIP procedures. The TLC technique increases the recovery of enriched material and improves the efficiency of ChIP using a simplified workflow. This approach excludes the use of beads by recruiting proximal binding of the antibody to specific adapters, followed by subsequent amplification of the ligated chromatin involving fewer steps and much shorter experimental times. This new application greatly reduces the methodological complexity by increasing the sensitivity of results obtained. These conditions have been optimized over time and more specific protocols have been developed.

2.1.1 ChIP in Microfluidic Droplets

This microfluidic-based method is another advanced tool able to overcome the limitations of standard ChIP described above. It involved for the first time the development of ChIP in droplets based on a microfluidic platform that combines nanolitre droplets, magnetic beads and tweezers. This innovative technique has many advantages including compartmentalization within nanolitre droplets, improved mixing, reduced consumption of samples and reagents, lower assay time (about 7 h), decreased number of cells required, lower cross contamination and very high sensitivity/specificity. The assay is very useful for the study of epigenetic processes/modifications. The protocol [39] was specifically designed to investigate four different histone modifications, with known epigenetic roles. This integrated ChIP procedure is performed using a Teflon tube platform. The system consists of a syringe pump which pushes sequential droplets across the tube toward a magnetic tweezer, whose activation/deactivation gives the extraction and dispersion of magnetic beads and droplets. The syringe pump permits droplet manipulation and processing by switching from injection to aspiration mode. The first train of droplets is made of magnetic beads associated with antibodies and chromatin, washing buffer, RNase and proteinase K. Magnetic beads are collected by magnetic tweezers and then extracted, washed in buffer droplets, transferred to the RNase droplet for RNA digestion and extracted after 5 min incubation. Subsequently, magnetic beads are dispersed in the proteinase K droplet for protein digestion and heated for 5 min at 70°C to release DNA from digested proteins. Magnetic beads are then discarded, and a new train of droplets is generated with charge switch magnetic beads, washing buffer and elution buffer droplets. These last beads are then transferred to the droplet containing DNA and digested proteins to capture the DNA, before it is extracted, washed and eluted and finally discarded before qPCR. The sensitivity of the droplet methodology is, however, highly dependent on an adequate quantity of magnetic beads as well as chromatin and specific antibodies. The choice of a well-defined range of starting material is therefore crucial to improve assay performance.

Recently, the droplet-based microfluidic system was also used by combining DNA barcoding and NGS in order to collect and compare orthogonal individual cell gene expression data [40]. This approach is crucial since epigenetic heterogeneity is often difficult to detect due to the complexity of chromatin organization and different patterns of variability. Assaf Rotem and colleagues identified the profiling of histone methylation marks (H3 lysine 4 trimethylation and demethylation) in a mixture of mouse populations of embryonic stem cells (ESCs), fibroblasts and haematopoietic progenitor cells by applying this method to discriminate and identify high-quality chromatin profiles for each cell present in the blend. Droplet-based ChIP assay thus provides a highly advantageous, rapid and cell-specific approach to epigenetic analyses.

2.1.2 G4 ChIP-Seq

G4 ChIP-seq is a novel method which can be used to determine G-quadruplex (G4) structure formation genome-wide in chromatin [41]. DNA sequences are able to form four-stranded G4 secondary structures that are involved in epigenetic regulation. This modified protocol applies standard ChIP-seq for the detection of secondary DNA structures using a specific antibody. As with any ChIP assay, the quality of the antibody used is critical. This technique uses the G4 structure-specific single-chain fragment variable antibody BG4, which is prepared using the expression vector pSANG10-3F-BG4.

Although G4 ChIP-seq is a highly innovative method that recognizes chromatin regions forming G4 DNA structures in a cell population, it presents technical limitations due to the detection of G4 structures at a given locus present in a heterogeneous cell population. In addition, this experimental approach is not able to detect the temporal organization of G4 structure and on which strand it is formed. The entire protocol requires less than 1 week to be completed and might be also combined with a microfluidic method (described in the subsection above) for single cells in order to avoid some of the issues linked to heterogeneous cell populations. Despite technical limits, the use of this method led to the identification of restricted regulatory regions of chromatin, focusing specifically on regions regulating cell fate and function [41] which are not often detectable by standard procedures. The next step will be to apply this technology in human tissues and other model organisms.

2.1.3 Standard Epigenome-Wide Association Studies (EWAS)

New high-throughput genome-wide epigenetic technologies such as ChIP-seq [42] and ChIP-on-chip [43] have recently been developed. ChIP-on-chip arrays are available from Affimetrix, Agilent and NimbleGen, in both whole-genome tiling array format for model organisms including human and mouse, as well as focused arrays for promoter regions and custom array designs. The main advantage of ChIP-on-chip is the huge amount of data generated which can be statically analysed minimizing the artefacts and improving the significance and resolution of results. EWAS are designed to identify a specific epigenetic mark [44] associated with a particular phenotype through the use of array and sequencing-based technologies.

One of the most studied epigenetic marks is DNA methylation of 5-methylcytosine (5mC) on CpGislands [45] since methylated DNA is the key component of epigenetic information in mammals [46]. A first step in studying DNA methylation is to quantify all global levels of DNA methylation. Massive parallel sequencing (MPS) [47] quantifies results after DNA pretreatment. Pretreatment may be affinity-based, which enriches methylated or unmethylated sites of DNA. Alternatively, pretreatment may be performed using methylation-sensitive restriction enzymes or bisulfite conversion of DNA. In the latter case, the gold standard approach is whole-genome bisulfite sequencing [48], which provides quantitative information about cytosines throughout the genome. However, this is a very expensive assay.

MPS approaches can be divided into two different groups: (1) reduced representation bisulfite sequencing (RRBS), which uses a measuring range of generated fragments from digestion with restriction enzymes for targeting deep sequencing at specific loci, and (2) HELP-tagging assays using digestion with methylation-sensitive restriction enzymes in order to generate tags at these sites, which represent the DNA methylation level. MPS provides data that cannot be obtained by microarray such as single-nucleotide polymorphisms. As microarrays are designed to recognize 5mC, in the context of CG, in some cases, levels of CHG and CHH (where H equals A, T or C) may not be detected or might interfere with restriction enzyme digestion. In bisulfite sequencing (BS-seq), cytosine that is not converted by bisulfite can be both 5mC and 5-hydroxymethylcytosine (5hmC). In order to overcome this issue, two different arrays are therefore necessary. Four different modifications including 5hmC, 5mC, 5-formylcytosine (5fC) and 5-carboxylcytosine (5caC) are read by bisulfite as unmethylated cytosine, which could require more specific assays [49], described in the following subsections. Although these assays have been improved to be specific for each type of modification, considering the huge amount of (epi)-genetic modifications underpinning different areas of the genome, they inevitably present both advantages and disadvantages (Table 4).

2.1.4 Tet-Assisted Bisulfite Sequencing (TAB-Seq)

TAB-Seq is a novel method using bisulfite conversion and Tet proteins to study 5hmC [50]. Specifically, genomic DNA is first treated consecutively with T4 phage β-glucosyltransferase (β-GT) and recombinant mouse Tet1 to convert 5hmC to 5gmC and 5mC to 5caC. β-GT catalyses the attachment of β-D-glucosyl residues from uridine diphosphoglucose to the hydroxyl group of 5hmC. DNA bisulfite treatment leaves 5hmC unchanged, while 5mC and unmethylated cytosines are converted to uracil. This assay provides an accurate representation of 5hmC localization in the genome and clearly differentiates between 5hmC and 5mC, specifically

Table 4 Advantages and disadvantages of different ChIP-seq applications

Method	Advantages	Disadvantages
TAB-seq	Accurate representation of 5hmC localization	Detection limit for the quantification of 5hmC
EnIGMA	To decipher the sequence converted by bisulphite back to the original sequence	DNMT1 enzyme does not methylate hemi-hydroxymethylated CpGs and non-methylated CpGs
CAB-seq	Detection of 5caC with single-base resolution in DNA	Low protection rate of 5caC deamination
fCAB-seq	Identification of 5fC using BS-seq	Useful for base-resolution mapping
scBS-seq	To avoid the loss of genomic DNA in individual cells during DNA purification	DNA degradation after bisulfite step
scRRBS	To provide genome-wide coverage of CpGs in islands at single-base resolution and of dense areas in CpG methylation	Biased sequence selection
snmC-seq	Gain insights into epigenetic alterations and regulation	Broader applications

identifying 5hmC. Further, CpG and non-CpG hydroxymethylation throughout the genome is covered at single-base resolution. This is a very accurate method, but large amounts of DNA starting material are required and the detection limit is insufficient for the quantification of 5hmC with extremely low 5hmC levels. To date, many other methods have been developed and improved such as those for the direct identification of mC, hmC and unmodified cytosine (C) at single-base resolution.

2.1.5 Enzyme-Assisted Identification of Genome Modification Assay (EnIGMA)

EnIGMA is a highly efficient and reliable analytical method based on the specificity of the enzyme DNMT1 [51]. Specifically, the enzyme methylates the cytosine of hemi-methylated CpGs but does not methylate hemi-hydroxymethylated CpGs and non-methylated CpGs. The technique includes genomic DNA digestion with the appropriate restriction enzyme, followed by digested DNA end-repairing and A-tailing. Ligation of hairpin-shaped adaptor DNA with deoxyuridine is then followed by "USER" enzyme digestion. Alternatively, digested DNA is dephosphorylated and directly ligated with hairpin-shaped adaptor DNA using the cohesive end of the restriction enzyme cutting site. DNA is then treated with DNA polymerase and DNMT1 enzyme. Finally, the method includes bisulfite treatment and PCR. Sequencing and CpG comparison are necessary to determine whether the cytosines in the original DNA were mC, hmC or unmodified C. The procedure can be applied to many types of epigenetic studies including comprehensive genome-wide analysis of hmC using massive parallel sequencers. This method is also very useful to decipher the original sequence converted by bisulfite, eliminating the need for re-sequencing. The technique has recently been used in studying genes involved in mouse brain development such as Arhgap27 and Nhlrc1, which have a significant number of hmC associated with single-nucleotide polymorphisms. EnIGMA is also important in the study of ESC memory, since cytosine modifications control epigenetic status across the epi-(genome) [51].

2.1.6 Chemical Modification-Assisted Bisulfite Sequencing (CAB-Seq)

CAB-Seq can detect 5caC with single-base resolution in DNA. In this technique, 5caC is selectively labelled by 1-ethyl-3(3-dimethylaminoproyl)-carbodiimide hydrochloride (EDC) by using a xylene-based primary amine [52]. This specifically labels 5caC but no other modified cytosine. This modified 5caC does not undergo deamination by bisulfite as does unmodified 5caC and is therefore sequenced as a cytosine instead of a thymine. However, this technique has a low protection rate of 5caC deamination (about 50–60%) and requires the sequencing of a non-treated DNA control. The use of this technique involves mouse ESCs, which are characterized by a strongly decreased quantity of 5caC.

2.1.7 5-Formylcytosine Chemical Modification-Assisted Bisulfite Sequencing (fCAB-Seq)

fCAB-seq [53] is a chemical method to identify 5fC using BS-seq. The technique exploits the reaction of hydroxylamine with 5fC but is limited by the fact that it is only useful for base-resolution mapping. In fCAB-seq, 5fC is protected by bisulfite-mediated deamination by treatment with O-ethylhydroxylamine. Hydroxylamine-protected 5fC is therefore read as a cytosine instead of a thymine, as in BS-seq. By comparing data from BS-seq and fCAB-seq, the locations of 5fC from unmodified cytosine and 5CaC can be differentiated. RRBS for 5fC is based on the chemical selective reduction of 5fC and subsequent bisulfite conversion. After bisulfite treatment, 5fC is deformylated and subsequently deaminated in uracil, so that it is read as a thymine and identified as an unmodified cytosine in BS-seq. This method has also been used in a mouse ESC model in which genomic distribution and the DNA glycosylase TDG strongly depend on 5fC regulation. 5fC profiling is further associated with 5mC/5hmC oxidation at different gene regulatory elements. In addition, 5fC accumulation correlates with p300 binding at poised enhancers, thus regulating important epigenetic mechanisms [53].

2.1.8 Single-Cell Bisulfite Sequencing (scBS-Seq)

This genome-wide DNA methylation sequencing method can also be reproduced in single cells by integrating some of the steps in the standard RRBS protocol in order to avoid the loss of genomic DNA in individual cells during DNA purification [54]. This method is useful for the epigenetic study of heterogeneous and/or rare cell populations. The approach has been applied in mouse ESCs used as a model of study due to their versatility, allowing identification of epigenetic and cellular biological heterogeneity [55]. The disadvantage of this procedure is that DNA may undergo degradation after the bisulfite step. This method has been improved in combination with single-cell genome and transcriptome sequencing in order to explore novel correlations between epigenetic modifications and gene expression in embryonic stem cells. The protocol has been modified with particular regard to amplification steps. Further, this method was used to study methylated DNA heterogeneity in a cell population in order to understand the role of this epi-mark in a well-defined developmental process of disease compared to normal conditions [56].

2.1.9 Single-Cell Reduced Representation Bisulfite Sequencing (scRRBS)

In the scRRBS method, genomic DNA is previously digested with one or multiple restriction enzymes (e.g. MspI) that identify common sites on CpG islands (CCGG sites) to produce sequence-specific fragmentation [54]. The fragmented genomic DNA is then treated with bisulfite and sequenced. This approach has the advantage of providing genome-wide coverage of CpGs in islands at single-base resolution as well as coverage of dense areas in CpG methylation. The disadvantage of conventional RRBS is biased sequence selection due to the fact that restriction enzymes cut at specific sites. In addition, the assay measures 10–15% of all CpGs in the genome and is not able to discriminate between 5mC and 5hmC. Differences from the standard method include the biotinylated 5′ end of universal adapters and 3′ end of index adapters and the introduction of rescue steps. In this way, it is possible to increase DNA sequences by PCR.

2.1.10 Single-Nucleus Methylcytosine Sequencing (snmC-Seq)

snmC-seq is a high-throughput method based on a whole-genome bisulfite sequencing approach to identify differentially methylated regions across thousands of cells in order to elucidate cellular diversity of complex tissues [57]. This high-throughput single-cell sequencing technique can be useful to gain insights into epigenetic alterations and regulation. The method has been applied in human and mouse frontal cortex neurons to identify specific neuronal clusters through the use of specific neuronal markers followed by correlation analyses [57]. Each cluster revealed differentially methylated regions characterized by low mCG in defined neuronal populations. Technical application on single neuron epigenomic profile therefore identified distinct mC regions containing crucial information for specific marker regions and regulatory elements both in human and mouse.

2.1.11 Single-Cell Genome-Wide Methylome and Transcriptome Sequencing (scM&T-Seq)

scM&T-seq is a recently developed method for the study of parallel single-cell genome-wide methylome and transcriptome sequencing, which provides information on transcriptional and epigenetic heterogeneity [58]. scM&T-seq uses Smart-seq2 and scBS-Seq to determine DNA methylation patterns. This approach is useful because there is no need to mask coding sequences during analysis since DNA and RNA are independently amplified. However, the technique has two main drawbacks: (1) it is impossible to distinguish between 5mC and 5hmC and (2) Smart-seq2 is not strand-specific and is applicable to only poly(A)-RNA. This technique provides a very important approach to the analysis of cellular plasticity and heterogeneity. Applications involve the use of mouse ESCs characterized by a strong heterogeneity state and variability. Using scM&T-seq it was possible to study these conditions at single-cell level by NANOG-low and REX1-low cells in sorted populations. NANOG and REX1 are in fact transcriptional factors involved in ESC differentiation, pluripotency and renewal [58], and this type of approach is of crucial importance to detect methylated distal regulatory elements associated with transcription of these key pluripotency genes.

2.1.12 Three-Dimensional (3D) Chromatin Methods

The architecture of chromatin is characterized by various levels of complexity that define the fate of specific genes in terms of regulation and expression, as well as regulatory regions. Thus, understanding the association of chromatin with specific proteins such as histones through 3D interactions is highly desirable. Many cutting-edge techniques have been developed to decipher specific genome areas as a "photo shot". These technologies involve chromosome conformation capture (3C) and 3C-based methods which decode the 3D organization of the genome and its structural conformation at specific loci. The 3C and 3C-derivative methods [59] offer the advantage of studying the genome from specific areas up to its entirety both in terms of complexity and function. The steps of these methods include those already used in standard ChIP assays, but unlike ChIP, they are able to identify ligation junctions between genomic loci. The protocol involves chemical cross-linking with formaldehyde and subsequent chromatin purification. Subsequently, specific restriction enzymes are used that are able to recognize interactions in DNA. Once chromatin fragments are released, they are further ligated and cross-linking is reversed. The final step is real-time PCR analysis, which recognizes the ligase product from the previously isolated DNA, providing a quantitative response of the interaction frequency between two ligated DNA regions. One of the techniques deriving from 3C is 4C, which allows the detection of an unknown region of DNA that interacts with a specific locus of interest. Unlike the 3C method, 4C uses a second restriction enzyme, which recognizes a different sequence from the first in order to recirculate the DNA of interest linked to its interacting DNA. 4C technology is an advanced 3C-based approach, as it combines 3C with NGS technology. A further variant of 3C is 5C technology, which has the added benefit of studying millions of interactions in genomic areas by using universal primers in one single assay. 3C-based technologies include Chromatin Interaction Analysis by Paired-End Tag Sequencing (ChIA-PET), which combines the 3C method with ChIP and examines genomic sites associated with specific transcription factors, which can be proximal and distal [60]. In ChIA-PET, a specific antibody is used together with sonicated chromatin and the protein of interest, and only this immunoprecipitated material linked to the protein of interest is used to construct the ChIA-PET. Then, biotinylated oligonucleotides are used to ligate DNA proximal fragments, which are subsequently ligated on beads. Finally, DNA is digested by a specific restriction enzyme which cuts to release a tag-linker-tag region. After pull-down of biotinylated linkers by specific beads, DNA tags are amplified. Hi-ChIP is an advanced Chia-PET method [61] which provides a more accurate interpretation of information on genome conformation. In Hi-ChIP, DNA interactions are firstly established in situ in the nucleus before lysis procedure. In this way it is possible to avoid false-positive interactions.

3 Epigenetic Regulatory Role of miRNA: New Analytical Tools and Technologies

miRNAs are small non-coding RNAs whose role as actors and targets in epigenetic machinery is well established. They are efficient regulators of heterogeneous physiological and pathological functions, as is well described in almost every cancer model. The clinical potential of miRNA lies in their fundamental tumour-suppressive and oncogenic functions and in their ability to modulate the efficacy of cancer treatments, such as chemotherapy and radiotherapy. miRNAs have proven to be efficient clinical diagnostic and prognostic markers in several human diseases including cancer, as well as therapeutic targets of biological and chemical molecules. Studies on selective miRNA modulation through antisense inhibition, mature miRNA replacement, chemical modification for better delivery or chemical compounds could have a significant impact on the successful translation of miRNA-based or miRNA-targeted therapeutics from bench to bedside. In recent years, different experimental and computational approaches using methodologies associated with miRNA detection and bioanalysis have therefore experienced rapid growth. This section provides an overview of miRNA research methods and technologies, mainly focusing on the latest and most advanced miRNA analysis strategies.

3.1 miRNA Quantification Techniques

qPCR, microarray and NGS are the most commonly used methods for miRNA quantification and profiling. They provide sensitivity and reproducibility and have become more cost-effective and easier to perform as a result of improved strategies and an increase in their global application.

3.1.1 qPCR

qPCR is currently the gold standard for miRNA quantification. Basically, miRNAs are amplified with reverse transcription (RT)-PCR to generate cDNA, and the miRNA of interest is then amplified and quantitatively measured in real time using fluorescent probes. SYBR Green and TaqMan (ThermoFisher Scientific) are two widely used systems. Although TaqMan technology offers greater miRNA detection sensitivity and specificity, qPCR procedures should avoid primer dimers and increase the sensitivity of detection threshold. Stem-loop RT [62, 63] and locked nucleic acid (LNA) primers [64] were developed to overcome these limits. Stem-loop RT-PCR is a new real-time RT-PCR for miRNA quantification. It includes two steps: RT and real-time PCR. First, stem-loop RT primers bind to the 3′ portion of miRNAs and are reverse transcribed by reverse transcriptase. cDNA is then quantified using conventional TaqMan PCR, which includes miRNA-specific forward primer, reverse primer and dye-labelled TaqMan probes (Fig. 6a). In this way, miRNA quantification shows

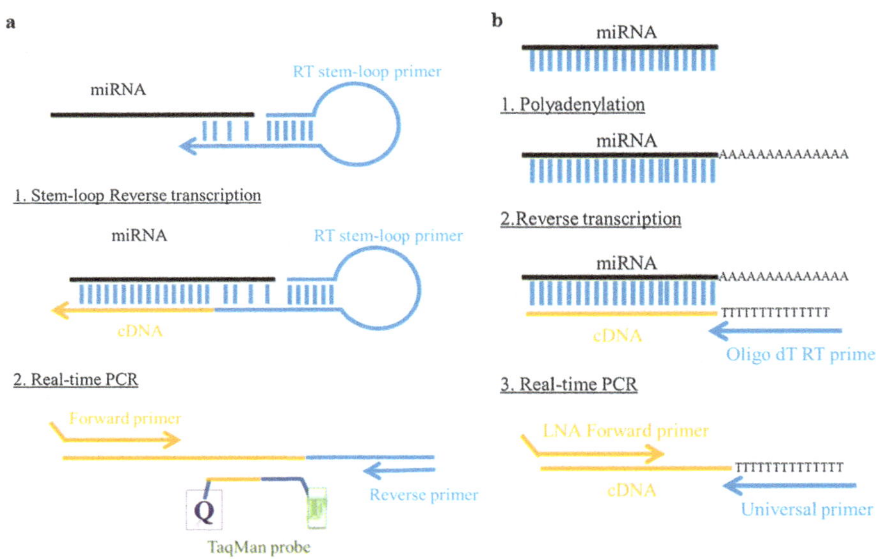

Fig. 6 Schematic representation of real-time RT-PCR for miRNA quantification by stem-loop RT (**a**) and LNA primers (**b**)

very high specificity, without false positives, and is sensitive for low amounts of these short RNA targets (such as 10 pg of total RNA). In the LNA assay, the RT-PCR reaction is performed with LNA primers. LNA is a modified RNA nucleotide in which the ribose ring is "locked" in the ideal conformation for Watson-Crick binding. LNA-enhanced oligonucleotides improve sensitivity and specificity of RT-PC, increasing the affinity for its DNA or RNA complementary oligos, mainly for small or highly similar targets. Thermal stability considerably enhances the Tm of the duplex, which increases by 2–8°C. In addition, LNA primers are efficient at discriminating single-nucleotide mismatches. In LNA RT-PCR, miRNAs are reverse transcribed from total RNA using miRNA-specific RT primers; reverse-transcribed miRNAs are then amplified using an LNA-enhanced PCR primer together with a universal PCR primer (Fig. 6b).

LNA oligos are also used in hybridization-based technologies including qRT-PCR platforms (e.g. Qiagen), microarray and in situ hybridization (ISH). For quantification of circulating miRNAs, new technologies such as droplet digital PCR (ddPCR), Ligo-miR and Two-tailed RT-qPCR have recently been introduced to overcome miRNA normalization in biofluids and to detect even single copies with high repeatability. The ddPCR method determines absolute miRNA expression and is based on partitioning of the sample into thousands of discrete, volumetrically defined, water-oil emulsion droplets. PCR amplification of the template molecules occurs in each individual droplet using reagents and workflows similar to those used for most standard TaqMan probe-based assays. Following PCR, each droplet is analysed by estimation of the number of molecules in the reaction under the assumption of a Poisson distribution. Results are expressed as target copies per microlitre of reaction [65, 66]. The high cost and complexity of ddPCR make it difficult to screen large numbers of samples across large panels of miRNA. Ligo-miR is a multiple assay technology to rapidly measure absolute miRNA copy numbers from a wide range of biological sources. Ligo-miR assay comprises two sequential ligation steps: capture and coding (Fig. 7). Firstly, a DNA universal adapter is ligated to the 3′ end of all miRNA molecules. Then in the coding ligation up to 26 miRNA-specific discrimination probes and an Alexa647-labelled common probe are hybridized to the miRNA templates to form a single-stranded DNA product. Thermal cycling is used to perform a 50X linear amplification by repeatedly generating Ligo-miR products from each miRNA template. The products can be identified and quantified using common DNA sizing methods such as electrophoresis (denaturing PAGE). Each band is a specific miRNA product where band intensity is proportional to quantity. Absolute copy numbers can be determined using the included spike-in controls and running reference samples. The sensitivity of Ligo-miR is as low as 20 copies per cell at 75 ng input levels [67]. Two-tailed RT-qPCR is a very recent system for the quantification of miRNAs (Fig. 8). In the RT step, two specific primers about 50 nucleotides long and complementary to the miRNA target are used, while the cDNA is amplified by qPCR using two target-specific primers and SYBR Green chemistry [68].

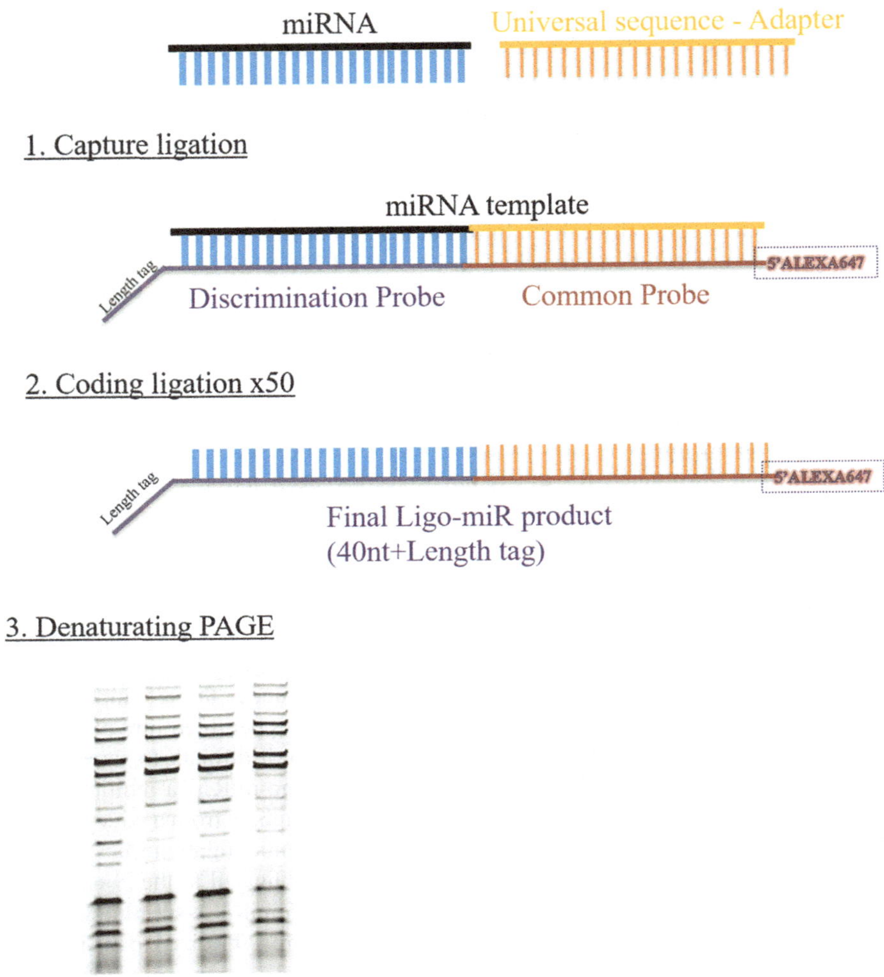

Fig. 7 Schematic representation of miRNA quantification by Ligo-miR

3.1.2 Microarray

Genome-wide analysis of miRNA expression can be performed by microarray. Microarray technology is a robust high-throughput method used to analyse simultaneously the expression of thousands of small non-coding RNAs, including miRNAs. Currently, a wide range of commercial platforms based on different technologies are available for global miRNA expression profiling, such as oligonucleotide microarray, LNA arrays (Exiqon), bead-based technology (Illumina) and microfluidic systems (Agilent, LC Biosciences). All these platforms have designed probes specific for mature miRNA sequences. The main differences are in the hybridization phase,

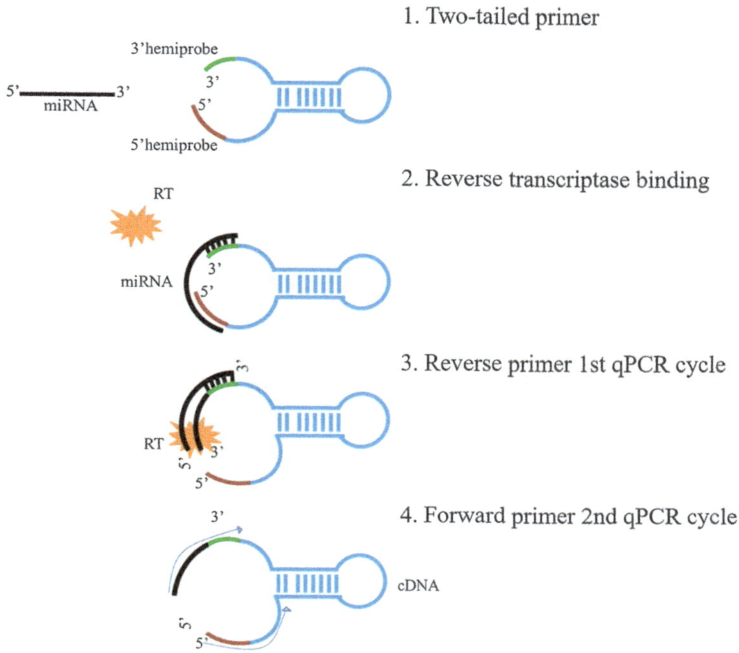

Fig. 8 Schematic representation of miRNA quantification by Two-tailed RT-qPCR

washing procedures and fluorescent dyes. The usefulness of these techniques in the clinic is limited owing to their high cost and poor reproducibility [69].

3.1.3 Next-Generation Sequencing

NGS, or second-generation sequencing, is characterized by high throughput and low cost. Currently, the most widely used sequencing systems for NGS are 454 (Roche), Ion Torrent sequencing platform (Life Technologies, Invitrogen), Illumina Genome Analyzer (Illumina) and SOLiD (Life Technologies Corporation). The basics of miRNA NGS are the same as in DNA or RNA sequencing. The NGS procedure for miRNA sequencing is divided into four phases: library preparation, sequencing, data analysis and biological interpretation. However, the various systems differ in some aspects of library preparation, such as enrichment by gel electrophoresis and choice of adapters/primers, and in data analysis. Sets of tools are available for the analysis of miRNA expression profiles, including CAP-miRSeq [70], mirTools 2.0 [71], sRNAtoolbox [72] and miARma-Seq [73]. In miRNA profiling, NGS technologies present additional advantages compared to microarray technologies, such as:

- High resolution: identification of single-base variants in the miRNA sequence.
- High throughout: de novo miRNA can be discovered.

- High accuracy: high-depth sequencing guarantees that every base is sequenced many times.

The development of high-throughput NGS methods to study miRNA expression profiles has exponentially increased the amount of data generated in the field of miRNA, bringing interesting applications in diagnostic and therapeutic approaches. Characterization of the entire miRNA spectrum by NGS provides an effective method to analyse limited sources, such as tumour biopsies from sites of difficult access, poor quality samples or circulating/exosomal miRNA from biological fluids [74–77].

3.1.4 Isothermal Exponential Amplification Reaction (EXPAR) and Advancements

EXPAR is an interesting miRNA analysis method based on isothermal amplification with high efficiency (10^6–10^9-fold). Galas and co-workers developed the EXPAR method using a combination of polymerase strand extension and single-strand nicking [78].

The basic reaction involves two steps: miRNAs hybridize with target primes and extend along the template in the presence of polymerase, and the nicking enzyme distinguishes the recognition site of double strands to cut off and generate a short DNA trigger. In the amplification phase, DNA polymerase extends the sequence to include a nicking enzyme recognition site. A nicking enzyme can then cut one of the double strands of DNA previously synthetized by DNA polymerase, which produces additional trigger sequences. This step is repeated several times usually using SYBR Green as the label. To date, different strategies have been proposed to improve EXPAR performance. Recently, an advanced miRNA assay based on the two-stage EXPAR and a single-quantum-dot (QD)-based nanosensor was developed. Importantly, EXPAR products were specifically hybridized with capture and reporter probes to form sandwich hybrids. These sandwich hybrids are collected on the surface of 605 nm emission QDs (605QDs), binding to reporter oligonucleotide/Cy5. Expression of the miRNA of interest can be analysed by Cy5 signals at excitation wavelength of 488 nm [79]. Another study proposed hairpin probes which are unfolded through catalysed hairpin assembly with the universal triggers as the primers [80]. Following a different approach, EXPAR was combined with high-resolution capillary electrophoresis single-strand conformation polymorphism (CE-SSCP) analysis. This modified EXPAR method involves the generation of signal barcodes labelled with a fluorescent dye to simultaneously analyse the amplified products via CE-SSCP. GeneMapper v4.1 (Applied Biosystems) is the leading software used to analyse the peak of interest obtained by electrophoretic mobility [81].

Table 5 Steps and materials required for miRNA detection by ISH

Phase	Standard	Alternative options
Fixation and permeabilization	Ethanol; methanol; paraformaldehyde	1-Ethyl-3-(3-dimethylaminopropyl) carbodiimide (EDC); no proteinase K
Hybridization	DNA; RNA-probes; formaldehyde buffer	Urea-based buffer. Probes: LNA/DNA; 2′OmeRNA; LNA/2′OmeRNA; morpholino; 2′F-RNA/DNA; padlock; circular; seal; Ultramar
Washing	Formamide; saline-sodium citrate (SSC)	Tetramethylammonium chloride (TMAC) + RNase A
Sequence amplification		Rolling circle amplification (RCA); oligo-fluorescence in situ hybridization (O-FISH); RT in situ PCR

3.1.5 In Situ Hybridization (ISH)

ISH is the only method able to provide insight into both miRNA level and localization in single cells. ISH can be divided into chromogenic enzyme-based in situ hybridization and fluorescent in situ hybridization (FISH). miRNA ISH is technically difficult due to miRNA features such as small size, sequence similarity and low expression levels. The standard ISH protocol has been variously modified to improve miRNA detection in several cell lines and tissues (Table 5). The first step is cell fixation and permeabilization. The former preserves the number and localization of small RNA molecules, while the latter improves cell and tissue diffusion of the probes. Permeabilization should not be strong as it can cause RNA loss. The second step is hybridization, with optimal hybridization temperatures depending on the probe. Probe design is a key step in ISH. There are linear probes directly labelled with fluorophore or ligand and probes that enable sequence amplification. Several modifications have been proposed to enhance their binding affinity to target sequences. The third step in the ISH protocol is the washing step, used to preserve the probe-target complex and eliminate off-target binding. Finally, signal detection is a critical step to detect miRNA subcellular distribution. Non-radioactive haptens, combined with probes, are commonly used and are detected by histochemical enzymatic reactions with enzyme-conjugated anti-hapten antibodies. Different controls (such as scrambled probes, and positive and negative controls) need to be used to guarantee specificity, adequate experimental conditions and good RNA quality [82].

3.2 Experimental miRNA Target Identification

To date, the majority of miRNA target identification approaches have been based on several experimental technologies, normally combined with overexpression/inhibition of the miRNA and followed by downstream gene expression or proteomic analysis [83]. Generally, miRNA overexpression can involve the transient transfection of a specific miRNA mimic or the stable introduction of a miRNA expression construct by a lentiviral vector. Conversely, miRNA inhibition can be achieved by chemically modified RNA analogues, such as anti-miRs [84], antagomiRs [85] and miRNA sponges [86], able to bind and block mature miRNAs. Reporter assay is used to validate individual miRNA:mRNA interactions, measuring the activity or expression of reporter protein [87]. Various computational and pull-down methods have been developed to improve the identification of the direct association of miRNA-target hybrids, including RNA immunoprecipitation of ectopically expressed components of the RNA-induced silencing complex or the affinity purification of synthetic miRNAs of interest, although the physiological significance of most identified miRNA-target associations is not still well understood [88]. Interestingly, newly developed techniques such as CLIP [89] and photoactivatable ribonucleoside-enhanced cross-linking and immunoprecipitation (PAR-CLIP) [90, 91] investigate the indirect relationships between miRNAs and their targets, combining the stabilization of protein-RNA complexes by ultraviolet (UV)-induced cross-linking with HTS (Fig. 9). Specifically, the CLIP method is able to define interaction sites by isolating and sequencing of small RNA segments cross-linked to RNA binding proteins (RBPs) [92]. However, this approach has some limitations, such as low RNA-protein cross-linking efficiency by exposure to 254 nm UV light and high background noise, requiring several control cross-linking experiments to correctly detect the isolated RNA fragment. Some of these problems were addressed by PAR-CLIP [90]. The protocol includes the incubation of cells with medium enriched with a photoactivatable nucleoside, such as 4-thiouridine and 6-thioguanosine, improving UV cross-linking [93, 94]. Irradiation of the cells by UV light of 365 nm, which is less harmful than 254 nm for in vivo experiments, triggers the cross-linking of photoreactive nucleoside-labelled cellular RNAs to interacting RBPs. Thus, PAR-CLIP is able to enhance RNA recovery and specificity of cross-linking, achieving single-nucleotide resolution of the binding site [95]. These approaches provide a large amount of molecular information on miRNA targets identified in RBP complexes, but the detailed protocol is time-consuming and costly and requires very specialized data analysis tools. In addition, the identification of miRNAs regulating specific mRNAs remains limited due to the imperfect complementarity of miRNA:mRNA transcripts. Interestingly, miRNA Trapping by RNA in vitro affinity purification (miR-TRAP) attempts to overcome this limitation, providing an evolved in vitro RNA affinity purification protocol for the rapid capture of the complex containing the trapped miRNA/target mRNA pair in the specific cellular context of choice [96]. The advanced miR-TRAP protocol includes photoactivatable ribonucleosides on transfected miRNA to allow only for complexes containing specific miRNAs and higher wavelength cross-linking, which is preferable for in vivo studies [96]. Specifically, the miRNA is conjugated to psoralen to produce a highly photoreactive probe. When the cells are exposed to UVA light, the Pso moiety of the miRNA reacts with uridine on target mRNAs. Pso-modified miRNA mimics act similarly to endogenous miRNAs, eliminating the use of antibodies, which can create non-specific background signals and complicate data interpretation. The cross-linking between the tagged miRNAs and target mRNAs is then stringently purified by streptavidin-coated beads,

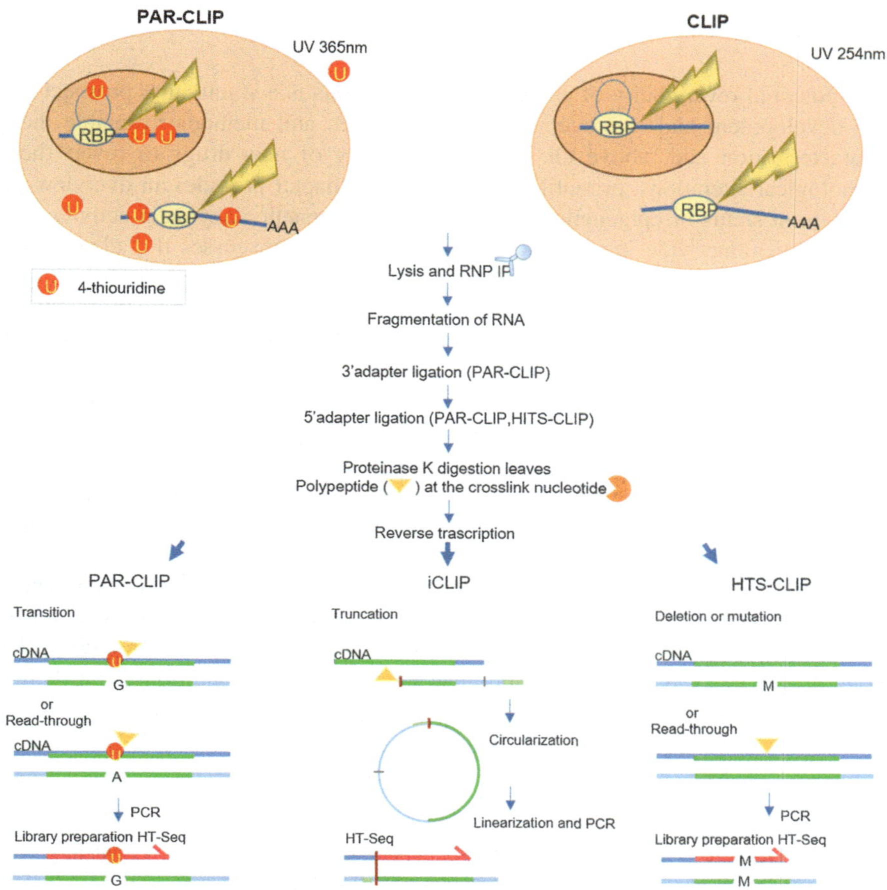

Fig. 9 Schematic representation of CLIP and PAR-CLIP procedures

minimizing the recovery of non-specific targets [96]. The pull-down method significantly enhances the enrichment of a specific target sequence for analysis via qRT-PCR. Most notably, miR-TRAP could be combined with HTS, providing the rapid identification of canonical but more importantly also of non-canonical and thus unpredictable regulatory miRNAs:mRNA networks in a cellular context [97]. This enhanced capacity has exceptional potential to discover novel mRNA targets for a miRNA of interest, including those that are transient or in low abundance. Accordingly, miR-TRAP is advantageous for the rapid identification of miRNAs at low false-positive and presumably low false-negative rates. Moreover, this approach is feasible and independent of genetic manipulations, allowing the analysis of several primary samples. Finally, miR-TRAP is a powerful, rapid, cost-effective and easy-to-handle tool that promotes the study of miRNA-dependent regulation in different diseases.

4 Conclusions

The potential reversibility of epigenetic modifications is a stimulating prospect for the development and improvement of technologies and methods to enable their characterization and, above all, to test the ability of new drugs to revert their pathological alterations, primarily in cancer. This chapter provides an overview of the current status of epigenetic assays based on epigenetic enzymatic activity and binding, genome-wide modifications and miRNAs and discusses the advantages, disadvantages and practical applications of each technique.

The growing amount of achievable information provided by innovative epigenetic approaches has prompted the parallel growth of so-called "computational epigenetics". A considerable number of computational approaches and tools are used in epigenetic studies for (1) epigenetic data repositories (e.g. International Human Epigenome Consortium Data Portal, Epigenome Atlas, Chromatin: 4DGenome, NIH Roadmap Epigenomics, miRbase, UCSC MethBase) and (2) statistical data analysis, prediction and visualization (e.g. Bioconductor (R) packages, epiGbs, MethPipe, MOABS, DaVIE, miRSystem) and data annotation, visualization and integration (e.g. DAVID, STRING, oPOSSUM, GeneMANIA, IPA).

Despite the plethora of accurate experimental and computational approaches, more efficient assays, bioinformatics methods and tools are still required to evaluate the complexity of epigenetic mechanisms and the sophisticated interaction networks governing gene expression in physiological and pathological conditions. Translating epigenetics into the clinic is undoubtedly a fundamental research perspective and the future direction of personalized medicine.

Compliance with Ethical Standards

Funding: The authors acknowledge AIRC17217; PON_0101227; VALERE: Vanvitelli per la Ricerca; Regione Campania lotta alle patologie oncologiche: iCURE (CUP B21C17000030007); and Regione Campania FASE2: IDEAL (CUP B53D18000080007). We thank C. Fisher for linguistic editing.

Conflict of Interest: The authors declare no competing interests.

Ethical Statement: This article does not contain any studies with human participants or animals performed by any of the authors.

References

1. Bernstein BE, Meissner A, Lander ES (2007) The mammalian epigenome. Cell 128 (4):669–681. https://doi.org/10.1016/j.cell.2007.01.033
2. Baylin SB, Jones PA (2011) A decade of exploring the cancer epigenome – biological and translational implications. Nat Rev Cancer 11(10):726–734. https://doi.org/10.1038/nrc3130
3. Cheuk IW, Shin VY, Kwong A (2017) Detection of methylated circulating DNA as noninvasive biomarkers for breast cancer diagnosis. J Breast Cancer 20(1):12–19. https://doi.org/10.4048/jbc.2017.20.1.12

4. Ho SM, Johnson A, Tarapore P, Janakiram V, Zhang X, Leung YK (2012) Environmental epigenetics and its implication on disease risk and health outcomes. ILAR J 53(3–4):289–305. https://doi.org/10.1093/ilar.53.3-4.289
5. Thomas ML, Marcato P (2018) Epigenetic modifications as biomarkers of tumor development, therapy response, and recurrence across the cancer care continuum. Cancers (Basel) 10(4). https://doi.org/10.3390/cancers10040101
6. Vardabasso C, Gaspar-Maia A, Hasson D, Punzeler S, Valle-Garcia D, Straub T, Keilhauer EC, Strub T, Dong J, Panda T, Chung CY, Yao JL, Singh R, Segura MF, Fontanals-Cirera B, Verma A, Mann M, Hernando E, Hake SB, Bernstein E (2015) Histone variant H2A.Z.2 mediates proliferation and drug sensitivity of malignant melanoma. Mol Cell 59(1):75–88. https://doi.org/10.1016/j.molcel.2015.05.009
7. Jia M, Jansen L, Walter V, Tagscherer K, Roth W, Herpel E, Kloor M, Blaker H, Chang-Claude J, Brenner H, Hoffmeister M (2016) No association of CpG island methylator phenotype and colorectal cancer survival: population-based study. Br J Cancer 115(11):1359–1366. https://doi.org/10.1038/bjc.2016.361
8. Ullman EF, Kirakossian H, Singh S, Wu ZP, Irvin BR, Pease JS, Switchenko AC, Irvine JD, Dafforn A, Skold CN et al (1994) Luminescent oxygen channeling immunoassay: measurement of particle binding kinetics by chemiluminescence. Proc Natl Acad Sci U S A 91 (12):5426–5430
9. Yasgar A, Jadhav A, Simeonov A, Coussens NP (2016) AlphaScreen-based assays: ultra-high-throughput screening for small-molecule inhibitors of challenging enzymes and protein-protein interactions. Methods Mol Biol 1439:77–98. https://doi.org/10.1007/978-1-4939-3673-1_5
10. Wigle TJ, Herold JM, Senisterra GA, Vedadi M, Kireev DB, Arrowsmith CH, Frye SV, Janzen WP (2010) Screening for inhibitors of low-affinity epigenetic peptide-protein interactions: an AlphaScreen-based assay for antagonists of methyl-lysine binding proteins. J Biomol Screen 15 (1):62–71. https://doi.org/10.1177/1087057109352902
11. Prabhu L, Chen L, Wei H, Demir O, Safa A, Zeng L, Amaro RE, O'Neil BH, Zhang ZY, Lu T (2017) Development of an AlphaLISA high throughput technique to screen for small molecule inhibitors targeting protein arginine methyltransferases. Mol Biosyst 13(12):2509–2520. https://doi.org/10.1039/c7mb00391a
12. Scarano S, Scuffi C, Mascini M, Minunni M (2011) Surface plasmon resonance imaging-based sensing for anti-bovine immunoglobulins detection in human milk and serum. Anal Chim Acta 707(1–2):178–183. https://doi.org/10.1016/j.aca.2011.09.012
13. Kim D, Lee IS, Jung JH, Yang SI (1999) Psammaplin A, a natural bromotyrosine derivative from a sponge, possesses the antibacterial activity against methicillin-resistant Staphylococcus aureus and the DNA gyrase-inhibitory activity. Arch Pharm Res 22(1):25–29
14. Duff MR Jr, Grubbs J, Howell EE (2011) Isothermal titration calorimetry for measuring macromolecule-ligand affinity. J Vis Exp (55). https://doi.org/10.3791/2796
15. Holdgate G (2009) Isothermal titration calorimetry and differential scanning calorimetry. Methods Mol Biol 572:101–133. https://doi.org/10.1007/978-1-60761-244-5_7
16. Jerabek-Willemsen M, Wienken CJ, Braun D, Baaske P, Duhr S (2011) Molecular interaction studies using microscale thermophoresis. Assay Drug Dev Technol 9(4):342–353. https://doi.org/10.1089/adt.2011.0380
17. Zillner K, Jerabek-Willemsen M, Duhr S, Braun D, Langst G, Baaske P (2012) Microscale thermophoresis as a sensitive method to quantify protein: nucleic acid interactions in solution. Methods Mol Biol 815:241–252. https://doi.org/10.1007/978-1-61779-424-7_18
18. Alpatov R, Lesch BJ, Nakamoto-Kinoshita M, Blanco A, Chen S, Stutzer A, Armache KJ, Simon MD, Xu C, Ali M, Murn J, Prisic S, Kutateladze TG, Vakoc CR, Min J, Kingston RE, Fischle W, Warren ST, Page DC, Shi Y (2014) A chromatin-dependent role of the fragile X mental retardation protein FMRP in the DNA damage response. Cell 157(4):869–881. https://doi.org/10.1016/j.cell.2014.03.040
19. Josling GA, Petter M, Oehring SC, Gupta AP, Dietz O, Wilson DW, Schubert T, Langst G, Gilson PR, Crabb BS, Moes S, Jenoe P, Lim SW, Brown GV, Bozdech Z, Voss TS, Duffy MF

(2015) A plasmodium falciparum bromodomain protein regulates invasion gene expression. Cell Host Microbe 17(6):741–751. https://doi.org/10.1016/j.chom.2015.05.009
20. Raha D, Hong M, Snyder M (2010) ChIP-seq: a method for global identification of regulatory elements in the genome. Curr Protoc Mol Biol Chapter 21:Unit 21 19 21-14. https://doi.org/10.1002/0471142727.mb2119s91
21. Wang CI, Alekseyenko AA, LeRoy G, Elia AE, Gorchakov AA, Britton LM, Elledge SJ, Kharchenko PV, Garcia BA, Kuroda MI (2013) Chromatin proteins captured by ChIP-mass spectrometry are linked to dosage compensation in Drosophila. Nat Struct Mol Biol 20 (2):202–209. https://doi.org/10.1038/nsmb.2477
22. Mohammed H, Taylor C, Brown GD, Papachristou EK, Carroll JS, D'Santos CS (2016) Rapid immunoprecipitation mass spectrometry of endogenous proteins (RIME) for analysis of chromatin complexes. Nat Protoc 11(2):316–326. https://doi.org/10.1038/nprot.2016.020
23. Cao Z, Lu C (2016) A microfluidic device with integrated sonication and immunoprecipitation for sensitive epigenetic assays. Anal Chem 88(3):1965–1972. https://doi.org/10.1021/acs.analchem.5b04707
24. Jafari R, Almqvist H, Axelsson H, Ignatushchenko M, Lundback T, Nordlund P, Martinez Molina D (2014) The cellular thermal shift assay for evaluating drug target interactions in cells. Nat Protoc 9(9):2100–2122. https://doi.org/10.1038/nprot.2014.138
25. Becher I, Werner T, Doce C, Zaal EA, Togel I, Khan CA, Rueger A, Muelbaier M, Salzer E, Berkers CR, Fitzpatrick PF, Bantscheff M, Savitski MM (2016) Thermal profiling reveals phenylalanine hydroxylase as an off-target of panobinostat. Nat Chem Biol 12(11):908–910. https://doi.org/10.1038/nchembio.2185
26. McNulty DE, Bonnette WG, Qi H, Wang L, Ho TF, Waszkiewicz A, Kallal LA, Nagarajan RP, Stern M, Quinn AM, Creasy CL, Su DS, Graves AP, Annan RS, Sweitzer SM, Holbert MA (2018) A high-throughput dose-response cellular thermal shift assay for rapid screening of drug target engagement in living cells, exemplified using SMYD3 and IDO1. SLAS Discov 23 (1):34–46. https://doi.org/10.1177/2472555217732014
27. Song Y, Madahar V, Liao J (2011) Development of FRET assay into quantitative and high-throughput screening technology platforms for protein-protein interactions. Ann Biomed Eng 39(4):1224–1234. https://doi.org/10.1007/s10439-010-0225-x
28. Alibhai D, Kelly DJ, Warren S, Kumar S, Margineau A, Serwa RA, Thinon E, Alexandrov Y, Murray EJ, Stuhmeier F, Tate EW, Neil MA, Dunsby C, French PM (2013) Automated fluorescence lifetime imaging plate reader and its application to Forster resonant energy transfer readout of Gag protein aggregation. J Biophotonics 6(5):398–408. https://doi.org/10.1002/jbio.201200185
29. Wade M, Mendez J, Coussens NP, Arkin MR, Glicksman MA (2004) Inhibition of protein-protein interactions: cell-based assays. In: Sittampalam GS, Coussens NP, Brimacombe K et al (eds) Assay guidance manual. Eli Lilly & Company and the National Center for Advancing Translational Sciences, Bethesda
30. Bacart J, Corbel C, Jockers R, Bach S, Couturier C (2008) The BRET technology and its application to screening assays. Biotechnol J 3(3):311–324. https://doi.org/10.1002/biot.200700222
31. Machleidt T, Woodroofe CC, Schwinn MK, Mendez J, Robers MB, Zimmerman K, Otto P, Daniels DL, Kirkland TA, Wood KV (2015) NanoBRET – a novel BRET platform for the analysis of protein-protein interactions. ACS Chem Biol 10(8):1797–1804. https://doi.org/10.1021/acschembio.5b00143
32. Hu F, Martin H, Martinez A, Everitt J, Erkanli A, Lee WT, Dewhirst M, Ramanujam N (2017) Distinct angiogenic changes during carcinogenesis defined by novel label-free dark-field imaging in a hamster cheek pouch model. Cancer Res 77(24):7109–7119. https://doi.org/10.1158/0008-5472.CAN-17-1058
33. Tollefsbol TO (2011) Advances in epigenetic technology. Methods Mol Biol 791:1–10. https://doi.org/10.1007/978-1-61779-316-5_1

34. Weinhold B (2006) Epigenetics: the science of change. Environ Health Perspect 114(3):A160–A167
35. Gasperskaja E, Kucinskas V (2017) The most common technologies and tools for functional genome analysis. Acta Med Litu 24(1):1–11. https://doi.org/10.6001/actamedica.v24i1.3457
36. Schwartzman O, Tanay A (2015) Single-cell epigenomics: techniques and emerging applications. Nat Rev Genet 16(12):716–726. https://doi.org/10.1038/nrg3980
37. Milne TA, Zhao K, Hess JL (2009) Chromatin immunoprecipitation (ChIP) for analysis of histone modifications and chromatin-associated proteins. Methods Mol Biol 538:409–423. https://doi.org/10.1007/978-1-59745-418-6_21
38. Zarnegar MA, Reinitz F, Newman AM, Clarke MF (2017) Targeted chromatin ligation, a robust epigenetic profiling technique for small cell numbers. Nucleic Acids Res 45(17):e153. https://doi.org/10.1093/nar/gkx648
39. Teste B, Champ J, Londono-Vallejo A, Descroix S, Malaquin L, Viovy JL, Draskovic I, Mottet G (2017) Chromatin immunoprecipitation in microfluidic droplets: towards fast and cheap analyses. Lab Chip 17(3):530–537. https://doi.org/10.1039/c6lc01535b
40. Rotem A, Ram O, Shoresh N, Sperling RA, Goren A, Weitz DA, Bernstein BE (2015) Single-cell ChIP-seq reveals cell subpopulations defined by chromatin state. Nat Biotechnol 33 (11):1165–1172. https://doi.org/10.1038/nbt.3383
41. Hansel-Hertsch R, Spiegel J, Marsico G, Tannahill D, Balasubramanian S (2018) Genome-wide mapping of endogenous G-quadruplex DNA structures by chromatin immunoprecipitation and high-throughput sequencing. Nat Protoc 13(3):551–564. https://doi.org/10.1038/nprot.2017.150
42. Gaasterland T, Oprea M (2001) Whole-genome analysis: annotations and updates. Curr Opin Struct Biol 11(3):377–381
43. Behjati S, Tarpey PS (2013) What is next generation sequencing? Arch Dis Child Educ Pract Ed 98(6):236–238. https://doi.org/10.1136/archdischild-2013-304340
44. Almouzni G, Cedar H (2016) Maintenance of epigenetic information. Cold Spring Harb Perspect Biol 8(5). https://doi.org/10.1101/cshperspect.a019372
45. Moore LD, Le T, Fan G (2013) DNA methylation and its basic function. Neuropsychopharmacology 38(1):23–38. https://doi.org/10.1038/npp.2012.112
46. Li E, Zhang Y (2014) DNA methylation in mammals. Cold Spring Harb Perspect Biol 6(5): a019133. https://doi.org/10.1101/cshperspect.a019133
47. Tucker T, Marra M, Friedman JM (2009) Massively parallel sequencing: the next big thing in genetic medicine. Am J Hum Genet 85(2):142–154. https://doi.org/10.1016/j.ajhg.2009.06.022
48. Li Q, Hermanson PJ, Springer NM (2018) Detection of DNA methylation by whole-genome bisulfite sequencing. Methods Mol Biol 1676:185–196. https://doi.org/10.1007/978-1-4939-7315-6_11
49. Lu X, Han D, Zhao BS, Song CX, Zhang LS, Dore LC, He C (2015) Base-resolution maps of 5-formylcytosine and 5-carboxylcytosine reveal genome-wide DNA demethylation dynamics. Cell Res 25(3):386–389. https://doi.org/10.1038/cr.2015.5
50. Yu M, Han D, Hon GC, He C (2018) Tet-assisted bisulfite sequencing (TAB-seq). Methods Mol Biol 1708:645–663. https://doi.org/10.1007/978-1-4939-7481-8_33
51. Kawasaki Y, Kuroda Y, Suetake I, Tajima S, Ishino F, Kohda T (2017) A novel method for the simultaneous identification of methylcytosine and hydroxymethylcytosine at a single base resolution. Nucleic Acids Res 45(4):e24. https://doi.org/10.1093/nar/gkw994
52. Lu X, Song CX, Szulwach K, Wang Z, Weidenbacher P, Jin P, He C (2013) Chemical modification-assisted bisulfite sequencing (CAB-seq) for 5-carboxylcytosine detection in DNA. J Am Chem Soc 135(25):9315–9317. https://doi.org/10.1021/ja4044856
53. Song CX, Szulwach KE, Dai Q, Fu Y, Mao SQ, Lin L, Street C, Li Y, Poidevin M, Wu H, Gao J, Liu P, Li L, Xu GL, Jin P, He C (2013) Genome-wide profiling of 5-formylcytosine reveals its roles in epigenetic priming. Cell 153(3):678–691. https://doi.org/10.1016/j.cell.2013.04.001

54. Guo H, Zhu P, Guo F, Li X, Wu X, Fan X, Wen L, Tang F (2015) Profiling DNA methylome landscapes of mammalian cells with single-cell reduced-representation bisulfite sequencing. Nat Protoc 10(5):645–659. https://doi.org/10.1038/nprot.2015.039
55. Smallwood SA, Lee HJ, Angermueller C, Krueger F, Saadeh H, Peat J, Andrews SR, Stegle O, Reik W, Kelsey G (2014) Single-cell genome-wide bisulfite sequencing for assessing epigenetic heterogeneity. Nat Methods 11(8):817–820. https://doi.org/10.1038/nmeth.3035
56. Clark SJ, Smallwood SA, Lee HJ, Krueger F, Reik W, Kelsey G (2017) Genome-wide base-resolution mapping of DNA methylation in single cells using single-cell bisulfite sequencing (scBS-seq). Nat Protoc 12(3):534–547. https://doi.org/10.1038/nprot.2016.187
57. Luo C, Keown CL, Kurihara L, Zhou J, He Y, Li J, Castanon R, Lucero J, Nery JR, Sandoval JP, Bui B, Sejnowski TJ, Harkins TT, Mukamel EA, Behrens MM, Ecker JR (2017) Single-cell methylomes identify neuronal subtypes and regulatory elements in mammalian cortex. Science 357(6351):600–604. https://doi.org/10.1126/science.aan3351
58. Angermueller C, Clark SJ, Lee HJ, Macaulay IC, Teng MJ, Hu TX, Krueger F, Smallwood S, Ponting CP, Voet T, Kelsey G, Stegle O, Reik W (2016) Parallel single-cell sequencing links transcriptional and epigenetic heterogeneity. Nat Methods 13(3):229–232. https://doi.org/10.1038/nmeth.3728
59. Han J, Zhang Z, Wang K (2018) 3C and 3C-based techniques: the powerful tools for spatial genome organization deciphering. Mol Cytogenet 11:21. https://doi.org/10.1186/s13039-018-0368-2
60. Li G, Cai L, Chang H, Hong P, Zhou Q, Kulakova EV, Kolchanov NA, Ruan Y (2014) Chromatin interaction analysis with paired-end tag (ChIA-PET) sequencing technology and application. BMC Genomics 15(Suppl 12):S11. https://doi.org/10.1186/1471-2164-15-S12-S11
61. Mumbach MR, Rubin AJ, Flynn RA, Dai C, Khavari PA, Greenleaf WJ, Chang HY (2016) HiChIP: efficient and sensitive analysis of protein-directed genome architecture. Nat Methods 13(11):919–922. https://doi.org/10.1038/nmeth.3999
62. Chen C, Ridzon DA, Broomer AJ, Zhou Z, Lee DH, Nguyen JT, Barbisin M, Xu NL, Mahuvakar VR, Andersen MR, Lao KQ, Livak KJ, Guegler KJ (2005) Real-time quantification of microRNAs by stem-loop RT-PCR. Nucleic Acids Res 33(20):e179. https://doi.org/10.1093/nar/gni178
63. Varkonyi-Gasic E, Wu R, Wood M, Walton EF, Hellens RP (2007) Protocol: a highly sensitive RT-PCR method for detection and quantification of microRNAs. Plant Methods 3:12. https://doi.org/10.1186/1746-4811-3-12
64. Jacobsen N, Andreasen D, Mouritzen P (2011) Profiling microRNAs by real-time PCR. Methods Mol Biol 732:39–54. https://doi.org/10.1007/978-1-61779-083-6_4
65. Campomenosi P, Gini E, Noonan DM, Poli A, D'Antona P, Rotolo N, Dominioni L, Imperatori A (2016) A comparison between quantitative PCR and droplet digital PCR technologies for circulating microRNA quantification in human lung cancer. BMC Biotechnol 16(1):60. https://doi.org/10.1186/s12896-016-0292-7
66. Hindson BJ, Ness KD, Masquelier DA, Belgrader P, Heredia NJ, Makarewicz AJ, Bright IJ, Lucero MY, Hiddessen AL, Legler TC, Kitano TK, Hodel MR, Petersen JF, Wyatt PW, Steenblock ER, Shah PH, Bousse LJ, Troup CB, Mellen JC, Wittmann DK, Erndt NG, Cauley TH, Koehler RT, So AP, Dube S, Rose KA, Montesclaros L, Wang S, Stumbo DP, Hodges SP, Romine S, Milanovich FP, White HE, Regan JF, Karlin-Neumann GA, Hindson CM, Saxonov S, Colston BW (2011) High-throughput droplet digital PCR system for absolute quantitation of DNA copy number. Anal Chem 83(22):8604–8610. https://doi.org/10.1021/ac202028g
67. Song Y, Kilburn D, Song JH, Cheng Y, Saeui CT, Cheung DG, Croce CM, Yarema KJ, Meltzer SJ, Liu KJ, Wang TH (2017) Determination of absolute expression profiles using multiplexed miRNA analysis. PLoS One 12(7):e0180988. https://doi.org/10.1371/journal.pone.0180988

68. Androvic P, Valihrach L, Elling J, Sjoback R, Kubista M (2017) Two-tailed RT-qPCR: a novel method for highly accurate miRNA quantification. Nucleic Acids Res 45(15):e144. https://doi.org/10.1093/nar/gkx588
69. Moody L, He H, Pan YX, Chen H (2017) Methods and novel technology for microRNA quantification in colorectal cancer screening. Clin Epigenetics 9:119. https://doi.org/10.1186/s13148-017-0420-9
70. Sun Z, Evans J, Bhagwate A, Middha S, Bockol M, Yan H, Kocher JP (2014) CAP-miRSeq: a comprehensive analysis pipeline for microRNA sequencing data. BMC Genomics 15:423. https://doi.org/10.1186/1471-2164-15-423
71. Wu J, Liu Q, Wang X, Zheng J, Wang T, You M, Sheng Sun Z, Shi Q (2013) mirTools 2.0 for non-coding RNA discovery, profiling, and functional annotation based on high-throughput sequencing. RNA Biol 10(7):1087–1092. https://doi.org/10.4161/rna.25193
72. Rueda A, Barturen G, Lebron R, Gomez-Martin C, Alganza A, Oliver JL, Hackenberg M (2015) sRNAtoolbox: an integrated collection of small RNA research tools. Nucleic Acids Res 43(W1):W467–W473. https://doi.org/10.1093/nar/gkv555
73. Andres-Leon E, Nunez-Torres R, Rojas AM (2016) miARma-seq: a comprehensive tool for miRNA, mRNA and circRNA analysis. Sci Rep 6:25749. https://doi.org/10.1038/srep25749
74. Garcia-Gimenez JL, Rubio-Belmar PA, Peiro-Chova L, Hervas D, Gonzalez-Rodriguez D, Ibanez-Cabellos JS, Bas-Hermida P, Mena-Molla S, Garcia-Lopez EM, Pallardo FV, Bas T (2018) Circulating miRNAs as diagnostic biomarkers for adolescent idiopathic scoliosis. Sci Rep 8(1):2646. https://doi.org/10.1038/s41598-018-21146-x
75. Gustafson D, Tyryshkin K, Renwick N (2016) microRNA-guided diagnostics in clinical samples. Best Pract Res Clin Endocrinol Metab 30(5):563–575. https://doi.org/10.1016/j.beem.2016.07.002
76. Rodriguez M, Bajo-Santos C, Hessvik NP, Lorenz S, Fromm B, Berge V, Sandvig K, Line A, Llorente A (2017) Identification of non-invasive miRNAs biomarkers for prostate cancer by deep sequencing analysis of urinary exosomes. Mol Cancer 16(1):156. https://doi.org/10.1186/s12943-017-0726-4
77. Buschmann D, Kirchner B, Hermann S, Marte M, Wurmser C, Brandes F, Kotschote S, Bonin M, Steinlein OK, Pfaffl MW, Schelling G, Reithmair M (2018) Evaluation of serum extracellular vesicle isolation methods for profiling miRNAs by next-generation sequencing. J Extracell Vesicles 7(1):1481321. https://doi.org/10.1080/20013078.2018.1481321
78. Van Ness J, Van Ness LK, Galas DJ (2003) Isothermal reactions for the amplification of oligonucleotides. Proc Natl Acad Sci U S A 100(8):4504–4509. https://doi.org/10.1073/pnas.0730811100
79. Zhang Y, Zhang CY (2012) Sensitive detection of microRNA with isothermal amplification and a single-quantum-dot-based nanosensor. Anal Chem 84(1):224–231. https://doi.org/10.1021/ac202405q
80. Liu H, Tian T, Zhang Y, Ding L, Yu J, Yan M (2017) Sensitive and rapid detection of microRNAs using hairpin probes-mediated exponential isothermal amplification. Biosens Bioelectron 89(Pt 2):710–714. https://doi.org/10.1016/j.bios.2016.10.099
81. Na J, Shin GW, Son HG, Lee SV, Jung GY (2017) Multiplex quantitative analysis of microRNA expression via exponential isothermal amplification and conformation-sensitive DNA separation. Sci Rep 7(1):11396. https://doi.org/10.1038/s41598-017-11895-6
82. Urbanek MO, Nawrocka AU, Krzyzosiak WJ (2015) Small RNA detection by in situ hybridization methods. Int J Mol Sci 16(6):13259–13286. https://doi.org/10.3390/ijms160613259
83. Thomas M, Lieberman J, Lal A (2010) Desperately seeking microRNA targets. Nat Struct Mol Biol 17(10):1169–1174. https://doi.org/10.1038/nsmb.1921
84. Elmen J, Lindow M, Silahtaroglu A, Bak M, Christensen M, Lind-Thomsen A, Hedtjarn M, Hansen JB, Hansen HF, Straarup EM, McCullagh K, Kearney P, Kauppinen S (2008) Antagonism of microRNA-122 in mice by systemically administered LNA-antimiR leads to up-regulation of a large set of predicted target mRNAs in the liver. Nucleic Acids Res 36 (4):1153–1162. https://doi.org/10.1093/nar/gkm1113

85. Krutzfeldt J, Rajewsky N, Braich R, Rajeev KG, Tuschl T, Manoharan M, Stoffel M (2005) Silencing of microRNAs in vivo with 'antagomirs'. Nature 438(7068):685–689. https://doi.org/10.1038/nature04303
86. Ebert MS, Sharp PA (2010) MicroRNA sponges: progress and possibilities. RNA 16 (11):2043–2050. https://doi.org/10.1261/rna.2414110
87. Kuhn DE, Martin MM, Feldman DS, Terry AV Jr, Nuovo GJ, Elton TS (2008) Experimental validation of miRNA targets. Methods 44(1):47–54. https://doi.org/10.1016/j.ymeth.2007.09.005
88. Beitzinger M, Peters L, Zhu JY, Kremmer E, Meister G (2007) Identification of human microRNA targets from isolated argonaute protein complexes. RNA Biol 4(2):76–84
89. Licatalosi DD, Mele A, Fak JJ, Ule J, Kayikci M, Chi SW, Clark TA, Schweitzer AC, Blume JE, Wang X, Darnell JC, Darnell RB (2008) HITS-CLIP yields genome-wide insights into brain alternative RNA processing. Nature 456(7221):464–469. https://doi.org/10.1038/nature07488
90. Hafner M, Landthaler M, Burger L, Khorshid M, Hausser J, Berninger P, Rothballer A, Ascano M Jr, Jungkamp AC, Munschauer M, Ulrich A, Wardle GS, Dewell S, Zavolan M, Tuschl T (2010) Transcriptome-wide identification of RNA-binding protein and microRNA target sites by PAR-CLIP. Cell 141(1):129–141. https://doi.org/10.1016/j.cell.2010.03.009
91. Danan C, Manickavel S, Hafner M (2016) PAR-CLIP: a method for transcriptome-wide identification of RNA binding protein interaction sites. Methods Mol Biol 1358:153–173. https://doi.org/10.1007/978-1-4939-3067-8_10
92. Ule J, Jensen KB, Ruggiu M, Mele A, Ule A, Darnell RB (2003) CLIP identifies Nova-regulated RNA networks in the brain. Science 302(5648):1212–1215. https://doi.org/10.1126/science.1090095
93. Favre A, Moreno G, Blondel MO, Kliber J, Vinzens F, Salet C (1986) 4-Thiouridine photosensitized RNA-protein crosslinking in mammalian cells. Biochem Biophys Res Commun 141(2):847–854
94. Bezerra R, Favre A (1990) In vivo incorporation of the intrinsic photolabel 4-thiouridine into Escherichia coli RNAs. Biochem Biophys Res Commun 166(1):29–37
95. Kishore S, Jaskiewicz L, Burger L, Hausser J, Khorshid M, Zavolan M (2011) A quantitative analysis of CLIP methods for identifying binding sites of RNA-binding proteins. Nat Methods 8 (7):559–564. https://doi.org/10.1038/nmeth.1608
96. Baigude H, Ahsanullah LZ, Zhou Y, Rana TM (2012) miR-TRAP: a benchtop chemical biology strategy to identify microRNA targets. Angew Chem Int Ed Engl 51(24):5880–5883. https://doi.org/10.1002/anie.201201512
97. Cambronne XA, Shen R, Auer PL, Goodman RH (2012) Capturing microRNA targets using an RNA-induced silencing complex (RISC)-trap approach. Proc Natl Acad Sci U S A 109 (50):20473–20478. https://doi.org/10.1073/pnas.1218887109

Index

A
A3, 69
A36, 166
A196, 146
A366, 128, 361, 362
A395, 134, 135
A485, 105, 111, 112
A893, 142
Abexinostat, 8, 9
AC-93253, 67
Acetylation, 2, 93–113, 173, 289, 340, 405, 429, 457, 527
Acetyl lysine, 4, 5, 17, 60, 103, 287, 292, 323
Acetyl-lysine posttranslational modifications (PTMs), 29
Activation, 6, 19, 45, 55, 76, 301, 431, 467, 492, 538
Acute myeloid leukaemia (AML), 67, 98, 180, 225, 262, 272, 303, 469
ADD domains, 340
S-Adenosyl-L-homocysteine (SAH/AdoHcy), 159, 163–175, 179, 260
S-Adenosyl-L-homocysteine hydrolase (SAHH), 164
Adipogenesis, 129
AdoHcy, 159, 163–175, 179, 260
AdoMet/SAM, 163, 179, 182, 258
ADS100380, 461
AGK2, 71
Ailanthus altissima, 78
AK-7, 71
ALDH1A1, 413
Allantodapsone, 466
AlphaScreen/AlphaLISA, 526
Alzheimer's disease (AD), 19, 57, 109, 203, 210
4-Amino-1-naphthol (4A1N), 111, 113
3-Amino pyridine carboxylic acid, 235
L-α-Aminosuberic hydroxamic acid (Asuha), 36
Aminothiazoles, 70
Amyotrophic lateral sclerosis, 57
Anacardic acid, 104, 107
2-Anilinobenzamide, 59
Ankyrin repeats, 340
Antibodies, 18, 272, 357, 415, 444, 494, 527, 552
Antibody-drug conjugates (ADCs), 211
Anticancer drugs, 1
Apicidin, 15, 39
Arginine demethylase, 162, 340
Arginine methylation, 159, 289, 340
Arginine methyltransferases (PRMTs), 126, 160–183, 265, 465
Aristoforin, 74, 76
AS1387392, 15
Astemizole, 133
Asthma, 102, 429–445, 492
Autism, 20, 129, 492
AZ-505, 142
Aza-AdoMet, 163
Azabenzimidazole, 77
5-Azacytidine (5azaC), 262, 492
Azumamide E, 15, 39

B
Base excision repair (BER) enzymes, 258
(S)-BAY-598, 142

BAY123807, 303
BAZ-PHD fingers, 348
BCL2, 18, 369, 498
Belinostat, 8, 11, 32
Benzodeazaoxaflavins, 66
BETs, 301–304, 320
BF1, 111
Binding free energy calculations, 455
Biophysical assays, 287
Bisbenzylidenebenzene-1,4-diamines, 72
Bisbenzylidenenaphthalene-1,4-diamines, 72
Bisindolylmaleimides (BIMs), 59
Bispiperazinecarbodithioic acid methyl ester, 75
Bisubstrate inhibitors, 103, 168, 176
Bisulfite sequencing, 505
BIX01294 (diazepin-quinazoline-amines), 127, 247
Bizine, 205
Bladder cancer, 144, 166, 169, 201, 203, 372, 374, 412
Bn-SNF, 143
brahma gene, 290
BRCA1, 101, 357
BRD4, 303–325, 535
BRD4770, 127, 129
BRD9539, 129
Breast carcinoma, 11, 60–71, 75, 98–101, 126, 135, 170, 201, 212, 237, 303, 312, 357, 372, 409–414, 492
Bromodomain, 3, 97, 287, 344, 348, 382, 417, 530, 535
(S)-8-Bromo-6-chloro-2-pentylchroman-4-one, 69
8-Bromo-NAD, 59
Burkitt's lymphoma, 66
Butein, 78
BxPC-3, 75

C
C107, 111, 112
Cambinol, 65, 66
Carbonic anhydrase, 7
5-Carboxycytosine (5caC), 257, 487
Cardiac hypertrophy, 20, 57, 102, 303, 419
CARM1 (PRMT4), 173
CBP/p300, 299
CBX6, 355
CBX7, 350
Chemical probes, 93
Chidamide, 14
Chlamydocin, 39, 41
4-Chlorophenyl pyrazole carboxamide, 68
Chromanones, 64, 69
Chromatin, 123, 181, 289
  domains, 339, 374
  modification, 95, 289, 401
  remodelling, 31, 95, 141, 146, 257, 277, 344, 431, 489, 526
Chromodomains, 350
Chronic lymphocytic leukaemia (CLL), 272, 411, 412
Chronic myelogenous leukemia, 144
Chronic myelomonocytic leukaemia (CMML), 262
Chronic obstructive pulmonary disease (COPD), 429
Cisplatin, 18, 410, 413–415
Citarinostat, 9
Citrulline, 162, 289
Colon cancer, 62, 65, 68, 169, 237, 261
Colorectal carcinoma, 65, 101, 178, 180, 213, 225, 378
Computational chemistry, 455
CoREST, 31
CP2, 246
Cpd5, 264
CPI-455, 242
CPI-1205, 132
CPTH2, 111
CPTH6, 111
CREB-binding protein (CBP), 299
CUDC-101, 11
Curcuma longa, 109
Curcumin, 107, 109
Cutaneous squamous cell carcinoma (CSCC), 101
Cyanidin, 78
Cyclic peptides, 29
Cyclin-dependent kinase inhibitor p21 (CIP1/WAF1), 18
cyl-1/2, 39
Cystic fibrosis, 20, 44, 261
Cytosine-5-methylenesulfonate (CMS), 504

D
DC_C01, 468
Deacetylase inhibitors (HDACi), 135, 429–445
Deacetylation, 55
3-Deazaneplanocin (DZNep), 131
Decitabine, 492
Demethylases, 197
Desethylamiodarone, 346
Diabetes, 20, 57, 78, 140, 166, 171

Diclone, 265
Dihydropyridines, 77
Dihydroquinazolinone, 300
Direct detection, 487
Disulfiram, 346
DNA damage response (DDR), 99, 135, 349, 412
DNA hydroxymethylation, 255
DNA methylation, 255, 259
DNA methyltransferases (DNMTs), 164, 255–261, 344, 358, 429, 440, 487
DNA repair, 99, 106, 140, 143, 163, 166, 258, 357
Disruptor of telomeric silencing 1-like (DOT1L), 123, 135
Drug design/development, 55
Drug discovery, structure-and ligand-based, 93
DU 145, 60
DZNep, 130

E

E11, 205
EBI-2511, 133
EED226, 134
EHMTs, 126
EI1, 131
EML425, 109
Enzyme-linked immunosorbent assay (ELISA), 494
Epidrugs, 417, 420, 431, 432
Epigenetic inhibitors, 263, 455–475
Epigenetic modulators, 123, 525
Epigenetic readers, 326, 339, 417
Epigenetic therapy, 289, 401, 417
Epstein-Barr virus (EBV), 19, 20, 179
EPZ-5676, 136
EPZ004777, 136
EPZ005687, 131
EPZ031686, 143
Esophageal squamous cell carcinoma (ESCC), 44
Euchromatin histone methyltransferases (EHMTs), 126
Ex-527, 66
EZH1/2, 123, 130

F

Fat mass and obesity-associated (FTO) proteins, 271
Fbox Only Protein 11 (FBXO11), 185
Filgastat, 248
FK228, 12
FL-411, 303
Flavanone 69, 265
Fluconazole, 20
Fluorescence polarization, 321
5-Fluorocytidine, 263
5-Formylcytosine (5fC), 257, 487
Fucoidan, 79
5-(3-(Furan-2-carboxamido)benzamido)-2-hydroxybenzoic acid, 74

G

G9a, 123
G9a/GLP inhibitors
  non-SAM-competitive, 127
  SAM-competitive, 129
GA001, 468
Garcinia cochinchinensis, 76
GATA4, 130
Gcn5-related *N*-acetyltransferase (GNAT), 95
Gemcitabine, 75, 411
Genitopatellar syndrome, 96, 102
Ginsenosides, 78
Givinostat, 8
GLIB-Seq, 504
Glioblastoma, 65, 101, 111, 135, 410–416
Globin disorders, 202
G-protein antagonists, 72
GSK-J1, 243
GSK-J2, 244
GSK126, 131
GSK343, 131
GSK-690, 205
GSK2807, 143
Guttiferone, 76
GW5054, 68, 69

H

H3K4, 441
HC-toxin, 39
Hemoglobin F (HbF), 126
Hepatitis B, 19, 20, 202
Hepatocellular carcinoma (HCC), 144, 167, 225, 410
Herpes simplex virus (HSV), 201, 203
Histone acetyltransferases (HATs), 93
Histone deacetylases (HDACs), 1–21, 29–47, 57, 289, 404, 455, 459
Histone demethylases (KDMs), 164, 197–213, 222–249, 441, 455, 469, 473
Histone H3 lysine 9 (H3K9), 63

Histone methyltransferases (HMTs), 123–144, 221, 247, 265, 346, 378, 405, 455, 465, 527
Histones, 197, 287
HIV, 19, 66, 130, 140, 168, 181, 202, 298, 341
HL-60, 38
Homology modeling, 455
Honokiol, 79
HOTAIR, 405, 406, 409–413
HPLC, 496
HR73, 65
HT-29, 68
Human anti-transferrin receptor (TfR), 415
Human silent information regulator type 1 (SIRT1), 55
Huntington's disease (HD), 20, 57, 129
Hydroxamic acid, 7, 29
5-Hydroxymethylcytosine (5hmC), 257, 487
5-Hydroxyuracil (5hmU), 267
Hyperacetylation, 31, 44, 99, 295
Hyperforin, 76
Hypoacetylation, 99, 105, 111

I
I-CBP112, 300
ICL-SIRT078, 73
Imidazothiazoles, 79
Inauhzin, 67
Indoles, 64
Inhibition, 55
Inhibitors, 197, 255
  discovery, 287
IOX1, 276
Ischemin, 300
Isogarcinol, 109

J
JARID1A, 345
JGB-1741, 64
JmjC histone demethylases, 221, 472
JMJD6, 162, 340
(+)-JQ1, 302

K
KDMs, 163, 221, 226, 341, 351, 440, 466
  KDM5A, 345
KDOAM-25, 242
Krüppel-like factor 2 (KLF2), 412

L
L3MBTL1, 364
Laccaic acid, 265
Largazole, 12, 13
lincRNA-p21, 412
LLY-507, 142
Long chain alkylidene malonates (LoCAMs), 108
Long noncoding RNA (lncRNAs), 401
LSD1 (KDM1A), 199, 469
LSD2 (KDM1B), 199
LTK-14, 109
LuCaP-77, 113
Lung cancer/carcinoma, 63, 101, 129, 181, 200, 261, 299, 341, 409, 429, 469
Luteolin, 78
Lyngbya majuscula, 76
Lys-CoA, 103
Lysine, 197
Lysine demethylases (KDMs), 163, 221, 226, 341, 351, 440, 466
Lysine methylation, 197, 222, 364, 374, 443
Lysine methyltransferases (KMTs), 123, 166, 176, 182, 199, 340, 350, 469
Lysine-specific histone demethylases 1/2 (LSD1/2), 197

M
Malaria, 20
MALAT1, 409, 415
Malignant brain tumor (MBT) domain, 339, 363
Manganese superoxide dismutase (Mn-SOD), 79
Marbostat-100, 10
Mass spectrometry, 487
Matrix metalloproteinases, 7
MC1946, 144
MC2141, 65
MCF-7, 38, 60
MCP-1, 140
Metalloenzymes, 4, 472
Methylation, 159, 197, 523
5-Methylcytosine (5mC), 257
3-(1-Methyl-4-phenylacetyl-1*H*-2-pyrrolyl)-*N*-hydroxy-2-propenamide (APHA), 34
Methyl-readers, 339
Methylthioadenosine (MTA), 164
MGCD290, 20
MI-1, 140

MIAT, 409
Microarray, 487
MiDAC, 31
MIND4, 72
miRNA, 523
Mithramycin, 130
MIV-6R, 140
Mixed lineage leukaemia (MLL), 169, 181, 267, 272
MLL-menin disruptors, 140
MLL1–4, 129, 138, 344, 373–378
MLL-AF9, 209, 213
MM-102, 138
MM-589, 138
Mocetinostat, 14
Molecular docking, 455
Molecular modeling, 455
Monomethylated arginine (MMA), 159, 161, 165, 170, 183
MS012, 128
MS049, 184
MS351, 354
MS453, 145
MS2126, 300
MS2177, 145
MS7972, 299
MS37452, 354
Muscular atrophy, 19, 44, 57, 341, 363, 418
Myelodysplastic syndrome (MDS), 209, 262, 272
Myeloid translocation protein-8, Nervy, and DEAF-1 (MYND) domain, 141
Myristoylethanolamide, 81
Myristoyllysine, 31
MYST, 98

N
Nahuoic acid A, 144
Nanatinostat, 8
Nanopore sequencing, 512
β-Naphthols, 64
Nasopharyngeal carcinoma, 96, 101, 134
Nexturastat A, 459
NgTET, 267
Nivolumab, 444
NK13650A/B, 110
Non-small cell lung carcinoma (NSCLC), 66, 101, 111, 201, 409, 412, 444
NSC663284, 146
Nucleoside analogues, 262, 277
NuRD, 31
Nusinersen (Spinraza), 419

O
Obesity, 20, 129, 171, 262
OICR-9429, 139
Olaparib, 75
Oleoylethanolamide, 81
OTX-015, 302
Ovarian carcinomas, 96, 101, 178, 358, 410, 414
Oxadiazole-carbonylaminothiourea, 72
Oxazolopyridines, 77
2-Oxoglutarate (2OG)-dependent oxygenases, 221

P
p53, 5, 44, 46, 60–67, 101, 126, 141–144, 165, 172, 178, 199, 201, 298–300, 357, 364, 412, 432
p300, 95–114, 292, 542
Paclitaxel, 130
Panax ginseng, 78
Pancreatic adenocarcinoma, 75
Panobinostat, 8, 9, 18, 19, 32, 213, 459, 460, 532
Pargyline, 204
Parkinson's disease (PD), 20, 57, 71, 378, 464
PCAF, 101, 298
PCI-34051, 11
PCPA, 206
Peptide probes, 29
PFI-2, 140
PFI-3, 295
PHD domain, 339
Phenelzine, 205
4-Phenyl-2-thiazolylhydrazones, 111
N-Phthaloyl-L-tryptophan, 264
Pinometostat, 136
Pivanex, 15
PKF118-310, 473
Plant homeodomain (PHD) fingers, 342
Plasmodium berghei, 20
Plasmodium falciparum, 38
PLZF, 130
Polycomb-like complex (PCL), 356
Polymyxin B, 205
Posttranslational modification (PTM), 429
Pracinostat, 8
PRC2, 130–135, 350, 356, 374–381, 404–419
  disruptors, 133
PRMT1, 165
PRMT2, 170
PRMT3, 172
PRMT4 (CARM1), 467

PRMT5, 178
PRMT6, 181
PRMT7, 183
PRMT8, 184
PRMT9, 185
Proline-tryptophan-tryptophan-proline domain (PWWP), 340, 371, 372
Prostate cancer, 60, 68, 101, 105, 112, 126, 166, 181, 186, 201, 207, 225, 272, 305, 341, 350, 354, 411
Protein arginine deiminases (PADs), 162
Protein arginine methyltransferase (PRMTs), 126, 159–183, 265, 465
Protein kinase C (PKC) inhibitors, 59
Protein lysine methyltransferases (PKMTs), 128, 135, 143, 465, 468
Pr-SNF, 143
Pseudomonas aeruginosa, 20
PU139, 110
Pyrazolo[1,5-*a*]pyrimidin-7(4*H*)-one, 241
Pyridoisothiazolones, 110
Pyrroloquinoxalines, 77, 81

Q
Quantitation, 487
Quercetin, 77
Quisinostat, 8

R
Reactive oxygen species (ROS), 79
Reduced representation bisulfite sequencing (RRBS), 505
Resminostat, 8
Resveratrol, 76–79
RG108, 264
RG119-1, 264
Ribonucleotide reductase, 7
Ricolinostat, 9, 10, 19, 20
Rituximab, 18
RNA therapy, 401
Romidepsin, 12
Royal Family domain, 339

S
Salermide, 64
Schistosoma mansoni, 20, 461
SDX437, 73
Selisistat, 67
Sequencing, 487
SET and MYND domain-containing (SMYD) family, 141
SETs, 123
  SETD1A/1B, 138
  SETD2 (KMT3A/SET2), 143
  SETD7 (SET9, SET7/9, KMT7), 140, 469
  SETD8, 144
  SETDB1 (ESET/KMT1E), 129
SGC110 (guadecitabine), 263
SGC0946, 136
SGF29, 349
SGI-1027, 264
Sickle cell disease (SCD), 19, 126, 202
SID49645275, 265
Sinefungin, 163
Single-molecule real-time (SMRT) sequencing, 511
SirReal2, 70
Sirtinol, 64
Sirtuins (SIRT1–7), 4, 30, 55, 432, 459, 463
  SIRT1, 55
smHDAC8, 21, 461, 462
SMRT/NCoR, 31
Solid-phase peptide synthesis (SPPS), 35
SP2509, 205
Spinal and bulbar muscular atrophy, 57
Spindlin-1, 361
Spiro-oxazolidinedione, 112
Splitomicin, 65
SRT1460, 79
SRT1720, 77, 79
Stilbamidine, 466
Structural biology, 287
Structure-based drug design, 455
Suberoylanilide hydroxamic acid (SAHA), 7, 32
Sulfobenzoic acids, 64, 71
Sulfonamides, 8, 323, 365–368
Suramin, 72, 353, 354, 465
SUVs, 123, 128
  SUV39H1/2, 130
  SUV420H1/2, 144

T
T5342129, 204
TAFII250, 293
Tanikolide, 76
α-TAT1, 98, 101
Tazemetostat (EPZ6438), 133
T-cell lymphoma, 7, 12, 14, 18, 44, 214
Tefinostat, 11

Temozolomide (TMZ), 415
Ten-eleven translocator (TET) enzymes, 255, 258, 440, 489
Tenovin, 73
Terpenylated coumarin, 77
Thailandepsin A, 12
Thalassemia, 202
Therapeutics, 159
Thiazolopyridine, 77
Thioacetyl-lysine, 61
Thiobarbiturate, 74, 75
Thiomyristoyl-lysine, 63
Thymidine-DNA glycosylase (TDG), 258
Tip60, 101
Tip3M15L, 276
TP53BP1, 412
Trapoxin A, 15, 36
TR-FRET, 321
Triazolophtalazines, 300
Trichostatin A (TSA), 8, 32
Trimethylamiodarone (WAG-003), 346
Tubastatin A, 462
α-Tubulin, 62
Tucidinostat, 14
Tudor domain, 339, 356
TUG1, 409, 416
Tumor necrosis factor alpha (TNFα), 63

U
UHRF1, 359
UNC0224, 127
UNC0321, 127
UNC0379, 145
UNC0638, 127
UNC0642, 128
UNC1999, 131
UNC2170, 359
UNC3567, 351
UNC3866, 351
UNC5114, 135
UNC5115, 135
Urease, 7

V
Valproic acid, 15
Varicella zoster virus (VZV), 201
Virtual screening, 455
Vorinostat, 7, 14, 20, 32

W
WD40, 340, 342, 363, 372–382
WDR5, 373
  inhibitors, 138
Wedelolactone, 133
WF3161, 39
Whole-genome bisulfite sequencing (WGBS), 505

X
XZ09, 470

Y
Y1467, 95

Z
Zebularine, 263, 492
Zinc metalloenzymes, 1

The manufacturer's authorised representative in the EU is Springer Nature Customer Service Centre GmbH, Europaplatz 3, 69115 Heidelberg, Germany. If you have any concerns regarding our products, please contact ProductSafety@springernature.com

Printed and bound by CPI Group (UK) Ltd, Croydon, CR0 4YY
22/09/2026
02220956-0016